AUTOMOTIVE

BRAKES, SUSPENSION, AND STEERING

OTHER BOOKS AND INSTRUCTIONAL MATERIALS BY WILLIAM H. CROUSE AND *DONALD L. ANGLIN

The Auto Book*
Auto Shop Workbook*
Auto Study Guide*
Auto Test Book*
Automotive Air Conditioning*
Workbook for Automotive Air Conditioning*
Automotive Body Repair and Refinishing*
Workbook for Automotive Body Repair and Refinishing*
Workbook for Automotive Brakes, Suspension, and Steering*
Automotive Dictionary*
Automotive Electronics and Electrical Equipment
Workbook for Automotive Electronics and Electrical Equipment*
Automotive Emission Control*
Workbook for Automotive Emission Control*
Automotive Engine Design
Automotive Engines*
Workbook for Automotive Engines*
Automotive Fuel, Lubricating, and Cooling Systems*
Workbook for Automotive Fuel, Lubricating, and Cooling Systems*
Automotive Mechanics
Study Guide for Automotive Mechanics*
Testbook for Automotive Mechanics*
Workbook for Automotive Mechanics*
Automotive Engines Sound Filmstrip Program
Automotive Service Business
Automotive Technician's Handbook*
Automotive Tools, Fasteners, and Measurements*
Automotive Manual Transmission and Power Trains*
Workbook for Automotive Manual Transmissions and Power Trains*
Automotive Automatic Transmissions*
Workbook for Automotive Automatic Transmissions*
Automotive Tuneup*
Workbook for Automotive Tuneup*
General Power Mechanics* (With Robert Worthington and Morton Margules)
Motor Vehicle Inspection*
Workbook for Motor Vehicle Inspection*
Motorcycle Mechanics*
Workbook for Motorcycle Mechanics*
Small Engine Mechanics*
Workbook for Small Engine Mechanics*

AUTOMOTIVE ROOM CHART SERIES

Automotive Brake Charts
Automotive Electrical Equipment Charts
Automotive Emission Control Charts
Automotive Engines Charts
Automotive Engine Cooling Systems, Heating, and Air Conditioning Charts
Automotive Fuel Systems Charts
Automotive Suspension, Steering, and Tires Charts
Automotive Transmissions and Power Trains Charts

AUTOMOTIVE TRANSPARENCIES BY WILLIAM H. CROUSE AND JAY D. HELSEL

Automotive Air Conditioning
Automotive Brakes
Automotive Electrical Systems
Automotive Emission Control
Automotive Engine Systems
Automotive Steering Systems
Automotive Suspension Systems
Automotive Transmissions and Power Trains
Engines and Fuel Systems

SIXTH EDITION

AUTOMOTIVE BRAKES, SUSPENSION, AND STEERING

WILLIAM H. CROUSE
DONALD L. ANGLIN

GREGG DIVISION/McGRAW-HILL BOOK COMPANY

New York ○ Atlanta ○ Dallas ○ St. Louis ○ San Francisco ○ Auckland ○ Bogotá ○ Guatemala
Hamburg ○ Lisbon ○ London ○ Madrid ○ Mexico ○ Montreal ○ New Delhi
Panama ○ Paris ○ San Juan ○ São Paulo ○ Singapore ○ Sydney ○ Tokyo ○ Toronto

ABOUT THE AUTHORS

William H. Crouse

Behind William H. Crouse's clear technical writing is a background of sound mechanical engineering training as well as a variety of practical industrial experience. After finishing high school, he spent a year working in a tinplate mill. Summers, while still in school, he worked in General Motors plants, and for three years he worked in the Delco-Remy division shops. Later he became director of field education in the Delco-Remy Division of General Motors Corporation, which gave him an opportunity to develop and use his writing talent in the preparation of service bulletins and educational literature.

Mr. Crouse became editor of technical education books for the McGraw-Hill Book Company and was the first editor-in-chief of McGraw-Hill's *Encyclopedia of Science and Technology.* He has contributed numerous articles to automotive and engineering magazines and has written many outstanding books.

William H. Crouse's outstanding work in the automotive field has earned for him membership in the Society of Automotive Engineers and in the American Society of Engineering Education.

Donald L. Anglin

Trained in the automotive and diesel service field, Donald L. Anglin has worked both as a mechanic and as a service manager. He has taught automotive courses and has also worked as curriculum supervisor and school administrator. Interested in all types of vehicle performance, he has served as a racing-car mechanic and as a consultant to truck fleets on maintenance problems.

Currently he devotes full time to technical writing, teaching, and visiting automotive instructors and service shops. Together with William H. Crouse he has co-authored magazine articles on automotive education and many automotive books published by McGraw-Hill.

Donald L. Anglin is a Certified General Automotive Mechanic, a Certified General Truck Mechanic, and holds many other licenses and certificates in automotive education service and related areas. His work in the automotive service field has earned for him membership in the American Society of Mechanical Engineers and the Society of Automotive Engineers. In addition, he is a member of the Board of Trustees of the National Automotive History Collection.

Sponsoring Editor: D. Eugene Gilmore
Editing Supervisor: Larry Goldberg
Design and Art Supervisor: Caryl Valerie Spinka
Production Supervisors: Kathleen Morrissey and S. Steven Canaris

Cover Designer: David R. Thurston
Cover Illustration: Fine Line, Inc.
Technical Studio: Technical Graphics

Library of Congress Cataloging in Publication Data

Crouse, William Henry (date)
Automotive brakes, suspension, and steering.

Rev. ed. of: Automotive chassis and body. 5th ed.
1976.
Includes index.
1. Automobiles—Brakes. 2. Automobiles—Springs and suspension. 3. Automobiles—Steering-gear. 4. Automobiles—Brakes—Maintenance and repair. 5. Automobiles—Springs and suspension—Maintenance and repair.
6. Automobiles—Steering-gear—Maintenance and repair.
I. Anglin, Donald L. II. Title.
TL269.C76 1983 629.2'4 82-17187

AUTOMOTIVE BRAKES, SUSPENSION, AND STEERING
Sixth Edition

7890 SEMSEM 89098

ISBN 0-07-014828-7

CONTENTS

PREFACE

Automotive Brakes, Suspension, and Steering is actually the sixth edition of *Automotive Chassis and Body*. When it came time to revise the book, the authors and publisher were faced with a tremendous task. This was to cover all the new developments in the field without producing an oversize and unwieldy book. In the meantime, however, the authors had produced two new books, *Automotive Body Repair and Refinishing* and *Automotive Air Conditioning*. For this reason, the chapters on these subjects were no longer needed in this book. This gave the authors many additional pages in which to cover the new developments.

At the same time, in addition to covering what was new in design, trouble diagnosis, and servicing, much of the original text was rewritten to improve readability. The result has been a still better readability level, as determined by the standard readability measurements.

There have been many new developments in brakes, suspension, and steering. The downsizing of cars into smaller, more fuel-efficient vehicles has brought about many changes. For example, many downsized front-wheel-drive cars have front-wheel disk brakes. The disks are only 12 inches [305 mm] in diameter and, because they are unventilated, the brake pads get very hot—up to 1000°F [538°C]. Organic pad linings, such as asbestos, cannot take such temperatures and give satisfactory life. Special semimetallic linings had to be developed.

Many other changes in design and materials have been made in recent years: the quick-takeup master cylinders for low-drag disk brakes; new types of truck and trailer brakes; modern suspension systems including MacPherson-strut front and rear suspensions, air and hydropneumatic suspensions, and electronic load levelers; the new Uniform Tire Quality Grading System (UTQGS); and more. This new edition covers these changes.

Trouble-diagnosis and servicing sections have been updated to include the latest techniques in such areas as brake servicing, wheel alignment, and balance.

During the preparation of this new edition, the latest technical and servicing literature issued by automotive manufacturers, both here and abroad, was analyzed to ensure complete coverage of the latest developments. Factory and dealer service facilities were visited to assess the latest service techniques so that the new developments could be incorporated into the new edition.

The metric system of measurements, introduced in the previous edition by dual dimensioning, has been continued in this edition. When a United States Customary measurement is used, it is followed by its metric equivalent in brackets; for example, 0.002 inch [0.051 mm].

A new edition of the *Workbook for Automotive Brakes, Suspension, and Steering* has also been prepared. It includes the basic brakes, suspension, and steering jobs as proposed in the latest recommendations of the Motor Vehicle Manufacturers Association—American Vocational Association Industry Planning Council. Used together, *Automotive Brakes, Suspension, and Steering* and its *Workbook* supply the student with the background information and hands-on experience needed to become a qualified and certified automotive brakes and wheel-alignment technician.

To assist the instructor, the *Instructor's Planning Guide* is available. This guide contains information on utilizing community resources, new laws affecting mechanics, and new methods and products. It also lists other materials, apart from the workbook, that are available for use with *Automotive Brakes, Suspension, and Steering*. These include the McGraw-Hill Automotive Transparencies and eight sets of Automotive Room Charts.

Used singly or together, these instructional materials ensure congruency between the school curriculum and the future needs of students entering the automotive service field. They will help the instructor to tailor the learning experience of the student around tested and proven competency-based objectives, and will help the student to meet the minimum standards of competence demanded of entry-level employees. The student will then be able to develop locally demanded career skills while mastering the necessary job competencies and performance indicators covered in *Automotive Brakes, Suspension, and Steering*.

William H. Crouse
Donald L. Anglin

ACKNOWLEDGMENTS

During the planning and preparation of this edition of *Automotive Brakes, Suspension, and Steering,* the authors and publishers had the advice and assistance of many people—educators, researchers, artists, editors, and automotive-industry service specialists. The authors gratefully acknowledge their indebtedness and offer their sincere thanks. All cooperated with the aim of providing accurate and complete information that would be useful in the training of automotive-service technicians.

Special thanks are owed to the following organizations for information and illustrations they supplied: AC-Delco Division of General Motors Corporation; American Motors Corporation; Ammco Tools, Inc.; ATW; Bear Automotive Service Equipment Company; Bendix Corporation; British Leyland, Limited; Buick Motor Division of General Motors Corporation; Cadillac Motor Car Division of General Motors Corporation; Chevrolet Motor Division of General Motors Corporation; Chrysler Corporation; Clayton Manufacturing Company; Delco Moraine Division of General Motors Corporation; FMC Corporation; Ford Motor Company; Ford of Europe, Inc.; General Motors Corporation; The General Tire & Rubber Company; Goerlich; B. F. Goodrich Company; Goodyear Tire and Rubber Company; Hunter Engineering Company; International Harvester Company; K-D Manufacturing Company; Lempco Industries, Inc.; Lisle Corporation; McQuay-Norris, Inc.; Maremont Corporation; Mazda Motors of America, Inc.; Mercedes-Benz of North America, Inc.; Monroe Auto Equipment Company; Moog Automotive, Inc.; Motor Vehicle Manufacturers Association; Oldsmobile Division of General Motors Corporation; Peugeot Motors of America, Inc.; Pontiac Motor Division of General Motors Corporation; Robert Bosch Corporation; Rubber Manufacturers Association; Shelby International, Inc.; Snap-on Tools Corporation; Stewart-Warner Alemite; Texaco, Inc.; Tire Industry Safety Council; Toyota Motor Sales Company, Limited; TRW, Inc.; Volkswagen of America, Inc.; Wagner Lockheed; Warner Electric Brake & Clutch Company; Weaver Manufacturing Company.

William H. Crouse
Donald L. Anglin

CHAPTER 1

INTRODUCTION TO SUSPENSION, STEERING, AND BRAKES

After studying this chapter, you should be able to:

1. Name the five basic parts, or components, of the car.
2. Describe the two types of frames cars use.
3. Explain why springs are needed between tires and the car body.
4. Describe how shock absorbers help provide a smoother ride.
5. Describe the basic operation of steering systems.
6. Define *hydraulic.*
7. Explain the difference between the two types of tires.

1-1 Components of the automobile Let us begin our studies of the suspension, steering, and brake system by taking a quick look at the complete automobile. It is made up of five basic parts, or components:

1. The power plant, or engine, which is the source of power that makes the wheels rotate and the car move. This includes the electric, fuel, lubricating, and cooling systems that are necessary to make the engine run.
2. The frame, which supports the car engine and body, and is supported by the wheels. In some cars, there is a complete one-piece frame. In others, the body itself serves as the central part of the frame, with two short, or stub, frames attached at front and back to the body.
3. The power train, which carries the power from the engine to the car wheels. It consists of the clutch (on cars with manual transmission), transmission, drive shaft or shafts, differential, and drive axle. In rear-wheel-drive cars, the power is carried from the transmission to the differential at the rear and from there to the rear-wheel axles. In front-wheel-drive cars, the differential is usually integral with the transmission in an assembly called the *transaxle*. Short shafts carry the power from the transaxle to the front wheels.
4. The car body.
5. The body accessories, which include the heater, air conditioner, lights, windshield wiper and washer, radio, and other such devices. These are not necessary to the basic operation of the vehicle. They add comfort and convenience, and permit the car to operate in rainy weather and at night.

Figure 1-1 shows, in phantom view, a car which has coil springs at front and rear. This is a rear-wheel-drive car with a drive shaft extending from the transmission at the front to the differential at the rear. Figure 1-2 is another rear-wheel-drive car with coil springs at the front and leaf springs at the rear. Figure 1-3 is a front-wheel-drive car using coil springs at both front and rear.

1-2 Frames The frame is the assembly of metal sections and structural parts that supports the engine and body, and that is supported by the wheels. There are two types of automotive frames:

1. The full frame, which supports the engine, power train, and body
2. The stub frame, which is used in cars that have the body itself serving as part of the frame

1-3 Full frames Figure 1-4 shows a typical full frame, or *perimeter frame,* for an automobile. The frame is made of channel, box, and tubular members (Fig. 1-5) carefully shaped and then welded or riveted together. *Crossmembers* reinforce the frame and support the engine and powertrain (Fig. 1-4). The frame is extremely solid and strong to withstand the shocks, twists, vibrations, and other strains it receives on the road. Brackets and pads are welded onto the frame at the proper places to support the body, engine, and other parts (Fig. 1-4).

Frames on cars today are designed to help absorb some of the crash forces from a front-end collision. During an impact, the frame collapses so that part of the shock of the collision is not carried to the passengers. In addition, the front of the car is equipped with some type of energy-

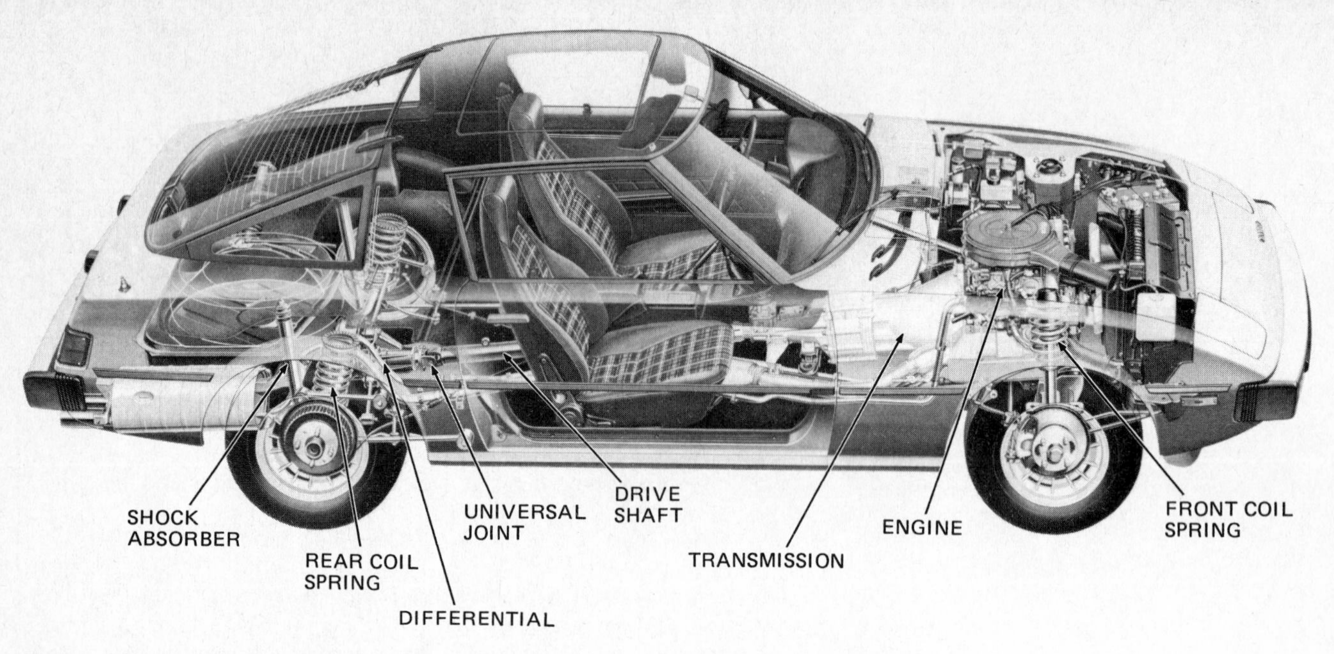

Fig. 1-1 Phantom view of an automobile, showing the suspension and drive train. This is a front-engine, rear-wheel-drive car. It uses coil springs front and rear. (*Mazda Motors of America, Inc.*)

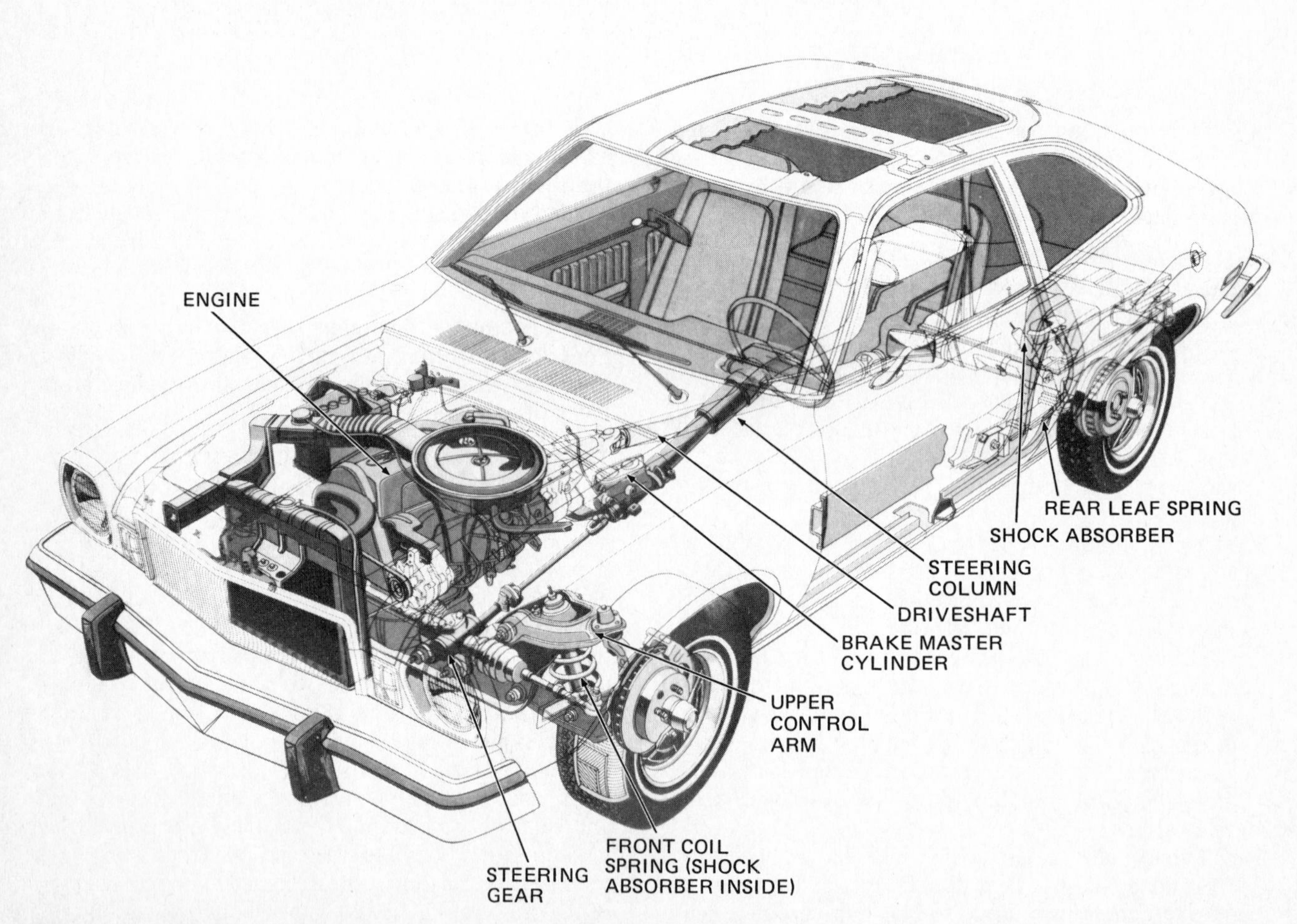

Fig. 1-2 Phantom view of an automobile, showing the suspension and steering systems. This is a front-engine, rear-wheel-drive car. It uses coil springs at the front and leaf springs at the rear. (*Ford Motor Company*)

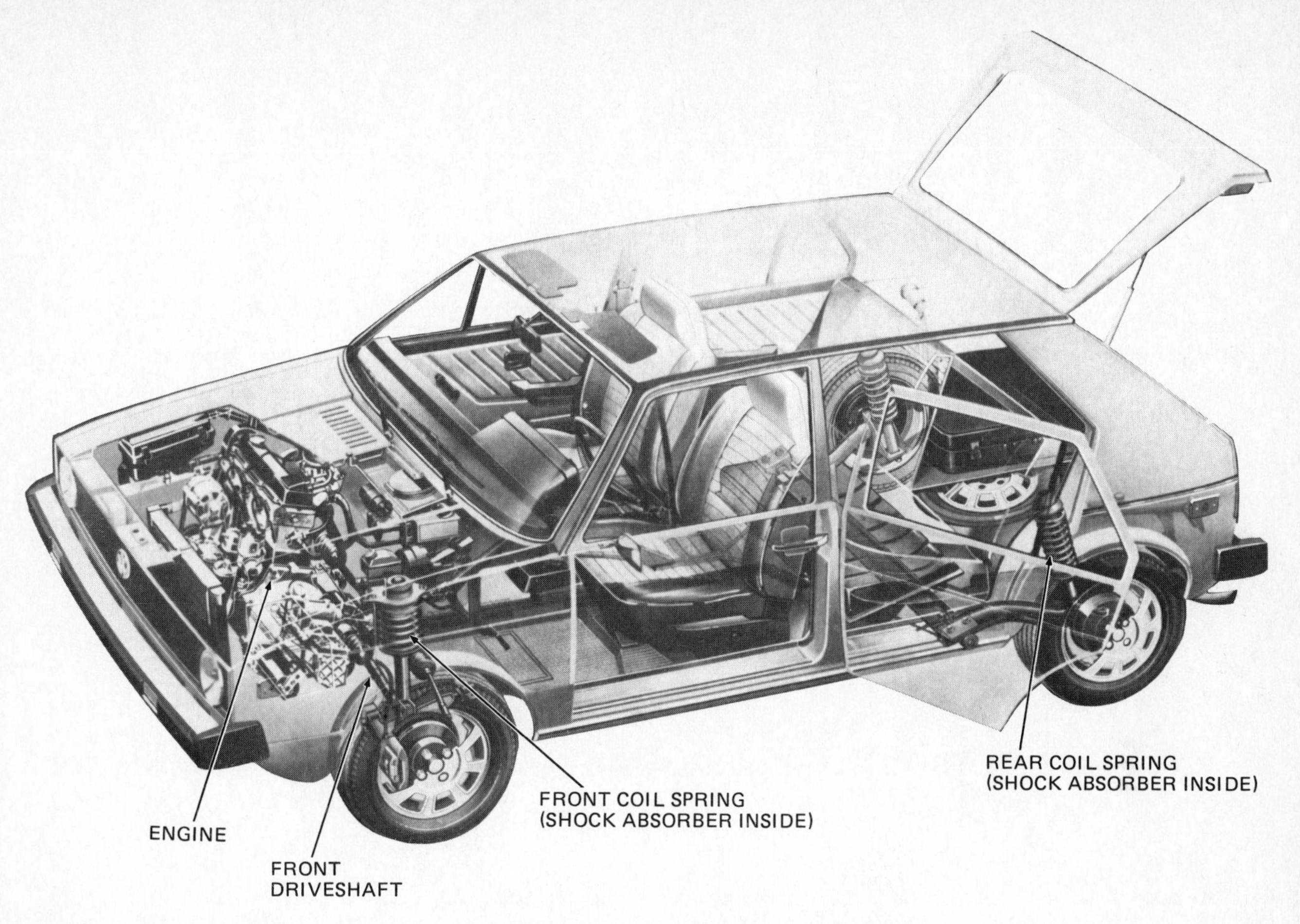

Fig. 1-3 Phantom view of an automobile, showing the suspension system. This is a front-engine, front-wheel-drive car. It uses coil springs front and rear. (*Volkswagen of America, Inc.*)

absorbing bumper. This also reduces the shock transmitted to the passengers. Figure 1-4 shows how front-bumper energy absorbers are an integral part of the front crossmember in one type of full frame.

BODY MOUNT BRACKET
TUBULAR CROSSMEMBER
ENERGY ABSORBER
BOX-SECTION REAR-FRAME ENDS
TUBULAR K BRACE
FRONT CROSSMEMBER WITH INTEGRAL ENERGY ABSORBERS

Fig. 1-4 Typical automobile frame. The frame curves upward at the rear (to the right) to provide room for the rear springs. The frame narrows at the front (to the left) to permit the front wheels to turn from side to side for steering. (*Ford Motor Company*)

Figure 1-6 shows how the car body is attached to the frame with *body bolts*. Every point of attachment includes rubber cushions. These insulate the body bolts and prevent any part of the metal from touching any body part. This prevents frame vibrations from traveling up through the body. The result is a smoother, quieter ride for the driver and passengers in the car.

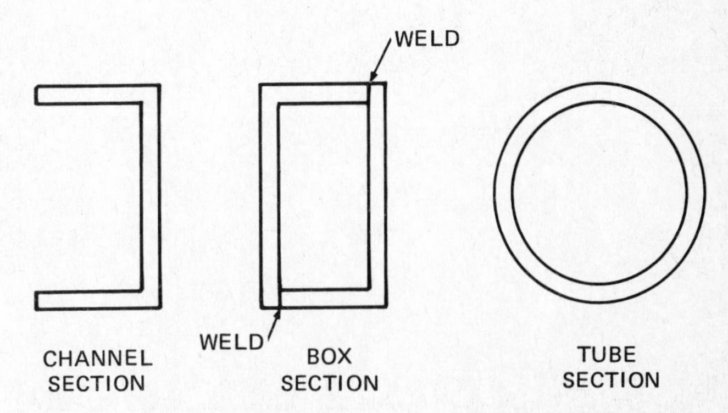

Fig. 1-5 End view of channel, box, and tube sections used in automotive frames.

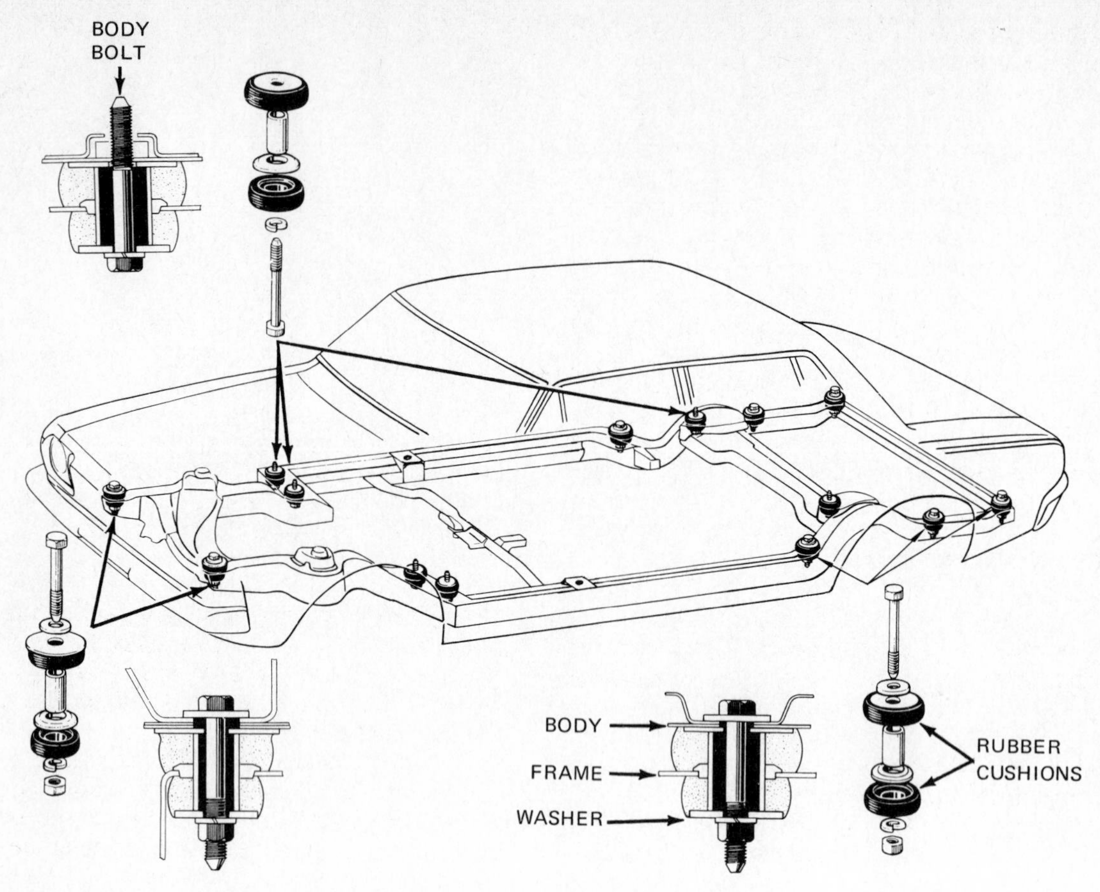

Fig. 1-6 Body bolts used to attach the body to the frame. All bolts are insulated by rubber cushions to prevent vibration from traveling from the frame to the body. (*Toyota Motor Sales, U.S.A., Inc.*)

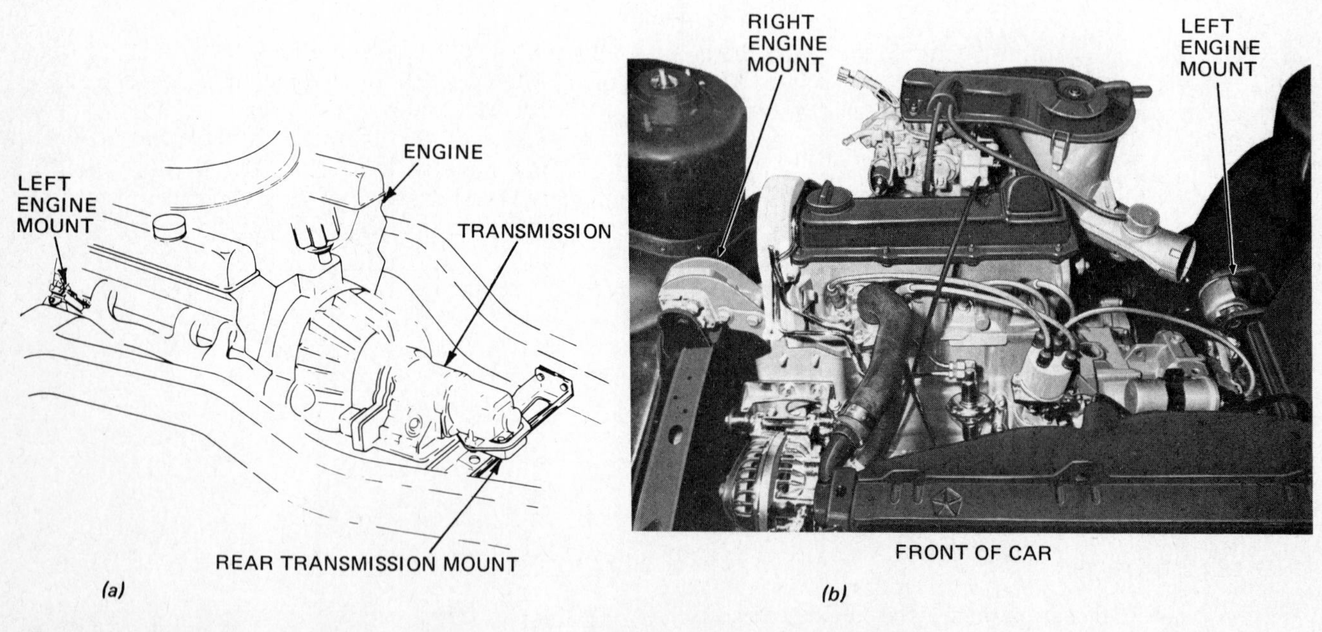

Fig. 1-7 (*a*) The engine-transmission assembly is usually supported by three mounts. The attachment points are cushioned by rubber to prevent engine vibration from reaching the frame. (*Buick Motor Division of General Motors Corporation*) (*b*) On a transverse engine, the weight is supported by a mount at each end of the engine. (*Chrysler Corporation*)

Noise and some vibration are part of engine operation. If the engine was bolted directly to the frame, the noise and vibration would travel from the engine to the frame, and from there to the occupants of the vehicle. To prevent this, the engine is always insulated from the frame.

The engine is mounted to the frame through flexible rubber cushions called *engine supports,* or *engine mounts.* The ends of the engine mounting brackets are supported in the rubber. Attaching bolts pass through the rubber mountings with no metal-to-metal contact. The rubber absorbs the engine noise and vibration so that they are not carried to the frame.

Generally, a three-point mounting system is used (Fig. 1-7*a*). A mount is placed on each side at the front of the engine. The third mount usually is placed at the rear of the transmission, between the transmission and the frame crossmember. Installing the rear mount in this location is possible because the transmission and engine are bolted together to form a rigid assembly.

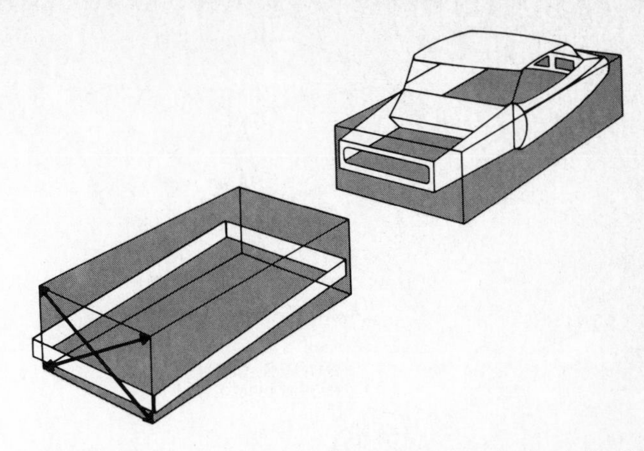

Fig. 1-8 With the unitized-body construction, the body itself becomes an integral part of the framework that supports the engine and other components of the automobile. (*Bear Division of Applied Power, Inc.*)

1-4 Stub frames Cars that are built with unitized-body construction are made with the frame and body welded together to form a single unit. The body itself becomes the structural assembly which is used as the supporting framework for the engine, suspension, and power train (Fig. 1-8). Today usually only larger, more expensive cars have separate body-and-frame construction. Almost all other cars use some form of unitized-body or integral construction.

A variety of arrangements are used to attach the engine, power train, and suspension components to the unitized body. Many of these arrangements have partial frames at the front and rear. These may be called *stub frames, miniframes,* or *cradles.*

The stub frame is only a half-frame, or less. Figure 1-9 shows how the front stub frame fits under the body. In this design, the body itself does not extend forward beyond the fire wall between the engine and the front passenger compartment. During final assembly of the

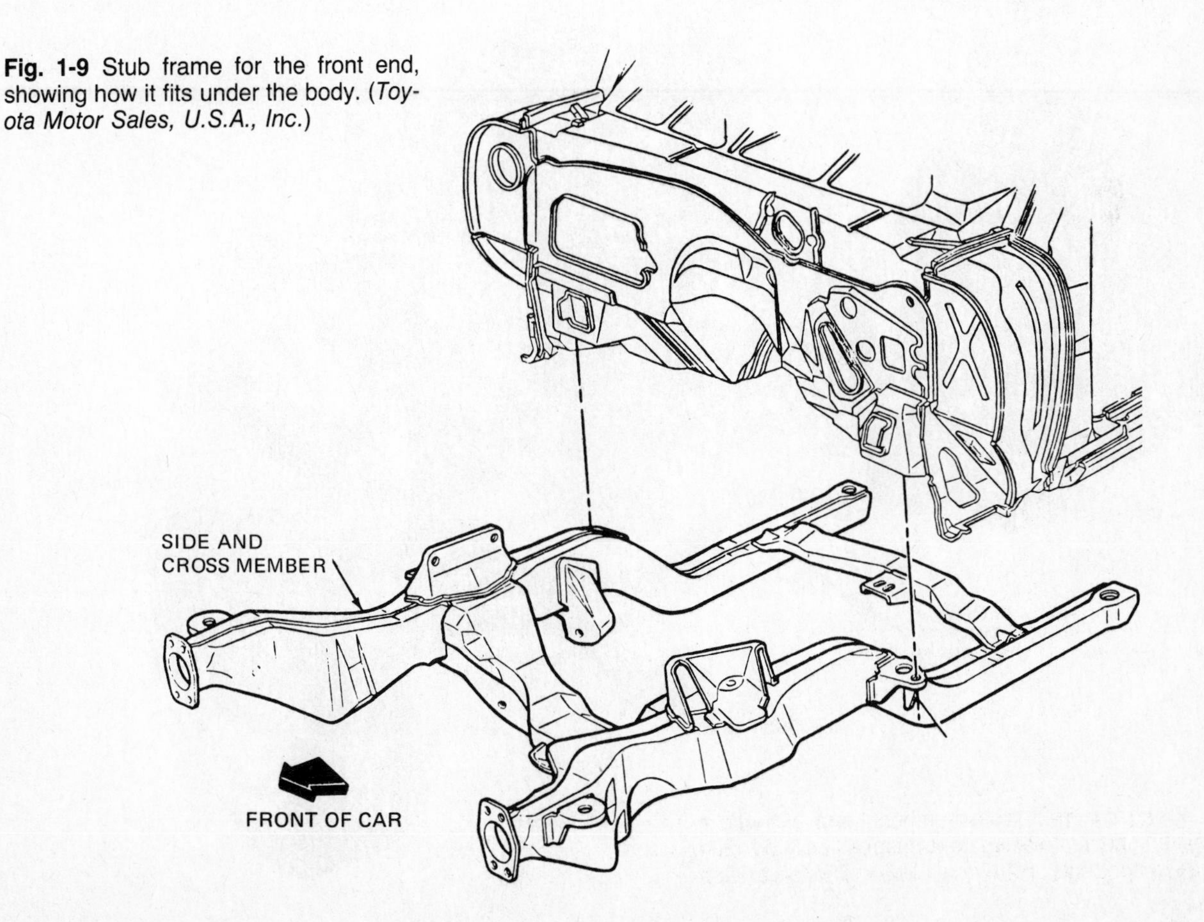

Fig. 1-9 Stub frame for the front end, showing how it fits under the body. (*Toyota Motor Sales, U.S.A., Inc.*)

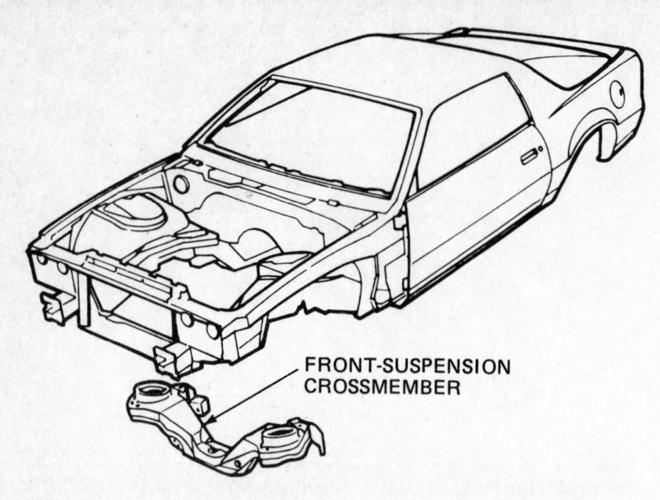

Fig. 1-10 The front-suspension crossmember is the only major bolt-on structural component of a car with full integral construction. (*Chevrolet Motor Division of General Motors Corporation*)

car, the front-end sheet metal is attached separately. In another design (Fig. 1-10), the body includes the internal front-end panels. Later, the hood, front fenders, doors, and trunk lid are added.

Cars that have transverse-mounted engines (Fig. 1-7*b*) usually are built with unitized-body construction. The engine mounting arrangement is similar to that for cars having a full frame. Both use a three-point mounting arrangement. A right and left mount carry the weight of the engine and attached transaxle (Fig. 1-11). However, the third mount does not always support weight. To control movement or "torquing" of the engine, the third mount acts as a stabilizer. It is attached to a reinforced section of the body and to the engine block. There the stabilizer limits unwanted engine movement, or torquing.

1-5 Front springs The car wheels are attached to the car frame through springs. Figures 1-1 to 1-3 show several arrangements. The springs support the weight of the vehicle. Figure 1-12 shows a front-suspension system using coil springs, separated from the rest of the vehicle so that its details can be seen. Note that the coil springs are positioned between the car frame at the top and a lower control arm. The steering knuckle connects to the outer ends of the upper and lower control arms through ball joints (Fig. 1-13). The ball joints allow the steering knuckles to swing from side to side so that the car can be steered. They also allow the two control arms to move up and down as the coil springs expand and compress.

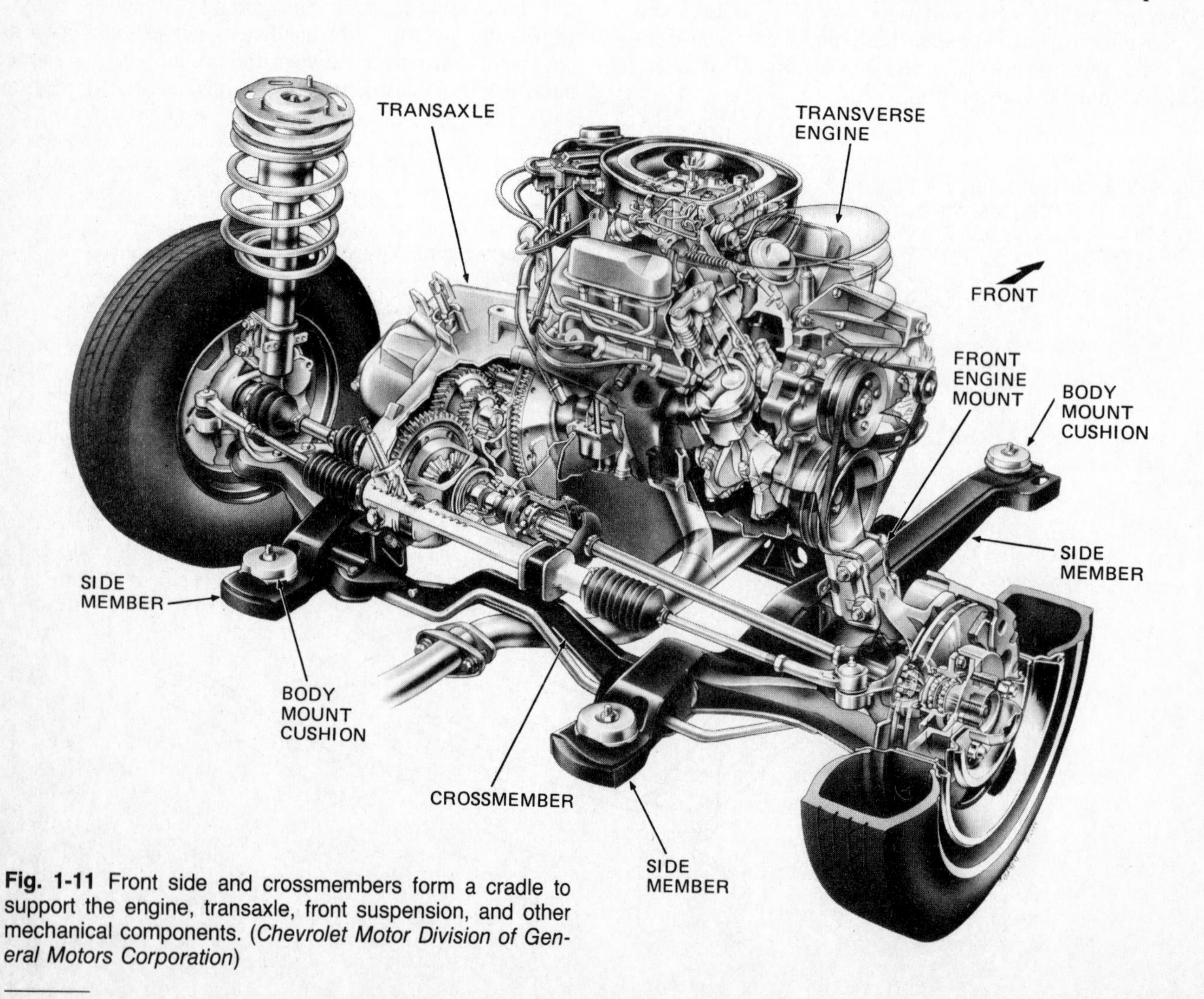

Fig. 1-11 Front side and crossmembers form a cradle to support the engine, transaxle, front suspension, and other mechanical components. (*Chevrolet Motor Division of General Motors Corporation*)

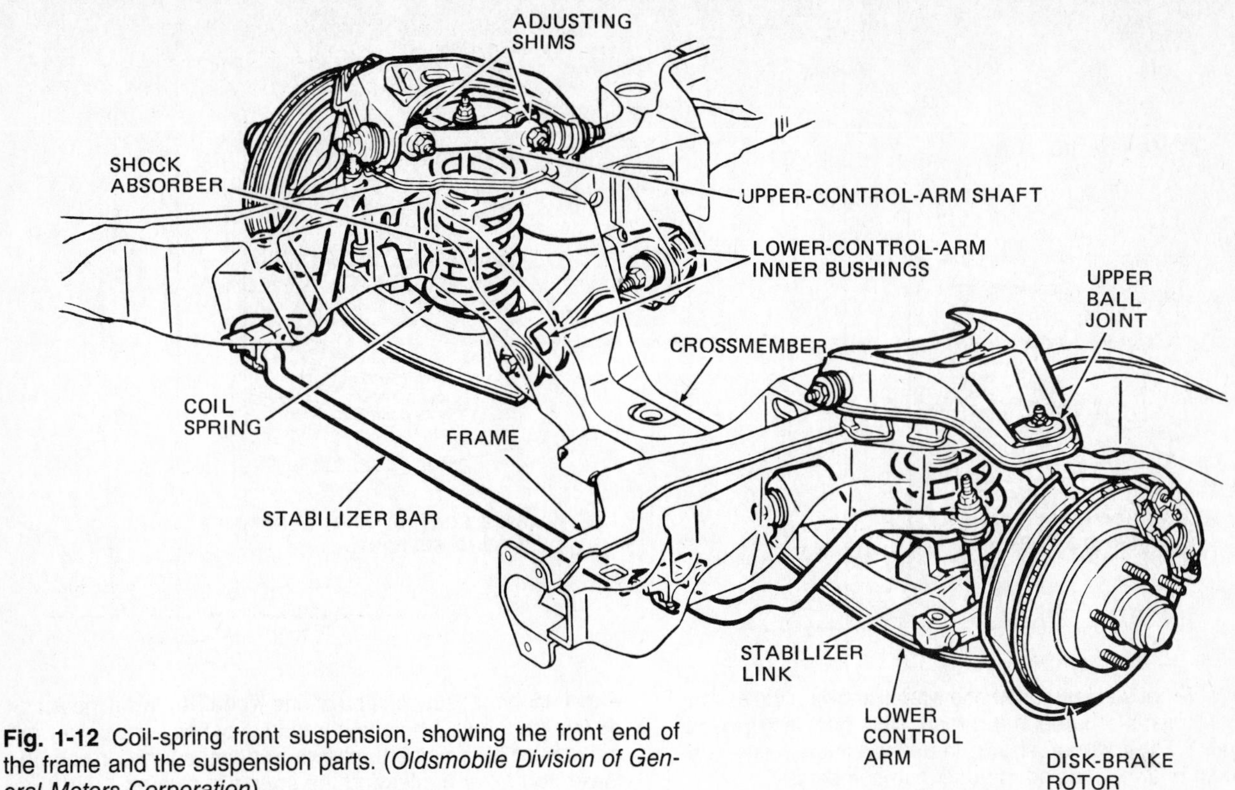

Fig. 1-12 Coil-spring front suspension, showing the front end of the frame and the suspension parts. (*Oldsmobile Division of General Motors Corporation*)

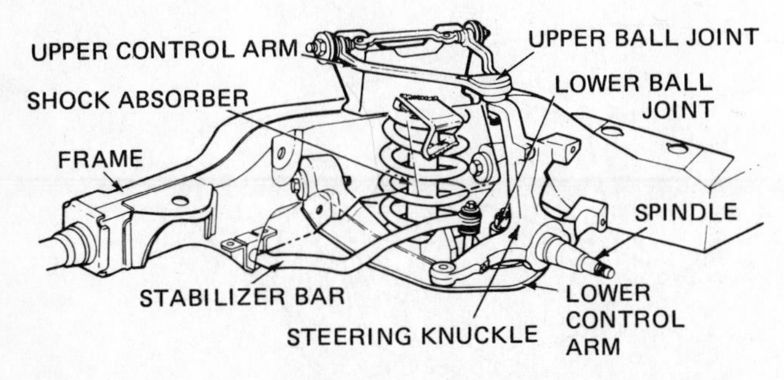

Fig. 1-13 Coil-spring front suspension for one wheel, showing how the wheel spindle and steering knuckle are attached to the upper and lower control arms through ball joints. (*Ford Motor Company*)

Figures 1-14 and 1-15 show the springing action. When a wheel meets a hole or bump in the road, the spring can expand or contract to absorb the wheel motion so that the motion is not carried up into the frame and car body. The coil spring is a heavy steel coil. The weight of the frame and body put an initial compression on the spring. If the wheel meets a bump in the road, the spring will further compress, as shown in Fig. 1-14. Note that the two control arms are attached to the car frame through pivots. The control arms pivot upward to allow the spring to compress and the wheel to move upward.

If the wheel drops into a hole in the road, the spring expands, as shown in Fig. 1-15. During this action, the control arms pivot downward to allow spring expansion and downward movement of the wheel. Several variations of coil-spring front suspension are covered in Chap. 8.

Some cars have a torsion-bar front-suspension system. In one system, a long, straight bar is fastened rigidly at the back to the car frame. It is fastened at the front to the lower control arm (Fig. 1-16). As the lower control arm moves up and down, the torsion bar twists more or less. This produces the springing action. In other torsion-bar front-suspension systems, the torsion bars are mounted transversely, or crosswise (Fig. 1-17), instead of from front to back. Torsion-bar suspension systems are covered in Chap. 8.

⚙ 1-6 Rear springs Rear springs are of two types, coil and leaf. Figures 1-1, 1-2, and 1-18 show rear-suspension systems using coil springs. In Fig. 1-18, note that two kinds of control arms are required. The lower control arms prevent the rear-axle housing from moving forward or backward. The upper control arms prevent the housing from moving from right to left. Both sets of control arms are pivoted so that they work together to allow the axle housing or wheel spindles to move up and down as the wheels encounter bumps or holes in the road. The springs compress or expand the wheels to move up and down, and thereby absorb the movement.

A leaf-spring rear-suspension system is shown in Figs. 1-2 and 1-17. It uses two leaf springs, one on each side. The leaf spring usually is made up of a series of flat plates, or leaves, or graduated length, one on top of another. The spring assembly acts as a flexible beam which is fastened at the two ends to the car frame and at the center to the drive axle housing. When the wheels encounter a bump, the spring bends upward to absorb the upward movement. When a wheel drops into a hole, the spring bends downward to permit the wheel to move

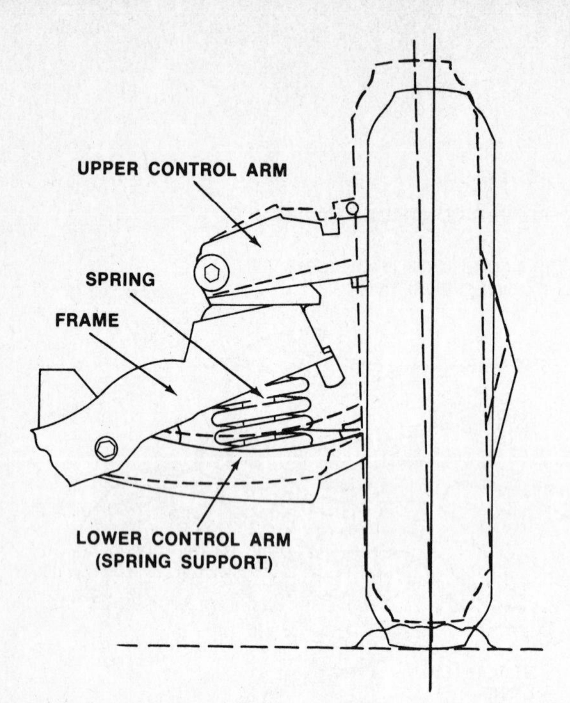

Fig. 1-14 Front suspension at one wheel, showing the action as the tire meets a bump in the road. Note how the upward movement of the wheel, shown in dashed lines, raises the lower control arm, causing the spring to compress.

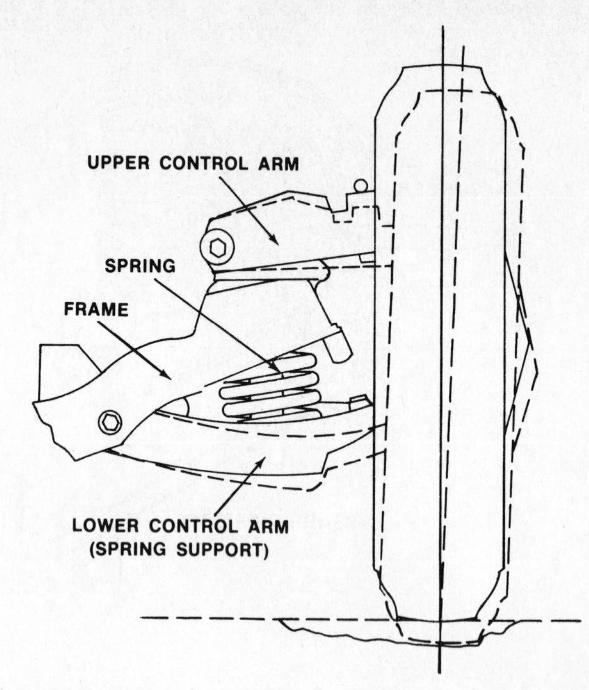

Fig. 1-15 Front suspension at one wheel, showing the action as the tire meets a hole in the road. Note how the downward movement of the wheel, shown in dashed lines, lowers the lower control arm, allowing the spring to expand.

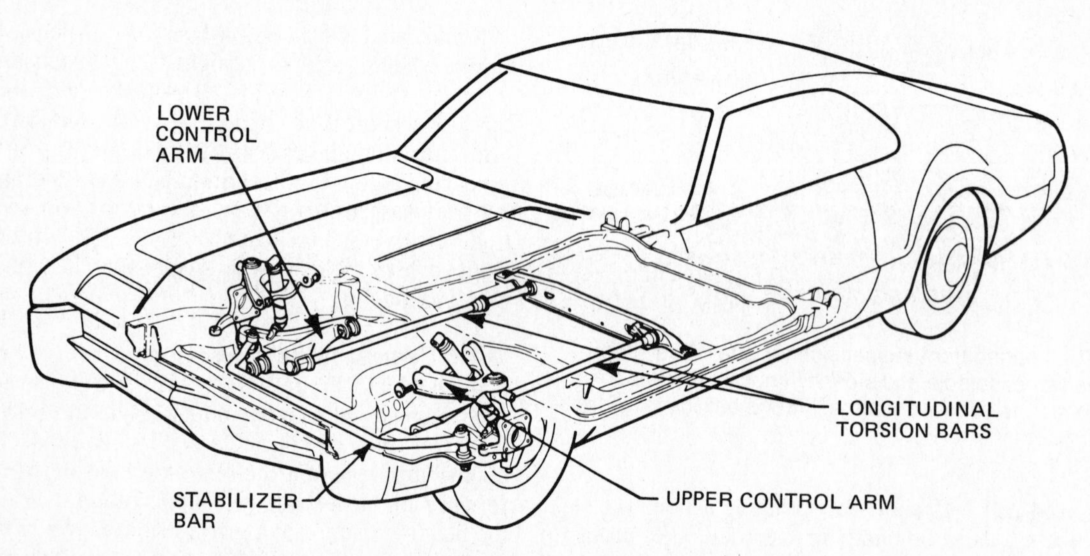

Fig. 1-16 Front suspension system of a car with front-wheel drive using longitudinal, or lengthwise, torsion bars. The torsion bars twist varying amounts as varying loads are applied. This allows the front wheels to move up and down. The action is similar to that for other springs. (*Oldsmobile Division of General Motors Corporation*)

down without carrying the motion up to the car frame and body.

1-7 Shock absorbers Springs alone cannot provide a satisfactorily smooth ride. They do absorb the wheel motions as the wheels meet holes and bumps. However, without some additional device, the springs would continue to compress and expand after the bump or hole had been passed.

When the wheel meets a bump, it moves up, compressing the spring (Fig. 1-14). After the bump has passed, the spring expands. However, it overrides its original position, expanding too much. This could cause the car frame to be thrown upward. Then, having expanded too much, the spring attempts to compress back to its original position. But it again overrides. This override in both directions would produce a very rough ride, with the tire and frame bouncing up and down.

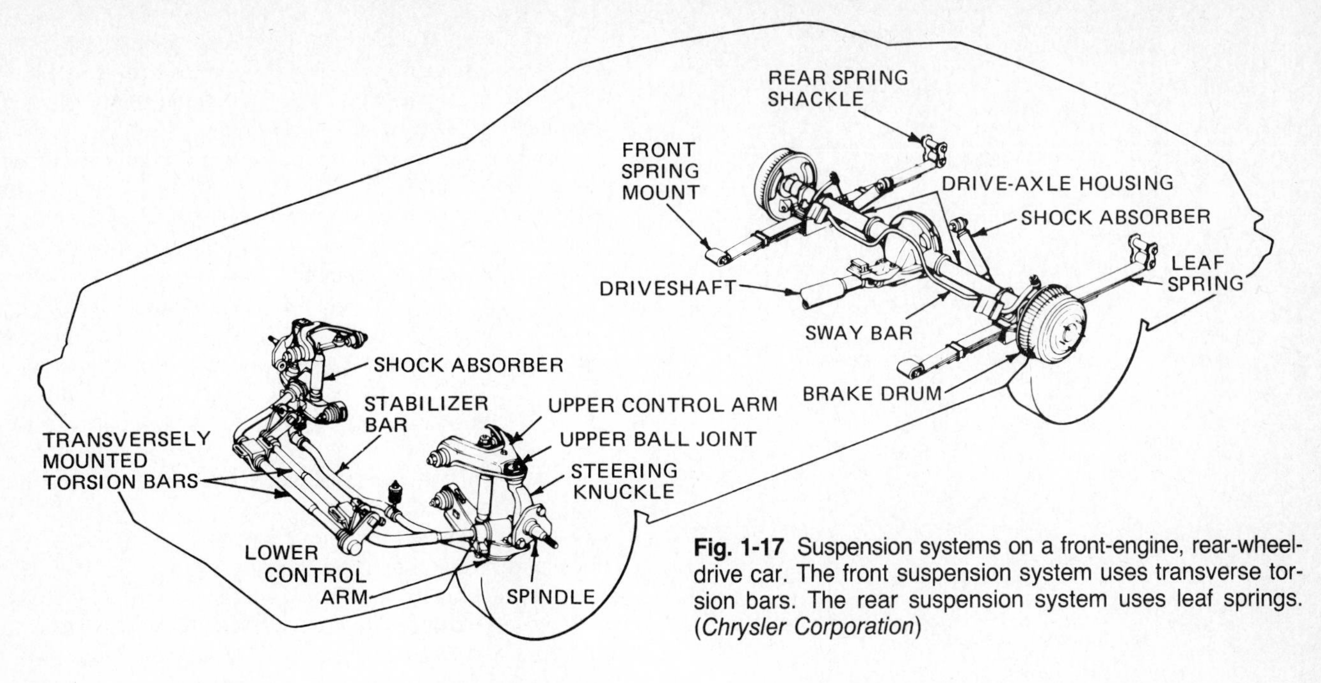

Fig. 1-17 Suspension systems on a front-engine, rear-wheel-drive car. The front suspension system uses transverse torsion bars. The rear suspension system uses leaf springs. (*Chrysler Corporation*)

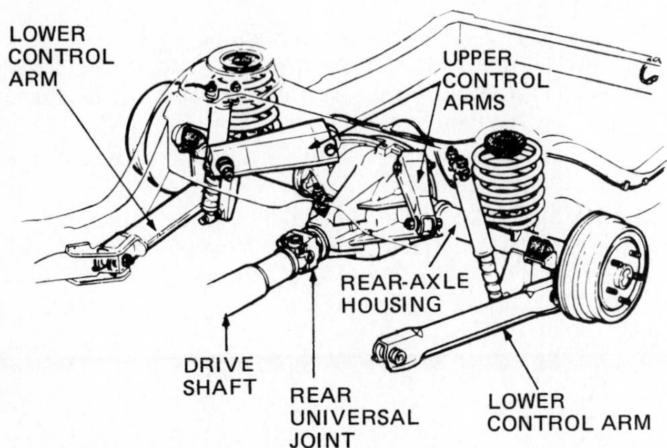

Fig. 1-18 Rear-suspension system using coil springs. (*Buick Motor Division of General Motors Corporation*)

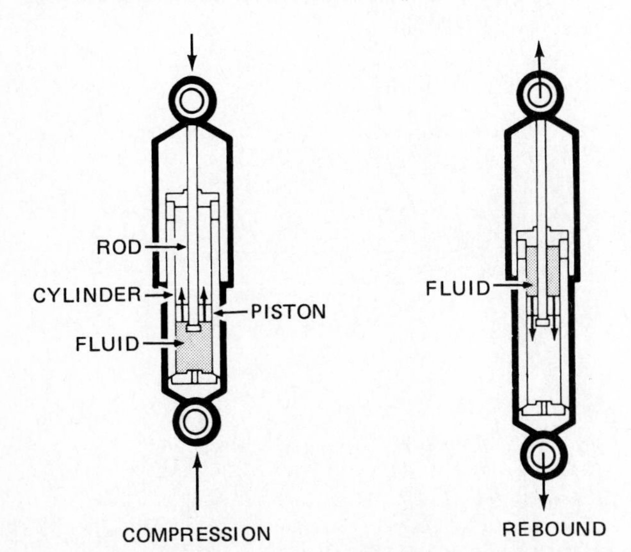

Fig. 1-19 Simplified drawing of shock absorbers in sectional view, showing the actions during compression and rebound. Fluid movement is shown by the arrows. (*Ford Motor Company*)

During the action, the wheel might be raised clear of the road. Although the spring quickly settles down after a single bump or hole, on a bumpy road the wheels would be in almost continuous bouncing motion. These bouncing motions might be so strong that the driver would lose control of the car, especially when rounding a curve.

To prevent such problems, cars are equipped with shock absorbers (Figs. 1-17 and 1-19). There is a shock absorber at each wheel and spring (Figs. 1-1 to 1-3). Figure 1-19 shows, in sectional view, two positions of a shock absorber which demonstrate the action. As the shock absorber shortens or lengthens, the piston moves inside the cylinder. This action displaces the fluid inside the unit. The fluid must then move through small openings in the piston, as shown by the arrows (Fig. 1-19). Since the fluid cannot move instantly through these openings, it restrains the up-and-down movement. In effect, it produces a dragging effect on the spring and wheel motion. The result is that after every bump or hole is passed, the spring quickly settles down again. The operation of shock absorbers is described further in Chap. 9.

❁ 1-8 Steering system To guide the car, the front wheels must be swung to one side or the other, away from straight ahead. The rest of the car then follows the direction in which the wheels are pointed. The front wheels are mounted on spindles which are attached to the control arms by ball joints (Fig. 1-17). The wheels can pivot on these ball joints. Steering arms, which are part of the spindles, are connected by rods, or linkage, to the steering gear (Fig. 1-20).

When the steering wheel is turned, the steering gear causes the pitman arm to swing from one side or the other. This motion is carried through linkage (tie rods and relay rods) to the steering arms. The steering arms, as they move, cause the wheels to pivot to one side or the other, so that the car is steered in the direction the driver selects.

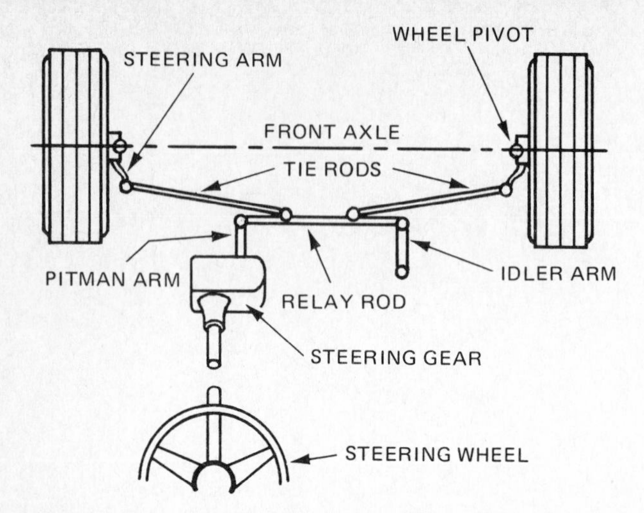

Fig. 1-20 Simplified drawing of a steering system, as seen from above.

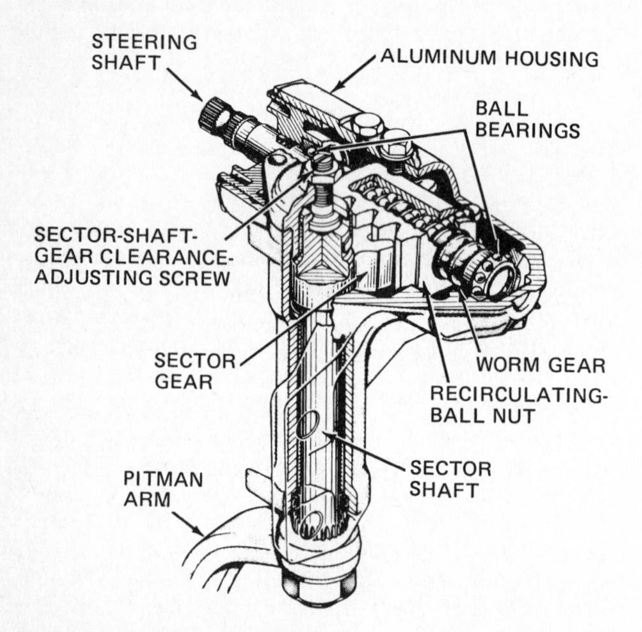

Fig. 1-21 Cutaway view of a recirculating-ball steering gear. (*Chrysler Corporation*)

There are several different types of steering gears. One type is shown in Fig. 1-21. The worm gear is attached to the end of the steering shaft. When the steering wheel and shaft turn, the worm turns. This causes the teeth in the sector gear to follow the worm teeth. The action causes the pitman-arm shaft to turn. As a result, the pitman arm, which is mounted on the end of the shaft, swings to one side or the other, according to the direction in which the steering shaft has turned.

There are several different designs of steering linkages and steering gears. In the rack-and-pinion steering gear (Fig. 1-22), the lower end of the steering shaft has a pinion (a small gear). This pinion is meshed with a rack, which is like a straight section of gear teeth. As the pinion turns, it causes the rack to move to one side or the other. This motion is carried by the tie rods to the steering arms.

Many steering systems have a power assist which provides an additional force to the steering system when the steering wheel is turned. This power assist supplies most of the force required to steer. Power steering, steering linkages, and various types of steering gears are discussed in later chapters.

⚙ 1-9 Brakes All automotive vehicles must have brakes (Fig. 1-23). Brakes slow or stop the car. Automotive brakes are operated by *hydraulic* pressure. Hydraulic pressure is pressure on a fluid. In the brake system, the fluid is a special brake fluid. It is not oil. When the driver applies the brakes by pushing down on the brake pedal, brake fluid is forced from a master cylinder through brake lines (tubes) to cylinders at each wheel.

As the fluid pressure increases in the brake cylinders, brake pads or shoes are forced against rotating disks or drums. The friction between the two causes the wheels to slow or stop. Many vehicles have power brakes. On these, when the driver pushes down on the brake pedal, a power booster (Fig. 1-23) acts to supply most of the effort required to apply the brakes. Later chapters cover various types of brakes, and their construction, operation, and servicing.

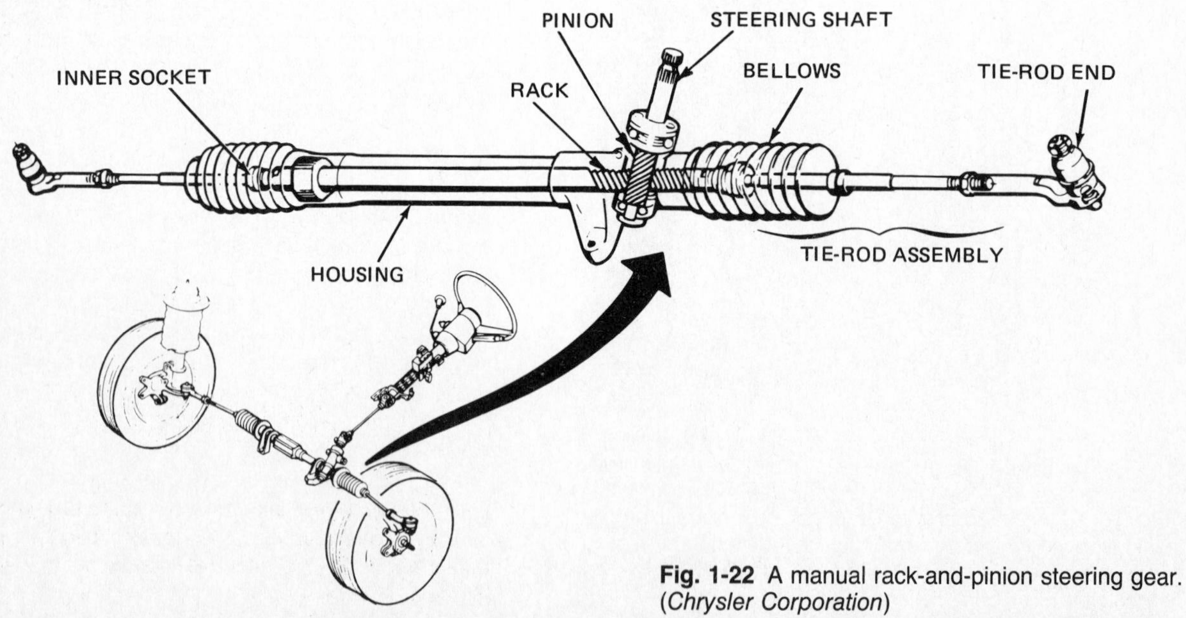

Fig. 1-22 A manual rack-and-pinion steering gear. (*Chrysler Corporation*)

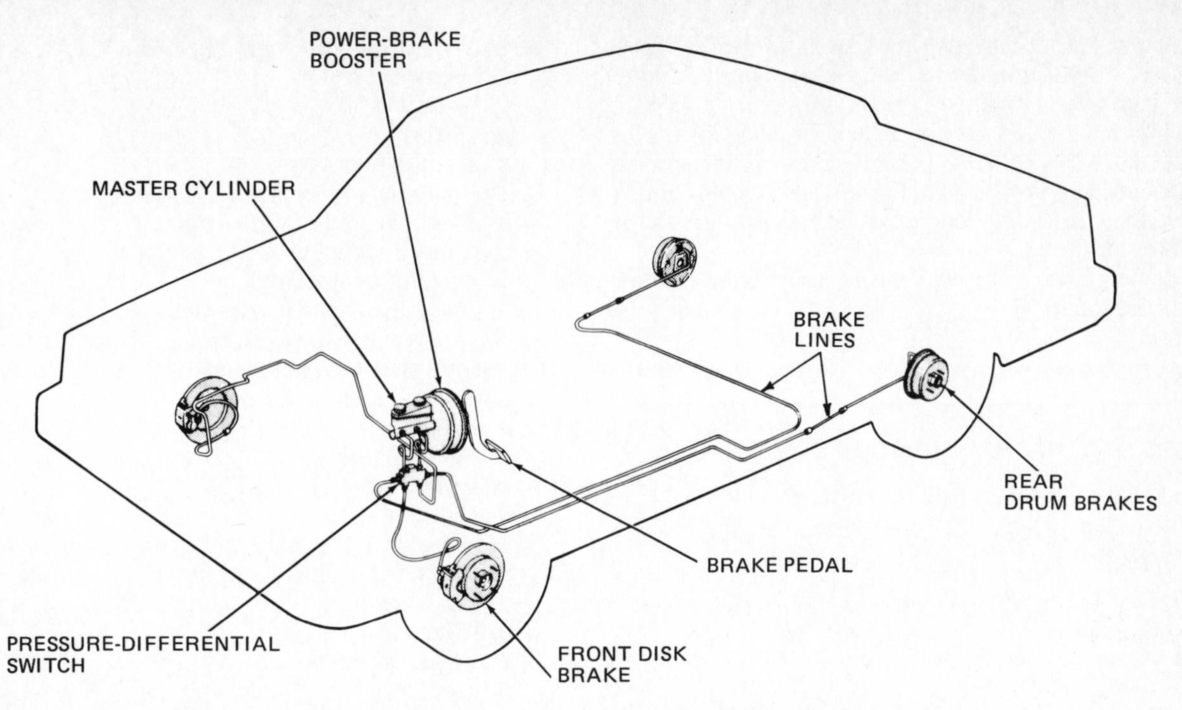

Fig. 1-23 Braking system on a car. The system is diagonally balanced, with floating-caliper front-disk brakes and rear-drum brakes operated by a power-brake booster. (*Chrysler Corporation*)

⚙ 1-10 Wheels and tires Automobiles use air-filled, or *pneumatic,* tires (Fig. 1-24) mounted on steel or aluminum wheel rims. Tires apply the driving power of the wheels to the road through frictional contact. Tires also absorb part of the road shock resulting from small bumps and holes the tires meet. The action prevents these shocks from being carried up to the frame and body of the car. As the tires roll over bumps, they flex. The outer surface, or tread, which is in contact with the road, bends inward against the cushion of air inside the tire.

Tires are of two types: those with inner tubes and those without. Tubeless tires are in general use today for automotive vehicles. Years ago, all pneumatic tires had inner tubes, and even today inner tubes are widely used in motorcycle, moped, and bicycle tires.

The tubeless tire mounts on the wheel rim so that the tire bead seals airtight against the rim flange (Fig. 1-25). In tires with inner tubes, the air is held in the inner tube itself.

Fig. 1-24 A typical automobile tire. (*The General Tire & Rubber Company*)

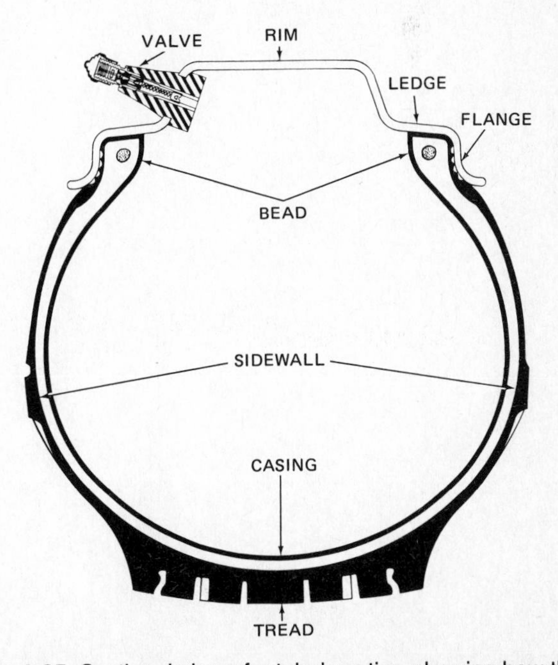

Fig. 1-25 Sectional view of a tubeless tire, showing how the tire bead rests between the ledges and flanges of the wheel rim to produce a good seal. (*Pontiac Motor Division of General Motors Corporation*)

The tire casing has an outer coating of rubber which is baked, or vulcanized, onto an inner structure of fabric, fiber glass, steel cord, or similar material. The tread, which is the thickest part of the outer rubber coating, is supplied in many different patterns. These different treads are designed to provide good traction and stopping ability for various operating conditions such as driving on the highway or in rain or snow.

Chapters 20 and 21 describe wheels and tires, and how to service them.

Chapter 1 review questions

Select the *one* correct, best, or most probable answer to each question. Then check your answers against the correct answers given at the end of the book.

1. At how many places is the engine ordinarily attached to the frame?
   a. one
   b. two
   c. three
   d. five
2. The frame provides support for the:
   a. engine
   b. body
   c. wheels
   d. all of the above
3. The purpose of the shock absorbers is to:
   a. attach the spring to the frame
   b. dampen spring oscillations
   c. tighten the mounting
   d. attach the frame to the wheel
4. As the steering wheel is turned, a worm gear on the steering shaft causes the end of the ______ to swing toward one side or the other of the car.
   a. kingpin
   b. steering gear
   c. pitman arm
   d. ball joint
5. Movement of the brake pedal forces brake fluid out of the master cylinder and into the:
   a. brake shoes
   b. brake cables
   c. wheel cylinders
   d. pedal rod

CHAPTER 2

FUNDAMENTAL PRINCIPLES

After studying this chapter, you should be able to:

1. Discuss atmospheric pressure and the effects of changes in it.
2. Define *vacuum.*
3. Explain the relationship between heat and pressure.
4. Discuss hydraulics and how it can be used to transmit force and motion.
5. Define *friction.*
6. List the three types of friction, and describe the characteristics of each.
7. Explain the difference between *friction bearings* and *antifriction bearings.*

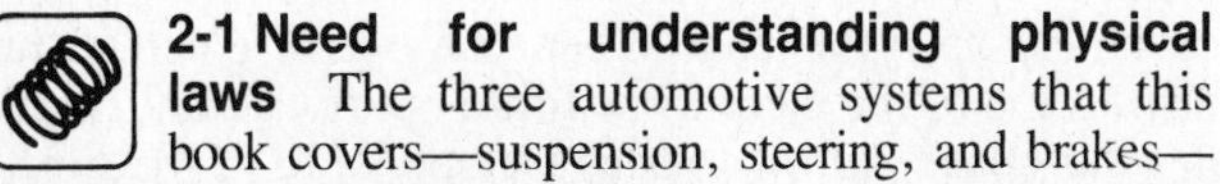

2-1 Need for understanding physical laws The three automotive systems that this book covers—suspension, steering, and brakes—work because of certain principles, or physical laws. For example, when we release a stone from our hand, the stone drops to the ground. When a vacuum exists in an engine cylinder, air rushes in as the intake valve opens.

When we step on the brake pedal, liquid is forced through tubes into the brake mechanisms at the wheel so that braking action takes place. But if air gets into the tubes, the braking action will be poor. These and many other things happen because of certain physical laws. For example, air acts one way and liquid another when pressure is applied. The principles discussed in this chapter explain why the three systems work.

2-2 Gravity Gravity makes the stone we release from our hand fall to the earth. Gravity is the attractive force that all objects have toward each other. The earth attracts the stone and pulls it toward the earth. When a car is driven up a hill, part of the engine power is used to overcome gravity so that the car can move up the hill. Likewise, a car can coast down a hill with the engine off because of the gravitational attraction of the earth on the car.

In the U.S. Customary System of measurement, gravitational attraction is usually measured in terms of *weight.* When we say that an object weighs 10 pounds, we mean that the object has enough mass for the earth to register that much pull on it. It is the gravitational attraction, or the pull of the earth, that gives an object its weight.

Weight or mass in the metric system of measurement is measured in *kilograms* (kg). For example, a 10-pound object weighs 4.54 kilograms. However, in the metric system, a pound of *force* is measured in units called *newtons* (N). One newton equals 4.448 pounds-force. Therefore, a force of 10 pounds exerts a push or pull of 44.48 newtons.

Pressure, vacuum, and heat

2-3 Atmospheric pressure Air has weight, just as any other material object has weight. Normally we do not think of air as having weight. This is because we cannot see it and because we have become accustomed to feeling its weight and movement—as, for example, when the wind (air) blows in our face. However, since air is an object, it does have weight. Air is pulled toward the earth by gravitational attraction.

At sea level, and at average temperature, a cubic foot [0.028 m^3 (cubic meter)] of air weighs about 0.08 pound [36 grams] (Fig. 2-1). This seems like a very small weight. However, the blanket of air (the atmosphere) that surrounds the earth is many miles thick. This means that there are, in effect, many thousands of cubic feet of air, piled one on top of another, all adding their weight (Fig.

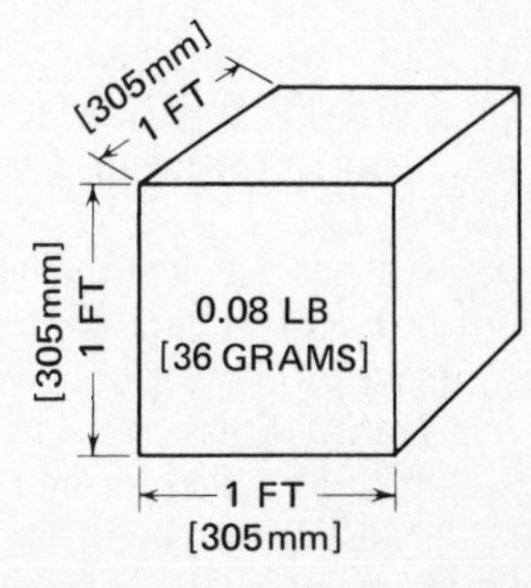

Fig. 2-1 One cubic foot of air at sea level and at average temperature weighs about 0.08 pound [36 grams].

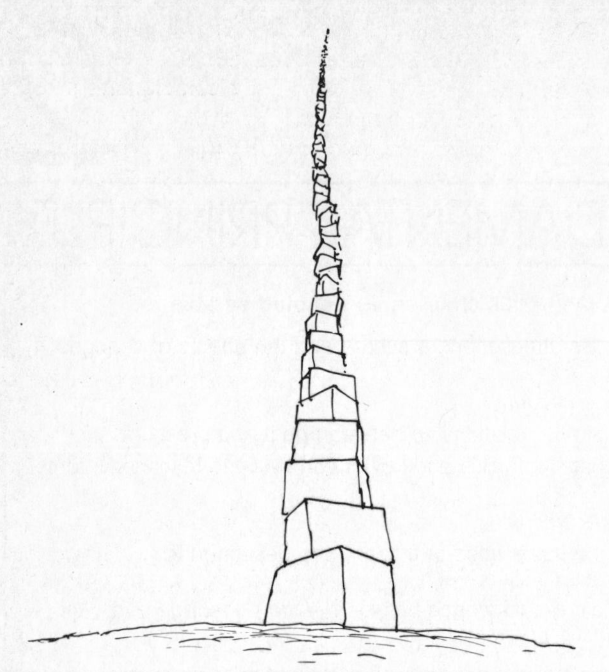

Fig. 2-2 Air pressure results from the miles of air stacked up above the earth.

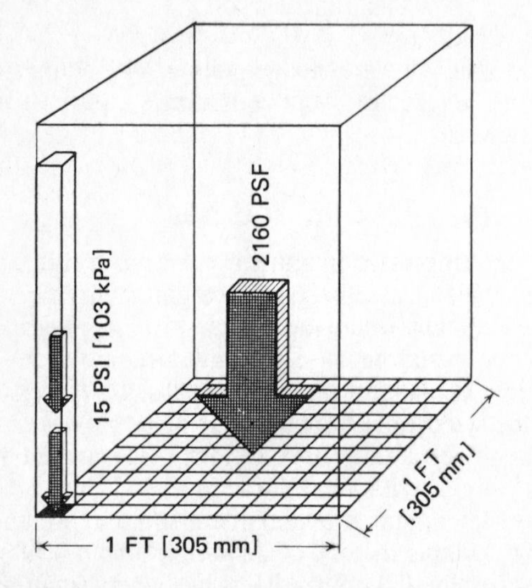

Fig. 2-3 A pressure of 15 psi [103 kPa] means 2160 pounds per square foot.

2-2). The actual weight of this air, or its downward push, at sea level is about 15 psi (pounds per square inch) [103 kPa (kilopascals)]. This is another way of saying that the atmospheric pressure (the force, or downward push, of the air) is 15 psi. And 15 psi means 2160 pounds per square foot (psf)[1] (Fig. 2-3).

Atmospheric pressure at sea level amounts to more than 2000 pounds on every square foot of the earth's surface or on any object on the earth at sea level. Since the human body has a surface area of several square feet, atmospheric pressure on the body amounts to several tons! The pressure doesn't crush you because the internal pressure of your body balances the external pressure of the air.

[1] 1 square foot = 144 square inches. $144 \times 15 = 2160$.

2-4 Changes in atmospheric pressure Atmospheric pressure is continually changing. It changes with the weather. It also changes as you move up from sea level by climbing a mountain or by flying in a plane. When you climb a mountain or fly upward in an airplane, atmospheric pressure is reduced. You put more and more of the air below you as you go up. There is less and less air above you to press down on you. At 30,000 feet [9144 m (meters)] above the earth's surface, the air pressure is less than 5 psi [34 kPa]. At 100,000 feet [30,480m], the air pressure is only about 0.15 psi [10 kPa]. You could not live at this height unless you wore a space suit which maintained a pressure around you and kept you supplied with enough air to breathe.

2-5 Vacuum Vacuum is the absence of air or other matter. When astronauts travel into space, they find no atmosphere at all. No air. At this distance from the earth there are only a very few particles of air, widely scattered. This is a vacuum.

But we do not need to leave the earth to find a vacuum. We can produce a vacuum anywhere on earth with a long glass tube closed at one end, plus a dish of mercury (a very heavy metal that is liquid at normal temperatures). To produce a vacuum, completely fill the tube with mercury and close the top end tightly. Next, turn the tube upside down, put the end into the dish of mercury, and open this end. When you open the end in the dish, some of the mercury will run out of the tube, leaving the upper part of the tube empty (Fig. 2-4). Since this upper part of the tube is closed, no air can get in. The upper part of the tube contains nothing. Therefore, it contains a vacuum.

The device shown in Fig. 2-5 is a barometer. It can be used to measure atmospheric pressure. All the mercury doesn't run out of the tube when it is turned upside down because the atmospheric pressure holds some mercury up in the tube. The atmospheric pressure presses down on the surface of the mercury in the dish. This push, transmitted through the mercury, holds the mercury up in the tube (Fig. 2-5).

The barometer indicates atmospheric pressure. When air pressure goes up, the air pushes harder on the mercury and forces it higher up in the tube. But when air pressure goes down, there is a weaker push and the mercury settles to a lower level in the tube.

2-6 Vacuum machines The barometer is one "machine" for producing vacuum. There are many devices (pumps of one sort or another) that produce vacuum. The automobile engine is, in one sense, a vacuum machine. A partial vacuum is created in the engine cylinders by the downward-moving pistons during intake strokes. As the pistons move down, atmospheric pressure pushes air toward the vacuum. The air passes through the carburetor, where it picks up a charge of fuel, and this air-fuel mixture moves on to the cylinders. Therefore, the vacuum causes a charge of air-fuel mixture to be delivered to the engine cylinders during the intake stroke.

The engine is also a compression machine. After the intake stroke, the piston moves up during the compression stroke and compresses the air-fuel mixture to one-eighth or less of its original volume.

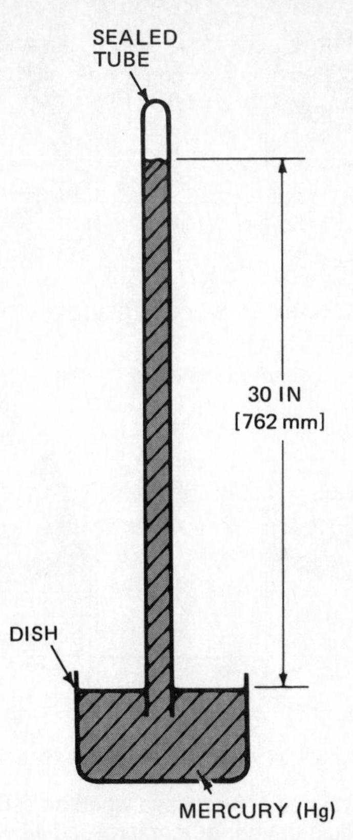

Fig. 2-4 A barometer. The mercury (Hg) in the tube will stand at about 30 inches [762 mm] above the surface of the mercury in the dish when the atmospheric pressure is 15 psi [103 kPa].

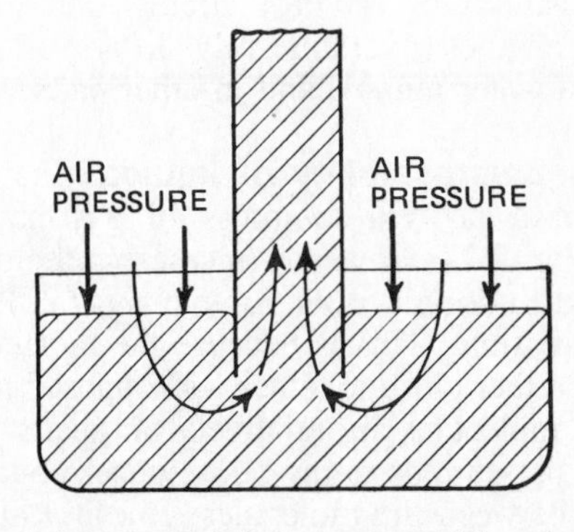

Fig. 2-5 In a barometer, the pressure of the air acts on the surface of the mercury and through the mercury to hold it up in the sealed tube.

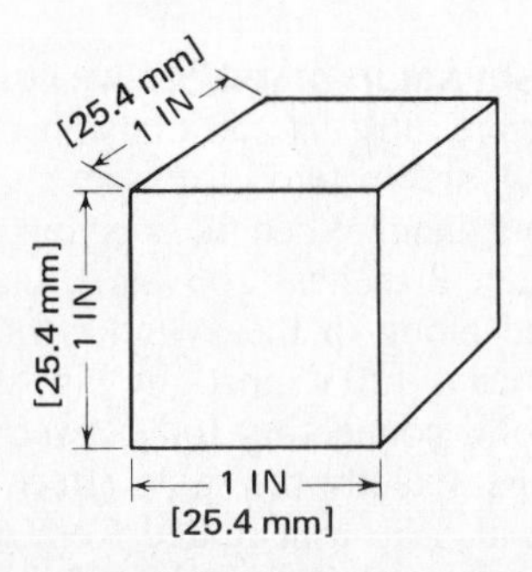

Fig. 2-6 One cubic inch of hydrogen gas at atmospheric pressure of 15 psi [103 kPa] and at a temperature of 32°F [0°C] contains about 880 billion billion atoms.

⚙ 2-7 Characteristics of air

Any liquid or gas is a *fluid,* which means it is a substance that flows easily. The fluid *air* is a gas, or vapor. Air can expand, or thin out. It can also be compressed, or packed into a smaller volume. Air is a mixture of several gases; about 20 percent is oxygen, and the rest is mostly nitrogen.

Air, or any gas or mixture of gases, is composed of tiny particles called *molecules* (combinations of atoms). These atoms or molecules are so tiny that there are literally billions upon billions of them in a cubic inch of gas. For example, in a cubic inch [16.4 cc (cubic centimeters)] of hydrogen gas at atmospheric pressure and 32°F (degrees Fahrenheit) [0°C (degrees Celsius)], there are about 880 billion billion atoms (Fig. 2-6). That can also be written as 880,000,000,000,000,000,000 atoms. Yet, the cubic inch of hydrogen is almost empty because the atoms are so very tiny.

We can prove that the cubic inch is almost empty by increasing the pressure on the gas. Suppose we put the cubic inch of hydrogen gas in a rigid box and fitted a square piston to the box. Then suppose we pushed down on the piston with a force of 150 pounds [667 N] (Fig. 2-7). The pressure would squeeze the atoms of hydrogen closer together so that the volume would be reduced to 1/10 cubic inch [1.6 cc]. As the volume of the gas decreased, its temperature would increase. This relationship between pressure and heat is discussed in ⚙ 2-10.

⚙ 2-8 Pressure

Pressure and force are closely related. A *force* is simply the push or pull on an object. The force produced by a spring when it is compressed or stretched is called *spring force*. *Pressure* is defined as the force divided by the surface area upon which the force acts, or *force per unit area*.

In the U.S. Customary System of measurement, a force such as spring force is given in pounds or ounces. Pressure is measured in *pounds per square inch,* which is usually abbreviated psi. In the metric system, force is measured in newtons (N). One pound of force equals 4.448 newtons. The metric unit for the measurement of pressure is the kilopascal (kPa). A pressure of one psi equals 6.9 kPa.

The atoms or molecules of gas are in constant motion. They have a relatively large space to move around in, and so they dart about in a constant turmoil. If the gas is enclosed in a container, molecules of the gas are constantly bumping into the inner sides of the container. There will be billions of these bumps every second in our cubic-inch container (Fig. 2-6). These billions of

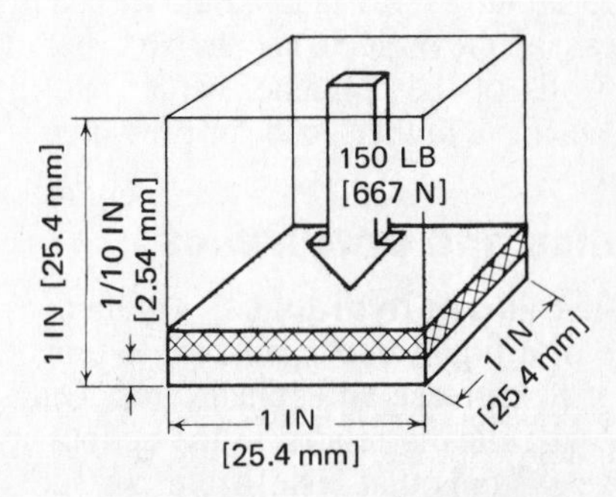

Fig. 2-7 Increasing the force to 150 pounds [667 N] decreases the volume of gas to 1/10 cubic inch [1.6 cc].

bumps—this constant bombardment—add up to the total force we know as pressure.

Now, when we compress the cubic inch into one-tenth of a cubic inch (Fig. 2-7), we have squeezed the gas molecules much closer together. Since the molecules are closer together and have much less room to move around in, they bump into the walls of the container more often. This means that the pressure has increased to 10 times as much. We started with a pressure of 15 psi [103 kPa] (Fig. 2-6) and increased it 10 times—to 150 psi [1034 kPa] (Fig. 2-7).

2-9 Heat Increasing the pressure on a gas decreases its volume. Increasing the pressure on a gas also increases its temperature. For actually, temperature, or heat, is nothing more than the speed of molecular motion. When the molecules move fast, the object is hot. When the molecules move slowly, the object is cold. In a piece of ice, the molecules are moving so slowly that they are more or less hanging together. The ice remains a solid. But if the ice is heated, it melts. The molecules begin to move a little faster, and they can no longer cling together to form a solid. Then the ice turns to water. If the water is heated, the molecules are set into even more rapid motion. Finally, they move so fast that they begin to jump clear of the liquid. The water boils, or turns to vapor.

When the cubic inch of gas is compressed to one-tenth of a cubic inch [1.6 cc], the molecules are pushed more closely together and set into more rapid motion. This is because of the more frequent collisions between molecules and more frequent collisions with the walls of the container. Since the gas molecules are in more rapid motion, the gas is hotter. The rapidly moving gas molecules, as they bombard the walls of the container, set the molecules of the container into more rapid motion. This is another way of saying that the container is heated. Outside molecules of air, as they pass the container or bump into the outer walls of the container, are then set into more rapid motion by the container molecules. Soon, the heat produced by compressing the gas has all dissipated to the container walls and then to the outside air.

2-10 Pressure increase with heat If we heat a container of air, the pressure inside the container increases. Suppose we start again with a cubic inch of air (Fig. 2-6). At 32°F [0°C], this cubic inch of air is at atmospheric pressure of 15 psi [103 kPa]. If we heated the container of gas to 100°F [37.8°C], its pressure would increase to about 17 psi [117 kPa]. In heating the gas, we have caused the molecules to move faster. They bombard the walls of the container harder and more often, thereby causing a higher push, or pressure.

Hydraulics and pneumatics

2-11 Meaning of hydraulics Hydraulics is defined as the use of a liquid under pressure to transfer force or motion, or to increase an applied force. Our special interest in hydraulics is related to the actions in the three automotive systems that result from pressure applied to a liquid. This is called *hydraulic pressure*. Hydraulic pressure is used in the brake system, in shock absorbers, and in power-steering systems. It is also used in automatic transmissions (control circuits and torque converter), in engines (hydraulic valve lifters, oil pump, fuel pump, and water pump), and in other parts.

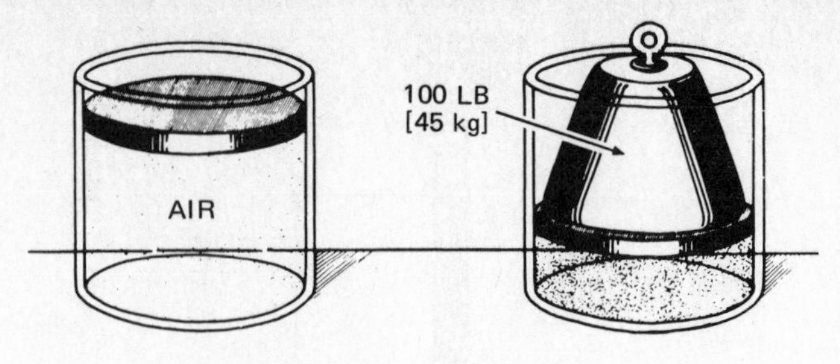

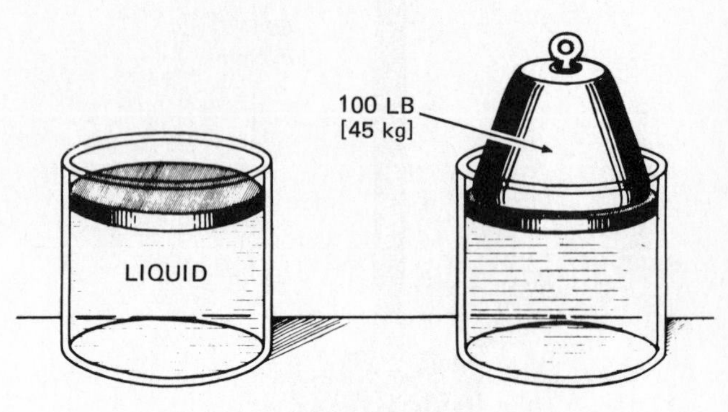

Fig. 2-8 Gas can be compressed when pressure is applied. However, a liquid cannot be compressed when pressure is applied. (*Pontiac Motor Division of General Motors Corporation*)

2-12 Incompressibility of liquids Increasing the pressure on a gas will compress the gas into a smaller volume (Fig. 2-7). However, increasing the pressure on a liquid will not reduce its volume (Fig. 2-8). A liquid is incompressible. This is because the molecules of the liquid are rather close together—as opposed to a gas, in which the molecules are relatively far apart.

Putting pressure on a gas can "squeeze out" some of the space between the molecules. But in a liquid, there is no extra space that can be squeezed out by the application of pressure. The molecules are already as close together as possible. Therefore, putting pressure on the molecules of a liquid cannot force them closer together.

2-13 Transmission of motion by liquid Since liquid is not compressible, it can transmit motion. For example, Fig. 2-9 shows two pistons in a cylinder, with a liquid separating them. When the applying piston is pushed into the cylinder 8 inches [203 mm], the output piston will be pushed along in the cylinder for the same distance, or 8 inches [203 mm]. In Fig. 2-9, you could substitute a solid connecting rod between piston A and piston B and get exactly the same effect.

Motion can also be transmitted from one cylinder to another by a tube or pipe. In Fig. 2-10, the applying piston is in cylinder A, and the output piston is in cyl-

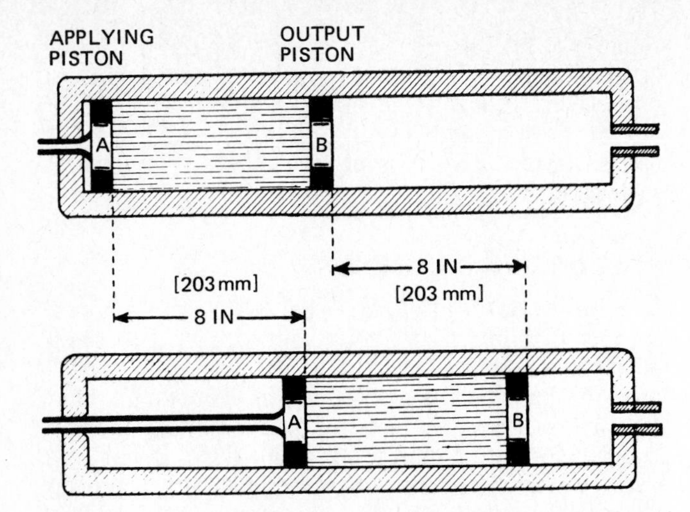

Fig. 2-9 Motion can be transmitted by liquid. When the applying piston is moved 8 inches [203 mm] in the cylinder, the output piston is also moved 8 inches [203 mm]. (*Pontiac Motor Division of General Motors Corporation*)

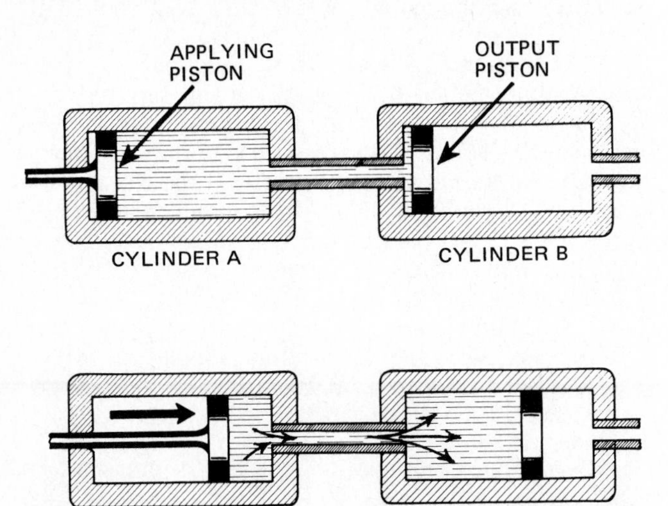

Fig. 2-10 Motion can be transmitted through a tube from one cylinder to another by a liquid, or hydraulic pressure. (*Pontiac Motor Division of General Motors Corporation*)

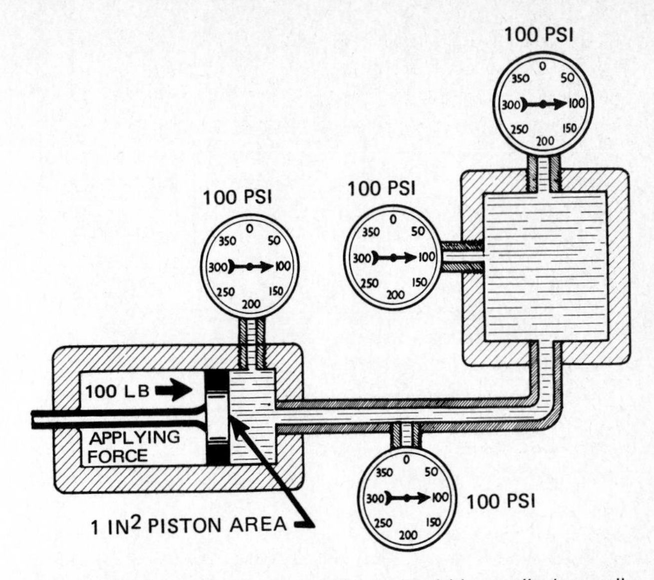

Fig. 2-11 The pressure applied to a liquid is applied equally in all directions. (*Pontiac Motor Division of General Motors Corporation*)

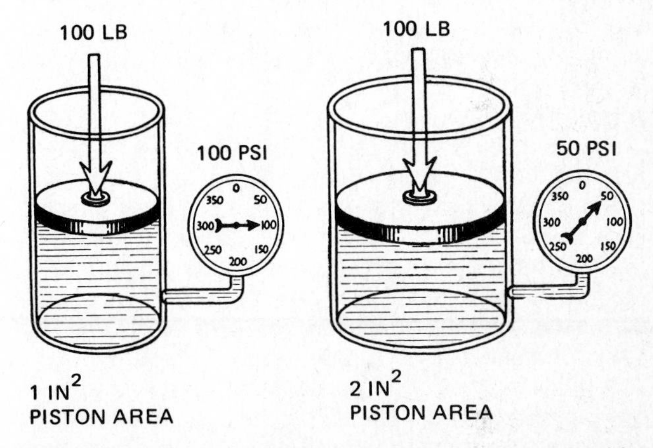

Fig. 2-12 Pressure in a hydraulic system is determined by dividing the applying force by the area of the applying piston. (*Pontiac Motor Division of General Motors Corporation*)

inder B. As the applying piston is moved into its cylinder, liquid is forced from cylinder A into cylinder B. This causes the output piston to be moved in its cylinder. If both pistons are the same size, the output piston will move the same distance as the applying piston.

2-14 Transmission of force by liquid

The force that is applied to a liquid is transmitted by the liquid in all directions and to every part of the liquid. For example, in Fig. 2-11 a piston with an area of 1 square inch [6.45 cm^2] is shown applying a force of 100 pounds [445 N]. This is a pressure of 100 psi [690 kPa]. Now suppose we attach pressure gauges to various parts of the system. We would find that the pressure on the liquid is the same at all points, regardless of where the measurement is taken.

When a piston is applying pressure to a liquid, we can calculate the pressure in pounds per square inch if we know the force being applied by the piston and the area of the piston in square inches. For example, in Fig. 2-12, when a 1-square-inch piston applies a force of 100 pounds, the pressure on the liquid is 100 psi. But if the area of the piston is 2 square inches and the piston is applying a force of 100 pounds, the pressure on the liquid is only 50 psi. Each square inch of the piston is applying a force of only 50 pounds. So the pressure is determined by dividing the applied force by the area of the piston in square inches.

In an input-output system (Fig. 2-13), the force of the output piston can also be determined. Output force is given by the equation

$$F = P \times A$$

where F = output force
P = pressure applied to piston
A = area of piston upon which pressure acts

To find the output force, multiply the applied pressure (in psi) times the area of the output piston (in square

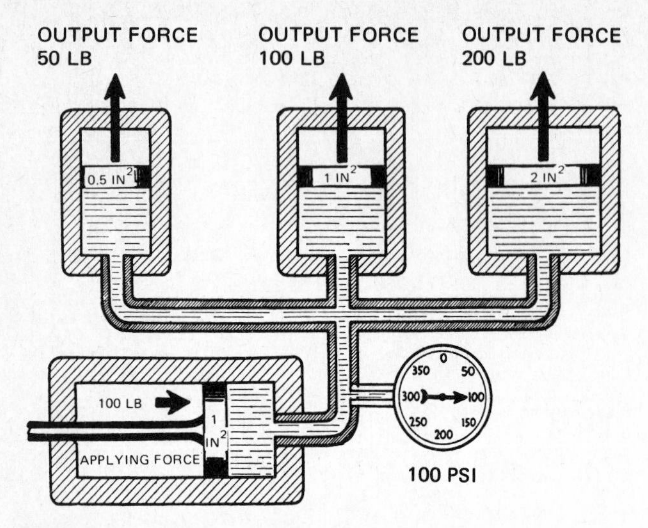

Fig. 2-13 The force applied to the output piston is the pressure in the system (in pounds per square inch) times the area of the output piston (in square inches). (*Pontiac Motor Division of General Motors Corporation*)

inches). For example, the hydraulic pressure shown in Fig. 2-13 is 100 psi [690 kPa]. The output piston to the left has an area of 0.5 square inch [3.23 cm^2]. Therefore, the output force of this piston is 100 times 0.5, or 50 pounds [222 N].

The center piston in Fig. 2-13 has an area of 1 square inch [6.45 cm^2]. Therefore its output force is 100 pounds [445 N]. The right piston has an area of 2 square inches [129 cm^2]. Its output force is 200 pounds (100 $\times$ 2) [890 N]. This shown that the bigger the output piston, the greater the output force. For example, if the area of the piston were 100 square inches [645 cm^2], then the output force would be 10,000 pounds [44,480 N].

Another way to increase the output force is to raise the applied pressure. The higher the hydraulic pressure, the greater the output force. If the hydraulic pressure on the 2-square-inch [12.9-cm^2] piston were to go up to 1000 psi [6900 kPa], then the output force on the piston would be 2000 pounds [8900 N].

In Fig. 2-13 and previous illustrations, a piston-cylinder arrangement has been shown as the means of producing the hydraulic pressure. However, any type of pump can be used. In automotive vehicles, the three most common types are gear, rotor, and vane.

❁ 2-15 Pneumatics Pneumatics means of or pertaining to air or other gases. Automobile tires are called *pneumatic tires* because they are filled with air. Any tool that is run by compressed air is a *pneumatic tool*.

Several types of pneumatic devices and systems have been used on motor vehicles—for example, air suspension (❁ 8-30), air power steering (❁ 10-25), and air brakes (Chap. 4). These devices are all operated by compressed air. Probably the most widely used pneumatic device on automobiles is the air shock absorber (❁ 9-7), which is also part of the automatic-level-control system (❁ 9-8).

Friction and lubrication

❁ 2-16 Friction Friction is the resistance to motion between two objects in contact with each other. Friction prevents one object from sliding on another. In the automobile, many parts are sliding on or rotating within other parts. Therefore, some of the power developed by the engine must be used to overcome this friction. This power is wasted because it does not contribute to moving the car.

However, friction is very valuable in the car brakes. It is the friction between the brake drums or disks and the brake shoes or pads that slows or stops the car when the brakes are applied.

❁ 2-17 Characteristics of friction Friction varies with the pressure applied between the sliding surfaces, the roughness of the surfaces, and the material of which the surfaces are made. For example, suppose that a platform with its load weighs 100 pounds [45 kg] and that it takes a force of 50 pounds [222.4 N] to pull it along the floor (Fig. 2-14). If you reduced the load so that the platform with load weighed only 10 pounds [4.5 kg], you could move it along the floor with a pull of only 5 pounds [22.2 N]. Friction varies with the load.

If you smoothed the floor and the sliding part of the platform with sandpaper, it would require less pull to move the platform on the floor. Friction varies with the roughness of the surfaces.

Friction varies with the type of material, too. For example, if you dragged a 100-pound [45-kg] bale of rubber across a concrete floor, it might require a pull of 70 pounds [311.4 N] (Fig. 2-15). But to drag a 100-pound [45-kg] cake of ice across the same floor might require a pull of only 2 pounds [8.9 N].

❁ 2-18 Coefficient of friction Engineers need a more exact way to express frictional differences than to say that one surface has a high friction and another surface a low friction. They therefore developed the idea of the coefficient of friction. This is simply a number that ac-

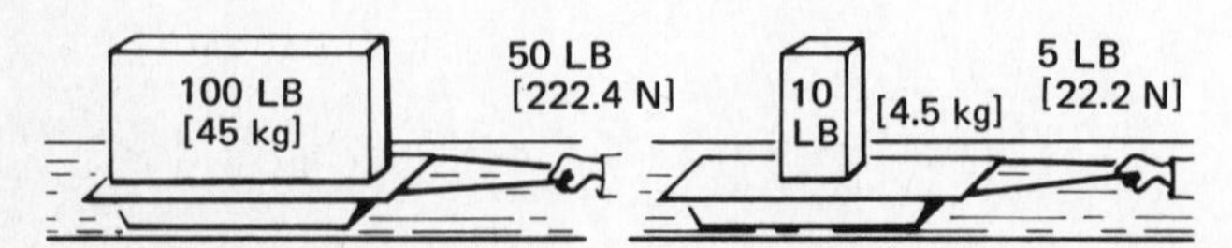

Fig. 2-14 Friction varies with the load applied between the sliding surfaces.

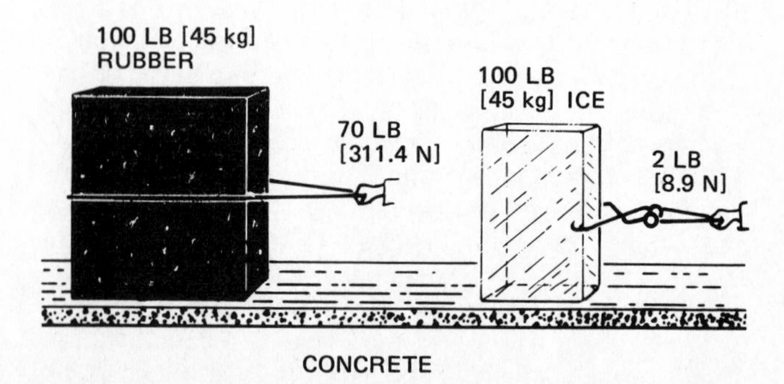

Fig. 2-15 Friction varies with the type of material.

curately states how much friction there is between two surfaces. For example, wood dragged on cast iron has a fairly high friction. A 100-pound (45.3-kg) block of wood might require a pull of 50 pounds [222.4 N] to be moved over a cast-iron slab (Fig. 2-16). To state this in terms of the coefficient of friction, divide the pull by the weight, or 50 divided by 100. This gives a coefficient of 0.5.

Knowing the coefficient, you could then determine the force required to pull a wood block of any weight over cast iron. All you have to do is multiply the weight by the coefficient. For example, a 250-pound block would require a pull of 125 pounds (25 × 0.5 = 125) [556 N].

The other example in Fig. 2-16 is that of a 100-pound [45.3-kg] block of bronze being dragged over a cast-iron slab. Here, the pull required is 20 pounds [90 N]. This gives a coefficient of friction of 0.2. If the bronze block weighed 40 pounds [18.1 kg], you could determine how much pull would be required to move it, knowing the coefficient of friction. It would be 40 × 0.2, or 8 pounds [35.6 N].

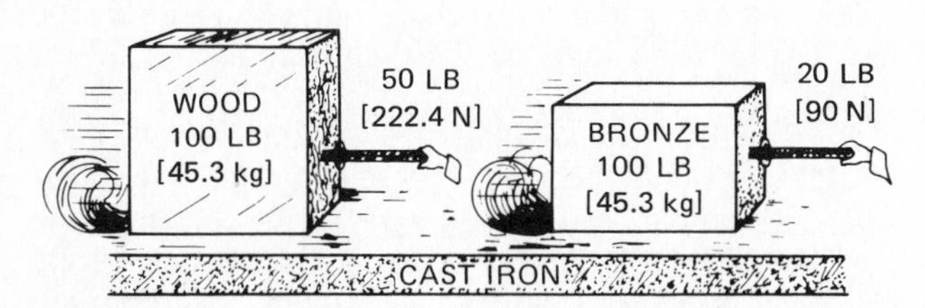

Fig. 2-16 The *coefficient of friction* is the force required to move an object divided by the weight of the object.

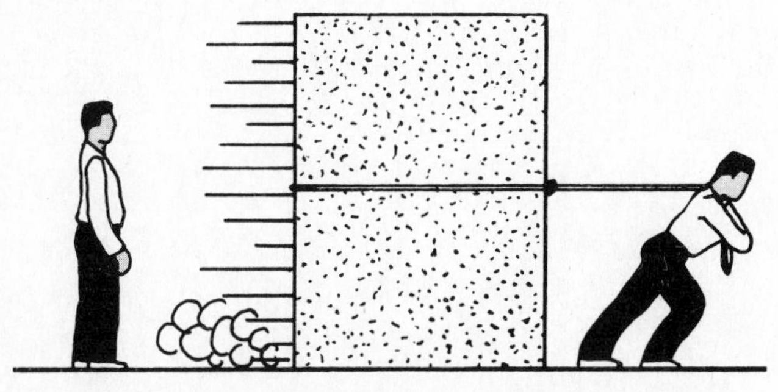

Fig. 2-17 Friction of rest is greater than friction of motion. In the example shown, it takes two people to overcome the friction of rest. But one person can keep the object moving by overcoming the friction of motion.

2-19 Friction of rest and motion

More force is required to start an object moving than to keep it in motion (Fig. 2-17). In the example shown, two people are needed to start the object moving. But once it is moving, one person can keep it moving. Therefore, the friction of an object at rest is greater than its friction in motion.

Engineers do not usually refer to these two kinds of friction as friction of rest and friction of motion. Instead, they call them *static friction* and *kinetic friction*. "Static" means at rest; "kinetic" means in motion, or moving. Therefore, static friction is friction of rest, and kinetic friction is friction of motion.

2-20 Causes of friction

One explanation of friction is that it is caused by surface irregularities. High spots on two surfaces in contact tend to catch on each other and hinder the motion between the two objects. When the surfaces are smoothed off, the high spots are cut down and there is less tendency for them to catch on each other. Friction is reduced. However, if the force between the two surfaces is increased, the high spots are pressed harder against each other, so that the friction is increased.

The fact that static friction is greater than kinetic friction can be explained similarly. When the surfaces are at rest, the force or weight between them tends to press the high spots of one surface into the other surface. Then it takes considerable force to move all the high spots of one surface up and out of the low spots of the other surface. But once moving, the high spots do not have a chance to "settle" into the opposing surface. Less force is required to keep the surfaces moving. Therefore, kinetic friction is less than static friction.

2-21 Friction in the car brakes

Friction is used in the car brake system. The friction between the brake drums or disks and brake shoes, or pads, slows or stops the car. This friction slows the rotation of the wheels. Then friction between the tires and road slows the motion of the car.

Note that it is the friction between the tires and the road that results in the car's stopping. The car does not stop more quickly if the wheels are locked so that the tires skid on the road. If the brakes are applied so hard that the wheels lock, the friction between the tires and road is kinetic friction (friction of motion as the tires skid on road). When the brakes are applied a little less hard, so that the wheels are permitted to continue rotating, the friction between the tires and the road is static friction. The tire surface is not skidding on the road but is rolling on it. Since this produces static friction between the road and tires, there is considerably greater braking effect.

A car will stop more quickly if the brakes are applied just hard enough to get maximum static friction between the tires and the road. If the brakes are applied harder than this, the wheels will lock. Then the tires will slide, and lower kinetic friction will result.

2-22 Classes of friction Up to now, we have been discussing *dry friction,* or the friction between two dry surfaces. There are two other classes of friction: greasy friction and viscous friction. These are the ones found most often in the moving parts of the automobile.

1. Greasy friction If we thinly coat the moving surfaces of two objects with a lubricant, such as grease or oil, the friction is greatly reduced. It is assumed that this thin coat tends to fill in the low spots of the surfaces. Therefore, the surface irregularities have less tendency to catch on each other. However, high spots will still catch and wear as the two surfaces move over each other. In automobile engines, greasy friction may occur when the automobile is starting. At this time, most of the lubricating oil has drained away from bearings, piston rings, and cylinder walls. When these surfaces first start to slide over each other, there is only a thin film to protect them against wear. However, the lubricating system quickly starts to pump oil to the moving surfaces to provide additional protective lubrication.

2. Viscous friction Viscosity refers to the tendency of liquids, such as oil, to resist flowing. A heavy oil is thicker, or more viscous, than a light oil. The heavy oil flows more slowly (has higher viscosity, or higher resistance to flowing). Viscous friction is the friction, or resistance to motion, between adjacent layers of liquid. As applied to machines, viscous friction occurs during relative motion between two lubricated surfaces. Figure 2-18 illustrates, in greatly exaggerated view, an object moving over a stationary object, the two being separated by lubricating oil. The oil is shown in five layers, A to E, for simplicity.

Layer A (in Fig. 2-18) adheres to the moving object and moves at the same speed. A layer of oil E adheres to the stationary object and is therefore stationary. However, there must be relative motion between the layers of oil A and E. This can be pictured as a slippage between many layers of oil between A and E. The nearer a layer is to the stationary layer E, the less it moves. This is shown by the progressively shorter arrows in layers B, C, and D. There is slippage between these layers. But there is resistance to this slippage, and this is called *viscous friction.*

2-23 Friction and wear When dry or greasy friction exists, the moving parts are in contact with each other. This means that the high spots are interfering with each other. They catch on each other, and particles of the material are torn off. Wear takes place. These tiny particles then add to the wear by scratching and gouging the moving surfaces. Soon, the roughness of the surfaces is increased, and wear goes on at a progressively swifter pace. To prevent this sort of wear, moving parts in machines are coated with oil or grease. The lubricant holds the moving surfaces apart so that only viscous friction results. Wear is kept to a minimum.

Bearings

2-24 Bearings A *bearing* is the part that transmits the load to a support, and in so doing absorbs the friction of the moving parts. To reduce friction and wear, the bearings must be supplied with lubricant such as oil or grease.

Machine bearings are classified as either *friction bearings* or *antifriction bearings.* These two names are somewhat misleading, since they seem to indicate that one type has friction while the other has not. Actually, the friction bearing does have greater friction, but both are low-friction devices. Figure 2-19 shows the difference between friction and antifriction bearings. In the friction bearing, one body slides over another. The load is supported on layers of oil, as shown in Fig. 2-18. In the antifriction bearing, the surfaces are separated by balls or rollers. The result is rolling friction between the two surfaces and the balls or rollers.

2-25 Friction bearings Friction bearings have sliding contact between the moving surfaces. The load is supported by layers of oil. There are three types of friction bearing (Fig. 2-20): journal, guide, and thrust. The journal type is symbolized by two hands holding a turning shaft, as shown in the upper left of Fig. 2-20. The hands support the shaft just as the surrounding bearing supports a shaft journal in an engine. Crankshaft, connecting-rod, camshaft, and piston-pin bearings are examples of this type of bearing used in the engine.

The bearing surface between the piston and cylinder wall is of the guide type (center, Fig. 2-20).

The thrust type of friction bearing checks endwise movement of the shaft (right, Fig. 2-20). The flats on the ends of the bearing are parallel to flats at the end of

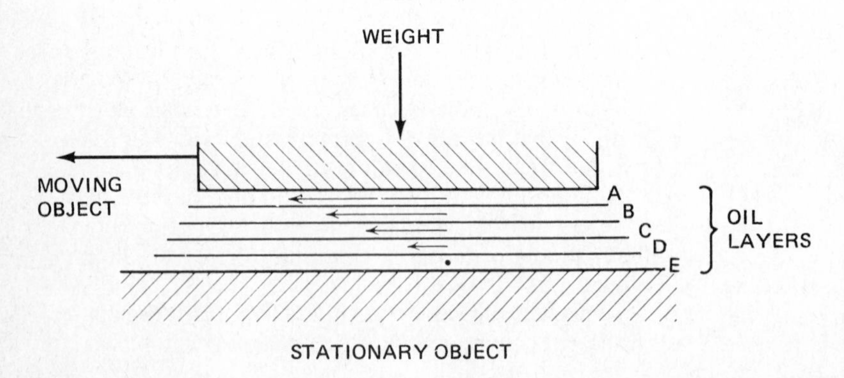

Fig. 2-18 Viscous friction is the friction between layers of liquid moving at different speeds. In the illustration, viscous friction is the friction between layers A, B, C, D, and E.

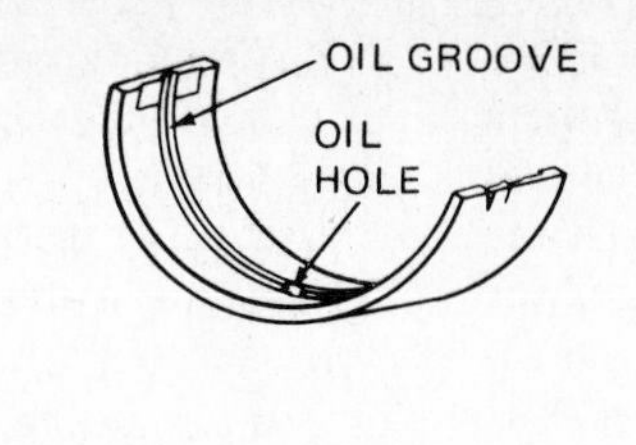

FRICTION (SLIDING) BEARING

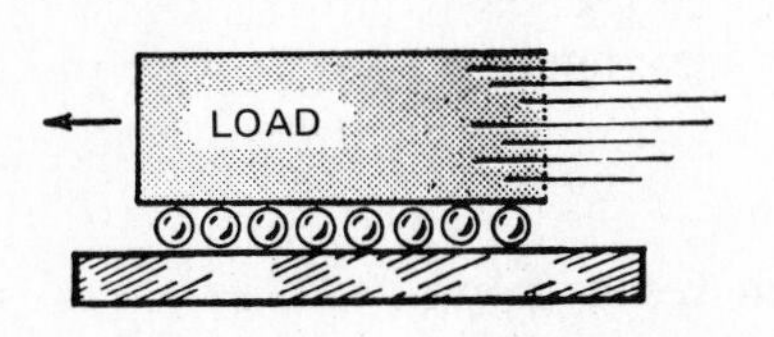

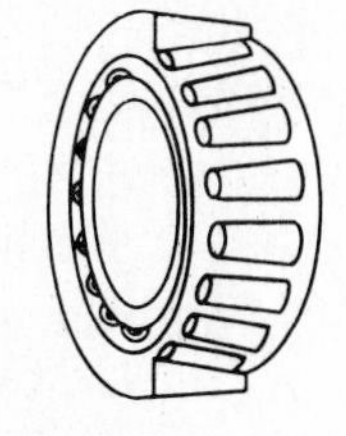

ANTIFRICTION (ROLLING) BEARING

Fig. 2-19 Friction and antifriction bearings. The friction bearing has sliding contact between its surfaces. The antifriction bearing has rolling contact.

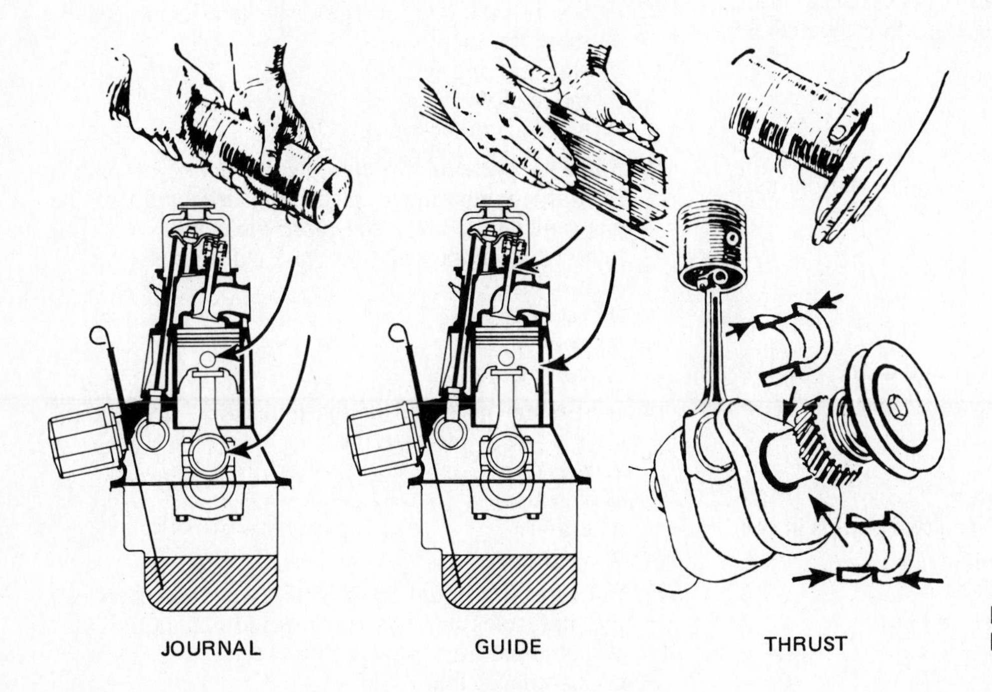

Fig. 2-20 Three types of friction-bearing surfaces.

the shaft journal. As the shaft attempts to move endwise, the flats, which are called *thrust faces,* prevent it.

2-26 Antifriction bearings Figure 2-21 shows three types of antifriction bearing: the ball, roller, and tapered roller. There are other types, including a spherical roller, thrust, and double-row ball. But all operate on the same principle by having a rolling object between the moving surfaces.

The ball bearing has an inner and an outer race in which grooves have been cut. Balls roll in these two race grooves and are held apart by a spacer assembly. When one race is held stationary (for example, by mounting it in a housing) and the other rotates (as it might when on a shaft), the balls roll in the two races to permit low-friction rotation.

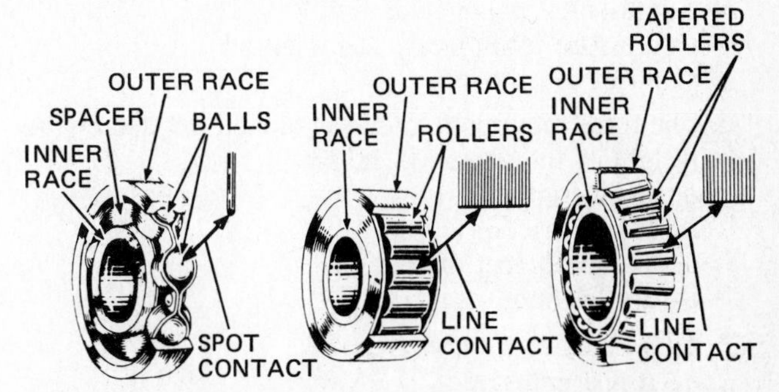

Fig. 2-21 Three types of antifriction bearings. These have rolling contact between the surfaces.

The roller bearing is similar to the ball bearing. However, the roller bearing has either plain or tapered rollers which roll between the inner and outer races. In the ball bearing, there is spot contact between the balls and the races. The roller bearing has line contact (the length of the roller) between the rollers and the races.

Antifriction bearings are usually lubricated by grease, which is essentially oil mixed with a solidifying agent (called soap). The solidifying agent does not contribute directly to the lubricating of the balls or rollers. But it does hold the oil in the bearing so that the bearing receives proper lubrication.

Chapter 2 review questions

Select the *one* correct, best, or most probable answer to each question. Then check your answers against the correct answers given at the end of the book.

1. Air is compressible but liquid is not because, in the liquid, the molecules are about as close together as:
   a. in the air
   b. they can be
   c. in a vacuum
   d. none of the above
2. Since liquids are incompressible, they can be used to transmit:
   a. force and motion
   b. pressure and friction
   c. static and kinetic friction
   d. all of the above
3. When a pressure is applied to a liquid, the pressure:
   a. increases with distance
   b. is reduced with distance
   c. is the same at all points
   d. all of the above
4. The resistance to motion between two objects in contact with each other is called:
   a. friction
   b. braking
   c. coefficient
   d. hydraulics
5. Friction between two surfaces varies with the:
   a. thickness and pressure
   b. pressure, weight, and pull
   c. pressure, roughness, and material
   d. all of the above
6. The force required to move an object, divided by the weight of the object, is called:
   a. static friction
   b. kinetic friction
   c. the coefficient of friction
   d. dry friction
7. Two kinds of friction are:
   a. at rest and static
   b. moving and kinetic
   c. static and kinetic
   d. all of the above
8. Three classes of friction are:
   a. dry, greasy, and viscous
   b. greasy, thin, and thick
   c. dry, wet, and viscous
   d. all of the above
9. In a comparison of friction and antifriction bearings, the one with the lower friction is the:
   a. friction bearing
   b. antifriction bearing
10. Two types of antifriction bearings are:
    a. guide and thrust
    b. ball and sleeve
    c. ball and roller
    d. none of the above
11. Atmospheric pressure results from:
    a. absence of vacuum
    b. compression of gas
    c. gravity
    d. none of the above
12. As a solid, a liquid, or a gas is heated, its molecules:
    a. move faster
    b. stop moving
    c. move slower
    d. none of the above
13. Since the pressure in a closed container of gas results from the bombardment of the container walls by the gas molecules, crowding more molecules in a container (by compressing the gas) will increase the:
    a. pressure
    b. vacuum
    c. gravity
    d. volume
14. Increasing the speed of the molecules—by heating the gas—will increase the:
    a. pressure
    b. vacuum
    c. gravity
    d. volume
15. Because in a liquid there is no extra space between the molecules that can be squeezed out, liquid is:
    a. compressible
    b. incompressible
    c. incompatible
    d. solid
16. The pressure applied to a liquid by a piston, in pounds per square inch, is the force on the piston divided by the:
    a. distance it moves
    b. piston area in square inches
    c. piston diameter
    d. none of the above
17. Comparing the two kinds of friction (static and kinetic), other things being equal, static friction is always:
    a. less
    b. equal
    c. greater
    d. none of the above

18. When you apply the brakes on your car so hard that you lock the wheels so they do not turn and the tires skid, the friction between the tires and road is:
    a. static friction
    b. kinetic friction
    c. both *a* and *b*
    d. neither *a* nor *b*
19. In a comparison of the three classes of friction, other things being equal, the class that offers the lowest friction is:
    a. dry friction
    b. greasy friction
    c. viscous friction
    d. wet friction
20. In comparing the actions of friction and antifriction bearings, the main difference is between:
    a. sliding and slipping contact
    b. spot and line contact
    c. sliding and rolling contact
    d. none of the above

CHAPTER 3
AUTOMOTIVE BRAKES

After studying this chapter, you should be able to:

1. List four types of parking brakes.
2. Explain the difference between the service brake and the parking brake.
3. Describe the basic construction and operation of the drum brake.
4. Describe the basic construction and operation of the disk brake.
5. Describe the basic operation of the hydraulic-brake system.
6. Name the three types of disk brakes, and describe the construction of each.
7. List the valves used on a car with front-disk and rear-drum brakes, and describe the operation of each.
8. Explain the construction and operation of the antilock brake system.
9. Discuss the types of brake fluid, and the differences among the types.
10. Explain why duo-servo drum brakes are self-energizing, and why disk brakes are not.

3-1 Automobile braking systems A brake is an energy-conversion device that is used to slow, stop, or hold the car. In operation, the brake uses friction to change the kinetic energy of motion into useless heat energy.

Figure 3-1 shows the complete layout on the chassis of the automobile braking system. It is made up of four major parts. First, there must be a source of energy to provide the energy required for braking. On vehicles, brakes may be operated by mechanical, hydraulic, pneumatic, or electrical devices. Second, there must be a control device to apply and release the brake. The foot pedal, operated by the driver, is the control for the service brake. The parking brake has a separate control. It may be either a hand lever or a foot pedal which latches until it is later released.

The third part that a brake system must have is some way to transmit the desired force and motion from the control device to the brake mechanism. Figure 3-1 shows the two methods used on cars. The service brake is operated hydraulically by forcing fluid through a tube or hose. This applies force to the friction elements at the brake. The parking brake is mechanically operated. Force and motion are transmitted from the control to the brake through rods and cables.

The fourth part of the brake system is the brake mechanism itself. On the car, a brake usually is found at each wheel. The brake contains the friction elements which are acted upon to oppose the motion of the vehicle by retarding the movement of the wheel.

Two completely independent braking systems are used on the car (Fig. 3-1). They are the *service brake* and the *parking brake*.

1. Service brake. The service brakes act to slow, stop, or hold the vehicle during normal driving as required by the driver. They are foot-operated by the driver depressing and releasing the brake pedal.
2. Parking brake. The primary purpose of the parking brake is to hold and maintain the vehicle in a stationary position while it is unattended. The parking brake is mechanically operated by the driver when a separate parking-brake foot pedal or hand lever is set.

Basically, all car brakes are friction brakes (Fig. 3-2). When the driver applies the brake, the control device forces brake shoes or pads against the rotating brake drums or disks at the wheels. Friction (❁ 2-16 to 2-21) between the shoes or pads and the drums or disks then slows or stops the wheels so that the car is braked.

❁ 3-2 Types of friction brakes In a friction brake (❁ 3-1), the brake lining wears away during normal operation. This is because parts which do not rotate (shoes and pads in Fig. 3-2) are forced against parts which rotate with the wheels (drum and disk in Fig. 3-2). Friction between these two parts causes the car to slow in proportion to the force applied.

Two basic types of friction brakes are used on automobiles. These are the drum brake and the disk brake. Both types are illustrated in Fig. 3-2. Older cars had

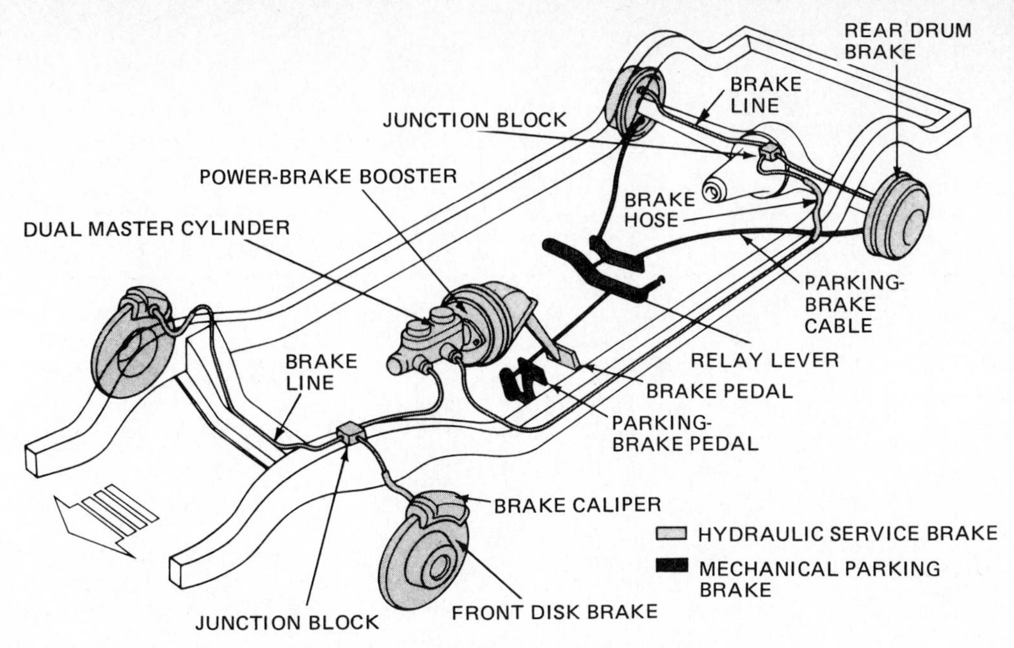

Fig. 3-1 Layout of the complete brake system on an automobile chassis. (*Texaco, Inc.*)

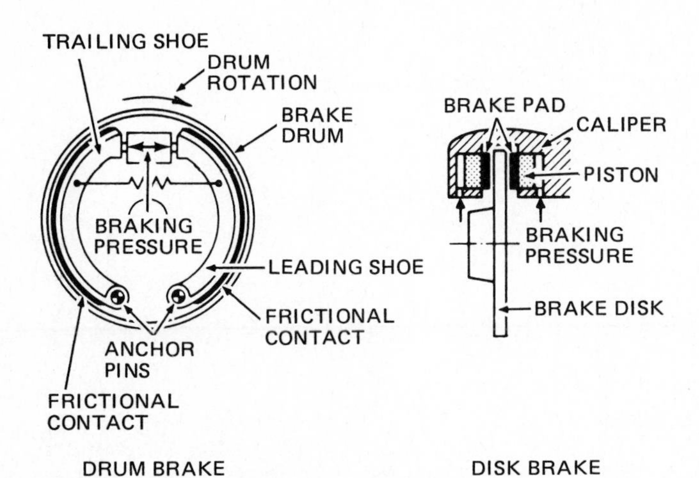

Fig. 3-2 Two basic types of friction brakes used on automobiles. (*Robert Bosch Corporation*)

drum brakes at all four wheels. Later, after disk brakes were introduced, most automotive manufacturers began using the combination drum-and-disk system shown in Fig. 3-1.

In the combination system, disk brakes are used at the front wheels and drum brakes at the rear. By continuing the use of rear drum brakes, the parking brake was also easily provided. Now many cars have disk brakes at all four wheels. The various types of parking brakes used with the different braking systems are described in ⚙ 3-5.

1. **Drum brakes** The drum brake has two curved shoes positioned inside a drum (Fig. 3-3). When the brakes are applied, the brake shoes are forced outward (Fig. 3-2) and into frictional contact with the rotating brake drum. The wheel mounts on the drum, so the wheel and

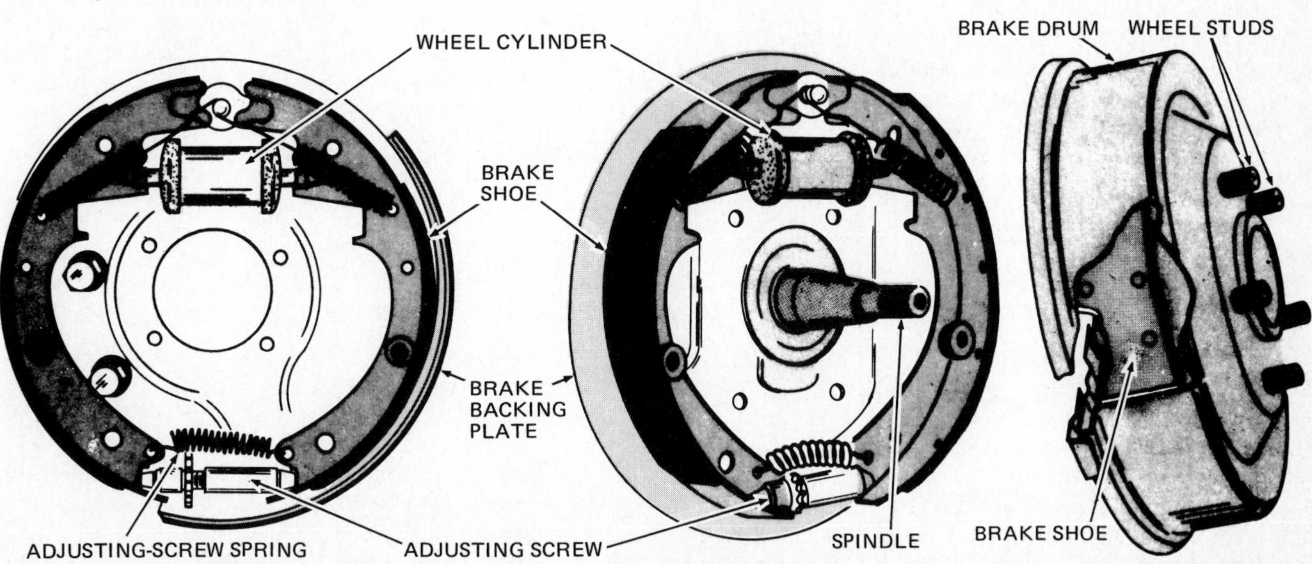

Fig. 3-3 Basic construction of the drum brake. Right, the brake drum has been partly cut away to show the shoe inside.

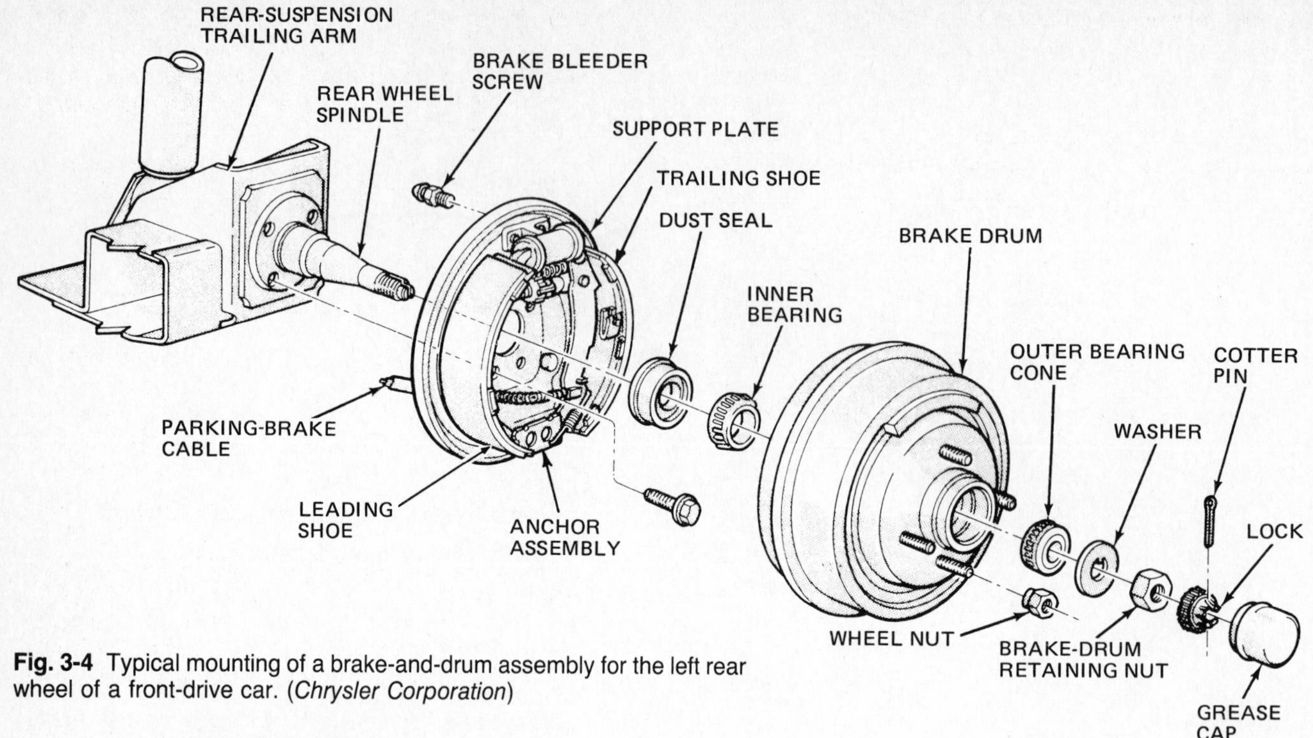

Fig. 3-4 Typical mounting of a brake-and-drum assembly for the left rear wheel of a front-drive car. (*Chrysler Corporation*)

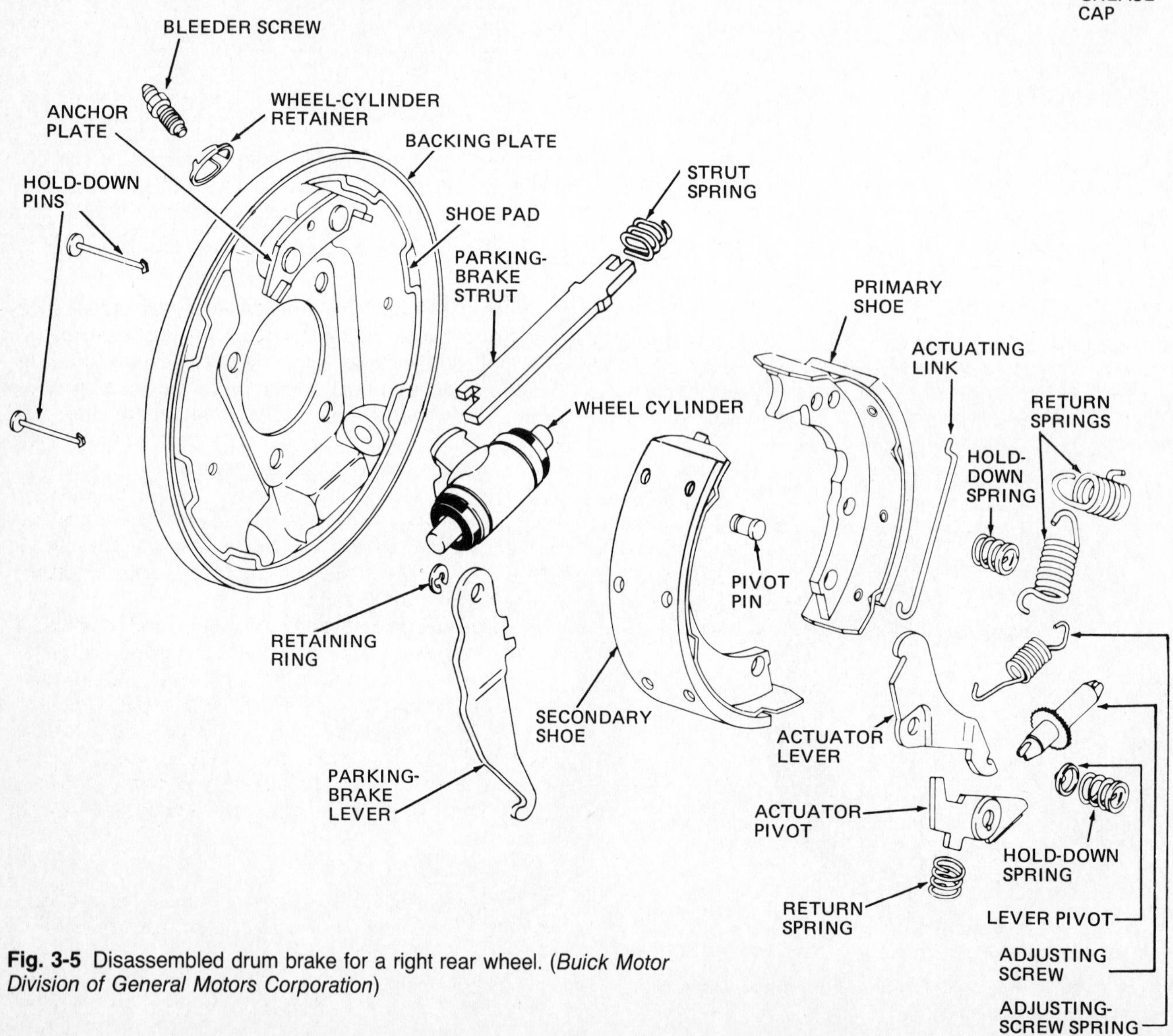

Fig. 3-5 Disassembled drum brake for a right rear wheel. (*Buick Motor Division of General Motors Corporation*)

drum rotate together. When the brake shoes are forced against the rotating drum, the friction between the two causes the drum and wheel to slow or stop. The same action takes place at all wheels, so that the car is braked.

Figure 3-4 shows how the drum brake attaches to the suspension at the left rear wheel. When the brake is assembled, the drum fits around the shoes as shown in Fig. 3-3. Figure 3-5 shows the disassembled drum brake for a right rear wheel.

2. Disk brakes In the disk brake, a disk, or rotor, is attached to the wheel. Brake pads (also called *linings* or *shoes*) are positioned on each side of the disk (Figs. 3-2 and 3-6). When the brakes are applied, the pads press against the disk as shown to the right in Fig. 3-6. This clamps the disk between the pads. Friction between the pads and the disk slows or stops the wheel, thereby providing braking action.

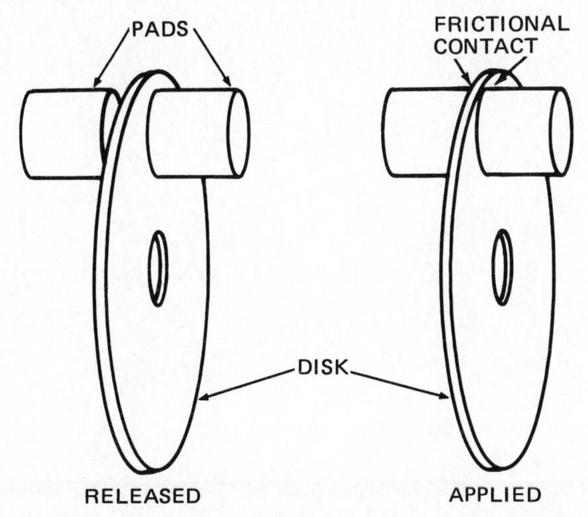

Fig. 3-6 Principle of disk brakes. Left, two brake shoes, or pads, are positioned on the two sides of the rotating disk. When the brakes are applied, the pads are moved into frictional contact with the disk, producing the braking action.

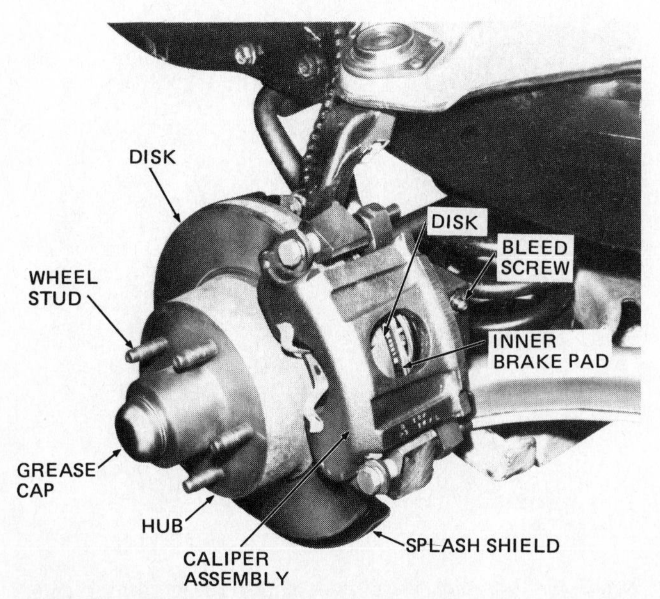

Fig. 3-7 Disk brake on the left front wheel of a rear-drive car. (*Ford Motor Company*)

Figure 3-7 shows a disk brake on the left front wheel of a rear-wheel-drive car. Notice that the thickness of the pad, or lining, can be seen through the hole in the center of the caliper. Figure 3-8 shows a disassembled front-wheel disk brake.

The construction and operation of the various types of disk brakes are described in ✪ 3-20. Drum brakes are discussed further in ✪ 3-16 to 3-19.

Brake-system components

✪ 3-3 Drum-brake lining A typical drum-brake shoe is shown in Fig. 3-9. The lining for this type of shoe is a thick, curved strip of asbestos or similar friction material. It is attached to the brake shoe either by rivets or by bonding with a special cement (Fig. 3-10). The friction material can withstand the heat and dragging effect imposed when the shoes are forced against the brake drum. During hard braking, the shoe may be forced against the drum by a pressure of 1000 psi [6895 kPa] or higher. Since friction increases as the pressure increases (✪ 2-7), a strong frictional drag is produced on the brake drum. Therefore a strong braking effects results at the wheel.

A large quantity of heat is produced by the frictional contact between the brake shoes and the drum. Under heavy braking conditions, drum-brake temperatures may reach 500°F [260°C]. Some heat flows through the brake linings to the shoes and backing plate, where the heat is carried away by the surrounding air. But most of the heat is absorbed by the brake drum. Some brake drums have cooling fins that provide additional surface for dissipating the heat more readily (Fig. 3-11). To help provide drum cooling, some brake linings are grooved at the center.

Excessive temperature may damage brakes and burn the brake linings. When the linings and drums are excessively hot, the coefficient of friction changes and less effective braking action results. This is the reason that brakes ''fade'' when they are used continuously for relatively long periods of time, such as when coming down a mountain or a long, steep downgrade.

Brake fade is the temporary reduction of braking effectiveness due to a loss of friction between the braking surfaces (the linings and the drums), resulting from heat. Water in the brakes, between the linings and the drum, may also cause brake fade.

To help overcome the problem of fade, brakes on some special-purpose vehicles have been equipped with metallic linings. Instead of asbestos or similar friction materials, the shoes have a series of metallic pads attached (Fig. 3-12). Metallic linings can withstand more severe braking and higher temperatures, with less tendency to fade. However, metallic linings tend to wear the drums quickly, to apply harshly under some conditions, and to be noisy.

✪ 3-4 Disk-brake lining In disk brakes, the disk disposes of most of the heat. Only a small part of the disk is clamped between the brake pads, and this is where the heat develops. The rest of the disk is circulating in open air and can dispose of its heat easily to the passing cooler

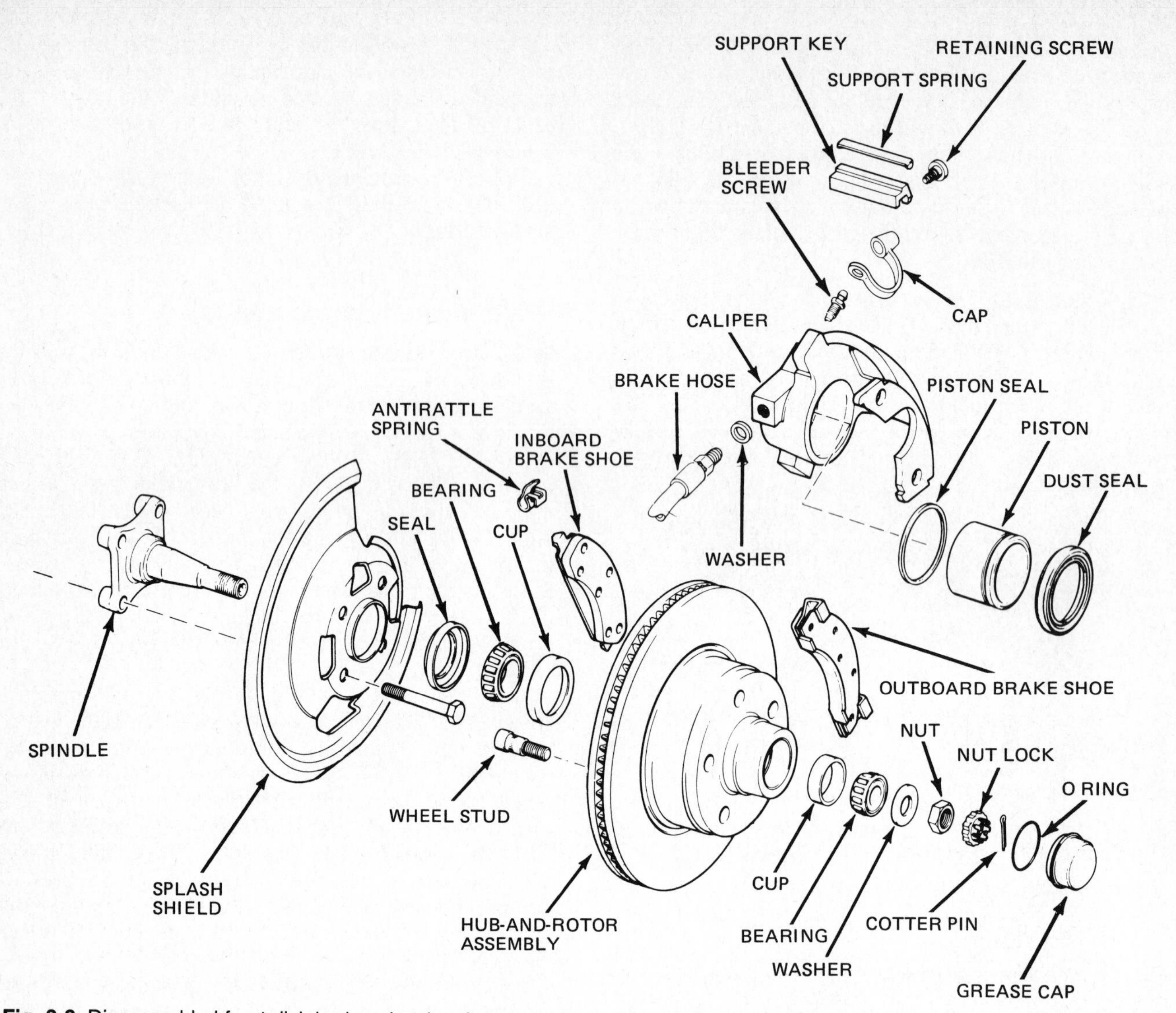

Fig. 3-8 Disassembled front disk brake, showing the components. (*American Motors Corporation*)

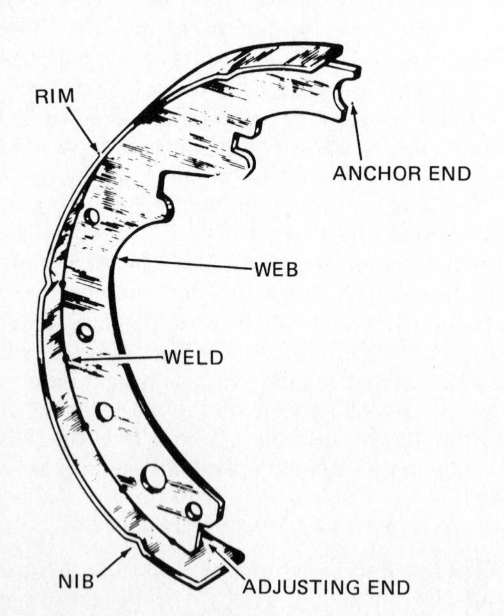

Fig. 3-9 Typical drum-brake shoe, without lining. (*Bendix Corporation*)

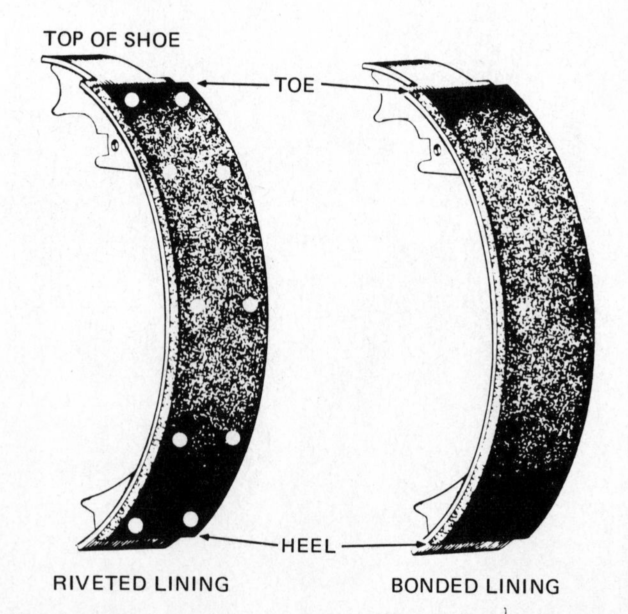

Fig. 3-10 Two methods of attaching the lining to the shoe. (*Bendix Corporation*)

Fig. 3-11 Brake drum with cooling fins that provide additional surface for dissipating heat. (*Oldsmobile Division of General Motors Corporation*)

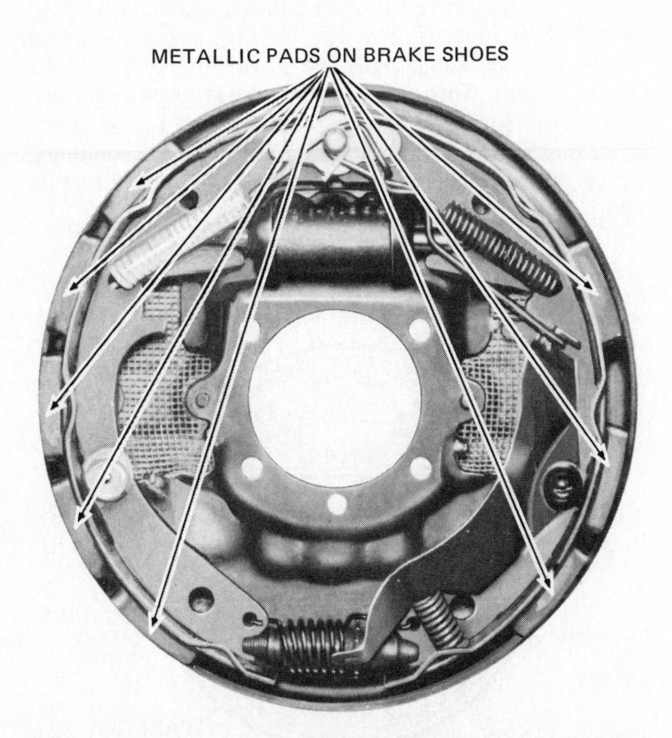

Fig. 3-12 Drum-brake assembly using metallic pads instead of brake linings. (*Chevrolet Motor Division of General Motors Corporation*)

air. Some disks have air-cooling passages, or *louvers,* which aid in dissipating the heat (Fig. 3-13).

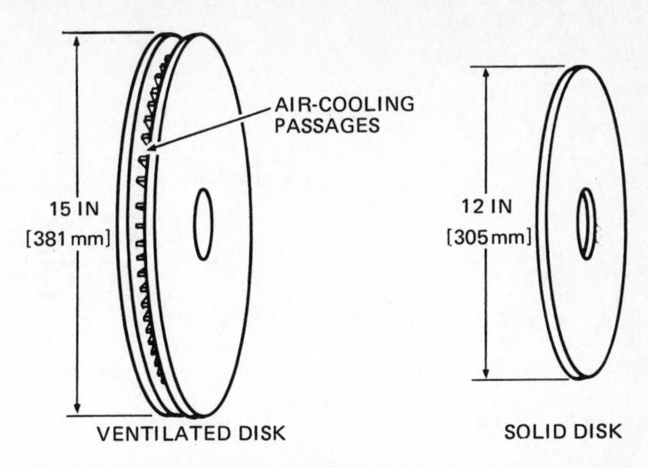

Fig. 3-13 For larger cars, the brake disks are 15 inches [381 mm] in diameter. Smaller cars use a 12-inch [305-mm] disk. (*ATW*)

Until recent years, brake pads for disk brakes had linings of organic (nonmetallic) material, such as asbestos. However, this material is not satisfactory for the latest disk-brake designs. Today disk-brake pad temperatures may reach 1000°F [538°C].

Earlier disk brakes have disks 15 inches [381 mm] in diameter, with air-cooling passages (Fig. 3-13). The disks used on late-model downsized cars are 12 inches [305 mm] in diamater and are solid (Fig. 3-13). This reduces the heat-dissipating ability of the disk. Also, many of the new small cars have front-wheel drive. With the engine and drive train at the front of the car, more of the car weight transfers to the front wheels during braking. This places a greater load on the smaller, unventilated front-disk brakes.

To provide effective braking, linings of semimetallic material are used. These materials can hold up under the higher temperatures, while not showing the undesirable characteristics of metallic linings (⚙ 3-3). Whenever disk-brake pads are replaced, the correct lining must be used. An organic pad installed in a disk brake for which semimetallic lining is specified may have less than one-third the life of the original lining.

⚙ 3-5 Parking brakes Figure 3-1 shows the typical layout of the complete automobile braking system on the chassis. In addition to the hydraulic foot-operated service brake, the car has a separate mechanical *parking brake.* It may be operated by a foot pedal (Fig. 3-1) or by a hand lever (Fig. 3-14).

The parking brake holds the car stationary while it is parked. Since the parking brake is independent of the service brakes, it can be used as an emergency brake if the services brakes fail. When the parking brake is operated by a hand lever, some manufacturers call it the *hand brake.*

A parking brake is designed to hold one or more brakes continuously in an applied position. There are two basic types of parking brakes. These are known as *integral* and *independent.*

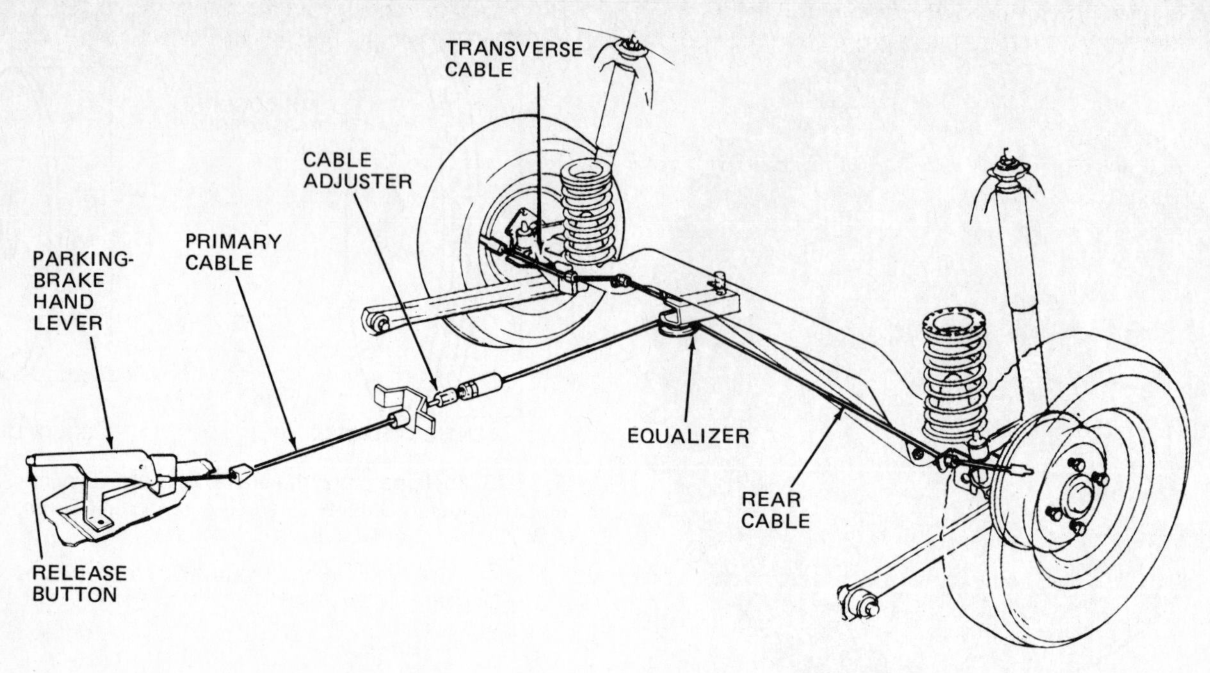

Fig. 3-14 Typical parking-brake system operated by a hand lever. (*Ford Motor Company*)

1. **Integral parking brakes** Integral parking brakes have parts which are common to both the parking-brake system and the service-brake system. For example, applying the parking brake may force the rear-wheel brake shoes or pads against the drum or rotor, just as the service brake does. However, the service brake operates the shoes or pads hydraulically, while the parking brake uses a separate mechanical system to transmit the braking force. Most cars built today have an integral parking brake.

Figures 3-5 and 3-15 show the parking-brake components for the right rear wheel of a drum brake. The parking-brake operating lever is located in back of the secondary shoe, and is attached to the shoe by a pivot pin at the upper end. A strut rod, located below the pivot pin, extends forward from the brake lever to the primary shoe.

The parking-brake cable is connected to the lower end of the brake lever (Fig. 3-15). When the parking brake is applied, the cable pulls the lower end of the brake lever forward. This causes the strut rod to push the primary shoe forward, while the upper end of the brake lever pushes the secondary shoe rearward. The combined action of the brake lever and the strut rod forces the shoes apart and into contact with the drum.

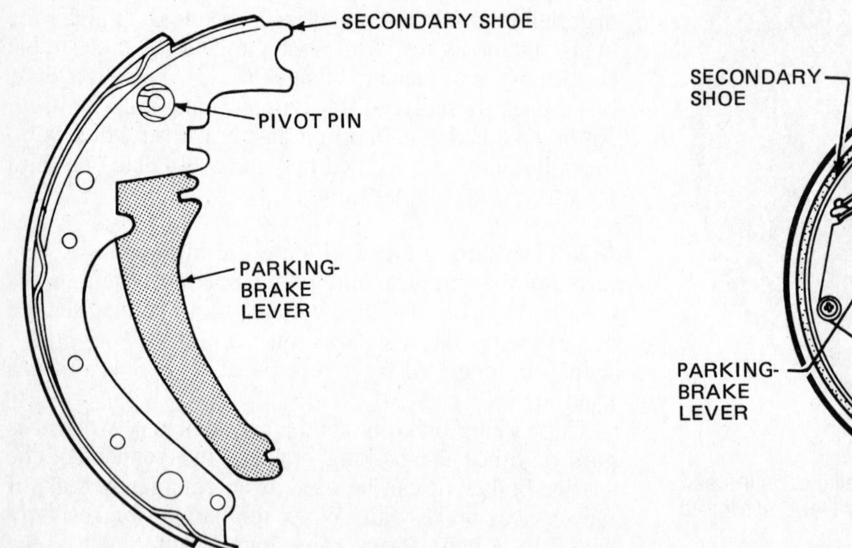

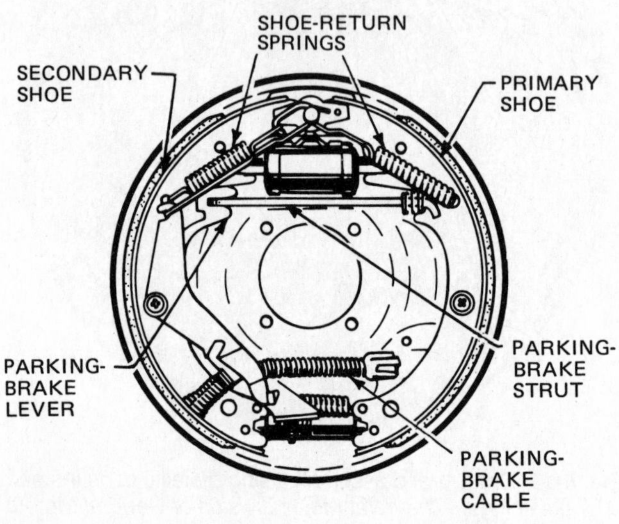

Fig. 3-15 Parking-brake components in a drum brake. Left, attachment of the parking-brake lever to the secondary shoe. Right, parking-brake mechanism in an assembled wheel brake. (*Ford Motor Company; Delco Moraine Division of General Motors Corporation*)

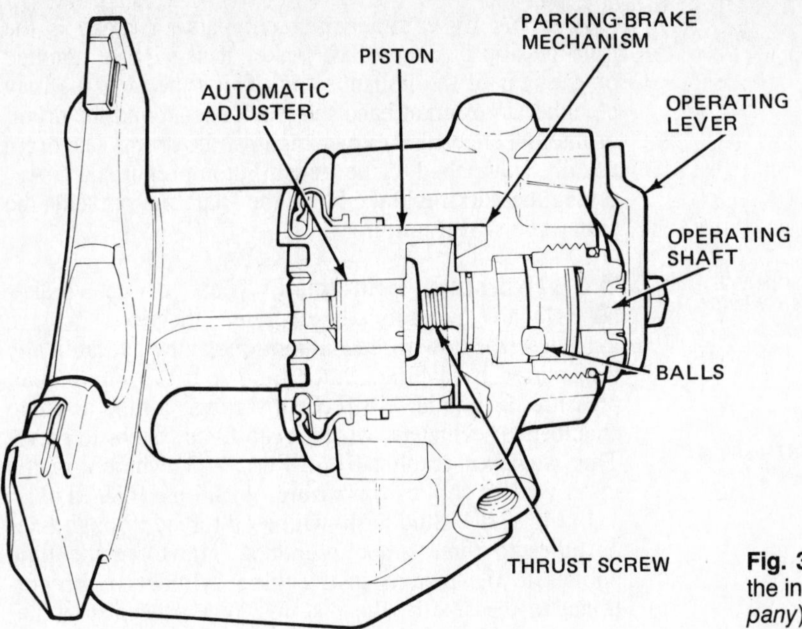

Fig. 3-16 Disk-brake caliper assembly, showing how the integral parking brake is built in. (*Ford Motor Company*)

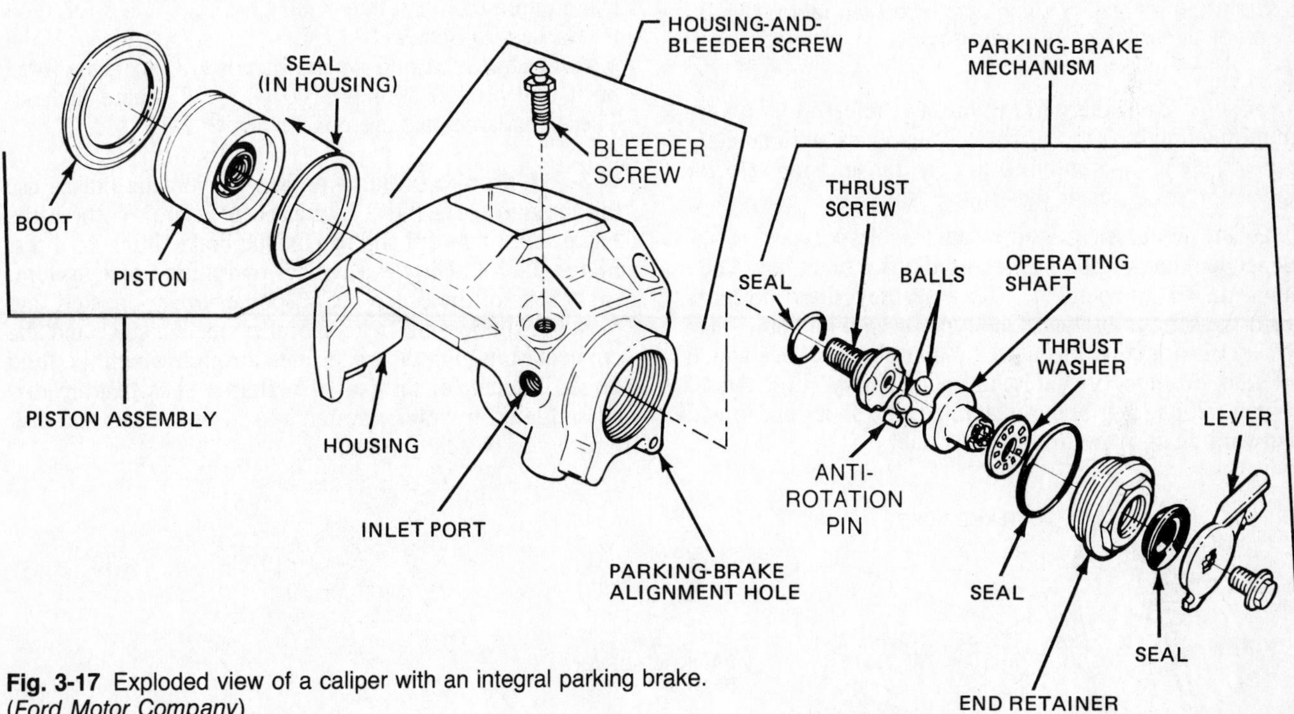

Fig. 3-17 Exploded view of a caliper with an integral parking brake. (*Ford Motor Company*)

On some cars with four-wheel disk brakes, an integral parking brake is built into each rear-wheel caliper (Fig. 3-16). When the driver applies the parking brake, the brake cables pull on the operating lever attached to the back of each caliper. As the lever moves, it turns an actuator screw which is threaded into the caliper piston (Fig. 3-17). This mechanically forces the piston to move outward and the caliper to slide inward, thereby forcing the disk-brake pads against the rotor. The piston assembly in the caliper also contains an automatic adjuster for the parking brake.

Some type of latching mechanism is built into the parking-brake control to keep the brake applied. When the parking brake is operated by a foot pedal (Fig. 3-1), a separate release lever must also be used. On cars with a hand lever, a release button usually is mounted in the outer end of the lever, as shown in Fig. 3-14. Pulling the release lever or depressing the release button unlatches the parking brake and allows the brake shoes or pads to retract. Return springs at the wheels of drum brakes pull the shoes away from the brake drums (Fig. 3-15).

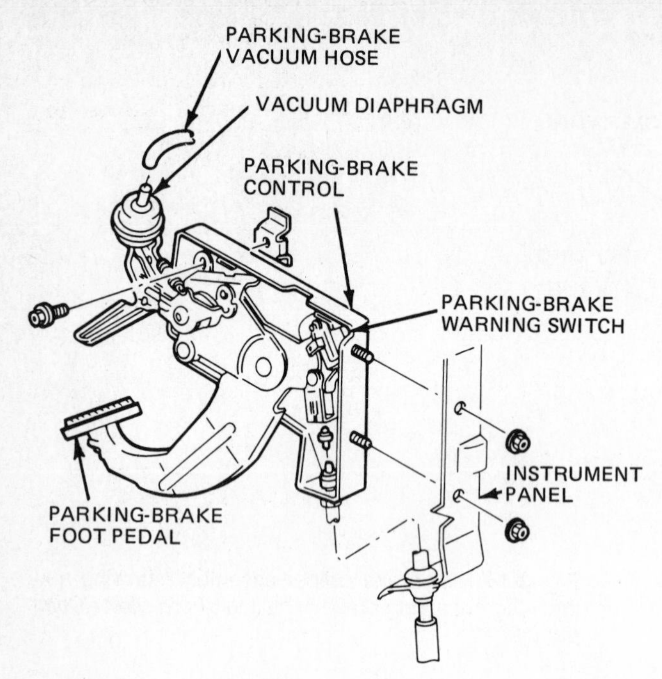

Fig. 3-18 A parking-brake foot pedal that is automatically released by a vacuum diaphragm. (*Cadillac Motor Car Division of General Motors Corporation*)

In some cars, the parking brake is released by a vacuum diaphragm (Fig. 3-18). This happens automatically when the engine is running and the transmission selector lever is moved out of the PARK position.

2. Independent parking brakes Two types of independent parking brakes are used on automobiles. They are called "independent" because they share no parts with the service brakes. One type, used with some rear-wheel disk brakes, has a set of small brake shoes which fit into the hub of a disk-drum assembly (Fig. 3-19). When the parking brake is applied, the shoes are forced outward against the drum in the hub.

A second type of independent parking brake is the transmission or drive-shaft brake. It usually is mounted on the rear of the transmission. This type of brake may be either an external band that contracts around the drum, or internal shoes that expand against the drum. The drum usually is attached to the transmission output shaft. Applying the parking brake holds the shaft and prevents the rear wheels from turning.

❂ 3-6 Hydraulic principles The service brakes (❂ 3-1) are hydraulically operated. Sections 2-11 to 2-14 describe how force and motion can be hydraulically transmitted by a fluid. Since fluid is not compressible, pressure on a fluid will force it through a tube and into chambers or cylinders, where it can force pistons to move. This is shown graphically in Fig. 2-13, where a piston in a cylinder applies a pressure of 100 psi [689 kPa].

In Fig. 2-13, fluid is shown being forced through lines or tubes to three other cylinders. The force the fluid applies to the pistons in the three cylinders is proportional to the size of the pistons. When the pistons has an area of 1 square inch [6.5 cm^2], there will be a force of 100 pounds [44.5 N] (or 100 psi) on it. If the piston has an area of 0.5 square inch [3.23 cm^2], the force on it will be 50 pounds (100 psi $\times$ 0.5 square inch). If the piston has an area of 2 square inches [13 cm^2], the force on it will be 200 pounds (100 psi $\times$ 2 square inches). These fundamentals are covered in ❂ 2-11 to 2-14.

❂ 3-7 Hydraulic braking action Hydraulic brakes use the pressure of a fluid (hydraulic pressure) to force the brake shoes outward and against the brake drums or disks. Figure 1-23 illustrates a typical hydraulic-brake system. It consists of the brake pedal, power-brake booster, and master cylinder; the wheel brake mechanisms; and the connecting tubing or brake lines through which the fluid flows. Except for the use of certain valves that are discussed later, hydraulic systems for drum brakes and disk brakes are basically the same.

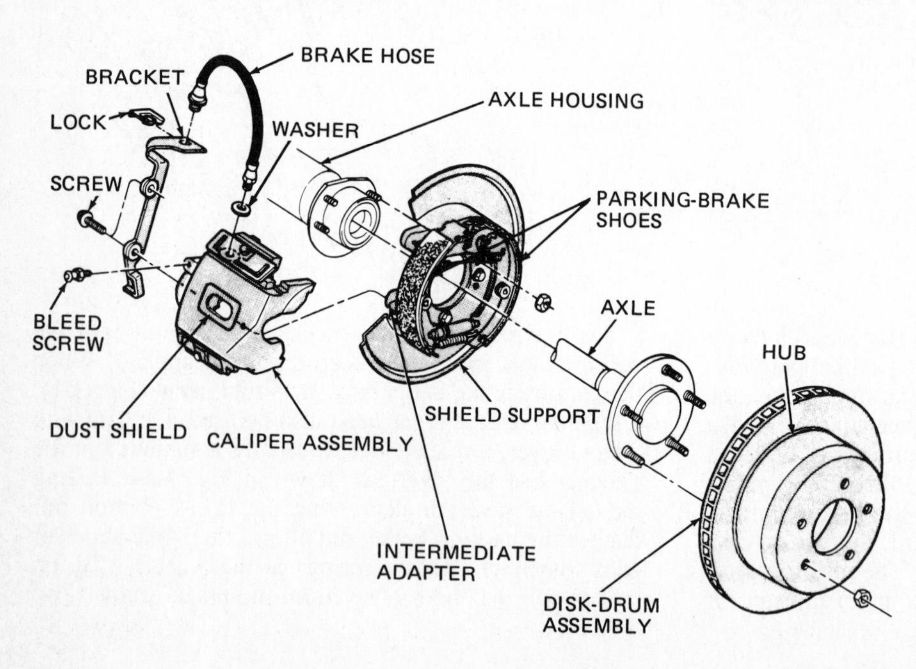

Fig. 3-19 A disassembled sliding-caliper disk brake which includes a small drum-type parking brake. (*Chrysler Corporation*)

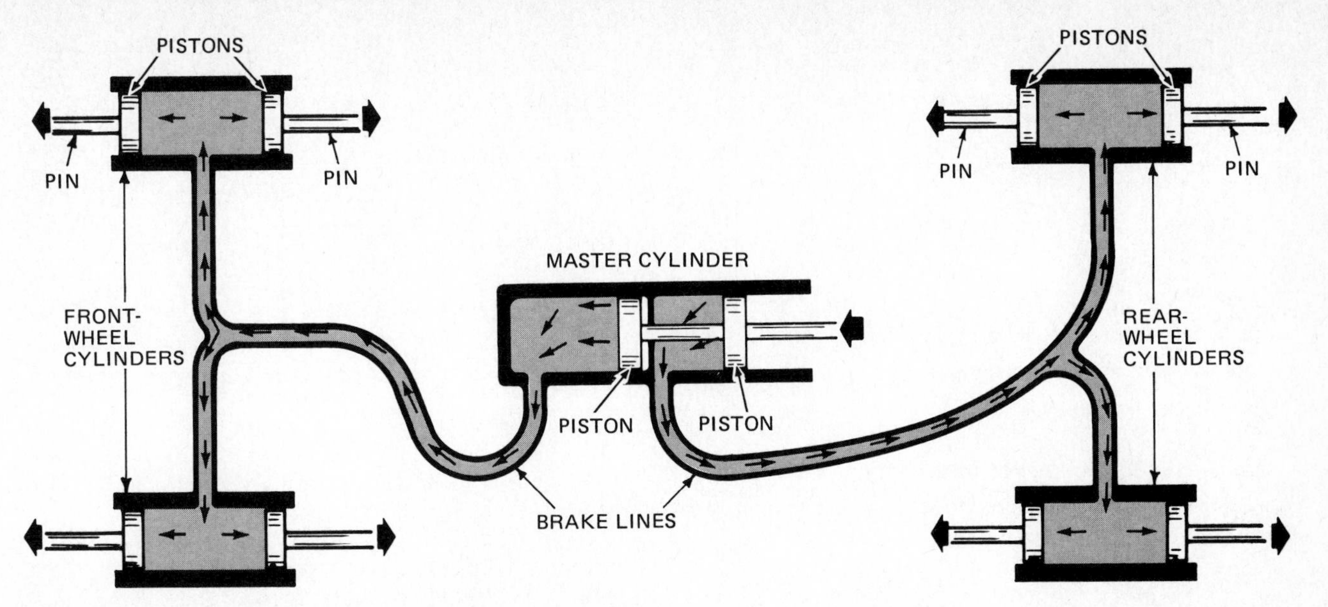

Fig. 3-20 Flow of brake fluid to the four wheel cylinders when the pistons are pushed into the master cylinder.

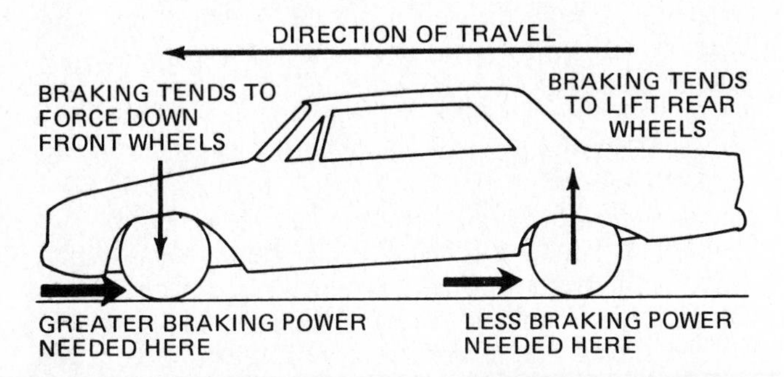

Fig. 3-21 During braking, the front wheels are forced down and the rear wheels up. (*Ammco Tools, Inc.*)

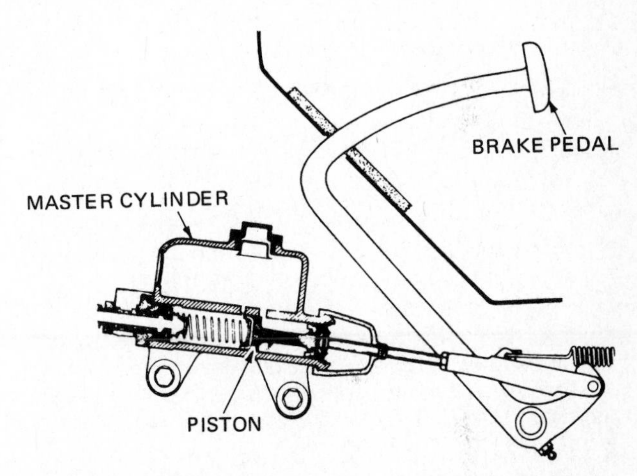

Fig. 3-22 Relationship between the brake pedal and the single-piston master cylinder. (*Pontiac Motor Division of General Motors Corporation*)

In the operation of the basic automotive hydraulic braking system, movement of the brake pedal forces pistons to move in the master cylinder (Fig. 3-20). This, in turn, forces the fluid ahead of the pistons through the brake lines to the wheel brakes.

Figure 3-20 shows the flow of brake fluid from the master cylinder to the wheels in a system with four-wheel drum brakes. At each wheel there is a wheel cylinder which has two pistons. Each piston is connected to one of the brake shoes by a pin or link (Fig. 3-3). When the master cylinder forces more fluid into the wheel cylinders, the two wheel-cylinder pistons are pushed outward. This outward movement forces the brake shoes outward and into contact with the brake drum.

When four-wheel drum brakes are used, the wheel-cylinder pistons are usually larger at the front wheels. This is because when the brakes are applied, more weight is transferred to the front wheels (Fig. 3-21). Braking places a greater load on the front-wheel brakes, and so they have greater braking power.

⚙ 3-8 Dual-brake system In older cars with four-wheel drum brakes, the master cylinder contained only one piston (Fig. 3-22). Its movement forced brake fluid to all four wheel cylinders. Now the hydraulic system is divided into two sections, a primary section and a secondary section (Fig. 3-23). This arrangement is called a *split-braking system* or a *dual-braking system*.

Figure 3-23 shows the two basic layouts of the split-braking system, the front/rear split and the diagonal split. With either arrangement, if one section fails due to damage or leakage, the other section will provide braking.

The dual-braking system includes an instrument-panel warning light that comes on to alert the driver when one section has failed. Figure 3-24 is a schematic view of the light system, which is controlled by the pressure-differential valve (⚙ 3-9).

⚙ 3-9 Warning light In the dual-braking system, a pressure-differential valve is used to operate a warning-light switch (Fig. 3-24). The valve has a piston that is

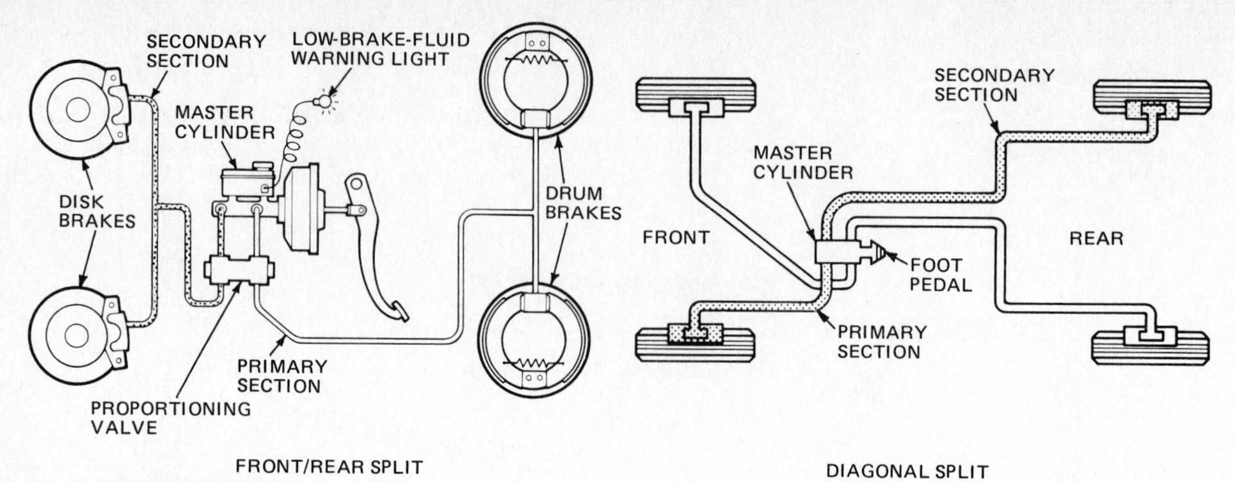

Fig. 3-23 Two basic layouts of the split-braking system. (*Mazda Motors of America, Inc.; Ford Motor Company*)

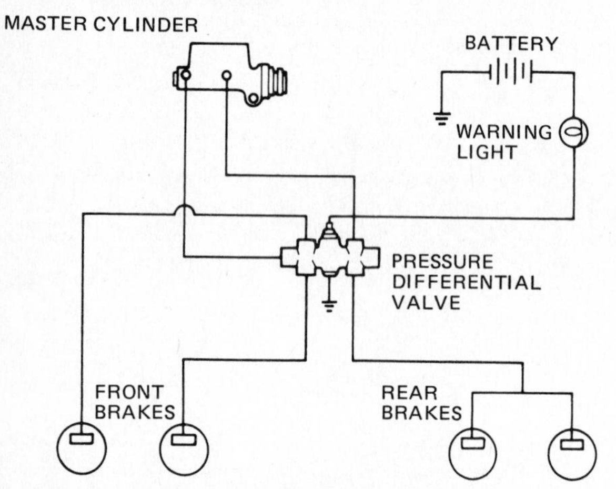

Fig. 3-24 Schematic view of the dual-brake system, showing the pressure-differential switch that operates the brake warning light. (*Chevrolet Motor Division of General Motors Corporation*)

centered when both sections of the hydraulic system are operating normally (Fig. 3-25a). However, if one section should fail, there will be low pressure on one side of the piston. The high pressure from the normally operating section will then move the piston and cause it to push the switch plunger upward (Fig. 3-25b). This closes contacts, which turns on the warning light on the instrument panel. Therefore, the driver is warned that one of the brake hydraulic systems has failed. Figure 3-25 shows the arrangement for a front/rear split. However, the system works similarly for brakes that are diagonally split. Even if the brakes are not used again, the light remains on until the trouble is fixed and the switch is reset.

The warning light does two jobs. First, it alerts the driver if either section of the service-brake hydraulic system has failed. In addition, as soon as the ignition switch is turned on, a separate parking-brake switch causes the light to come on if the parking brake is applied. By using

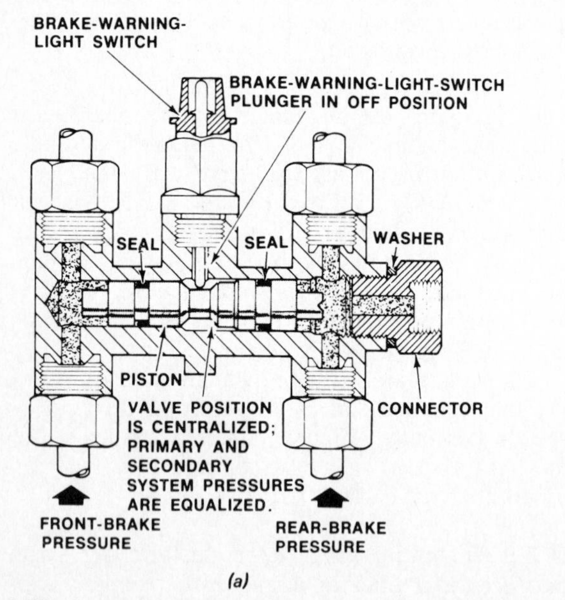

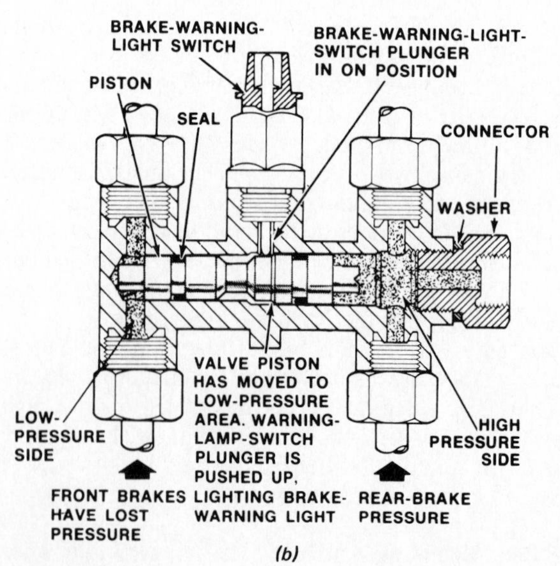

Fig. 3-25 The two positions of the pressure-differential valve: (*a*) normal operation; (*b*) with front brake system failed. (*Ford Motor Company*)

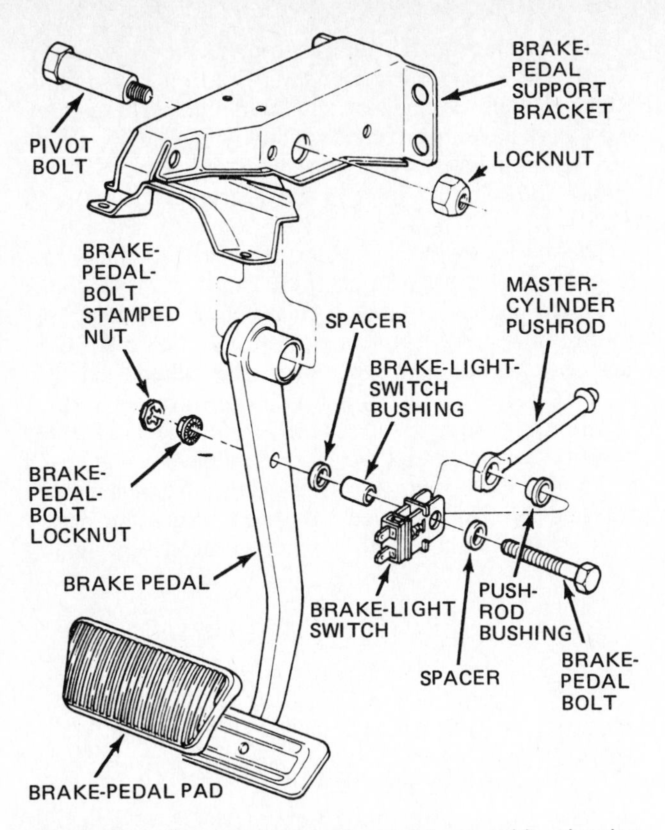

Fig. 3-26 A disassembled brake-pedal assembly, showing the stoplight, or brake-light, switch. (*American Motors Corporation*)

one bulb for both functions, the bulb is tested every time the parking brake is applied with the ignition switch on.

3-10 Stoplight switch All cars have brake lights or stoplights at the rear. These lights come on every time the driver depresses the brake pedal. This alerts the driver of any car following that the car in front is slowing or stopping.

A *stoplight switch,* or *brake-light switch,* is used to complete the electric circuit to the stoplights (Fig. 3-26). Until the introduction of the dual-brake system, most stoplight switches were hydraulic. They contained a small diaphragm that was moved by hydraulic pressure when the brakes were applied. This action closed a switch which connected the stoplights to the battery.

With the dual-brake system (3-8), the hydraulic stoplight switch could not be used. If the hydraulic switch were connected into one system and that system failed, the car would have not stoplights even though the other system was still working and stopping the car. Therefore the mechanical stoplight switch is used.

Figure 3-27 shows the operation of one type of mechanical stoplight switch. The switch is fastened to the brake pedal. When the brakes are not applied, contacts in the switch are open and the brake lights are off. When the pedal is pushed for braking, it carries the switch contacts with it (to the left in Fig. 3-27*b*). The switch actuating pin closes the contacts in the switch so that the stoplights come on.

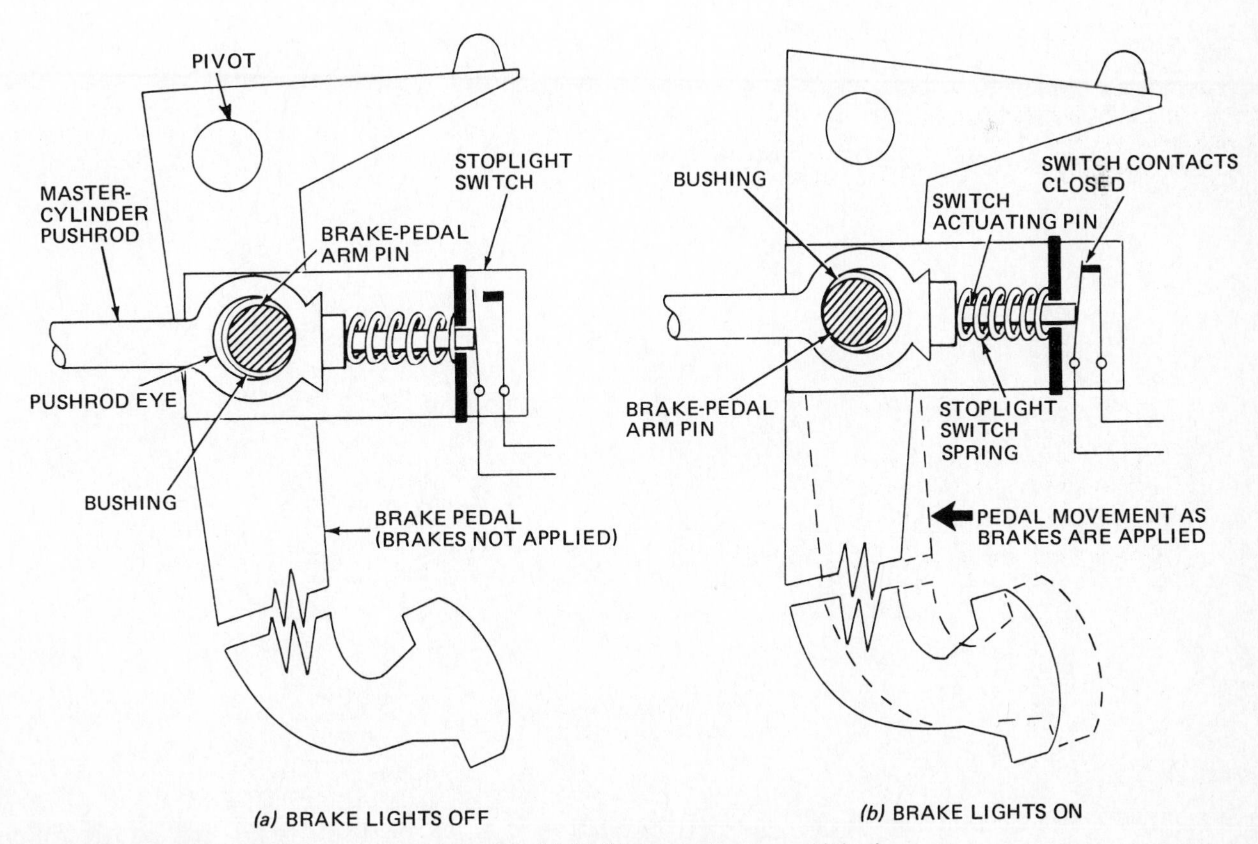

Fig. 3-27 Operation of a mechanical stoplight switch. (*a*) Switch contacts open, with brakes not applied. (*b*) Switch contacts closed, with brakes applied. (*Ford Motor Company*)

⚙ 3-11 Brake pedal In most cars, the brake pedal is suspended from a bracket under the dash (Figs. 3-26 and 3-27). The master-cylinder pushrod attaches to the brake pedal a short distance below the brake-pedal pivot. When the driver depresses the brake pedal, it acts as a lever to increase the applied force. This is called the *pedal ratio*. In one system, for example, a force of 100 pounds [445 N] applied to the brake pedal produces a force of 750 pounds [3336 N] on the pushrod.

⚙ 3-12 Master cylinders There are two different types of master-cylinder construction. However, both types operate similarly. The older type, which is still widely used, has the reservoirs integrally cast with the cylinder to form a single-piece master-cylinder body (Fig. 3-28). It usually is made of cast iron. The later type is sometimes called a *composite* master cylinder. It has reservoirs formed from sheet metal or plastic, then attached to the cylinder with rubber grommets or seals (Fig. 3-29). The separate cylinder to which the reservoirs are attached often is made of cast aluminum. These reservoirs may have translucent windows for visually checking the fluid level. Other master cylinders have a low-brake-fluid warning light in the instrument panel (Fig. 3-23).

The master cylinder in older cars had only one piston (Fig. 3-22). The dual-brake system has a master cylinder with two pistons in tandem (one behind the other) operating in a single cylinder bore (Figs. 3-28 and 3-29). The operation of both types of master cylinder has two separate sections functioning independently (⚙ 3-8).

The two pistons in the dual master cylinder are called the *primary piston* and the *secondary piston* (Figs. 3-28 and 3-29). The primary piston is the piston that is directly actuated by the brake-pedal pushrod. When the master cylinder is mounted in the engine compartment against

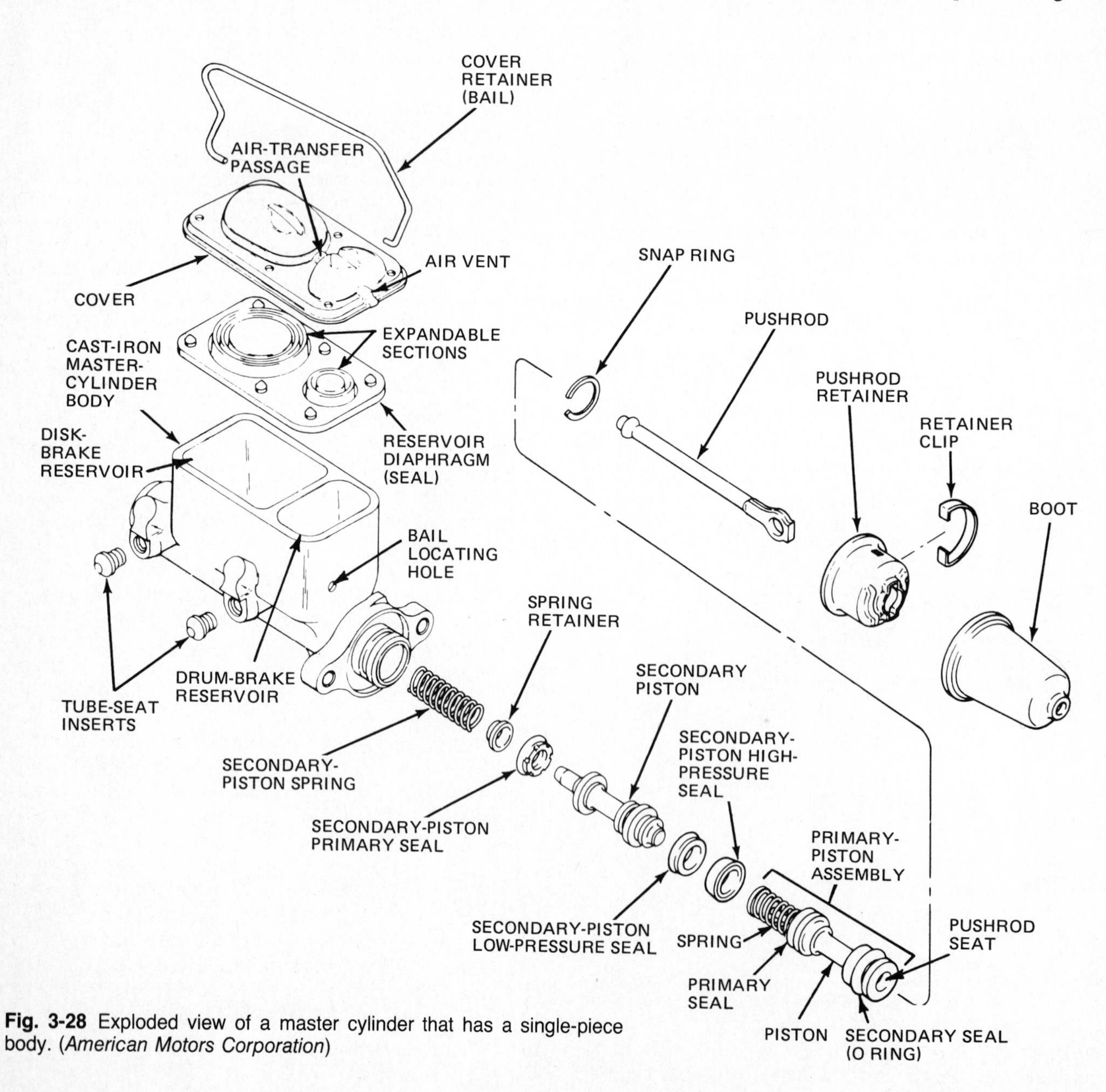

Fig. 3-28 Exploded view of a master cylinder that has a single-piece body. (*American Motors Corporation*)

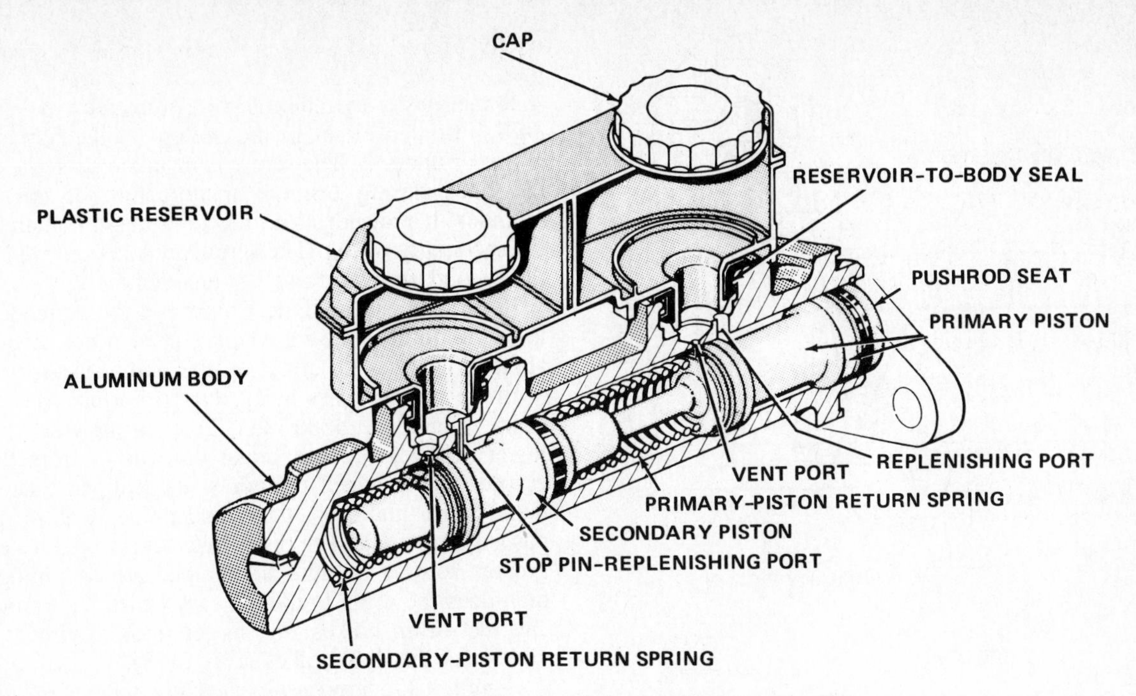

Fig. 3-29 Sectional view of a composite master cylinder, which has separate reservoirs attached to the body. (*Chrysler Corporation*)

the fire wall, the primary piston is the rearward piston. This is the one closest to the fire wall. The secondary piston is at the forward end of the master cylinder.

In Fig. 3-30, note that there are two holes in the bottom of each reservoir. The small hole toward the front of each reservoir (in the direction of piston travel) is the *bypass hole*. A typical bypass hole has a diameter of about 0.030 inch [0.76 mm]. The larger hole is the *compensating port*. In a typical master cylinder, the compensating port has a diameter of about ⅛ inch [3.2 mm].

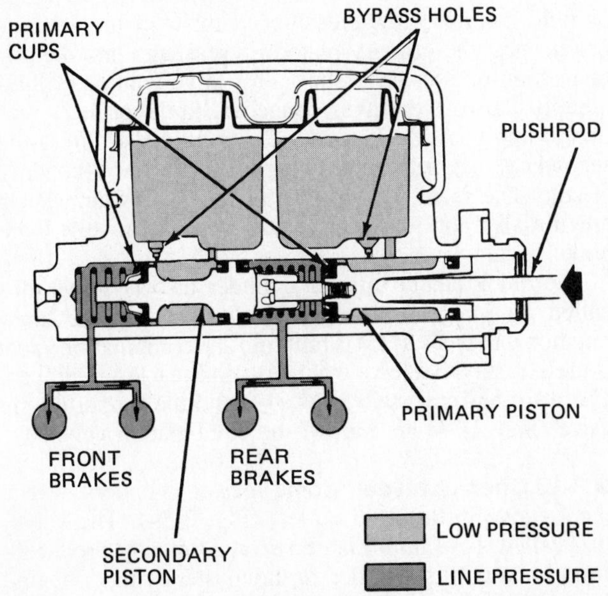

Fig. 3-30 Dual master cylinder, showing the low-pressure areas and the high- or line-pressure areas with the brakes applied. (*Delco Moraine Division of General Motors Corporation*)

When the brake pedal is depressed, the master-cylinder pushrod forces the primary piston forward (Fig. 3-30). Under normal conditions, the combination of hydraulic pressure and the force of the primary-piston spring moves the secondary piston forward at the same time. As the primary seals, or cups, on the pistons move past the bypass holes in each section of the cylinder bore, the fluid is trapped ahead of each piston. This condition is shown in Fig. 3-30. Pressure rises rapidly, and is transmitted through the brake lines to the front and rear brakes, causing the brakes to apply.

Most seals used in master cylinders (and wheel cylinders, discussed in ⚙ 3-17) are cup-type seals (Figs. 3-28 to 3-30). The cup is installed with the lip facing the fluid pressure, which forces the lip against the cylinder bore to make the seal. Some master cylinders have an O-ring seal installed as the secondary or low-pressure seal on the primary piston. This type of seal is shown next to the pushrod seat on the primary piston in Fig. 3-28.

When the brake pedal is released, the master-cylinder springs force the pistons to return to their released positions faster than fluid can flow back into the master cylinder (Fig. 3-31). This tends to create a momentary vacuum. To avoid the vacuum, fluid flows from the reservoirs through relatively large compensating ports into the cylinder low-pressure chambers. From there, the fluid passes through small compensating holes in the pistons, around the primary cups, and into the high-pressure chambers. Some pistons do not have compensating holes. In these, additional piston clearance is allowed so that the compensating-fluid flow is around the outside diameter of the piston seal.

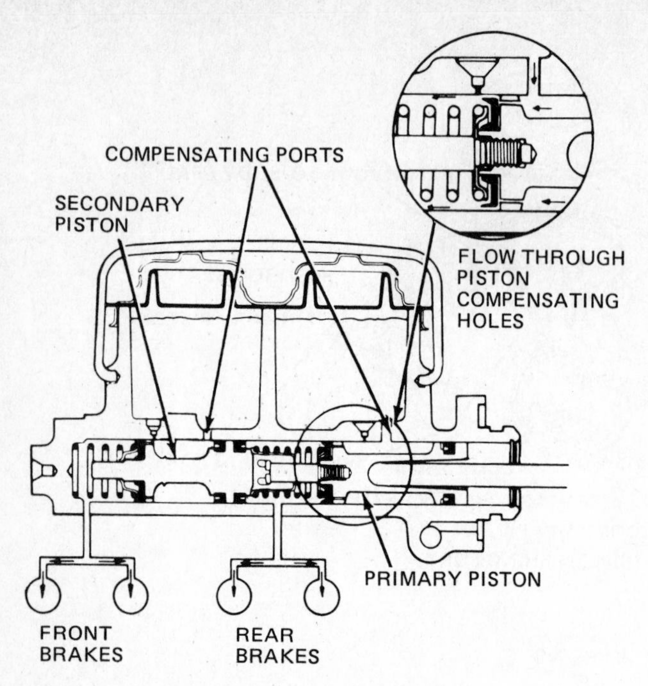

Fig. 3-31 Conditions in the master cylinder as the brakes start to release. (*Delco Moraine Division of General Motors Corporation*)

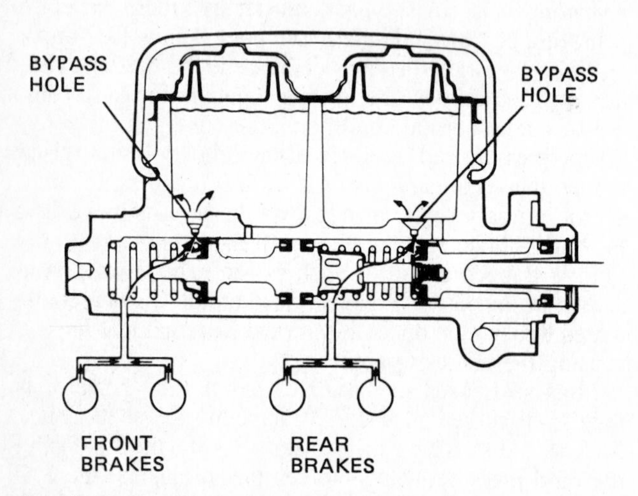

Fig. 3-32 Conditions in the master cylinder as the brake release ends. (*Delco Moraine Division of General Motors Corporation*)

As the pistons return to their released positions, line pressure is still higher than in the fluid reservoirs. As soon as the piston seals uncover the bypass holes, some fluid flows through them back into the reservoirs (Fig. 3-32). This backflow continues until the pressures have equalized.

To prevent an air lock from forming as the fluid level in the master cylinder gets lower, the reservoirs are vented to the outside air (Fig. 3-28). However, to prevent the air from reaching the brake fluid and adding moisture to it (❁ 3-15), a seal with expandable sections fits between the cover and the reservoirs. As the fluid level lowers, the diaphragm sections expand to apply air pressure to the top of the fluid while preventing the air from touching it.

Some master cylinders have a stop screw or pin (Fig. 3-29) installed either in the bottom of the forward reservoir or under it. The screw is installed on some assembly lines during original manufacture of the master cylinder. It protrudes into the low-pressure chamber for the secondary piston. The screw must be removed before the secondary piston can be removed.

In some master cylinders there is only a threaded hole in the bottom of the reservoir in which no screw is installed. This has no effect on the brake system.

Brake lining wears away during normal use. As the lining wears, the brake-fluid level in the reservoir gets lower. This is because brake fluid flows from the reservoirs to the wheel cylinders and calipers. The additional brake fluid keeps the cylinders full, with the pistons pushed further out. The disk-brake caliper requires a much greater volume of fluid to compensate for lining wear than does the drum brake. In cars which have front-disk and rear-drum brakes, the larger master-cylinder reservoir is for the disk brakes (Fig. 3-28).

Figure 3-33 shows a *quick-takeup master cylinder*. It is designed for use with low-drag disk brakes, which pull the brake pads away from the disk when the brake pedal is released. This moves a fairly large amount of fluid out of the caliper. As a result, when the brake pedal is depressed the next time, a large volume of fluid must be quickly returned to the caliper to move the brake pads back into the contact with the disk. Only then will further application of the brake pedal cause rapid pressure buildup and normal braking.

In the quick-takeup master cylinder (Fig. 3-33), the secondary section works basically the same as in other units. However, the primary piston works in a step bore, which has two different diameters. When the brake pedal is depressed, the initial movement forces a large volume of fluid from the low-pressure chamber of the primary piston, past the primary-piston lip seal, and into the primary high-pressure chamber. This large volume of fluid quickly causes the brakes connected to the primary section to apply. At the same time, the fluid also forces the secondary piston to travel quickly to fill the secondary circuit. The result is that the brakes at all four wheels are initially applied with relatively little travel of the brake pedal.

The quick-takeup master cylinder also is sometimes called an *integral master cylinder* because the combination valve (❁ 3-25) is built in. The combination valve is used in cars that have front-disk and rear-drum brakes. The warning-light switch (❁ 3-9) and the proportioning valve (Fig. 3-33) are part of the combination valve.

❁ 3-13 Check valves Some master cylinders have a check valve in the fluid outlets (Fig. 3-34). The valve, also called a *residual check valve,* is mostly used with drum brakes. It is installed in the master cylinder behind the tube-seat insert (Fig. 3-28). Fluid must pass through the check valve to leave the master cylinder.

When the brake is released, fluid tries to flow back into the master cylinder through the check valve. The

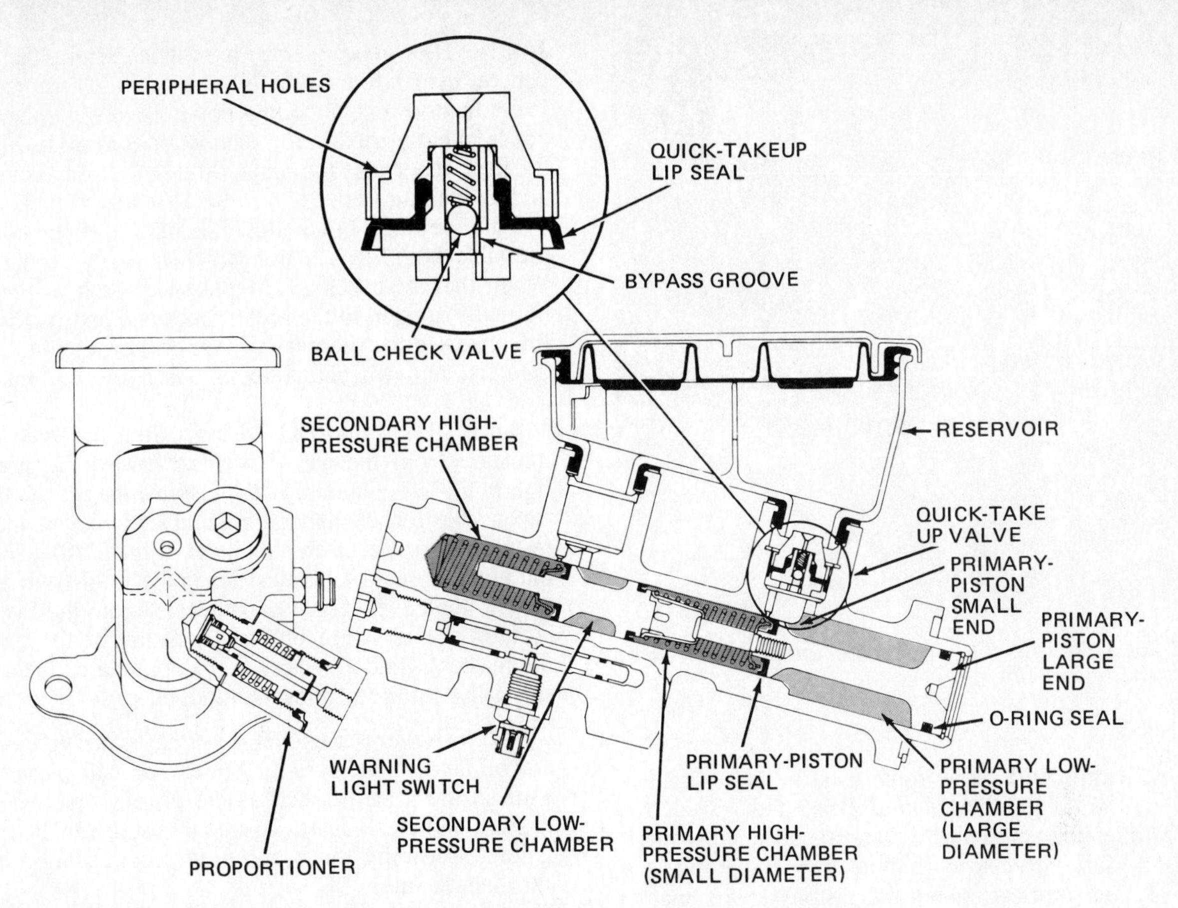

Fig. 3-33 A quick-takeup master cylinder which has a step-bore cylinder and an integral combination valve. (*Delco Moraine Division of General Motors Corporation*)

pressure overcomes the spring force and causes the valve to unseat. Now fluid flows around the valve and into the master cylinder.

As the pressure drops to about 6 to 18 psi [41 to 124 kPa], the spring forces the valve closed. This traps a slight pressure (*residual line pressure*) ahead of the check valve in the hydraulic system. With drum brakes, the pressure keeps the cups in the wheel cylinders securely against the bores. This helps prevent fluid from leaking out and air from leaking into the hydraulic system.

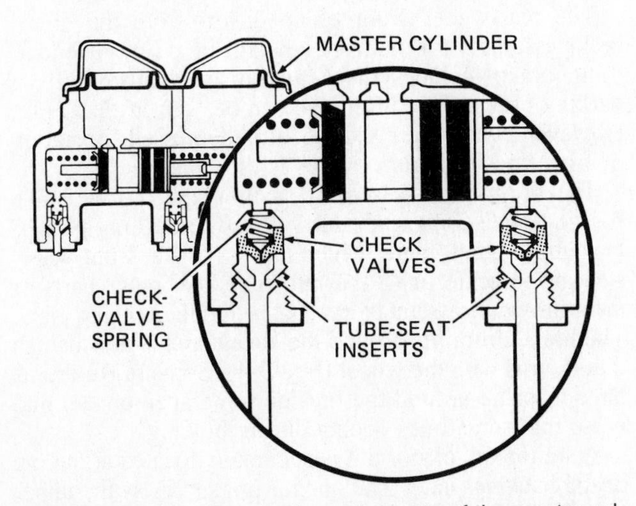

Fig. 3-34 Check valves in the outlet ports of the master cylinder, which maintain a slight residual line pressure. (*Ford Motor Company*)

NOTE: General Motors and some other manufacturers have discontinued the use of check valves in all dual master cylinders for passenger cars. However, not all manufacturers have discontinued their use.

❁ 3-14 Brake lines Brake fluid is carried by steel pipes called *brake lines* from the master cylinder to the various valves and then to the brakes at the wheels (Figs. 3-1 and 3-35). Brake lines usually pass under the floor pan of the car, where they are exposed to stones and other road damage. Because of this, brake lines often are wrapped with a wire "armor" to protect them against damage.

A short, flexible *brake hose* (or "flex hose") is used to connect the stationary brake lines to the brake assemblies that move up and down and swing with each front wheel for steering (Fig. 3-35). Steel tubing will crack if it vibrates or is flexed too much. With the use of a T-fitting junction block, only a single flexible hose is needed at the rear axle to take care of the up-and-down movement of the housing (Fig. 3-1).

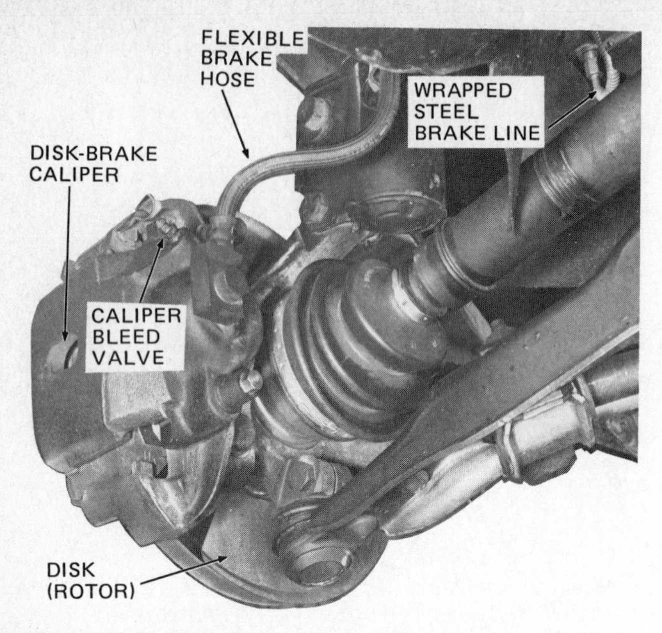

Fig. 3-35 Brake lines for a front wheel on a front-drive car. (*Chrysler Corporation*)

3-15 Brake fluid The liquid used in the hydraulic-brake system is called *brake fluid*. Brake fluid must have very definite characteristics. It must be chemically inert; it must be little affected by high or low temperatures; it must provide lubrication for the pistons in the master cylinder, wheel cylinders, and calipers; and it must not attack the metal, plastic, and rubber parts in the braking system. Therefore, only the brake fluid recommended by the car manufacturer must be used when the addition of brake fluid becomes necessary.

Three types of hydraulic-brake fluid are used in automobiles. These are classified by the Department of Transportation (DOT) as DOT 3, DOT 4, and DOT 5. This classification must appear on the brake-fluid container label.

DOT 3 brake fluid is the type most widely used. Its *wet boiling point* must be at least 284°F [140°C]. This is a test for the minimum boiling temperature of a brake fluid after a certain amount of water is added. New DOT 3 and DOT 4 brake fluids have a clear to amber color.

DOT 4 brake fluid has a higher wet boiling point (311°F [155°C]) than DOT 3 fluid. It was developed for use in disk-brake hydraulic systems, and these reach higher temperatures than drum brakes. DOT 4 fluid absorbs moisture at a slower rate than DOT 3 fluid.

DOT 5 brake fluid is a newer type that is not widely used in cars. It is usually a silicone fluid with additives, which absorbs practically no water when compared with DOT 3 and DOT 4 brake fluids. DOT 5 brake fluid has a minimum wet boiling point of 356°F [180°C]. It has about half the viscosity of DOT 3 and DOT 4 brake fluids at −40°F [−42°C]. New DOT 5 brake fluid has a purple color.

Moisture must be prevented from getting into any hydraulic brake system and from getting into the brake fluid. Some brake fluids (DOT 3 and DOT 4) are *hygroscopic*. They readily absorb moisture from the atmosphere, even while in the brake system. Moisture in the brake fluid lowers its boiling point. Then when the brakes are frequently used hard, such as on a long downgrade, the heat may cause the brake fluid in the wheel cylinders and calipers to boil.

With boiling, brake fluid changes from a noncompressible liquid to a vapor which is easily compressed. When the brake pedal is depressed, much of the force normally applied to the brake shoes and pads is lost. The force is used to compress the vapor bubbles in the lines, calipers, and wheel cylinders. As a result, braking is lost.

Since DOT 3 and DOT 4 brake fluids readily absorb moisture from the air, they must always be stored in tightly closed containers. This is also the reason that an expandable rubber diaphragm completely covers the fluid reservoirs in the master cylinder (Fig. 3-28). The diaphragm, or seal, separates the brake fluid from the air vent in the cover. Once moisture gets into the hydraulic system, the moisture promotes rusting of the bores in the master cylinder, wheel cylinders, and calipers. Rust can cause failure of these components.

Careful! Never put engine oil or any other type of mineral oil in the brake system. Mineral oil will cause rubber parts in the system, such as the piston cups and seals, to swell and break apart. This could cause complete brake failure. Use only the brake fluid recommended by the car manufacturer.

Drum brakes

3-16 Drum-brake construction A car may be equipped with drum brakes at all four wheels, or with disk brakes on the front and drum brakes on the rear (Fig. 3-23). Figure 3-3 shows a basic front-wheel drum brake. Figure 3-4 shows how the drum-and-brake assembly attaches to the rear wheel of a front-wheel-drive car. Figure 3-5 shows the rear-wheel drum brake completely disassembled.

The rear-wheel drum brake differs from the drum-brake assembly for the front wheel (Fig. 3-36). The rear-drum brake includes the actuating mechanism for the parking brake. The difference can be seen in Fig. 3-36. However, when the brake pedal is depressed, operation of front and rear drum brakes is the same.

The drum-brake assembly is mounted on the brake backing plate (Figs. 3-5 and 3-36). The backing plate is a stamped-steel plate which bolts to the front-wheel steering knuckle or to the rear axle. All other parts of the drum-brake assembly are attached to the backing plate. The brake drum fits around the shoes, and is attached so that it turns with the wheel (Fig. 3-4). Some brake drums have a spring around the outside to dampen chatter and noise that sometimes occurs during braking.

A stationary anchor pin is accurately located at the top of the backing plate. The anchor pin serves as the upper pivot point for the brake shoes, and is located above the wheel cylinder. In some drum brakes, the anchor pin attaches the top of the backing plate to the steering knuckle.

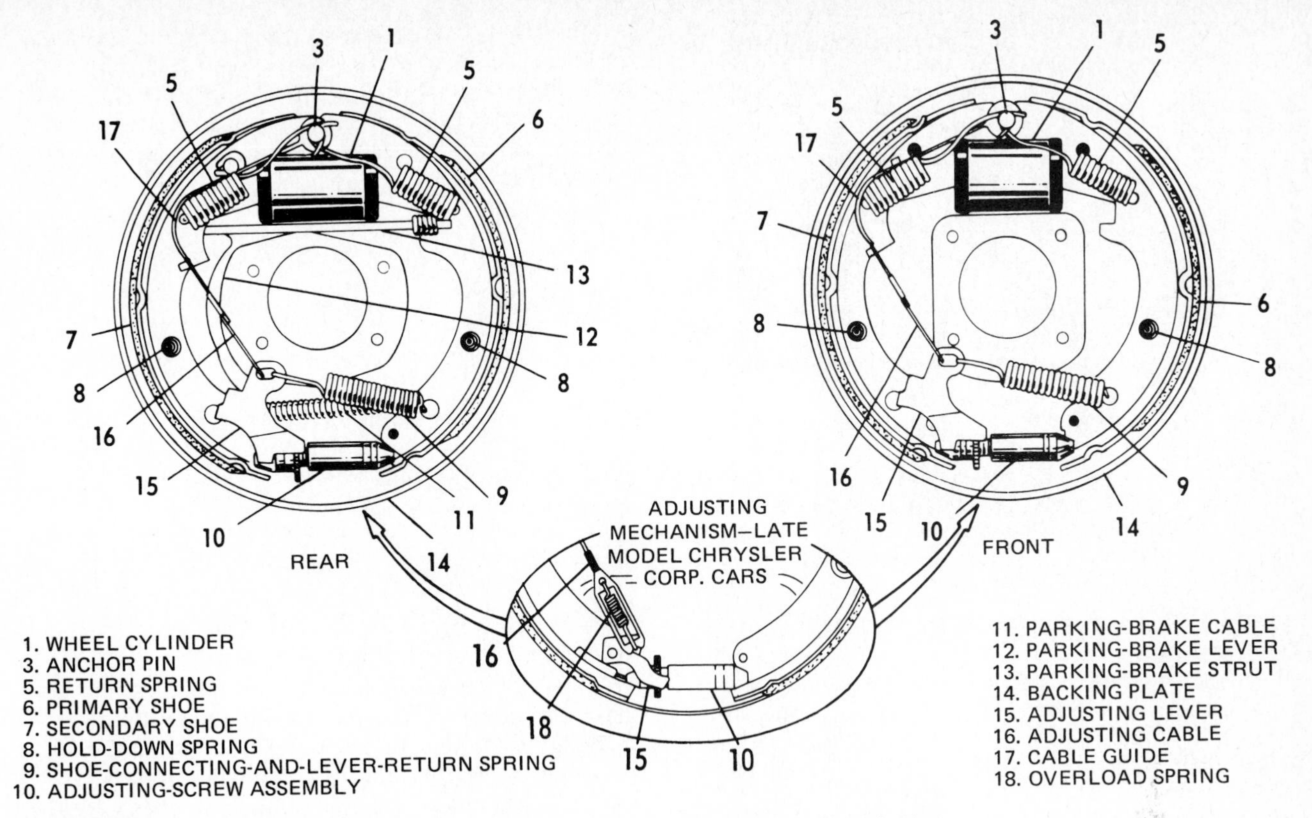

Fig. 3-36 Typical duo-servo type of drum brake for the front and rear wheels. Each brake assembly has a self-adjusting mechanism. (*Delco Moraine Division of General Motors Corporation*)

In others, the anchor pin is fastened through the backing plate to a flange on the axle housing, or tube.

Each drum brake has two brake shoes. The shoe installed toward the front of the car is called the *primary shoe*. The shoe installed toward the rear is the *secondary shoe*. Linings of a friction material such as asbestos are bonded or riveted to the metal brake shoes (⚙ 3-3).

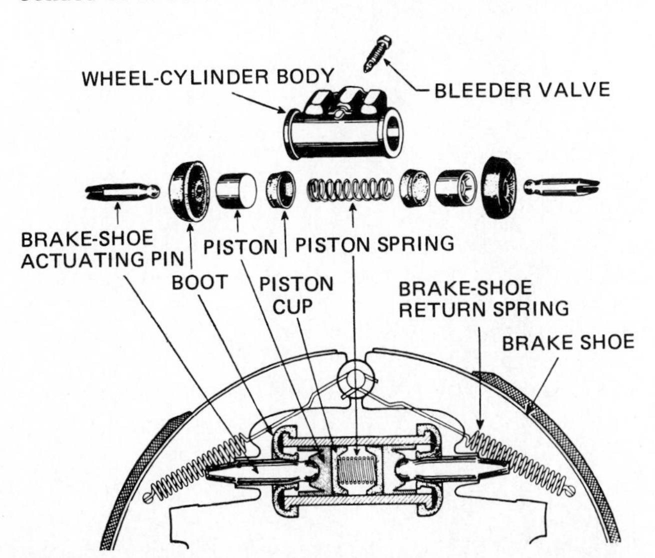

Fig. 3-37 Disassembled and sectional views of a wheel cylinder. (*Pontiac Motor Division of General Motors Corporation*)

The tops of the shoes are fitted to the anchor pins and secured by brake-shoe *return*, or *retracting, springs* (Fig. 3-36). At the bottom, the shoes fit into grooves at each end of the adjusting-screw (or *star-wheel*) assembly. A spring may be installed above the adjusting screw, connecting the shoes (Fig. 3-3). Each shoe is also attached to the backing plate by a hold-down spring and retainer cup.

The drum brake described in this section is sometimes called the *duo-servo brake*. However, many brake-shoe arrangements and combinations have been used, along with a variety of wheel cylinders and various numbers of anchor pins. Another drum brake, found on some cars today, is the *leading-trailing-shoe* brake (Figs. 3-3 and 3-4). In this brake, each shoe is anchored at the bottom by a separate anchor pin.

For many years, drum brakes have been equipped with self-adjusters. Self-adjusting brake-shoe mechanisms are shown in Figs. 3-5 and 3-36. Operation of self-adjusters is described in ⚙ 3-19.

⚙ 3-17 Wheel cylinders Figure 3-37 shows the construction of a drum-brake wheel cylinder. It has two pistons separated by a spring. Most automotive wheel cylinders are of this type.

The pistons often are made of powdered metal with lubricant added, to resist corrosion and sticking. Each piston is connected to a brake shoe by a solid connecting link that passes through a rubber boot on the end of the

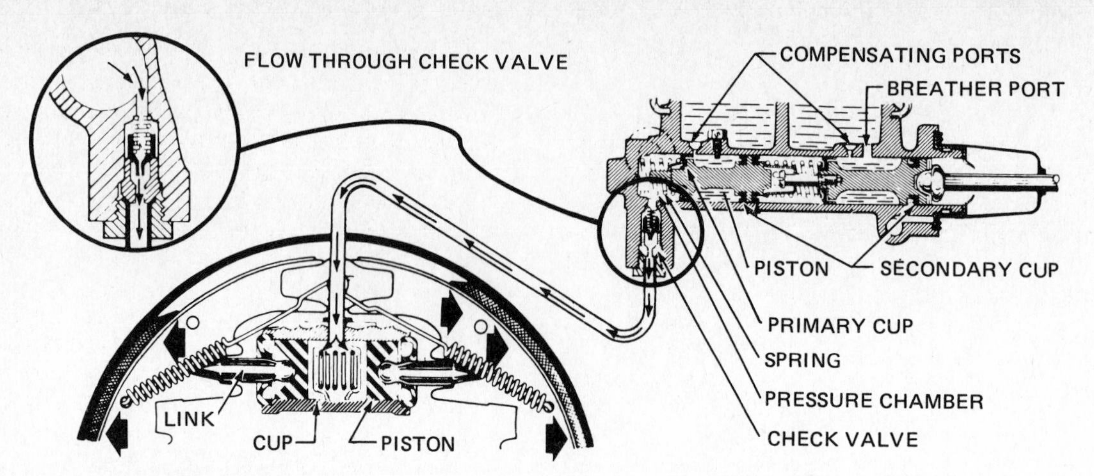

Fig. 3-38 Conditions in a drum-brake system while the brakes are being applied. Brake fluid flows from the master cylinder to the wheel cylinder, as shown by the arrows. The fluid forces the wheel-cylinder pistons outward, applying the brakes.

cylinder. The boots may be either internal (Fig. 3-15, right) or external (Fig. 3-37).

A piston cup on the inner side of each piston provides a seal for the brake fluid. The piston cups are shaped so that the hydraulic pressure forces them tightly against the cylinder wall of the wheel cylinder. This produces an effective sealing action that holds the fluid in the cylinder.

When the brake pedal is depressed, the increasing pressure on the brake fluid acts against the piston cups and forces the pistons outward. This movement is transferred by pins or links to the brake shoes, and the shoes are forced against the brake drums.

⚙ 3-18 Drum-brake operation

The most commonly used drum brake is the duo-serve or *self-energizing* type (Fig. 3-36). It is called self-energizing because the shoes have a tendency to wrap themselves into the drum during braking. However, some cars use the leading-trailing-shoe brake. The operation of both types of drum brakes is described below.

1. Duo-servo (self-energizing) type When the brakes are applied, as shown in Fig. 3-38, the wheel cylinder pushes the brake shoes away from the anchor pin and toward the rotating drum. This action forces the shoes outward at the top. As the shoes move outward, they

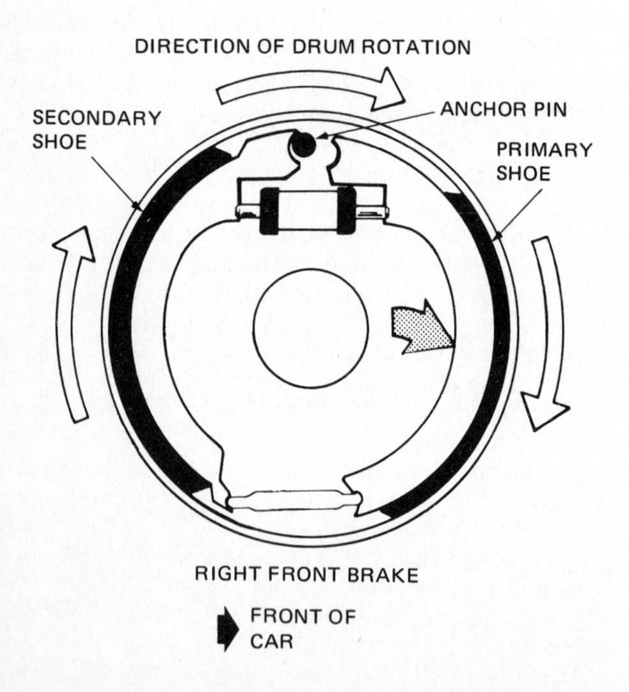

Fig. 3-39 Friction of the rotating drum against the top of the primary shoe forces the shoe to shift downward. This forces the whole shoe against the drum. (*Ford Motor Company*)

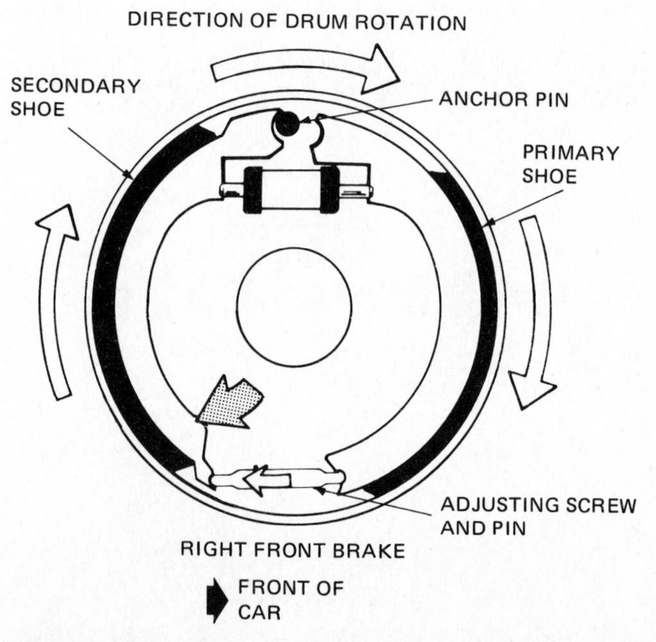

Fig. 3-40 The shifting of the primary shoe downward pushes the adjusting screw and pin to the left so that the bottom of the secondary shoe is forced against the drum. (*Ford Motor Company*)

stretch the shoe-return springs, which later retract the shoes to their normal rest position.

As the shoes contact the drum, friction causes the shoes to try to rotate with it (Fig. 3-39). This tends to pull the top of the primary shoe away from the anchor pin. It also pushes the bottom of the primary shoe against the adjusting screw with a rearward force (Fig. 3-40). As a result, the primary shoe is forced tightly against the rotating drum, increasing braking action.

When the bottom of the primary shoe pushes against the adjusting screw (Fig. 3-40), the adjusting screw shifts slightly in the dirction of drum rotation. This rotates the secondary shoe upward until the shoe web contacts the anchor pin.

Now, the secondary shoe is pushed against the drum by two forces. These are the apply force and the friction force of the primary shoe. This combination of forces acting on the secondary shoe provides the self-energizing action that gives this type of drum brake its name. The self-energizing action applies an increased force to the brake shoes with less pedal effort than other types of drum brakes. As a result, the secondary shoe provides about twice as much braking effect as the primary shoe. For this reason, the secondary-shoe lining area usually is larger than that of the primary shoe (Fig. 3-41).

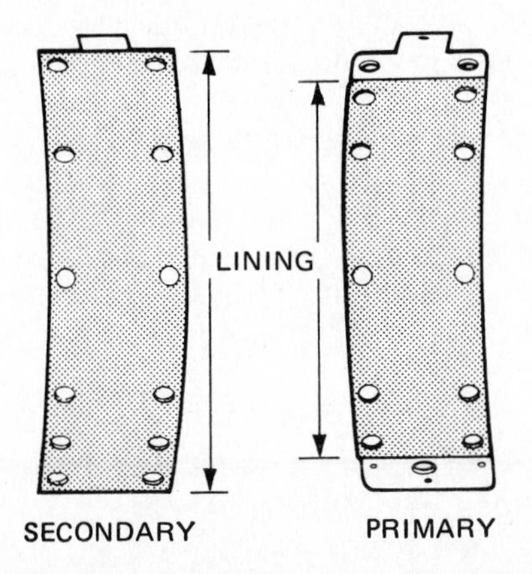

Fig. 3-41 The primary-shoe lining is smaller, and the primary shoe is installed toward the front of the car. The secondary-shoe lining is larger, and the secondary shoe is installed toward the rear of the car. (*Ford Motor Company*)

If the brakes are applied with the vehicle moving in reverse, the rear shoe becomes, in effect, the primary shoe. The self-energizing action now would be applied to the front shoe, and the rear shoe would move away from the anchor pin.

When the brake pedal is released, spring force acting on the linkage and against the master-cylinder pistons (Fig. 3-42) pushes the pistons back to their normal rest positions. This relieves the pressure in the lines and wheel cylinders. Now the brake-shoe return springs (Fig. 3-42) pull the brake shoes away from the drums. This forces the wheel-cylinder pistons inward. A small amount of brake fluid flows through the lines to the master cylinder, as shown by the arrows in Fig. 3-42. If check valves (⚙ 3-13) are used in the master cylinder (as shown in Fig. 3-42), a slight residual line pressure remains in the lines.

2. Leading-trailing-shoe type Figures 3-3 and 3-4 show a leading-trailing-shoe type of drum brake. When the brakes are applied as the car is moving forward, the wheel cylinder forces the tops of the shoes outward against the drum. Friction between the drum and the lining causes the leading shoe to try to rotate with the drum. This self-energizing action of the leading shoe forces the bottom, or heel, of the shoe against the bottom anchor assembly. The action is similar to that of the duo-servo brake described in item 1 above. However, in the leading-trailing-shoe brake, the bottom anchors are fixed. No force is transferred from the leading shoe to the trailing shoe.

With the brakes applied, the trailing shoe is not self-energized. This is because the friction force from the shoe contact with the drum opposes the input or apply force from the wheel cylinder. Therefore, the leading shoe does the greater proportion of the braking. When the car is driven in reverse and the brakes applied, the trailing shoe becomes the energized shoe.

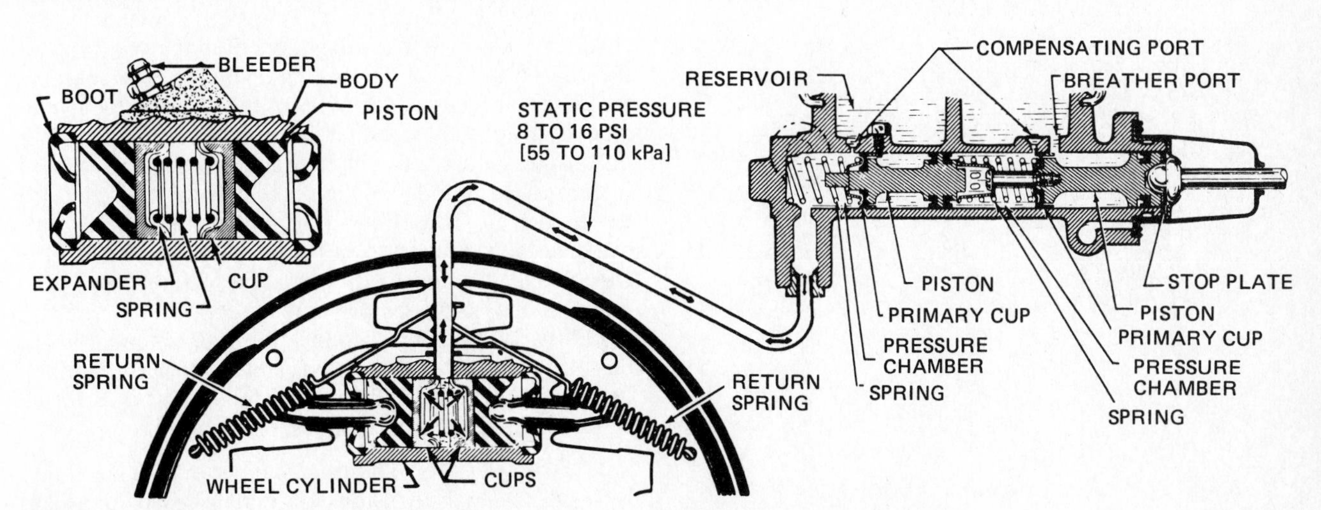

Fig. 3-42 Conditions in the drum-brake system when the brakes are released. Brake fluid flows back to the master cylinder, as shown by the arrows. (*Buick Motor Division of General Motors Corporation*)

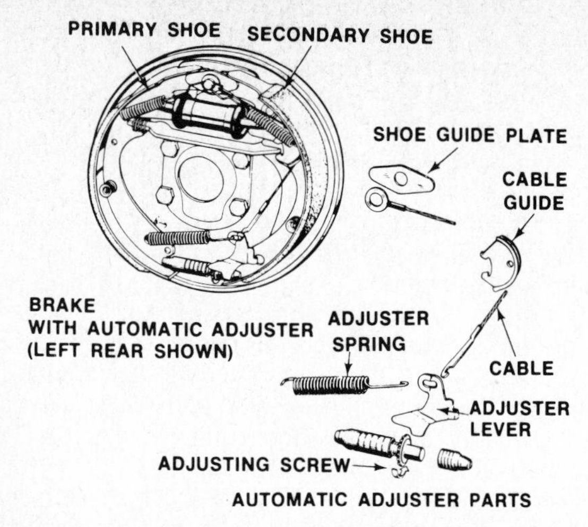

Fig. 3-43 Drum-brake assembly with automatic self-adjuster, showing the self-adjuster parts disassembled. (*Bendix Corporation*)

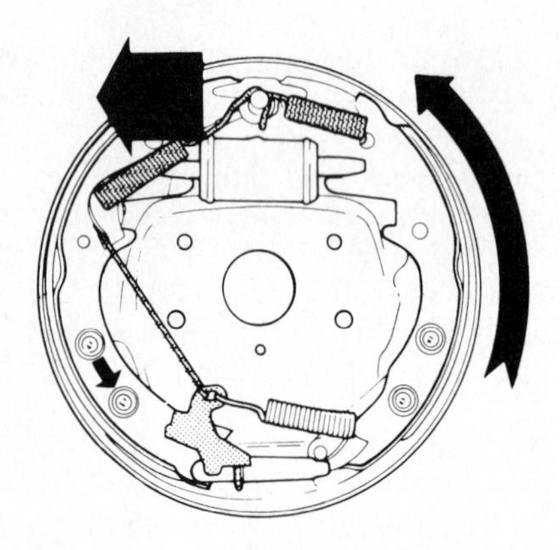

Fig. 3-44 Operation of the self-adjuster. When the brakes are applied as the car is being backed, the shifting of the brake shoes actuates the adjusting lever. (*Ford Motor Company*)

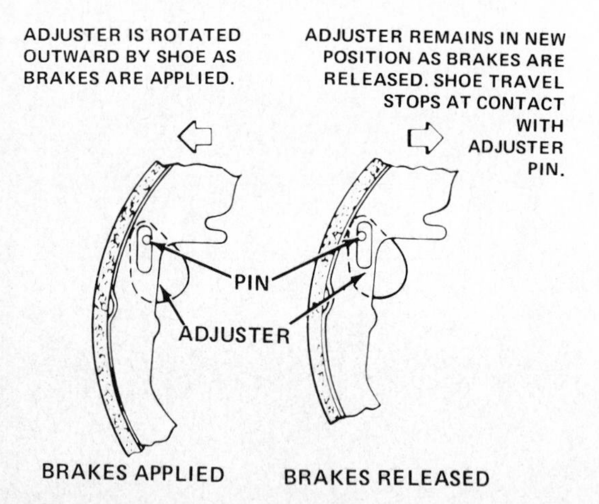

Fig. 3-45 Self-adjusting mechanism used with some types of leading-trailing-shoe drum brakes. Adjustment takes place whenever the brakes are applied. (*Delco Moraine Division of General Motors Corporation*)

⚙ 3-19 Self-adjusting drum brakes Most automotive drum brakes have a self-adjusting mechanism that automatically adjusts the brakes to compensate for brake-lining wear. Figures 3-43 and 3-44 illustrate a typical arrangement used with the duo-servo brake (⚙ 3-18). The self-adjuster is attached to the secondary brake shoe. Adjustment takes place when the brakes are applied as the car is moving rearward. Then an adjustment is made only when the brake linings have worn enough to make adjustment necessary.

As the brakes are applied with the car moving backward, friction between the primary shoe and the brake drum forces the primary shoe against the anchor pin. The wheel cylinder forces the upper end of the secondary shoe away from the anchor pin and downward (Fig. 3-44). This causes the adjuster lever to pivot on the secondary shoe so that the lower end of the lever is forced against the star wheel on the adjusting-screw assembly. If the brake linings have worn enough, the adjuster screw will be turned a full tooth. This spreads the lower ends of the brake shoes a few thousandths of an inch—enough to compensate for lining wear.

On some cars, the self-adjustment mechanism operates with the car moving forward when the brakes are applied.

On cars with leading-trailing-shoe brakes (⚙ 3-18), adjustment takes place, if needed, whenever the brakes are applied (Fig. 3-45). The car may be standing still or moving in either direction. Other cars have a special parking-brake rod-and-strut assembly that adjusts the brake shoes when the parking brake is applied.

Disk brakes

⚙ 3-20 Types of disk brakes The disk brake has a metal disk, or rotor, instead of a drum, and a pair of pads, or flat shoes, instead of the curved shoes used with drum brakes (Figs. 3-2 and 3-6). There are three general types: fixed-caliper, sliding-caliper, and floating-caliper.

The caliper is the assembly which straddles the disk and in which the brake pads are held. It contains the flat shoes or pads, which are positioned on the two sides of the disk (Figs. 3-2 and 3-46). In operation, the shoes are forced inward against the sides of the disks by the movement of pistons in the caliper assembly (Fig. 3-47). These pistons are actuated by the hydraulic pressure developed in the master cylinder as the brake pedal is pushed down by the driver. The effect is to clamp the rotating disk between the stationary shoes (Fig. 3-6). As a result, the shoes apply friction to the disk and attempt to stop its rotation. This provides the braking action.

Figure 3-47 shows a connecting tube through which fluid flows to force the pistons inward toward the disk. In many disk-brake calipers, the fluid passage is in the caliper housing, not through an external tube.

1. Fixed caliper The fixed-caliper disk brake (Figs. 3-46 and 3-48) has pistons on both sides of the disk. Some fixed-caliper disk brakes have two pistons, one on each side (Fig. 3-49). Others have four pistons, two on each side (Fig. 3-50). The caliper is rigidly attached to a stationary car part. If the caliper is at the front of the

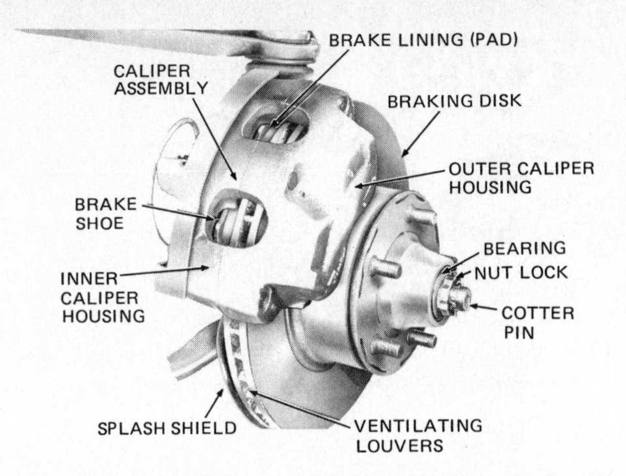

Fig. 3-46 A fixed-caliper disk brake. (*Chrysler Corporation*)

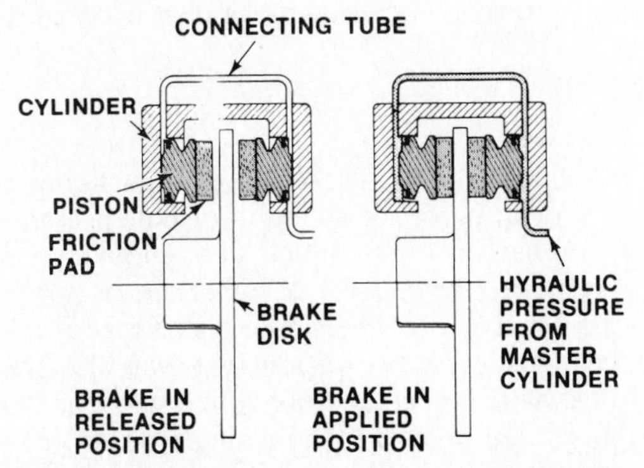

Fig. 3-47 Sectional views showing how hydraulic pressure forces the friction pads inward against the brake disk to produce braking action.

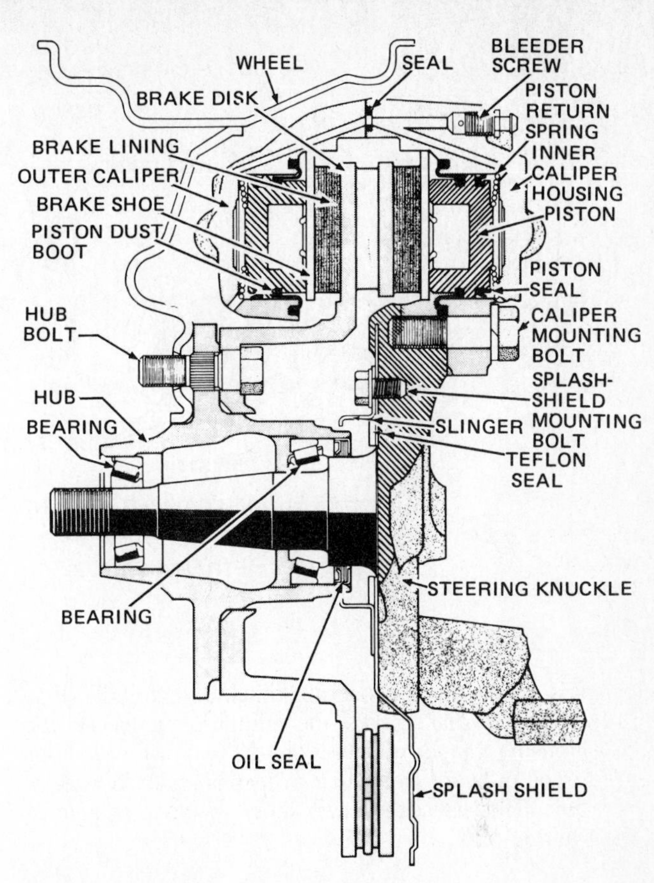

Fig. 3-48 Sectional view of a fixed-caliper disk brake. (*Chrysler Corporation*)

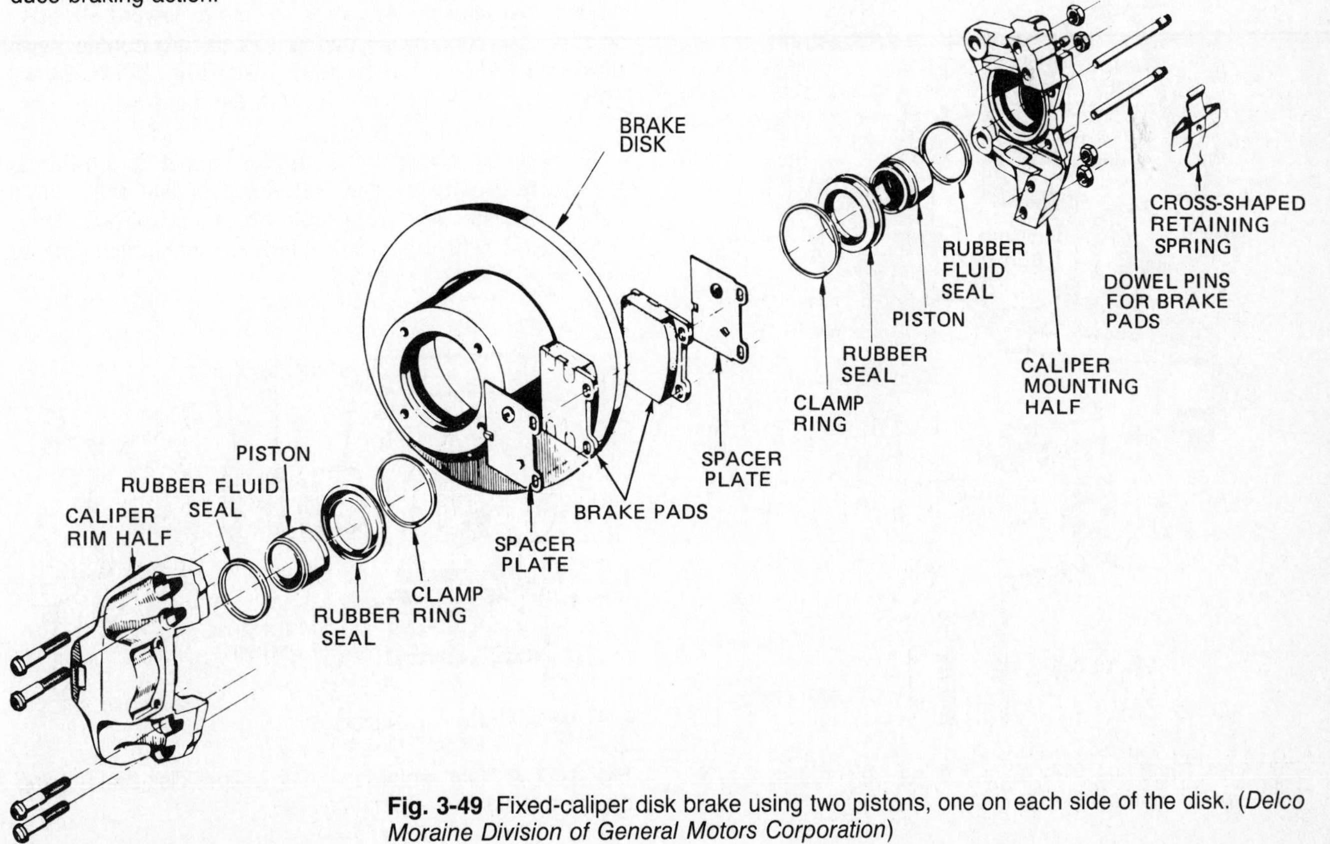

Fig. 3-49 Fixed-caliper disk brake using two pistons, one on each side of the disk. (*Delco Moraine Division of General Motors Corporation*)

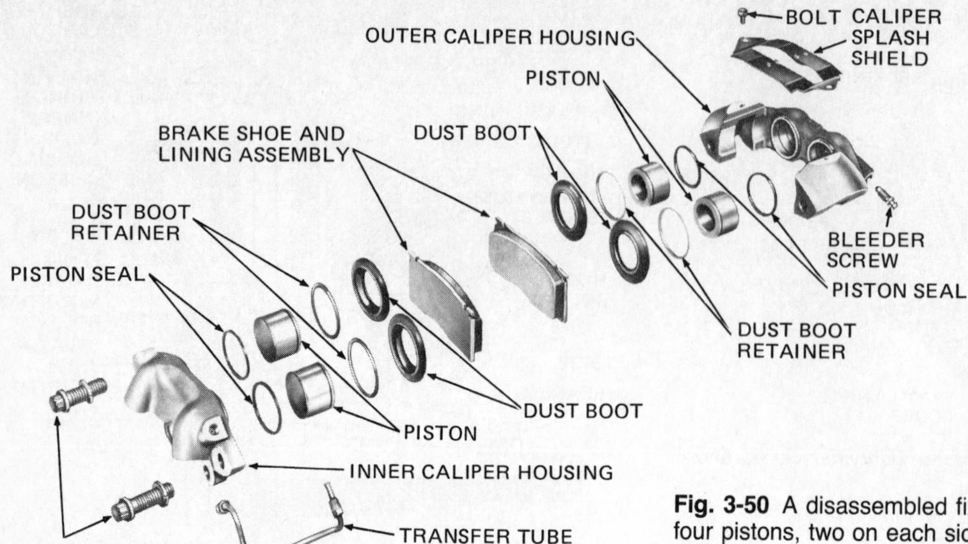

Fig. 3-50 A disassembled fixed-caliper assembly that uses four pistons, two on each side of the disk. (*Chrysler Corporation*)

car, it is attached to the steering knuckle, or spindle (Fig. 3-48). On rear disk brakes, the caliper is attached to the axle-housing flange. In operation, the two or four pistons are forced outward from their caliper bores by hydraulic pressure. This causes the two shoes to move in against the rotating disk.

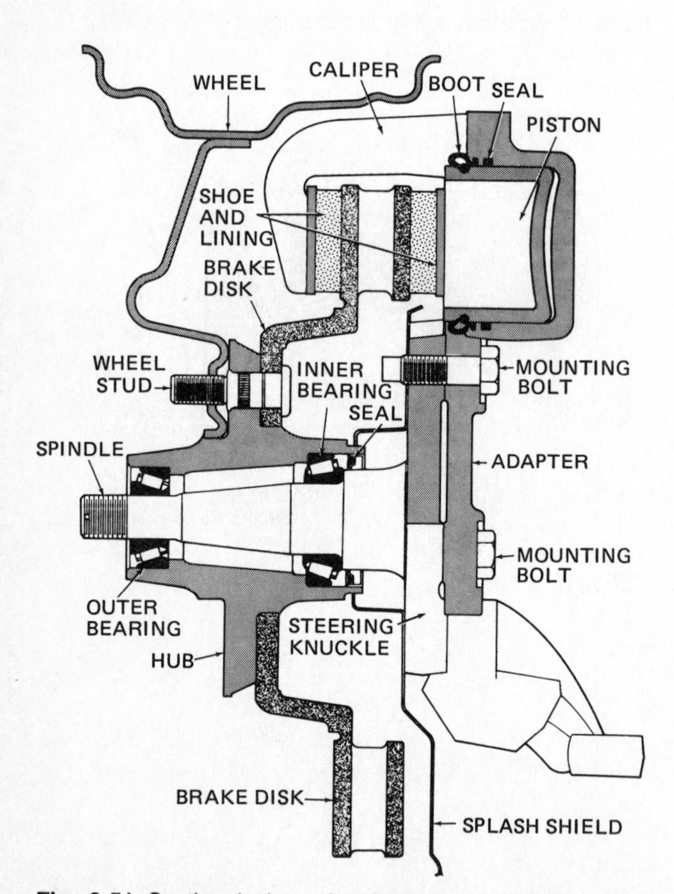

Fig. 3-51 Sectional view of a floating-caliper disk brake. (*Chrysler Corporation*)

2. Floating caliper The floating, or swinging caliper (Fig. 3-51) can pivot, or swing in or out. It is mounted through rubber bushings which give enough to permit this movement. The caliper has either one or two pistons (Figs. 3-52 and 3-53). The single-piston type is used for passenger cars. The two-piston type was designed for light trucks which need additional braking. The two pistons apply more pressure than a single piston.

In operation, hydraulic pressure back of the piston forces the brake pad on the piston side against the rotating disk. This produces a reaction force against the caliper that causes the caliper to move inward slightly, so that the other brake pad is forced into contact with the other side of the rotating disk (Fig. 3-54). Now, braking action is the same as with the fixed-caliper type.

3. Sliding caliper The sliding caliper is similar to the floating caliper. The difference is that the sliding caliper is suspended from rubber bushings on bolts (Fig. 3-55). Sliding calipers are supplied with either one or

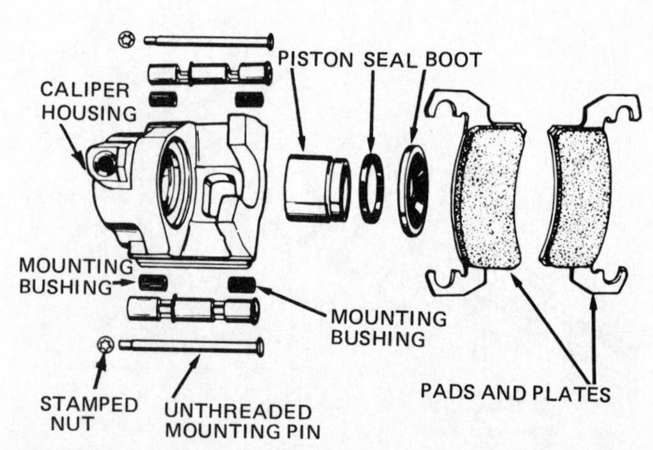

Fig. 3-52 A disassembled floating-caliper disk brake using one piston. (*Bendix Corporation*)

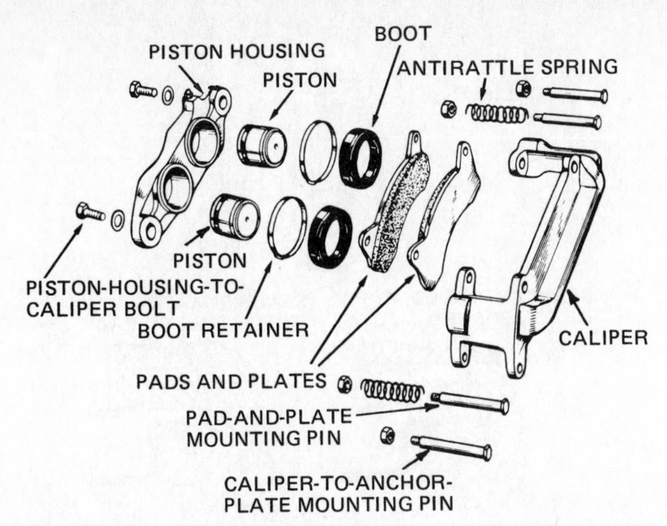

Fig. 3-53 A disassembled floating-caliper disk brake using two pistons. (*Bendix Corporation*)

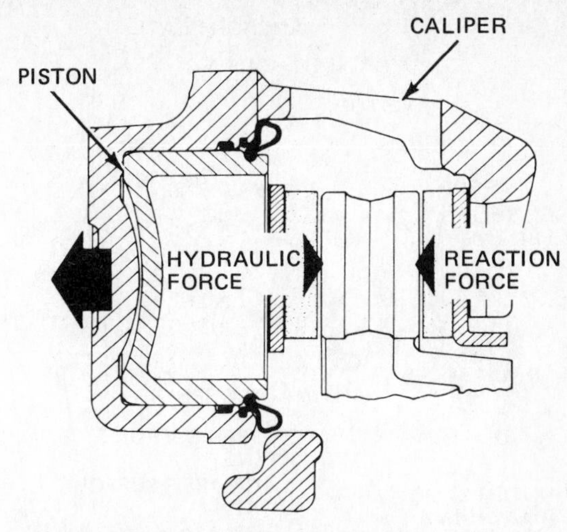

Fig. 3-54 Action of the floating-caliper disk brake. The hydraulic force in back of the piston pushes the caliper to the left while the piston is moving to the right. (*Ford Motor Company*)

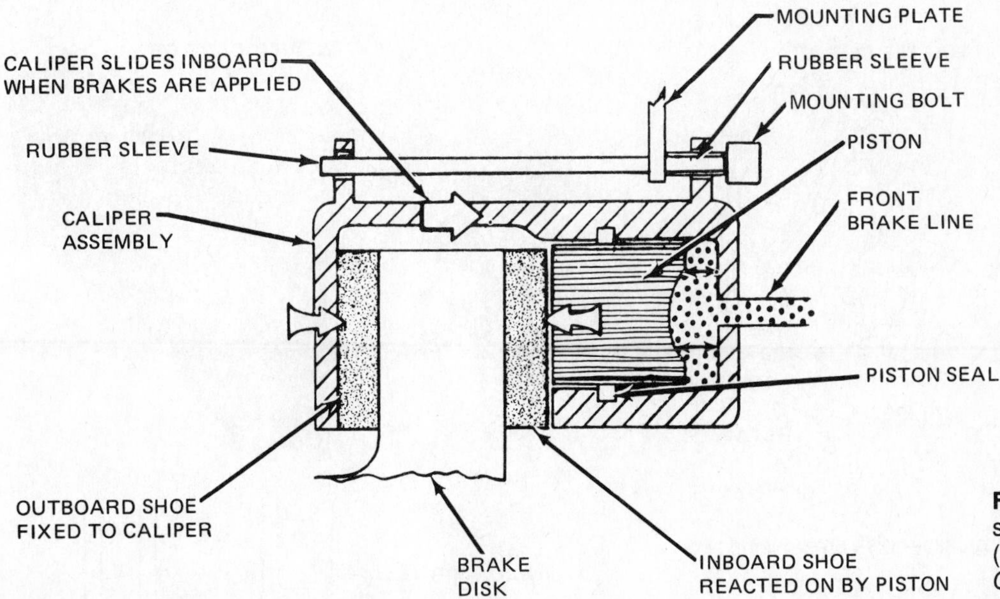

Fig. 3-55 Operation of the sliding-caliper disk brake. (*Chevrolet Motor Division of General Motors Corporation*)

two pistons (Figs. 3-56 to 3-58). The caliper shown in Fig. 3-56 is for a passenger car. The caliper shown in Fig. 3-57 is for motor homes. The caliper shown in Fig. 3-58 is for heavy-duty trucks. It is similar to the two-piston floating caliper shown in Fig. 3-53 except for the mounting arrangement.

When hydraulic pressure is applied back of the piston or pistons, the piston or pistons move and push the inboard shoe into hard contact with the rotating disk. At the same time, the hydraulic pressure on the inner end of the piston cylinder or cylinders causes the caliper to slide inward (Fig. 3-55). This brings the outboard shoe into hard contact with the outer surface of the rotating disk. The pressure of the shoes on the two sides of the disk provide the braking effort.

NOTE: In many disk brakes, the brake linings on the brake pads or shoes contact the brake disk at all times. This is not a heavy contact (except when braking). It is a light sliding contact that keeps the disk wiped clean of foreign matter. Because of this ''zero'' clearance, relatively little brake-pedal movement is required to produce braking action.

⚙ 3-21 Self-adjustment of disk brakes Disk brakes are self-adjusting. On some types, there is a piston-return spring back of each piston. This is a light spring that keeps the piston in the forward position and holds the brake lining lightly against the disk when the brakes are released. In this position, only a small additional movement of the pistons will force the lining against the disk for quick braking.

Another system provides automatic self-adjustment by means of the piston seals. When the brakes are applied, the piston slides toward the disk, distorting the piston

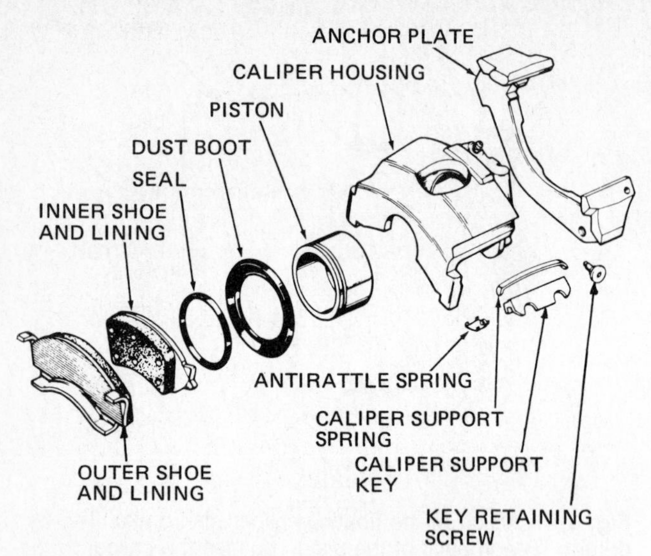

Fig. 3-56 Disassembled sliding-caliper disk brake. (*Bendix Corporation*)

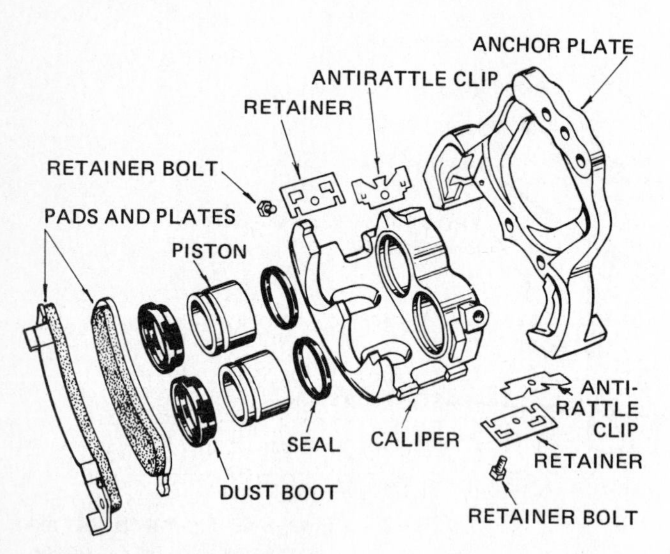

Fig. 3-57 Disassembled sliding-caliper disk brake using two pistons. (*Bendix Corporation*)

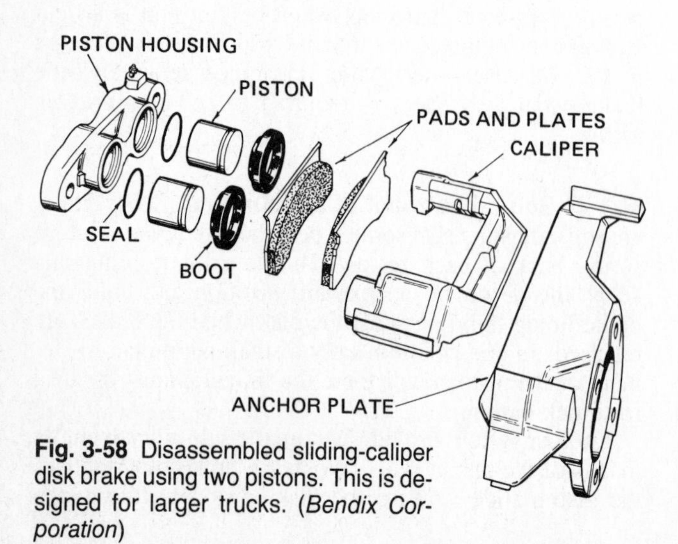

Fig. 3-58 Disassembled sliding-caliper disk brake using two pistons. This is designed for larger trucks. (*Bendix Corporation*)

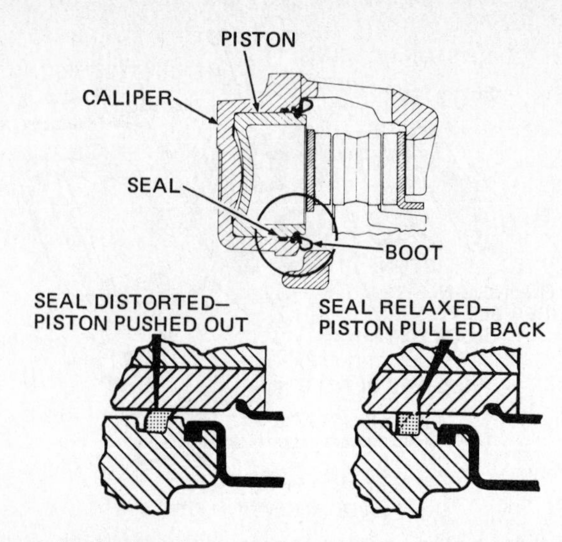

Fig. 3-59 Seal action, providing the self-adjusting of disk brakes as the piston moves out and back. (*Ford Motor Company*)

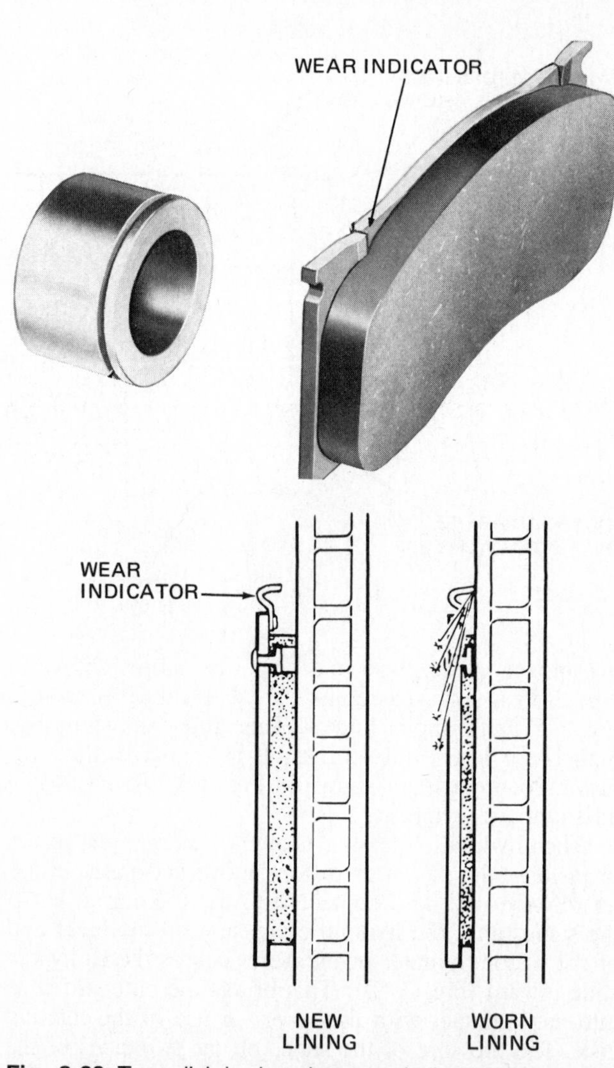

Fig. 3-60 Top, disk-brake piston and shoe-and-lining assembly, showing one type of wear indicator. Bottom, when the lining wears down, the wear indicator scrapes on the disk and makes a warning noise. (*Chrysler Corporation; Chevrolet Motor Division of General Motors Corporation*)

seal, as shown in Fig. 3-59. Then, when the brakes are released, relaxation of the seal draws the piston slightly away from the disk.

As the brake linings wear, piston travel tends to exceed the seal deflection limit. Then the piston slides outward through the seal to compensate for lining wear.

3-22 Wear indicators Many brake shoes for disk brakes have a wear indicator made into the shoe (Fig. 3-60, top) or attached to it (Fig. 3-60, bottom). When the brake linings are worn down to where they should be replaced, the wear-indicating tabs, or sensors, rub against the disk. This makes a loud scraping noise that warns the driver that the brake shoes must be replaced.

3-23 Metering valve Some cars equipped with front-disk and rear-drum brakes include a metering valve in the hydraulic line to the front brakes (Figs. 3-61 and 3-62). This valve works to improve front-to-rear braking balance during light braking. It prevents the front disk brakes from applying until after the rear brake shoes overcome the shoe return springs and the linings contact the drums. If the front brakes applied first, the rear wheels would lockup and skid. The metering valve is sometimes called a *hold-off valve*.

3-24 Proportioning valve A proportioning valve is used in the rear brake line of some cars with front-disk and rear-drum brakes (Figs. 3-61 and 3-62). The proportioning valve improves front-to-rear braking balance during hard braking. This is when more of the car weight is transferred to the front wheels (Fig. 3-21). As a result, more braking is needed at the front wheels and less at the rear wheels. If normal braking continued, the rear wheels could lockup and skid.

The proportioning valve reduces the pressure to the rear-wheel brakes when hard braking and high fluid pressures develop.

3-25 Combination valve Many cars equipped with front-disk and rear-drum brakes have a combination valve in the hydraulic system (Fig. 3-63). In the combination valve, the pressure-differential valve (3-9), metering valve (3-23), and proportioning valve (3-24) are combined into a single unit.

All combination valves include the failure warning switch, but not all contain the metering valve or proportioning valve. These valves are not used in all systems. However, all combination valves look the same from the outside, and all serve as the front junction block.

Antilock brake systems

3-26 Antilock brake systems About 40 percent of all car accidents involve skidding. The most efficient braking takes place when the wheels are still revolving (2-21). If the brakes lock the wheels so that the tires skid, kinetic friction results, and braking is much less effective. To prevent skidding and provide maximum effective braking, several antilock devices have been developed. Some provide skid control at the rear wheels only. Others provide control at all four wheels.

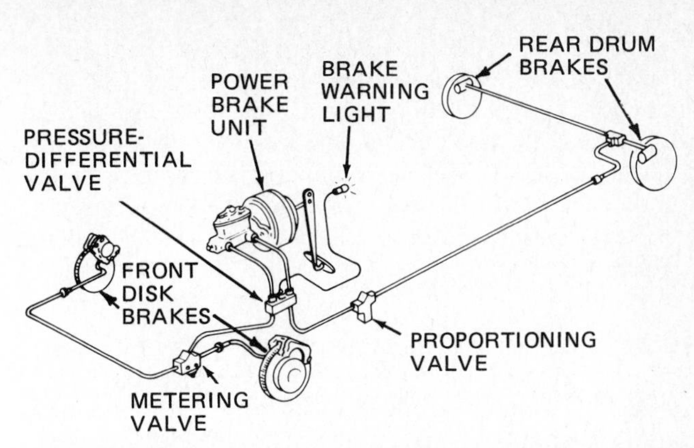

Fig. 3-61 Valves used in the hydraulic systems of a car with front-disk and rear-drum brakes. (*Wagner Lockheed*)

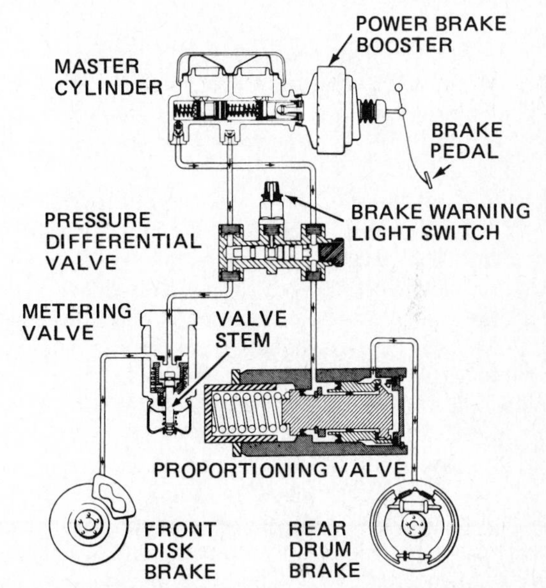

Fig. 3-62 Sectional view of the valves used in the hydraulic systems of a car with front-disk and rear-drum brakes. (*Ford Motor Company*)

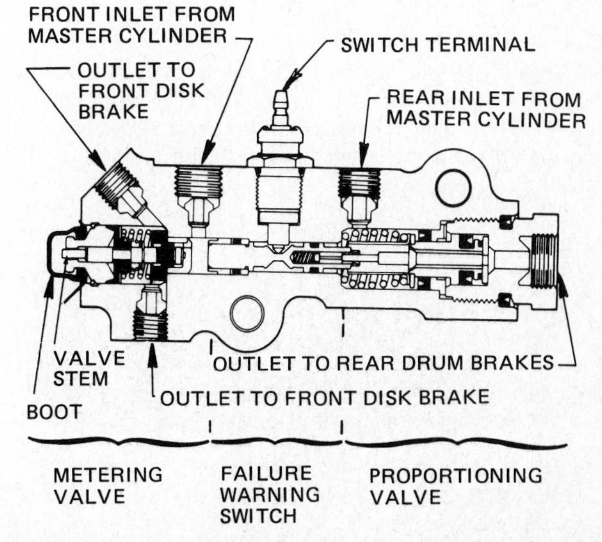

Fig. 3-63 Combination valve with warning-light switch, metering valve, and proportioning valve all in the same assembly. (*Delco Moraine Division of General Motors Corporation*)

"Control" means that as long as the wheels are rotating, the antilock device permits normal application of the brakes. But if the brakes are applied so hard that the wheels tend to stop turning, and a skid starts to develop, the device comes into operation. It partly releases the brakes so that the wheels continue to rotate. However, braking continues. But it is held to just below the point where a skid would start. The result is maximum braking effect.

NOTE: Antilock systems were offered at one time as optional equipment on various car models. The public apparently was not interested, and therefore the option was dropped for most passenger cars. However, antilock braking systems are used in some heavy-duty trucks and buses. The following section describes a typical passenger-car antilock brake system.

⚙ 3-27 Chrysler Sure-Brake The Chrysler antilock brake system is designed to prevent wheel lockup when the brakes are applied while the car is moving at above about 5 mph [8 km/h]. Figure 3-64 shows the antilock system layout on the car.

At the front wheel there is a magnetic wheel which acts as a speed sensor, attached to the brake disk. As the wheel and disk revolve, the magnetic wheel produces alternating current (ac) in the sensor. The sensor is a coil of wire, or a winding. A similar action takes place at the other wheels. These ac signals from the car wheels are fed into a logic control unit, located in the trunk.

When the brakes are applied, the logic control compares the ac signals from the wheels. The frequency of the ac increases with speed. As long as the frequency of the ac from all wheels is about the same, normal braking will take place. However, if the ac from any wheel is slowing down too rapidly, it means that the wheel is also slowing down too rapidly. The tire is beginning to skid.

When the logic control unit senses a rapid drop in the frequency of the ac, it signals modulators at the front of the car. Figure 3-64 shows the locations of the front-wheel and rear-wheel modulators. The hydraulic pressure from the master cylinder to the wheel cylinder or calipers passes through these modulators. When the logic control unit senses that a wheel is about to skid, it signals the modulator to reduce the hydraulic pressure to the brake for that wheel. When the pressure is reduced, the braking effect at that wheel is reduced, so the skid is prevented.

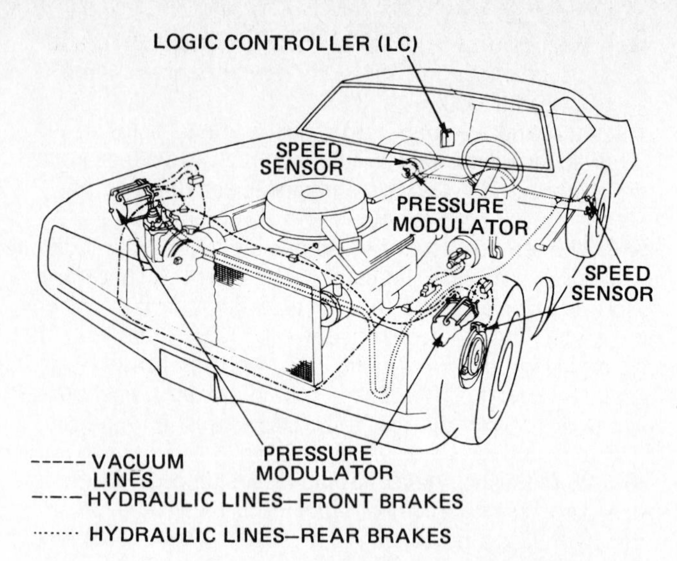

Fig. 3-64 Antilock-brake system used by Chrysler, showing the locations of the components and their connections to the vacuum and hydraulic lines. (*Chrysler Corporation*)

Chapter 3 review questions

Select the *one* correct, best, or most probable answer to each question. Then check your answers against the correct answers given at the end of the book.

1. The service brakes are operated:
   a. electrically
   b. hydraulically
   c. by air pressure
   d. by vacuum
2. During braking, the brake shoe is moved outward to force the lining against the:
   a. wheel piston or cylinder
   b. anchor pin
   c. brake drum or disk
   d. wheel rim or axle
3. A piston with an area of 2.5 square inches is packed into a liquid-filled cylinder with a force of 500 pounds. The pressure on the liquid is:
   a. 200 psi
   b. 500 psi
   c. 1250 psi
   d. 5000 psi
4. With a pressure of 800 psi acting on it, the force on a piston with an area of 0.4 square inch would be:
   a. 200 pounds
   b. 320 pounds
   c. 500 pounds
   d. 5000 pounds
5. If the piston in the master cylinder has an area of 0.8 square inches and the push applied on it is 800 pounds, the hydraulic pressure in the brake fluid is:
   a. 100 psi
   b. 640 psi
   c. 1000 psi
   d. 6400 psi
6. In most drum brakes, each wheel cylinder contains:
   a. one piston
   b. two pistons
   c. three pistons
   d. four pistons
7. When comparing the front and rear wheel-cylinder pistons, the pistons in the front wheel cylinders usually are:
   a. larger in diameter
   b. smaller in diameter
   c. the same size
   d. none of the above

8. In the dual-brake system, the master cylinder has:
   a. one piston
   b. two pistons
   c. three pistons
   d. four pistons
9. There are three types of disk brake:
   a. fixed-caliper, tab-action, and two-piston
   b. fixed-caliper, sliding-caliper, and floating-caliper
   c. floating-caliper, swinging-caliper, and proportioning
   d. all of the above
10. The antilock control system called Sure-Brake uses a vacuum-powered actuator, an electronic control module, and:
    a. wheel sensors
    b. a load-sensing valve
    c. a governor
    d. all of the above
11. When used, the purpose of the residual check valve is:
    a. to allow the driver to pump up the brakes
    b. to prevent air from entering the hydraulic system
    c. to prevent wheel lockup by reducing the hydraulic pressure
    d. all of the above
12. The brake warning light warns the driver of:
    a. low fluid level in the master cylinder
    b. air in the hydraulic system
    c. failure of the primary or secondary section of the hydraulic system
    d. power-brake failure
13. Mechanic A says that most self-adjuster mechanisms are attached to the secondary brake shoe. Mechanic B says that self-adjuster mechanisms are attached to the primary brake shoe. Who is right?
    a. A only
    b. B only
    c. both A and B
    d. neither A nor B
14. The purpose of the spring that is sometimes found around the outside of a brake drum is to:
    a. strengthen the drum
    b. dissipate heat
    c. prevent bellmouthing
    d. dampen chatter and noise
15. On a drum brake with a cable-type self-adjuster attached to the secondary shoe, what action must be taken to operate the self-adjuster?
    a. Apply the brake with the car moving forward.
    b. Apply the brake with the car moving backward.
    c. Apply the parking brake.
    d. Release the parking brake.
16. The brake warning light is turned on by the:
    a. metering valve
    b. proportioning valve
    c. pressure-differential valve
    d. none of the above
17. On a car with front-disk and rear-drum brakes, the front brakes grab when light pedal pressure is applied. This problem could be caused by a defective:
    a. proportioning valve
    b. pressure-differential valve
    c. metering valve
    d. check valve
18. On a car with front-disk and rear-drum brakes, the rear brakes lockup when the brake pedal is depressed normally. Which of these could cause this problem?
    I. Leaking front-caliper seals
    II. Defective proportioning valve
    a. I only
    b. II only
    c. either I or II
    d. neither I nor II
19. In the dual master cylinder, the primary piston is the piston that is:
    a. directly actuated by the brake-pedal pushrod
    b. closest to the fire wall
    c. toward the rear of the car
    d. all of the above
20. One cause of a low brake-fluid level in the master cylinder could be:
    a. normal wearing away of the brake lining
    b. a defective brake drum
    c. a defective rotor
    d. excessively hard brake lining

CHAPTER 4

POWER BRAKES

After studying this chapter, you should be able to:

1. Explain how atmospheric pressure and vacuum can be used to move a piston in a cylinder.
2. List and describe the types of vacuum-assisted power brakes.
3. Describe the operation of the hydraulic-assisted power brake.
4. Discuss the construction and operation of a basic air-brake system.
5. Explain the operation of an electric brake.

4-1 Power brakes For hard braking and fast stops, a considerable pressure must be exerted on the brake pedal with the brake system described in Chap. 3. Also, the heavier the vehicle, the greater the braking effort required. For many years, buses and trucks have used special equipment that assists the driver to brake the vehicle. This equipment may use either compressed air or vacuum. When the driver applies the brake, the compressed air or vacuum supplies most of the effort required for braking. Other systems use electric and hydraulic devices for braking.

In recent years, more than 80 percent of all U.S.-built passenger cars have been supplied with *power brakes*. These are vacuum- or hydraulic-assist systems. Essentially, they all operate in a similar manner. When the brake pedal is moved to apply the brakes, a valving arrangement is actuated. With the vacuum-assist system, the valves admit atmospheric pressure on one side of a piston or diaphragm and apply vacuum to the other side. The piston or diaphragm then moves toward the vacuum side. This movement transmits most of the hydraulic pressure, through the brake fluid, to the wheel cylinders or calipers. Operation of the hydraulic-assist system is described in later sections.

⚙ 4-2 Atmospheric pressure and vacuum The atmospheric pressure is about 15 psi [103 kPa] at sea level (⚙ 2-3 to 2-6). Vacuum is an absence of air. If we arranged a simple cylinder and piston (as shown in Fig. 4-1) and then applied atmospheric pressure to one side and vacuum to the other, the piston would move toward the vacuum side. If we held the piston stationary, we could calculate the pressure, or push, being exerted on it, provided we knew the area of the piston, the atmospheric pressure, and the amount of vacuum. Suppose the piston had an area of 50 square inches (about 8 inches in diameter). We'll assume also that the atmospheric pressure is 15 psi [103 kPa] and that the vacuum is great enough to have brought the pressure down to only 5 psi [34 kPa]. With 15 psi [103 kPa] on one side, and only 5 psi [34 kPa] on the other, the difference in pressure is 10 psi [69 kPa]. There is an effective pressure of 10 pounds on every square inch of the piston area. Since there are 50 square inches of piston area, the force on the piston, pushing it to the vacuum side, is 500 pounds (50×10). It is this effective pressure, or push, that is utilized in power brakes. The vacuum is supplied by the automobile engine. In one way, the engine is a vacuum pump (⚙ 2-6). With every intake stroke, the downward-moving piston produces a partial vacuum in the cylinder

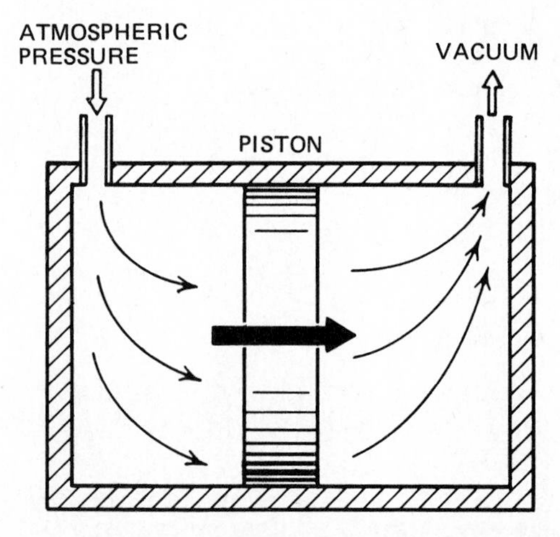

Fig. 4-1 If atmospheric pressure is applied to one side of a piston and vacuum to the other side, the piston will move toward the vacuum side.

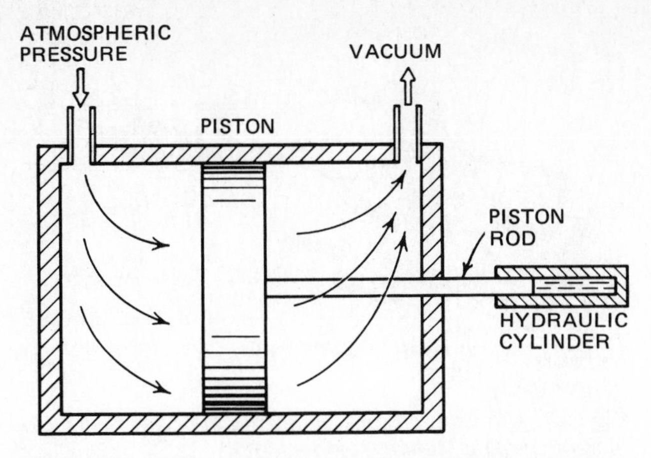

Fig. 4-2 If the piston rod is placed in a hydraulic cylinder, the pressure on the piston in the atmospheric-pressure cylinder will be converted into hydraulic force.

and therefore in the intake manifold. The vacuum side of the power-brake cylinder (Fig. 4-1) is connected to the intake manifold so that it can utilize intake-manifold vacuum.

⚙ 4-3 Putting the vacuum to work If we add a hydraulic cylinder to the cylinder and piston of Fig. 4-1 as shown in Fig 4-2, we can utilize the push on the piston to produce hydraulic pressure. All the pressure on the piston is carried through the piston rod and into the hydraulic cylinder. Therefore, in the example described above, the piston rod would push into the hydraulic cylinder with a 500-pound force. If the end of the piston rod had an area of 0.5 square inch, the pressure in the hydraulic fluid would be 1000 psi [6895 kPa] (or 500 pounds divided by the area, 0.5 square inch). The hydraulic pressure can be altered by changing the size of either the piston or the rod (with the same pressure differential acting on the piston). For example, a piston with an area of 100 square inches and a rod of 0.2 square inch area would produce a hydraulic pressure of 5000 psi [34,474 kPa] (or 1000 pounds divided by 0.2 square inch).

If the hydraulic cylinder is connected to the wheel cylinders (as shown in Fig. 4-3), the hydraulic pressure produced will result in braking action. Note that even though the hydraulic cylinder has been increased in diameter, the piston rod entering it (Fig. 4-3) still displaces liquid. It still produces the same pressure increase as it would if the cylinder were the same size as the rod (as shown in Fig. 4-2).

Automotive power brakes

⚙ 4-4 Types of vacuum-assisted power brakes Vacuum-assisted power brakes can be divided into two general categories: vacuum-suspended and atmospheric-suspended. In the vacuum-suspended type, intake-manifold vacuum is applied to both sides of a piston or diaphragm in the power-brake unit when no braking action is taking place. To produce braking, atmospheric pressure is admitted to one side of the piston or diaphragm.

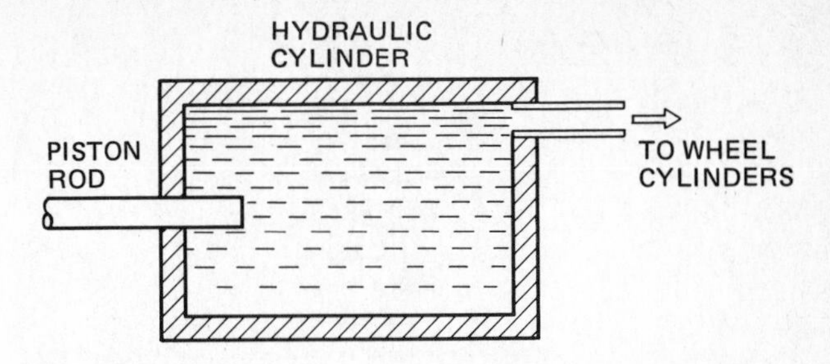

Fig. 4-3 If the hydraulic cylinder is connected by tubes to the wheel cylinders, movement of the piston rod into the hydraulic cylinder will produce braking action.

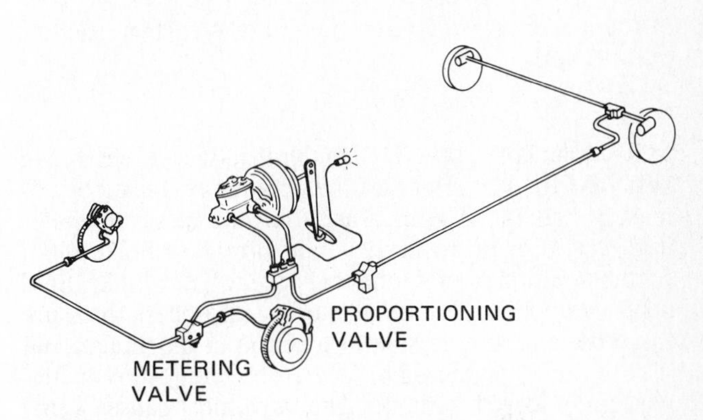

Fig. 4-4 An integral type of power-brake system. (*Wagner Lockheed*)

The difference in pressures causes the piston or diaphragm to move, producing the braking action. Most automotive power-brake units are of this type.

In the atmospheric-suspended type, atmospheric pressure is applied to both sides of the diaphragm or piston. To produce braking action, one side must be connected to a source of vacuum, such as the intake manifold. However, this type of power brake will not operate if the engine is not running, unless there is a reserve vacuum supply. In some applications, a small vacuum tank is included to provide enough vacuum for several brake applications after the engine has stopped.

Both types of power brake are "fail safe." They can be operated even though the power assistance is not operating. If the power diaphragm or piston does not operate, braking can still be achieved. However, a much heavier brake-pedal application is necessary.

The vacuum-assisted power brakes can be classified in another way, as the integral type, the multiplier type, and the assist type.

1. Integral type The integral-type power-brake system (Fig. 4-4) has the brake master cylinder as an integral part of the power-brake assembly. When the brake pedal is operated, it actuates a valve in the power-brake assembly that applies atmospheric pressure to one side of a piston or diaphragm and vacuum to the other side. This causes the piston or diaphragm to move. The movement forces a piston to move into the master cylinder and therefore apply the brakes. An integral-type system is described in ⚙ 4-6. Most passenger cars and light trucks use this type of system.

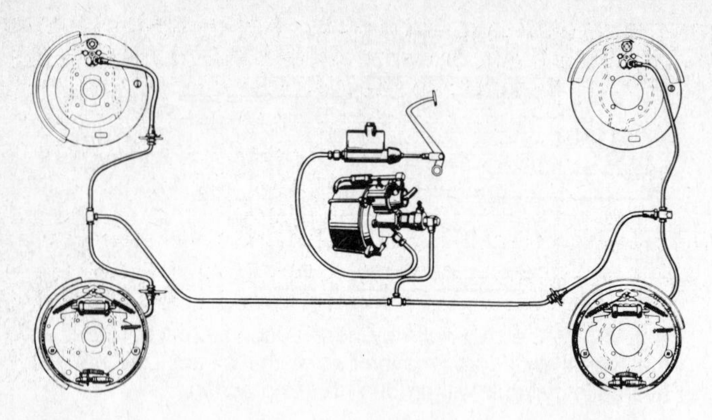

Fig. 4-5 A multiplier type of power-brake system. (*Bendix Corporation*)

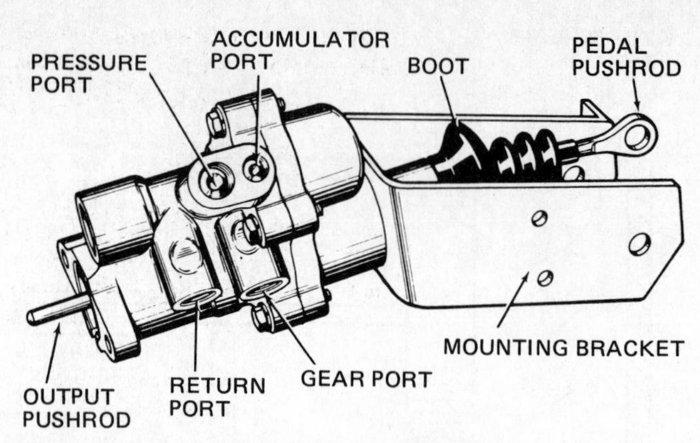

Fig. 4-7 A hydraulic-brake booster. (*Chevrolet Motor Division of General Motors Corporation*)

2. Multiplier type The multiplier-type power-brake system (Fig. 4-5) multiples the pressure produced by the master cylinder. The pressure from the master cylinder is directed to the multiplier unit through a brake tube. In the multiplier unit, the pressure of the brake fluid actuates a valve. The valve causes atmospheric pressure to be directed to one side of a piston or diaphragm and vacuum to the other side. Therefore, the piston or diaphragm is forced to move. The movement causes a piston to move in a hydraulic cylinder that is part of the multiplier unit. This produces a high hydraulic pressure which is carried by brake lines to the wheel cylinders. In this system, the relatively low pressure from the master cylinder, produced by brake-pedal movement, is multiplied to a high pressure by the multiplier unit. Therefore, a relatively light brake-pedal pressure produces heavy braking action. A multiplier-type power-brake system is described in ⚙ 4-8.

3. Assist type The assist-type power-brake system (Fig. 4-6) has a power-cylinder assembly that assists in applying the brakes through a mechanical linkage. When the brake pedal is moved, linkage to the power cylinder is actuated, causing valve action and therefore movement of a diaphragm or bellows within the cylinder. This movement is carried through linkage to the master cylinder, thereby increasing the total force being applied, which in turn increases the braking action.

⚙ 4-5 Hydraulic-assisted brake booster The hydraulic-assisted brake booster (Fig. 4-7) uses hydraulic pressure supplied by the power-steering pump to assist in applying the brakes. Power steering is covered in a later chapter. Figure 4-8 shows a dissassembled booster unit. Figure 4-9 is a sectional view. Figure 4-10 shows the system schematically. The system shown in Fig. 4-10, which is for a truck, includes an oil cooler. Higher pressures and temperatures are developed in the power-steering fluid because of the greater steering and braking demands of vehicle operation. The fluid flows, as shown by the arrows in Fig. 4-10, from the pump to the steering gear, hydraulic booster, oil cooler, and then to the pump reservoir. Not all hydraulic-assist systems use an oil cooler.

When the power-steering pump is running, there is always pressure in the smaller cylinder in the brake booster. As the brakes are applied, the pressure through

Fig. 4-6 An assist type of power-brake system. (*Bendix Corporation*)

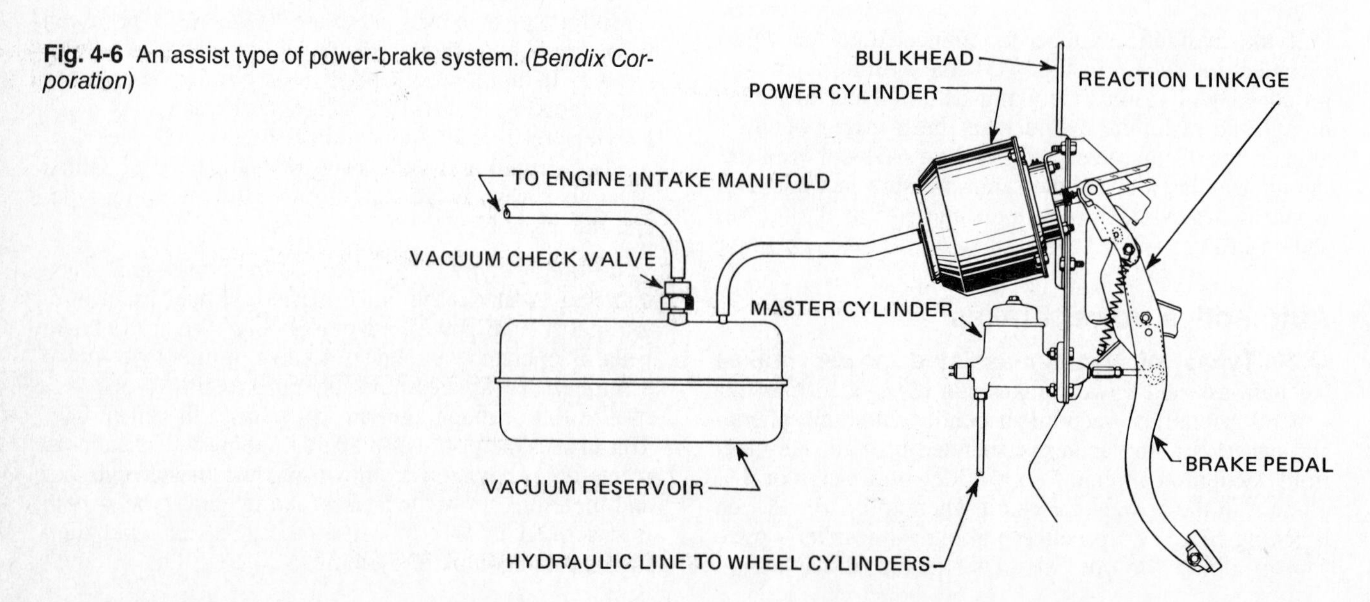

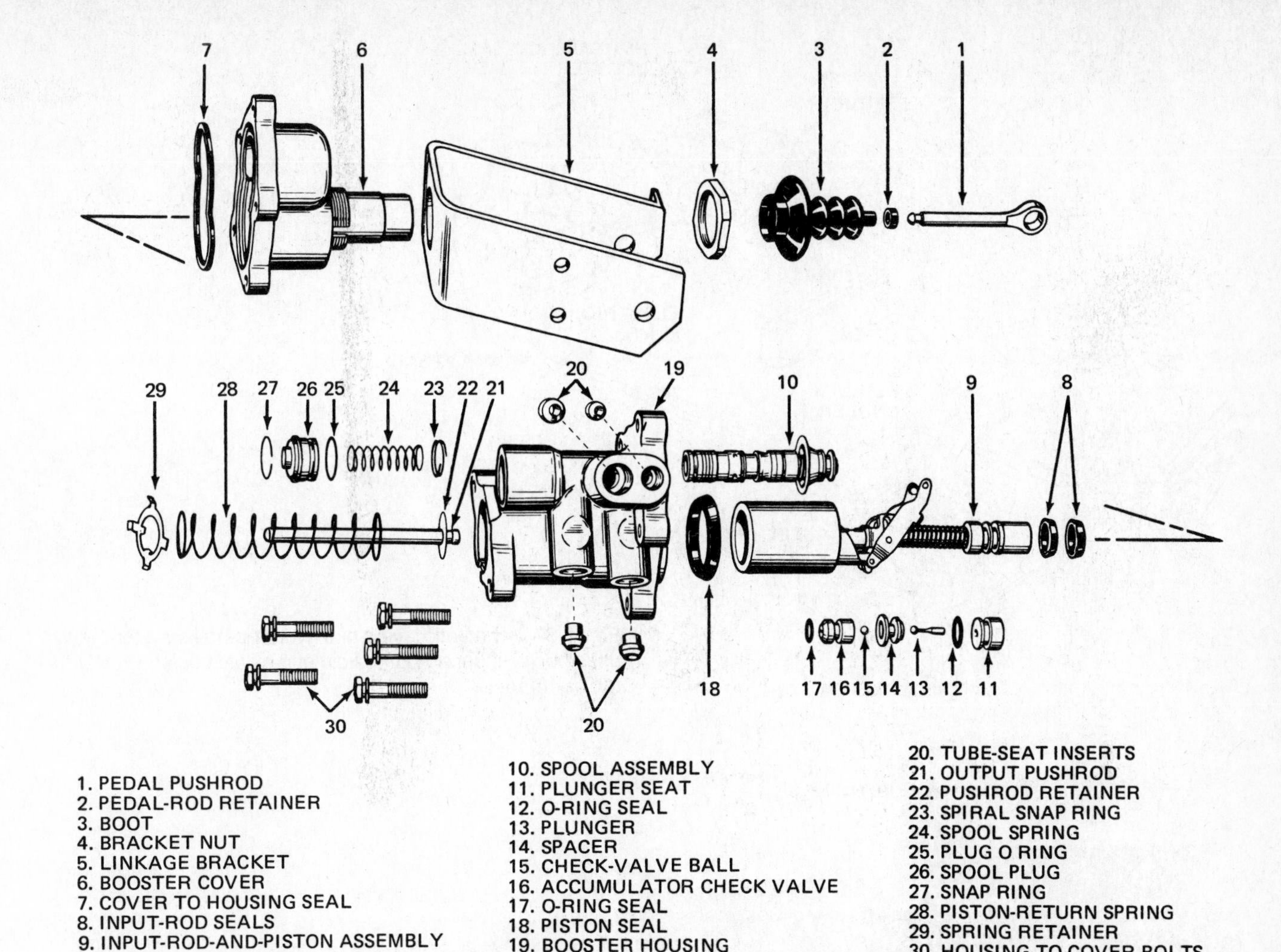

Fig. 4-8 Disassembled view of a hydraulic-brake booster. (*Chevrolet Motor Division of General Motors Corporation*)

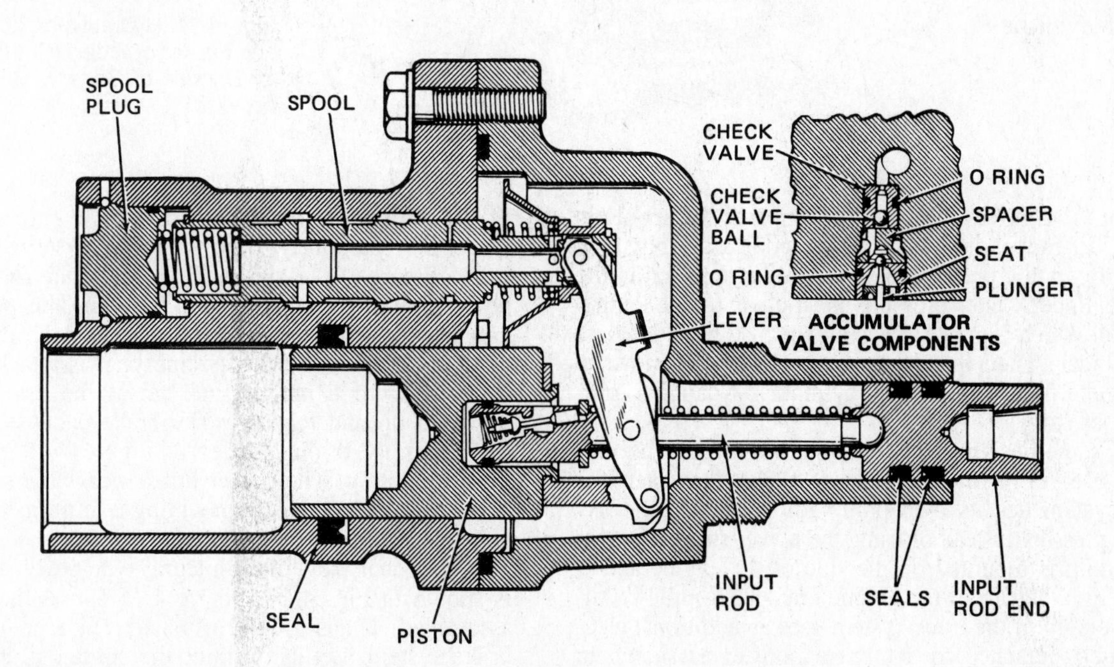

Fig. 4-9 Sectional view of a hydraulic-brake booster. (*Chevrolet Motor Division of General Motors Corporation*)

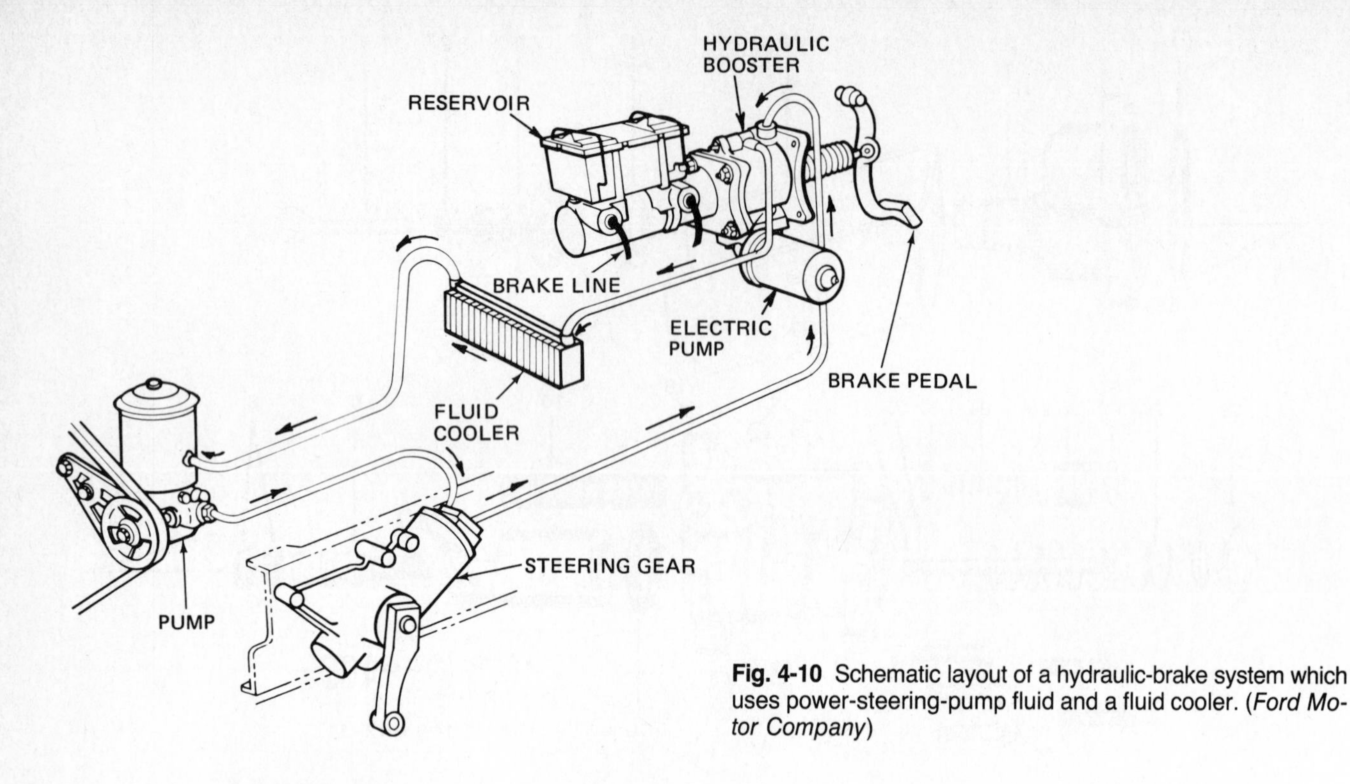

Fig. 4-10 Schematic layout of a hydraulic-brake system which uses power-steering-pump fluid and a fluid cooler. (*Ford Motor Company*)

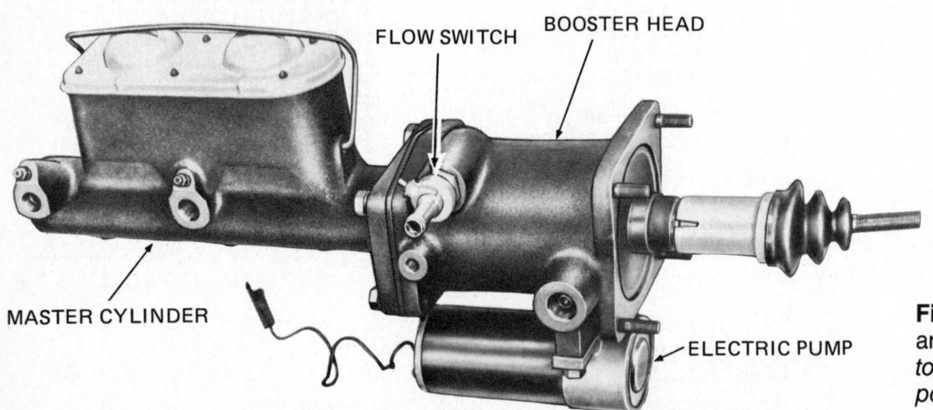

Fig. 4-11 Hydraulic-brake booster and master cylinder. (*Chevrolet Motor Division of General Motors Corporation*)

the input rod moves the lever. This forces the spool assembly to move off center. The spool assembly now admits hydraulic pressure back of the large piston in the large cylinder. This pressure is applied to the output pushrod, which is pressing against the pistons in the master cylinder. Then the master-cylinder pistons move to send brake fluid to the wheel cylinders or calipers, producing braking.

This same design of hydraulic-assisted brake booster is also used in medium and heavy trucks. However, the truck system has its own pump for producing the hydraulic pressure instead of using the power-steering pump. This pump is mounted on the side of the engine and is driven by a belt from the engine crankshaft pulley. Another version of the truck system uses an additional electric pump, attached to the brake booster as shown in Figs. 4-10 and 4-11. The electric pump assures adequate hydraulic pressure for the truck brakes.

❁ 4-6 Integral-type power brake Figure 4-4 illustrates the integral-type power-brake system schematically. Figures 4-12 and 4-13 are external views of the power-brake assembly, which includes the power unit and the dual master cylinder. Figure 4-12 shows one method of mounting the assembly. It is attached to the fire wall with a bracket that raises the assembly and moves it up and forward. The brake pedal is linked to the assembly through a lever. Figure 4-13 shows the flush mounting. The power brake works the same way. The only difference is in mounting it to fit into the available space in the engine compartment.

A sectional view of an integral-type power-brake unit is shown in Fig. 4-14. Figure 4-15 shows the unit disassembled. It has a tandem master cylinder for a dual-brake system. Let us examine the operation of a typical integral-power-brake assembly. Figure 4-16 shows the assembly in the released position, with the brakes not

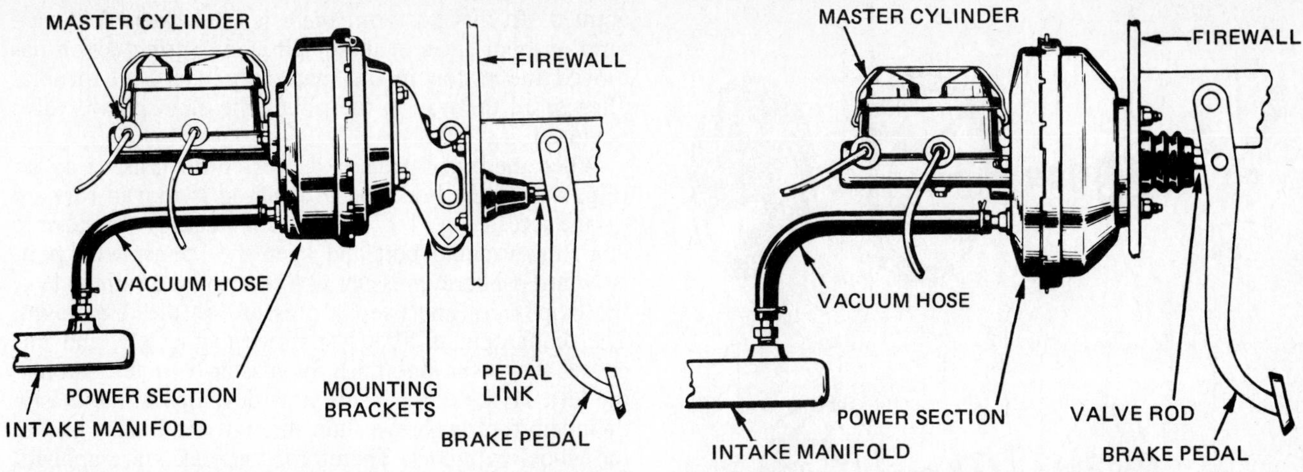

Fig. 4-12 Bracket-mounted power-brake unit. (*Bendix Corporation*)

Fig. 4-13 Flush-mounted power-brake unit. (*Bendix Corporation*)

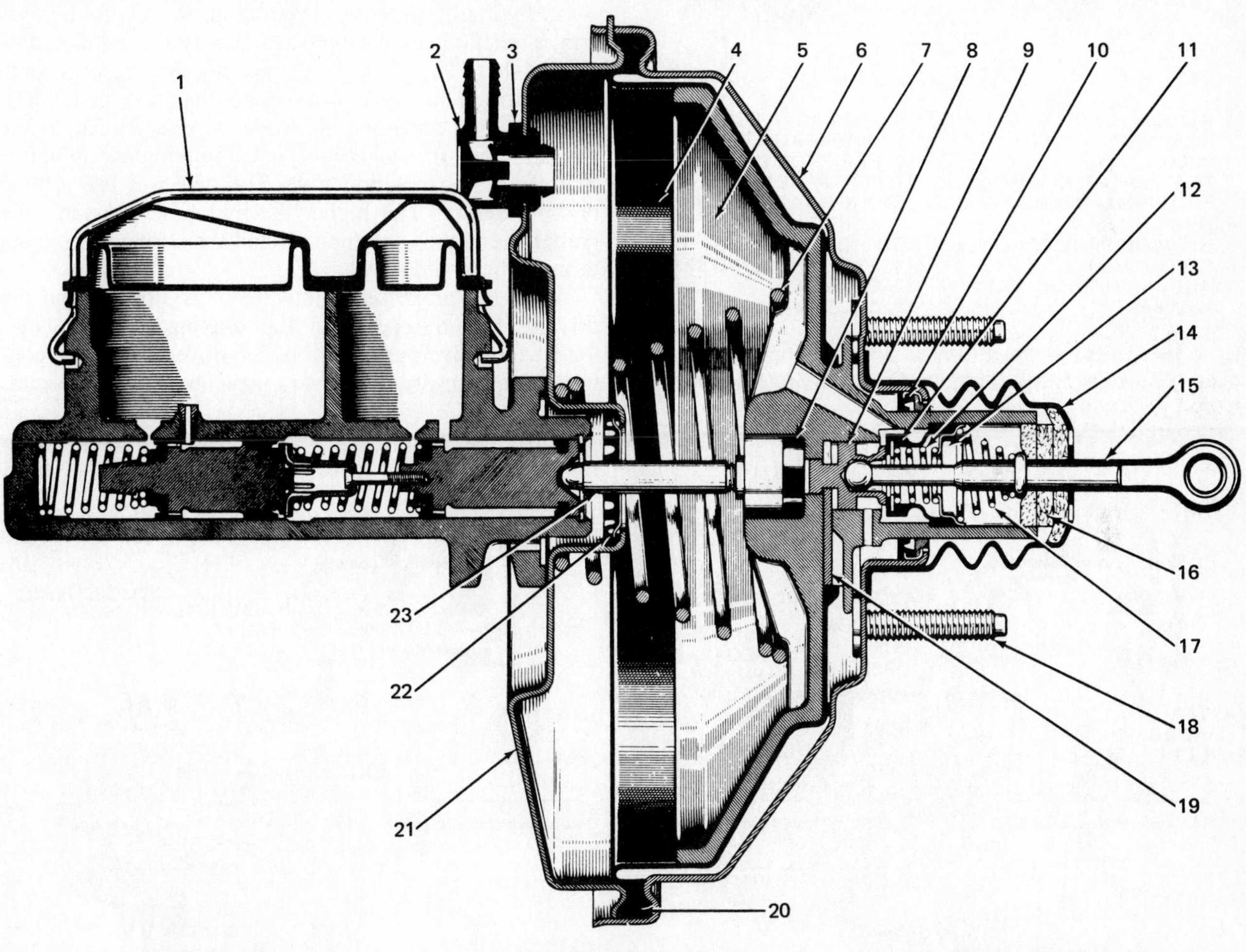

Fig. 4-14 Sectional view of a Bendix single-diaphragm power brake. (*Chevrolet Motor Division of General Motors Corporation*)

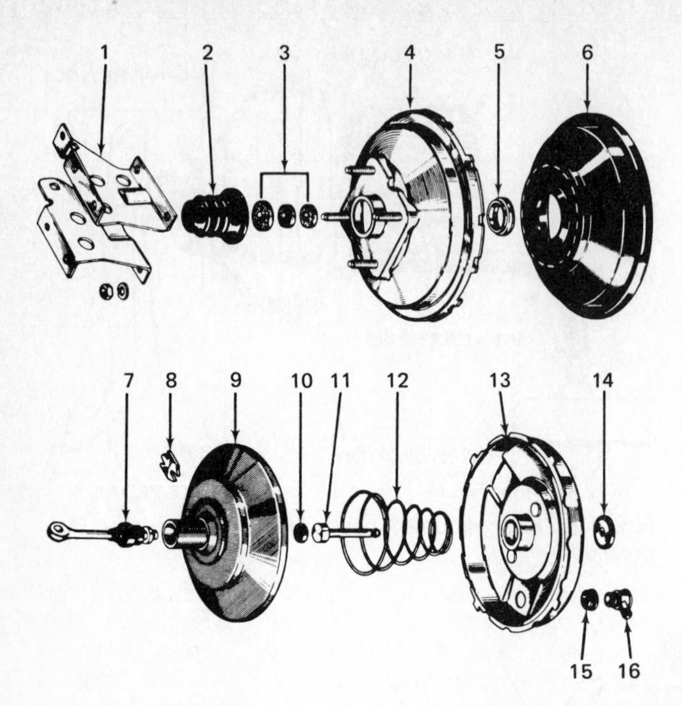

1. REAR-HOUSING MOUNTING BRACKETS (TRUCK)
2. PUSHROD BOOT
3. FOAM AND FELT AIR-FILTER SILENCERS
4. REAR HOUSING
5. REAR-HOUSING SEAL
6. DIAPHRAGM
7. AIR-VALVE PUSHROD ASSEMBLY
8. AIR-VALVE LOCK
9. DIAPHRAGM PLATE
10. REACTION DISK
11. PISTON ROD
12. DIAPHRAGM RETURN SPRING
13. FRONT HOUSING
14. FRONT-HOUSING SEAL
15. GROMMET
16. CHECK VALVE

Fig. 4-15 A disassembled Bendix single-diaphragm power brake. (*Chevrolet Motor Division of General Motors Corporation*)

applied. In this position, there is intake-manifold vacuum on both sides of the diaphragm. Spring action has moved the piston in the master cylinder and the diaphragm all the way to the right. The atmospheric valve is closed.

When the brake pedal is depressed for braking action (Fig. 4-17), the brake-pedal pushrod is moved forward (to the left in Fig. 4-17). This action causes the valve to close the vacuum port and open the atmospheric port. Now atmospheric pressure can enter on the right side of the diaphragm and exert a pressure on the diaphragm. The diaphragm is forced to move to the left, and this causes the pushrod to push the master-cylinder piston to the left. Hydraulic pressure now develops in the master cylinder, forcing brake fluid through the brake lines to the wheel cylinders. Therefore, the brakes are applied. The harder the driver presses on the brake pedal, the wider the atmospheric port is opened and the harder the diaphragm presses on the pushrod to produce braking.

As hydraulic pressure develops in the hydraulic system, a reaction counterforce acts against the reaction disk (Fig. 4-14). This disk then transmits the reaction force back through the valve pushrod and the brake pedal. The reaction, or ''push-back,'' force is proportional to the hydraulic pressure and to the actual braking taking place. Therefore, the reaction gives the driver a feel of the braking action. The higher the hydraulic pressure, the harder the brakes are applied, and the stronger the reaction on the brake pedal.

When the brake-pedal movement is stopped and the driver holds the pedal in the braking position (Fig. 4-18), the valve pushrod stops its movement of the control-valve plunger. However, the unbalanced pressures

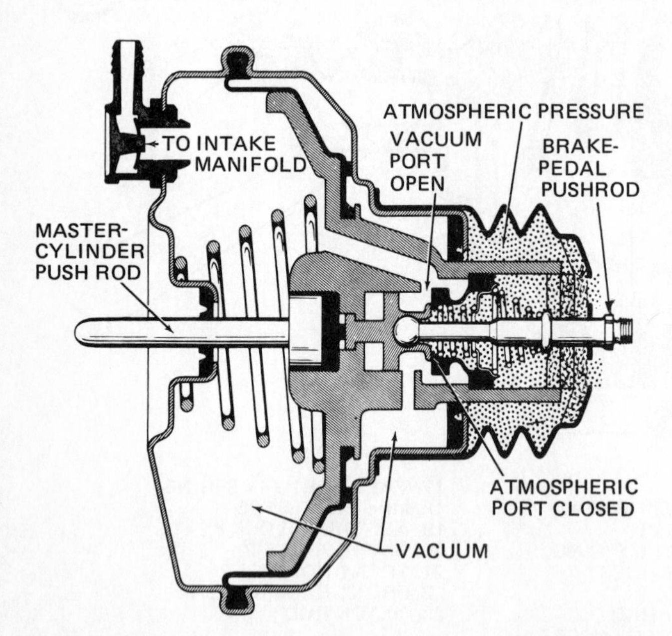

Fig. 4-16 Sectional view of a Bendix single-diaphragm power brake in the released position. (*Chevrolet Motor Division of General Motors Corporation*)

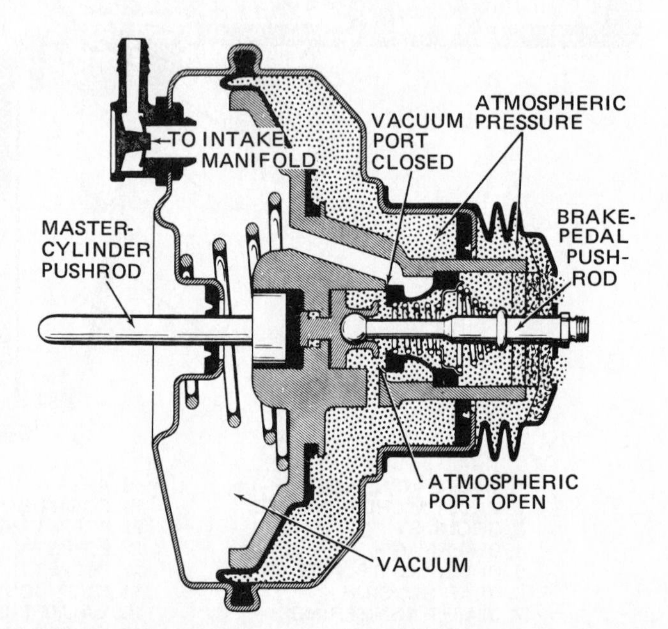

Fig. 4-17 Sectional view of a Bendix single-diaphragm power brake in the applying position. (*Chevrolet Motor Division of General Motors Corporation*)

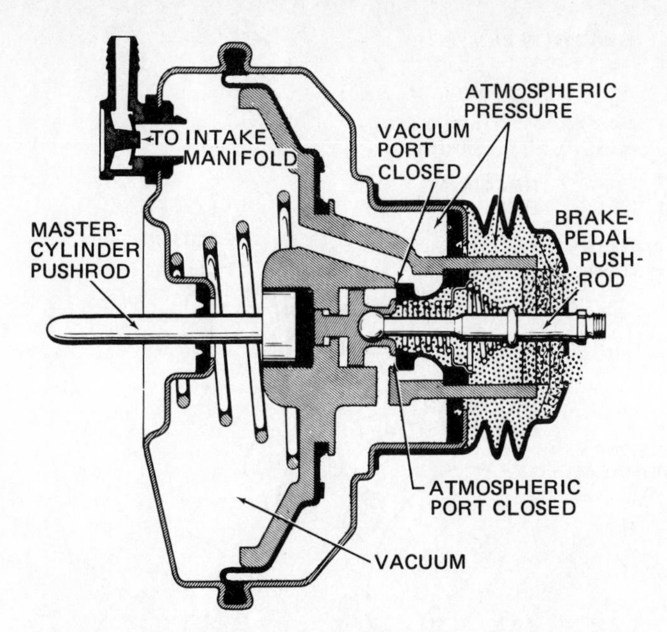

Fig. 4-18 Sectional view of a Bendix single-diaphragm power brake in the holding position. (*Chevrolet Motor Division of General Motors Corporation*)

on the two sides of the diaphragm will continue to move the outer sleeve of the control-valve plunger forward, keeping the vacuum port closed. At the same time, the reaction force acting on the reaction disk will tend to move the atmospheric valve to the closed position. When these forces reach a balance, the vacuum port will remain closed. The atmospheric valve will cut off any further passage of atmospheric pressure to the right side of the diaphragm. Therefore, the hydraulic pressure will be maintained at a constant value so that the braking effect continues.

When the brake pedal is released (Fig. 4-16), the spring action closes the atmospheric port and opens the vacuum port so that vacuum is applied to both sides of the diaphragm. Therefore, the brakes are released.

⚙ 4-7 Dual-diaphragm power brake The dual-diaphragm assembly (Fig. 4-19) is an integral type power brake similar to the unit discussed in ⚙ 4-6 above. However, it has a second diaphragm and plate. A disassembled view of a very similar construction is shown in Fig. 4-20. The purpose of the secondary diaphragm is to provide additional braking power. The unit works in the same manner as the power brake described in ⚙ 4-6.

⚙ 4-8 Multiplier-type power brake Figure 4-5 shows a multiplier-type brake system schematically. Figures 4-21 to 4-23 show a brake assembly of this type. In operation, when the brakes are applied, the hydraulic pressure from the master cylinder is applied to the control valve. This action causes the valve to admit atmospheric pressure to one side of the power diaphragm. With intake-manifold vacuum on the other side of the power diaphragm, the diaphragm is forced to move. This motion forces the piston into the hydraulic cylinder so that the brake fluid is forced from the hydraulic cylinder to the wheel cylinders. Then braking takes place. The relatively light brake-pedal force and the hydraulic force in the master cylinder are multiplied several times by the power brake.

Truck air brakes

⚙ 4-9 Air brakes Many medium-duty and all heavy-duty trucks and highway tractors are equipped with air brakes. The air-brake system uses compressed air to apply the braking force to the brake shoes. A typical truck air-brake system is shown in Fig. 4-24. The air compressor, air-reservoir tank, brake chamber, and wheel

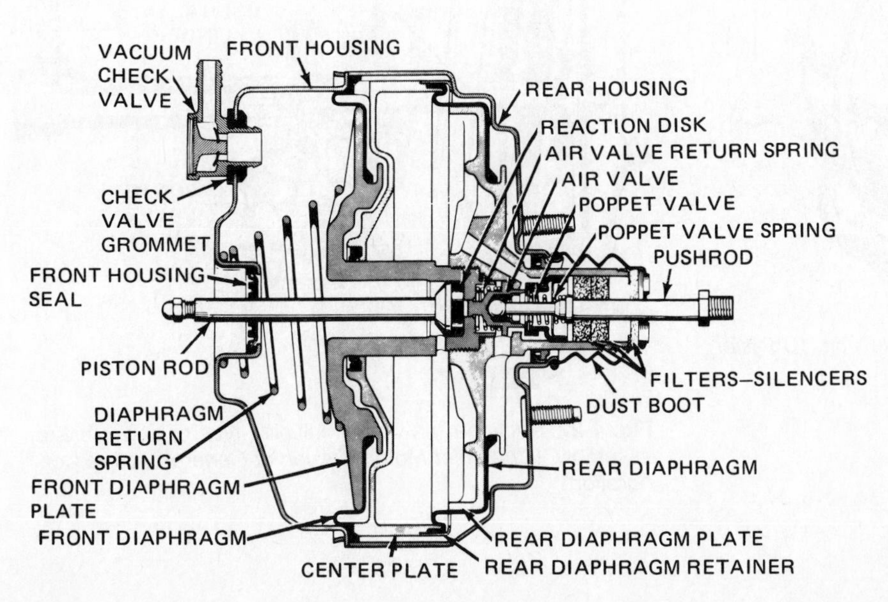

Fig. 4-19 Sectional view of a Bendix dual-diaphragm power brake. (*Chevrolet Motor Division of General Motors Corporation*)

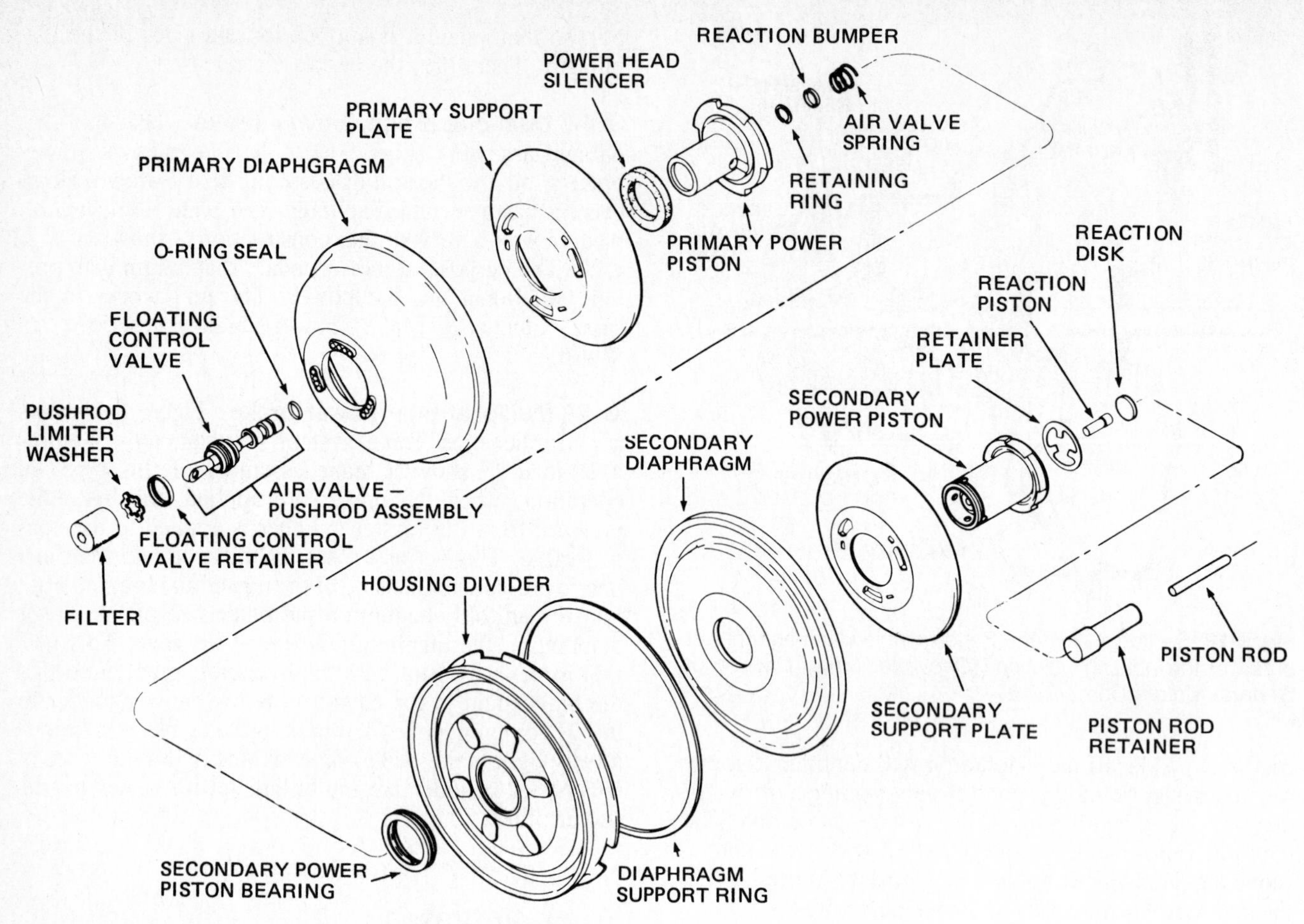

Fig. 4-20 Disassembled Bendix dual-diaphragm power brake. (*Buick Motor Division of General Motors Corporation*)

Fig. 4-21 A multiplier type of power-brake unit. (*Chrysler Corporation*)

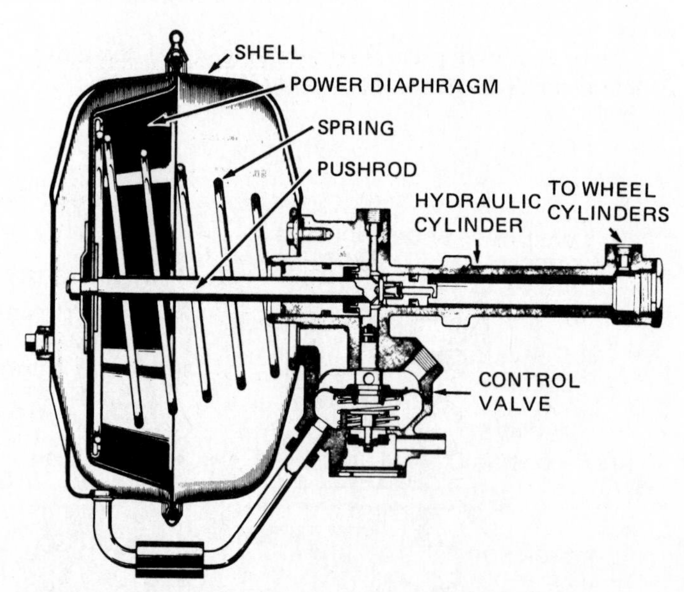

Fig. 4-22 Sectional view of a multiplier type of power-brake assembly. (*Chevrolet Motor Division of General Motors Corporation*)

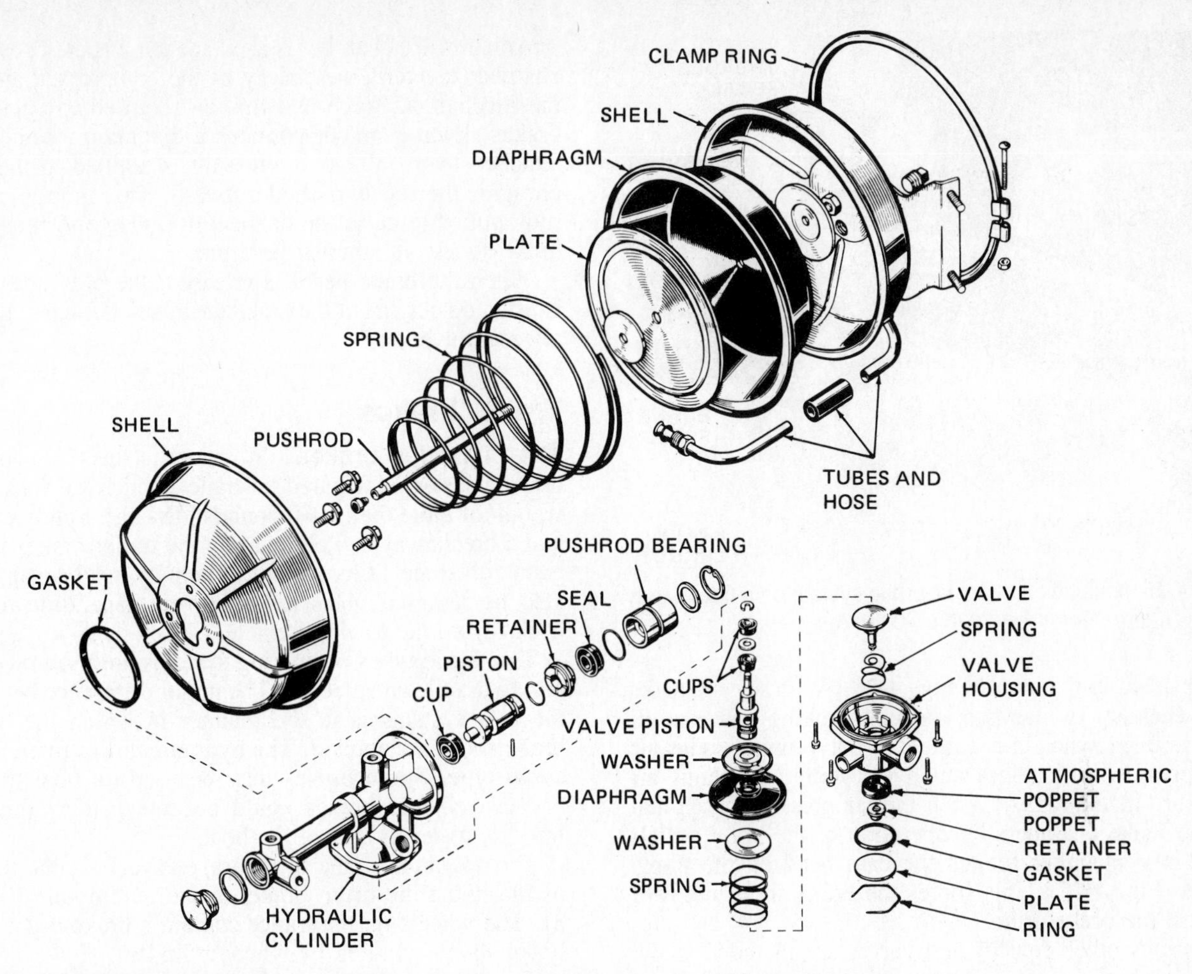

Fig. 4-23 Disassembled multiplier type of power-brake assembly (*Chevrolet Motor Division of General Motors Corporation*)

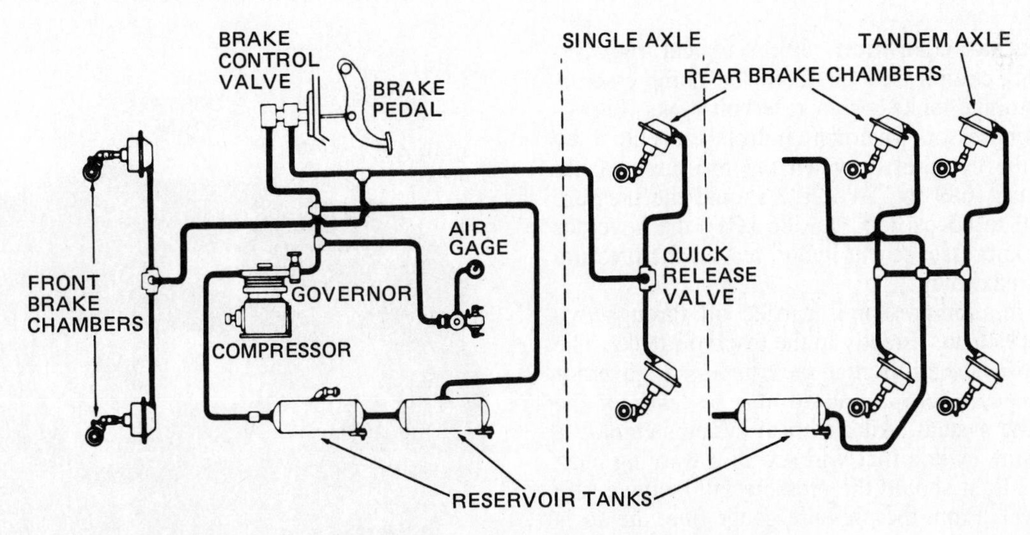

Fig. 4-24 A truck air-brake system. (*Chevrolet Motor Division of General Motors Corporation*)

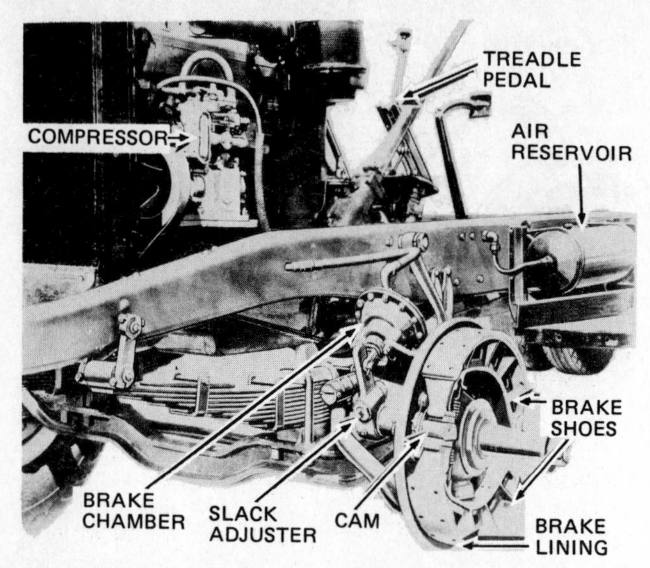

Fig. 4-25 Installation of an air-brake system on a heavy-duty truck. (*International Harvester Company*)

mechanism are shown in Fig. 4-25. Air-reservoir tanks are necessary to maintain adequate braking power at all times, even when the engine is not running. The air compressor, which is a small air pump, maintains air pressure in the tanks. When the air pressure is applied to the brake chambers by operation of the brake pedal, the brake chamber rotates the cam between the brake shoes (Fig. 4-25). This forces the brake shoes outward against the brake drum.

Various other devices and valves are included in the truck air-brake system. These help provide the braking force, and also act as designed-in safety devices. Actions in the air-brake system during a brake application are described in the following section.

❁ 4-10 Air-brake operation In the typical truck air-brake system, compressed air from the compressor is sent to the storage tanks, or air-reservoir tanks (Fig. 4-24). The compressor governor controls the output by cutting out the compressor when the pressure reaches 100 to 105 psi [689 to 724 kPa]. Should the pressure fall below 80 to 85 psi [552 to 586 kPa], the governor allows the compressor to cut in and restore the pressure to its preset maximum.

Air from the compressor is carried off through two lines. One lines leads directly to the reservoir tanks. The other line carries the air through the compressor governor to an air gauge, where a reading may be taken of the pressure in the system. All air-brake systems employ a low-air-pressure switch that will set up a warning buzz or flash a red light should the pressure fall below a safe operating level. From the pressure-gauge line, the air is carried to an outlet from the second tank and then piped directly to the brake control valve. The pressure at the brake control valve is the same as the reading on the pressure gauge.

When the brake pedal is depressed, the brake control valve is opened, and full pressure is allowed to pass to the front brake chambers. Another smaller line carried full tank pressure to the rear brake chambers.

An air brake chamber is used at each wheel. The brake chamber converts the energy of the compressed air into the mechanical force and motion required to apply the brakes. Each chamber contains a diaphragm with a rod attached to it. When air pressure is applied to the diaphragm, the rod is pushed outward. This motion of the rod applies force to the connecting linkage to move the brake shoes out against the drum.

When the brake pedal is released, the pressure is exhausted quickly from the brake chambers through a quick-release valve.

Trailer brakes

❁ 4-11 Trailer brakes In many states, a separate braking system is required for trailers which have a loaded weight of more than 1000 pounds [454 kg]. Safety chains and a breakaway switch to apply the trailer brakes in the event of some failure in the hitching mechanism may also be required. This helps prevent separation of the trailer from the towing vehicle.

Two basic types of braking systems are used on trailers, hydraulic and electric. The major difference between the two systems is in the manner in which the trailer brake shoes are applied. The hydraulic brake often is the surge type. Surge brakes may be used on boat trailers because electric brakes could be damaged by backing into the water to unload the boat.

The operation of the surge brake is very similar to that of the hydraulic drum brake on the automobile (Chap. 3). The wheels on the trailer contain a brake drum with brake shoes operated by wheel cylinders. However, the braking systems on the towing vehicle and the trailer are not interconnected. The master cylinder for the surge brake is located on the tongue of the trailer. When the driver applies the car brakes to slow down, the trailer

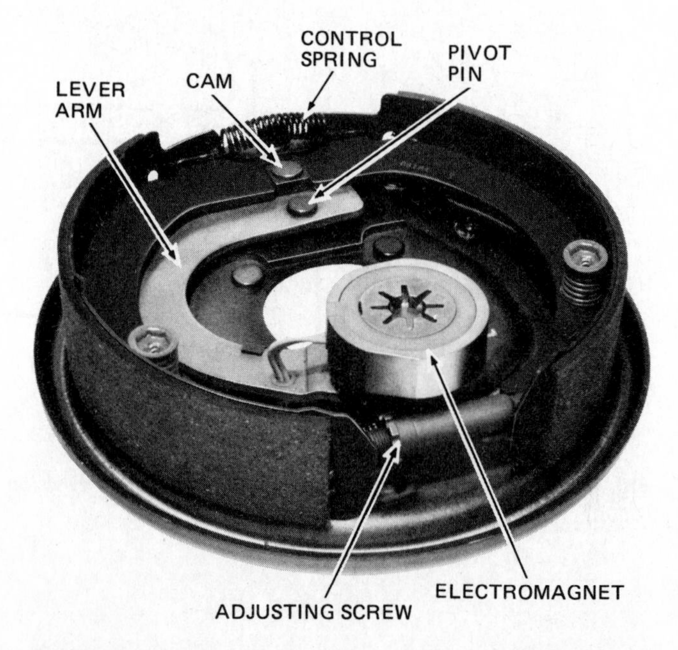

Fig. 4-26 Construction of an electric brake. (*Warner Electric Brake & Clutch Company*)

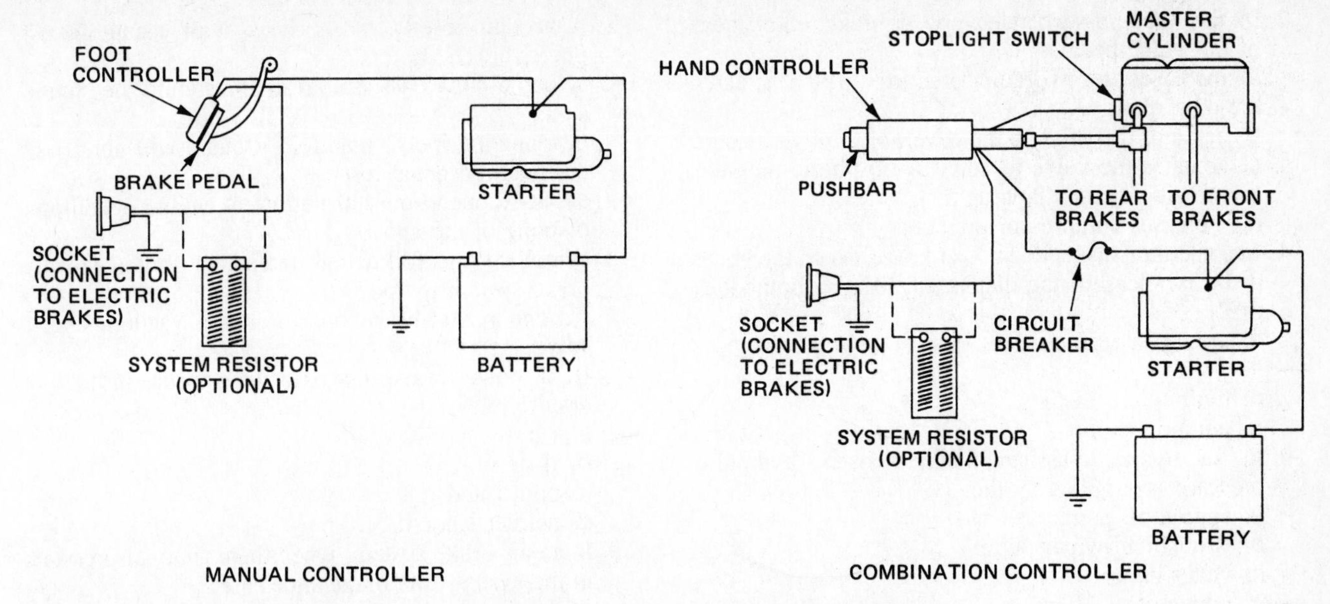

Fig. 4-27 Two types of controllers for electric brakes. Left, a manual controller which is foot-operated by applying the service brake. Right, a combination controller which operates automatically as the rear brakes are applied. This system can also be manually controlled by pressing the pushbar. (*Warner Electric Brake & Clutch Company*)

tongue pushes against the hitch ball on the towing car. As a result, the piston is forced into the master cylinder, which transmits pressure through the system to apply the trailer brakes. No action of the driver is required to apply a surge brake.

Surge brakes do not permit simulaneous braking of the car and the trailer. In addition, there is no way to separately apply the trailer brakes. For these reasons, electric brakes (❁ 4-12) are the most popular type used on trailers.

❁ 4-12 Electric brakes Many trailers that are pulled by cars and light trucks are equipped with electric brakes. The construction of the electric brake is similar to that of the drum brake (Chap. 3). However, electric brakes use electromagnets to force the brake shoes out against the drum (Fig. 4-26).

Each wheel brake contains a semistationary electromagnet and an armature disk that revolves with the wheel. The electromagnet operates on current from the battery. A *controller* (Fig. 4-27) connects the electromagnet to the battery and then varies the current flow through the electromagnet as needed. This, in turn, varies the braking action.

There are three types of controllers: automatic, manual, and combination. Figure 4-27 (left) shows a manual controller and how it is wired into the vehicle's electrical system. Figure 4-27 (right) shows a combination controller, which is an automatic controller with manual override when the pushbar is pressed.

In the electric brake shown in Fig. 4-26, the magnet is attached to one end of a lever arm. The other end of the lever arm has a cam fastened to it. When the driver applies the brakes, the controller allows current to flow through the magnet. This builds up a magnetic field, causing magnetic attraction between the semistationary electromagnet and the rotating armature disk.

The electromagnet tries to rotate with the disk. This shifts the brake lever outward in the direction of wheel rotation. As the brake lever pivots outward on the pivot pin, the cam turns, forcing the brake shoes into the drum. The greater the current flowing through the electromagnet, the greater the braking effort.

Chapter 4 review questions

Select the *one* correct, best, or most probable answer to each question. Then check your answers against the correct answers given at the end of the book.

1. The two types of power-brake boosters used on automobiles are:
   a. air and vacuum
   b. compressed air and electric
   c. hydraulic and atmospheric
   d. vacuum and hydraulic
2. To operate, a vacuum-assisted power-brake booster makes use of the pressure differential between vacuum from the intake manifold and:
   a. venturi vacuum
   b. compressed air
   c. atmospheric pressure
   d. none of the above

3. In the vacuum-suspended power brake, movement of the brake pedal:
a. increases the hydraulic pressure, which actuates the control valve
b. actuates the valve in the governor through linkage
c. actuates the valve to admit atmospheric pressure to one side of the diaphragm
d. increases compressor pressure

4. In a vacuum-suspended power brake, when the brake pedal is released, the diaphragm has, on both sides of it:
a. atmospheric pressure
b. a vacuum
c. hydraulic pressure
d. compressed air

5. In the hydraulic-assisted brake booster, hydraulic pressure is supplied by the:
a. engine oil pump
b. windshield-wiper pump
c. water pump
d. power-steering pump

6. The purpose of the dual-diaphragm power brake is to:
a. provide additional braking power
b. provide a safety feature in case one diaphragm fails
c. reduce the pedal force required to activate the power-brake booster
d. all of the above

7. A vacuum-reserve tank is used with vacuum-assist power brakes so that:
a. some power assist is available should the engine stall
b. changing intake-manifold vacuum will not affect the required pedal force
c. brake operation will not affect engine operation
d. none of the above

8. Which of the following statements abut a power-brake system is true?
I. The power-brake booster has a vacuum check valve.
II. A power brake increases the pressure in the hydraulic lines.
a. I only
b. II only
c. both I and II
d. neither I nor II

9. In an air-brake system, when there is no air pressure in the system, the brake shoes are:
a. applied
b. released
c. forced into contact with the drum
d. none of the above

10. Electric brakes on a trailer are activated by the:
a. electromagnet
b. brake springs
c. surge brakes
d. controller

CHAPTER 5

BRAKE TROUBLE DIAGNOSIS, TESTING, AND INSPECTION

After studying this chapter, and with proper instruction and equipment, you should be able to:

1. Diagnose troubles in drum brakes.
2. Diagnose troubles in disk brakes.
3. Test brakes using a platform brake tester.
4. Test brakes using a dynamic brake analyzer.
5. Check vehicle stopping distance.
6. Road-test the brakes.
7. Make a visual inspection of brake-system components.
8. Check the operation of the brakes in the shop.

CAUTION: **Breathing dust containing asbestos fibers can cause serious bodily harm. The United States Surgeon General has determined that ingesting asbestos fibers, either by breathing or through the mouth, can cause serious physical diseases. You should not sand or grind brake linings or clean brake parts with a dry brush or compressed air. Instead, wipe parts off with a cloth dampened with water.**

5-1 Brake trouble diagnosis The charts and sections that follow give you a means of tracing troubles in the brake system back to their causes. This permits quick location of the causes and quick correction of the troubles. If the cause is known, the troubles can usually be readily corrected. There are two trouble-diagnosis charts, one for drum brakes (5-2) and one for disk brakes (5-15). Following the trouble-diagnosis charts are sections that cover the adjustment and repair procedures for different types of automotive brakes.

Trouble diagnosis is more than following a series of steps in an attempt to find the solution to a problem. It is a way of looking at systems that are not working properly. Here are the basic rules:

1. Know the system. This means that you should know how that parts go together, how they work together as a system, and what happens if some part goes bad or the parts fail to work together as they should.
2. Know the history of the system in the car. How old is the system? What sort of treatment has it had? What is its service history? Has it been serviced before for the same problem? The answers to these questions might save you a lot of time.
3. Know the history of the condition causing the driver's complaint. Did it start all at once? Did the trouble come on gradually? Was it related to some other condition, like the accident shown in Fig. 5-1 or to a previous service problem?
4. Know the odds. Some troubles happen more often than others. Be aware of what happens frequently and what happens rarely. A trouble such as "the engine cranks normally but does not start" is more likely to be caused by an empty fuel tank than by a broken timing chain.

Fig. 5-1 Know the history of the condition causing the driver's complaint. (*ATW*)

5. Don't cure the symptom and leave the cause. Charging a run-down battery may be a temporary fix. But, if the trouble is really a defective alternator, you have not eliminated the problem.
6. Be sure you have fixed the basic condition that caused the trouble.
7. Be sure you get all the information (items 1, 2, and 3) that you can from the driver. This information may greatly simplify your search for the cause of the trouble.

There are several ways to study the troubleshooting or trouble-diagnosis chart. One way is to go through this chapter page by page. Another way is to study the complaints listed in the chart one at a time. Read through the possible causes and checks or corrections for each complaint. Then read the section of the chapter that discusses the complaint. The section will give you more detailed information on the causes and the checks or corrections to be made. Knowing how to troubleshoot the brake system is important, so refer to the troubleshooting charts as often as needed. To help you learn the complaints, causes, and corrections, copy the information on 3 × 5 inch [76 × 128 mm] cards. Then carry the cards around with you. When you have time, study each card until you know the complaints and their causes.

Drum-brake trouble diagnosis

✪ 5-2 Drum-brake trouble-diagnosis chart A variety of braking problems bring the driver to the mechanic. Few drivers will know exactly what is causing a trouble. The chart that follows lists possible troubles in drum-brake systems, their possible causes, and checks or corrections to be made. Following sections describe the troubles and causes or corrections in detail. The chart in ✪ 5-15 covers possible disk-brake troubles.

Drum-Brake Trouble-Diagnosis Chart

(See ✪ 5-3 to 5-14 for detailed explanation of the trouble causes and corrections listed in the chart).

COMPLAINT	POSSIBLE CAUSE	CHECK OR CORRECTION
1. Brake pedal goes to floorboard (✪ 5-3)	a. Linkage or shoes out of adjustment	Adjust
	b. Brake linings worn	Replace
	c. Lack of brake fluid	Add fluid; bleed system (see item 10 below)
	d. Air in system	Add fluid; bleed system (see item 9 below)
	e. Worn master cylinder	Repair
2. One brake drags (✪ 5-4)	a. Shoes out of adjustment	Adjust
	b. Clogged brake line	Clear or replace line
	c. Wheel cylinder defective	Repair or replace
	d. Weak or broken return spring	Replace
	e. Loose wheel bearing	Adjust bearing
3. All brakes drag (✪ 5-5)	a. Incorrect linkage adjustment	Adjust
	b. Trouble in master cylinder	Repair or replace
	c. Mineral oil in system	Replace damaged rubber parts; use only recommended brake fluid
4. Car pulls to one side when braking (✪ 5-6)	a. Brake linings soaked with oil	Replace linings and oil seals; avoid overlubrication
	b. Brake linings soaked with brake fluid	Replace linings; repair or replace wheel cylinder
	c. Brake shoes out of adjustment	Adjust
	d. Tires not uniformly inflated	Inflate correctly
	e. Brake line clogged	Clear or replace line
	f. Defective wheel cylinder	Repair or replace
	g. Brake backing plate loose	Tighten
	h. Mismatched linings	Use same linings all around
5. Soft or spongy pedal (✪ 5-7)	a. Air in system	Add brake fluid; bleed system (see item 9 below)
	b. Brake shoes out of adjustment	Adjust
6. Poor braking action requiring excessive pedal force (✪ 5-8)	a. Brake linings soaked with water	Will be all right when dried out
	b. Shoes out of adjustment	Adjust
	c. Brake linings hot	Allow to cool
	d. Brake linings burned	Replace
	e. Brake drum glazed	Turn or grind drum
	f. Power-brake assembly not operating	Overhaul or replace

Drum-Brake Trouble-Diagnosis Chart (*Continued*)

(See ❁ 5-3 to 5-14 for detailed explanation of the trouble causes and corrections listed in the chart).

COMPLAINT	POSSIBLE CAUSE	CHECK OR CORRECTION
7. Brakes too sensitive or grab (❁ 5-9)	a. Shoes out of adjustment	Adjust
	b. Wrong linings	Install correct linings
	c. Brake linings greasy	Replace; check oil seals; avoid overlubrication
	d. Drums scored	Turn or grind drums
	e. Backing plates loose	Tighten
	f. Power-brake assembly malfunctioning	Overhaul or replace
	g. Brake linings soaked with oil	Replace linings and oil seals; avoid overlubrication
	h. Brake linings soaked with brake fluid	Replace linings; repair or replace wheel cylinders
8. Noisy brakes (❁ 5-10)	a. Linings worn	Replace
	b. Shoes warped	Replace
	c. Shoe rivets loose	Replace shoe or lining
	d. Drums worn or rough	Turn or grind drums
	e. Loose parts	Tighten
9. Air in system (❁ 5-11)	a. Defective master cylinder	Repair or replace
	b. Loose connections, damaged tube	Tighten connections; replace tube
	c. Brake fluid lost	See item 10 below
10. Loss of brake fluid (❁ 5-12)	a. Master cylinder leaks	Repair or replace
	b. Wheel cylinder leaks	Repair or replace
	c. Loose connections, damaged tube	Tighten connections; replace tube
Note: After repair, add brake fluid and bleed system.		
11. Brakes do not self-adjust (❁ 5-13)	a. Adjustment screw stuck	Free and clean up
	b. Adjustment lever does not engage star wheel	Repair; free up or replace adjuster
	c. Adjuster incorrectly installed	Install correctly
12. Warning light comes on when braking (dual system) (❁ 5-14)	a. One section has failed	Check both sections for braking action; repair defective section
	b. Pressure-differential valve defective	Replace

❁ 5-3 Brake pedal goes to floorboard When the brake pedal goes to the floorboard, there is no pedal reserve (Fig. 5-2). Full pedal movement does not produce adequate braking. This would be a very unlikely situation with a dual-brake system. One section might fail, but it would be rare for both to fail at the same time. It is possible that the driver has continued to operate the car with one section out. (Either the driver ignored the warning light or the light or the pressure-differential valve has failed.) Causes of failure could be linkage or brake shoes out of adjustment linings worn, air in the system, lack of brake fluid, or a worn master cylinder.

❁ 5-4 One brake drags If one brake drags, this means that the brake shoes are not moving away from the brake drum when the brakes are released. This trouble could be caused by incorrect shoe adjustment or a clogged brake line which does not release pressure from the wheel cylinder. It could also be due to sticking pistons in the wheel cylinder, to weak or broken brake-shoe return springs, or to a loose wheel bearing which permits the wheel to wobble so that the brake drum comes in contact with the brake shoes even though they are retracted.

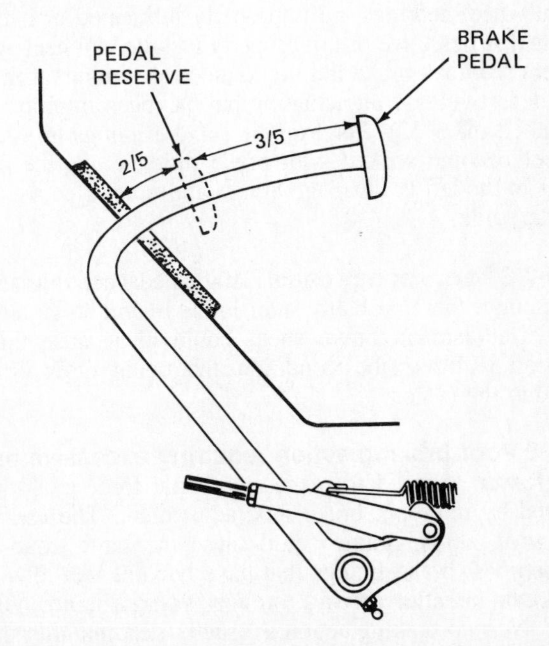

Fig. 5-2 Pedal reserve is the additional pedal travel available after the brakes are firmly applied.

⚙ 5-5 All brakes drag When all brakes drag, the brake pedal may not have sufficient free travel. In that case, the pistons in the master cylinder do not fully retract. This would prevent the lip of the piston cup from clearing the compensating port, and hydraulic pressure would not be relieved as it should be. As a result, the wheel cylinders would not release the brake shoes.

A similar condition could result if engine oil was added to the system. Engine oil will cause the piston cups to swell. If they swelled enough, they would not clear the compensating ports even with the piston in the "fully retracted" position. A clogged compensating port would have the same result. Do not use a wire or drill to clear the port. This might produce a burr that would cut the piston cup. Instead, clear the port with alcohol and compressed air.

Clogging of the reservoir vent might also cause dragging brakes by pressurizing the fluid in the reservoir. This would prevent relase of pressure on the fluid in the lines. However, a clogged vent could also cause leakage of air into the system (⚙ 5-11 below).

⚙ 5-6 Car pulls to one side If the car pulls to one side when the brakes are applied, more braking force is being applied to one side than to the other. This happens if some of the brake linings have become soaked in oil or brake fluid, if brake shoes are unevenly or improperly adjusted, if tires are not evenly inflated, or if defective wheel cylinders or clogged brake lines are preventing uniform braking action at all wheels. A loose brake backing plate or the use of two different types of brake lining will cause the car to pull to one side when the brakes are applied. A misaligned front end or a broken spring could also cause this problem.

Linings will become soaked with oil if the lubricant level in the differential and drive axle is too high. This usually causes leakage past the oil seal (Fig. 5-3). At the front wheel, brake linings may become oil-soaked if the front-wheel bearings are improperly lubricated or if the oil seal is defective or not properly installed. Wheel cylinders will leak brake fluid onto the brake linings if they are defective or if an actuating pin has been improperly installed (⚙ 5-12). For example, if the linings at a left wheel become soaked with brake fluid or oil, the car pulls to the left because the brakes are more effective on the left side.

⚙ 5-7 Soft or spongy pedal If the pedal action is soft or spongy, there probably is air in the hydraulic system. Out-of-adjustment brake shoes could also cause this. Section 5-11 describes conditions that could allow air to get into the system.

⚙ 5-8 Poor braking action requiring excessive pedal force A need for excessive pedal force could be caused by improper brake-shoe adjustment. The use of the wrong brake lining could cause the same trouble. Sometimes, brake linings that have become wet after a hard rain or after driving through water will not hold well. Normal braking action is usually restored after the brake linings have dried out.

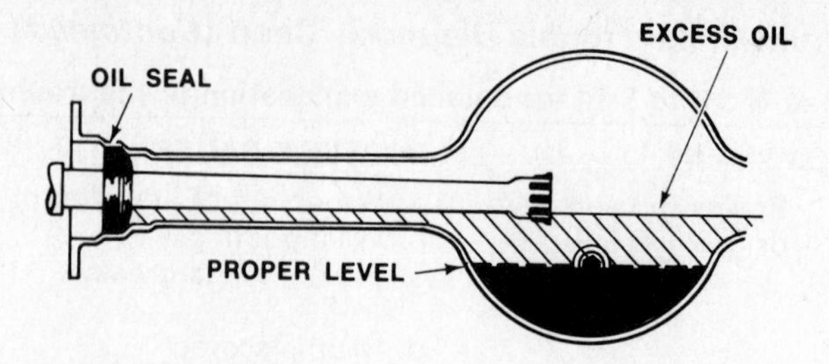

Fig. 5-3 A high lubricant level in the differential and axle housing may cause leakage past the oil seat. This would result in oil-soaked brake linings. (*Pontiac Motor Division of General Motors Corporation*)

Another possible cause of poor braking action is excessive temperature. After the brakes have been applied for long periods, as in coming down a long hill, they begin to overheat. This overheating reduces braking effectiveness so that the brakes "fade." Often, if brakes are allowed to cool, braking efficiency is restored. However, excessively long periods of braking at high temperature may char the brake linings so that they must be replaced. Further, this overheating may glaze the brake drum so that it becomes too smooth for effective braking action. Then the drum must be ground or turned to remove the glaze. Glazing can also take place even though the brakes are not overheated. Failure of the power-brake assembly will noticeably increase the force on the foot pedal required to produce braking.

⚙ 5-9 Brakes too sensitive or grab If linings are greasy, or soaked with oil or brake fluid, the brakes tend to grab with slight pedal force. Then the linings must be replaced. If the brake shoes are out of adjustment, if the wrong linings are used, or if drums are scored or rough (Fig. 5-4), grabbing may result. A loose backing plate may cause the same condition. As the linings come into contact with the drum, the backing plate shifts to give hard braking. A defective power-brake assembly can also cause grabbing.

⚙ 5-10 Noisy brakes Brakes become noisy if the brake linings wear so much that the rivets come into contact with the brake drum; if the shoes become warped so that pressure on the drum is not uniform; if shoe rivets become loose so that they contact the drum; or if the drum becomes rough or worn (Fig 5-4). Any of these conditions may cause a squeak or squeal when the brakes are applied. Loose parts, such as the brake backing plate, also may rattle.

⚙ 5-11 Air in system If air gets into the hydraulic system, poor braking and a spongy pedal will result. Air can get into the system if the filler vent becomes plugged (Fig. 5-5), since this may tend to create a partial vacuum in the system on the return stroke of the piston. Air could then bypass the rear piston cup, as shown by the arrows in Fig 5-5, and enter the system. It is possible accidentally to plug the vent when the filler plug or cover is removed. Always check the vent and clean it when the plug or cover is removed. Air can also get into the hy-

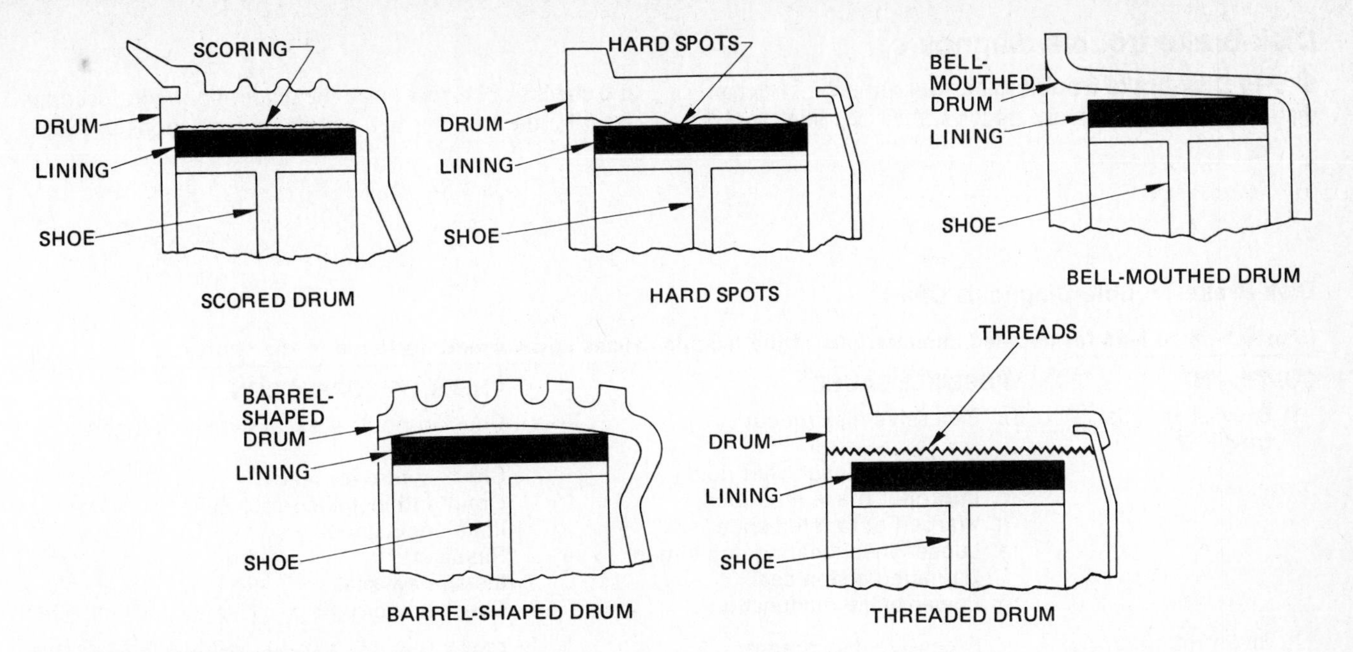

Fig. 5-4 Various types of brake-drum defects that require drum service. (*Bear Manufacturing Company*)

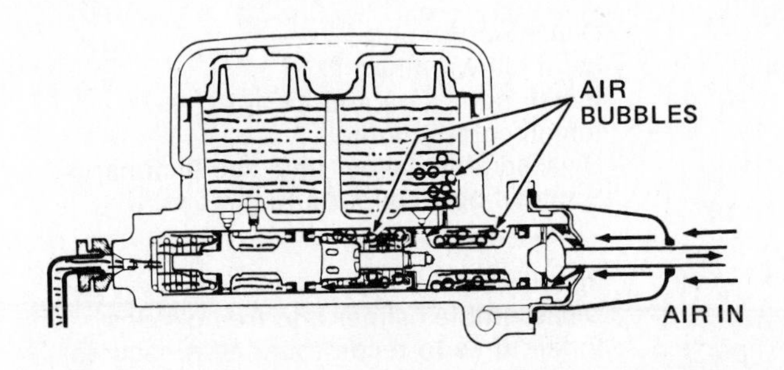

Fig. 5-5 If the air vent in the master-cylinder cover becomes clogged, air may be drawn into the system. The air gets past the low-pressure seal on the primary piston during the return stroke of the piston. (*Pontiac Motor Division of General Motors Corporation*).

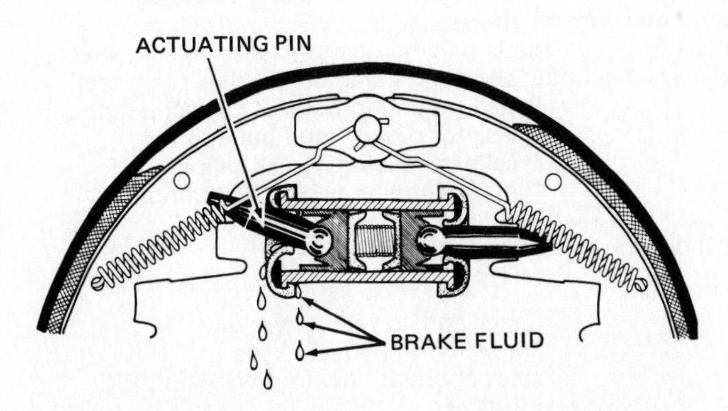

Fig. 5-6 Incorrect installation of the pin in the wheel cylinder will cause a side thrust on the piston which permits brake fluid to leak out past the cup. The pin must align with the notch in the brake shoe. (*Pontiac Motor Division of General Motors Corporation*)

draulic system if the master-cylinder valve is leaky and does not hold pressure in the system. A leak could allow air to seep in around the wheel-cylinder piston cups, since there would be no pressure holding the cups tight against the cylinder walls. Probably the most common cause of air in the brake system is insufficient brake fluid in the master cylinder. If the brake fluid drops below the compensating port, the hydraulic system will draw air in as the piston moves forward when braking. Air in the system must be removed by adding brake fluid and bleeding the system, as described in Chap. 6.

5-12 Loss of brake fluid Brake fluid can be lost if the master cylinder leaks, if the wheel cylinder leaks, if the line connections are loose, or if the line is damaged. One possible cause of wheel-cylinder leakage is incorrect installation of the actuating pin (Fig. 5-6). If the pin is cocked, the side thrust on the piston may permit leakage past the piston. Leakage from other causes at the master cylinder or wheel cylinder requires removal and repair or replacement of the defective parts.

5-13 Brakes do not self-adjust Brakes do not self-adjust if the adustment screw is stuck, the adjustment lever does not engage the star wheel, or the adjuster was incorrectly installed. It is necessary to inspect the brake to find and correct the trouble.

5-14 Warning light comes on when braking If the warning light comes on when braking, it means that one of the two braking sections has failed. Both sections should be checked so that the trouble can be found and eliminated. It is dangerous to drive with this condition. Even though the car slows, only half the wheels are being braked.

Disk-brake trouble diagnosis

✽ 5-15 Disk-brake trouble-diagnosis chart The chart that follows lists disk-brake troubles, their possible causes, and checks or corrections to be made. Following sections describe the troubles and causes or corrections in detail.

Disk-Brake Trouble-Diagnosis Chart

(See ✽ 5-16 to 5-24 for detailed explanations of the trouble causes and corrections listed in the chart.)

COMPLAINT	POSSIBLE CAUSE	CHECK OR CORRECTION
1. Excessive pedal travel (✽ 5-16)	a. Excessive disk runout	Check runout; if excessive, install new disk
	b. Air leak, or insuficient fluid	Check sytem for leaks
	c. Improper brake fluid (boil)	Drain and install correct fluid
	d. Warped or tapered shoe	Install new shoe
	e. Loose wheel-bearing adjustment	Readjust
	f. Damaged piston seal	Install new seal
	g. Power-brake malfunction	Check power unit
2. Brake roughness or chatter (pedal pulsation) (✽ 5-17)	a. Excessive disk runout	Check runout; if excessive, install new disk
	b. Disk out of parallel	Check runout; if excessive, install new disk
	c. Loose wheel bearing	Readjust
3. Excessive pedal force, grabbing, or uneven braking action (✽ 5-18)	a. Power-brake malfunction	Check power unit
	b. Brake fluid or grease on linings	Install new linings
	c. Lining worn	Install new shoe and linings
	d. Incorrect lining	Install correct lining
	e. Frozen or seized pistons	Disassemble caliper and free pistons; install new caliper and pistons, if necessary.
4. Car pulls to one side (✽ 5-19)	a. Brake fluid or grease on linings	Install new linings
	b. Frozen or seized pistons	Disassemble caliper and free pistons
	c. Incorrect tire pressure	Inflate tires to recommended pressures
	d. Distorted brake shoes	Install new brake shoes
	e. Front end out of alignment	Check and align front end
	f. Broken rear spring	Install new rear spring
	g. Restricted hose or line	Check hoses and lines and correct as necessary
	h. Unmatched linings	Install correct lining
5. Noise (✽ 5-20): Groan	Brake noise when slowly releasing brakes (creep-groan). Not detrimental to function of disk brakes—no corrective action required. This noise may be eliminated by slightly increasing or decreasing brake-pedal efforts.	
Rattle	Brake noise or rattle at low speeds on rough roads may be due to excessive clearance between the shoe and the caliper. Install new shoe and lining assemblies to correct.	
Scraping	a. Mounting bolts too long	Install mounting bolts of correct length
	b. Disk rubbing housing	Check for rust or mud buildup on caliper housing; check caliper mounting and bridge bolt tightness
	c. Loose wheel bearings	Readjust
	d. Linings worn, allowing wear indicator to scrape on disk	Replace linings
6. Brakes heat up during driving and fail to release (✽ 5-21)	a. Power-brake malfunction	Check and correct power unit
	b. Sticking pedal linkage	Free sticking pedal linkage
	c. Driver riding brake pedal	Instruct driver how to drive with disk brakes
	d. Frozen or seized piston	Dissasemble caliper, clean cylinder bore, clean seal groove, and install new pistons, seals, and boots
	e. Residual check valve in master cylinder	Remove valve from cylinder
	f. Incorrect linkage adjustment	Adjust linkge

Disk-Brake Trouble-Diagnosis Chart (*Continued*)

(See ⚙ 5-16 to 5-24 for detailed explanations of the trouble causes and corrections listed in the chart.)

COMPLAINT	POSSIBLE CAUSE	CHECK OR CORRECTION
7. Leaky caliper cylinder (⚙ 5-22)	a. Damaged or worn piston seal	Install new seal
	b Scores or corrosion on surface of piston or caliper bore	Disassemble caliper, clean cylinder bore; if necessary, install new pistons or replace caliper
8. Brake pedal can be depressed without braking action (⚙ 5-23)	a. Piston pushed back in cylinder bores during servicing of caliper (and lining not properly positioned)	Reposition brake shoe and lining assemblies. Depress pedal a second time and if condition persists, look for the following causes:
	b. Leak in system or caliper	Check for leak, repair as required
	c. Damaged piston seal in one or more cylinders	Disassemble caliper and replace piston seals as required
	d. Air in hydraulic system, or improper bleeding procedure	Bleed system
	e. Bleeder screw opens	Close bleeder screw and bleed entire system
	f. Leak past primary cup in master cylinder	Recondition master cylinder
9. Fluid level low in master cylinder (⚙ 5-24)	a. Leaks in system or caliper	Check for leak, repair as required
	b. Worn brake-shoe linings	Replace shoes
10. Warning light comes on when braking (dual system) (⚙ 5-14)	a. One section has failed	Check both sections for braking action; repair defective section
	b Pressure-differential valve defective	Replace

⚙ 5-16 Excessive pedal travel Anything that requires excessive movement of the caliper pistons will require excessive pedal travel. For example, if the disk has excessive runout, it will force the pistons farther back in their bores when the brakes are released. Therefore, additional pedal travel is required when the brakes are applied. Warped or tapered shoes, a damaged piston seal, or a loose wheel bearing could cause the same problem. In addition, air in the brake lines, insufficient fluid in the system, or incorrect fluid, which boils, will cause a spongy pedal and excessive pedal travel. If the power brake is defective, it could also cause greater than normal pedal travel.

⚙ 5-17 Brake-pedal pulsation Brake-pedal pulsation (Fig. 5-7) is probably due to a disk with excessive runout. The problem also may be caused by a loose wheel bearing.

⚙ 5-18 Excessive pedal force If excessive pedal force is required to obtain normal braking action, the power brake may be defective. In addition, if the linkages are worn, hot, or water-soaked, they will not brake properly. Neither will a caliper that has a piston jammed in it. All these conditions require an excessively hard push on the brake pedal to provide braking action.

⚙ 5-19 Car pulls to one side Pulling to one side is due to uneven braking action. It could be caused by incorrect front-wheel alignment, uneven tire inflation, or a broken or weak suspension spring. Within the braking system itself, such things as brake fluid on the linings, unmatched linings, warped brake shoes, jammed pistons, floating or sliding caliper seized, or restrictions in the brake lines could cause the car to pull to one side when braking.

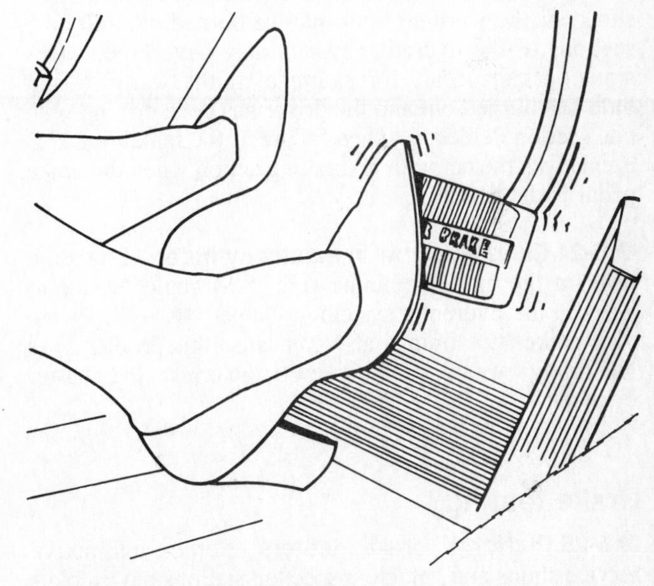

Fig. 5-7 Brake-pedal pulsation is probably due to a disk with excessive runout or to a loose wheel bearing. (*Ford Motor Company*)

⚙ 5-20 Noise The chart covers various noises and their causes. Refer to it for details.

⚙ 5-21 Brakes fail to release Brake-release failure could result from anything from a sticking pedal linkage or malfunctioning power brake to pistons stuck in the calipers. It could also be due to the driver's riding the brake pedal or to failure of the master cylinder to release the pressure when the brakes are released.

⚙ 5-22 Leaky caliper cylinder A leaky caliper cylinder could be due to a damaged or worn piston seal. The leak would also be caused by roughness on the surface of the piston as a result of scores, scratches, or corrosion.

⚙ 5-23 Brake pedal can be depressed without braking action If the brake calipers have been serviced, the pistons may be pushed back so far in their bores that a single full movement of the brake pedal will not produce braking. After any service on disk brakes, the brake pedal should be pumped several times. Then the master-cylinder reservoir should be properly filled before the car is moved. Pumping the pedal moves the pistons into position so that normal brake-pedal application causes braking.

Other conditions can prevent braking action when the pedal is depressed. These include leaks or air in the hydraulic system. Leaks can occur at the piston seals, bleeder screws, or brake-line connections, or in the master cylinder. Also, the pressure-differential valve may be stuck, or the warning light may be burned out and both sections of the hydraulic system may have failed. This situation can occur, for example, if the warning-light bulb has burned out and the driver has been driving with one section defective. Then failure of the remaining section leaves the car with no braking action when the brake pedal is depressed.

⚙ 5-24 Fluid level low in master cylinder Low fluid level in the master cylinder (Fig. 5-8) could be due to leaks in the hydraulic system or caliper (⚙ 5-23). Worn disk-brake-shoe linings also can cause this problem. As the linings wear, the fluid level lowers in the master cylinder.

Brake testing

⚙ 5-25 Platform brake testers Some automotive service shops and vehicle inspection stations have a platform-type brake tester (Fig. 5-9). These units can check brake action at each wheel in only a few seconds, without removing the wheels or brake drums. Visual inspection may reveal some brake troubles. However, actual brake performance at each wheel can be quickly checked by the use of the platform brake tester.

With the tester shown in Fig. 5-9, the car is driven onto the four tread plates at about 5 mph [8 km/h], and then the brakes are applied hard. The braking effort at each wheel registers on one of four dials or in one of four glass columns (which contain colored liquid that rises in proportion to the braking effort). The four tread plates are on rollers and are spring loaded in the horizontal direction. When a rolling wheel on a tread plate is suddenly stopped by a brake application, the plate moves forward against the spring force. This movement registers the amount of braking action on the dial or in the glass column.

The vehicle needs further brake inspection and service if:

1. The readings are lower than local laws or inspection limits allow.
2. The reading on any one wheel is less than 75 percent of the reading on the other wheel on the same axle. This means that the two front wheels should read within 75 percent of each other, and the same applies for the two rear wheels.

⚙ 5-26 Dynamic brake analyzers Some brake testers use motored rollers to spin both wheels on the front or rear end of the car at the same time. Then, when the

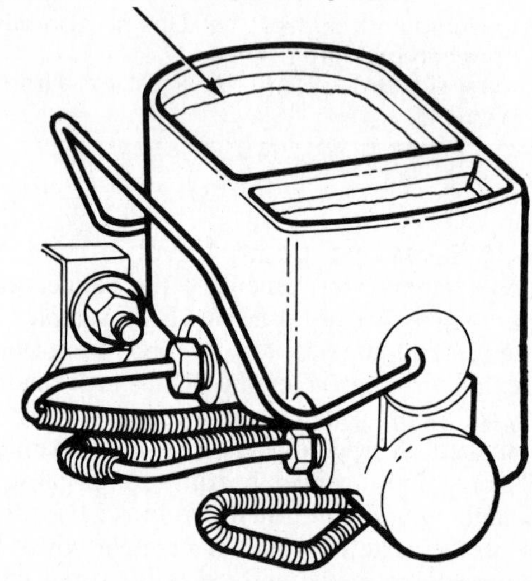

Fig. 5-8 Low brake-fluid level in the master cylinder may be caused by a leak or by worn disk-brake pads. (*Ford Motor Company*)

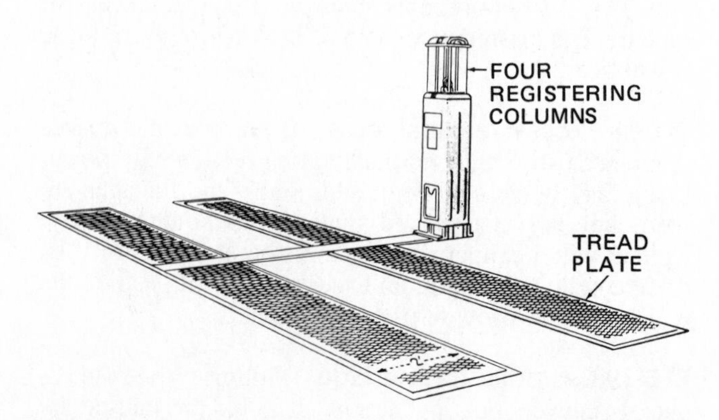

Fig. 5-9 A platform type of brake tester. (*Weaver Manufacturing Company*)

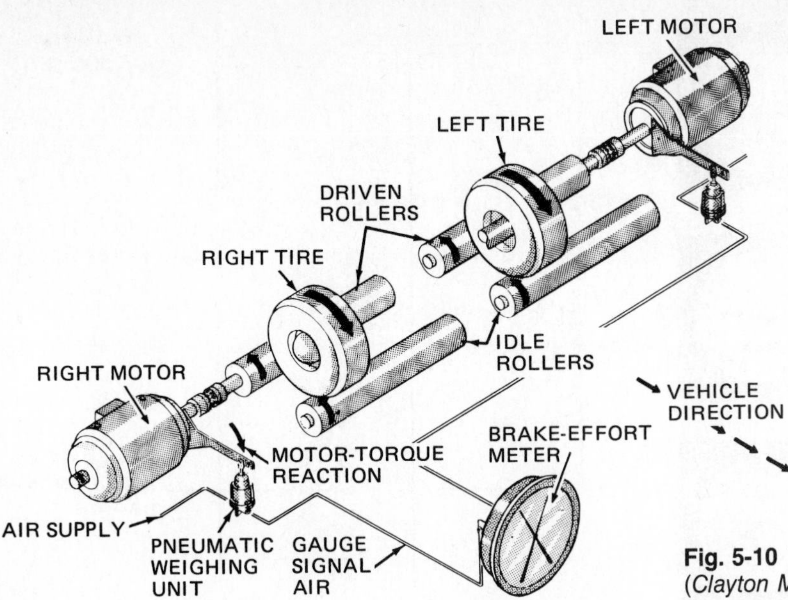

Fig. 5-10 The operating parts of a dynamic brake analyzer. (*Clayton Manufacturing Company*)

brakes are applied, the braking action at each of the wheels can be checked. Testers of this type are called roller-type brake testers, or *dynamic brake analyzers*. Some chassis dynamometers can also perform this type of brake test.

Figure 5-10 shows the construction of a dynamic brake analyzer. It has two pairs of motor-driven rollers on which either the front or rear wheels of the vehicle are placed (Fig. 5-11). Then the tester operator turns on the electric motors. This causes the rollers and the car wheels to spin at a predetermined medium-to-high speed. To test the brakes, the motors are shut off. Then the tester operator depresses the brake pedal. As the brakes apply, the wheels slow, thereby slowing the roller under each tire at the same rate. The braking effort at each wheel registers on a dial (Fig. 5-12), on meters, or on a number display.

On the meter dial shown in Fig. 5-12, there are two hands, one for each wheel. This shows up any imbalance between the left and right brakes. Ideally, both brakes should exert the same stopping force. However, if one brake applies normally and the other does not, the car will tend to be thrown sideways. This could send the vehicle into another lane of traffic, and an accident could result.

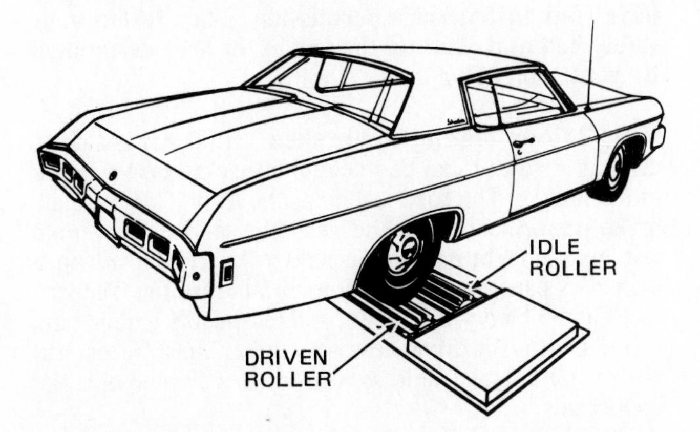

Fig. 5-11 The drive wheels of the vehicle are placed on the rollers of the dynamic brake analyzer. (*Clayton Manufacturing Company*)

The vehicle needs further brake inspection and service if:

1. The readings are lower than local laws or inspection limits allow.
2. One wheel shows less than 80 percent of the braking effort of the other wheel on the same axle.

⚙ 5-27 Checking stopping distance When platform testers (⚙ 5-25) and dynamic brake analyzers (⚙ 5-26) are not available, a check of the vehicle stopping distance can be made. A safe area, preferably with a premarked lane 12 feet [3.7 m] wide, is required (Fig. 5-13). This a type of *road test,* and it should be conducted on level, dry, smooth pavement that is free from oil, grease, and loose dirt. Before starting the test, the tires on the car must be checked and properly inflated.

CAUTION: **Before driving or road testing any car, always press the brake pedal to make sure there is ample pedal reserve (⚙ 5-3) to stop the car. If there is, then make a series of very slow-speed stops to determine if the brakes are safe for driving.**

Drive the car toward the marked lane at a speed of 20 mph [32 km/h]. Upon entering the lane, apply the foot brake firmly. The vehicle should come to a smooth stop within the specified distance and without pulling excessively to the right or left. Passenger vehicles built after 1971 should stop within 20 feet [6.1 m]. Older cars should stop within 25 feet [7.6 m].

The vehicle needs further brake inspection and service if:

1. It swerves enough for any part of the car to leave the 12-foot [3.7-m] lane.
2. It fails to stop within the specified distance.

In the first, or rolling resistance test, the technician does not touch the brake pedal. The solid hand records action of left brake, while the dash indicates operation of right. A wide spread of the hands indicates presence of a dragging brake or uneven tire pressures.

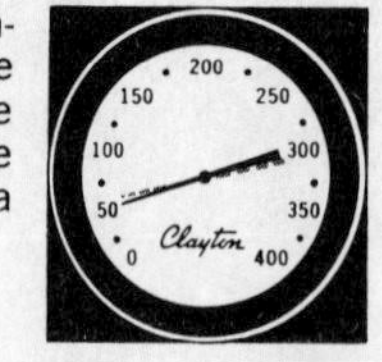

This test determines mechanical application of the brakes. A light application is made to observe first contact of brake shoes with the drums. Often revealed are such things as sticking wheel cylinders, corrosion or ledges between shoes and mounting plates, and/or unhooked or broken return springs.

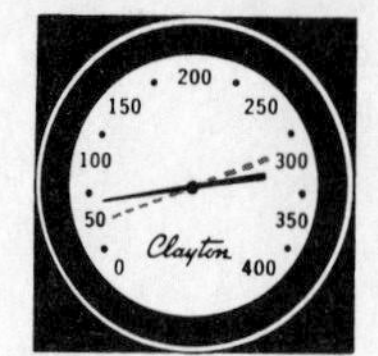

Hydraulic application (the difference in braking power between the two wheels under test) is found by making two short applications of the brakes. Slow application—shown here—reveals brake balance at moderate pressures.

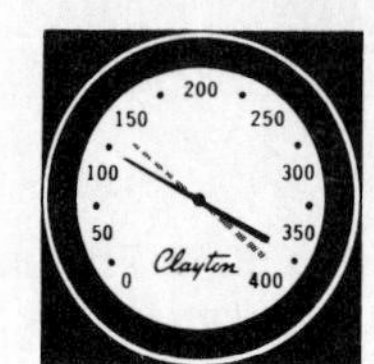

A rapid application is shown here. Any hydraulic restrictions are indicated by a lag in one hand or the other. Hydraulic lag creates momentary imbalance that may cause a car to react wildly during an emergency stop at freeway speeds.

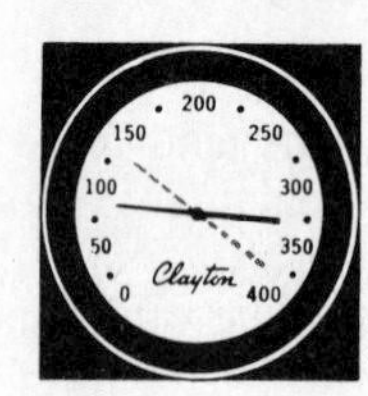

A high effort application on very good brakes. Note closeness of hands.

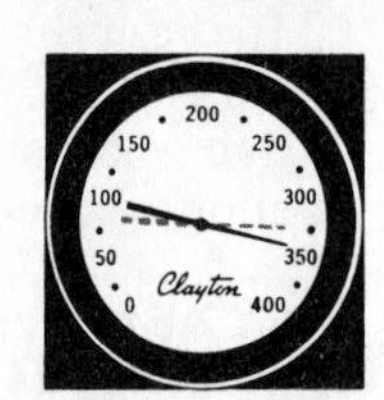

The same hydraulic test on brakes with modest imbalance, but within acceptable limits.

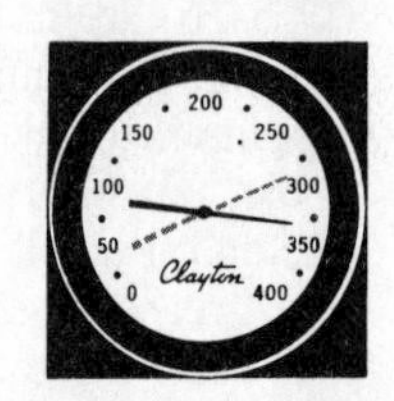

A high effort application or simulated stop from high speeds. These brakes are marginal and can be dangerous in a panic stop. Condition could be improved by repair.

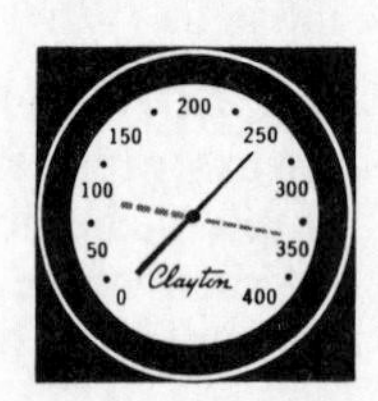

These brakes could cause a fatal accident if not repaired. Note wide spread of the hands. With hydraulic imbalance like this, a car could be literally jerked into another lane of traffic . . . a divider . . . or into a head-on collision. This car is unsafe in any driving situation.

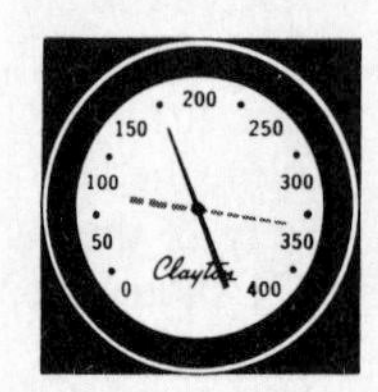

Fig. 5-12 Various meter readings and their meanings for one type of dynamic brake analyzer. The solid needle shows the action of the left brake. The dashed needle shows the action of the right brake. (*Clayton Manufacturing Company*)

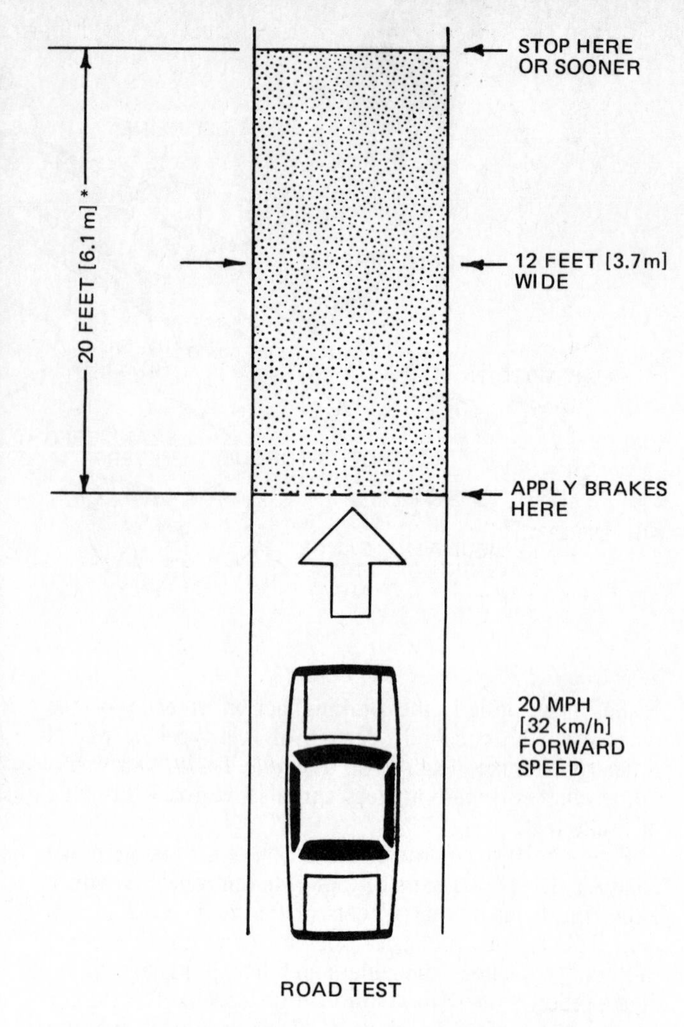

Fig. 5-13 Using a premarked lane to check vehicle stopping distance. (*Motor Vehicle Manufacturers Association*)

CAUTION: **Do not drive a car unless you have a valid driver's license, and do not make this test unless you have your instructor's permission. Then fasten your safety belt and conduct the test in the area designated by your instructor.**

❁ 5-28 Road testing the brakes If the car is safe to drive, a road test can be made to check the performance of the brakes. The road test helps the technician evaluate brake performance and the need for service. Any road test involving brake testing should be conducted on a stretch of pavement that is free of other traffic. The surface should be dry and clean, and reasonably smooth and level. During the road test, the driver must be careful not to cause brake fade as a result of continuous brake applictions.

CAUTION: **Before driving or road testing any car, always press the brake pedal to make sure there is ample pedal reserve (❁ 5-3) to stop the car. If there**

is, then make a series of very slow-speed stops to determine if the brakes are safe for driving.

CAUTION: **Do not drive a car unless you have a valid driver's license, and do not make this test unless you have your instructor's permission. Then fasten your safety belt and conduct the test in the area designated by your instructor.**

A. Low-speed test for effectiveness, pulls, and noise

1. Make a series of light and medium stops at from 10 to 15 mph [16 to 24 km/h]. Bring the car to a complete stop each time. Notice the force you must apply with your foot against the brake pedal to make each stop. It should not be too light, produce brake grabbing, or be too hard.
2. Check for pulling. On cars with rear-wheel drive, unequal front brake action will cause the car to pull in the direction of the brake doing the most work. Unequal rear brake action may not cause noticeable pulling during slow-speed stops.
3. Check for noise. Open the car windows and turn off the radio and all other accessories. Then listen carefully to determine the type of noise and the wheel from which it is coming. Driving the car alongside a building or wall will make noises easier to hear when the brakes are applied.

B. High-speed test for roughness and pulsations

1. Make several light stops from the maximum legal speed.
2. Check for any roughness and pulsations in the brake pedal during braking by noticing the pedal feel and car vibration.

C. High-speed test for effectiveness, pull, and noise

1. Make several hard stops from the maximum legal speed, bringing the car to a complete stop each time. Do not repeat hard stops within 2 miles [3.2 km] of each other. This will avoid high brake temperatures and fade.
2. Notice the force you must apply with your foot against the brake pedal to make each stop. It should not be too light, produce brake grabbing, or be too hard.
3. Check for pulling. Unequal brake action will cause the car to pull in the direction of the brake doing the most work.
4. Check for noise. If there is any, determine the type and the wheel from which it is coming.

D. Other factors affecting brake performance Four conditions outside of the brake system may affect brake performance during a road test. These are:

1. Wheel bearings. A loose wheel bearing may permit the drum to tilt during braking. This will produce spotty brake contact, causing erratic brake action. Make sure that the wheel bearings are properly adjusted (❁ 12-7).
2. Front-wheel alignment. Any misalignment of the front wheels can cause brake action to appear unequal from side to side. Alignment should be checked against the manufacturer's specifications (Chap. 13). At the same time, the tire wear patterns should be noted.
3. Shock absorbers. During quick stops, the feel of erratic or too severe braking may be caused by defective or ineffective front shock absorbers. The shock absorbers should be checked (❁ 9-10), and replaced if necessary.
4. Tires. Before the brakes are road-tested, check that the tires are properly inflated. Tires on the same axle should carry the same pressure and have identical or equivalent treads. Radial tires and bias tires must not be mixed on the car. The tires at all four wheels should be either radials or bias-ply. Any of these improper tire conditions may cause unequal braking.

Brake inspection

❁ 5-29 Inspecting brakes Many conditions require the technician to make a visual inspection and then check the operation of the brakes while the car is in the shop. The brake system can be divided into three components: hydraulic elements, mechanical elements, and friction elements. All of these must be checked during a brake inspection. The following procedure is recommended by General Motors for inspecting and checking brakes in the shop.

A. Pedal checks

1. Apply and release the foot brake several times while checking for friction and noise. The engine should be running if the vehicle is equipped with power brakes. Pedal movement should be smooth, the pedal should return fast, and no squeaks should be heard from the pedal or brakes.
2. Apply heavy force with your foot and check the pedal for sponginess. The engine should be running if the car has power brakes. The pedal should feel firm, not springy. Then measure pedal reserve. The pedal should be more than 2 inches [51 mm] from the floor for manual brakes and high-pedal-type power brakes. For low-pedal-type power brakes, the pedal should be more than 1 inch [25 mm] from the floor.
3. Check for hydraulic leaks. Hold a light force against the pedal with your foot for 15 seconds. The engine should be off for power brakes. There should be no pedal movement. Then repeat the test with a heavy foot force. The engine should be on for power brakes. Again, there should be no pedal movement.
4. Apply and release light pedal force. The engine should be running for power brakes. Check that the stop lights go on and off.
5. Check the power brakes for proper operation. With the engine stopped, apply and release the brake pedal several times to eliminate all vacuum from the system (vacuum brake booster). Apply the brakes and, while maintaining medium force on the pedal with your foot, start the engine. If the vacuum brake booster is functioning properly, the brake pedal will noticeably sink toward the floor as the engine vacuum is restored to the system.

B. Master-cylinder and brake-fluid inspection

1. Remove the master-cylinder reservoir cover. Make sure the cover diaphragm is in place and in good condition. Then check that the cover vent holes are open.
2. Check that the brake fluid is at the proper level in both sections of the master cylinder, and that the fluid is clean. Master cylinders with nylon reservoirs have fluid-level marks in the reservoir. Most cast-iron resrvoirs shoul be filled to within ¼ inch [6.35 mm] of the top. Add the proper fluid, if necessary.
3. Check for external hydraulic leaks. Look for dampness around the master-cylinder body, fittings, brake pipes, and at the mating surface between the master cylinder and the power-brake booster. On manual brakes, look for leakage at the pushrod end of the master cylinder.

C. Hose, pipe, and wheel inspection

1. Under the hood, check the brake hose, pipes, and connections for leaks. Then raise the car on a lift, and make the same check under the car.
2. Check the brake backing plates and wheels for traces of brake fluid or grease leaks.
3. Make sure the brake pipes are free from dents and other damage. Check that the brake hose is flexible and free from cracks, cuts (Fig. 5-14), and bulges.

D. Parking-brake check

1. Set the parking brake. Check that the lever or pedal moves no more than two-thirds of full travel. Then check that the rear wheels are locked.
2. Raise the car on a lift or with a jack, and then place safety stands under the car. Release the parking brake, and check that all wheels turn freely without drag.

E. Brake-lining inspection

1. Whenever brake-lining wear is suspected, or at the intervals recommended by the manufacturer, remove the brake drums and inspect the linings. On cars with drum brakes, remove one of the front drums. On cars with front-disk brakes and rear-drum brakes, remove one of the rear drums.
2. If less than ¹⁄₃₂ inch [0.8 mm] of usable lining remains (above the rivet heads on shoes with riveted lining), replace the shoes and linings. If further usage of the lining is questionable, remove the other brake drums and inspect the other linings.
3. Replace the shoes and linings (and refinish or replace the drums) if the lining wear is uneven. Linings that are oil-soaked, have loose rivets, or have foreign material embedded in them should be replaced.

F. Brake mechanism inspection

1. If you are uncertain about the condition of the brakes, or when the brakes at one or more wheels are suspected of causing trouble, pull and inspect the brake drums. With the drums off, visually inspect the brake mechanisms.
2. Check the linings for being loose on the shoes. Then look for cracks, unusual wear, and foreign material embedded in the lining. Make sure that the linings are not contaminated with oil, grease, or brake fluid.
3. Inspect the brake shoes for cracks, distortion, and broken welds.
4. Make sure that all springs are properly installed. Check for cracks, distortion, and discoloration.
5. Check that the shoe hold-down parts—pins, springs, and cups or clips—are properly installed and undamaged.
6. Check the brake backing plates for distortion and cracks. Make sure that the bolts and anchors are tight. Carefully pry each shoe away from the backing plate, and check the backing-plate pads for grooving or other damage. Make sure the pads are properly lubricated. Check that the shoes retract fully to contact the anchor pin.
7. Inspect the wheel-cylinder boots for cuts, cracks, hardening, and other signs of deterioration. Make sure there is no sign of brake-fluid leakage. If any leakage is noted—more than a drop—overhaul or replace the wheel cylinder.
8. Check that all parts of the self-adjuster are properly installed. Check the parts for any damage or corrosion that would prevent proper operation. Check the adjusting screw for freedom of rotation.

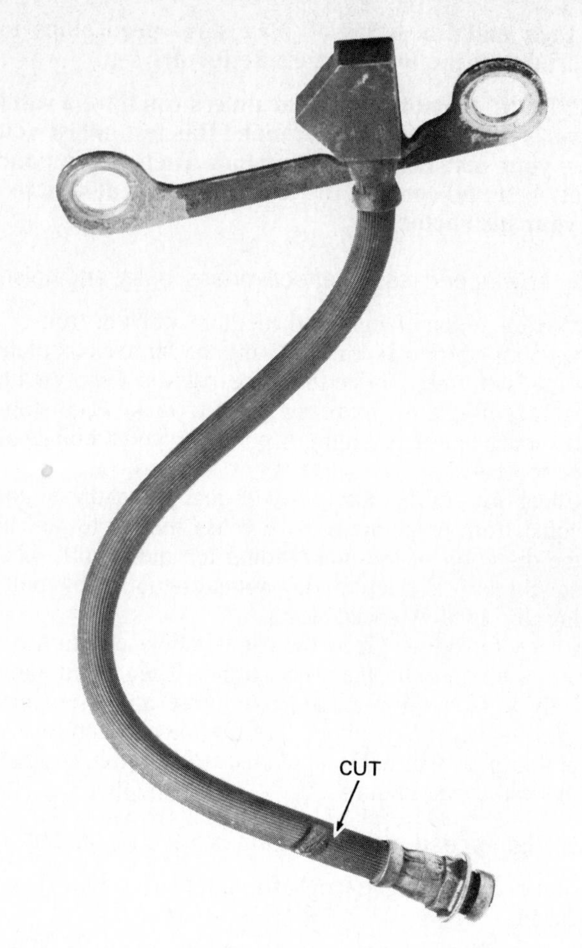

Fig. 5-14 A flexible brake hose that has been cut by rubbing against the edge of the car frame. (*ATW*)

9. Check for foreign objects or accumulations of dirt. Brake mechanisms must be reasonably clean to operate properly.

Chapter 5 review questions

Select the *one* correct, best, or most probable answer to each question. The check your answers against the correct answers given at the end of the book.

1. The driver of a car with front-disk and rear-drum brakes complains that the brake pedal moves slowly to the floor while the pedal is depressed at a traffic light. This problem could be caused by:
   a. a leaking primary piston cup in the master cylinder
   b. a leaking power-brake booster
   c. a leaking residual check valve in the master cylinder
   d. an internal leak in the combination valve
2. On a car with single-piston floating-caliper disk brakes, the brake pad between the piston and the disk is badly worn. The other brake pad is slightly worn. Mechanic A says that too much rotor runout could be the cause. Mechanic B says that the piston is binding in the caliper. Who is right?
   a. A only
   b. B only
   c. either A or B
   d. neither A nor B
3. Which of these problems could be caused by putting a residual check valve in the master-cylinder outlet for the front-disk brakes?
   a. reduced brake-pedal travel
   b. reduced brake-system pressure
   c. increased brake-pad wear
   d. increased rotor runout
4. The front end of a car with front-disk and rear-drum brakes dips too much when the brakes are lightly applied. The most probable cause is:
   a. loose front-wheel bearings
   b. air in the front brake lines
   c. binding caliper-piston seals
   d. defective metering valve
5. The owner of a car with four-wheel disk brakes says that the brake pedal pulsates when the brakes are applied. Any of the following could be the cause *except:*
   a. a deeply scored or grooved rotor
   b. excessive rotor runout or wobble
   c. uneven rotor thickness
   d. a loose wheel bearing
6. On a car with front-disk and rear-drum brakes, the rear wheels lockup with normal brake-pedal application. Which of these could cause this problem?
   I. A defective pressure-differential valve
   II. A defective proprotioning valve
   a. I only
   b. II only
   c. either I or II
   d. neither I nor II
7. The disk-brake reservoir on a dual master cylinder is low on fluid. Mechanic A says the brake pads may be worn. Mechanic B says the residual check valve may be faulty. Who is right?
   a. A only
   b. B only
   c. both A and B
   d. neither A nor B
8. A car pulls to the right when the brakes are applied. Mechanic A says that the cause could be a defective master cylinder. Mechanic B says that the cause could be a defective metering combination valve. Who is right?
   a. A only
   b. B only
   c. either A or B
   d. neither A nor B
9. In the master cylinder, the brake-fluid level should usually be:
   a. not much more than ⅛ inch [3.2 mm] below the top
   b. level with the top
   c. ¼ to ½ inch [6.4 to 12.7 mm] below the top
   d. ½ to ¾ inch [12.7 to 19 mm] below the top
10. Dragging brakes can be caused by:
    a. a blocked compensating port
    b. improper brake-pedal pushrod adjustment
    c. restricted or kinked brake hose
    d. all of the above

CHAPTER 6

BRAKE ADJUSTMENTS AND SERVICE

After studying this chapter, and with proper instruction and equipment, you should be able to:

1. Adjust drum brakes.
2. Adjust parking brakes.
3. Service disk brakes.
4. Service drum brakes.
5. Service the master cylinder.
6. Repair a damaged brake tube.
7. Replace a defective brake hose.
8. Flush, fill, and bleed the hydraulic system.

Adjusting brakes

6-1 Drum-brake adjustments Any complaint of faulty braking should be diagnosed to determine its cause. Sometimes all that is required (on drum brakes without self-adjusters) is a brake-shoe adjustment to compensate for lining wear. On self-adjusting drum brakes (Figs. 3-43 to 3-45), the brake shoes are automatically adjusted.

Drum brakes are usually adjusted without removing the wheels and brake drums from the car. However, self-adjusting brakes require a preliminary adjustment after the brake shoes are replaced. Measurements are taken and initial adjustments made before the drums are installed on the car.

❁ 6-2 Manual adjustment of drum brakes Self-adjusting drum brakes require adjustment only after replacement of brake shoes, grinding of the brake drums, or other service in which disassembly of the brakes was performed. This adjustment is called the *manual adjustment,* or *preliminary adjustment.* (Disk brakes require no adjustment.) Typical manual adjustment procedures for drum brakes are given below.

1. Drums on the car A Bendix self-adjusting drum brake and its automatic adjuster parts are shown in Fig. 3-43. To adjust the brakes with the drums in place, remove the adjustment-hole cover from the brake drum or backing plate (Fig. 6-1). Insert the adjusting tool or screwdriver and move it up or down as required to expand the shoes. Stop when the shoes drag heavily against the drum as you turn it. Then insert a wire hook or thin-blade screwdriver into the adjustment hole to hold the adjuster lever away from the adjustment screw. Back off the adjustment screw approximately 20 notches and check that there is no drag on the drum. Repeat the procedure for each drum brake being adjusted. Then install the adjustment-hole cover.

Careful! After making the preliminary adjustment, press firmly on the brake pedal. There must be adequate pedal before you move the car.

To complete the preliminary adjustment of self-adjusting brakes, make several alternating reverse and forward stops. Press firmly against the brake pedal. Repeat this procedure until adequate pedal reserve has built up. If making the alternating reverse and forward stops does not restore the pedal to normal height, the automatic adjusters are not working. Remove the drums and check the brakes for defective parts or assembly procedures.

When normal pedal reserve has been obtained, adjust the parking brake (❁ 6-3). Then road-test the car and check for normal brake operation (❁ 5-28).

2. Drums off the car If the brake drums are off the car, the manual adjustment can be made quickly and easily by using the brake-shoe adjusting gauge (Fig. 6-2). First, set the gauge to the inside diameter of the drum and then tighten the gauge lock screw. Fit the other end of the gauge over the brake shoes. Turn the adjusting screw to expand the shoes until the gauge just slides over the linings. Rotate the gauge around the lining surface

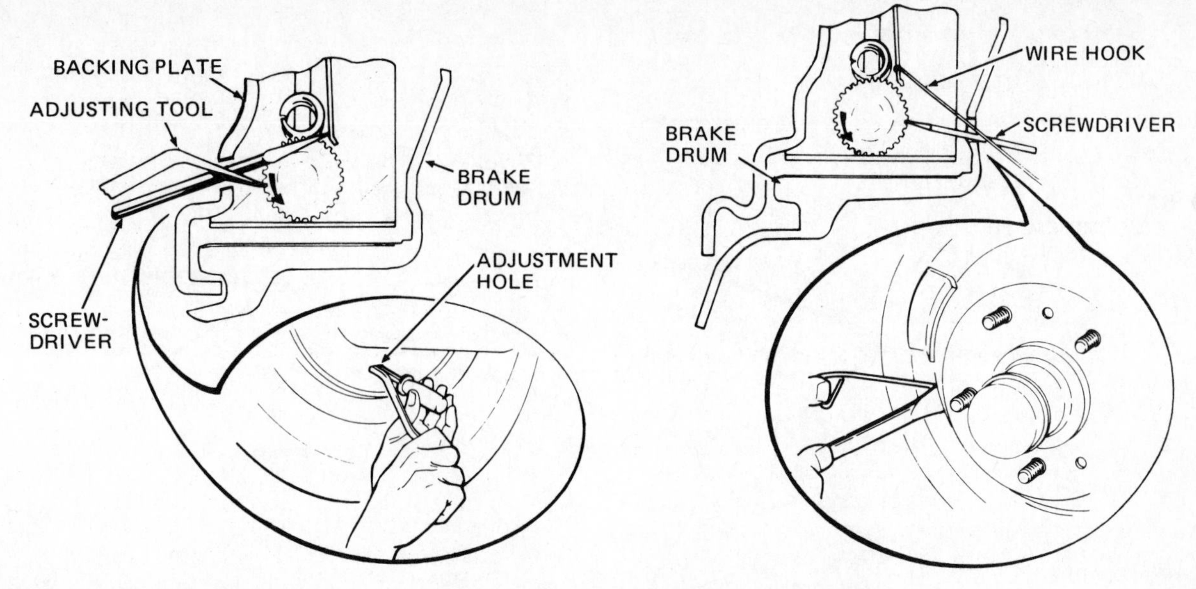

Fig. 6-1 Holding the adjusting lever off the adjusting screw while turning it. Left, moving the tool down expands the shoes. Right, moving the screwdriver up expands the shoes. (*Delco Moraine Division of General Motors Corporation*)

to ensure proper clearance. When the gauge will just fit over the linings at all points, the preliminary clearance between the linings and drum has been set. Install the drums and wheels.

Careful! After making the preliminary adjustment, press firmly on the brake pedal. There must be adequate pedal before you move the car.

Complete the preliminary adjustment by making several alternating reverse and forward stops. Press firmly against the brake pedal for each stop. Repeat this procedure until adequate pedal reserve has built up. If normal pedal height is not restored, then the self-adjusters are not working. Pull the drum and check the brakes for defective parts or assembly procedures.

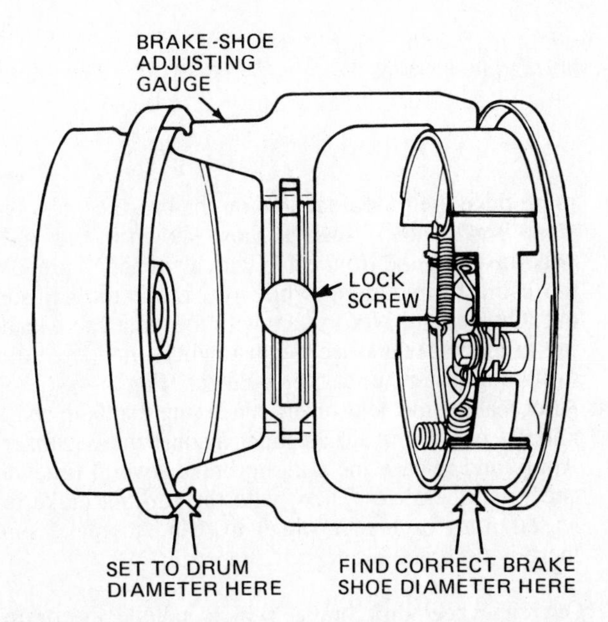

Fig. 6-2 Using the brake-shoe adjusting gauge to set the preliminary clearance between the linings and the drum. (*Ford Motor Company*)

NOTE: Some cars may have a different type of drum brake. For manual adjustment procedures, refer to the manufacturer's service manual.

❁ 6-3 Parking-brake adjustment The parking brake should be adjusted anytime the rear brakes have been disconnected, and when the parking-brake foot pedal or hand lever travels too far before applying the brakes. The parking brake should be adjusted so that adequate holding action is ensured. On many cars, there should be some pedal or lever reserve. It should not be necessary to move the lever or pedal through its full travel to get full braking. However, the adjustment should not be so tight that in the released position the parking brake causes the lining to drag.

Rear-wheel drum brakes use the same shoes for both service brake and parking brake. Any adjustment of the shoe adjusting wheel will affect both brake systems. For this reason, the parking brake should be checked and adjusted anytime a manual brake adjustment (❁ 6-2) is made on the rear brakes.

Basically all parking brakes are adjusted in some way so that proper tension is restored to the brake cables. This compensates for wear of the brake lining and for stretching of the cables. Typical procedures for adjusting parking brakes are given below. Refer to the manufacturer's service manual for the required specifications.

1. Release or set the parking brake, as required. Then raise the car on a lift.
2. On most cars with rear-wheel drum brakes, the parking-brake pedal or lever is connected to a short front

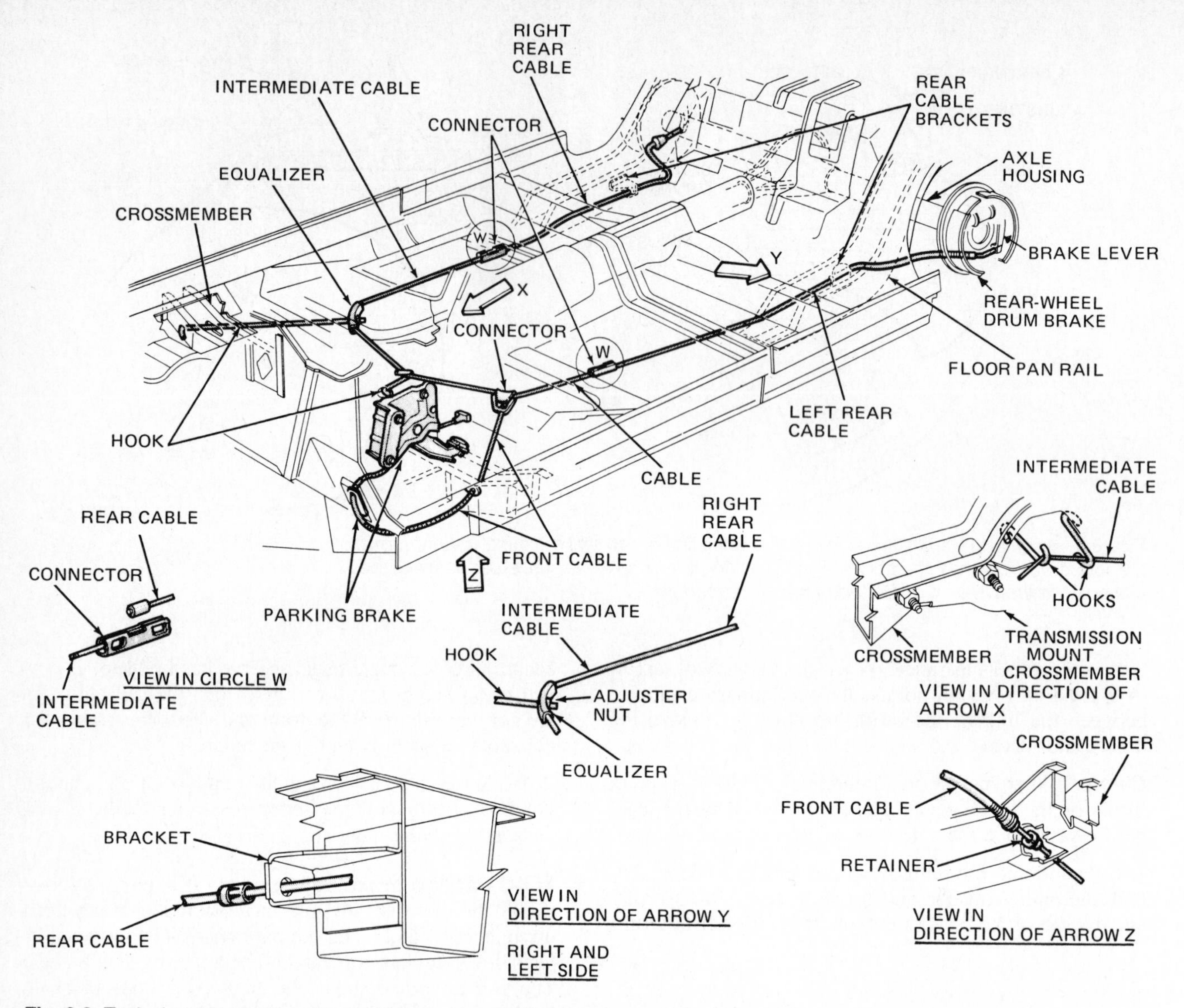

Fig. 6-3 Typical parking-brake cable routing on a rear-wheel-drive car with rear-drum brakes. (*Chrysler Corporation*)

cable (Fig. 6-3). The cable usually has an *equalizer* attached so that an even pull is applied to each of the rear brakes. On some cars, the front cable extends back to where individual cables branch off to the two rear brakes (Fig. 6-3). On other cars, the front cable is routed along the left side of the car through the equalizer, to which the cable is attached (Fig. 6-4). From the equalizer, the cable continues on as the left rear cable. The cable for the right rear brake is connected to the equalizer and routed around the axle housing to enter the brake assembly from the rear. Notice in Fig. 6-4 that the left cable enters the brake assembly from the front.

3. Adjustment of most parking brakes usually is made by tightening or loosening the equalizer adjuster nut (Fig. 6-4). Another method is by alternately loosening and tightening the check nuts on either side of the equalizer. This moves the equalizer forward or back to produce the proper tension on the brake cables. On some brake cables, such as those shown in Fig. 6-3, you must hold the front cable with a wrench to prevent the cable from turning while you move the adjuster nut. On General Motors cars with the front cable leading to the left rear wheel, adjustment is made by turning the adjuster nut at the equalizer (Fig. 6-4).
4. After making the adjustment, make sure that both check nuts (if two are used) are tight against the equalizer. Apply and release the parking brake several times to check its operation. Then, with the parking brake released, turn each rear wheel by hand to make sure there is no brake drag.

On rear-wheel disk brakes which include the drum-type parking brake, the parking-brake shoes may require periodic adjustment. The adjusting tool is inserted through the backing plate, as shown in Fig. 6-5. Then the ad-

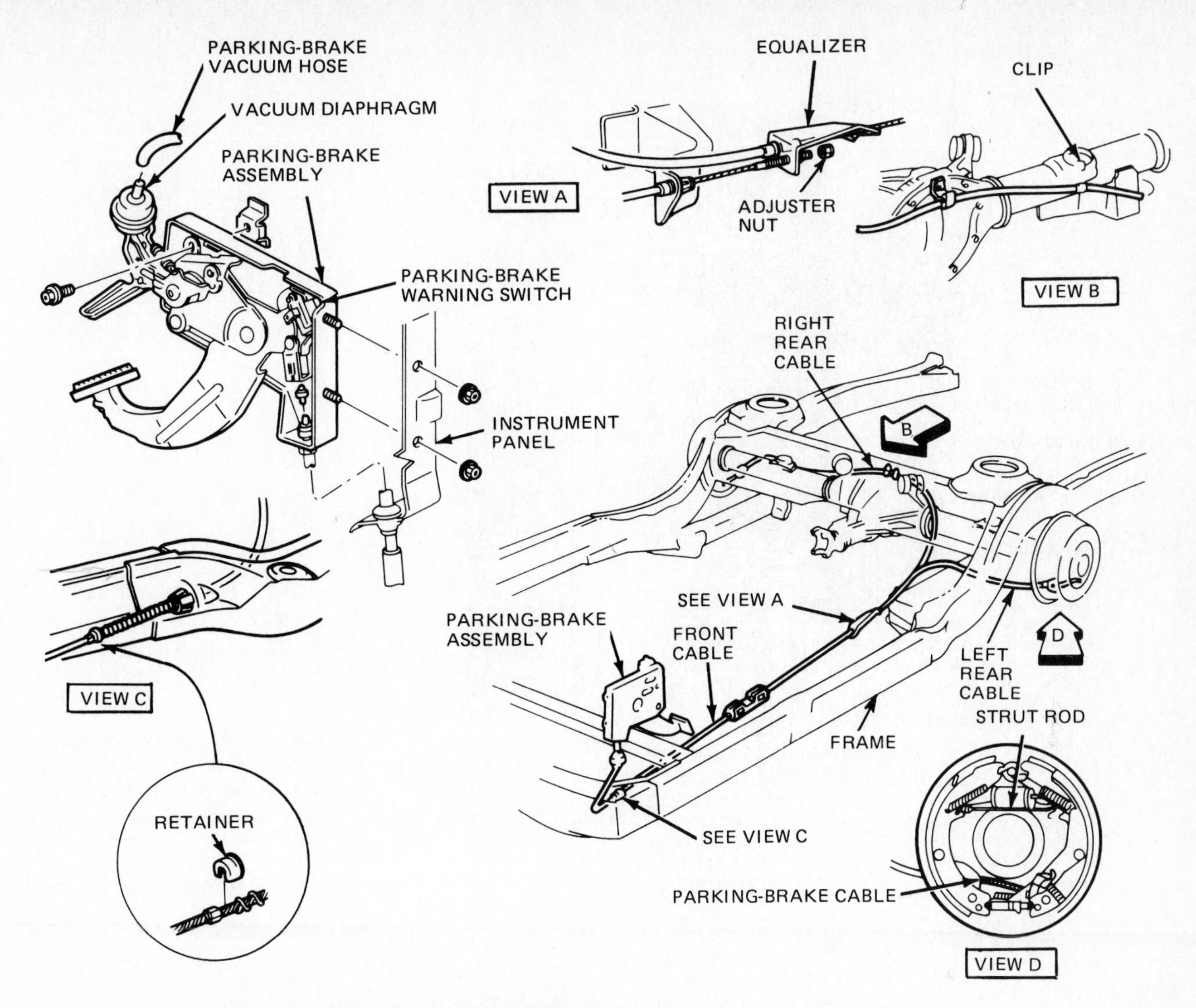

Fig. 6-4 On some cars, the parking-brake cable enters the brake assembly from the rear. (*Cadillac Motor Car Division of General Motors Corporation*)

justing tool is pivoted either on the splash shield or on the clamp to turn the adjuster star wheel one way or the other, as required.

Rear-wheel disk brakes which have an integral parking brake (❁ 3-5) do not require periodic adjustment. The piston assembly in the caliper contains a self-adjusting mechanism for the parking brake.

Disk-brake service

❁ 6-4 Fixed-caliper disk-brake service The chart in ❁ 5-15 lists troubles in disk-brake systems and their possible causes. Although all disk brakes work similarly, their design and construction varies. There are three general types, as described in ❁ 3-20: fixed-caliper, floating-caliper, and sliding-caliper. This section describes the servicing of the fixed-caliper disk brake (Figs. 3-46 to 3-50 and 6-6). Following sections discuss servicing of the other two types.

1. Removing the brake shoes Raise the vehicle on a lift or on a jack. Then support the car on safety stands. Remove the wheel covers and the wheel-and-tire assembly. Remove the bolts holding the splash shield, and remove the shield and antirattle spring (Fig. 6-6). Using two pliers to grip the tabs on the shoe, pull the shoe out (Fig. 6-7). If the shoe hangs on the pistons, push the pistons in with slip-joint pliers as shown in Fig. 6-7. Watch for master-cylinder reservoir overflow when pushing the pistons in. Mark the shoes so you can return them to the same side of the caliper from which you removed them.

On disk brakes that have spring-loaded pistons, a special piston-compressing tool is used to push the pistons back into their bores and hold them in that position during the time that the brake shoe is out.

With the brake shoes out, check the caliper and pistons for possible leaks and wipe all parts clean. If brake fluid is present, remove the caliper as explained below for replacement of piston seals.

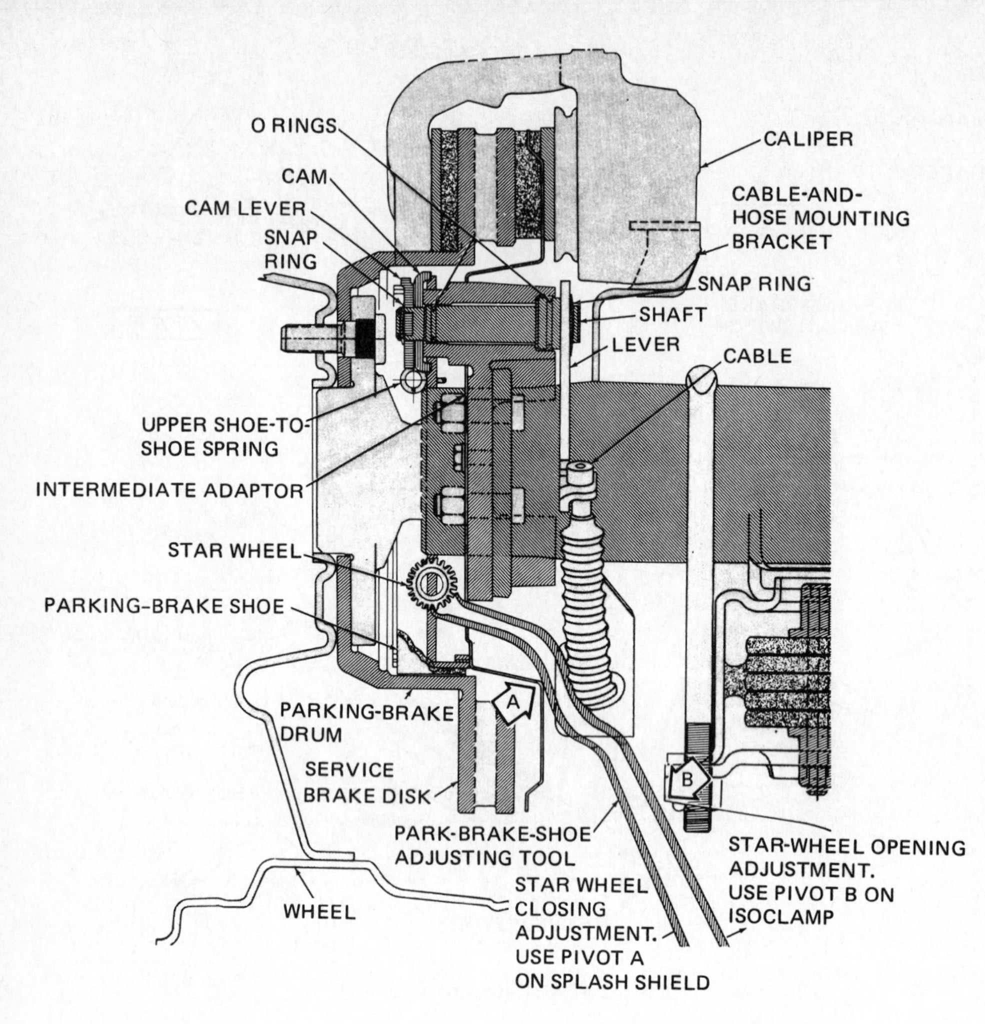

Fig. 6-5 Adjusting the parking-brake shoes on a sliding-caliper disk brake with an independent drum-type parking brake. The two positions of the parking-brake-shoe adjusting tool are shown at A and B. (*Chrysler Corporation*)

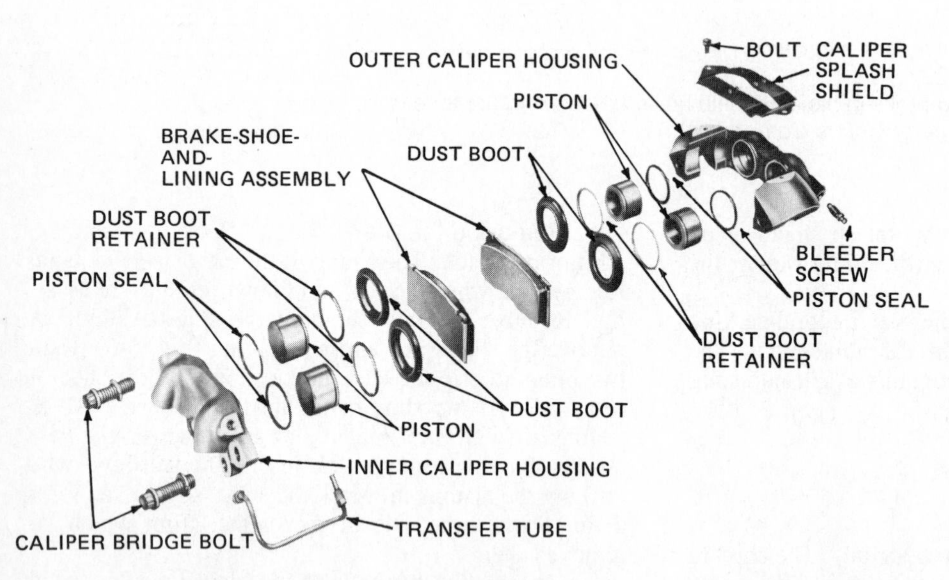

Fig. 6-6 Disassembled view of a four-piston fixed-caliper assembly. (*Chrysler Corporation*)

2. Installing the brake shoes Push the pistons back into their bores to allow room for installation of new thicker shoes and linings. Then slide the new shoe into the caliper with the ears of the shoe resting on the bridges of the caliper. Be sure the shoe is fully seated, with the lining facing the disk. Install the other shoe. Install the caliper splash shield and antirattle spring assembly. Pump the foot pedal several times until a firm pedal is obtained. This ensures proper seating of the shoes. Install the wheel and tire. Check and refill the master cylinder if necessary.

With your instructor's permission, road-test the car. Make several heavy braking stops (*not* skids) from about 40 mph [64 km/h] to ensure good seating of the brake

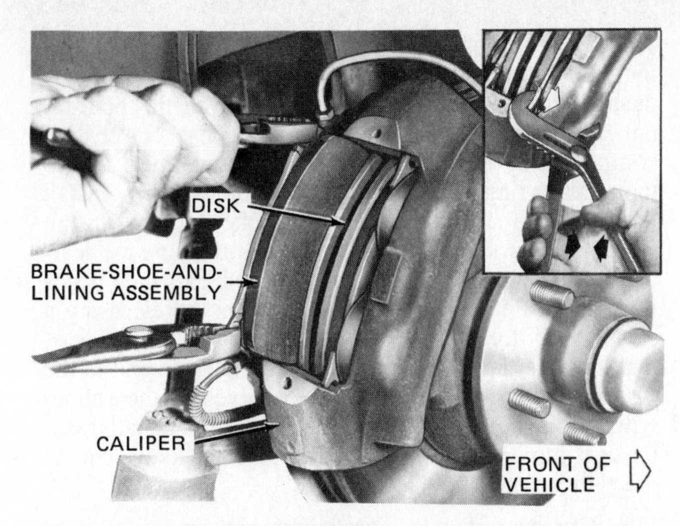

Fig. 6-7 Removing the brake shoe and lining from the disk-brake assembly. Upper right, forcing the piston into the bore with slip-joint pliers. (*Chrysler Corporation*)

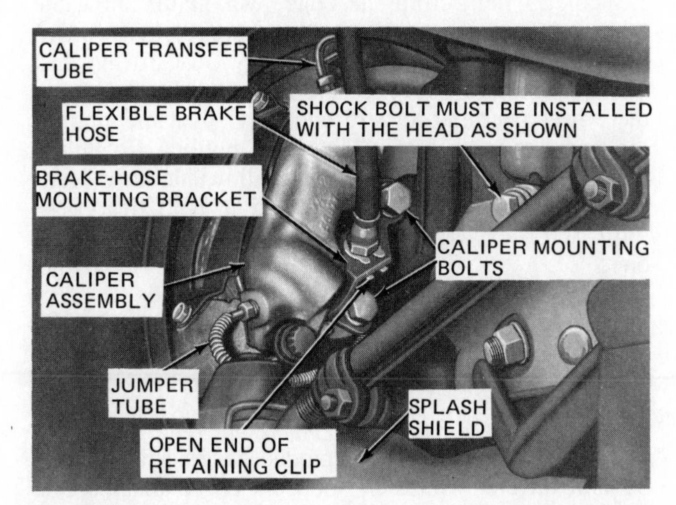

Fig. 6-8 Mounting of the disk-brake caliper at a front wheel. (*Chrysler Corporation*)

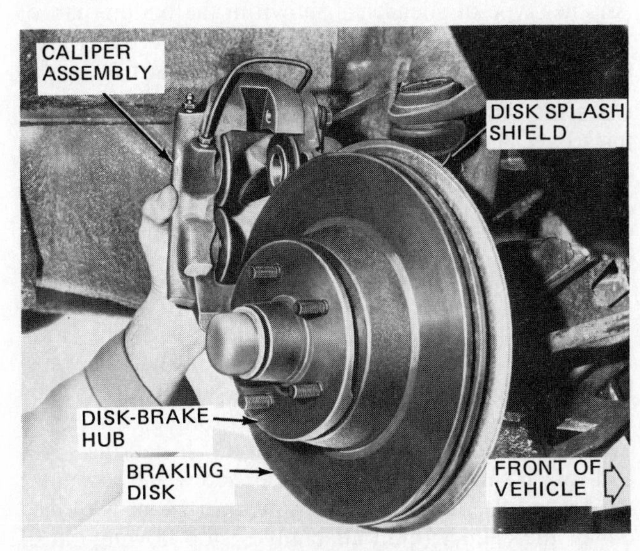

Fig. 6-9 Removing the disk-brake caliper. (*Chrysler Corporation*)

lining on the disk. Unless this is done, initial braking may cause the car to pull to one side or the other.

3. **Removing the caliper** Refer to Fig. 6-8 and proceed as follows. Raise the car on a lift or on a jack. Then support the car on safety stands and remove the wheel cover and wheel-and-tire assembly. Disconnect the brake flexible hose from the brake tube at the frame mounting bracket. Plug the brake tube to prevent loss of fluid. Remove the attaching bolts and lift the caliper assembly up and out (Fig. 6-9).

4. **Disassembling the caliper** Refer to Fig. 6-6 and proceed as follows. Remove the splash shield and anti-rattle spring assembly. Clamp the caliper mounting lugs in the soft jaws of a vise and remove the transfer tube and armored jumper tube. Remove shoes. Separate the two halves of the caliper by removing two bolts. Peel dust boots off the caliper and pistons. Use a special tool (Fig. 6-10) to remove the pistons, being careful to avoid scratching the pistons and bores. Then use a small pointed wood or plastic stick to remove piston seals from grooves in the piston bores. Do not scratch the bores!

5. **Cleaning caliper parts** Clean all parts, except shoes and linings, in brake fluid and wipe dry with lint-free towels. Blow out drilled passages and bores with compressed air. Discard old piston seals. Inspect the piston for pitting, scoring, corrosion, and surface damage. Discard dust boots and pistons that appear damaged in any way. If piston bores are scratched, clean them with crocus cloth or a special hone. However, if more than about 0.002 inch [0.05 mm] must be removed to clean up deep scratches, discard the caliper. Carefully clean the caliper to remove all traces of dust or dirt.

NOTE: Some manufacturers recommend cleaning brake parts in dentured alcohol. Others specify that dentured alcohol should *not* be used. Follow the manufacturer's recommendations for the car you are servicing.

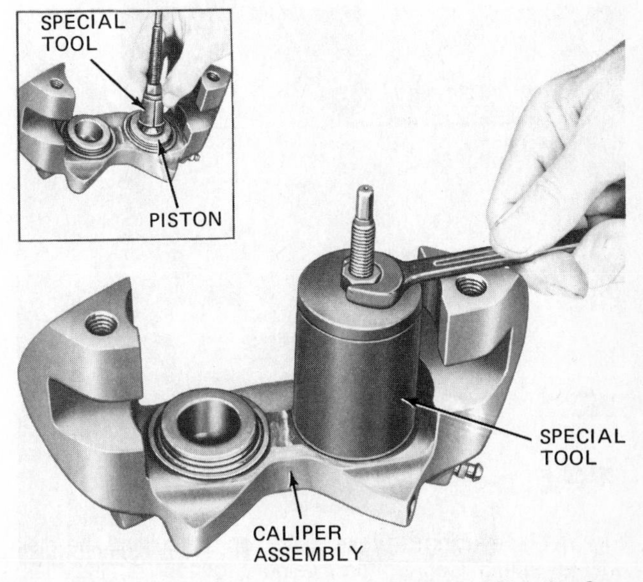

Fig. 6-10 Removing pistons from the caliper. (*Chrysler Corporation*)

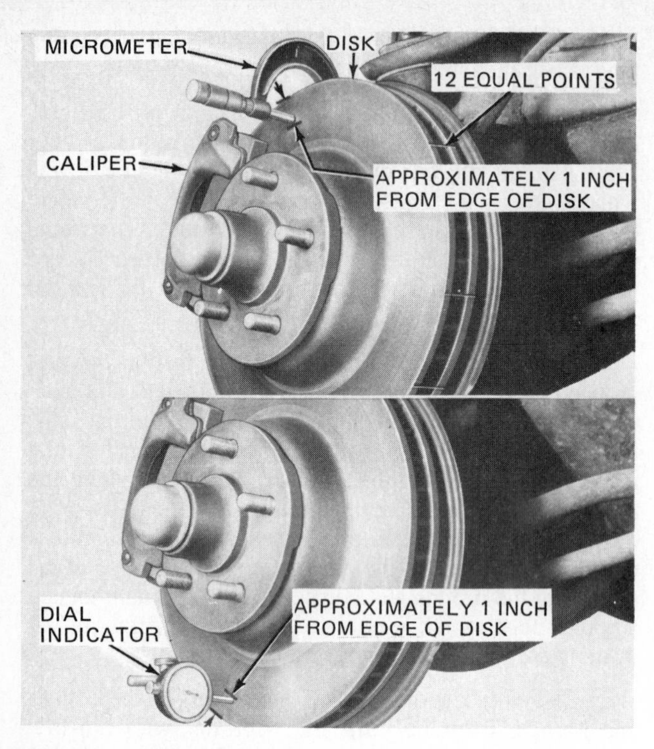

Fig. 6-11 Checking the thickness and runout of the disk. (*Chrysler Corporation*)

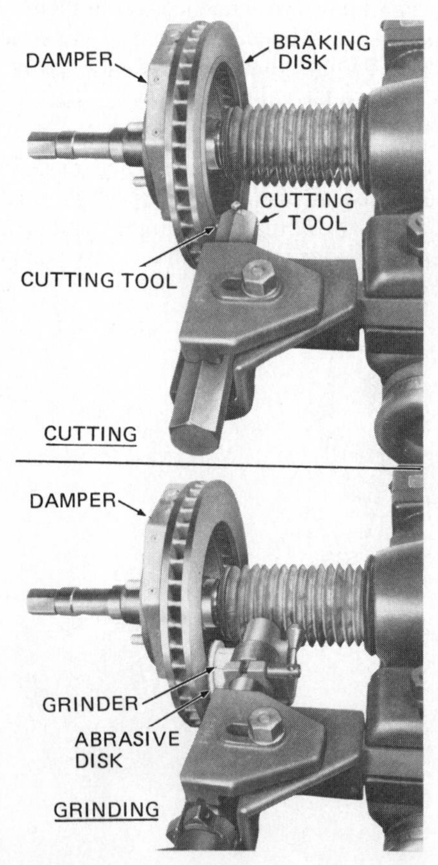

Fig. 6-12 Refacing a brake disk. Top, machining the disk with a cutting tool. Bottom, using a grinder on the disk. (*Chrysler Corporation*)

6. Reassembling the caliper Clamp the inner half, by the mounting lugs, in the soft jaws of a vise. If a new dust-boot retainer ring is necessary, clean the boot-retainer groove in the caliper. Then apply a special sealing compound to the retainer groove in the caliper, and to the retainer ring where it will seat in the housing. Install the ring. Dip new piston seals in brake fluid and install them in the grooves in the piston bores. Be sure the seals are not twisted or rolled. Coat the outside of the pistons with brake fluid and slide them into the bores, using a slow, steady force from your hand. Install new dust boots, making sure they seat over the retaining rings and in the piston grooves. Reattach the two halves of the caliper with special bolts, torqued to specifications. Install transfer and jumper tubes. Install the bleeder screw but do not tighten it.

7. Checking disk for parallelism and runout Before reinstalling the caliper, check the disk for runout and thickness, as follows. Use a micrometer [Fig. 6-11 (top)] and measure thickness at four or more equal points about 1 inch [25 mm] from the edge. Maximum allowable thickness variation usually is about 0.0005 inch [0.013 mm]. If thickness varies excessively, discard the disk and install a new one. Measure runout by first adjusting the wheel bearning to zero end play, then mounting a dial indictor as shown in Fig. 6-11, bottom. Rotate the disk and check runout. Maximum allowable lateral runout usually is about 0.004 to 0.005 inch [0.10 to 0.13 mm]. If it is excessive, discard the disk and install a new one.

Careful! Readjust wheel bearings after the check!

Light scores and wear of the disk are okay, but if the scores are fairly deep, the disk should be refinished (Fig. 6-12).

NOTE: Machining brake disks produces considerable noise. Several different kinds of silencers are available. One type is shown attached to the disk in Fig. 6-12. Another view of this silencer, or damper, is shown in the top illustration of Fig. 6-13. It is held in place by magnets. Another type of silencer is shown in the bottom part of Fig. 6-13. This type is a heavy or weighted rubber band that is wrapped around the disk before it is machined.

Careful! All vehicles built since 1971 (and some before that) have the specification for the minimum allowable disk thickness cast into the disk (Fig. 6-14). This measurement is the minimum thickness to which the disk can be refinished. If it is necessary to refinish the disk so that its thickness is less than that specified, discard it. The disk is too thin to use safely.

8. Installing the caliper After the disk has been reinstalled (if it was removed for service) and the wheel bearings adjusted, install the caliper assembly and tighten the mounting bolts to the specified torque. Check the disk for lateral runout (item 7, above). Install the shoes and splash shield. Open the bleeder screw and reconnect the brake line. Allow fluid to flow until all air is pushed out of the caliper (until air bubbles stop flowing out of the bleeder valve). Tighten the bleeder screw. Refill the

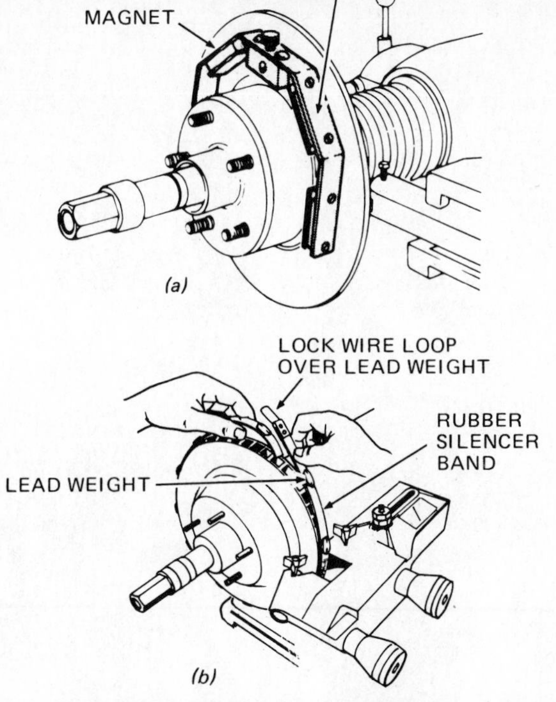

Fig. 6-13 Two types of dampers or silencers used on disk-brake rotors while machining the rotor faces. (*Ammco Tools, Inc.*)

master cylinder to the proper level with the specified type of brake fluid.

6-5 Floating-caliper disk-brake service The chart in 5-15 lists troubles in the disk-brake systems and their possible causes. This section describes the servicing of the floating-caliper disk brake. The operation of this brake is described in 3-20 and illustrated in Figs. 3-51 to 3-54. Figure 6-15 shows a disassembled floating-caliper assembly.

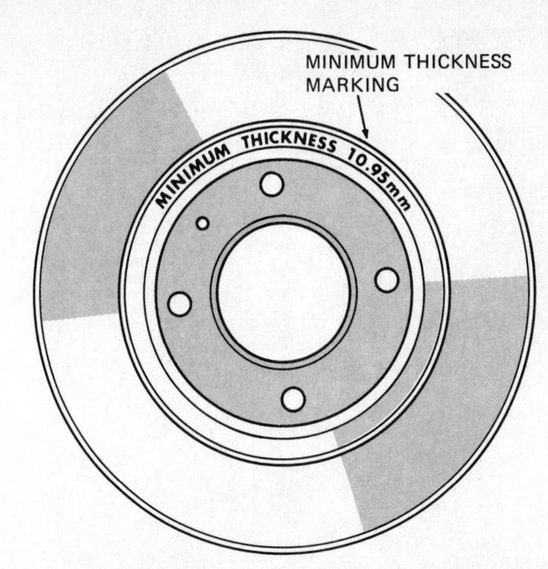

Fig. 6-14 Typical minimum thickness marking, cast into the disk-brake rotor. (*Chrysler Corporation*)

1. Removing the brake shoes Raise the car on a lift or safety stands and remove the front-wheel covers and the wheel-and-tire assemblies. Remove the caliper guide pins and positioners that attach the caliper to the adapter. Now, you can slide the caliper up and away from the disk (Fig. 6-16). Support the caliper firmly so that you don't damage the brake hose.

Slide the outboard and inboard shoe-and-lining assemblies out (Fig. 6-17). Mark the shoes so that you can put them back in the same places in the caliper. Push the outer bushings from the caliper with a wooden or plastic stick (Fig. 6-18). Throw the bushings away. Slide the inner bushings off the guide pins and discard them.

2. Inspecting caliper parts Check for piston-seal leaks (brake fluid in or around the boot area and inboard lining). If the boot is damaged or fluid has leaked, disassemble the caliper to install new parts as explained in

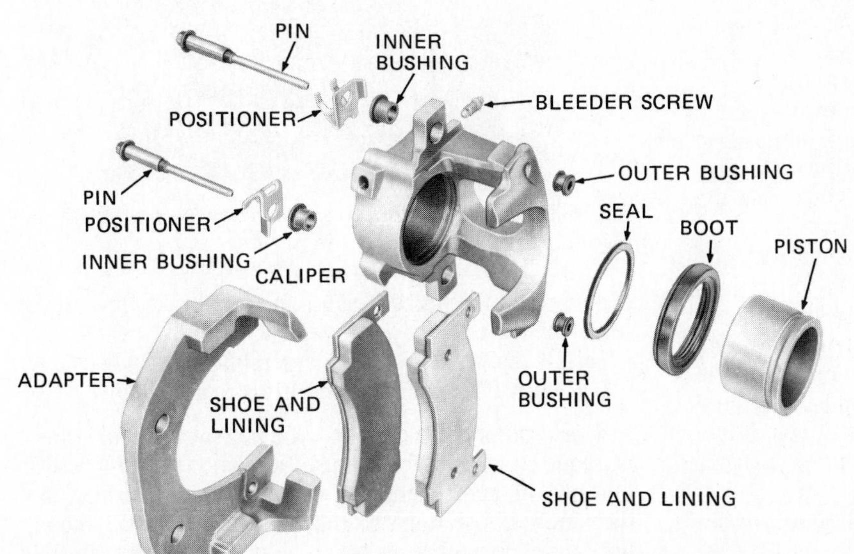

Fig. 6-15 Disassembled view of a floating caliper. (*Chrysler Corporation*)

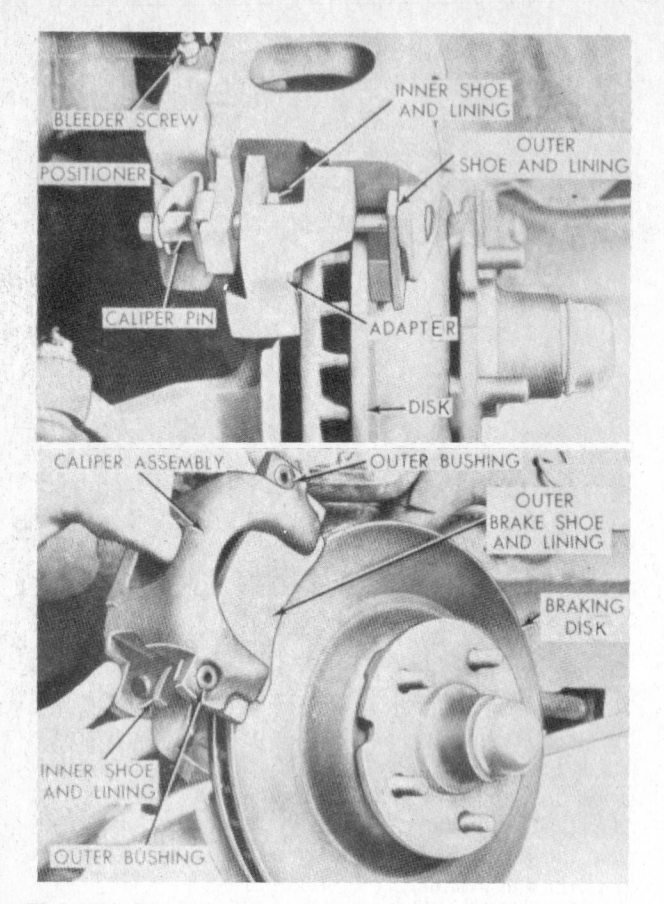

Fig. 6-16 Removing or installing calipers. (*Chrysler Corporation*)

Fig. 6-17 Removing or installing brake shoes and linings. (*Chrysler Corporation*)

item 5 below. Inspect the brake disk and service it if necessary (⚙ 6-4, item 7).

3. Installing brake shoes New positioners and inner and outer bushings will be required (Fig. 6-15). Slowly and carefully push the piston back into the bore until it bottoms. Watch for possible fluid overflow at the brake master cylinder. Install new bushings, noting their proper relationship as shown in Fig. 6-15. Slide new shoe-and-lining assemblies into position (Fig. 6-17). Make sure the metal part of the shoe is fully in the recess of the caliper and adapter. Hold the outboard lining in position and slide the caliper down into place over the disk (Fig. 6-16). Align the guide-pin holes of the adapter, and the inboard and outboard shoes (Fig. 6-19).

Install new positioners over the guide pins with the open ends toward the outside and the stamped arrows pointing upward. Install each guide pin through the bushing, caliper, adapter, inboard shoe, outboard shoe, and outer bushing (Fig. 6-19). Press in on the end of the guide pin and thread the pin into the adapter. Use care to avoid cross-threading. Tighten to specifications. Make sure tabs of positioners are over the machined surfaces of the caliper.

Depress the brake pedal several times until a firm pedal has been obtained. Check and refill the master-cylinder reservoir if necessary. If you cannot get a firm pedal, bleed the brake system and add brake fluid to the reservoir (⚙ 6-14).

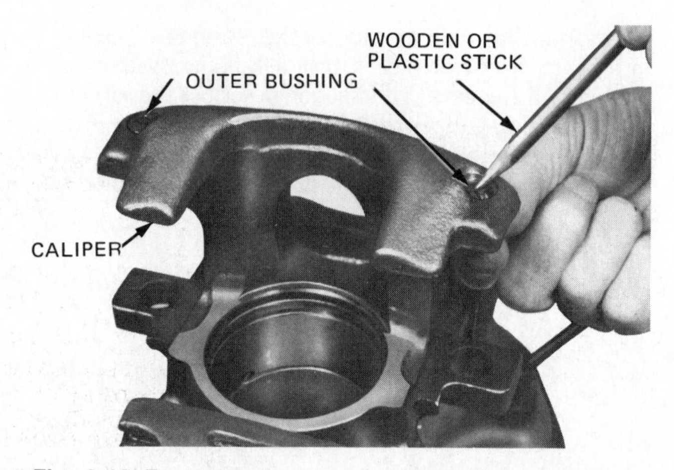

Fig. 6-18 Removing the outer bushings. (*Chrysler Corporation*)

Install the wheel-and-tire assemblies and wheel covers. Remove the car from the lift or safety stands.

4. Removing the caliper If a new piston seal or boot is required, the caliper must be removed. Proceed as outlined in item 1 above, with the additional step that the flexible hose must be disconnected from the tube at the frame. The tube must then be plugged to prevent loss of fluid.

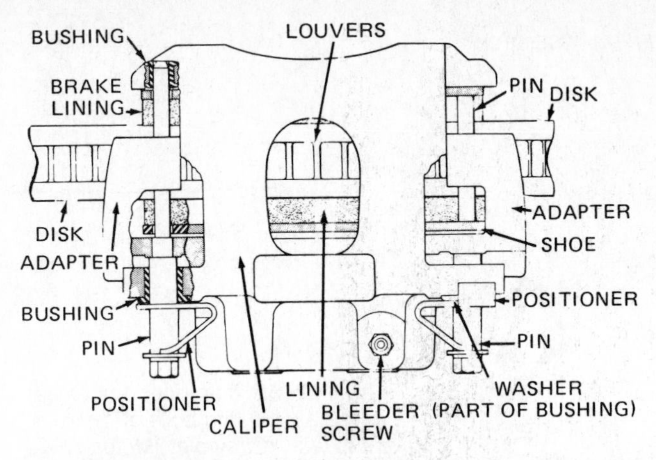

Fig. 6-19 Sectional view of a floating-caliper assembly, showing the positions of the adapter, pins, bushings, and positioners. (*Chrysler Corporation*)

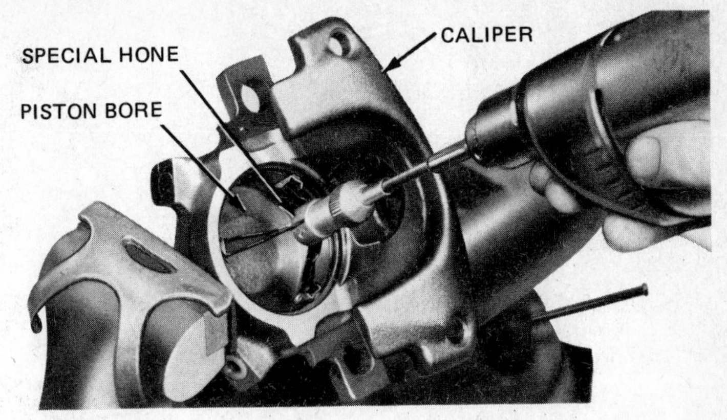

Fig. 6-21 Honing the piston bore in the caliper. (*Chrysler Corporation*)

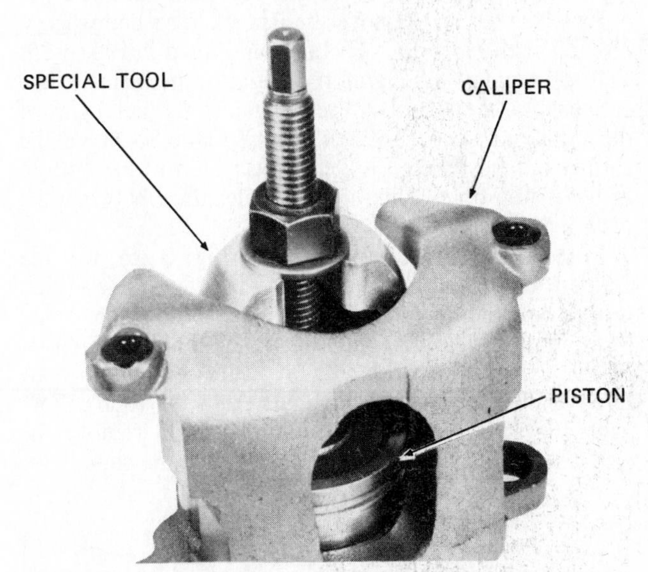

Fig. 6-20 Using the special tool to remove the piston from the caliper. (*Chrysler Corporation*)

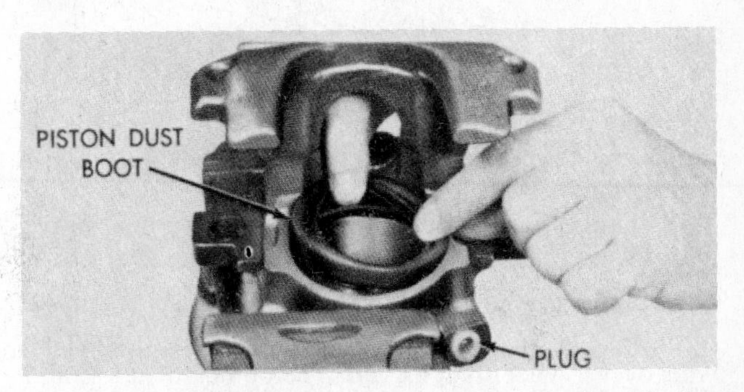

Fig. 6-22 Installing the dust boot in the piston bore in the caliper. (*Chrysler Corporation*)

5. Disassembling the caliper Clamp the caliper lightly in the soft jaws of a vise, and remove the dust boot. Use a special tool (Fig. 6-20) to remove the piston from the caliper.

Use a pointed wood stick to work the seal out of its groove in the piston bore. Never use a screwdriver or metal tool. It could scratch the bore or burr the edge of the seal groove. This would ruin the caliper.

6. Cleaning and inspecting caliper parts Clean parts with denatured alcohol or clean solvent and blow dry with compressed air. Inspect the bore for pits or scoring. Install a new piston if the old one is pitted or scored or if the plating is worn off. Light score marks in the bore can be cleaned off with crocus cloth. Deeper scores require honing (Fig. 6-21) provided no more than 0.002 inch [0.05 mm] is removed. If the bore does not clean up, discard the old caliper.

Careful! After using crocus cloth or honing, clean the caliper thoroughly with brake fluid, including drilled passages. Wipe the bore with a clean, lintless cloth. Continue wiping until the cloth shows no sign of dirt.

7. Assembling the caliper Clamp the caliper in the soft jaws of a vise and install the new piston seal in the groove in the bore. (Never reuse the old seal!) Lubricate the seal with special lubricant supplied in the service kit. Position the seal in one area and carefully work it into the groove, using clean fingers. Make sure the seal is not twisted or rolled.

Coat the new piston boot with lubricant, leaving plenty on the inside circumference. Install the boot in the caliper (Fig. 6-22), working it into place using only your clean fingers. Temporarily plug the fluid inlet to the caliper and bleeder-screw hole. Coat the piston with a generous amount of lubricant. Spread the boot with the fingers of one hand and push the piston straight down into the bore. The trapped air under the piston will force the boot around the piston and into its groove as the piston is pushed down. Remove the plug and push the piston down until it bottoms. Apply force uniformly all around the piston to keep it from cocking.

Reinstall the caliper. Then reconnect the flexible brake hose.

8. Checking the disk for parallelism and runout This procedure is covered in ❁ 6-4, item 7.

❁ 6-6 Sliding-caliper disk-brake service The chart in ❁ 5-15 lists troubles in disk-brake systems and their

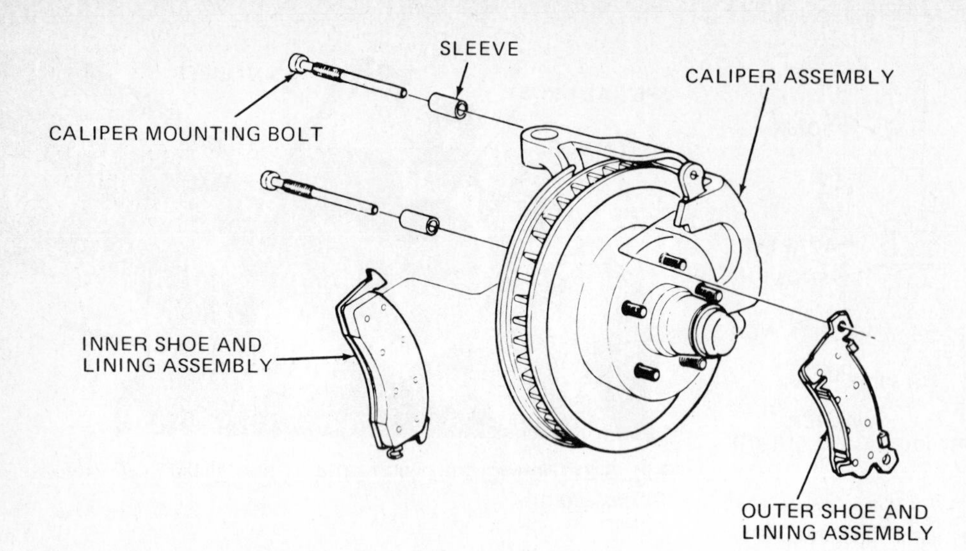

Fig. 6-23 Sliding-caliper disk brake with the mounting bolts and shoes removed. (*Chevrolet Motor Division of General Motors Corporation*)

possible causes. This section describes the servicing of the sliding-caliper disk brake. The operation of this brake is described in ❁ 3-20 and illustrated in Figs. 3-55 to 3-58. Figure 6-23 shows the sliding-caliper disk brake with the mounting bolts and shoes removed.

1. Inspecting shoes and linings Linings should be inspected for wear every 6000 miles [9656 km] and also whenever a wheel is removed. The outboard shoe could be checked at both ends, as shown by the arrows in Fig. 6-24. The inboard lining can be checked through the inspection hole, as shown by the single arrow. If a lining is worn to within 0.020 inch [0.56 mm] of the rivet at either end, all shoe-and-lining assemblies should be replaced.

2. Removing shoes and linings Remove two-thirds of the brake fluid from the master-cylinder section feeding the disk brakes. Discard the fluid. Do not reuse it.

Raise the car and remove the wheel covers and wheels. Use a 7-inch [178 mm] C clamp as shown in Fig. 6-25. The solid end of the clamp rests against the inside of the caliper, and the screw end rests against the metal part of the outboard shoe. Tighten the C clamp to move the caliper out far enough to push the piston to the bottom of the piston bore. This produces clearance between the disk and the shoes.

Remove the two mounting bolts (Fig. 6-26). Lift the caliper off the disk. Support it with a wire hook so that it does not hang from the brake hose. Remove the shoes. Mark the shoes so you can return them on the same side of the caliper from which they were removed. Next, remove the sleeves and bushings from the four caliper ears (Fig. 6-27). A special tool is used to remove the sleeves. The bushings fit into grooves in the ears.

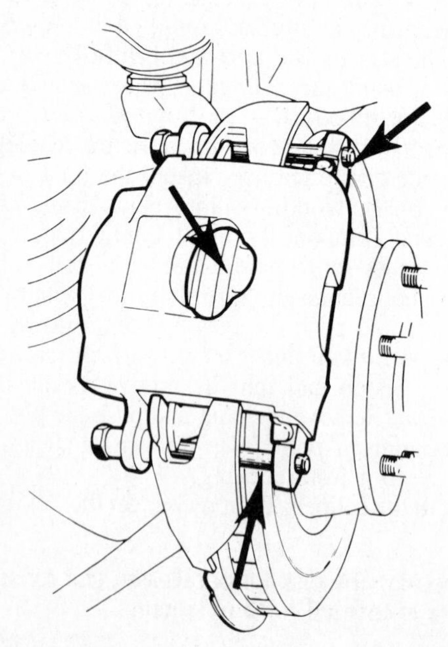

Fig. 6-24 Arrows indicate the shoe-lining inspection points. (*Buick Motor Division of General Motors Corporation*)

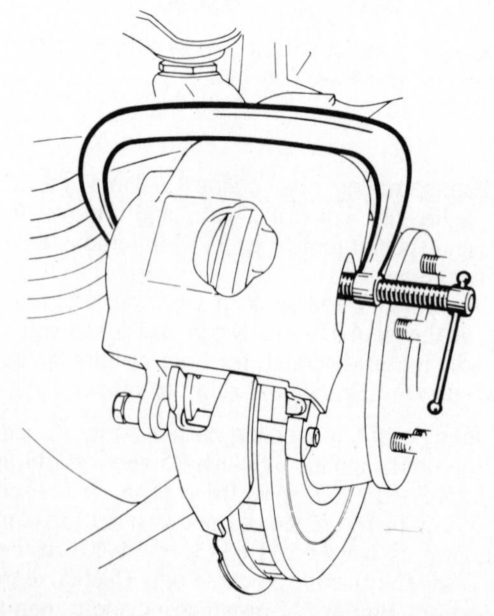

Fig. 6-25 Using a C clamp to force the piston into the caliper bore. (*Buick Motor Division of General Motors Corporation*)

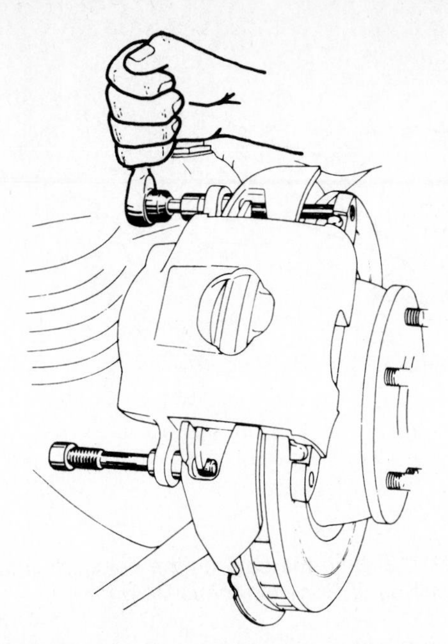

Fig. 6-26 Using a ratchet and socket to remove the caliper mounting bolts. (*Buick Motor Division of General Motors Corporation*)

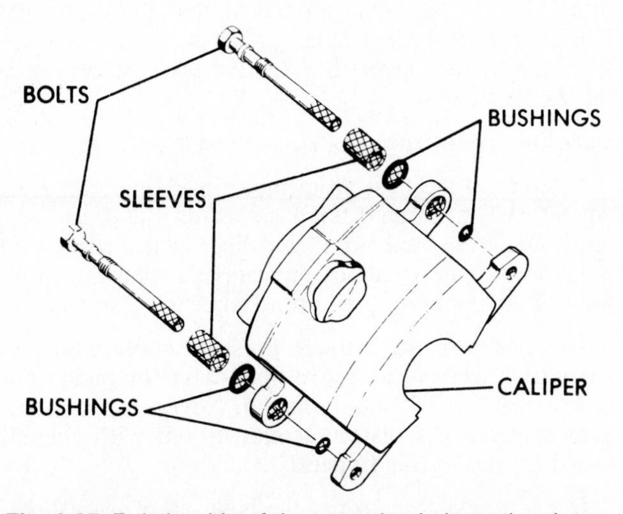

Fig. 6-27 Relationship of the mounting bolts to the sleeves and bushings. The areas to be lubricated with silicone lubricant are indicated. (*Buick Motor Division of General Motors Corporation*)

3. Cleaning and inspection Clean the holes and grooves in the caliper ears and wipe dirt from the mounting bolts. Replace the bolts if they are corroded or damaged. Wipe the inside of the caliper clean while inspecting it for brake-fluid leakage. If leakage is noted, remove the caliper for overhaul (see item 5 below). Make sure the dust boot is in good condition and is properly installed in the piston and caliper (Fig. 6-28). Check the rotor for wear and runout (⚙ 6-4, item 7). If it needs service, remove it.

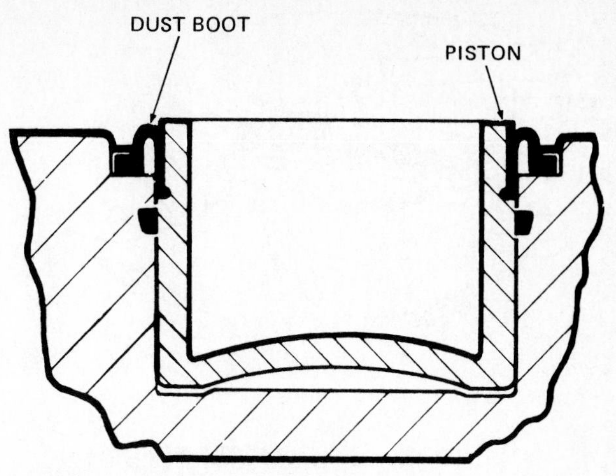

Fig. 6-28 Installation of the dust boot. (*Buick Motor Division of General Motors Corporation*)

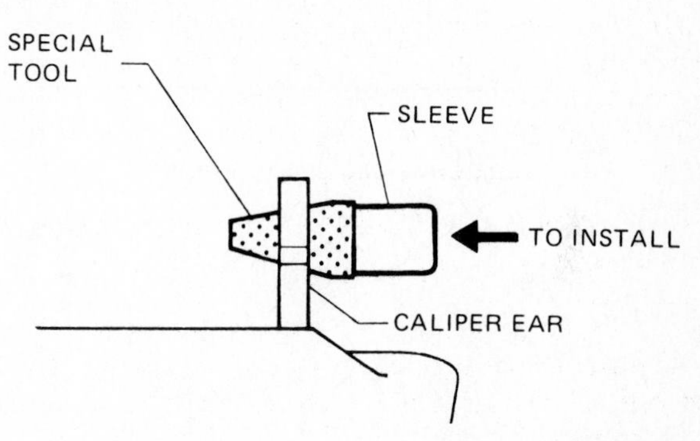

Fig. 6-29 Using the special tool to install the sleeve in the caliper ear. (*Buick Motor Division of General Motors Corporation*)

4. Installing shoes and caliper Apply special silicone lubricant to the new sleeves and bushings, the bolts, and the bushing holes and grooves in the caliper ears (Fig. 6-27).

Careful! Always use new sleeves and bushings, properly lubricated, to ensure easy sliding of the caliper.

Install the four bushings in the caliper ears. Use the special tool as shown in Fig. 6-29 to install the sleeves. The outer ends of the sleeves should be flush with the surface of the ears.

Install the shoe support spring and the inboard shoe in the center of the piston cavity as shown in Fig. 6-30. Install the outboard shoe as shown in Fig. 6-31.

Now position the caliper over the disk, making sure that the brake hose is not twisted or kinked. Start the bolts through the sleeves, making sure that the bolts pass under the retaining ears on the inboard shoes (Fig. 6-32). Push the bolts on through, making sure that they go through the holes in the outboard shoe and the ears in the caliper. Screw the bolts into the mounting holes and tighten them to the proper torque.

Add fresh brake fluid to the reservoir and depress the brake pedal several times to seat the linings against the disk. Now clinch the upper ears of the outboard shoe

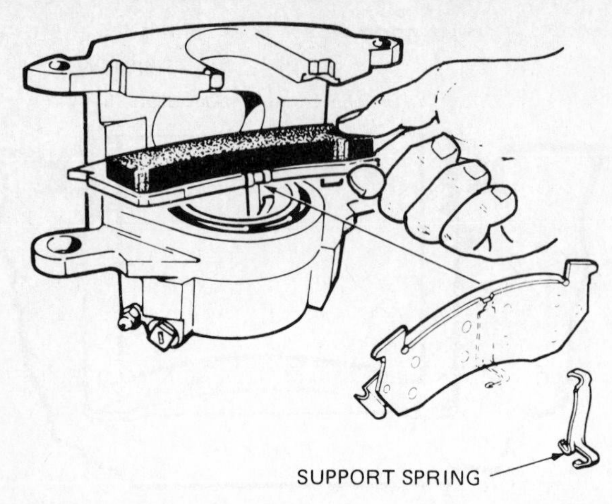

Fig. 6-30 Installing the inboard shoe. Note the location of the support spring. (*Chevrolet Motor Division of General Motors Corporation*)

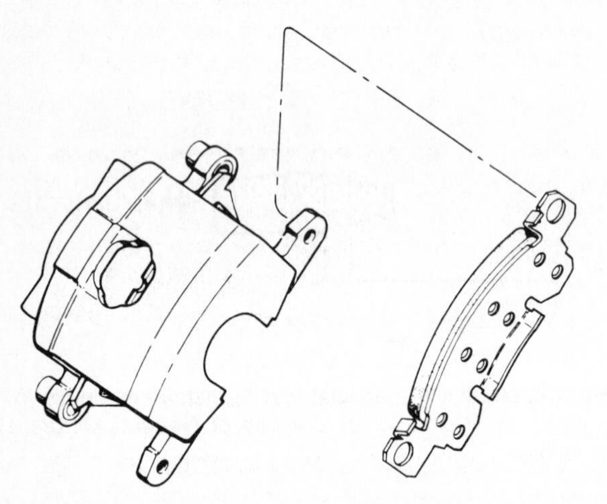

Fig. 6-31 Position of the outboard shoe in the caliper. (*Buick Motor Division of General Motors Corporation*)

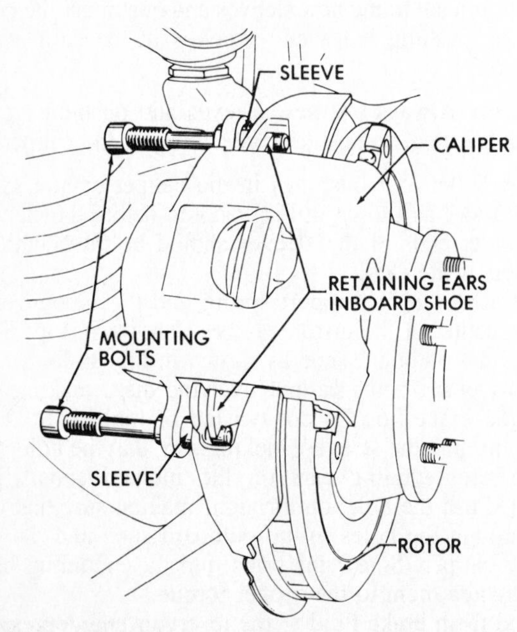

Fig. 6-32 Installation of the caliper mounting bolts. (*Buick Motor Division of General Motors Corporation*)

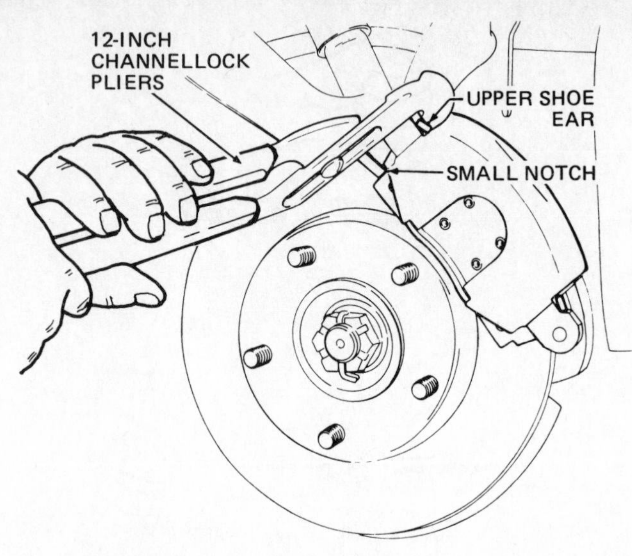

Fig. 6-33 Clinching the ears on the outboard shoe. (*Buick Motor Division of General Motors Corporation*)

with Channellock pliers as shown in Fig. 6-33. After clinching, the ears should be flat against the caliper.

5. Overhauling the caliper To remove the caliper, first detach the caliper and remove the shoes as described in item 2. Then disconnect the hose from the steel brake line and cap the fittings to keep dirt out. Detach the hose from the frame support bracket and take the caliper to the workbench.

Careful! Your bench, tools, and hands must be clean!

Disconnect the hose from the caliper, discarding the copper gasket. Discard the hose also if it appears damaged. Drain the fluid from the caliper. Use a clean shop towel to pad the inside of the caliper as shown in Fig. 6-34. Then apply air pressure to force the piston out.

CAUTION: **Use only enough pressure to force the piston out! Excessive pressure may drive the piston out so hard that it will be damaged. Never use your fingers to catch the piston. It can fly out with enough speed to mash your fingers!**

Use a screwdriver to pry the boot out of the caliper. Do not scratch the bore! Use a plastic toothpick to remove the seal from the groove in the bore. Do not use a metal tool! It could scratch the bore. Remove the bleeder valve.

Discard the boot, piston seal, rubber bushings, and sleeves. Use new parts on reassembly. Clean the piston and caliper with brake cleaner. Blow out all passages with compressed air.

NOTE: Lubricated shop air can leave a film of oil on the metal parts. This oil will damage the rubber parts. Be sure you are using only dry, filtered compressed air.

Discard the piston if it has any nicks, scratches, or worn spots. Examine the bore in the caliper. Minor corrosion or stains can be polished away with crocus cloth (not emery cloth). If the bore cannot be cleaned, discard the caliper.

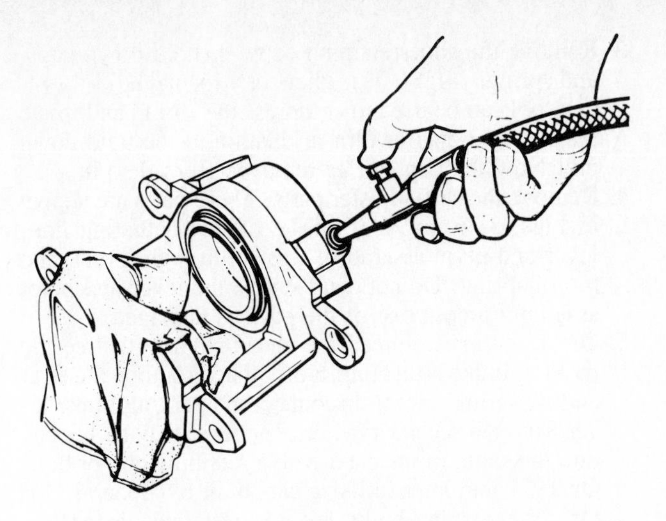

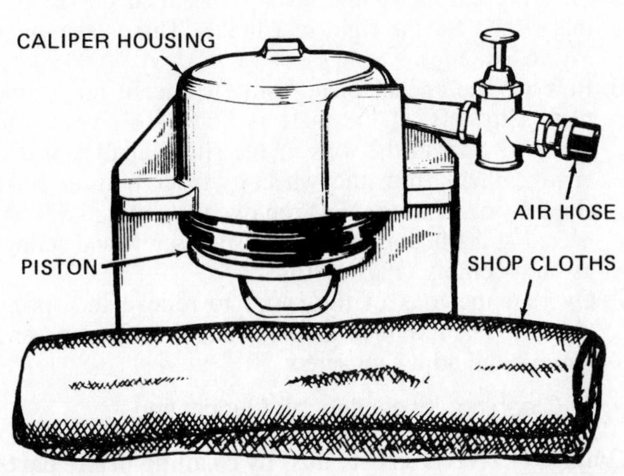

Fig. 6-34 Using air pressure to remove the piston from the caliper. (*Buick Motor Division of General Motors Corporation; Bendix Corporation*)

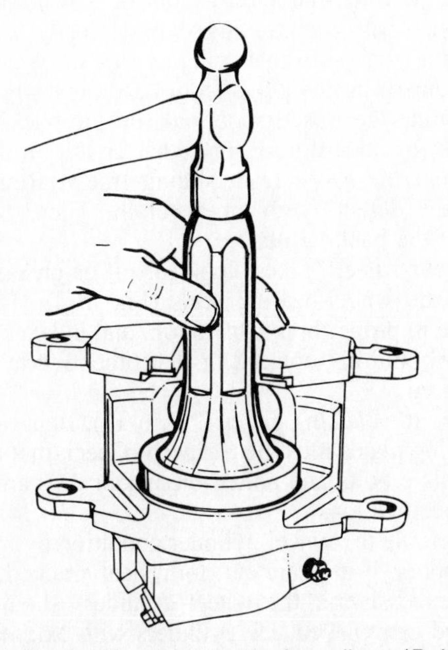

Fig. 6-35 Seating the dust boot in the caliper. (*Buick Motor Division of General Motors Corporation*)

To install the piston, first lubricate the bore and the new piston seal with brake fluid. Install the seal in the groove. Lubricate the piston and assemble a new boot into the piston groove. Install the piston in the bore, being careful not to unseat the seal. Push the piston to the bottom of the bore. This will require a strong push. Put the outside of the boot into the caliper counterbore and seat it with the special tool as shown in Fig. 6-35.

Now reconnect the brake hose, using a new copper gasket. Install the caliper, attach the hose to the frame bracket, and reconnect it to the steel tube.

Finish installation as described in item 4 above. Bleed the system (❁ 6-14).

Drum-brake service

❁ 6-7 Servicing drum brakes The chart in ❁ 5-2 lists troubles in drum-brake systems and their possible causes. Although all drum brakes work similarly, their design and construction varies. Following sections cover servicing of typical drum brakes. Later sections cover servicing of the hydraulic system, including master-cylinder and brake-line service.

❁ 6-8 Replacing drum-brake shoes With the car on a lift or on safety stands, remove the wheels and brake drums (❁ 6-9). Figure 3-5 shows a disassembled drum brake. Figure 6-36 shows a rear brake with the drum removed and all parts named.

A. Removing the brake shoes To remove the brake shoes, proceed as follows:

1. If required, install a wheel-cylinder clamp on each wheel cylinder (Fig. 6-37). The clamp prevents fluid leakage and keeps air from getting into the hydraulic system while the shoes are removed. However, not all wheel cylinders require the use of the clamp. Some wheel cylinders have stops which will hold the wheel-cylinder parts in place.

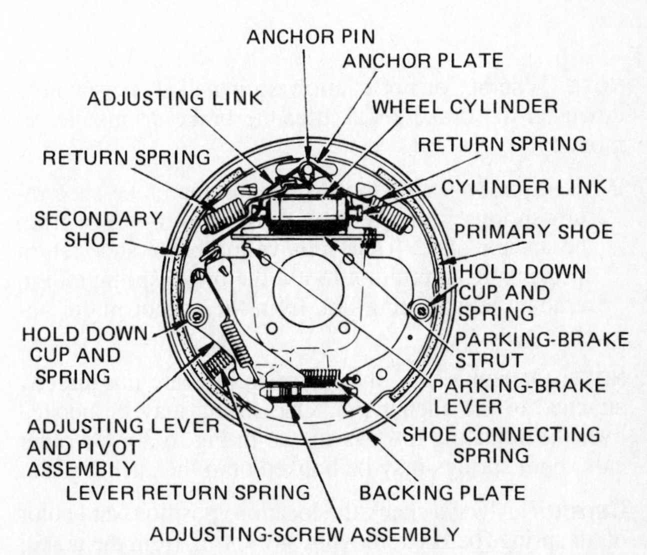

Fig. 6-36 A rear-drum brake with the drum removed and all parts named. (*Delco Moraine Division of General Motors Corporation*)

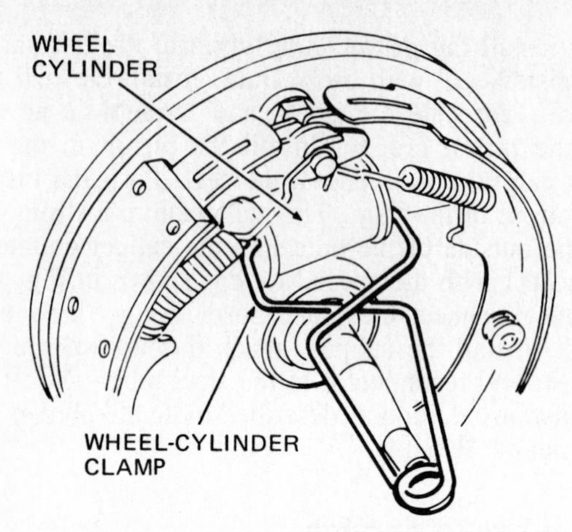

Fig. 6-37 Before removing the brake shoes, place a wheel-cylinder clamp on each wheel cylinder. (*Delco Moraine Division of General Motors Corporation*)

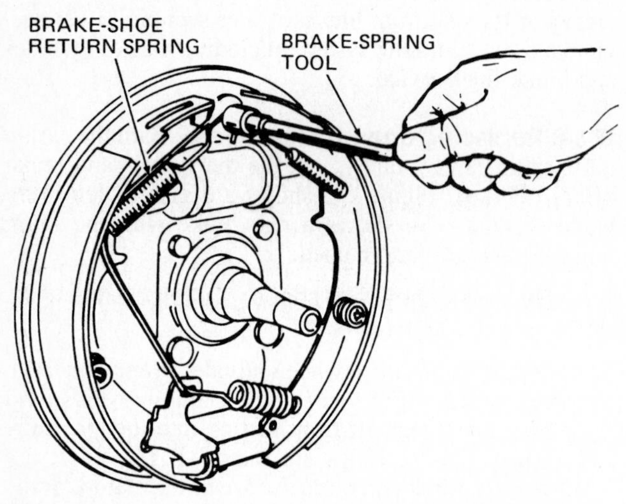

Fig. 6-38 Using a brake-spring tool to unhook the return springs from the anchor pin. (*Delco Moraine Division of General Motors Corporation*)

NOTE: Whether or not clamps are installed, never press down on the brake pedal after the brake drums are removed.

2. Use a brake-spring tool to unhook the brake-shoe return springs from the anchor pin (Fig. 6-38) or from the anchor plate. Then remove the brake-shoe return springs. Be careful not to use the brake-spring tool to remove the adjusting link from the anchor pin or anchor plate.

NOTE: On some cars, the return springs are not directly attached to the anchor pin. One spring may be hooked over the adjusting link, as shown in Fig. 6-36. On other cars, both springs may be hooked onto the anchor plate.

Careful! Always check the location, position, and color of all springs before removing any spring from the brake. This will enable you to reinstall each spring in its original location and position.

3. Remove the shoe-retaining cups, or hold-down cups, and springs (Fig. 6-39). Pliers or a special hold-down-cup tool can be used to compress the spring and rotate the cup one-quarter turn in relation to the hold-down pin. Now the cup can be removed from the pin.
4. Remove the self-adjuster parts. These parts are shown in Figs. 3-43 and 6-36. Take off the actuating link, lever-and-pivot assembly, sleeve (through lever), and return spring. Do not disassemble the lever-and-pivot assembly unless one of the parts is damaged.
5. On rear brakes, spread the shoes slightly to free the parking-brake strut (Fig. 6-36). Then remove the strut and its spring. Next, disconnect the parking-brake lever from the secondary shoe. The lever may be hooked into the shoe, or attached with a retaining clip or bolt. On 1977 and later full-size cars built by General Motors, the parking-brake lever is installed on the primary shoe for the right rear brake. This is shown in Fig. 6-36.
6. Remove the anchor plate from the anchor pin, if the plate slips off. If the plate is bolted or riveted on, leave it. Spread the tops of the shoes apart and disengage them from the wheel-cylinder pins or links (Fig. 6-36), if used. Now remove the shoes (still connected at the bottom by the adjusting screw and spring) as an assembly (Fig. 6-40).
7. Overlap the tops of the shoes to relieve the spring tension. Then unhook the adjusting-screw spring and remove the adjusting screw.

B. Cleaning, inspecting, and lubricating

CAUTION: **Never create dust by cleaning brake parts with a dry brush or compressed air. This could cause asbestos fibers to get into the air. Inhaling dust containing asbestos fibers may cause bodily harm.**

1. Clean, inspect, refinish, or replace the brake drums as necessary (⚙ 6-9).
2. Using a water-dampened cloth or a water-based solution, clean the backing plates, struts, levers, and other metal parts that will be reused. Wet cleaning prevents asbestos fibers from becoming airborne.
3. Examine the raised shoe pads on the backing plate. Look for corrosion or any other condition that could prevent the shoes from sliding freely. Remove any surface defects with emery cloth. Then thoroughly clean the backing plate.
4. On rear-wheel brakes, look for oil or grease leakage past the wheel-bearing seals (Fig. 5-3). This could cause improper brake operation, and indicates that the seal should be replaced or that other service work is required.
5. Check the backing plates (Fig. 6-41). Bent or cracked backing plates must be replaced. Check that the backing-plate bolts and bolted-on anchor pins are torqued to specifications.
6. Determine the wheel-cylinder condition by inspecting the boots. If they are cut, torn, heat-cracked, or leaking excessively, the wheel cylinders should be repaired or replaced. On cylinders with external boots, pull back the lower edge of the boot (Fig. 6-42). If

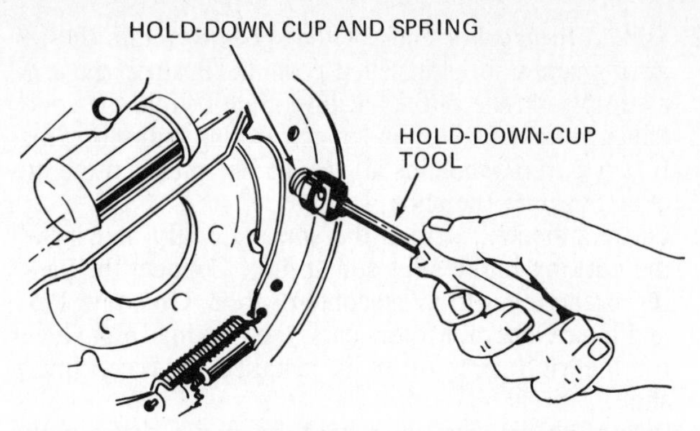

Fig. 6-39 A hold-down-cup tool is used to compress the spring and rotate the cup one-quarter turn so that the cup and spring can be removed from the hold-down pin.

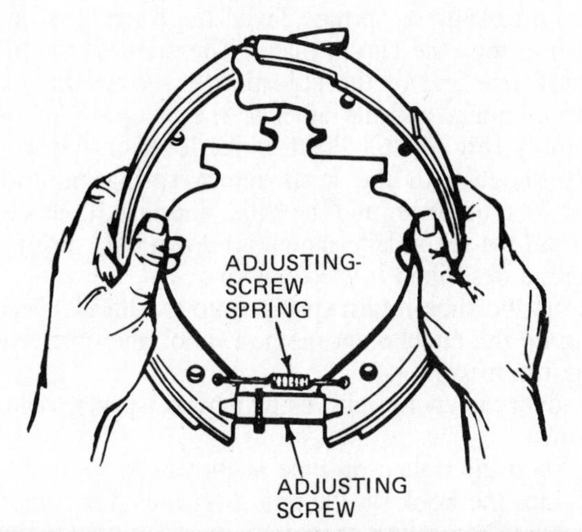

Fig. 6-40 Remove the shoes from the backing plate with the adjusting-screw spring and the adjusting screw still engaged. (*Delco Moraine Division of General Motors Corporation*)

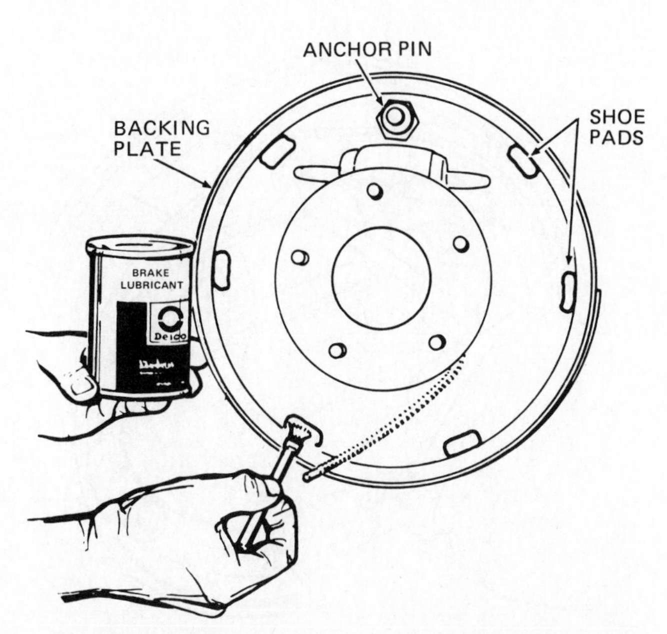

Fig. 6-41 Lubricating the brake pads on the backing plate. (*Delco Moraine Division of General Motors Corporation*)

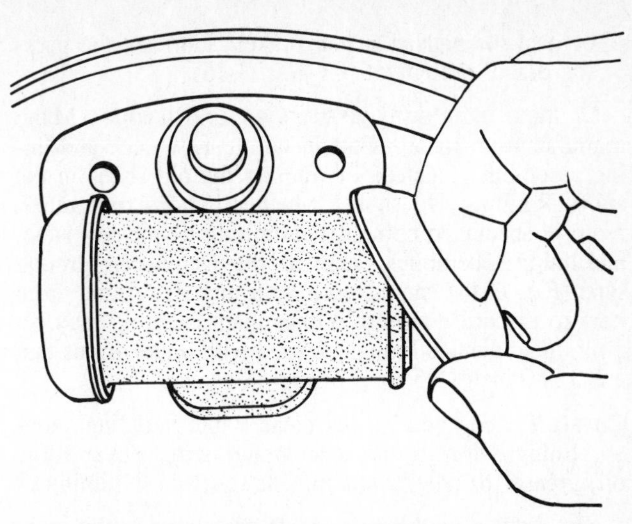

Fig. 6-42 Check the condition of the wheel cylinder by pulling back the boot. (*ATW*)

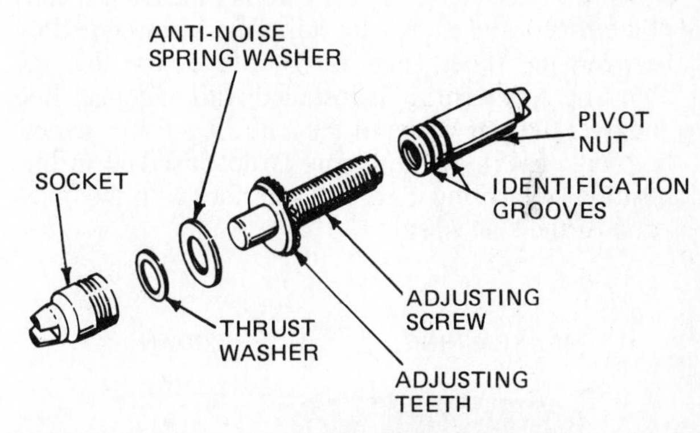

Fig. 6-43 A disassembled adjusting-screw assembly. (*Delco Moraine Division of General Motors Corporation*)

more than a drop of fluid runs out, the leakage is excessive and the wheel cylinder should be serviced. However, a small amount or trace of fluid behind the boot is normal. This serves to lubricate the pistons.

On wheel cylinders with internal boots, remove one of the wheel-cylinder connecting links to check for leakage.

7. Disassemble the adjusting-screw assembly (Fig. 6-43) and clean the parts. Check that the adjusting screw will thread completely into the pivot nut without sticking or binding. None of the adjusting-screw teeth can be damaged. This would interfere with the operation of the self-adjuster. Lubricate the adjusting-screw threads with brake lubricant and reassemble the adjusting-screw assembly. Be careful not to get lubricant on the adjusting teeth or star wheel. Most adjusting-screw assemblies have a thrust washer between the adjusting screw and the socket (Fig. 6-43). Some also have an antinoise spring washer. Thread the adjusting screw into the pivot nut as far as the screw will go.
8. Apply a thin film of brake lubricant to the raised shoe pads on the backing plate (Fig 6-41). Before installing the brake shoes (as described below), check that there are no burrs on the edges of the shoes where they

contact the pads. On rear brakes, lubricate the parking-brake-lever pivot pin (Fig. 3-15).

C. Installing new brake shoes and linings Many manufacturers recommend installing new shoe-and-lining assemblies, instead of relining the old shoes in the shop. Relining old shoes, whether bonded or riveted, requires special equipment and training. Some new shoe-and-lining assemblies, such as the Delco cam-ground type (Fig. 6-44) can be used with any size drum, from new to discard diameter. Other manufacturers may require that oversize linings be used with brake drums that are 0.060 inch [1.5 mm] or more oversize.

Careful! Keep your hands clean while handling shoes and linings, drums, and other brake parts. Never allow oil, grease, or other contaminants to touch the linings.

1. On a clean workbench, place the shoes in the same relative positions as they are to be installed on the backing plates. Hook the adjusting-screw spring between the shoes (Fig. 6-40). Overlap the anchor ends of the shoes, and engage the adjusting-screw assembly between the shoes (Fig. 6-45). Make sure that the adjusting-screw spring is installed with its longer free end over the star wheel of the adjusting-screw assembly (Fig. 6-45). If the spring is not installed in this position, the spring coils will interfere with the operation of the star wheel.

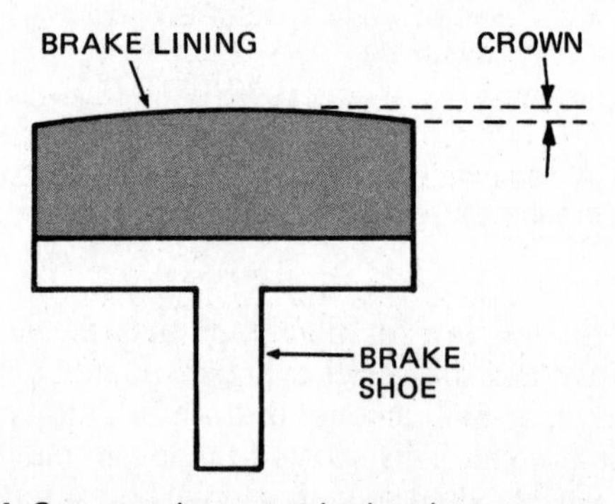

Fig. 6-44 Some replacement brake shoes are ground with a slight crown so that they can be used with any size drum, from new to discard diameter. (*AC-Delco Division of General Motors Corporation*)

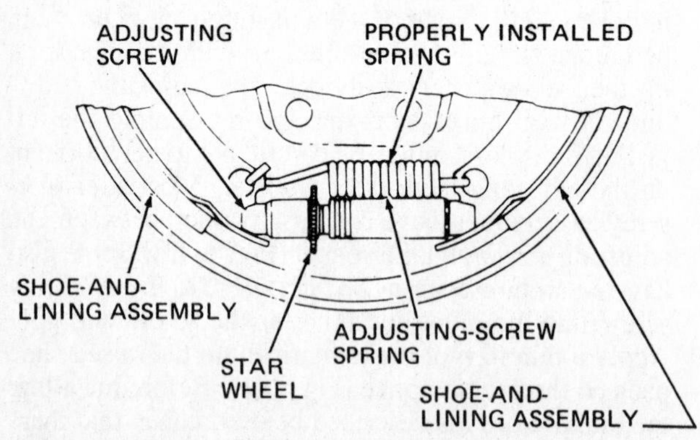

Fig. 6-45 Proper installation of the adjusting-screw spring. (*Delco Moraine Division of General Motors Corporation*)

2. Spread the anchor ends of the shoes to retain the adjusting screw in its installed position. Position the shoe assembly on the brake backing plate so that the shoes engage the wheel-cylinder connecting links or pins. If it was removed, install the anchor plate on the anchor pin over the shoe webs.
3. On rear brakes, spread the shoes slightly and install the parking-brake strut and spring. Connect the parking-brake lever to the secondary shoe. On some 1977 and later General Motors cars, the parking-brake lever for the right rear brake is installed on the primary shoe.
4. Insert the primary-shoe hold-down pin through the holes in the backing plate and shoe, and hold the pin in position. Place the hold-down spring on the pin. Grasp the cup with pliers and position the cup in front of the pin. Then force the cup into the spring pin, compressing the spring. Twist the pliers one-quarter turn to lock the cup in place. Then release the pliers.
5. Hook one end of the adjusting link over the anchor pin or plate, and the other to the lever-and-pivot assembly (Fig. 6-36). Position the lever on the secondary shoe, with the lever return spring in position between the lever and the shoe, and install the sleeve. Install the secondary-shoe hold-down pin, spring, and cup as described in step 4 above.
6. Hook the shoe-return springs into the shoes. Depending on the attachment method, hook the other end of the return spring:
   a. over the anchor pin, using a brake-spring tool (Fig. 6-46).
   b. onto the bolted-on-type anchor plate.
   c. into the hook on the adjusting link (for some secondary-shoe return springs).

Careful! Never use a brake-spring tool to remove or install the adjusting link to the anchor pin.

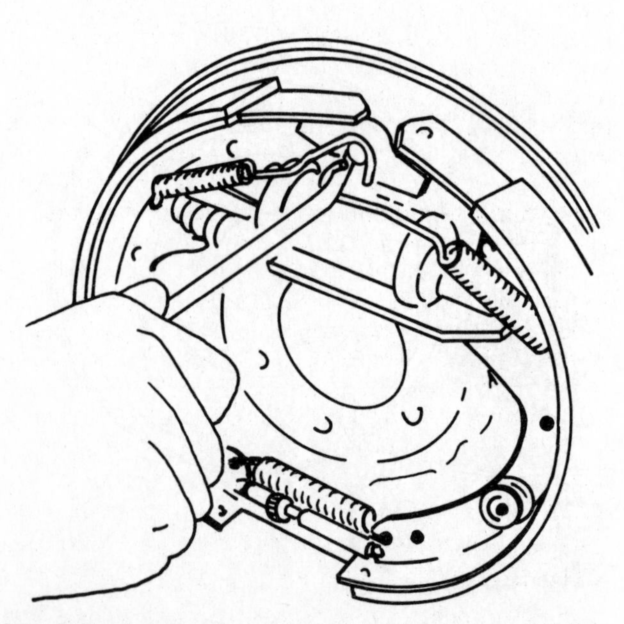

Fig. 6-46 Use a brake-spring tool to hook one end of the shoe-return spring over the anchor pin. (*K-D Manufacturing Company*)

7. Make the preliminary manual adjustment of the self-adjusting brakes (❁ 6-2). Install the drums and wheels. Then make the final adjustments to the service brakes and to the parking brakes (❁ 6-3).

❁ 6-9 Brake-drum service Before any inspection or service can be performed on the brake drum, it must first be removed. Then the drum can be inspected, and refinished (if necessary), discarded, or reinstalled. Or a defective brake drum can be replaced with a new one. Wheel bolts pass through the brake drum on some front brakes. If any wheel bolts are damaged, they can be replaced while the drum is off the car. These services are described below.

A. Removing the brake drum

1. Release the parking brake. Then raise the car on a lift or on safety stands, and remove the wheels.
2. Relieve all tension on the parking-brake cables. On most cars this is done by loosening or removing the adjusting nut at the equalizer (❁ 6-3). Then back off the brake-shoe adjustment as necessary so that the drums can be removed (❁ 6-2).
3. On most cars, the front wheels have integral hub-and-drum assemblies (Fig. 6-47). On these cars, remove the dust or grease cap, and pull the cotter pin. Then remove the castellated nut, or nut and nut lock. Now remove the thrust washer and outer bearing. Pull the drum-and-hub assembly from the spindle. Do not drop the bearing or allow dirt to get in it. Some cars have front drums which can be removed with the hub in place. On these, the front wheel bearing is not disturbed by removing the drum.
4. On rear drums, check to see if the drum has a locating tang. If you cannot find one, mark one of the studs and its hole in the drum so that the drum can be reinstalled in the same position. Remove any speed nuts that may be retaining the drum. Then pull the drum off the axle flange. If the drum does not come off easily, make sure that the brakes are fully released. Then try a sharp rap on the drum with a plastic mallet to break the drum loose.

B. Inspecting brake drums

1. With the drum removed, visually inspect the brake linings (❁ 5-29). Many times the condition of the brake linings will indicate defects in the drums. For example, if the linings on one wheel are worn more than others, this may indicate a rough drum. Uneven lining wear from side to side on any one set of brake shoes may be caused by a tapered drum. If some linings are worn badly at the toe or heel, it may indicate an out-of-round brake drum.

CAUTION: **Never create dust by cleaning brake parts with a dry brush or compressed air. This could cause asbestos fibers to get into the air. Inhaling dust containing asbestos fibers may cause bodily harm.**

2. Thoroughly clean the brake drums using a water-dampened cloth or a water-based cleaning solution. Wet cleaning methods must be used to prevent asbestos fibers from becoming airborne. If the drums have oil or grease on them, thoroughly clean the drums with a non-oil-base solvent (carburetor cleaner or lacquer thinner) after washing to remove any dust and dirt. Determine the source of the oil or grease, and correct the problem before the drums are reinstalled.
3. Visually inspect the drums for cracks or other defects in the friction surfaces. The most common surface defects are shown in Fig. 5-4. Some cars have composite rear brake drums (Fig. 6-48). These are made of aluminum with a cast-iron liner providing the friction surface. On composite drums, check that the liner has not separated from the drum.

Careful! Cracked or separated drums must be replaced. never attempt to weld or repair a drum with this type of damage.

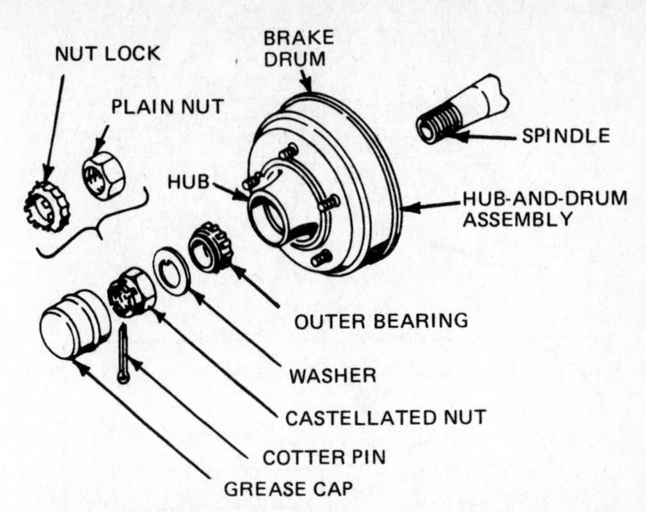

Fig. 6-47 A disassembled front drum-and-hub assembly, showing the hub and the bearing. (*Delco Moraine Division of General Motors Corporation*)

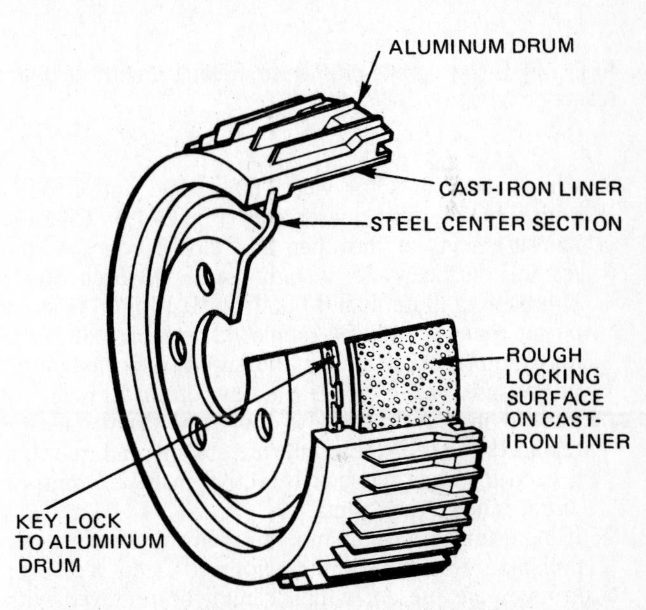

Fig. 6-48 A composite brake drum. The finned drum is made of aluminum with a cast-iron liner on the inside to provide the friction surface for the brake lining. (*Ford Motor Company*)

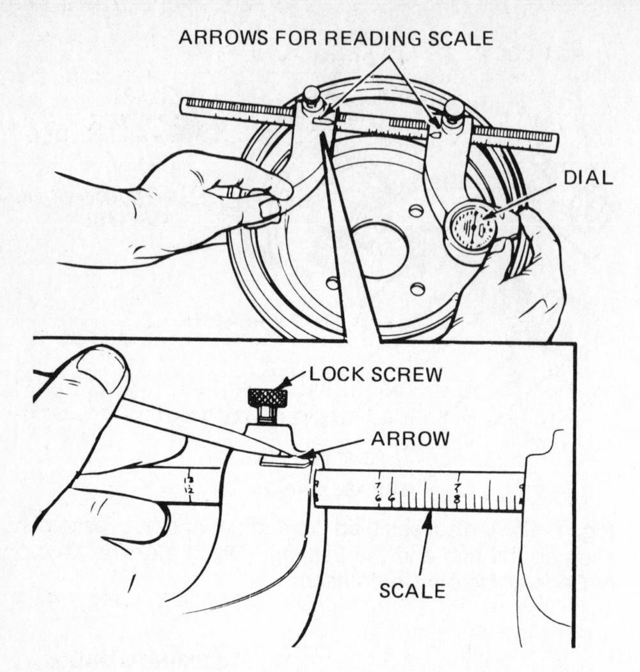

Fig. 6-49 Using a brake-drum micrometer to determine drum diameter. (*Ammco Tools, Inc.*)

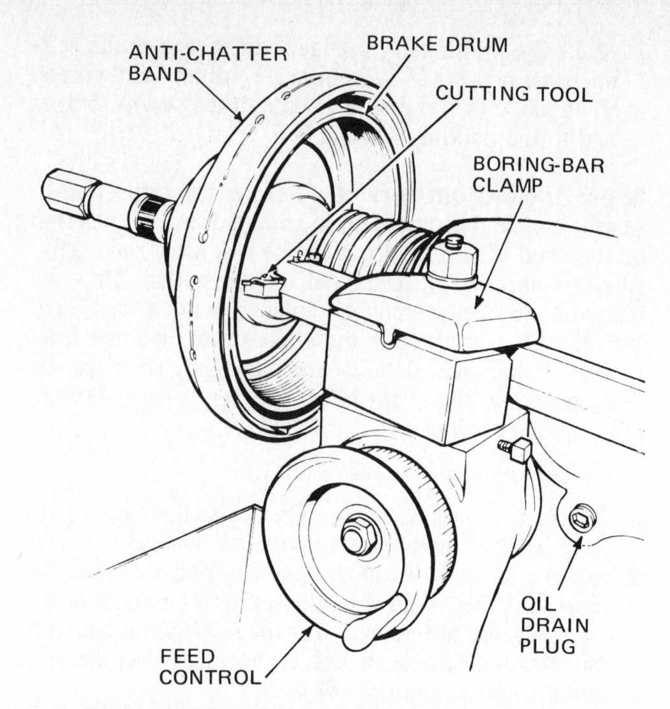

Fig. 6-50 Refinishing a brake drum on a brake-drum lathe using a cutting tool. (*Ammco Tools, Inc.*)

4. Check the drums for wear, taper, and out-of-round with a brake-drum micrometer (Fig. 6-49). Take the measurements at the open and closed edges of the friction surface, and at right angles to each other. Drums with more than 0.006 inch [0.15 mm] taper or out-of-round should be refinished or replaced. If the maximum drum diameter (measured from the bottom of any grooves) exceeds the new drum diameter by more than 0.060 inch [1.5 mm], the drum must be replaced. Also, if the drums are smooth and true, but exceed the new diameter by 0.090 inch [2.3 mm] or more, replace the drums.
5. If the drums are true, smooth up any slight scores by polishing with fine emery cloth. If deep scores or grooves are present which cannot be removed with emery cloth, then the drum must be refinished or replaced.

C. Refinishing brake drums

1. Brake drums can be refinished on a brake lathe either by turning using a cutting tool (Fig. 6-50), or by grinding with a grinding wheel (Fig. 6-51). Some manufacturers recommend that the drums be turned, using a very light finish cut. Follow the procedure in the operating instructions for the brake lathe you are using.
2. When refinishing a brake drum, only enough metal should be removed to obtain a true, smooth friction surface. However, if one drum is refinished, the other drum on the same axle should also be refinished by the same method and to the same diameter.
3. Brake drums manufactured in 1971 and later have the discard diameter on them (Fig. 6-52). This is the maximum diameter allowable, and not the allowable machining dimension. After refinishing, there must be 0.030 inch [0.76 mm] left for wear. Typical drum diameters for new and refinished drums, along with their discard diameters, are shown in Fig. 6-53. These specifications apply to all drums, including those without a stamped discard diameter.
4. Replacement drums are available as semifinished and as full-finished. Semifinished drums may require additional machining to obtain the proper diameter and surface finish. Full-finished brake drums do not require additional refinishing unless it is necessary to match the diameter of an old drum on the same axle. New drums have a rustproofing coating which must be cleaned off the friction surfaces. A non-oil-base solvent such as carburetor cleaner or lacquer thinner should be used to remove any oil or grease.

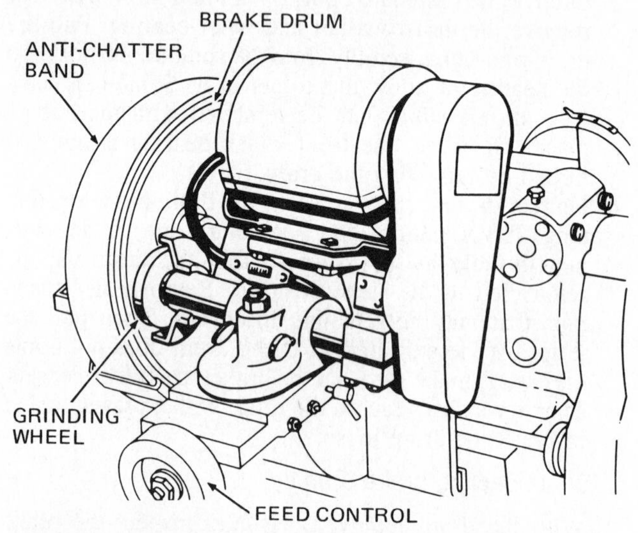

Fig. 6-51 Refinishing a brake drum on a brake-drum lathe using a grinding wheel. (*Ford Motor Company*)

Fig. 6-52 Discard diameter cast into the outside of a brake drum. Note that this drum has the diameter given in millimeters. (*Chrysler Corporation*)

DIMENSIONS—INCHES		
NEW	MAXIMUM REFINISHED	DISCARD DIAMETER
7.874 [200 mm]	7.899 [200.64 mm]	7.929 [201.40 mm]
9.000	9.060	9.090
9.500	9.560	9.590
10.000	10.060	10.090
11.000	11.060	11.090
12.000	12.060	12.090

Fig. 6-53 Typical drum diameters for new and refinished drums. (*Delco Moraine Division of General Motors Corporation*)

D. Replacing wheel bolts The following procedure may be used to replace one or more damaged wheel bolts. Or, if the drum must be replaced and the wheel hub is to be reused, this procedure may be used to replace all wheel bolts and the drum. Figure 6-54 shows various methods of mounting brake drums.

1. Secure the hub-and-drum assembly. Mark the center of the bolt head with a center punch. Drill a ⅛-inch pilot hole in the bolt head. Redrill the bolt head using a 9/16-inch drill. Cut off any remaining part of the bolt head with a chisel. Use a drift punch to drive out the bolt in the direction shown by the heavy black arrow in Fig. 6-55.
2. Use the procedure in step 1 to remove any other damaged bolts.
3. If the drum is to be replaced, follow the procedure in step 1 to remove all bolts, and replace the drum.
4. Press new bolt or bolts into the drum and hub.
5. Inspect the drums as described earlier in this section. Refinish the drums as necessary.

E. Installing brake drums

1. Rear brake drums. Install the rear brake drums over the wheel studs (Fig. 6-54). If the drum has an alignment tang or a hole for a locating screw, make sure that this is aligned with the hole in the hub flange. Install any retaining clips or speed nuts used to hold the drum in position.

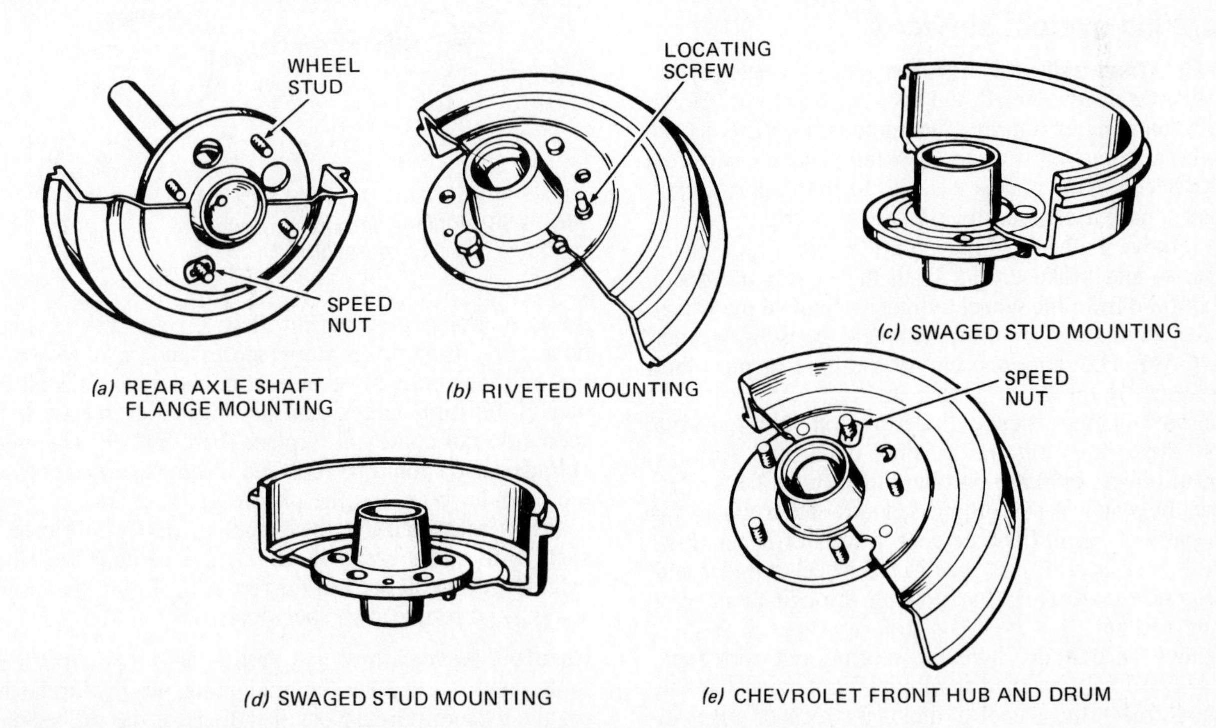

Fig. 6-54 Various brake-drum mounting arrangements. (*Bendix Corporation*)

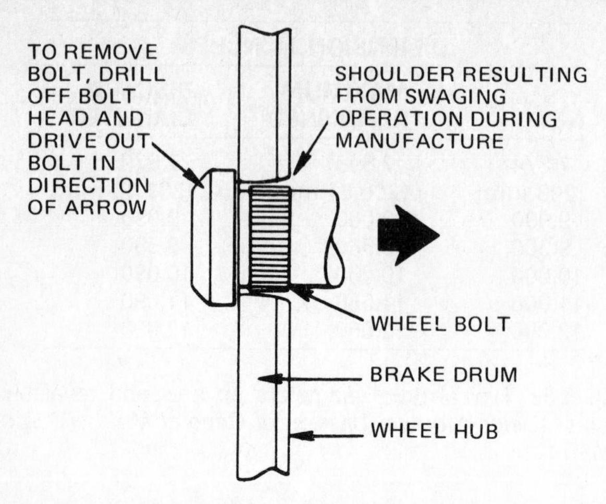

Fig. 6-55 Procedure for replacing a front-wheel bolt. (*Delco Moraine Division of General Motors Corporation*)

2. Front brake drums. At the front wheels, first install the inner bearing and grease seal (if they were removed) in the hub. Then install the hub-and-drum assembly, outer bearing, thrust washer, and nut on the spindle (Fig. 6-42). Adjust the front-wheel bearing following the manufacturer's recommended procedure. Spin the wheel and tighten the spindle nut to the specified torque with the wheel turning. Install the nut lock (if used) and cotter pin.

   Install the wheels. Adjust the service brake (⚙ 6-2) and parking brake (⚙ 6-3) as required. Then road-test the car (⚙ 5-28).

Careful! After making the preliminary brake adjustments, press firmly on the brake pedal to be sure that there is adequate brake pedal before moving the car.

Hydraulic-system service

⚙ 6-10 Wheel-cylinder service Most wheel cylinders can be disassembled and rebuilt on the car. However, many manufacturers recommend that the wheel cylinder be removed from the backing plate and serviced on the bench. This makes it easier to thoroughly clean, inspect, and reassemble the cylinder properly.

To remove a wheel cylinder from the car, first remove the wheel and brake drum. Then disconnect the brake hose or tube from the wheel cylinder. Remove the wheel cylinder by taking out the attachment bolts or retainer (Fig. 6-56). Then tape the end of the hose or pipe shut to prevent any dirt from getting in.

Disassemble the wheel cylinder by first pulling off the boots. Then push out the pistons, cups, and springs. Clean all wheel-cylinder parts in clean brake fluid.

Dry the parts with compressed air. Then place the dried parts on clean lint-free shop towels or paper (Fig. 6-57). Check that all passages in the wheel cylinder and bleeder screw are clear by blowing through them with compressed air.

Inspect the cylinder bore for scoring and corrosion. Crocus cloth may be used to remove light corrosion and stains. Replace the wheel cylinder if crocus cloth does not remove the corrosion, or if the bore is pitted or scored.

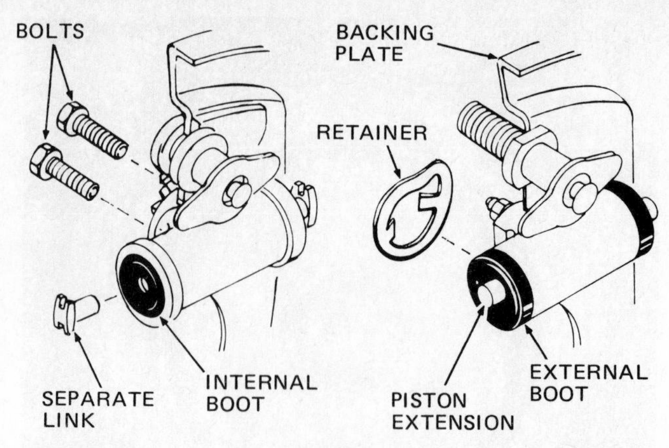

Fig. 6-56 Two methods of attaching the wheel cylinder to the backing plate. (*Delco Moraine Division of General Motors Corporation*)

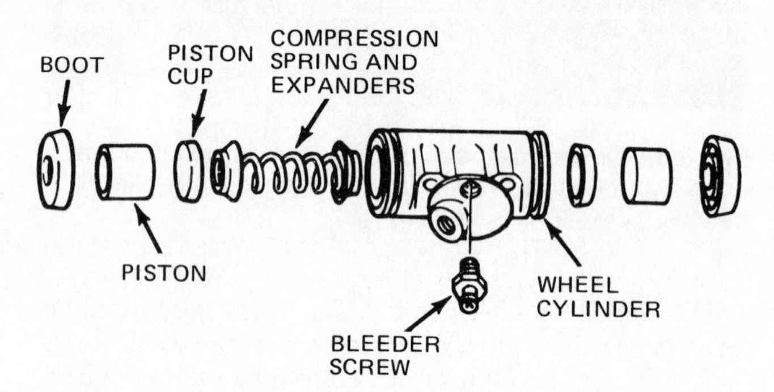

Fig. 6-57 A disassembled wheel cylinder. (*American Motors Corporation*)

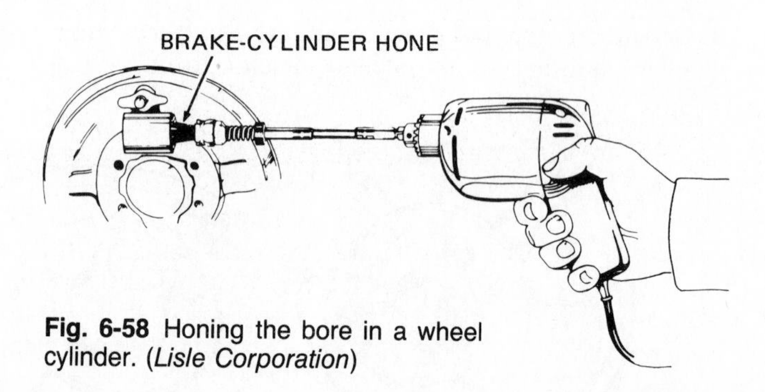

Fig. 6-58 Honing the bore in a wheel cylinder. (*Lisle Corporation*)

Some manufacturers permit the use of a brake-cylinder hone (Fig. 6-58) to remove scores and rust. However, the cylinder bore must not be honed more than 0.003 inch [0.08 mm] larger than its original diameter. If the scores do not come out, replace the cylinder. The wheel cylinder also should be replaced if the clearance between the cylinder bore and the pistons is excessive.

When reassembling the wheel cylinder, lubricate all parts with clean brake fluid. Then assemble the wheel cylinder, using all parts in the repair kit. Install the bleeder screw, and torque it to specifications.

Careful! Never allow any grease or oil to contact the rubber parts, or other internal parts, of the brake hydraulic system. Grease or oil will cause the rubber parts to swell, which may lead to brake failure.

⚙ 6-11 Master-cylinder service The service procedures for master cylinders used with disk brakes and drum brakes are very similar. One difference is that with disk brakes, a larger brake-fluid reservoir is required. Figure 3-28 is a disassembled master cylinder used with a braking system that has front-wheel disk brakes and rear-wheel drum brakes. Note that one of the fluid reservoirs is larger than the other. A typical servicing procedure follows.

1. Disassembly (Fig. 3-28) Clean the outside of the master cylinder. Then remove the cover and seal, and pour out any remaining brake fluid. Mount the master cylinder in a vise. If the car has a manual brake system, slide the boot to the rear, remove the retainer clip, and then remove the retainer, pushrod, and boot. Use the pushrod to force the primary piston inward, and remove the snap ring from the groove in the piston bore. Then remove the primary-piston assembly and discard it. The repair kit contains a complete new primary-piston assembly.

Remove the secondary-piston stop screw, if so equipped. Using the shop air hose, apply slight air pressure through the compensating port at the bottom of the reservoir. This will force out the secondary-piston assembly. Remove the piston seals from the secondary piston.

The outlet-tube seats, check valves, and springs must not be removed from some master cylinders. They are permanent parts of the master cylinder. However, these parts may be removed from other master cylinders. Follow the procedures in the manufacturers service manuals.

NOTE: Most disk-brake hydraulic systems do not use check valves.

2. Inspection and repair Clean all parts in brake fluid or brake-cleaning solvent only. Blow dry with filtered compressed air. Blow out all passages and ports to be sure they are clear. If the master cylinder is scored, corroded, pitted, cracked, porous, or otherwise damaged, replace it. However, some manufacturers permit slight honing of the master-cylinder bore. But if the pits or scoring are deep, a new master cylinder must be installed.

3. Reassembly Dip all parts into brake fluid (except the master cylinder itself). Insert the complete secondary-piston assembly, with return spring, into the master-cylinder bore. Install the secondary-piston stop screw, if so equipped. Put the primary-piston assembly into the bore. Depress the primary piston and install the snap ring in the bore groove. Install the pushrod, boot, and retainer on the pushrod, if so equipped. Install the pushrod assembly into the primary piston. Make sure the retainer is properly seated and holding the pushrod securely.

Careful! Never add a stop screw to a master cylinder that does not have one, even though it has a threaded hole for the screw. If the stop screw is installed in a master cylinder of this type, the master cylinder may not function properly.

Position the inner end of the pushrod boot (if so equipped) in the retaining groove in the master cylinder. Put the seal into the cover, and install it on the master cylinder. Secure the cover with the retainer.

NOTE: The master cylinder should be bled (⚙ 6-14) before it is installed on the car. The procedure involves filling the master cylinder with brake fluid, then working the pistons back and forth to get rid of any trapped air.

⚙ 6-12 Hydraulic-brake tubing repair Most hydraulic-brake tubing is made of double-walled, welded-steel tubing which is coated to resist rust. Only the tubing specified by the automotive manufacturer should be used. When replacing a tube, use the old tube as a pattern to form a new tube. Do not kink the tubing or make sharp bends.

Brake tubing must be cut off square with a special tube cutter. Do not use a jaw-type cutter or a hacksaw to cut brake tubing. Either of these can distort the tubing and leave heavy burrs that would prevent normal flaring of the tube. After the tube has been cut off, a special flaring tool must be used to double-flare the end of the tube (Fig. 6-59). Double flaring is a three-step operation, as shown in Fig. 6-59. First, install the fitting (tube nut) on the tube. Next, dip the end of the tube in brake fluid (for lubrication during flaring). Then perform the three steps as shown in Fig. 6-59. A different set of tools to do the same job is shown in Fig. 6-60.

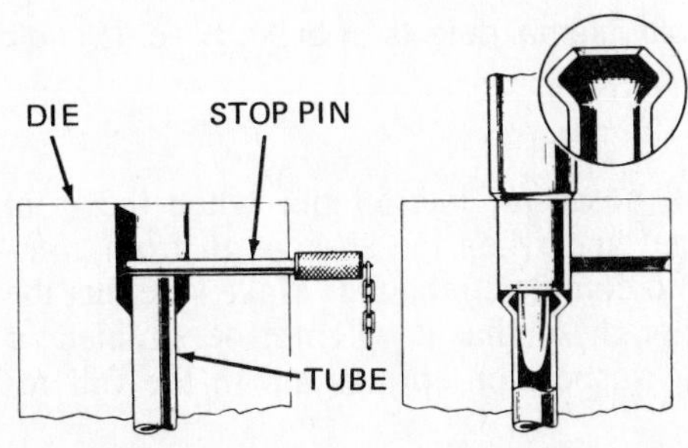

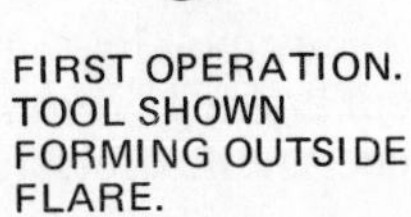

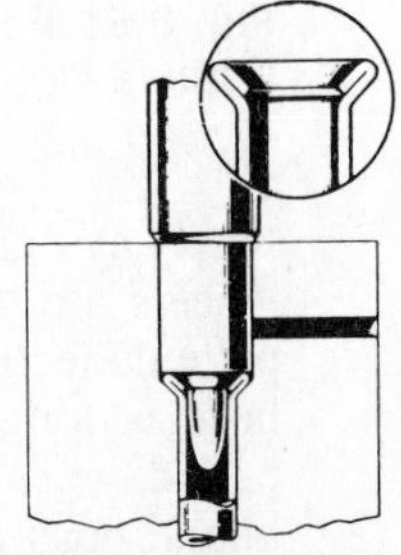

SECTIONAL VIEW OF DIE SHOWING TUBING LOCATED AGAINST STOP PIN.

FIRST OPERATION. TOOL SHOWN FORMING OUTSIDE FLARE.

SECOND OPERATION. TOOL SHOWN FORMING INSIDE FLARE AND SEAT. COMPLETED DOUBLE-LAP FLARE SHOWN IN INSERT.

Fig. 6-59 The three steps required in double-flaring hydraulic-brake tubing. (*Ford Motor Company*)

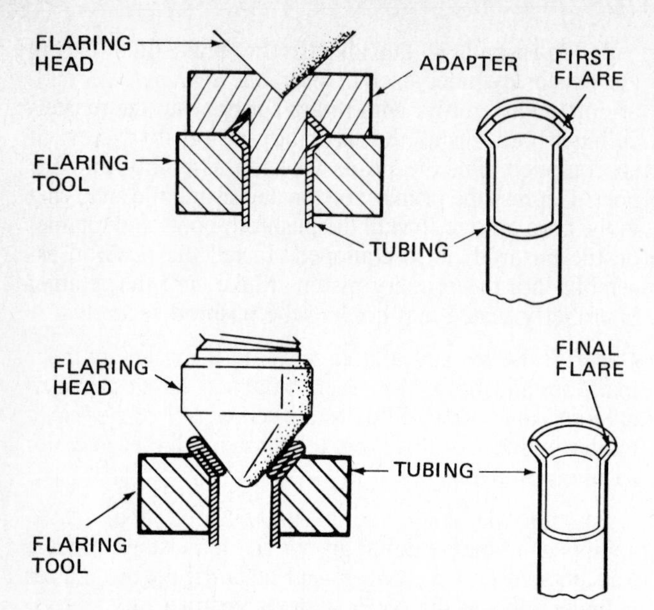

Fig. 6-60 Flaring hydraulic-brake tubing. Top, using an adapter to make the first flare. Bottom, making the second flare in the double-flare procedure.

Some tubing has a chamfer-type flare instead of a double flare. Figure 6-61 shows the two types.

When removing or installing tubing, using the correct tube-nut wrench (Fig. 6-62). Also, screw a spare tube nut into a female fitting to avoid distortion (Fig. 6-62).

Always use new copper gaskets on fittings that require gaskets. Once compressed, copper gaskets should not be reused because they may not provide a leakproof seal.

Careful! After flaring, blow compressed air through the tube to blow out any metal chips or particles of dirt.

6-13 Hydraulic-brake-hose replacement Hydraulic hoses form the flexible link between the tubing and the brakes or axle housing. The tubing is attached firmly to the car frame. However, the wheels and axle housings move up and down in relation to the car frame. Therefore, flexible hoses must be used to complete the hydraulic line to the brakes.

Brake hoses are supplied in different diameters and lengths, with a variety of end fittings (Fig. 6-63). When installing a hose, be careful not to twist it. Hold the hose-end fitting securely while attaching a tube nut to a hose. If a clip is used to hold a hose-end fitting to a support, make sure the clip is properly installed.

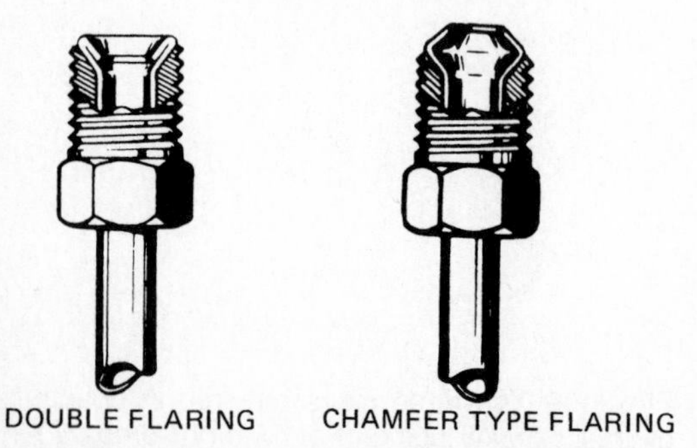

Fig. 6-61 Two types of flares. (*Bendix Corporation*)

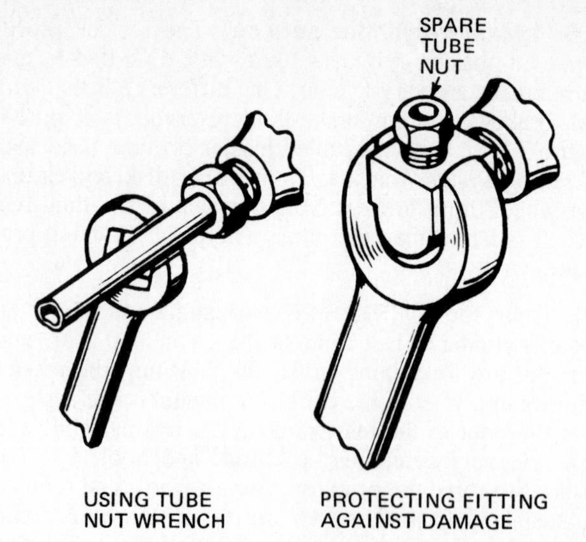

Fig. 6-62 Tightening hydraulic fittings. (*Bendix Corporation*)

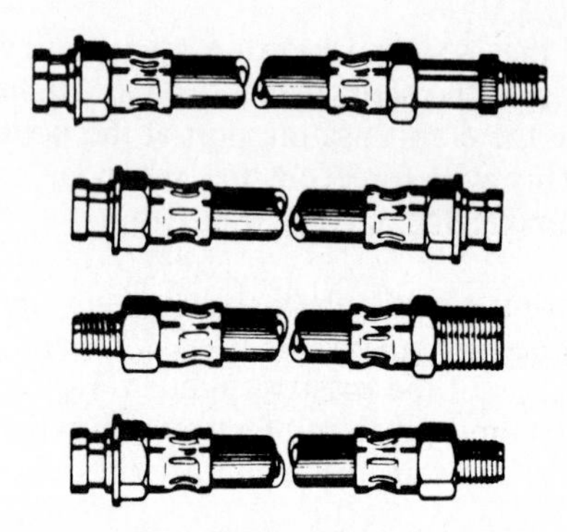

Fig. 6-63 Typical brake-hose fittings. (*Bendix Corporation*)

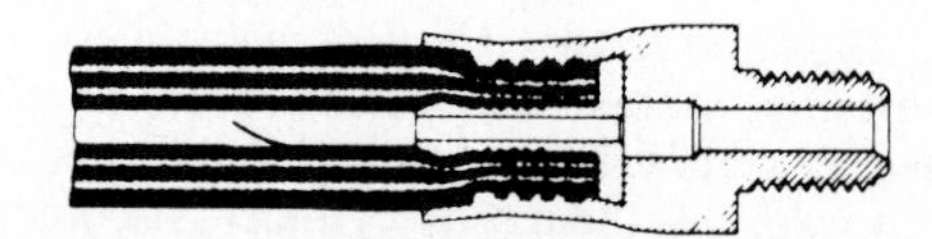

Fig. 6-64 Possible internal defects of brake hose. (*Bendix Corporation*)

Always check hoses for interference when the front wheels are turned and when the springs go from complete deflection to complete rebound. Make sure that the hose is long enough so that it will not be strained or pulled when the suspension springs are in the full rebound position.

Figure 6-64 shows possible internal defects of hoses that could prevent braking. Torn inner lining or a leaking fitting can impede the flow of brake fluid.

6-14 Flushing, filling and bleeding the hydraulic system The hydraulic system must be free of any con-

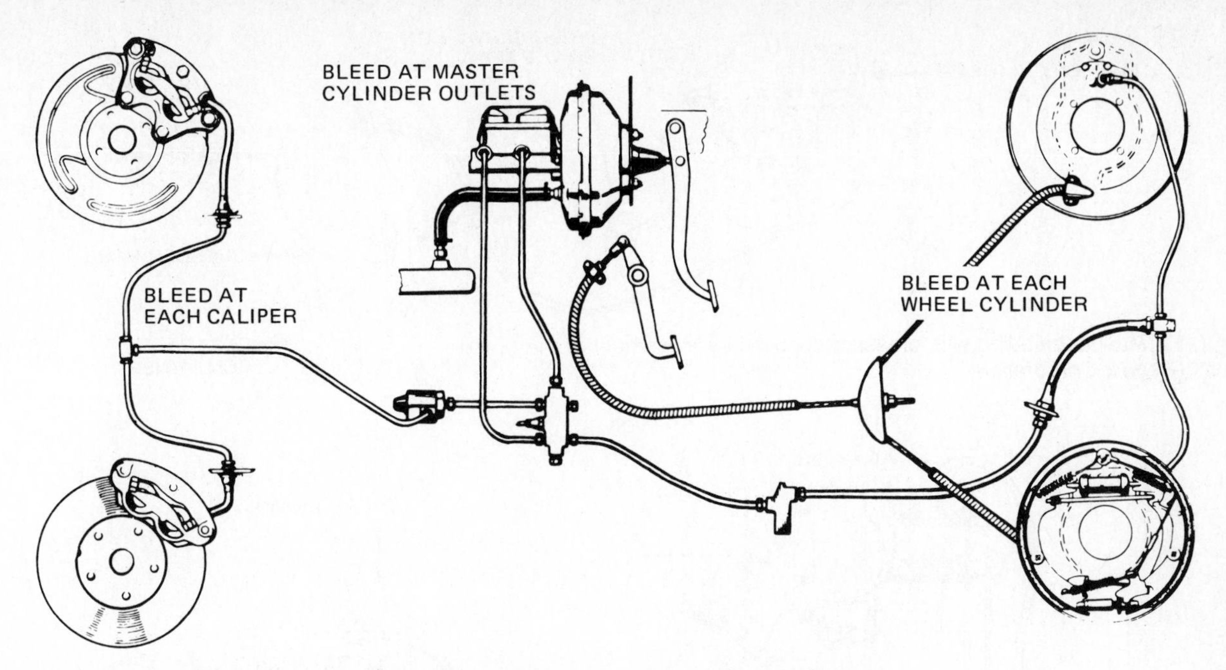

Fig. 6-65 Typical hydraulic-brake system, showing the bleeder points. (*Bendix Corporation*)

taminant, such as air, dirt, or mineral oil. If air gets into a brake line or wheel cylinder or caliper, the brakes will not work properly. The brake pedal will be spongy. Instead of the pedal going down firmly to produce normal braking, it will be springy. The air will compress, and braking will be erratic.

Air can enter the system if any brake lines or system components are disconnected. Dirt can also enter, especially if components are disconnected without the connections being wiped clean first. Mineral oil (such as engine oil) must never be added to the brake system. Mineral oil will attack the rubber parts in the system, causing them to swell and disintegrate. The result can be complete failure of the braking system.

The process of getting rid of air in a brake line or component is called *bleeding*. The process of purging the system of all the old contaminated brake fluid is called *flushing*. Flushing is the same as bleeding except that a much greater quantity of new brake fluid must be used to flush out all the old brake fluid.

1. **Bleeding and filling** In the bleeding process, brake fluid is forced through the brake line or component that is contaminated with air. The brake fluid is forced from the master cylinder to the wheel cylinder or caliper. Figure 6-65 shows the bleeder points in a typical brake system. Each wheel cylinder or caliper has a bleeder screw. Figure 6-66 shows the location of a bleeder screw on a wheel cylinder.

Figure 6-66 also shows the bleeding procedure. A bleeder hose is attached at the wheel-cylinder bleeder connection, and the bleeder screw is backed off to open the outlet. Then pressure is applied to the brake fluid at the master cylinder.

Careful! Clean away dirt and grease from around the bleeder screw to avoid getting any dirt into the system. Any dirt at a bleeder connection could be drawn into the cylinder if the pressure on the brake fluid at the master cylinder is released. The dirt could then cause brake failure at the wheel.

When pressure is applied to the brake fluid at the master cylinder, brake fluid flows from the master cylinder through the brake line to the wheel cylinder. This pushes the air out ahead of it. The lower end of the bleeder hose is put into a clear container partly filled with brake fluid, as shown in Fig. 6-66. The bleeding procedure is continued until no more bubbles show up in the fluid flowing into the container.

During bleeding, the end of the hose must be kept immersed in the brake fluid. This is so that any bubbles of air that come out can be seen. It also prevents any air from being pulled back into the system if the pressure on the brake fluid is released before the bleeder screw is tightened.

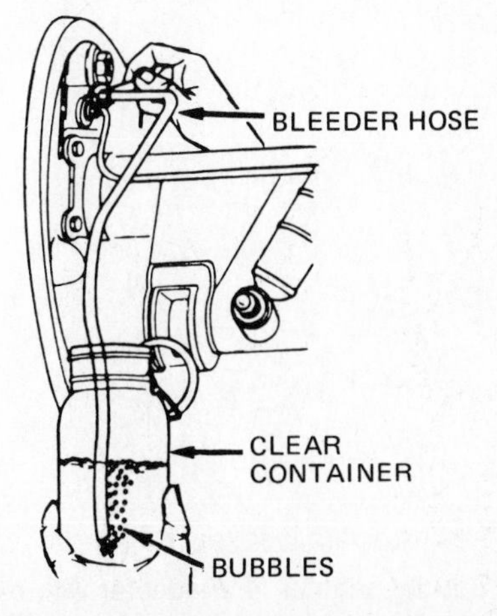

Fig. 6-66 Bleeding at a wheel cylinder. (*Bendix Corporation*)

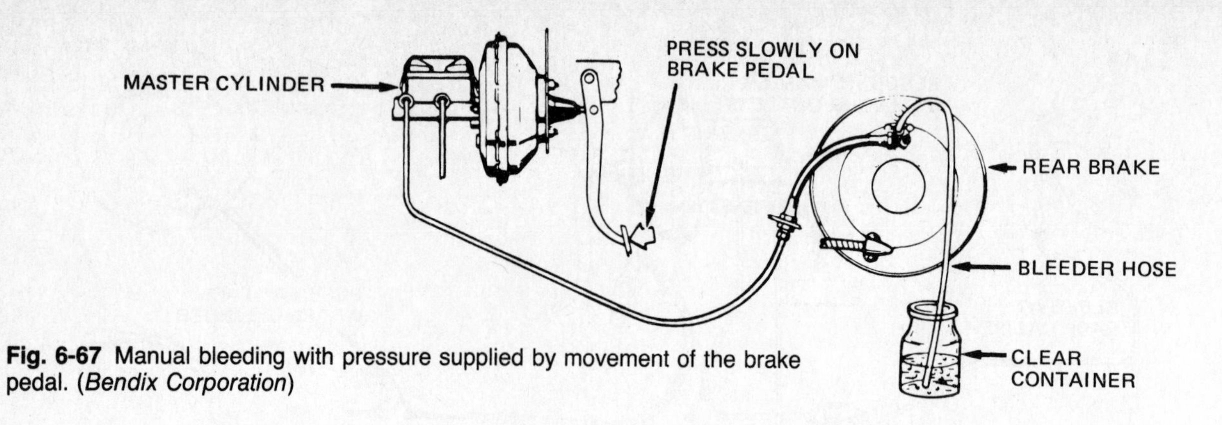

Fig. 6-67 Manual bleeding with pressure supplied by movement of the brake pedal. (*Bendix Corporation*)

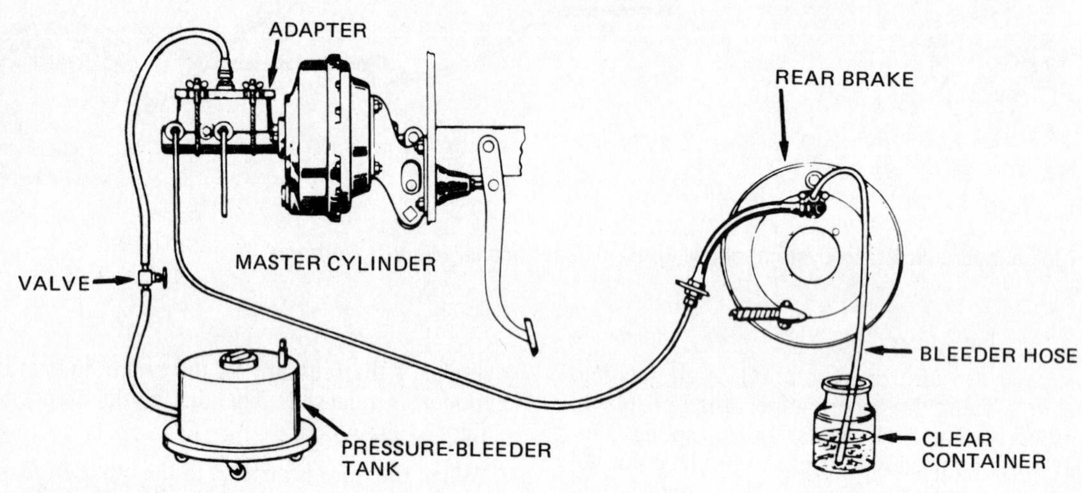

Fig. 6-68 Pressure-bleeder adapter in place on master cylinder. (*Bendix Corporation*)

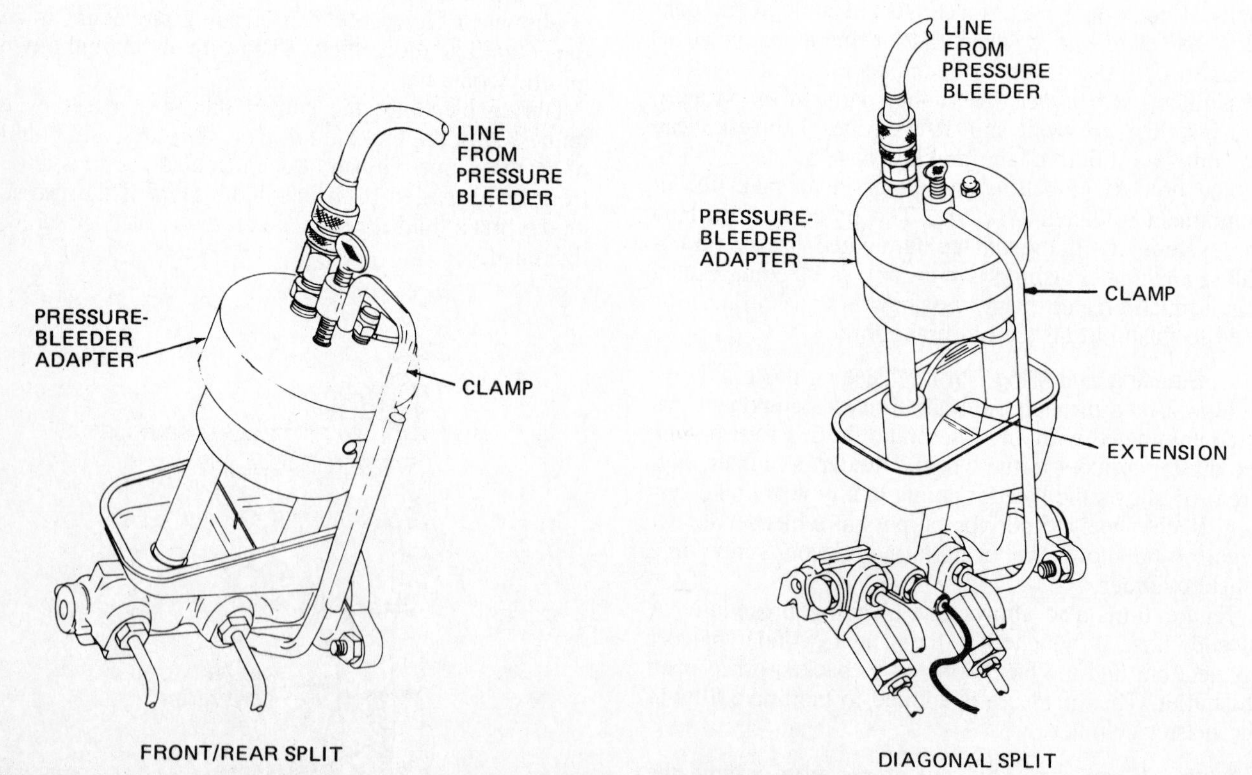

Fig. 6-69 Special adapter, and adapter with extension, needed to bleed different types of composite master cylinders. (*Delco Moraine Division of General Motors Corporation*)

2. Applying pressure The pressure is applied at the master cylinder either manually or with a pressure tank called a *brake bleeder*. The manual bleeding procedure requires an assistant who applies even pressure to the brake pedal (Fig. 6-67). The master-cylinder reservoir must be filled to within one-half inch [12.7 mm] of the top with clean brake fluid. During the bleeding procedure, more brake fluid must be added to the reservoir if the level falls to less than half full. When the fluid coming out of the bleeder hose is free of air bubbles, close the bleeder screw and release the brake pedal. Repeat the procedure at the other bleeder valves.

Pressure bleeding requires a pressure-bleeder tank and an adapter to fit in or over the master-cylinder reservoir. No assistant is needed. Figure 6-68 shows a typical bleeder adapter installed on the reservoir of a cast-iron master cylinder. Figure 6-69 shows the special fixture needed to bleed composite master cylinders, and the extension that must be added to bleed General Motors diagonal-split master cylinders.

Be sure the pressure bleeder tank has enough brake fluid in it. Then charge the tank with compressed air to about 20 to 25 psi [138 to 172 kPa]. Fill the reservoir to the top with clean brake fluid. Install the adaptor (Fig. 6-69).

Cars with front-disk and rear-drum brakes have either a metering valve (⚙ 3-23) or a combination valve (⚙ 3-25) in the hydraulic system. To allow fluid to flow to the front-disk brakes during pressure bleeding, it may be necessary to hold the valve stem open manually. To hold the metering valve (Fig. 3-63) or combination valve (Fig. 3-63) open, either push the valve stem in or pull it out. The action to take depends on the valve type, so use only a light force to avoid damaging the valve. For GM cars, a special tool is available to hold the valve stem in. For AMC, Chrysler, and Ford cars, a special tool is available to hold the valve stem in. Open the valve on the bleeder tank to allow pressurized fluid to flow from the tank to the reservoir. When the fluid coming out of the bleeder hose is free of air bubbles, close the bleeder screw. Repeat the procedure at the other bleeder valves (Fig. 5-79).

3. Filling As a final step in bleeding, make sure that the master-cylinder reservoir is filled to the specified level with brake fluid.

4. Bleeder-screw locations In addition to the bleeder locations at the vehicle wheels, bleeder screws may also be found in many cars next to the outlet ports of the master cylinder. They are also found on some combination valves used with front-disk and rear-drum brakes.

5. Bleeding sequence First, bleed at the master-cylinder bleeder screws, if the cylinder is so equipped. Next, bleed at the combination valve if it is equipped with a bleeder screw. Next, bleed the wheel cylinder or caliper that is farthest from the master cylinder. Then bleed the next farthest, and so on. On drum brakes with two wheel cylinders, bleed the upper cylinder first.

NOTE: On some American Motors, International Harvester, and Jeep vehicles, the switch terminal plug should be removed from the brake warning-light switch before bleeding. This protects the switch terminal from damage. See the manufacturer's shop manual before bleeding the brake hydraulic system.

6. Bench-bleeding master cylinders Before installing a new or rebuilt master cylinder, bleed it. This saves bleeding time on the vehicle. When the master cylinder is first filled with brake fluid, air will be trapped ahead of the pistons in the master cylinder. This air can be eliminated before the master cylinder is installed.

Install bleeder tubes (Fig. 6-70), or plug the outlets (Fig. 6-71). Push in and release the master-cylinder pistons several times. This removes the air trapped in the cylinder. Repeat until no more air bubbles show up. Keep your face away from the reservoir to avoid spraying brake fluid.

7. Flushing the system Flushing the system is the same as bleeding except that more new brake fluid must be used to make sure that all contaminated brake fluid is purged from the system. The basic procedure is the same. However, flushing must be continued until all the old brake fluid has been flushed out.

Some manufacturers recommend the use of a special flushing fluid. This fluid is used instead of new brake fluid during the flushing operation. Flushing is continued until all the old brake fluid has been flushed out. Then the flushing fluid is purged by applying clean, dry air through the master cylinder to blow the fluid out. Do not use too much air pressure. After all flushing fluid is out,

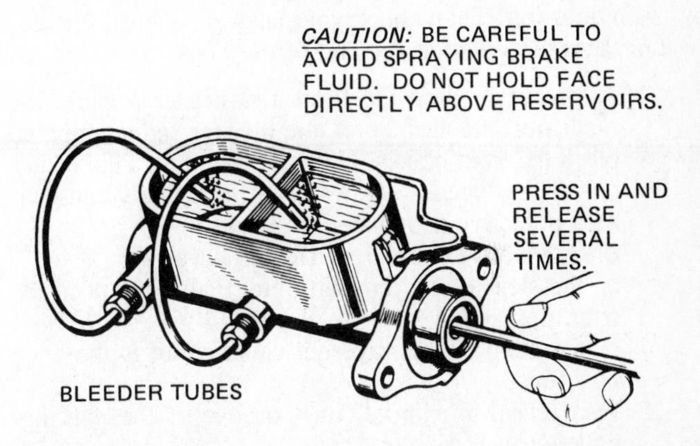

Fig. 6-70 Using bleeder tubes to bleed the master cylinder. (*Bendix Corporation*)

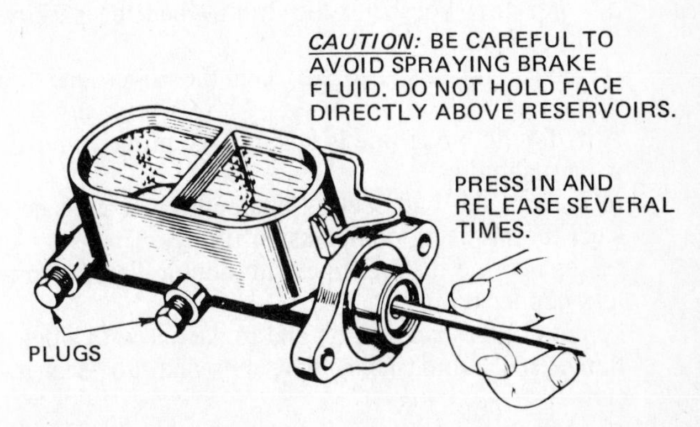

Fig. 6-71 Plugging outlets to bleed the master cylinder. (*Bendix Corporation*)

fill the master-cylinder resrvoir with new brake fluid and bleed the system as explained above.

8. Mineral-oil damage If mineral oil such as engine oil or automatic-transmission fluid has been added to the brake system, all rubber parts in the system must be replaced. In addition, the system must be thoroughly flushed to remove all traces of the oil.

9. Changing brake fluid Some vehicle manufacturers recommend changing the brake fluid periodically. This is because DOT 3 and DOT 4 brake fluids absorb moisture. The recommended change interval varies from 1 to 4 years. The correct brake fluid to use is the fluid specified in the owner's manual of the manufacturer's service manual for the car you are servicing.

While DOT 4 fluid can be used to fill systems that originally were filled with DOT 3, fluid classed as DOT 5 must not be added to either. To properly fill the system with DOT 5 fluid, the system must be flushed and all traces of the old fluid removed. When changing to DOT 5 fluid, wheel cylinders should be rebuilt or replaced. DOT 5 fluid must not be added to a system containing DOT 3 or DOT 4 fluid.

Chapter 6 review questions

Select the *one* correct, best, or most probable answer to each question. Then check your answers against the correct answers given at the end of the book.

1. After overhauling the front disk-brake calipers, the mechanic attaches a pressure bleeder and attempts to bleed the front brakes. However, no brake fluid comes out of the bleeder screws. The most likely cause is:
   a. a defective proportioning valve
   b. a defective pressure-differential valve
   c. the stem of the proportioning valve is not in the open position
   d. the stem of the metering valve is not in the open position
2. In a rear-drum brake, the purpose of the parking-brake strut rod is to:
   a. equalize the braking force between the shoes during normal braking
   b. center the shoes after each brake application
   c. force the shoes into the drum when the parking brake is applied
   d. reduce the distance between the shoes and the drum
3. A hydraulic brake line is leaking. To properly repair it, you should:
   a. cut out the bad section and replace it with new steel tubing using compression fittings
   b. replace the leaking line with double-flared seamless copper tubing
   c. cut out the bad section and replace it with single-flared steel tubing using flare nuts and tube-seat inserts
   d. replace the leaking line with double-flared steel tubing
4. A drum-brake hydraulic system has been flushed with a pressure bleeder, using new brake fluid. Mechanic A says that the brake fluid remaining in the system should be removed by applying air to the wheel cylinder. Mechanic B says that the brake fluid remaining in the system should be removed by applying air to the master cylinder. Who is right?
   a. A only
   b. B only
   c. both A and B
   d. neither A nor B
5. Disk-brake rotors are checked for each of the following conditions *except:*
   a. bellmouthing or taper
   b. parallelism
   c. scoring
   d. runout
6. Mechanic A says that the drum of a drum brake can be warped, causing braking problems, if the lug nuts are installed incorrectly. Mechanic B says that the disk-brake rotor can be warped, causing braking problems, if the lug nuts are installed incorrectly. Who is right?
   a. A only
   b. B only
   c. both A and B
   d. neither A nor B
7. Before starting to work on disk brakes, removing some brake fluid from the disk-brake reservoir in the master cylinder will:
   a. make bleeding easier
   b. prevent fluid from spilling out when the pistons are pushed into the calipers
   c. allow the pressure to the front and rear axles to equalize
   d. prevent the brake warning light from coming on
8. A stuck piston in a disk-brake caliper can be removed by all of the following *except:*
   a. a special puller
   b. compressed air
   c. a sharp chisel
   d. hydraulic pressure
9. Brake drums should be checked for all of the following *except:*
   a. scoring
   b. runout
   c. bellmouthing
   d. rust on outer surfaces
10. Brake drums should not be refinished to larger than the original standard drum diameter by more than:
    a. 0.020 inch [0.5 mm]
    b. 0.040 inch [1 mm]
    c. 0.060 inch [1.5 mm]
    d. 0.090 inch [2.3 mm]
11. The maximum allowable thickness variation of the disk (rotor parallelism) usually is about:
    a. 0.005 to 0.007 inch [0.13 to 0.18 mm]
    b. 0.0000 to 0.0003 inch [0.00 to 0.008 mm]
    c. 0.002 to 0.005 inch [0.05 to 0.13 mm]
    d. 0.0005 to 0.0007 inch [0.013 to 0.018 mm]

12. The surface of the piston that slides past the seal must be inspected for:
    a. pitting
    b. scoring
    c. corrosion
    d. all of the above
13. After the caliper is removed from the disk, you may do all of the following *except:*
    a. disconnect the brake hose from the brake line
    b. allow the caliper to hang by the brake hose
    c. rest the caliper on the suspension
    d. hang the caliper on a wire hanger or hook from the suspension
14. To install the piston in the caliper, use a slow, steady force:
    a. with your hand
    b. with a large hammer
    c. with compressed air
    d. with the shop press
15. In the hydraulic system, rubber seals and O rings should be lubricated before installation, but *never* with:
    a. brake fluid
    b. engine oil
    c. special hydraulic-brake assembly lubricant
    d. silicone grease
16. After removing the brake drum, the dust and dirt can be removed from the brake assembly by:
    a. blowing it off with compressed air
    b. brushing it off with a stiff brush
    c. wiping it off with a water-dampened cloth
    d. none of the above
17. Proper bleeding of the hydraulic system usually results in:
    a. a firmer brake pedal
    b. removal of all air from the hydraulic system
    c. both A and B
    d. neither A nor B
18. When using a pressure bleeder on a car with drum brakes, the tank should be charged with compressed air to about:
    a. 2 to 8 psi [14 to 55 kPa]
    b. 15 to 35 psi [103 to 241 kPa]
    c. 40 to 65 psi [276 to 448 kPa]
    d. 60 to 80 psi [413 to 552 kPa]
19. When assembling a disk-brake caliper, a special tool is used to:
    a. seat the bolt in the caliper counterbore
    b. install the piston
    c. both A and B
    d. Neither A nor B
20. When assembling a drum brake, the shoe pads on the backing plate should be lubricated. Mechanic A says to use SAE 30 oil. Mechanic B says to use petroleum jelly. Who is right?
    a. A only
    b. B only
    c. both A and B
    d. neither A nor B

CHAPTER 7

POWER-BRAKE SERVICE

After studying this chpter, and with proper instruction and equipment, you should be able to:

1. Diagnose troubles in power-brake systems.
2. Overhaul a dual-diaphragm power-brake unit.
3. Make a pushrod adjustment on a power-brake unit.

7-1 Introduction to power brake service

This chapter discusses the trouble diagnosis, adjustment, removal, repair, and installation of power brakes. As a typical servicing procedure, the overhaul instructions are given for the Bendix power-brake booster. This unit is widely used on General Motors and other manufacturers' passenger cars. However, Chrysler and Ford do not provide overhaul instructions for power-brake units in their service manuals. These manufacturers recommend replacement of defective units.

Power-Brake Trouble-Diagnosis Chart

(See ⚙ 7-3 and 7-4 for details of checks and corrections listed. Not all the possible causes and checks or corrections listed apply to all models of power brakes.)

COMPLAINT	POSSIBLE CAUSE	CHECK OR CORRECTION
1. Excessive brake-pedal force	a. Defective vacuum check valve	Free or replace
	b. Hose collapsed	Replace
	c. Vacuum fitting plugged	Clear, replace
	d. Binding pedal linkage	Free
	e. Air inlet clogged	Clear
	f. Faulty piston seal	Replace
	g. Stuck piston	Clear, replace damaged parts
	h. Faulty diaphragm	Replace (applies to diaphragm type only)
	i. Causes listed under item 6 in chart in ⚙ 5-2 or under item 3 in chart in ⚙ 5-15	
2. Brakes grab	a. Reaction, or "brake-feel," mechanism damaged	Replace damaged parts
	b. Air-vacuum valve sticking	Free, replace damaged parts
	c. Causes listed under item 7 in chart in ⚙ 5-2 or item 3 in chart in ⚙ 5-15	
3. Pedal goes to floorboard	a. Hydraulic-plunger seal leaking	Replace
	b. Compensating valve not closing	Replace valve
	c. Causes listed under item 1 in chart in ⚙ 5-2 or item 8 in chart in ⚙ 5-15	
4. Brakes fail to release	a. Pedal linkage binding	Free up
	b. Faulty check-valve action	Free, replace damaged parts
	c. Compensator port plugged	Clean port
	d. Hydraulic-plunger seal sticking	Replace seal
	e. Piston sticking	Lubricate, replace damaged parts as necessary
	f. Broken return spring	Replace
	g. Causes listed under item 3 in chart in ⚙ 5-2 or item 6 in chart in ⚙ 5-15	
5. Loss of brake fluid	a. Worn or damaged seals in hydraulic section	Replace, fill and bleed system
	b. Loose line connections	Tighten, replace seals
	c. Causes listed under item 10 in chart in ⚙ 5-2 or items 7 and 9 in chart in ⚙ 5-15	

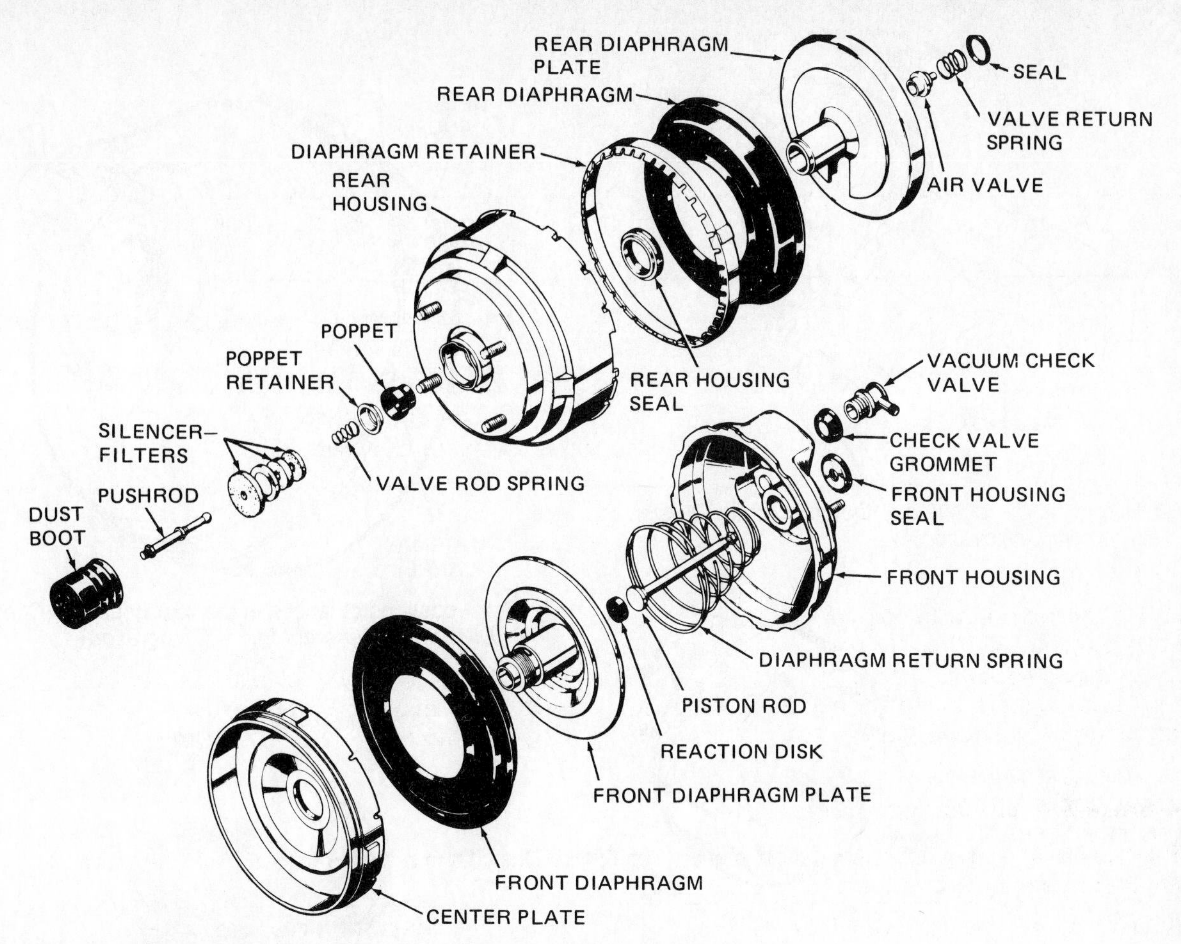

Fig. 7-1 Disassembled view of a Bendix dual-diaphragm power-brake booster. (*Chevrolet Motor Division of General Motors Corporation*)

⚙ 7-2 Power-brake trouble diagnosis chart The chart on the previous page relates various power-brake troubles to their possible causes and corrections. This chart gives you a means of logically tracing troubles to their actual causes. Its use permits quick location of causes and their rapid correction. The chart and the sections that follow pertain to power-brake units only. Generally, the trouble-diagnosis charts in ⚙ 5-2 and 5-15, which cover hydraulic brakes, also apply to power-brake systems. Therefore, the troubles listed in the charts, as well as the trouble corrections described in Chap. 5, also apply to power brakes.

⚙ 7-3 Servicing power-brake units Even though the different types of power-brake units operate in a similar manner and have a similar exterior apearance, each model requires a special disassembly and reassembly procedure. Before you attempt to service a specific model, refer to the shop manual covering that model.

Keep your workbench and tools clean. Small particles of dirt in the valves could cause malfunctioning of the power brakes. Examine the rubber parts as the unit is disassembled. Discard any part that is cracked, cut, or worn. Rubber seals and other parts must be in good condition for normal valve and power-brake action. Replace any that are in questionable condition. Usually, the manufacturer's instructions require replacement of all old seals during an overhaul.

⚙ 7-4 Overhauling Bendix tandem power-brake units The Bendix tandem, or dual-diaphragm, power-brake unit (Fig. 7-1) has been widely used by General Motors and other car manufacturers. Procedures for removal, disassembly, cleaning, reassembly, installation, and adjustment of this unit follow.

1. Removal To remove the unit, disconnect the pushrod clevis from the brake-pedal arm. If the clevis will not pass through the hole in the fire wall, take the clevis off the rod, first noting its approximate location. Disconnect the vacuum hose from the power unit and the hydraulic lines from the master cylinder. Cap the lines to keep dirt out. Remove the nuts and the lock washers that attach the power-brake assembly to the fire wall, and take the assembly out of the engine compartment.

2. Disassembly Take the master cylinder off the power unit and lay it aside. Master-cylinder service was described in ⚙ 6-11.

Scribe lines across the flanges of the front and rear housings, in line with the master-cylinder cover, to provide guidelines for reassembly. Pull the piston rod from the front housing. (Figures 4-19 and 7-1 show the locations and appearance of the parts.) The seal will come off with the piston rod. Pull the vacuum check valve out. Discard the valve and rubber grommet.

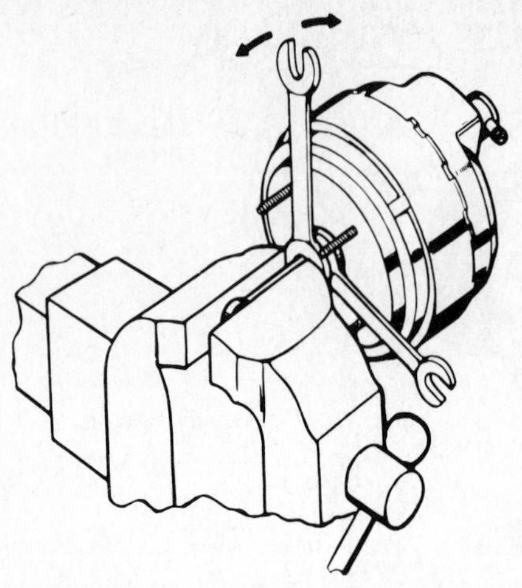

Fig. 7-2 Removing the pushrod. (*Chevrolet Motor Division of General Motors Corporation*)

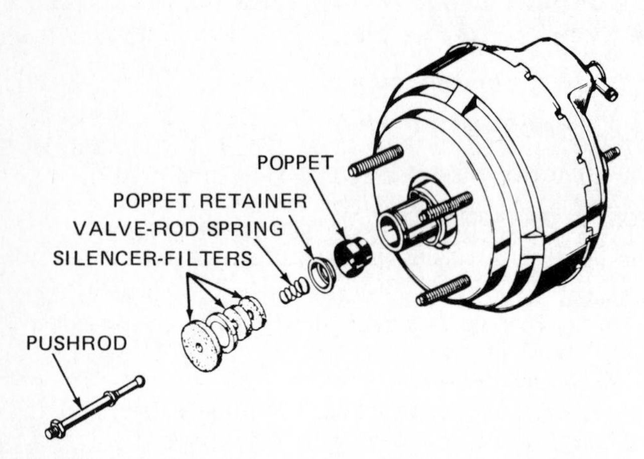

Fig. 7-3 Locations of the pushrod, silencer filters, and poppet parts. (*Chevrolet Motor Division of General Motors Corporation*)

If the pushrod has a clevis, remove the clevis. Unseat the dust boot from the housing and remove it and the silencer. Use a thin-bladed screwdriver to pry the silencer retainer off the end of the hub of the rear diaphragm plate. Do not chip the plastic. Squirt denatured alcohol down the pushrod to lubricate the rubber grommet in the air valve.

Clamp the end of the pushrod in a vise, leaving enough room to position two open-end wrenches between the vise and the retainer on the hub of the rear plate (Fig. 7-2). Using the wrench nearest the vise as a pry, force the air valve off the ball end of the pushrod. Do not damage the plastic hub or allow the power unit to fall to the floor.

Slide the air filter and air silencer from the pushrod. Remove the poppet spring, retainer, and poppet (Fig. 7-3).

Figure 7-4 shows the two types of lances on the edge of the rear housing. Four are the deep type. The metal that forms these must be partly straightened out so the lances will clear the cutouts on the front housing. If the metal tabs break, the housing must be replaced. After straightening the lances, attach a holding fixture to the front housing with nuts and washers drawn tight to eliminate bending of studs. Put the holding fixture in an arbor press (Fig. 7-5) with rear housing up. Use a 1½-inch (38 mm) wrench, as shown, to keep the lower unit from turning. It will turn a little, but when the wrench comes up tight against the arbor press, the unit cannot turn further.

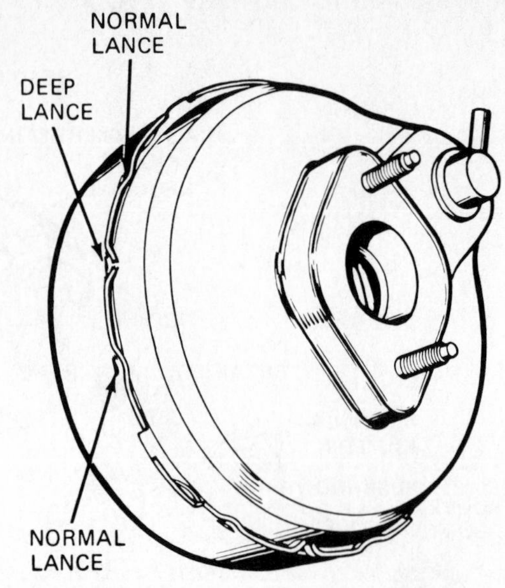

Fig. 7-4 Locations of lances in the rear housing. (*Chevrolet Motor Division of General Motors Corporation*)

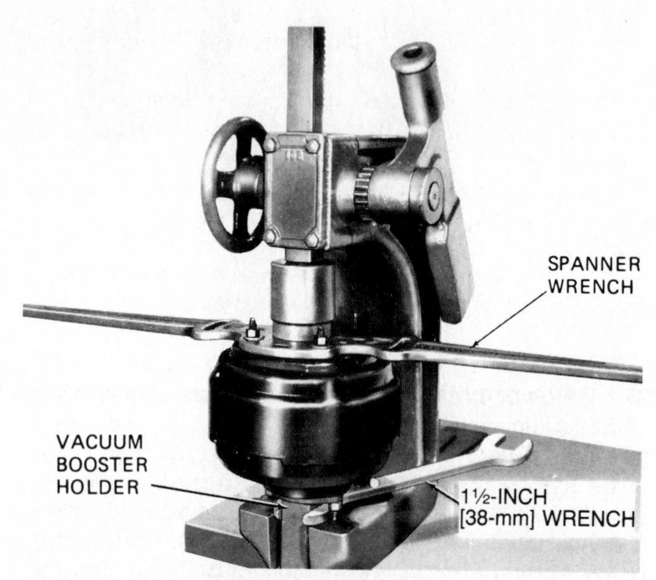

Fig. 7-5 Using special fixtures to hold and separate the housings in an arbor press. (*Chevrolet Motor Division of General Motors Corporation*)

Fasten the special spanner wrench to studs on the rear shell with nuts and lock washers. Place a piece of 2-inch [51-mm] pipe about 3 inches [76 mm] long over the plastic hub of the diaphragm. Put a piece of flat steel stock over the end of the pipe and press the housing down with the arbor press to relieve the spring force. Rotate the spanner counterclockwise to unlock the shells.

Release the arbor press and remove the diaphragm return spring. Detach the spanner and the holding fixture. Work the edges of the front diaphragm from under the

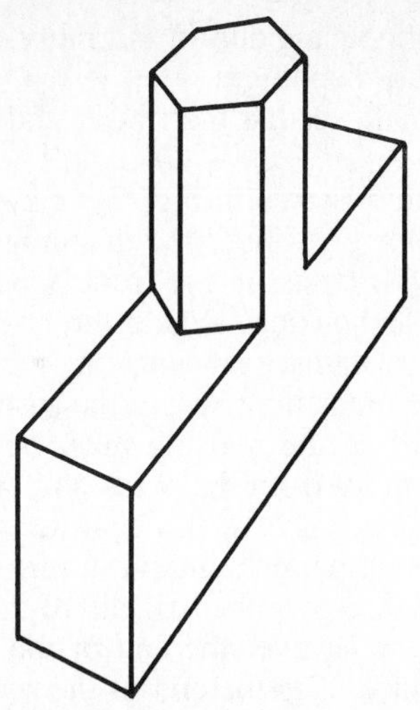

Fig. 7-6 Special tool to hold the front plate. The hex head fits into the hex openings of the front plate. (*Chevrolet Motor Division of General Motors Corporation*)

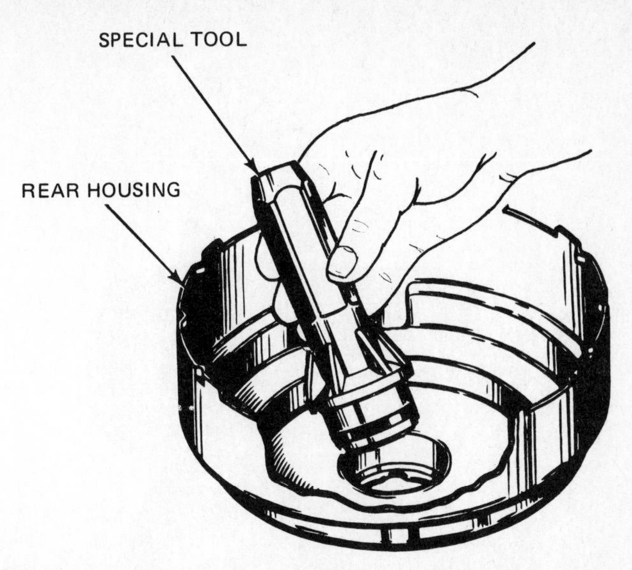

Fig. 7-7 Installing the rear-housing seal with the special tool. (*Chevrolet Motor Division of General Motors Corporation*)

lances of the rear housing and remove the complete vacuum assembly. Bosses on the center plate (Fig. 7-1) must align with cutouts in the rear housing to permit removal.

Wet the rear diaphragm retainer with denatured alcohol and remove it with your fingers only. Do not use any tool.

Clamp the special tool (Fig. 7-6) in a vise, hex head up. Put the diaphragm-and-plate assembly on the tool with the tool seated in the hex opening in the front plate. Twist the rear diaphragm plate counterclockwise, using hand leverage on the outer edge of the plate. Remove the plates from the tool and place them, front plate down, on a bench. Unscrew the rear plate completely and lift it off, catching the air valve and return spring as the parts are separated.

Remove the square ring seal from the shoulder of the front-plate hub. Remove the reaction disk from inside the front diaphragm plate. The vacuum seal may stay in front of the center diaphragm plate. If the seal assembly is defective, the center-plate-and-seal assembly must be replaced as a unit.

Remove the diaphragms from the plates. If the rear-housing seal requires replacement, use a blunt punch or 1¼-inch [32-mm] socket to drive the seal out.

3. Cleaning and inspection Clean all parts with denatured alcohol. Blow out all passages and holes with compressed air. Air-dry parts. If slight rust is found on the inside surface of the power-cylinder housing, polish it with crocus cloth and clean with denatured alcohol.

CAUTION: **Never use gasoline, kerosene, or other solvent to clean power-brake parts. These liquids will damage the rubber parts.**

Rubber parts must be in good condition, or the brake will not work properly. Replace them if there is the slightest trace of damage.

4. Reassembly Be sure all parts are clean. Rewash them before reassembly if there is any doubt about their being clean. Lubricate rubber, plastic, and metal friction points with special silicone lubricant.

If the rear housing seal was removed, press a new seal into place with the special tool (Fig. 7-7). Install the reaction disk in the front-plate hub, small tip side first. Use a rounded rod to seat it.

Clamp the special tool (Fig. 7-6) in a vise. Put the front plate on the tool with hex head of tool in front plate. Put the front diaphragm on the front plate with the long fold of the diaphragm facing down. Install the seal protector over the threads on the front-plate hub (Fig. 7-8). Apply a light film of silicone lubrication on the seal, and then guide the center plate, seal first, onto the front-plate hub. Remove the seal protector.

Apply a light film of silicone lubricant to the front and rear bearing surfaces of the air valve, but not to the rubber grommet inside the valve. Install the square ring

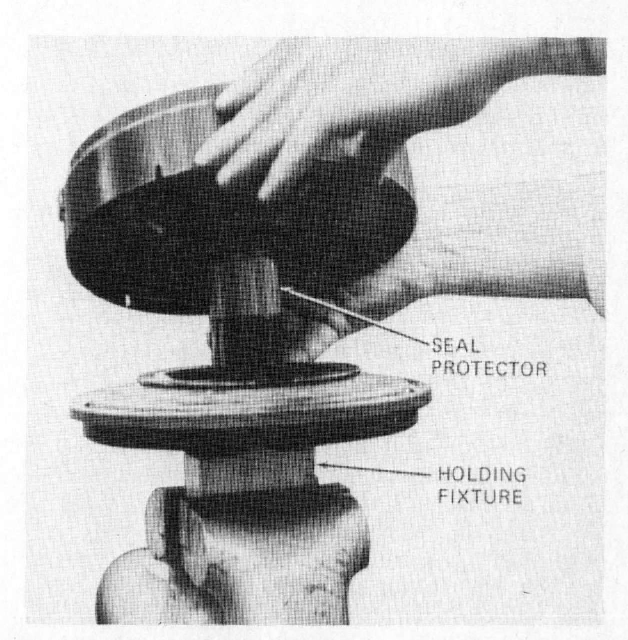

Fig. 7-8 Seal protector installed to protect the seal from threads on the front-plate hub. (*Chevrolet Motor Division of General Motors Corporation*)

Fig. 7-9 Installing the air-valve assembly. (*Chevrolet Motor Division of General Motors Corporation*)

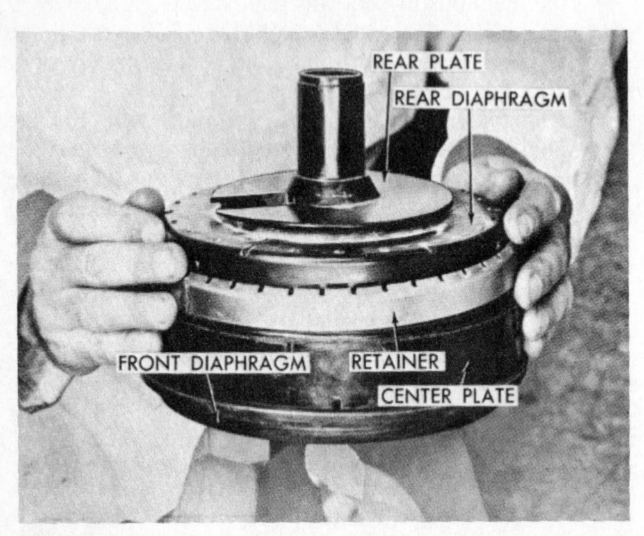

Fig. 7-10 Installing the rear-diaphragm retainer. (*Chevrolet Motor Division of General Motors Corporation*)

seal on the front-plate hub. Then install the return spring and air valve in the base of the front-plate hub (Fig. 7-9).

Set the rear plate over the hub of the front plate. Use your hands only and screw the plate onto the hub, making sure that the valve and spring are properly aligned. Use your index finger to check the travel of the valve plunger. It should be free. Plates should be tight, but do not overtorque.

Assemble the rear diaphragm to the rear plate, and put the lip of the diaphragm in the groove in the rear plate. Install the diaphragm retainer, using your fingers to press the retainer until it seats on the shoulder of the center plate (Fig. 7-10).

Apply talcum powder to the inside wall of the rear housing and silicone lubricant to the scalloped cutouts of the front housing and to the seal in the rear housing. Assemble the diaphragm-and-plate assembly into the rear housing. Bosses on the center plate must align with cutouts in the rear housing during assembly. Work the outer rim of the front diaphragm into the rear housing with a screwdriver blade so that the rim is under the lances in the housing.

With the setup shown in Fig. 7-5 compress the housings in the arbor press until the diaphragm edge is fully compressed with tangs on the front housing against the slots in the rear housing. Rotate the spanner clockwise until tangs butt against rear-housing stops.

Bend lanced areas in to secure the assembly. (If tangs break, that half of the housing must be replaced.) Remove the assembly from the press and detach tools.

Wet the poppet valve with denatured alcohol, install retainer inside the poppet, and put it into the hub. Install silencers and filters over the ball end of the pushrod (Fig. 7-3). Put the spring over the end of the rod, then push the rod into place. Tap the end of the rod with a plastic hammer to seat the ball in the poppet. Seat the filters and silencers into the hub, and install the retainer on the end of the hub. Assemble the silencer in the dust boot, wet the dust-boot opening with denatured alcohol, and assemble over the plate hub and rear-housing hub.

If the pushrod has a clevis, attach it. Dip a new check-valve grommet in denatured alcohol and install it in the front housing. Dip a new check valve in denatured alcohol and install it in the grommet.

Apply silicone lubricant to the piston end of the piston rod and insert the rod into the front plate. Twist it to eliminate air bubbles between it and the reaction disk. Assemble the seal over the rod and press it into the recess in the front housing.

To adjust the piston rod, use the special piston-rod gauge as shown in Fig. 7-11. To adjust, grasp the serrated end of the piston rod with pliers and turn the adjusting screw either in or out as necessary. The adjustment screw is self-locking.

Install the master cylinder and then install the assembly in the car. Bleed the hydraulic system after connecting the tubes to the master cylinder (⚙ 6-14).

⚙ 7-5 Adjustments of other power-brake units

American Motors Corporation, Chrysler Corporation, and Ford Motor Company no longer provide overhaul instructions on power-brake units. They specify that if trouble occurs in the unit, replace it. Many power-brake units have a pushrod adjustment as shown in Fig. 7-11. However, some pushrods are not adjustable (Fig. 7-12). On these, if the height is not correct, the complete power-brake unit must be replaced.

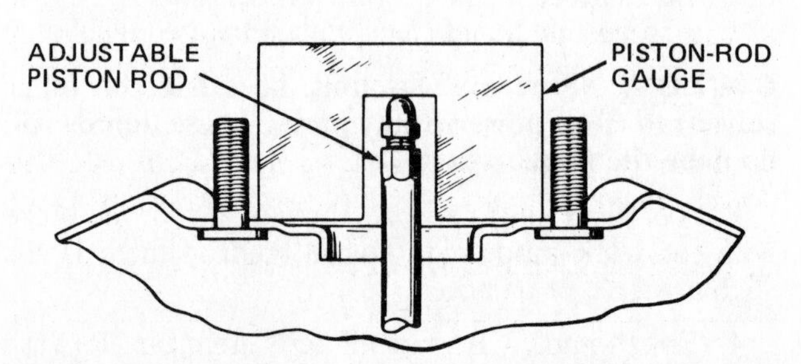

Fig. 7-11 Checking the adjustment of the piston rod with the piston-rod gauge. (*Chevrolet Motor Division of General Motors Corporation*)

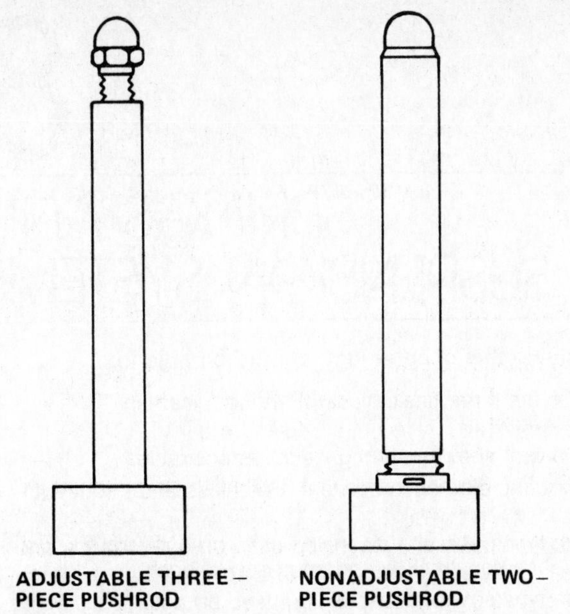

Fig. 7-12 Some piston rods are adjustable, others are not. (*American Motors Corporation*)

Chapter 7 review questions

Select the *one* correct, best, or most probable answer to each question. Then check your answers against the correct answers given at the end of the book.

1. Which of the following could cause excessive brake-pedal force to be applied for braking on a car with power brakes?
   a. defective vacuum check valve
   b. plugged vacuum fitting
   c. stuck piston
   —*d.* all of the above
2. On a car with power brakes, which of the following could cause the brakes to fail to release?
   —*a.* binding pedal linkage
   b. clogged air inlet
   c. both *a* and *b*
   d. neither *a* nor *b*
3. When disassembling the power-brake booster, you should scribe lines across the:
   —*a.* flanges of the front and rear housings
   b. master-cylinder cover and body
   c. fire wall and brake-pedal pushrod
   d. all of the above
4. To disassemble the Bendix dual-diaphragm power-brake booster, you need a:
   a. large hammer
   b. drill press
   —*c.* arbor press
   d. brake-drum lathe
5. The pushrod adjustment cannot be made on some power-brake units because:
   a. the pushrod is adjustable
   —*b.* the pushrod is not adjustable
   c. the pushrod cannot be bent
   d. the pushrod can be bent too easily

CHAPTER 8

SPRINGS AND SUSPENSION SYSTEMS

After studying this chapter, you should be able to:

1. Discuss the three basic types of springs used on automobiles.
2. Explain why a low unsprung weight is desirable.
3. Describe the difference between Hotchkiss and torque-tube drive.
4. List the types of rear suspension used on automobiles, and discuss the construction and operation of each.
5. List the types of front suspension used on automobiles, and discuss the contruction and operation of each.
6. Discuss the basic operation of an air-suspension system.
7. Describe the construction and actions in a hydropneumatic suspension system.

Springs

8-1 Purpose of springs The car frame supports the weight of the engine, power train, body, and passengers. The frame, in turn, is supported by the springs. Figures 1-1 and 1-3 show various suspension systems using coil and leaf springs. Figures 1-16 and 1-17 show front-suspension systems using torsion bars. Regardless of the type of spring, all work the same way. The weight of the frame, body, and load applies an initial compression to the springs. Then the springs will further compress or expand as the car wheels meet bumps or holes in the road. The wheels can move up and down somewhat independently of the frame and car body. This allows the springs to absorb a large part of the up-and-down motion of the wheels, instead of this motion being transmitted to the car frame and from it to the passengers.

Figures 1-14 and 1-15 show how coil springs compress and expand as the wheels encounter bumps or holes in the road. The illustrations show springs at a front wheel. However, the principle of spring action is the same at the rear wheels.

⚙ 8-2 Types of springs The automobile uses three basic types of springs: coil, leaf, and torsion-bar. Air suspension was offered at one time as optional equipment for cars. Today it is used on some trucks and buses. A modification of the air-suspension system, used on some cars, is designed to keep the rear of the car level even though the load changes. This system is called *automatic level control* and is covered in Chap. 9.

Most automobiles use either coil springs or torsion bars in the front suspension. Some cars have coil springs in the rear suspension. Others use leaf springs at the rear wheels. A few cars, and many trucks, use leaf springs at the front wheels also.

⚙ 8-3 Coil springs The coil spring is made of a length of round spring-steel rod or wire wound in the shape of a coil (Fig. 8-1). The spring is formed while the steel is white-hot. Then it is cooled and heat-treated to give it the proper characteristics of elasticity and "springiness." Spring characteristics are discussed in ⚙ 8-7.

Many coil springs are wound with tapered wire, which produces a spring that has a larger diameter at the center than at the two ends (Fig. 8-2). This gives the coil spring a variable *spring rate* (⚙ 8-7). As the spring is compressed, its resistance to further compression increases. The result is a more uniform ride.

⚙ 8-4 Leaf springs There are two types of leaf spring: the multileaf and the single-leaf. The latter is called a *tapered-plate spring* by the manufacturer.

1. Multileaf spring The multileaf spring is made up of a series of flat steel plates of graduated length placed one on top of another (see Fig. 8-3). Figure 1-2 shows a car using leaf springs at the rear. The plates, or leaves, are held together at the center by a center bolt which passes through holes in the leaves. Clips placed at intervals along the spring keep the leaves in alignment (Fig. 8-3). These are called *rebound clips*. They prevent excessive leaf separation during rebound after the wheel has passed over a bump in the road.

Instead of clips, some leaf springs are sheathed in a metal cover. The longest, or master, leaf is rolled at both ends to form spring eyes through which bolts are placed to attach the spring ends. On some springs, the ends of

Fig. 8-1 Coil spring used in automotive suspension systems.

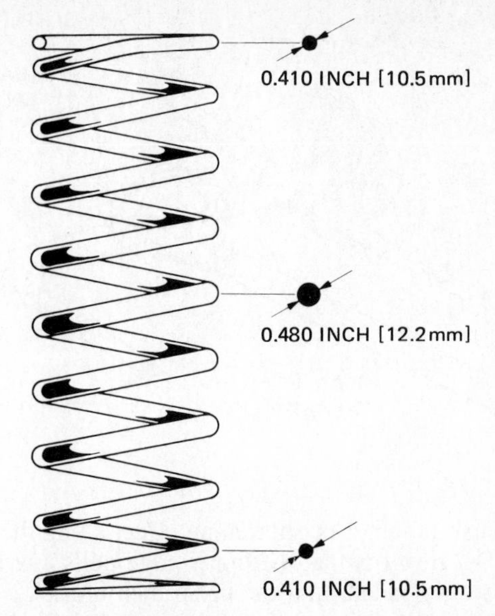

Fig. 8-2 Coil spring made from tapered wire, or rod. The wire is larger in diameter at the center of the coil. (*Mazda Motors of America, Inc.*)

the second leaf are also rolled partway around the two spring eyes to reinforce the master leaf.

In operation, the leaf spring acts like a flexible beam. An ordinary solid beam strong enough to support the car weight would not be very flexible. This is because as the beam bends, as shown in Fig. 8-4 (left), the top edge tries to become longer and the lower edge tries to become shorter. There is a pull-apart effect at the upper edge,

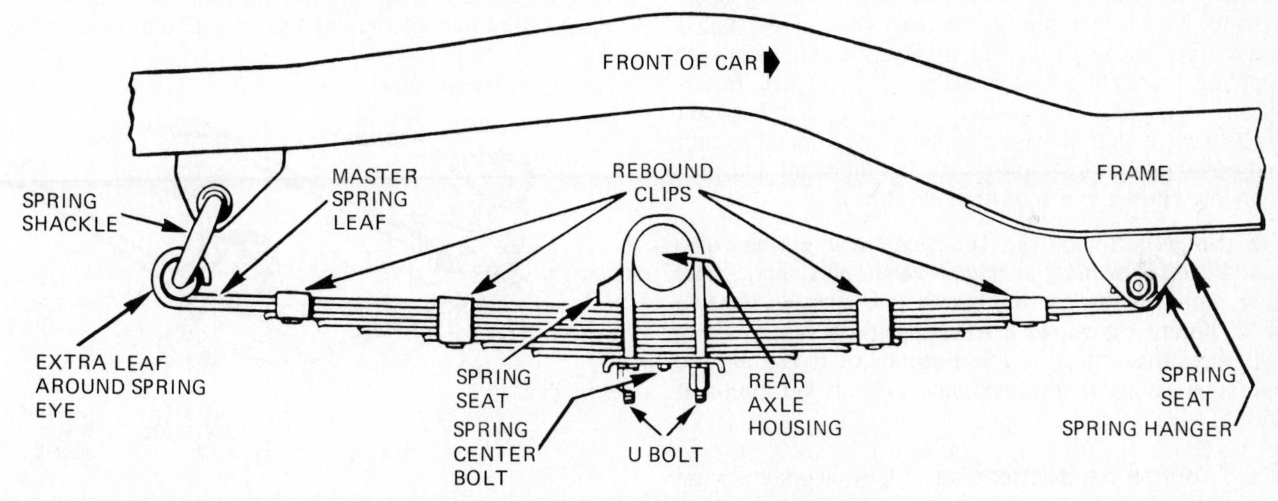

Fig. 8-3 Typical leaf spring, showing how it is attached to the frame and axle housing.

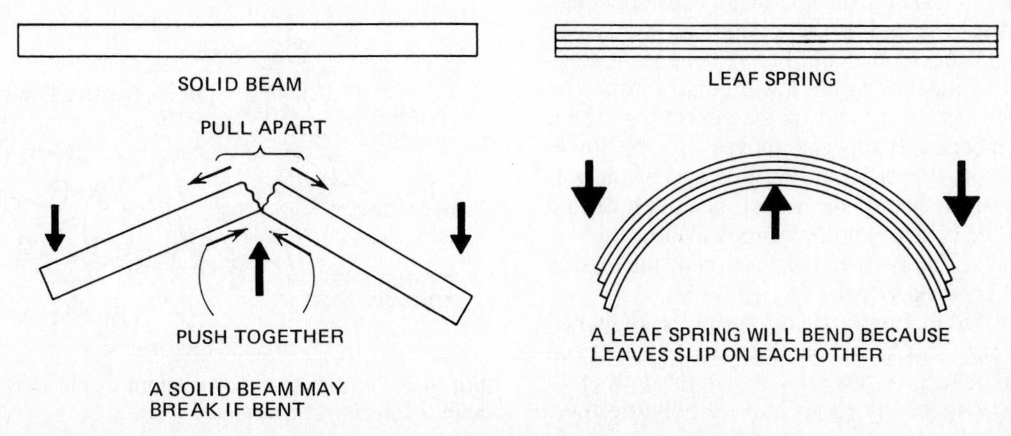

Fig. 8-4 The effect of bending a solid beam (left) and a leaf beam, or spring (right).

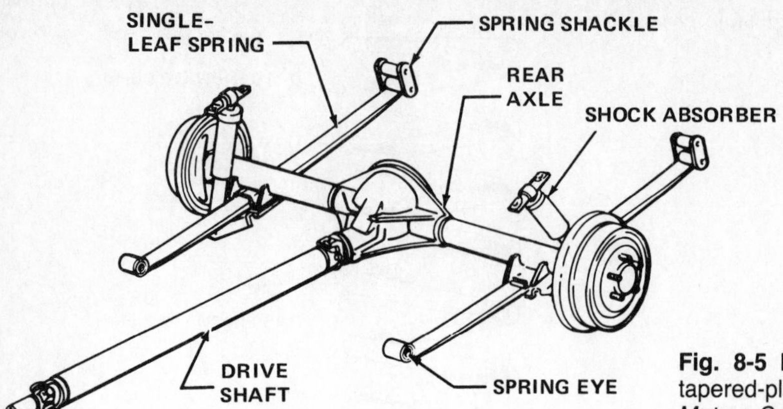

Fig. 8-5 Rear-suspension system using a single-leaf, or tapered-plate, spring. (*Chevrolet Motor Division of General Motors Corporation*)

and a push-together or shortening effect along the lower edge. The result is that the upper edge pulls apart if the beam is overloaded, and the beam then breaks.

However, there is a different action when the beam is made of a series of thin leaves, one on top of another, as shown in Fig. 8-4 (right). The leaves slip over each other to take care of the pull-apart and push-together tendencies of the two edges of the beam. Figure 8-4 (lower right) shows how much the leaves will slip over each other if the beam is sharply bent. All the leaves are the same length. The amount that the inner leaf projects beyond the outer leaves shows the amount of slippage.

In the actual leaf spring, the leaves are of graduated length. To permit the leaves to slip, various means of applying lubricant between the leaves are used. In addition, some leaf springs have special inserts between the leaves to permit easier slipping. The metal sheath that covers some leaf springs retains lubricant and prevents the entrance of moisture and dirt.

2. Single-leaf spring The single-leaf spring, also called a *tapered-plate spring,* is made of a single steel plate which is thick at the center and tapers to the two ends. Figure 8-5 shows a rear-suspension system using two single-leaf springs. The methods of mounting and operation are generally the same as with the multileaf spring.

⚙ 8-5 Torsion-bar suspension Many automobiles use torsion-bar front suspension. Figures 1-16 and 1-17 show this type of suspension system. Figure 8-6 is a view of a similar system, isolated from all other car components. In these systems, one end of the torsion bar is attached rigidly to the frame so that the bar is held stationary. The other end is attached to the lower control arm. The car weight places an initial twist on the torsion bar. Then, if the wheel meets a bump and moves up, the lower control arm pivots upward, causing that end of the torsion bar to twist further. If the wheel meets a hole and drops, so that the control arm pivots downward, the torsion bar untwists. The twist-untwist action of the torsion bar supplies the spring effect.

The torsion bar is simply a coil spring straightened out. When a coil spring is compressed, the rod from which the coil spring is made twists. It must twist to permit the coils of the spring to move closer together. This twisting provides the spring effect of the coil spring, just as the twisting effect does in the torsion bar.

In smaller cars space up front is limited, and so the torsion bars are placed from side to side, or transversely (Fig. 8-7). The up-and-down movement of the wheels and spindles causes the torsion bars to twist and untwist to provide the springing action. Height adjustment can be made by turning bolts located on the inner ends of the torsion bars, as shown in Fig. 8-7.

⚙ 8-6 Sprung and unsprung weight In the automobile, the terms *sprung weight* and *unsprung weight* refer to the part of the car that is supported on springs

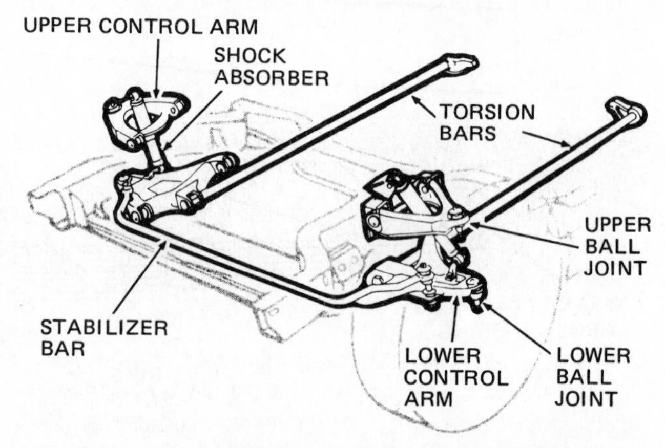

Fig. 8-6 Front-suspension system using longitudinal torsion bars. (*Moog Automotive, Inc.*)

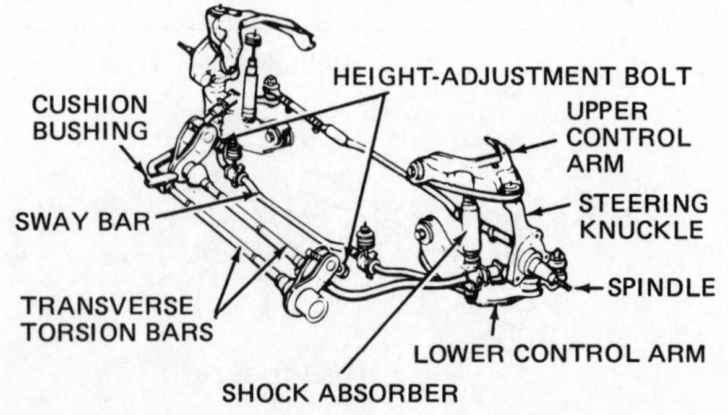

Fig. 8-7 Front-suspension system using transverse torsion bars. (*Chrysler Corporation*)

and the part of the car that is not. The frame and the parts attached to the frame are sprung. Their weight is supported on the car springs. However, the wheels, the wheel axles, and the drive axle are not supported on the springs. They represent unsprung weight.

Generally, unsprung weight should be kept as low as possible. This is because the roughness of the ride increases as unsprung weight increases. For example, consider a single wheel. If it is light, it can move up and down as road irregularities are met without causing much reaction to the car body. But if the weight of the wheel is increased, its movement will become more noticeable to the car occupants.

Suppose the unsprung weight at the wheel is equal to the sprung weight above the wheel. Then the sprung weight would tend to move almost as much as the unsprung weight. The unsprung weight, which must move up and down as road irregularities are met, would tend to cause a similar motion of the sprung weight. This is the reason why the unsprung weight should represent only a small portion of the total weight of the car.

❁ 8-7 Characteristics of springs and spring rate The ideal spring for an automotive suspension would absorb road shock rapidly and then return to its normal position slowly. However, such an ideal spring is not possible. An extremely flexible, or soft, spring allows too much movement, while a stiff, or hard, spring gives too rough a ride. Therefore, satisfactory riding qualities are attained by using a fairly soft spring combined with a shock absorber (Chap. 9). Suspension engineers carefully balance spring action with shock-absorber action to get the best handling and control.

Softness or hardness of a spring is referred to as its *rate*. The rate of a spring with uniform deflection is the weight required to deflect it 1 inch [25.4 mm]. Some automotive springs have almost constant rates through the operating range, or deflection, in the car. Here is an example of Hooke's law as applied to coil springs: The spring will compress in direct proportion to the weight applied. Therefore, if 600 pounds [272 kg] will compress the spring 3 inches [76 mm], 1200 pounds [544 kg] will compress the spring 6 inches [152 mm].

A spring that is designed not to have a constant rate is called a variable-rate spring (❁ 8-3). These are used on some cars to provide a smooth ride at slow speed and with light loads. Then, as speed and load increase, the spring stiffens to improve vehicle handling and control.

The variable-rate coil spring is wound from tapered wire. Figure 8-2 shows an example. The wire is larger in diameter at the center of the coil than at the two ends. This gives the coil spring a variable rate. The coil shown in Fig. 8-2 has a spring rate that varies from an initial 72 pounds per inch [1.29 kg per mm] to 163.5 pounds per inch [2.92 kg per mm]. This provides for a more uniform ride and better handling, especially at higher speeds.

❁ 8-8 Hotchkiss and torque-tube drives In rear-suspension systems, the springs may have an additional job to do besides supporting the car load. This is to absorb a reactive force known as *rear-end torque*.

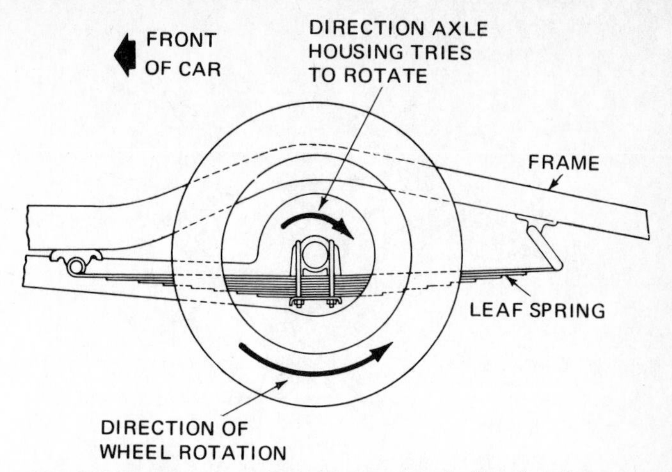

Fig. 8-8 The axle housing tries to rotate in the direction opposite to that of wheel rotation.

Whenever the rear wheel is being driven through the power train by the engine, the wheel rotates as shown in Fig. 8-8 (for forward car motion). A fundamental law of physics states that for every action there must be an equal and opposite reaction. Therefore, when the wheel rotates in one direction, the wheel-axle housing tries to rotate in the opposite direction, as shown in Fig. 8-8. The twisting motion applied to the axle housing is called rear-end torque. Two different rear-end designs have been used to combat this twisting motion of the axle. These are the Hotchkiss drive and the torque-tube drive (Fig. 8-9).

1. **Hotchkiss drive** In the Hotchkiss drive, the twisting effect, or torque, is taken by the springs. In Fig. 8-8, notice that the spring is firmly attached to the axle housing. The torque applied by the housing to the spring tends to lift the front end of the spring (on the left in Fig. 8-8 or on the right in Fig. 8-9). At the same time, the torque action tends to lower the rear end of the spring. The spring does flex a little to permit a slight amount of housing rotation. This absorbs the rear-end torque, or twisting effort.

2. **Torque-tube drive** The torque-tube drive is now seldom used on cars. It consists of a rigid tube (Fig. 8-9) that surrounds the drive shaft. (The drive shaft carries the engine power from the transmission to the drive-wheel axles.) The rigid tube is attached to the transmission at the front and to the axle housing—actually the differential housing—at the rear. The axle housing, in attempting to rotate, tries to bend the tube, but the tube resists this effort. Therefore, the twisting effort of the housing, or rear-end torque, is absorbed by the torque tube.

❁ 8-9 Rear-end torque and squat When the rear wheels are driving the car, the axle housing tries to rotate in a direction opposite to wheel rotation (Fig. 8-8). This motion is called rear-end torque. In a leaf-spring rear suspension, the leaf springs absorb the rear-end torque. On the coil-spring rear suspension, the control arms absorb the rear-end torque. Later sections cover these two types of rear-end suspension systems.

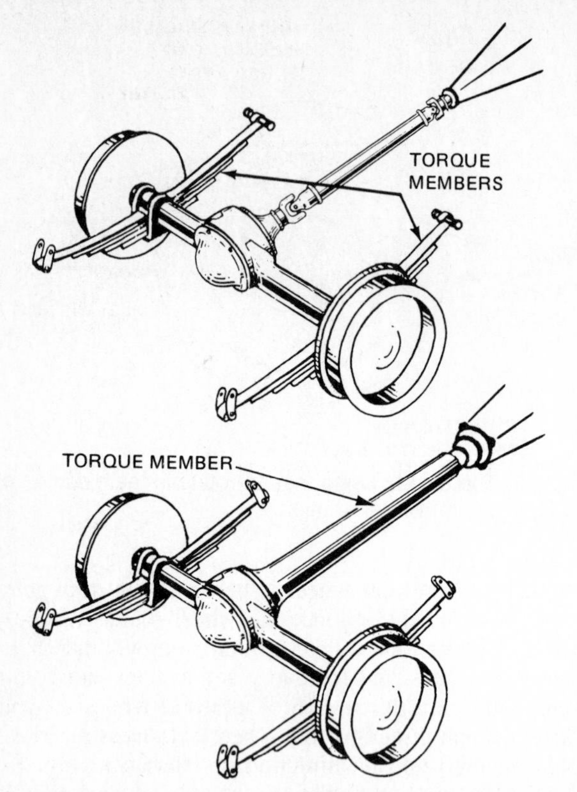

Fig. 8-9 Hotchkiss drive (top) compared with torque-tube drive (bottom).

One effect of rear-end torque is rear-end "squat" on acceleration (Fig. 8-10). When a car is accelerated from a standing start, the drive pinion in the differential tries to climb the teeth of the differential ring gear. As a result, the drive pinion and differential pivot upward, and the springs are twisted and compressed by the differential action. The effect is that the rear end of the car moves down, or squats, when the car is accelerated, and the front end moves up. On deceleration, or braking, the rear of the car moves upward, because of the inertia of the car.

Rear-suspension systems

⚙ 8-10 Coil-spring rear suspension In some rear-suspension systems using coil springs, the springs are placed between spring housings in the car frame and brackets on the rear-axle housing (Figs. 1-1 and 8-11). Figure 8-12 is a close-up view of one coil spring in a similar suspension design. The spring fits between a circular depression in the frame and a bracket mounted on the axle housing.

Figure 8-13 shows the rear-suspension system of the Chevrolet Chevette. It uses torque-tube drive (⚙ 8-8) and coil springs. The purpose of the control arms and track bar is to hold the rear-axle housing in proper alignment with the frame. This prevents the tires from rubbing the body during turns. The axle housing must be permitted to move up and down in relation to the frame. However, the housing must not be allowed to move excessively forward, backward, or sideways with respect to the car frame. The rear stabilizer bar (in Fig. 8-13) prevents excess body roll, or lean-out, on turns.

Control arms permit the rear-axle housing to move up and down as the springs compress or expand. At the same time, the control arms keep the axle in proper alignment with the frame. On some cars, control arms are also used to prevent sideways movement of the axle housing. For example, in Fig. 1-18 the lower control

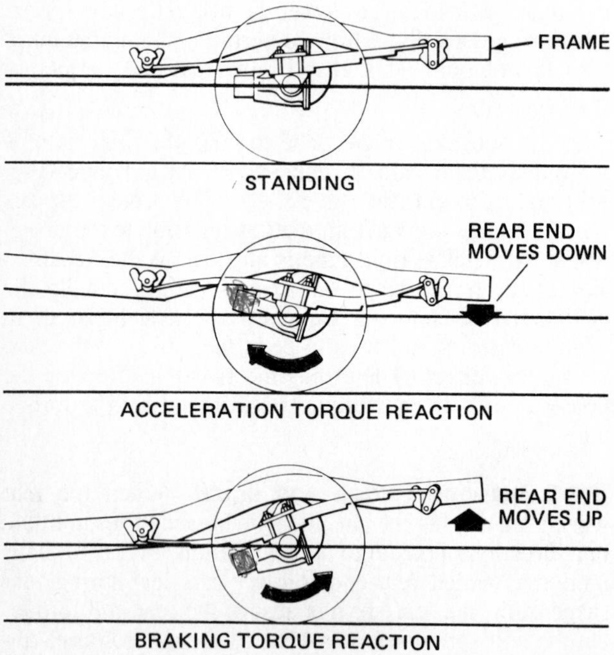

FRAME
STANDING
REAR END MOVES DOWN
ACCELERATION TORQUE REACTION
REAR END MOVES UP
BRAKING TORQUE REACTION

Fig. 8-10 Actions of the spring and rear end when the car is standing (top), accelerated (center), and braked (bottom). (*Ford Motor Company*)

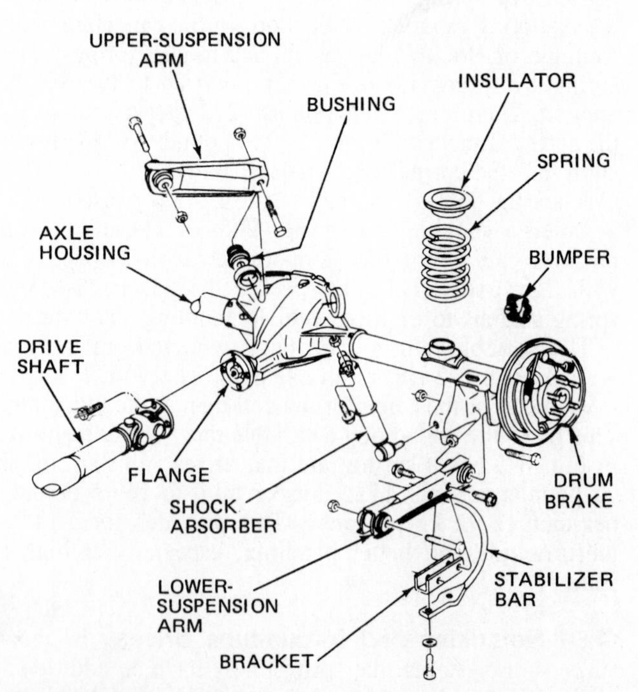

Fig. 8-11 Disassembled left half of a coil-spring rear-suspension system. (*Cadillac Motor Car Division of General Motors Corporation*).

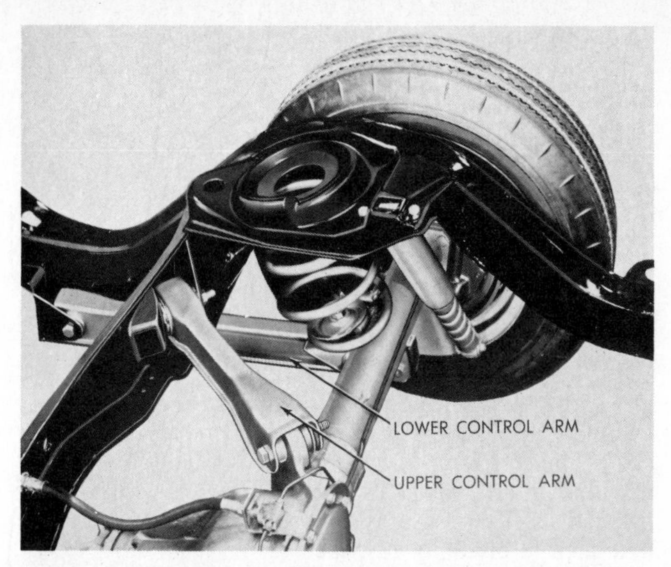

Fig. 8-12 Coil-spring suspension system for the right rear wheel. (*Pontiac Motor Division of General Motors Corporation*)

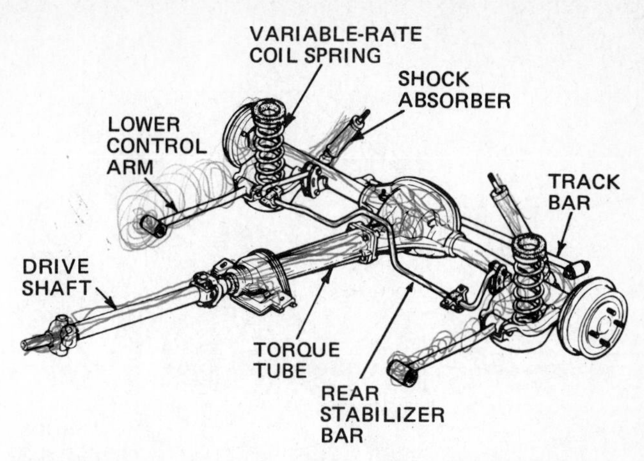

Fig. 8-13 Rear-suspension system using torque-tube drive and coil springs. (*Chevrolet Motor Division of General Motors Corporation*)

arms prevent backward and forward movement. The upper control arms prevent sideways movement.

A variety of coil-spring rear-suspension systems have been used, besides the type shown in Figs. 1-1, 8-12, and 8-13. The design shown in Fig. 8-14 has two sets of control arms and a track bar, with the coil springs positioned between the body and the lower-control arms. In the design in Fig. 8-15, the shock absorbers are centered under the coil springs. The assembly, called a *MacPherson strut,* is positioned between the body and the trailing control arms. In Fig. 8-16, the shock absorbers are centered in the coil springs. The springs are situated between the lower suspension arms and the suspension crossmember, which is part of the car frame.

The coil-spring rear-suspension systems in Figs. 8-13 to 8-16 show the variety of designs used in modern automobiles. Regardless of type, the coil springs must be located between members that move up and down with the wheels, and stationary body or frame members.

The rear-suspension systems shown in Figs. 8-11 to 8-14 are used on front-engine, rear-wheel-drive cars. The drive shaft carries engine power through the differential and axles to the wheels. The differential and wheel axles are enclosed in the axle housing. With this arrangement, the two wheels cannot move independently of each other. If one rear wheel meets a hole or bump, it will move down or up. This causes the rear-axle housing to also move up and down. As a result, some of the up-and-down motion is transmitted through the axle housing to the other wheel. Because of this, these rear-suspension systems are called *semi-independent suspensions*.

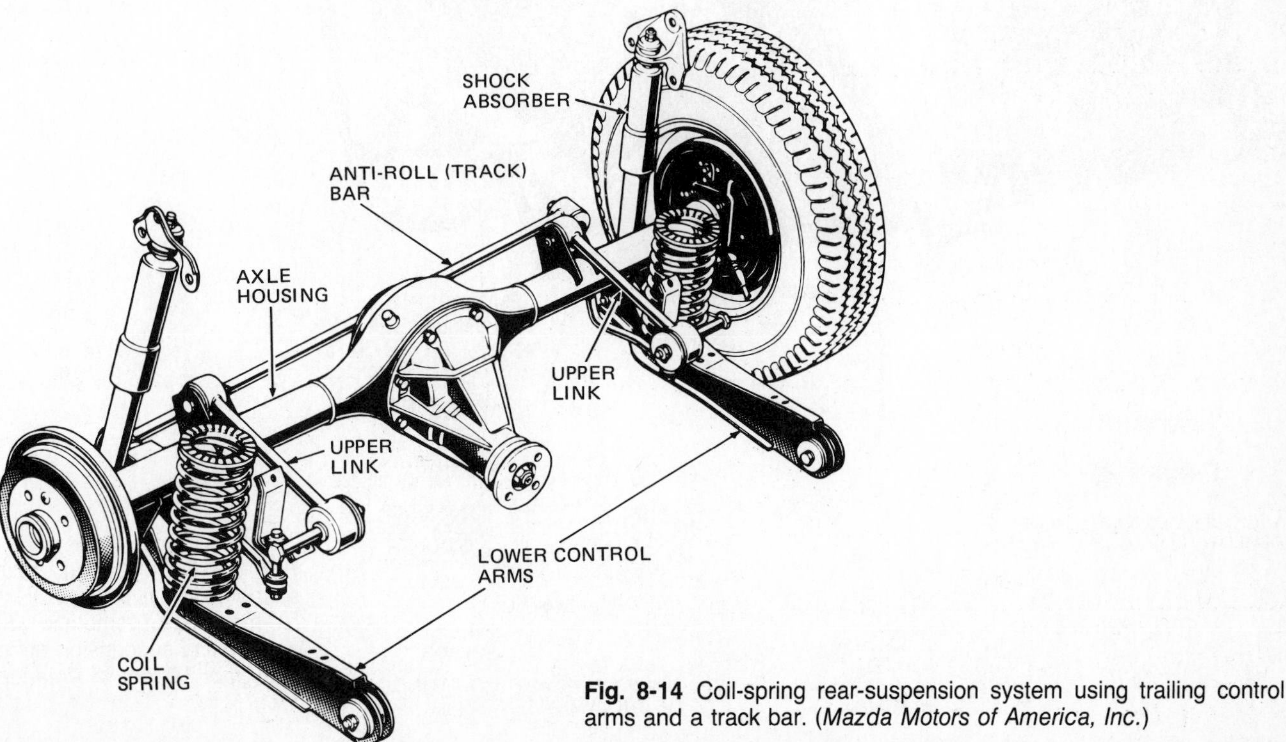

Fig. 8-14 Coil-spring rear-suspension system using trailing control arms and a track bar. (*Mazda Motors of America, Inc.*)

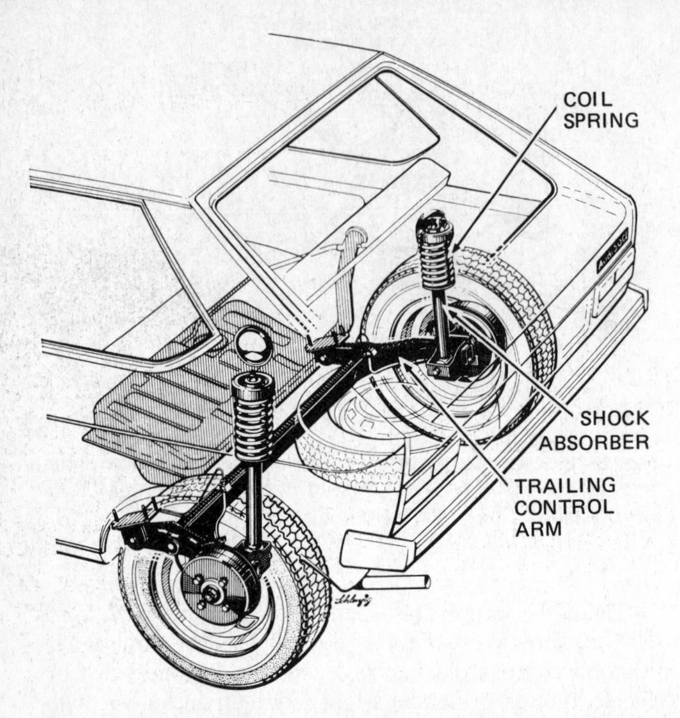

Fig. 8-15 MacPherson-strut rear-suspension system. (*Volkswagen of America, Inc.*)

Figures 8-15 and 8-16 show rear-suspension systems that are fully independent. One rear wheel can move up and down without greatly influencing the other. On these, each wheel has its own suspension arrangement. The system shown in Fig. 8-15 is for a front-wheel-drive car. The system shown in Fig. 8-16 is for a rear-wheel-drive car. The drive shafts that connect the differential to the wheel hubs are flexible. They have universal joints that transmit torque to the wheels even though they are moving up and down. Figure 8-17 shows this arrangement.

8-11 Leaf-spring rear suspension A variety of leaf-spring rear-suspension systems have been used in automobiles. The leaf spring most commonly used is a semielliptical spring. It has the shape of half an ellipse, and that is the reason for its name. In the following illustrations of rear-suspension systems using leaf springs, note their semi-elliptical shape.

Figure 8-3 shows how the leaf spring is installed. The usual method is to attach the spring to the axle housing with two U bolts (Fig. 8-3). The spring is, in effect, hanging from the axle housing. A spring plate or straps are used at the bottom of the spring, as shown in Fig. 8-18. Some installations include insulating strips or pads of rubber to reduce noise transfer from the axle housing to the spring.

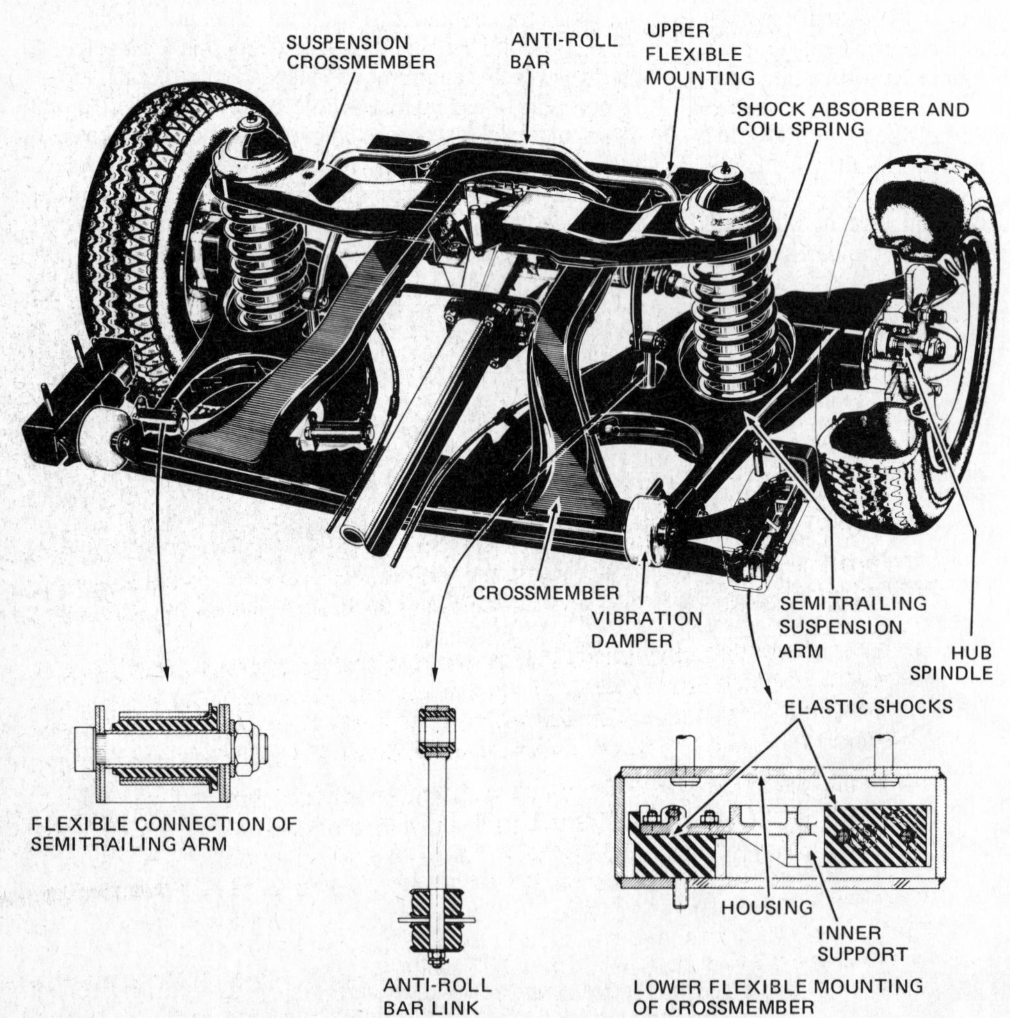

Fig. 8-16 Coil-spring rear-suspension system with flexible-mounted crossmember and semitrailing suspension arms. (*Peugeot Motors of America, Inc.*)

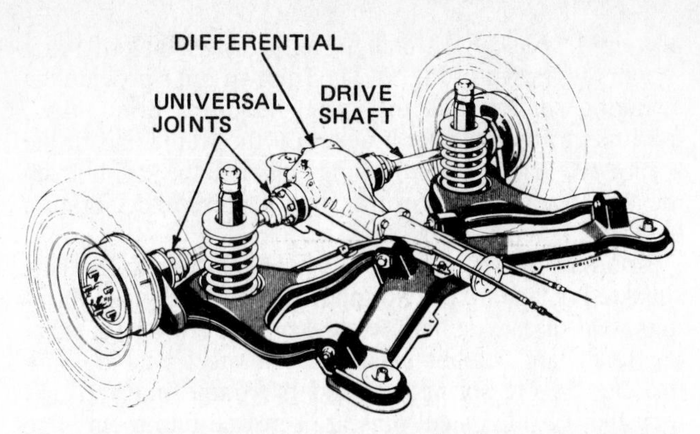

Fig. 8-17 Independent rear-suspension system using coil springs mounted on semitrailing arms. The drive shafts each have two universal joints. (*Ford of Europe, Inc.*)

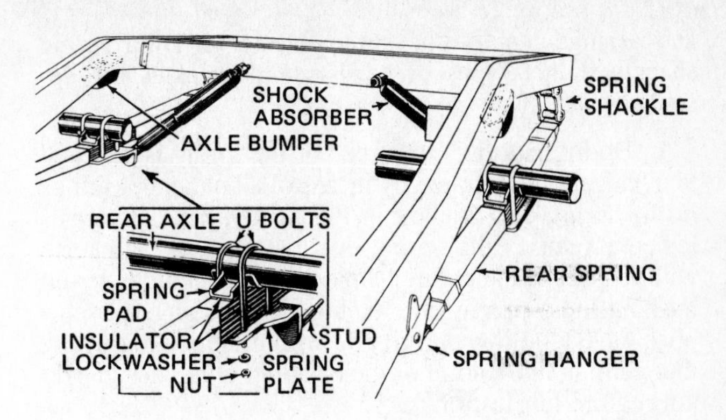

Fig. 8-18 Leaf-spring rear-suspension system showing attachment of the spring to the axle housing.

Leaf-spring rear-suspension systems do not require control arms, as do coil-spring rear-suspension systems. The leaf springs absorb rear-end torque and side thrust that occur when the car is rounding a corner. The exception to this is the transverse leaf spring, such as that used on the Chevrolet Corvette (see ❁ 8-12 for more information).

On some vehicles, the spring is placed on top of the axle housing, rather than under it. This is a common arrangement in trucks (❁ 8-13).

The two ends of the spring are attached to the frame by a spring hanger at the front and by a spring shackle at the rear. A typical rear leaf-spring installation is shown disassembled in Fig. 8-19. Examine the construction

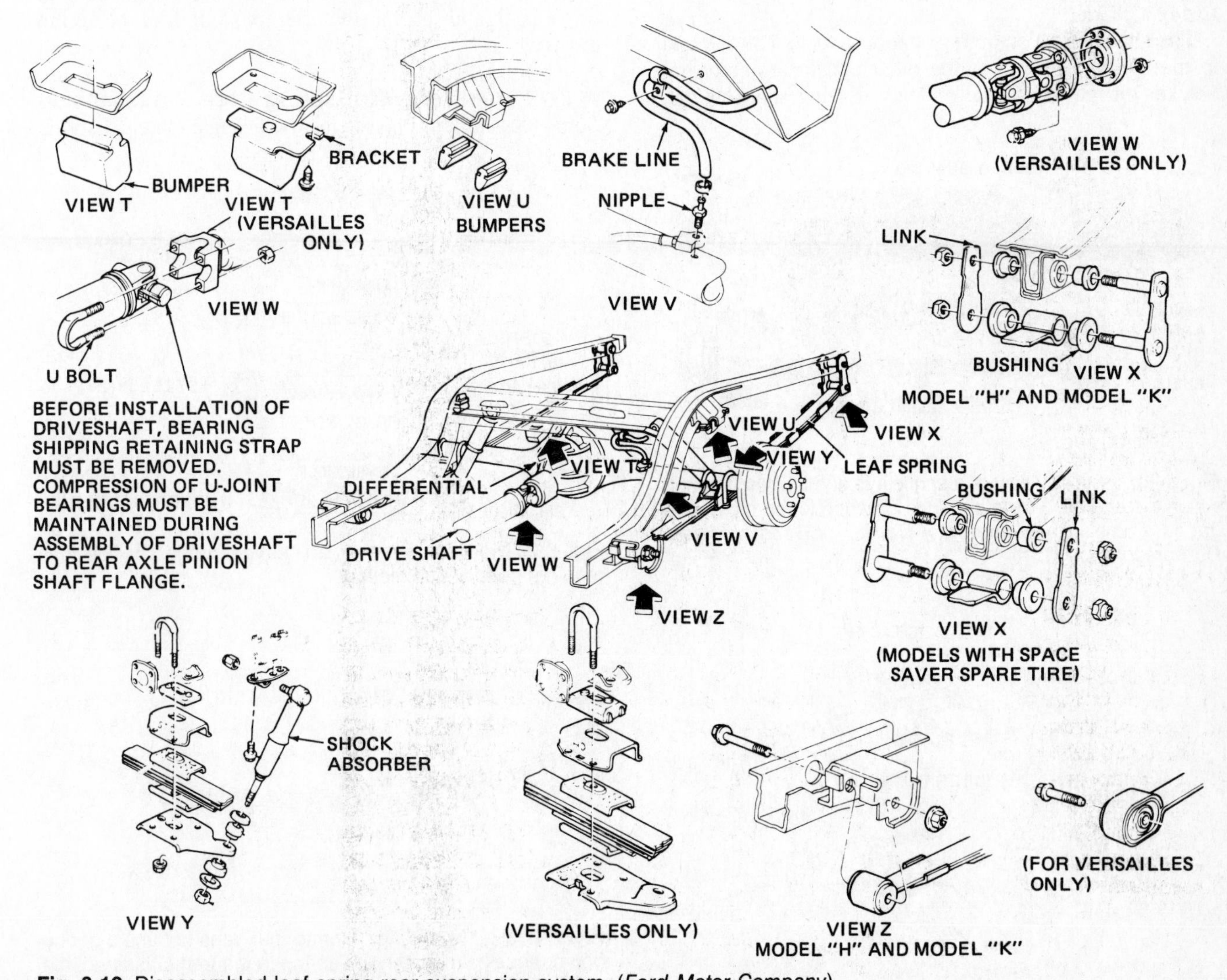

Fig. 8-19 Disassembled leaf-spring rear-suspension system. (*Ford Motor Company*)

and arrangement of the spring hanger and the spring shackle. Other types of hangers and shackles are described later.

1. Spring Hanger One end of the spring is attached to a hanger on the frame by means of a bolt and bushings in the spring eye (view z in Fig. 8-19). The spring, as it bends, causes the spring eye to turn back and forth with respect to the spring hanger. The attaching bolt and bushing must permit this rotation. Some hanger assemblies have a hollow spring bolt with a lubrication fitting that permits lubrication of the bushing. Other designs do not require lubrication.

Some hangers have a bushing made up of an inner and an outer metal shell. Between these two shells is a molded rubber bushing that carries the weight. The rubber also acts to dampen vibration and noise, and prevents them from reaching the car body. Figure 8-20 shows one type of rubber-bushed mounting. No lubrication is required.

2. Spring shackle As the spring bends, the distance between the two spring eyes change. If both ends of the spring were fastened rigidly to the frame, the spring would not be able to bend. To permit bending, the spring is fastened at one end to the frame through a link called a *spring shackle*.

The shackle is a swinging support attached at one end to the spring eye and at the other end to a supporting bracket on the car frame or body. Spring shackles can be seen in Figs. 8-18 and 8-19. A disassembled spring shackle is shown in Fig. 8-21. The two links provide the swinging support that the spring requires. Bolts attach the links to the shackle bracket on the frame and to the spring eye. The rubber bushings insulate the spring from the frame and prevent a transfer of noise and vibration between the two.

Another type of shackle is shown in Fig. 8-22. This shackle is made up of two internally threaded steel bushings; two threaded, hollow-steel pins; a draw bolt; shackle links; and rubber washers. The steel bushings are installed in the spring eye and the frame bracket, and then the threaded steel pins are screwed into them. The shackle links are held on the pins by the draw bolt. The washers will protect the threaded pins from dirt and retain the lubricant in the shackle. Lubricant fittings are provided for each bushing.

While shackles with lubrication fittings require periodic lubrication, rubber-bushed shackles (Fig. 8-21) should not be lubricated. Oil or grease on the bushings may cause them to soften and deteriorate.

In some trucks, no spring shackle or hanger attachment is used. Instead, the two ends of the spring are straight, and they ride on hangers located on the frame, as shown in Fig. 8-23. This permits the spring ends to move back and forth as the effective length of the spring changes.

8-12 Transverse-leaf-spring rear suspension

Some rear-suspension systems use a single multileaf spring

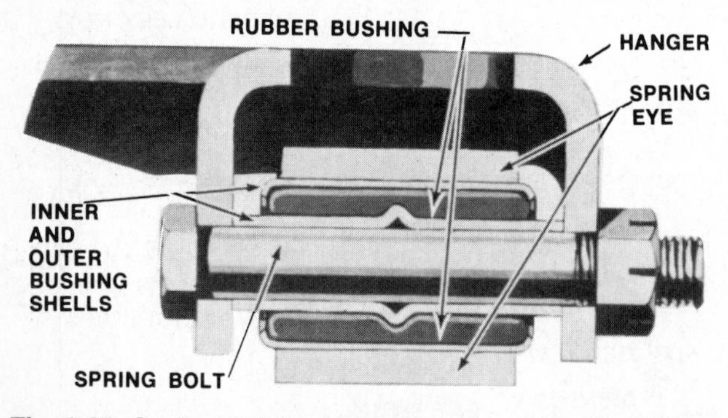

Fig. 8-20 Sectional view of a spring eye and bushing through which the end of the leaf spring is attached to the hanger on the car frame. (*Chevrolet Motor Division of General Motors Corporation*)

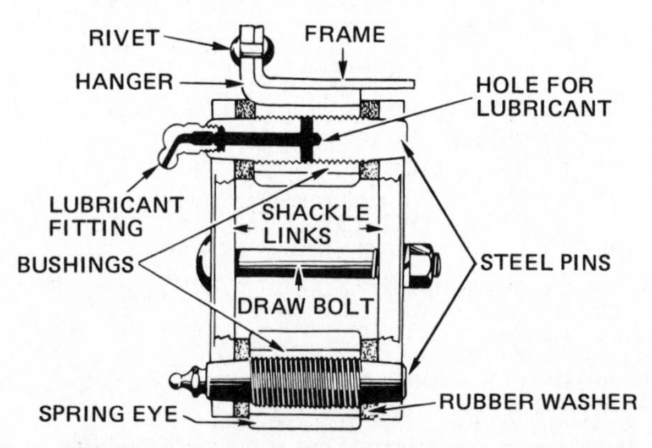

Fig. 8-22 Sectional view of a link-type spring shackle that requires periodic lubrication.

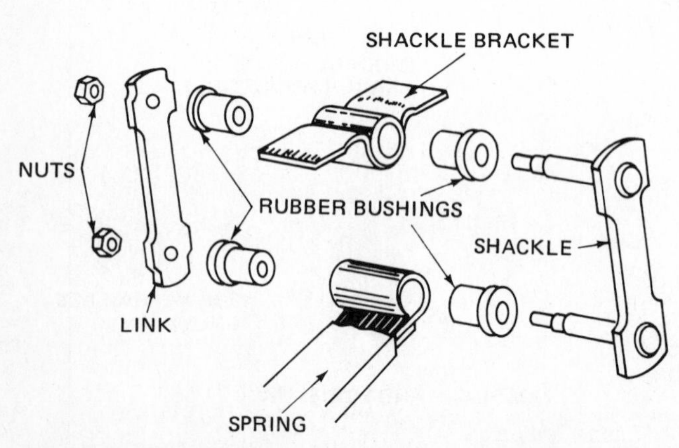

Fig. 8-21 Disassembled rubber-bushed spring shackle.

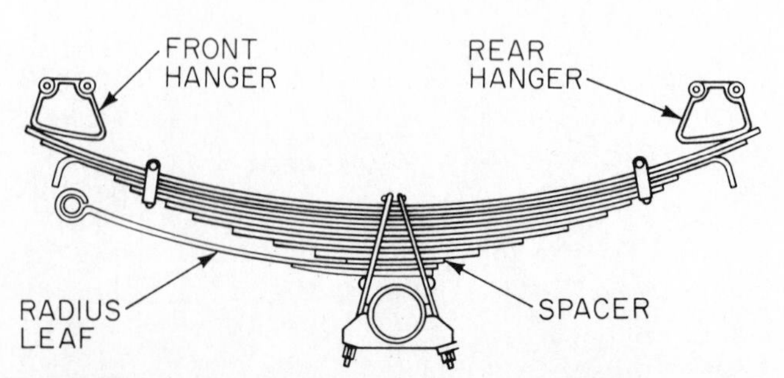

Fig. 8-23 Heavy-duty leaf spring that does not use a spring bolt or shackle to attach the spring to the frame. Instead, the top leaf rides on hangers at the front and rear. (*Chevrolet Motor Division of General Motors Corporation*)

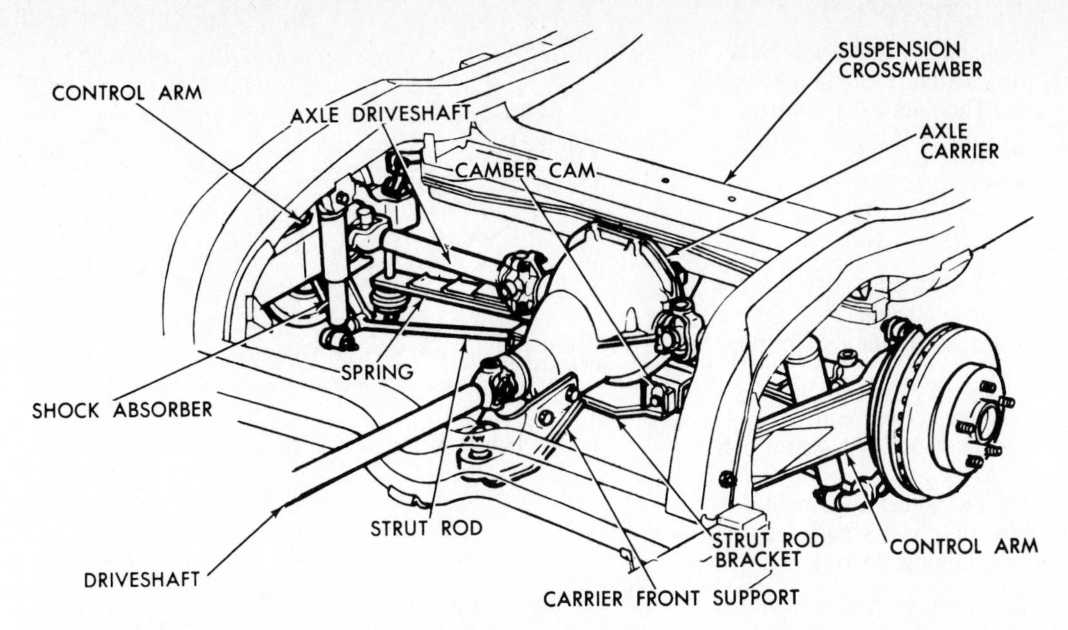

Fig. 8-24 Rear-suspension and power-train components in the Chevrolet Corvette. Note the transverse leaf spring and the drive shafts, which each have two universal joints. (*Chevrolet Motor Division of General Motors Corporation*)

Fig. 8-25 Attachment of the rear leaf spring to the frame and axle housing on a heavy-duty truck. The upper spring is an auxiliary, or helper, spring that comes into use when heavy loads are applied. (*International Harvester Company*)

that is mounted transversely (Fig. 8-24). Each wheel is independently suspended by one end of the spring. In the system shown in Fig. 8-24, each rear wheel is driven by a separate shaft, which includes two universal joints. These universal joints are necessary to permit the power from the engine to be carried through the differential and the shafts to the rear wheels.

❁ 8-13 Heavy-duty rear suspension Figures 8-25 and 8-26 show the spring arrangement used at the rear of heavy-duty trucks. The spring is carried above the axle housing, not slung below it as in most automobile suspension systems. An auxiliary or helper spring is mounted above the main spring.

The helper spring comes into action only when the truck is heavily loaded or when the wheel encounters a large bump. Then, as the main spring goes through a large deflection, the ends of the helper spring meet the two bumpers on the frame. The auxiliary spring deflects and adds its tension to the tension of the main spring.

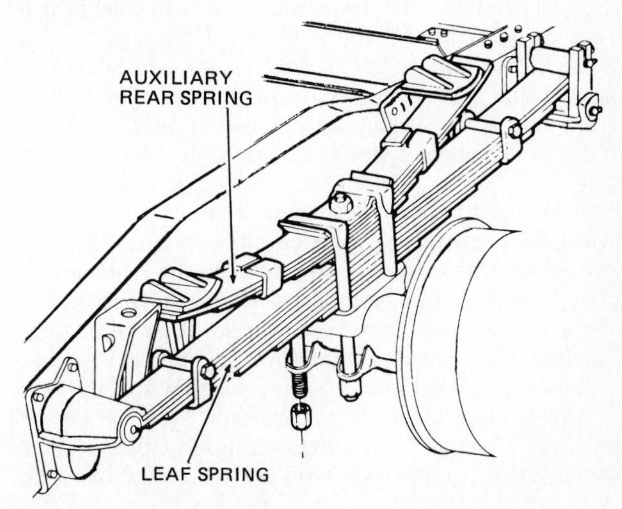

Fig. 8-26 Rear leaf-spring suspension system using an auxiliary spring for a heavy-duty truck. (*Chevrolet Motor Division of General Motors Corporation*)

Some heavy-duty leaf springs are not attached at either end (Fig. 8-23). Instead, the two ends bear on spring hangers attached to the frame. The radius leaf maintains the forward-and-back relationship of the axle with the frame.

Front-suspension systems

⚙ 8-14 Front suspension The suspension system for the front wheels is more complicated than for the rear wheels. This is because the front-suspension system must perform four jobs:

1. It supports the weight of the car's front end.
2. It absorbs road shocks and cushions the passengers against them.
3. It provides steering control and wheel alignment.
4. It reduces the major strains on chassis parts caused by severe braking while maintaining steering control.

Today, all cars use independent front suspension. In this system, each front wheel is independently supported by either a coil spring or a torsion bar. The coil-spring arrangement is the most common. With either type, movement of one front wheel cannot cause a movement of the other front wheel.

To permit the front wheels to swing to one side or the other for steering, each wheel is supported on a spindle which is part of a steering knuckle. The steering knuckle is supported, through ball joints, by upper and lower control arms which are attached to the car frame.

A coil spring can be used with two control arms, or with a strut that includes a built-in shock absorber and only a lower control arm. Because of the way the strut attaches to the body, no upper control arm is needed. This design is called the *MacPherson-strut suspension system* (⚙ 8-23).

The various types of front-suspension systems used on modern cars are discussed in following sections. Chapter 10 covers steering systems.

⚙ 8-15 Types of front-suspension systems There are three types of front-suspension systems. These are the independent front suspension, the twin I-beam front suspension, and the solid-axle front suspension (Fig. 8-27). Today, all cars use some type of independent front suspension. The twin I-beam front suspension and the solid-axle front suspension are widely used in trucks. Each of the three types is described in following sections.

With any type of independent front suspension, each wheel is a separate suspension unit. As a result, it has the ability to conform to variations in the road surface. Also, it tends to maintain a level position regardless of the action of the other front wheel. In addition, independent front suspension improves the ride and handling qualities of the vehicle, while prolonging tire life.

Most independent front-suspension systems use coil springs. There are three basic designs. Other types of independent front-suspension systems use torsion bars.

⚙ 8-16 Independent front-suspension systems with coil springs The three basic designs of independent front-suspension systems with coil springs are:

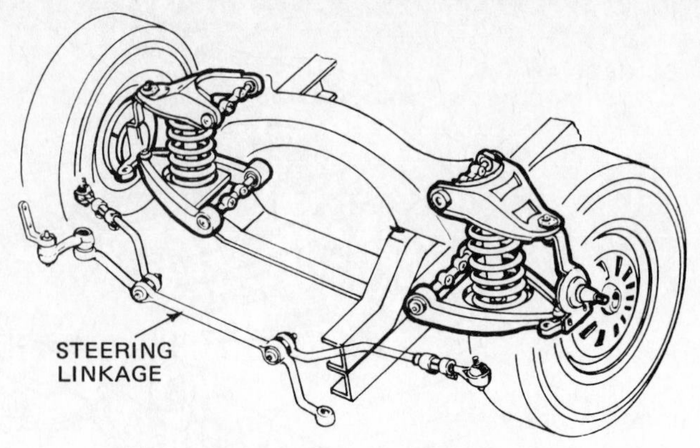

(a) INDEPENDENT FRONT SUSPENSION

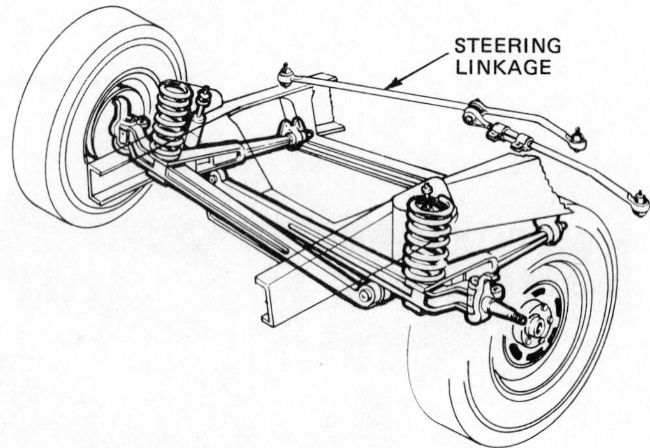

(b) TWIN I-BEAM FRONT SUSPENSION

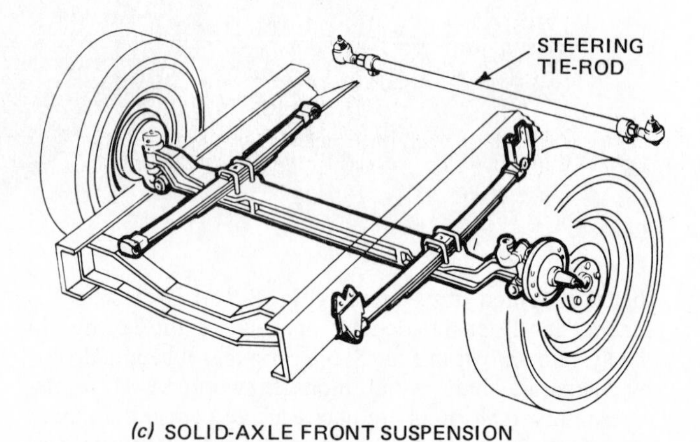

(c) SOLID-AXLE FRONT SUSPENSION

Fig. 8-27 Types of front-suspension systems. (*Moog Automotive, Inc.*)

1. Coil spring between the upper and lower control arms.
2. Coil spring above the upper control arm.
3. MacPherson-strut suspension, which does not use an upper control arm.

The following sections discuss each of these.

⚙ 8-17 Front-suspension system with coil spring between control arms This system is shown in Figs. 8-28 and 8-29. In both of these, the coil spring is between the control arms. The upper end of the coil spring

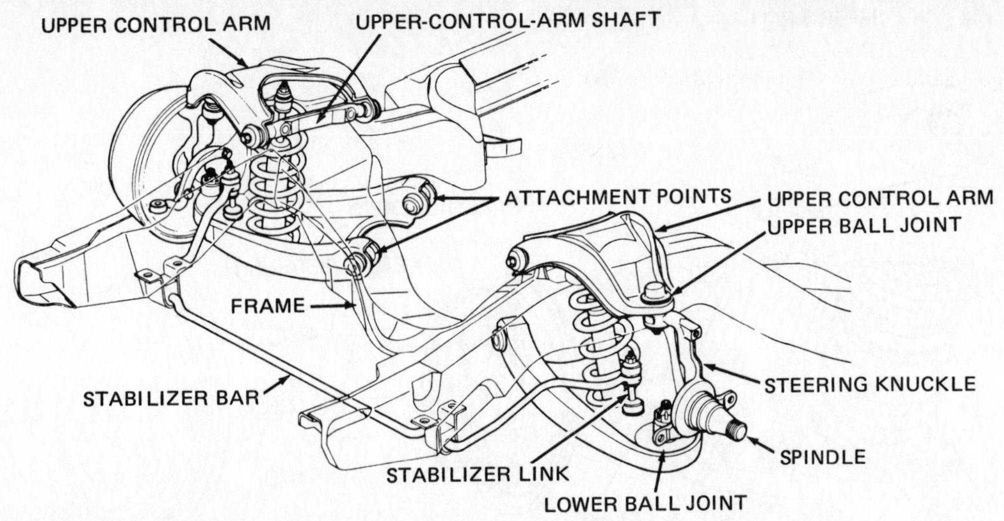

Fig. 8-28 Front-suspension system with the coil springs between the upper and lower control arms. The lower control arm has two pivoting attachment points to the frame. (*Chevrolet Motor Division of General Motors Corporation*)

rests in a pocket in the car frame. The lower end of the coil spring rests on the lower control arm. As the wheel moves up and down, the two control arms pivot up and down as the spring shortens and extends. The shock absorbers, located inside the coil springs, dampen spring oscillation. Shock absorbers are discussed in Chap. 9.

There is a difference between the two front-suspension systems shown in Figs. 8-28 and 8-29. In the system shown in Fig. 8-28, the lower control arm is the wishbone, or "A," type. It is attached to the frame at two points. It does not need any added bracing to keep the outer end from moving forward or back.

In contrast, the lower control arm in Fig. 8-29 is attached at a single point. This control arm is sometimes called a beam-type arm because of its single attachment point to the frame. The design requires extra bracing to keep the outer end of the control arm from swinging forward or backward. The bracing consists of a strut, or brake-reaction rod, which is attached to the outer end of the lower control arm and to the car frame. The frame attachment is flexible so that it does not hinder the up-and-down movement of the lower control arm. But it does prevent the lower control arm from swinging forward or backward. During wheel alignment, caster (⚙ 10-6) is adjusted by changing the length of the strut rod.

The brake assembly is mounted on the spindle, or steering knuckle, as shown in Fig. 8-30. The brake drum and wheel are mounted, by bearings, on the tapered spindle shaft. The wheel can turn freely on the bearings. At the same time, the steering knuckle can swing back and forth on the two ball joints at the outer ends of the control arms. This turns the attached wheel in or out so that the car can be steered. Chapter 10 covers steering systems.

UPPER CONTROL ARM
LOWER-ARM ATTACHMENT POINT
STABILIZER BAR
STEERING KNUCKLE
SPINDLE
STRUT
LOWER CONTROL ARM

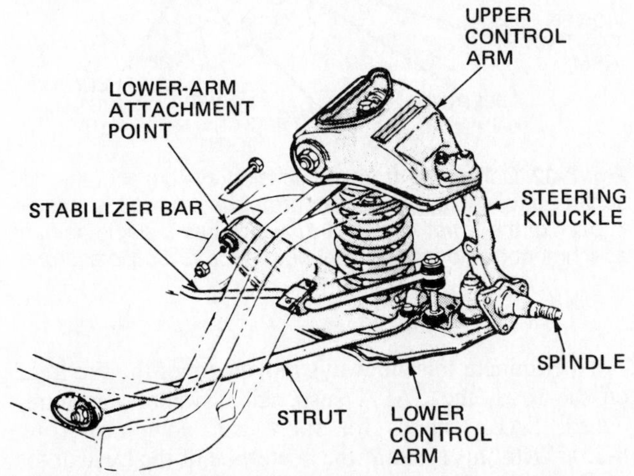

Fig. 8-29 Front-suspension system with the coil spring between the upper and lower control arms. A beam-type lower control arm is used, which is attached to the frame at only one pivoting point. A strut connects the lower control arm to the frame. (*Ford Motor Company*)

⚙ 8-18 Front-suspension system with coil spring above upper control arm Figure 8-31 shows the independent front suspension system which has the coil spring above the upper control arm. The upper end of the spring rests in a spring tower or housing that is part of the body sheet metal. Figure 8-32 shows a similar example of this arrangement. In Fig. 8-32, the strut is almost hidden behind the stabilizer-bar link that attaches the lower control arm to the bar.

The action is the same as with the system with the spring between the two control arms. As the wheels move up and down, the control arms pivot up and down, and the spring shortens and lengthens. The shock absorber, located inside the coil springs, controls wheel and spring action as explained in Chap. 9.

⚙ 8-19 Parallelogram vs. SALA suspension Early independent front-suspension systems had two control arms of about the same length (Fig. 8-33). They were called parallelogram systems because the lines through the four pivot points formed a parallelogram. While this system did permit the two wheels to move up and down

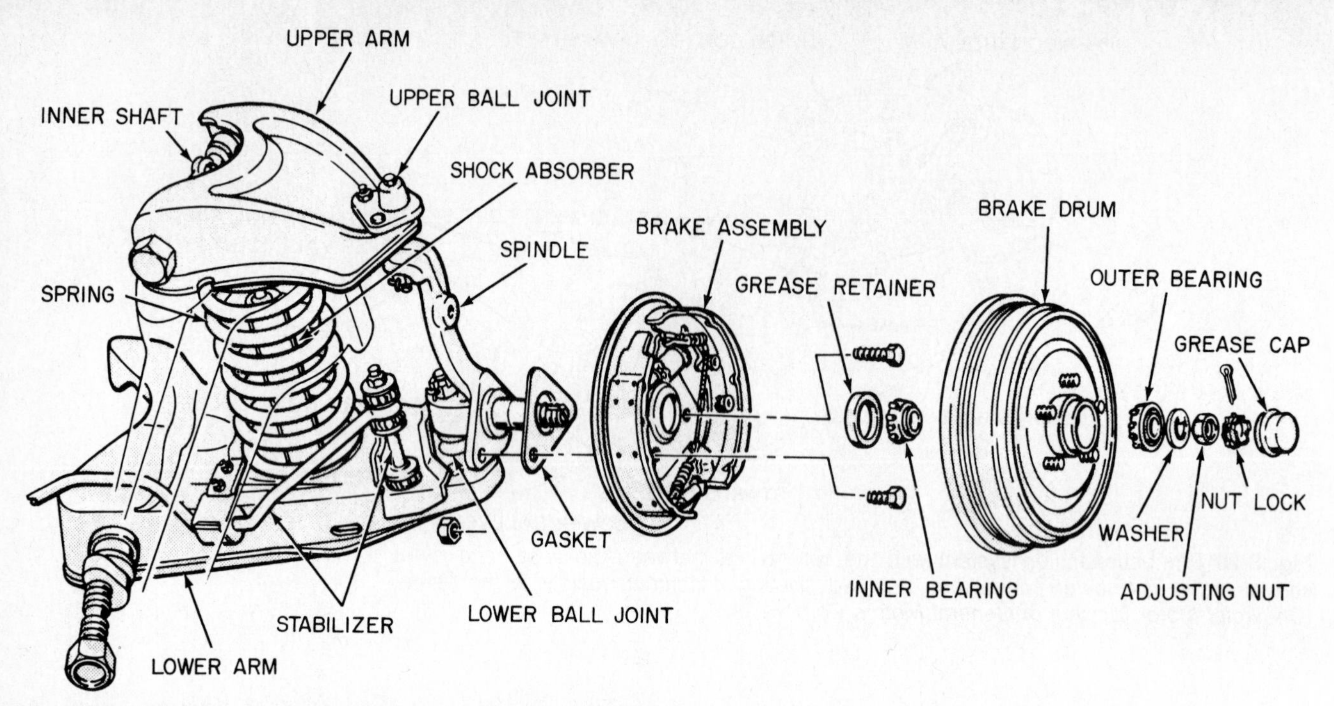

Fig. 8-30 Coil-spring front suspension at one wheel, showing the brake assembly and drum attachment to the spindle, or *steering knuckle.* (*Ford Motor Company*)

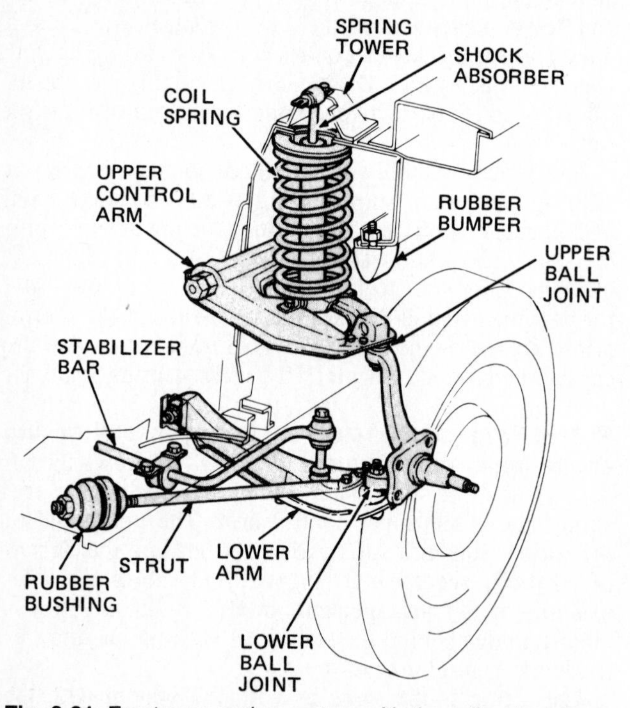

Fig. 8-31 Front-suspension system with the coil spring above the upper control arm. The lower control arm has a single point of attachment to the frame and requires the use of a strut. (*Ford Motor Company*)

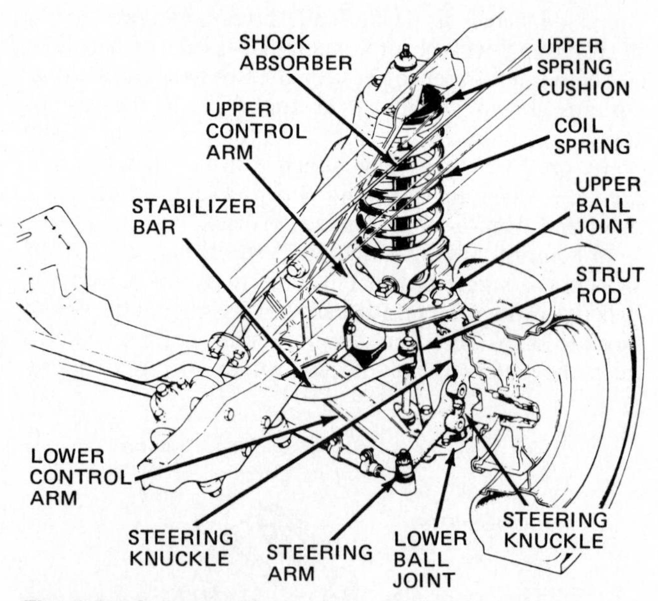

Fig. 8-32 Coil-spring front-suspension system with the coil spring above the upper control arm. The strut rod is located in back of the lower control arm, which has a single point of attachment to the frame. (*American Motors Corporation*)

independently, as shown in Fig. 8-34, there was a drawback. Whenever a tire encountered a bump in the road, the wheel and tire moved upward and inward (Fig. 8-34). This inward movement would drag the tire sideways and cause rapid wear of the tire tread.

To eliminate this sideways movement of the tire tread on the road, the SALA suspension system was introduced. SALA stands for *s*hort *a*rm, *l*ong *a*rm (Fig. 8-35). With this system, the centerline of the tread at the road remains the same as the tire moves up and down (Fig. 8-36). Therefore, there is no sideways sliding of the tread on the road as the tire meets bumps and holes.

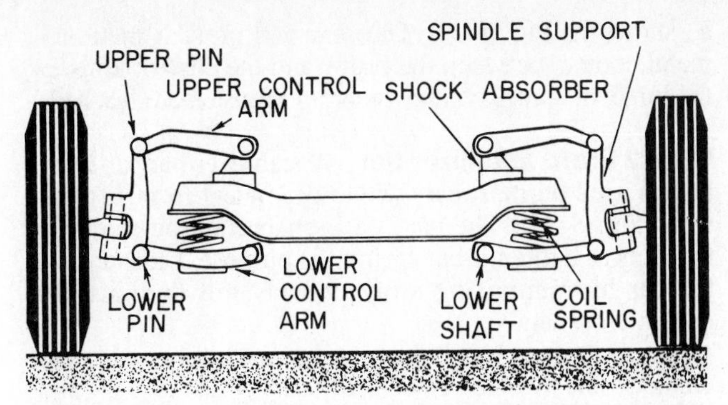

Fig. 8-33 Front-suspension system in which both upper and lower control arms are the same length. (*Ammco Tools, Inc.*)

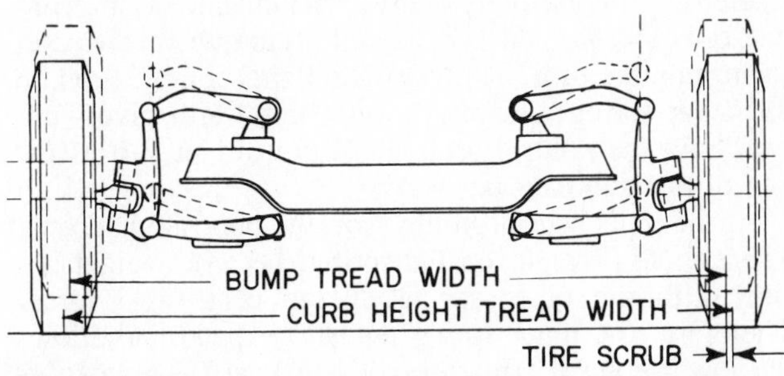

Fig. 8-34 When both control arms are the same length, upward movement of the wheel also causes the wheel to move inward. This forces the tire to slide sideways on the road. (*Ammco Tools, Inc.*)

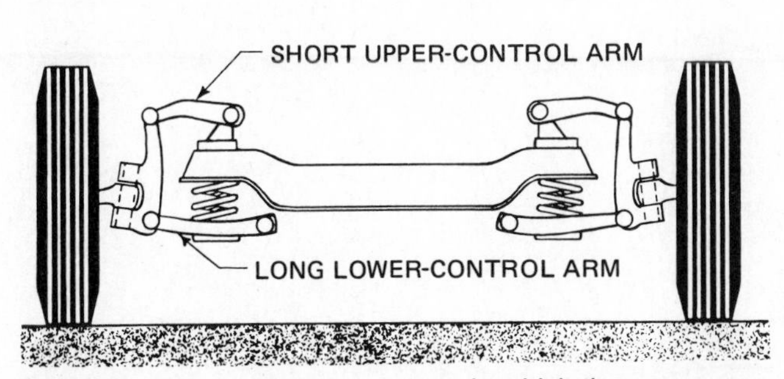

Fig. 8-35 Front-suspension system in which the upper control arm is shorter than the lower control arm. (*Ammco Tools, Inc.*)

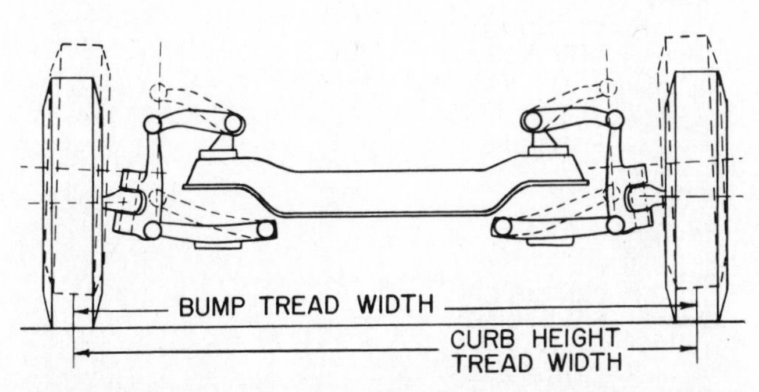

Fig. 8-36 When the upper control arm is shorter, as shown, the tread width remains the same even though the wheel moves up and down. However, the top of the tire moves inward as it moves up. (*Ammco Tools, Inc.*)

Another effect is also important. The top of the tire moves inward whenever the wheel moves up or down (Fig. 8-36). When the wheel tips inward, the camber of the wheel changes. Camber is the inward or outward tilt of the wheel (viewed from the front of the car). A wheel that is perfectly vertical (not tilted inward or outward) has zero camber. When the top of the wheel tilts inward, the wheel has negative camber. If it tilts outward, it has positive camber. The SALA suspension system shown in Fig. 8-35 gives the wheel negative camber as it moves up or down. There is more on camber in Chap. 13.

8-20 Moving instant center outside car The instant center is the point of intersection of lines drawn through the control-arm attachment points for any one position of the wheel. Earlier-model cars were designed so that the instant center was inside the car (Fig. 8-37). In this design, the wheel takes on negative camber by tilting inward as it moves upward.

A later design places the instant center outside the car. In this design, the wheel takes on a positive camber and tilts outward as it moves upward. Figure 8-38 shows how this affects car stability. When a front wheel meets a bump, the later design causes the wheel to tilt outward as it moves up (Fig. 8-38). The effect of this is to impart an outward thrust at the road which reduces the side effect of the bump (which tends to push the car to the left in Fig. 8-38).

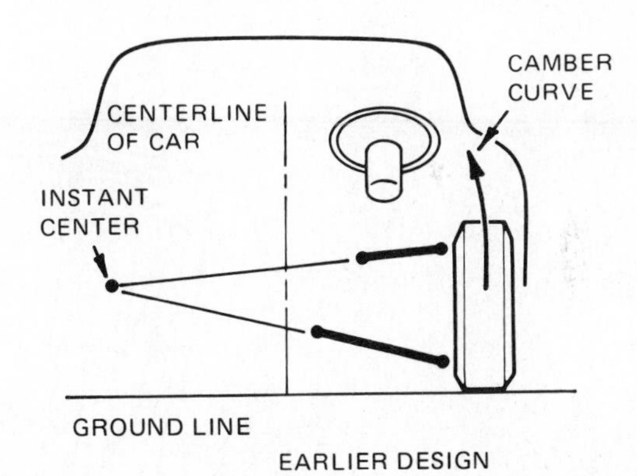

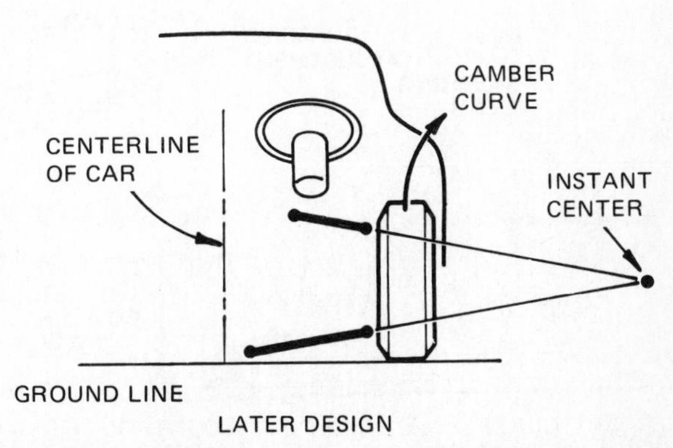

Fig. 8-37 Comparison of earlier and later front-suspension systems. In the later design, the inner attachment points to the control arms have been moved apart so that the instant center is outside the car. (*Buick Motor Division of General Motors Corporation*)

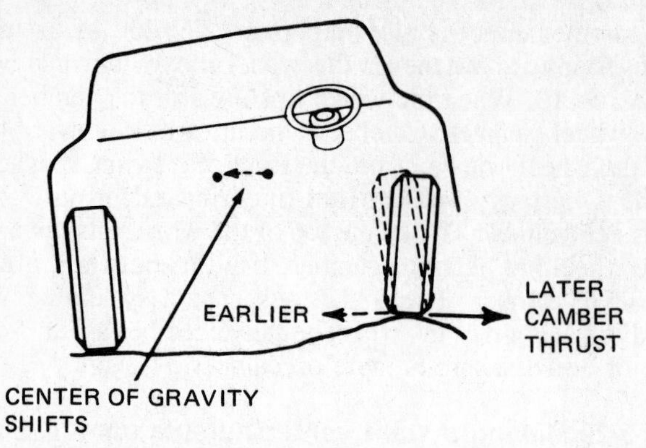

Fig. 8-38 With the instant center outside the car, the center of gravity shifts as the wheel moves up. (*Buick Motor Division of General Motors Corporation*)

Stability is improved as the car rounds a curve. Centrifugal force throws more of the weight on the outer wheel, so that this wheel moves up. As it does so, it tilts outward as shown in Fig. 8-38. The tire opposes the centrifugal force that is trying to push the car sideways. In addition, the contact point of the tire with the road shifts inward as the tire moves up. Also, the center of gravity of the car shifts into the curve to help counteract the centrifugal force pushing the car outward.

8-21 Rubber bumpers Rubber bumpers are attached to the frame or crossmember and a control arm, as shown in Fig. 8-31. The bumpers prevent metal-to-metal contact between the frame and the control arms as the limits of spring compression or expansion are reached.

8-22 Front stabilizer bar A stabilizer bar, or sway bar, is used on most cars with independent front suspension (Fig. 8-31). The bar is a U-shaped spring-steel rod that acts as a torsion-bar spring. Its purpose is to stabilize the car by minimizing sway or body roll during turns and in crosswinds.

The stabilizer bar is mounted transversely so that it connects the two lower control arms. Figures 8-28 and 8-31 show how the bar is attached to the lower control arms and pivoted at two points on the frame.

When the car is moving around a curve, centrifugal force tends to keep the car moving in a straight line. Therefore, the car body leans to the outside. With lean-out, or body roll, additional weight is thrown on the outer springs on the turn. This puts additional compression on the outer spring, and the lower control arm pivots upward. As the control arm pivots upward, it carries the end of the stabilizer bar with it.

At the inner wheel on the turn, the opposite happens. There is less weight on the spring, because weight has shifted to the outer spring due to centrifugal force. Therefore, the inner spring tends to expand and allows the lower control arm to pivot downward. This carries the end of the stabilizer bar downward.

During the turn, the outer end of the stabilizer bar is carried upward by the outer control arm. The inner end

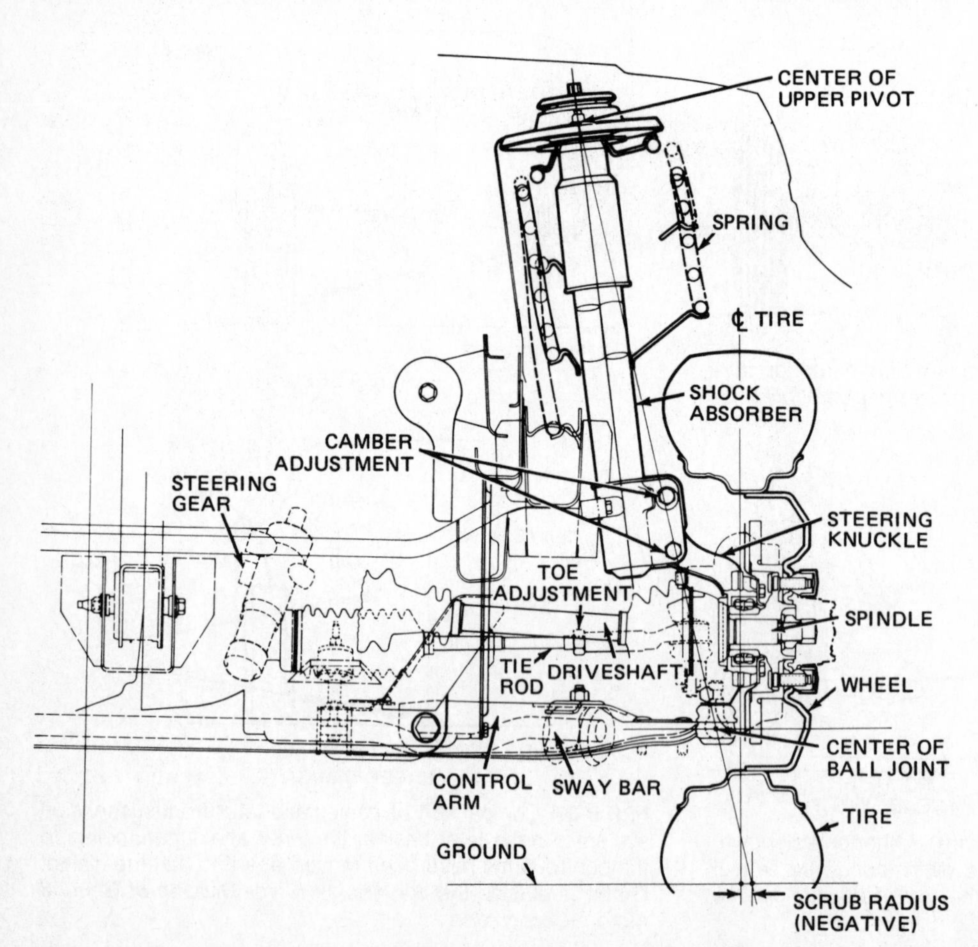

Fig. 8-39 Sectional view of the MacPherson-strut suspension system for the left front wheel. (*Chrysler Corporation*)

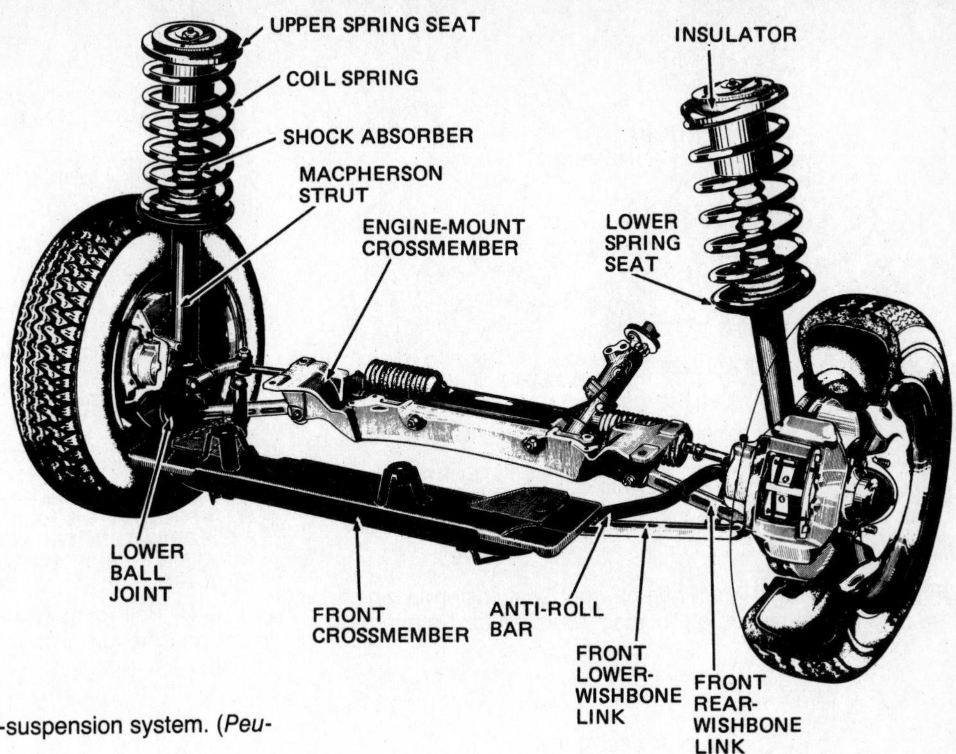

Fig. 8-40 MacPherson-strut front-suspension system. (*Peugeot Motors of America, Inc.*)

of the stabilizer bar is carried downward by the inner control arm. This combined action twists the stabilizer bar. The resistance of the bar to twisting fights the tendency of the car body to lean out on turns. Therefore, there is less body roll, or lean-out, than there would be without the stabilizer bar. The stabilizer bar works the same way as a torsion bar in a torsion-bar front-suspension system (❁ 8-25).

When both front springs expand or compress by the same amount, such as when the front end hits a dip in the road, the stabilizer bar is ineffective. It merely moves up and down with the lower control arms. Only when the left and right control arms move unequally does the stabilizer bar come into action.

❁ 8-23 MacPherson-strut front suspension The MacPherson-strut front suspension has become increasingly popular in smaller cars due to its simplicity, low weight and cost, and compactness. It has been used for years in many foreign cars. Figures 1-1, 1-3, and 8-39 to 8-41 show front-suspension systems using the MacPherson strut. This type of front suspension has no upper control arm. Instead, the top of the strut is fastened to the front-end sheet metal.

The strut combines the shock absorber, strut, and spindle into a single assembly unit (Fig. 8-40). The top of the strut mounts to the car body through an insulator or thrust bearing (Fig. 8-41). At the lower end of the strut, a ball joint attaches the spindle to the lower control arm.

A shock absorber is built into the upper section of the strut, and the coil spring fits around the shock absorber. The lower spring seat is welded to the lower section of the strut, which moves up and down with the lower control arm. The upper spring seat is fastened to the shock-absorber piston rod.

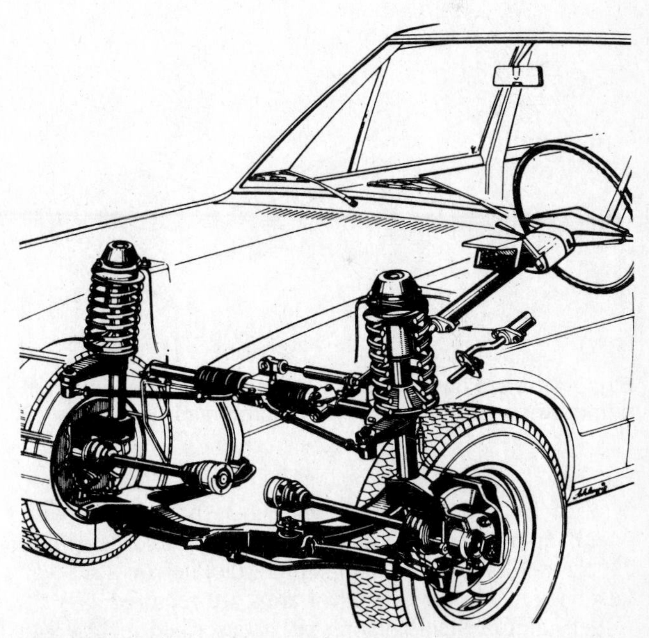

Fig. 8-41 Phantom view of the front end of a car, showing the MacPherson front-suspension and steering systems. (*Volkswagen of America, Inc.*)

Figure 8-42 shows a modified version of the MacPherson front suspension. The spring is located between the lower control arm and the body, instead of being mounted directly on the strut.

MacPherson struts are also used in rear-suspension systems in many small cars, which usually have the same

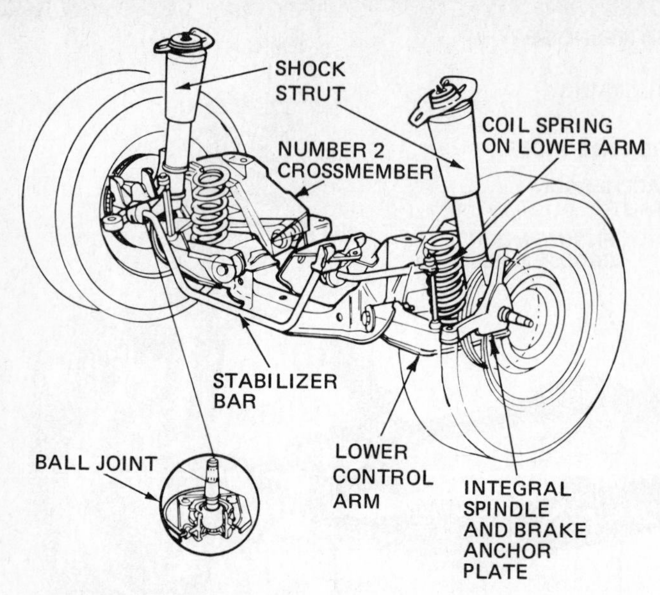

Fig. 8-42 Modified MacPherson front suspension in which the springs are separately located. (*Ford Motor Company*)

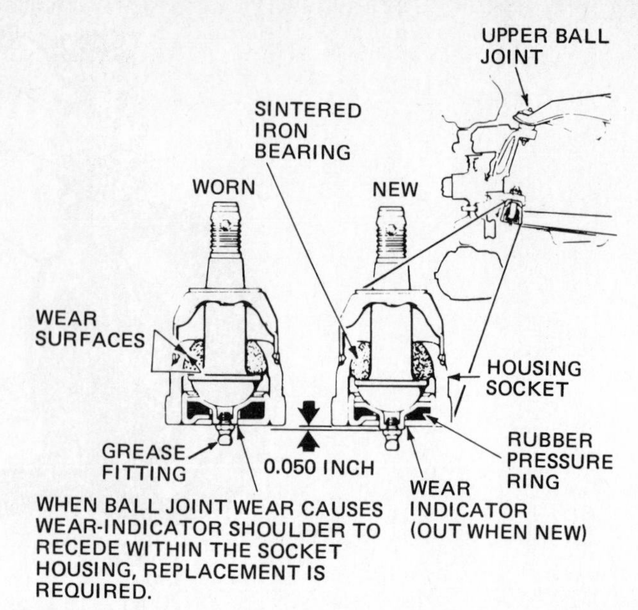

Fig. 8-44 Ball joint, showing how it attaches the steering knuckle to the lower control arm, and how to read the wear indicator.

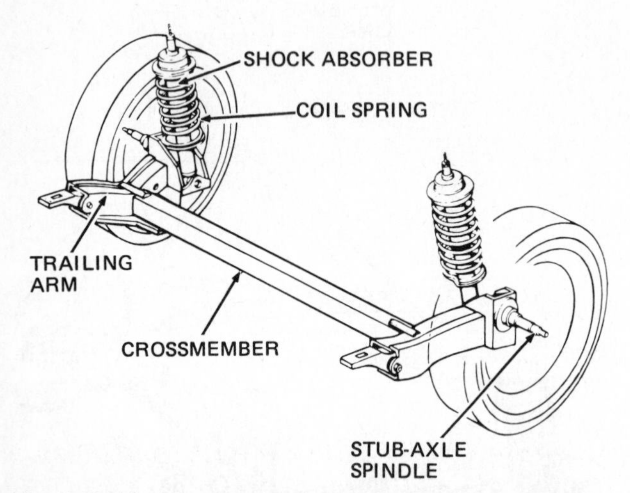

Fig. 8-43 MacPherson struts used in the rear suspension of a front-wheel-drive car. (*Chrysler Corporation*)

type of front suspension. Figure 1-3 shows a car with MacPherson struts in both the front and rear suspensions. Figure 8-43 shows the details of a MacPherson-strut rear suspension. No upper control arms are required.

Servicing a MacPherson strut is described in ✪ 9-14.

✪ 8-24 Ball joints A ball joint is a flexible pivot consisting basically of a ball and a socket (Fig. 8-44). Ball joints are shown attaching the spindle to the control arms in Figs. 8-40, 8-42, and 8-44. The use of ball joints allows the spindle to swing forward and backward for easy steering. They also pivot up and down slightly as the vehicle moves over the road, causing the control arms to move up and down.

In the ball joint, the ball is a heavy steel ball or half-ball formed on the end of a tapered stud. The ball is seated in a steel socket. There is a nylon bearing between the ball and the socket. This allows the ball to turn freely. In some ball joints, a spring keeps pressure on the ball to take up any play. The ball joint is sealed by a rubber dust cover that keeps out dust and water, and keeps in lubricant. On many cars, the ball joint has a grease fitting (Fig. 8-44) or plug that is removed so the ball joint can be lubricated.

Usually, the ball stud is bolted to the steering knuckle, and the socket is attached to the control arm. The socket may be pressed, welded, or riveted to the control arm. The ball joints must be tight. Any looseness will cause poor steering and handling. How to check ball joints for wear, and how to replace them if they are excessively worn are covered in later chapters.

✪ 8-25 Torsion-bar front suspension Torsion bars are attached firmly at one end to the body or frame, and at the other end to the lower control arm. As the wheel and arm move up and down, the control arm pivots and causes the torsion bar to twist or untwist. The twist-untwist action provides the springing.

There are two types of torsion-bar front-suspension systems: longitudinal and transverse. Longitudinal torsion bars run along the length of the car as shown in Fig. 8-6 and 8-45. Transverse torsion bars run across the car, from one side to the other, as shown in Figs. 1-17 and 8-7.

Figure 8-6 shows how longitudinal torsion bars fit into a front-wheel-drive car. Figure 8-45 is a similar partly disassembled torsion-bar front suspension system. In many of the new smaller cars, the torsion bars are installed transversely, or from one side to the other (Fig. 8-46). One end of the torsion bar is fastened rigidly to a frame member. The other end is fastened to the lower control arm. The action is the same as with longitudinal torsion bars. As the lower control arm pivots up and down, the torsion bar is twisted more or less.

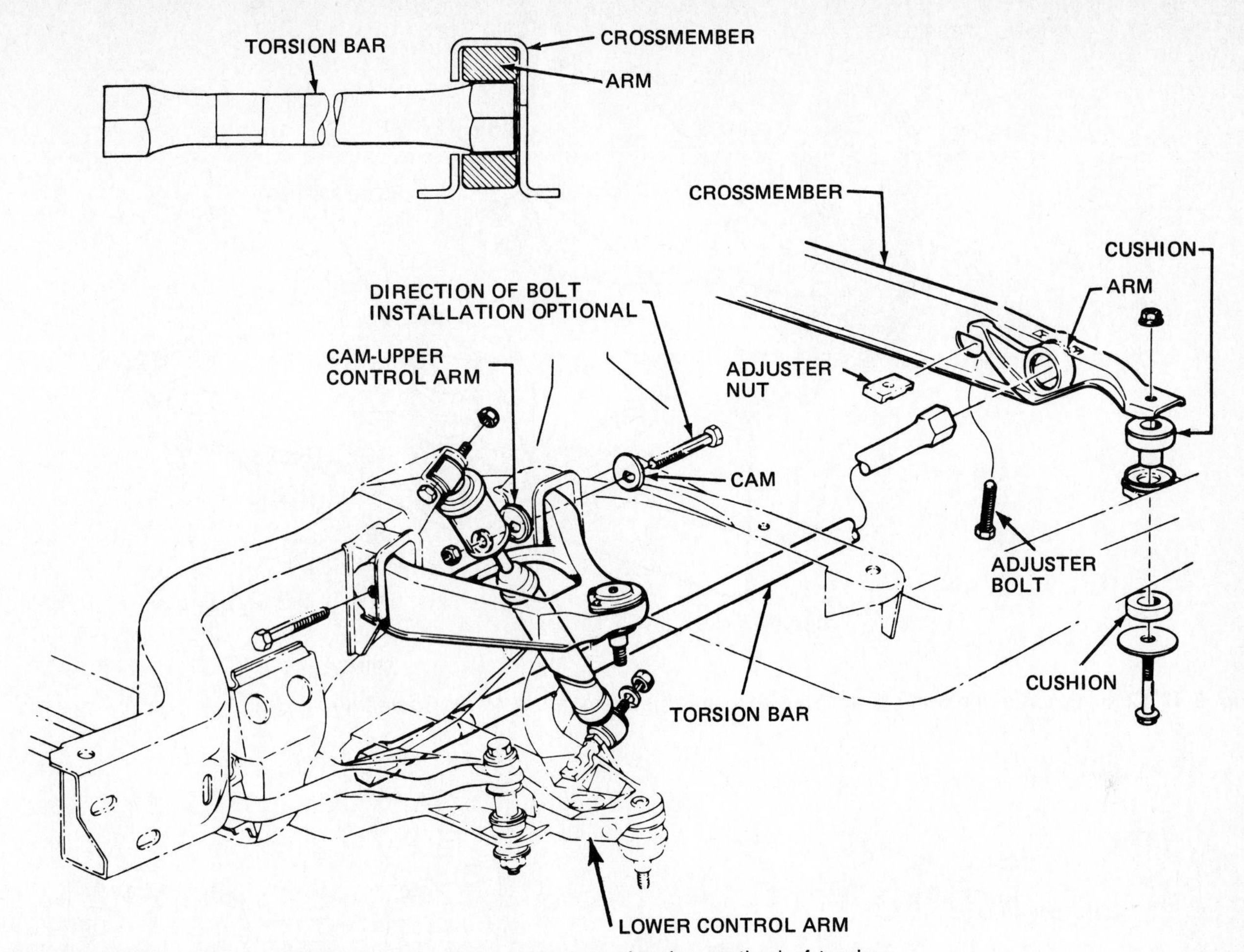

Fig. 8-45 Front suspension using a longitudinal torsion bar, showing method of torsion-bar attachment. (*Cadillac Motor Car Division of General Motors Corporation*)

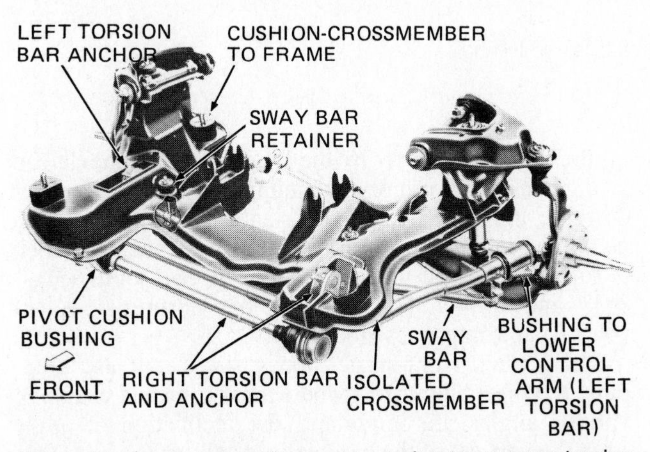

Fig. 8-46 Front-suspension system using transverse torsion bars. (*Chrysler Corporation*)

Torsion-bar suspension systems include a means of height adjustment. A sag or front-end height change can occur due to a change in the torsion-bar characteristics, or due to a change in the front-end load. If this happens, height correction can be made by turning an adjusting bolt. On some cars, the adjustment is made at the end of the torsion bar, where the bar is attached to the frame crossmember (Figs. 8-7 and 8-45). On other cars, the adjustment is made at the lower control arm, where the torsion bar is attached to the control arm.

8-26 Twin I-Beam front suspension Figures 8-27*b* and 8-47 show the twin I-beam front-suspension system used on many Ford trucks. Each front wheel is supported at the end of a separate I beam. The opposite ends of the I beams are attached to the frame through flexible pivots. The wheel ends of the two I beams are attached to the frame by radius arms which prevent backward or forward movement of the wheels. The arrangement provides adequate suspension flexibility with the added strength of the I-beam construction.

Some four-wheel-drive trucks combine the twin I-beam design with independent front suspension (Fig. 8-48). In this system, the front drive-axle housing is made in two parts with a pivot point to the right side of the differential. When one wheel strikes a bump, that side of the axle housing "bends up" around the pivot point. This reduces the interaction between the front wheels.

The differential is linked to the right front wheel with a drive shaft that has a universal joint at each end. This

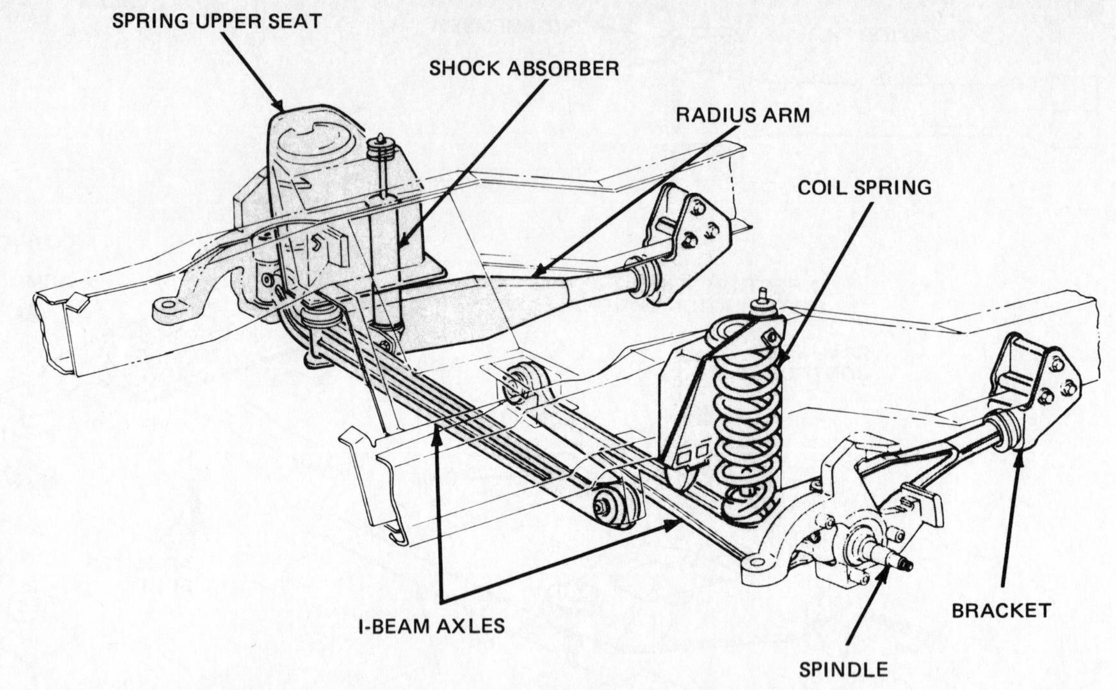

Fig. 8-47 Construction of the twin I-beam front-suspension system. (*Ford Motor Company*)

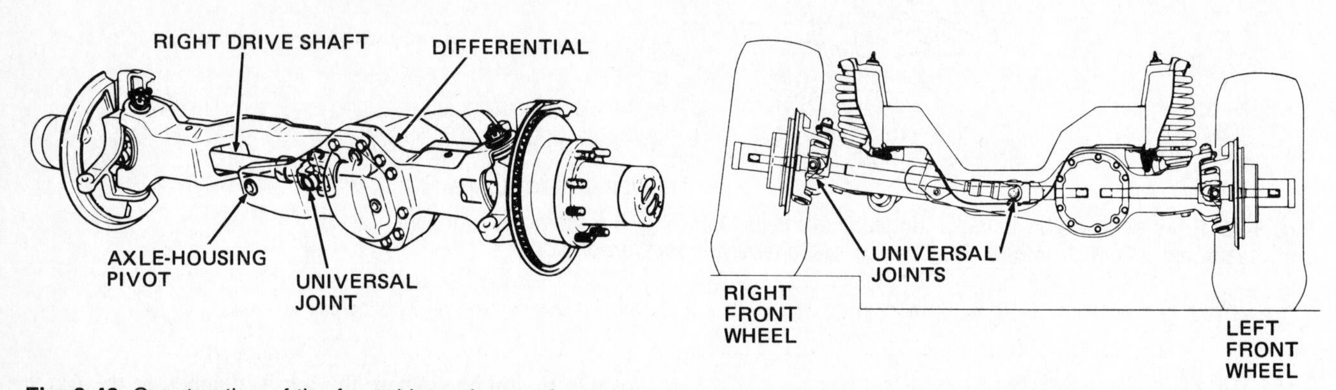

Fig. 8-48 Construction of the front drive axle on a four-wheel-drive truck which has independent front suspension. (*Ford Motor Company*)

permits the drive shaft to move up and down while driving the wheel.

⚙ 8-27 Mono-beam front suspension This arrangement, shown in Fig. 8-49, is used on some Ford four-wheel-drive vehicles. The system has two radius arms, similar to those used in the twin I-beam front suspension (Fig. 8-49). They maintain front-to-rear alignment of the front wheels. The coil springs permit the wheel axles and wheels to move up and down as the wheels meet irregularities in the road. The track bar is attached to the bracket on the left frame side member and to the right side of the axle housing. It maintains lateral stability and axle alignment by preventing the wheels from shifting from side to side.

⚙ 8-28 Kingpin Earlier passenger-car front-suspension systems used a kingpin (Fig. 8-50). In this design, the knuckle support is attached by pivots, top and bottom, to the two control arms. The knuckle is supported on the knuckle support by the kingpin. The knuckle can swing back and forth on the kingpin for steering. In later designs, the support, kingpin, and knuckle have been combined into a single part, the steering knuckle, which is supported at the top and bottom by ball joints (Figs. 8-28 and 8-44). This reduces the unsprung weight (⚙ 8-6) and improves ride quality.

Many truck front-suspension systems still use kingpins. In wheel alignment, where measuring and changing various angles are important, the inclination or angle from the vertical of the kingpin is usually measured. This is called *kingpin inclination* (KPI). It is the centerline around which the front wheel swings for steering.

When ball joints are used instead of a kingpin, the steering-axis inclination (SAI) is measured. The steering axis is an imaginary line drawn through the centers of the two ball joints. The front wheel swings around this line for steering.

Compare Fig. 8-30, which shows ball joints in the suspension system, with Fig. 8-50, which shows the use

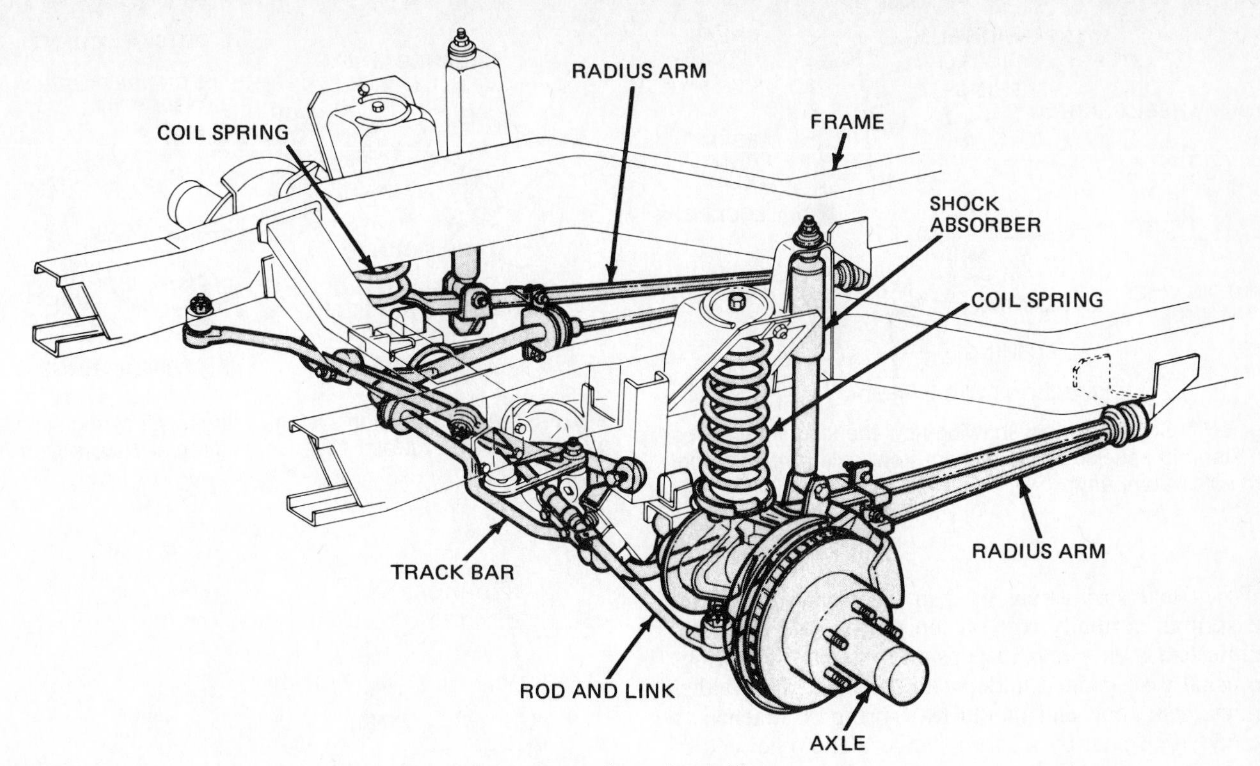

Fig. 8-49 Mono-beam front suspension, as used on some four-wheel-drive vehicles. (*Ford Motor Company*)

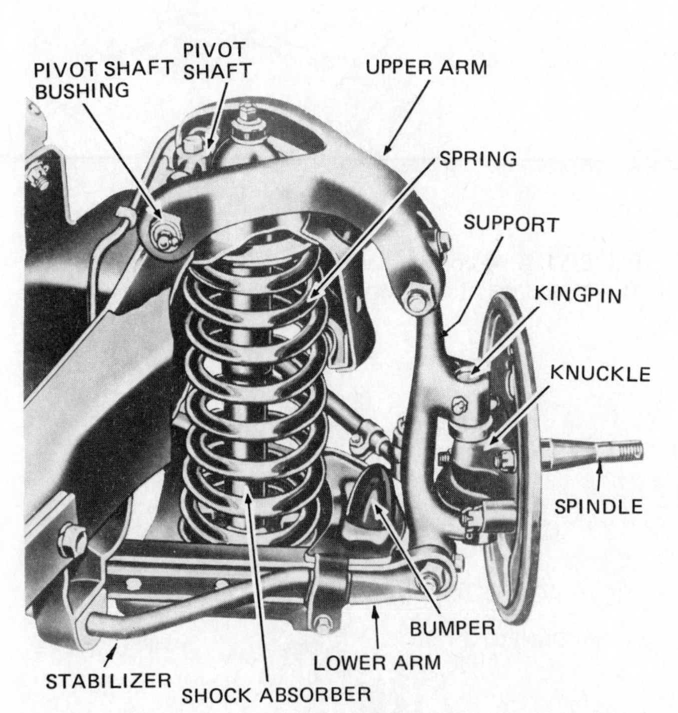

Fig. 8-50 Coil-spring front suspension using a kingpin. (*Ford Motor Company*)

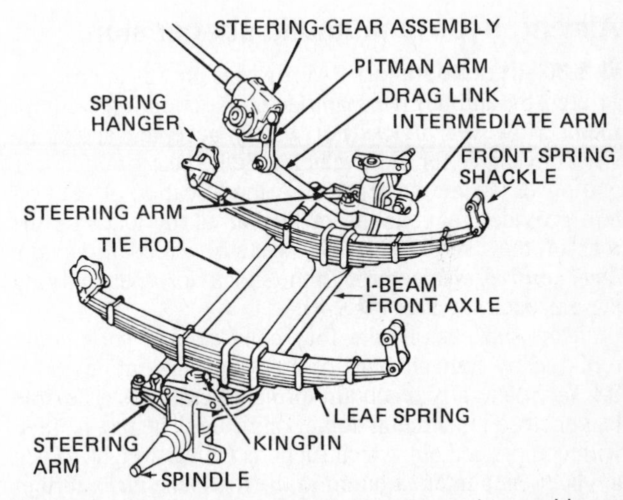

Fig. 8-51 Truck front-suspension system using an I-beam front axle and two leaf springs. (*Ford Motor Company*)

of a kingpin. In Fig. 8-50, notice how the only job of the kingpin is to allow the spindle to swing backward and forward on it. Two additional pivots are required at the top and bottom of the knuckle support to allow for the up-and-down movement of the control arms. A ball joint provides for both actions in a single flexible pivot (Fig. 8-44). As a result, a front suspension using ball joints is lighter than the type using kingpins. Unsprung weight at the wheel is less, which helps produce a smoother ride.

8-29 Leaf-spring front suspension Today almost all cars use either coil-spring or torsion-bar front-suspension systems. However, many trucks have leaf springs at the front (Fig. 8-51). There usually are two leaf springs, one at each wheel. Figure 8-52 shows how the steering knuckle is attached to the end of the axle by the kingpin.

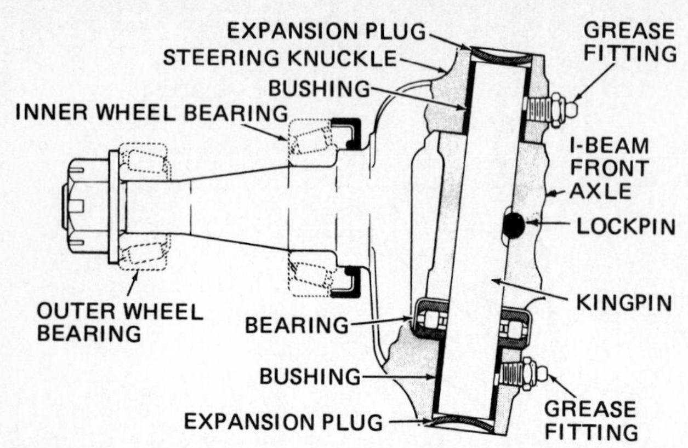

Fig. 8-52 Sectional view showing how the kingpin attaches the steering knuckle to the I-beam front axle. (*International Harvester Company*)

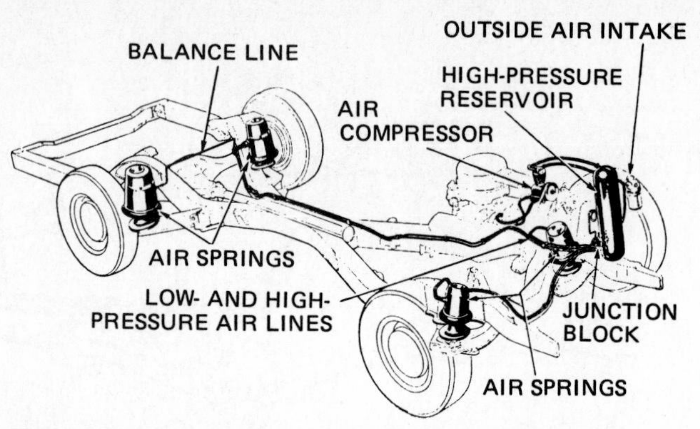

Fig. 8-53 Air-suspension system, using an air spring at each wheel. (*Chevrolet Motor Division of General Motors Corporation*)

When leaf springs are used in the front suspension, the springs normally rest on top of the axle. In automobile leaf-spring rear-suspension systems, the springs are usually suspended underneath the axle. With either arrangement, one end of the leaf spring is attached to the body or frame by a spring hanger. The other end of the spring is attached by a shackle. This permits the spring to move back and forth slightly as its effective length changes.

Air and hydropneumatic suspension

❁ 8-30 Air suspension Air suspension was at one time offered by automotive manufacturers as optional equipment. However, it was not widely accepted and is no longer available for most vehicles. Some heavy-duty trucks and buses use air suspension. A modification of the system provides for leveling of the car as the loads on the rear of the car change. The system, called *automatic level control,* works through the shock absorbers to bring the car back to level (❁ 9-8).

In air suspension, the four conventional springs are replaced by four air bags or air-spring assemblies (Fig. 8-53). Essentially, each air-spring assembly is a flexible bag enclosed in a metal dome, or girdle. The bag is filled with compressed air, which supports the car weight. When a wheel encounters a bump in the road, the air is further compressed and absorbs the shock.

An air compressor, or air pump, supplies air to the system. The compressor is driven by a belt from the crankshaft. Air is admitted to the four air bags through height-control or leveling valves at each wheel. When there is too little air in an air bag, that side of the car will ride low. This causes the linkage to move the leveling arm so that the valve is opened, admitting more air and leveling the car. When a passenger gets out of the car or a load is removed, the air bag goes high. Then the leveling valve releases air to lower the bag to the proper level.

The control and operation of the system are very similar to that of the automatic-level-control system (❁ 9-8). However, the air-suspension system provides springing for all four wheels, instead of only the two rear wheels as with automatic level control.

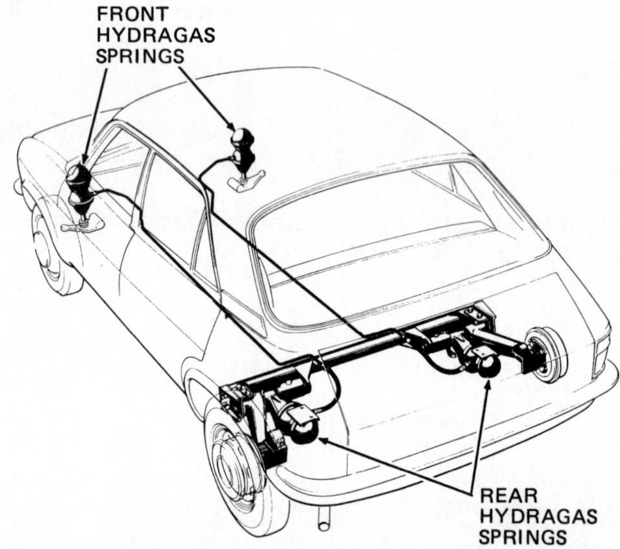

Fig. 8-54 A hydropneumatic suspension system, using Hydragas springs at each wheel. (*British Leyland*)

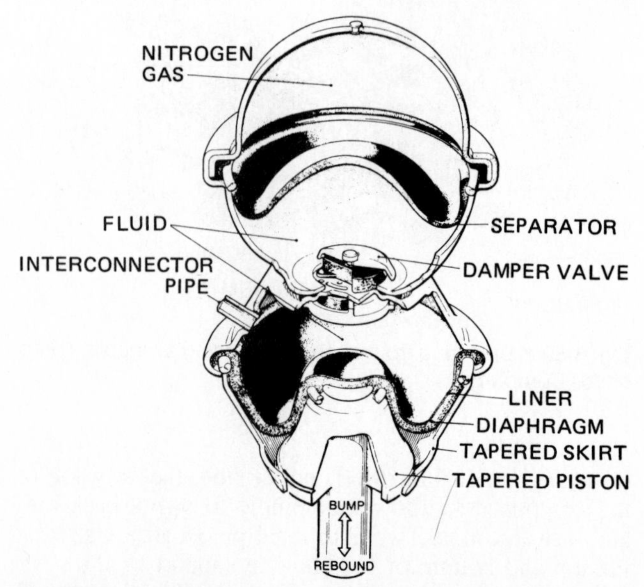

Fig. 8-55 A partially cut away Hydragas spring. (*British Leyland*)

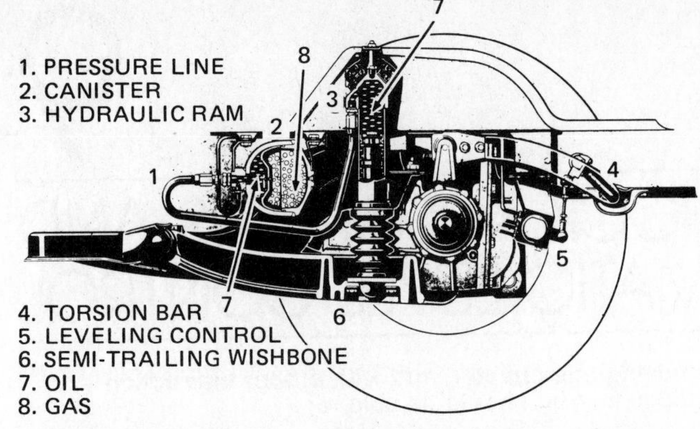

Fig. 8-56 Hydropneumatic suspension using an oil-filled strut connected to a canister filled with oil and a gas at each wheel. (*Mercedes-Benz of North America, Inc.*)

8-31 Hydropneumatic suspension Various suspension systems using air, gas, fluid, or a combination of these have been developed. One of the best-known is the Hydragas suspension system. It uses gas-filled spring units (called Hydragas springs), one at each wheel (Fig. 8-54). Each unit has a sealed chamber containing a quantity of nitrogen gas at high pressure. Below this chamber is a displacement chamber filled with water-based fluid (Fig. 8-55). When the wheel meets a bump, the fluid is pushed upward, compressing the gas. This action provides the springing effect.

In addition, the two units on each side of the car are interconnected front to back (Fig. 8-54). Therefore, when the left front wheel meets a bump, for example, part of the fluid from the left front unit is forced through a pipe to the left rear unit. This action raises the left rear wheel also. Therefore the shock is distributed between the left front and the left rear wheels. This improves the ride.

Figure 8-56 shows a different type of hydropneumatic suspension system used by Mercedes-Benz. The basic spring element is provided by an oil-filled strut at each wheel. Road shocks are absorbed by the strut and a canister which is filled with oil and a gas.

Chapter 8 review questions

Select the *one* correct, best, or most probable answer to each question. Then check your answers against the correct answers given at the end of the book.

1. A front stabilizer bar is used to:
   a. increase vehicle load-carrying capacity
   b. provide a softer ride
   c. control suspension movement and body roll
   d. all of the above
2. Which of the following statements is *true* about a MacPherson-strut front suspension?
   a. Upper and lower control arms are used.
   b. Two ball joints are used with each strut.
   c. Only an upper control arm is used.
   d. The shock absorber is built into the strut.
3. In the coil-spring rear-suspension system, the axle housing is kept in place by:
   a. U bolts
   b. the stabilizer bar
   c. a track bar and control arms
   d. none of the above
4. The bolt on the end of a torsion bar is used for:
   a. locating the control-arm end of the torsion bar
   b. holding the back end of the torsion bar to the chassis
   c. adjusting the ride height of the vehicle
   d. caster adjustment
5. Three types of springs used in automotive suspension systems are:
   a. coil, leaf, and torsion-bar
   b. coil, torsion-bar, and lever
   c. leaf, rubber, and gas
   d. all of the above
6. The rear-suspension system in which rear-end torque is absorbed by the springs is called the:
   a. torque-tube drive
   b. differential drive
   c. Hotchkiss drive
   d. Hooke's drive
7. The rubber bushing in the eye of a leaf spring:
   a. absorbs vibration
   b. should be oiled regularly
   c. can be left out when the spring is replaced
   d. all of the above
8. A strut rod, or brake-reaction rod, is used with:
   a. each MacPherson strut
   b. leaf springs
   c. lower control arms using a single frame-attachment point
   d. upper control arms having two frame-attachment points
9. The steering knuckle is attached to the upper and lower control arms by:
   a. the kingpin
   b. upper and lower ball joints
   c. the stabilizer bar
   d. the spindle
10. The front-suspension system that uses leaf springs and a solid front axle usually is found on:
    a. race cars
    b. passenger cars
    c. pickup trucks
    d. none of the above

CHAPTER 9

SHOCK ABSORBERS AND AUTOMATIC LEVEL CONTROL

After studying this chapter, and with proper instruction and equipment, you should be able to:

1. Describe the purpose and operation of shock absorbers.
2. Explain the construction and operation of automatic level control.
3. List six shock absorber troubles, and discuss the possible causes of each.
4. Perform a bounce test and a bench test on shock absorbers.
5. Replace a shock absorber.
6. Install an overload shock absorber.
7. Service a MacPherson strut.

Shock-absorber types

9-1 Purpose of shock absorbers Springs alone are not satisfactory for a car suspension system. This is because the spring must be a compromise between flexibility and stiffness. It must be flexible so that it can absorb road shock. But if it is too flexible, it will flex and rebound excessively and repeatedly, giving a rough ride. A stiff spring will not flex and rebound so much after a bump has been passed. However, it will give a hard ride because it will transmit too much of the road shock to the car. By using a relatively flexible, or soft, spring and a shock absorber at each wheel, a satisfactorily smooth ride will be achieved.

You can demonstrate to yourself why a spring alone is unsatisfactory for a car suspension. Hang a weight on a coil spring, as shown in Fig. 9-1. Then lift the weight and let it drop. It will expand the spring as it drops. Then it will rebound, or move up. The spring, as it expands and contracts, will keep the weight moving up and down, or oscillating, for some time.

On the car, a very similar action takes place with a flexible spring. The spring is under an initial compression because of the car weight. As the wheel passes over a bump, the spring is further compressed. After the bump is passed, the spring attempts to return to its original position, but it overrides the position and expands too much. This causes the car frame to be thrown upward. Now, having overexpanded, the spring compresses. Again it overrides and compresses too much. As this happens, the wheel may be raised clear off the road and the frame may drop. Then the spring expands again, so the oscillations continue, gradually dying out. But every time the wheel encounters a bump or hole in the road, the same series of oscillations will take place.

Such spring action on a vehicle produces a rough and unsatisfactory ride. On a bumpy road, and especially on a curve, the oscillations might become serious enough to cause the driver to lose control of the car. Therefore it is necessary to use some device to quickly dampen out the spring oscillations once the wheel has passed the hole or bump in the road. The shock absorber is the device universally used today.

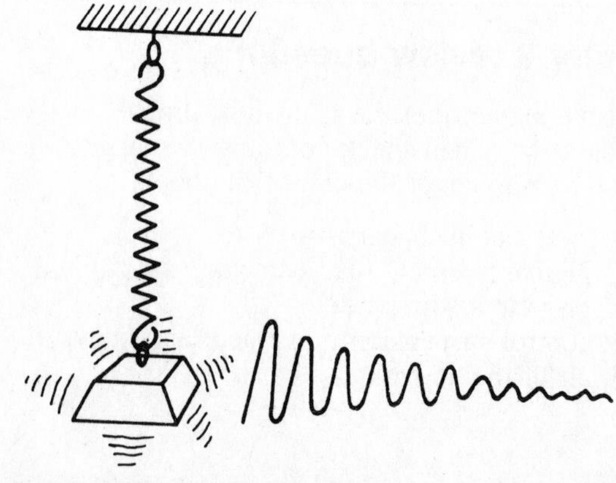

Fig. 9–1 If a weight hanging from a coil spring is set into up-and-down motion, it will oscillate for some time. The distance the weight moves up and down gradually shortens, as indicated by the curve to the right. Finally, the motion will die out.

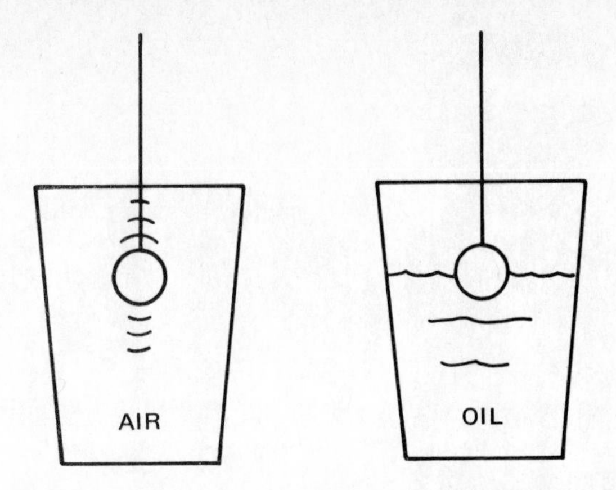

Fig. 9–2 A weight on a rubber band, if dropped into an empty glass (to left), will bounce up and down several times. But if the glass is partly filled with fluid (to right), the weight will meet with resistance and will quickly come to rest.

There have been many types of shock absorber used on cars. They have operated on friction, on compressed air, and hydraulically. The hydraulic shock absorber is the only type in common use today. It contains a fluid that is forced through restricting orifices as the shock absorber is operated by the flexing of the spring. The resistance to the movement of the fluid through the restricting orifices imposes a drag on spring movement. This quickly dampens out the spring oscillations.

A weight on a rubber band demonstrates how a hydraulic shock absorber works (Fig. 9-2). Drop the weight into an empy glass, and the weight will bounce up and down. But put some fluid in the glass (right, Fig. 9-2), and the weight will meet with resistance. Almost at once, the weight will settle down and not rebound.

Several designs of hydraulic shock absorbers have been used, including the parallel-cylinder, opposed-cylinder, and vane types. In the parallel- and opposed-cylinder types, there was one cylinder for compression and another for rebound. Suspension-arm movement, as the wheel moved up and down, caused pistons to be forced into one or the other of these cylinders. The piston movement forced fluid to flow through restricting orifices. This imposed restraint on the spring and wheel movement. The vane-type shock absorber had vanes that rotated in a cylindrical chamber filled with fluid. The most commonly used shock absorber is the direct-acting type, described in ✪ 9-2.

✪ 9-2 Shock-absorber construction The direct-acting, or telescopic, shock absorber is used on both front- and rear-suspension systems. Figures 9-3 and 9-4 show how shock absorbers are mounted at the front and rear of the car. Other methods of shock-absorber mounting at the front and rear wheels are shown in many of the illustrations in Chap. 8. The two ends of the shock absorber (which are the studs or eyes by which they are attached) are encased in rubber grommets or bushings. This provides a flexible mounting which absorbs road vibration and noise.

Regardless of the method of mounting, the shock absorber is attached so that it shortens and lengthens as the wheel moves up and down. However, the shock absorber imposes a restraint on this movement. Excessive wheel and spring movements are prevented, as are spring oscillations.

A shock absorber with its internal parts is shown in sectional view in Fig. 9-5. Figure 9-6 shows a cutaway view of an assembled shock absorber. Figure 9-7 shows a similar unit disassembled. This shock absorber consists

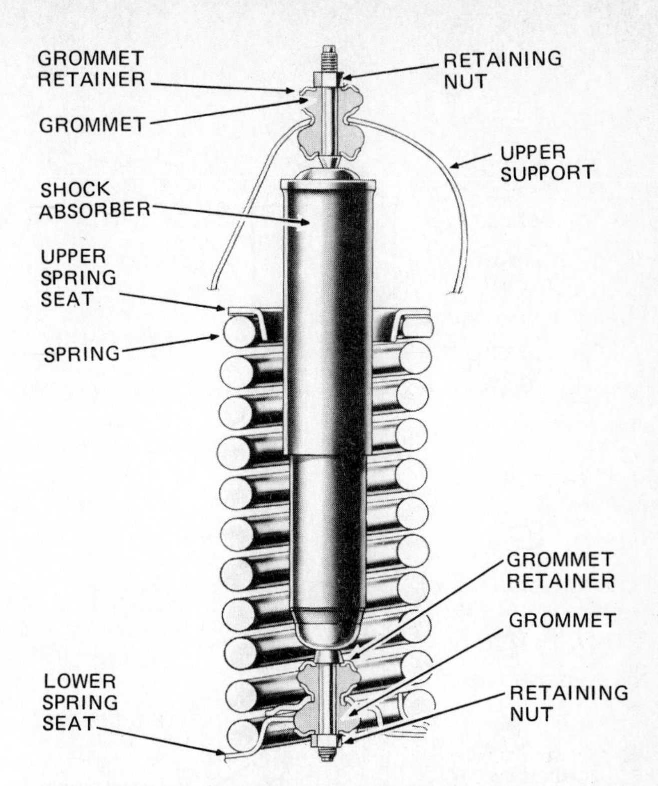

Fig. 9–3 One method of attaching a direct-acting shock absorber in a front-suspension system. (*Chevrolet Motor Division of General Motors Corporation*)

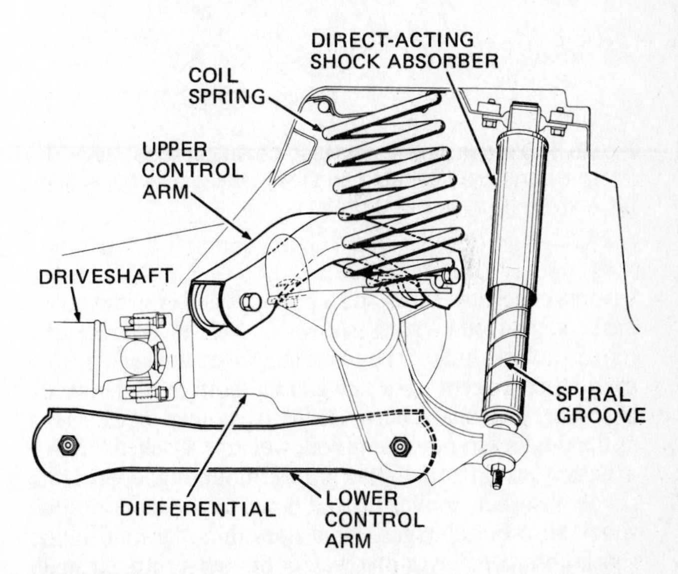

Fig. 9–4 Side view showing the shock-absorber mounting for the left rear wheel. (*Chevrolet Motor Division of General Motors Corporation*)

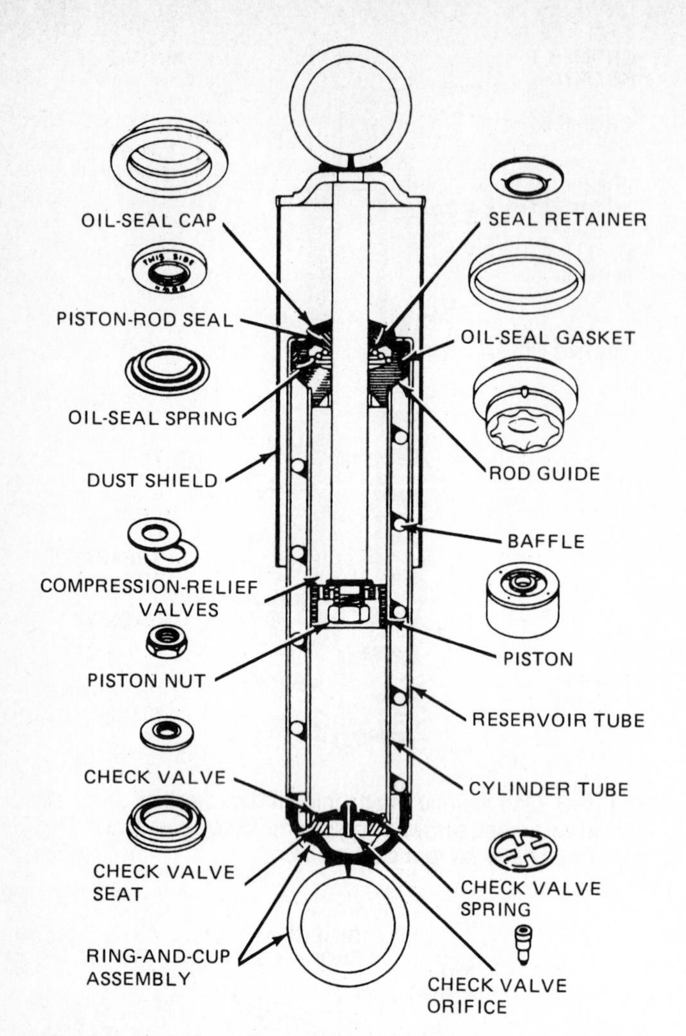

Fig. 9–5 Direct-acting shock absorber in sectional view. The internal parts are illustrated and their positions in the assembly shown. (*Chrysler Corporation*)

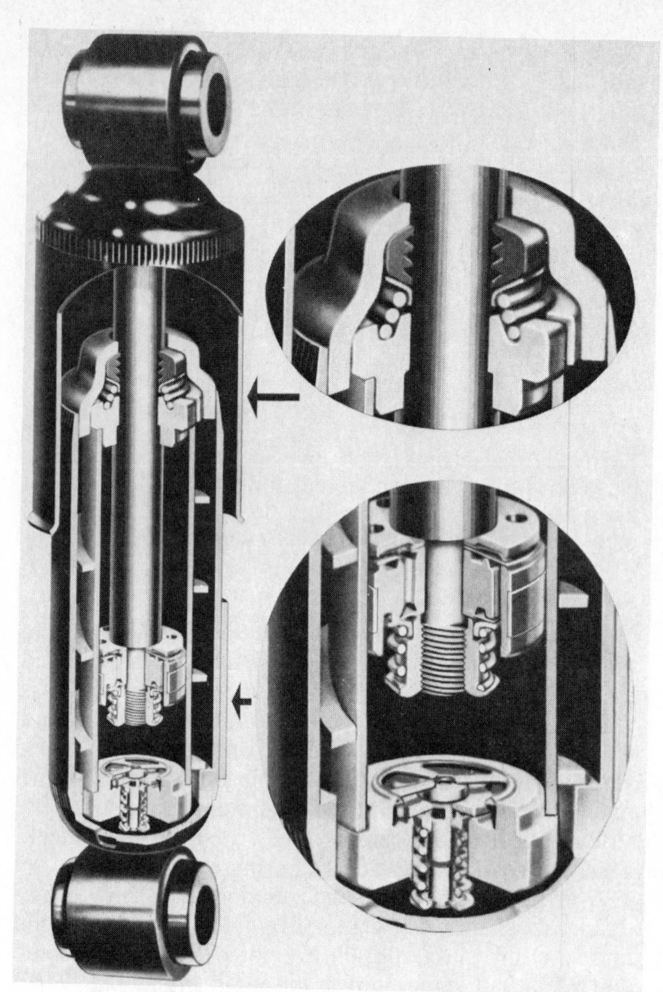

Fig. 9–6 Cutaway view of a shock absorber. (*Monroe Auto Equipment Company*)

of three concentric tubes and a piston, together with valves, gaskets, and other parts shown in Fig. 9-5. The outer tube is a dust shield. The two inner tubes are sealed from each other, except for a valve at the bottom of the shock absorber. The space between the two inner tubes serves as the fluid reservoir. The piston, which is attached through a heavy piston rod to the upper mounting eye of the shock absorber, moves up and down as the length of the shock absorber changes. As it does this, the fluid in the shock absorber moves one way or the other through small passages in the piston (Fig. 9-8). The effect this action has on spring movement is described in ✪ 9-4.

✪ 9-3 Shock-absorber operation The action taking place in the shock absorber when the wheel is moving up, compressing the spring and shock absorber, is shown to the left in Fig. 9-9. When the tire encounters a bump, causing the wheel to move up toward the frame, the spring compresses. At the same time, the shock absorber is telescoped, or shortened. This causes the piston to move downward in the inside cylinder, or tube.

Downward movement of the piston puts pressure on the fluid below the piston. At the same time, it creates a vacuum in the cylinder above the piston. The fluid is forced through the small openings, called *orifices,* in the piston and into the upper part of the cylinder. Meanwhile, fluid in the lower end of the cylinder flows out through the check-valve orifice and into the reservoir that surrounds the inner cylinder.

If the spring movement is very rapid—as it might be if a large bump in the road is encountered—the relief valves flex away from the upper face of the piston, permitting the opening of additional passages. Regardless of this, however, the orifices in the piston tend to restrict the movement of the liquid, which thereby slows the movement of the piston. This, in turn, places a restriction or damper on the spring action.

On rebound, when the wheel moves downward after passing a bump, or when it encounters a hole in the road, the shock absorber extends (Fig. 9-9, right). As this happens, the piston moves into the upper part of the cylinder. The fluid above the piston is forced to pass to the lower part of the cylinder through the small orifices in the piston. At the same time, the check valve in the bottom of the cylinder is lifted off its seat. This permits fluid to flow from the reservoir into the lower end of the cylinder.

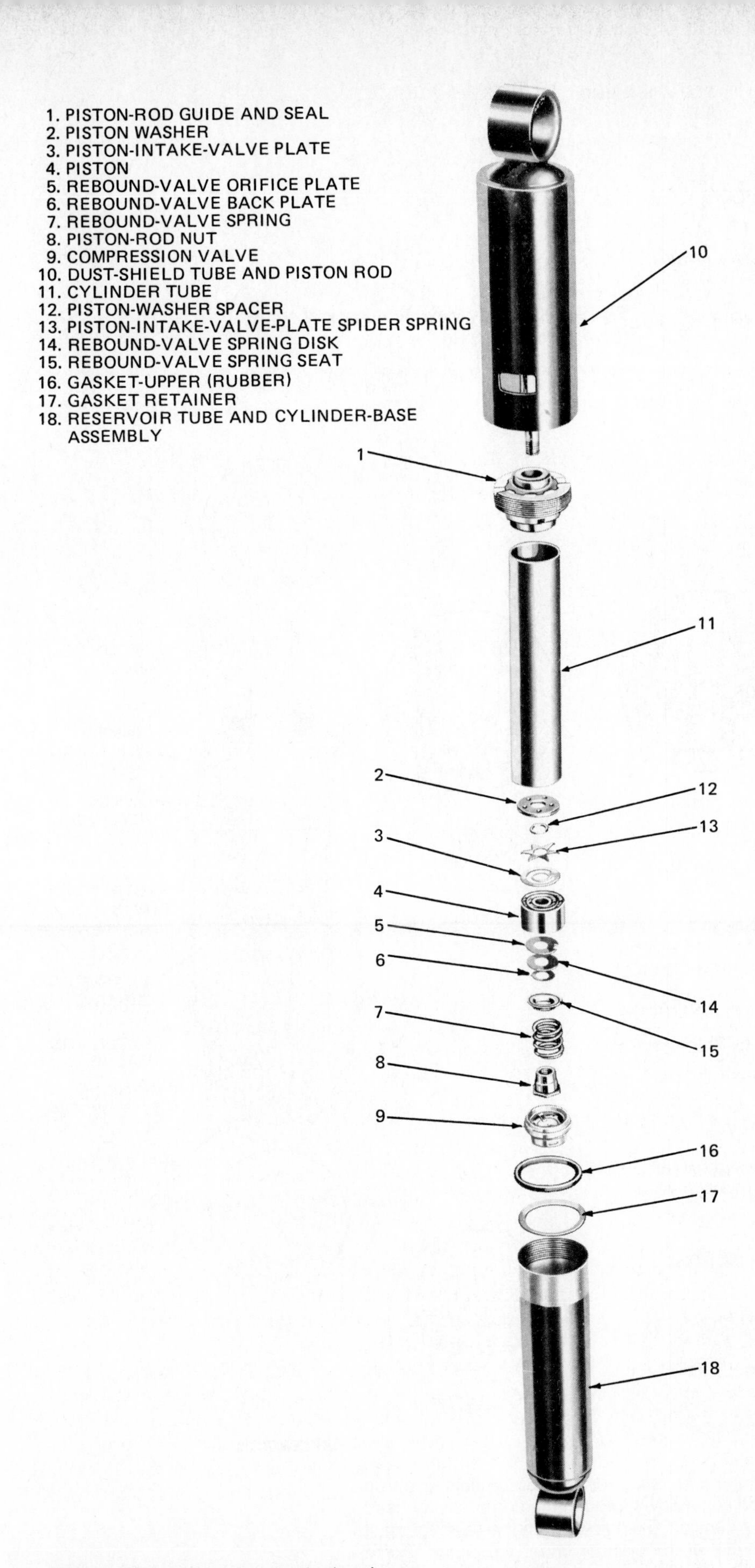

Fig. 9–7 Disassembled shock absorber. (*Chrysler Corporation*)

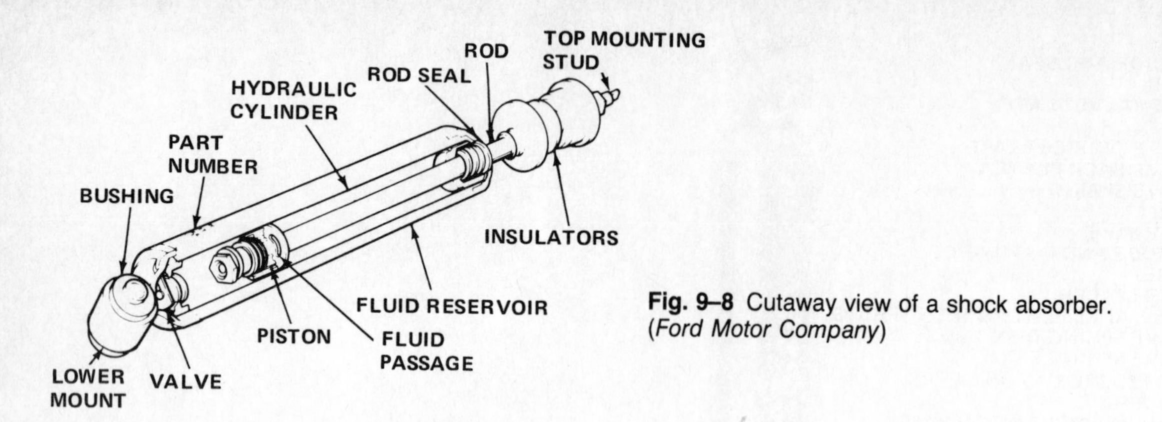

Fig. 9–8 Cutaway view of a shock absorber. (*Ford Motor Company*)

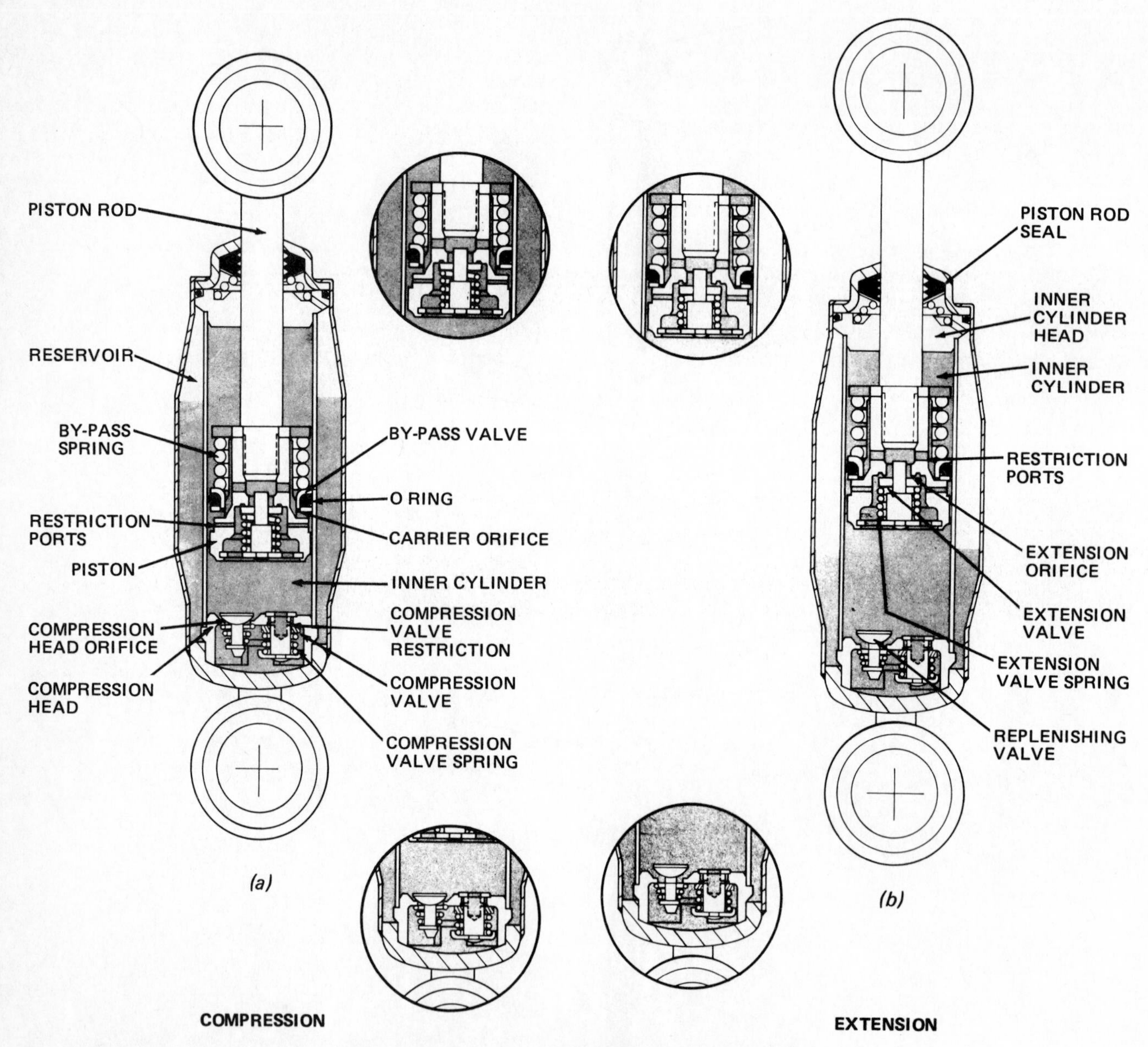

Fig. 9–9 Shock-absorber action. At left, action during compression. Some fluid flows up through the piston, past the bypass valve, and some down through the compression head and into the reservoir. At right, action during extension. Fluid flows from the reservoir past the replenishing valve and up into the lower part of the cylinder under the piston. Some fluid flows down through the extension valve and into the cylinder under the piston. (*Maremont Corporation*)

In both compression and rebound, the valves in the shock absorber open varying amounts, allowing varying speeds of liquid movement through the orifices. This permits rapid spring movements, while still imposing a restraining action. Also, this prevents the excessive pressure rise in the fluid that might otherwise occur when large bumps in the road are encountered. Shock absorbers that dampen spring action during both compression and rebound are called *double-acting*. Most passenger-car shock absorbers are this type. When the damping action occurs equally in both directions, the shock is said to be set for ''50/50'' damping.

⚙ 9-4 Gas-assisted shock absorbers The continuing rapid movement of the fluid between the chambers during rebound and compression can cause foaming, or aeration, of the fluid. Aeration means that free air mixes with the fluid. When this occurs, the shock absorber develops lag. Then the piston moves through an air pocket that offers little resistance before the piston hits fluid again. This reduces the effectiveness of the shock absorber.

Two methods of eliminating foaming are used. One uses a spiral groove in the reservoir tube, as shown in Fig. 9-4. The groove breaks up the air bubbles in the fluid. The other method uses a sealed gas-filled cell (Fig. 9-10), which is a plastic bag filled with Freon gas. A similar design uses a chamber filled with nitrogen gas. Either gas replaces the free air in the shock absorber, and acts as an air chamber. It expands and contracts as the piston moves up and down. Since there is no free air to mix with the fluid, aeration and foaming are prevented.

⚙ 9-5 Adjustable shock absorbers Some replacement shock absorbers are adjustable (Fig. 9-11). By turning the upper dust tube, in relation to the lower reservoir tube, the damping effect is changed. Therefore, turning the dust tube one way will produce a softer ride, and turning it the other way will produce a harder ride. This permits the selection of the ride best suited to the operating needs. For example, rough roads and off-road driving are best handled by a stiffer ride. A soft ride would result in excessive up-and-down motion.

⚙ 9-6 Spring-assisted shock absorbers Some shock absorbers have a variable-rate coil spring (⚙ 8-7) installed between the piston and the fluid-reservoir tube (Fig. 9-12). This type of shock absorber combines spring action with the functions of the shock absorber to help keep the vehicle level regardless of load. The spring may provide up to 700 pounds [318 kg] or extra support for the rear of the vehicle. The spring-assisted shock absorber often is installed on passenger cars that carry full passenger loads or a heavily loaded truck, or that tow trailers or boats. It can also be used to help return sagging suspensions to their original height.

When a vehicle is heavily loaded in the rear, the rear springs sag and the rear end may bottom. Also, there is poor steering, excessive sway, and abnormal front-tire wear. In addition, the headlights are pointed up, instead of ahead of the vehicle. Installation of spring-assisted

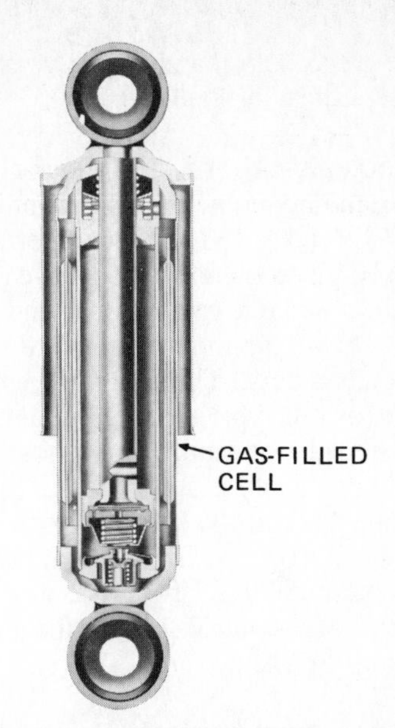

Fig. 9–10 Gas-filled shock absorber. (*Chevrolet Motor Division of General Motors Corporation*)

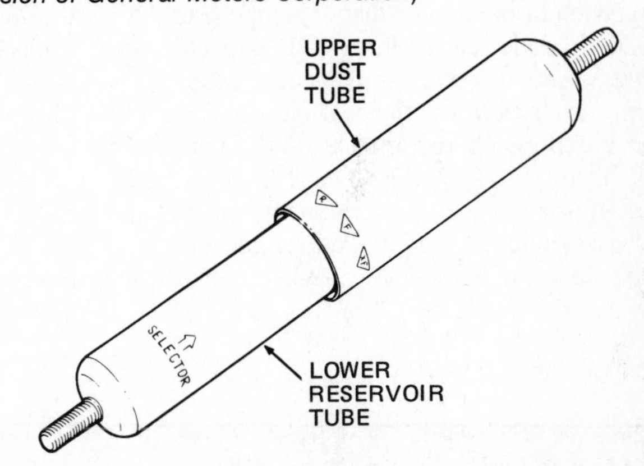

Fig. 9–11 An adjustable shock absorber. The regular, or *R*, setting gives the softest ride. Firm, or *F*, gives a stiffer ride. Extra firm, or *EF*, is the stiffest setting. It provides maximum control with heavy loads on rough roads. (*Maremont Corporation*)

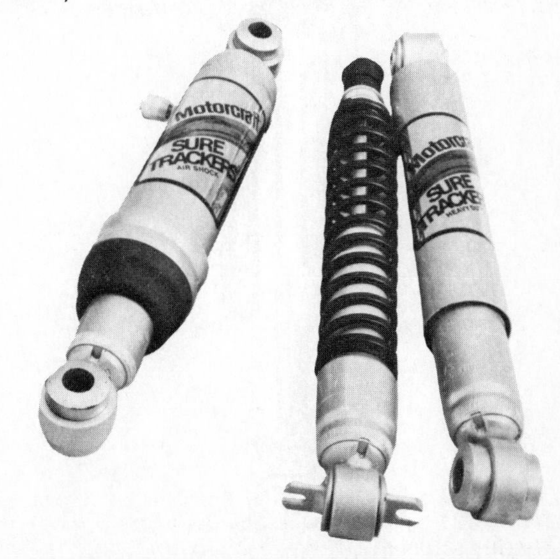

Fig. 9–12 Types of shock absorbers. Left, an air shock absorber. Center, a spring-assisted shock absorber. Right, a heavy-duty shock absorber. (*Ford Motor Company*)

shock absorbers can help maintain the vehicle at normal ride height, regardless of load.

⚙ 9-7 Air shock absorbers Air shock absorbers have an air cylinder and neoprene boot surrounding the shock itself (Figs. 9-12 and 9-13). This compartment is filled with compressed air to increase the load-carrying capacity of the vehicle without causing rear-end sag.

Figure 9-14 shows the mounting and air line between two air shock absorbers. This is the arrangement used with a Scrader (or tire-type) air valve mounted inside the fuel-fill door. To add air, apply an air hose to the valve. In some installations, air flows into the first shock absorber and from there to the second shock absorber.

The air shock absorber is also used with the automatic-level-control system (⚙ 9-8). The basic difference is that, in the automatic-level-control system, the air comes from a compressor on the car.

⚙ 9-8 Automatic level control Automatic level control, which is standard equipment on some cars and optional equipment on others, compensates for variations in load in the rear of the car (Fig. 9-15). When a heavy load is added to the trunk or rear seat, the springs will compress and allow the rear end to lower. This changes the handling characteristics of the car and also causes the headlights to point upward. Automatic level control prevents all this by automatically raising the rear back to level when a load is added, and automatically lowering the rear back to level when the load is removed.

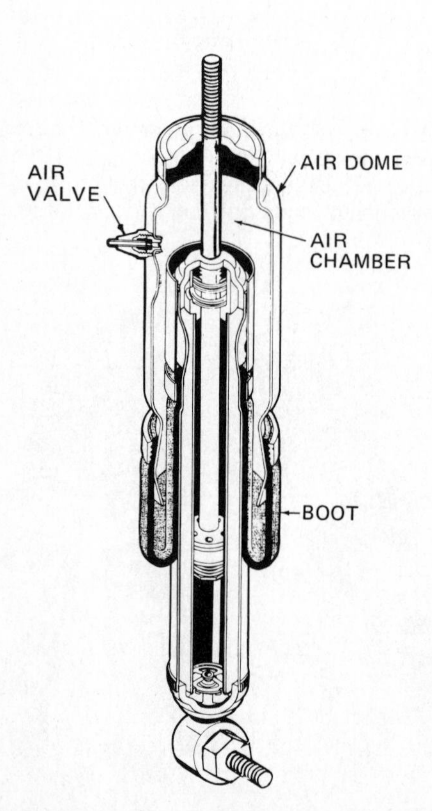

Fig. 9–13 Sectional view of an air shock absorber. (*Buick Motor Division of General Motors Corporation*)

NOTE: At one time, there were two variations of the automatic-level-control system. In one, two rubber air cylinders were installed in the rear-suspension coil springs. In the other, the air cylinders were incorporated in the shock absorbers. This latter type is the only system used today and is the one described here.

The automatic-level-control system (Fig. 9-16) includes a compressor, an air-reserve tank, a height-control valve, and two air shock absorbers (⚙ 9-7). The compressor is operated either by engine intake-manifold vacuum (Fig. 9-16) or by an electric motor (Fig. 9-17).

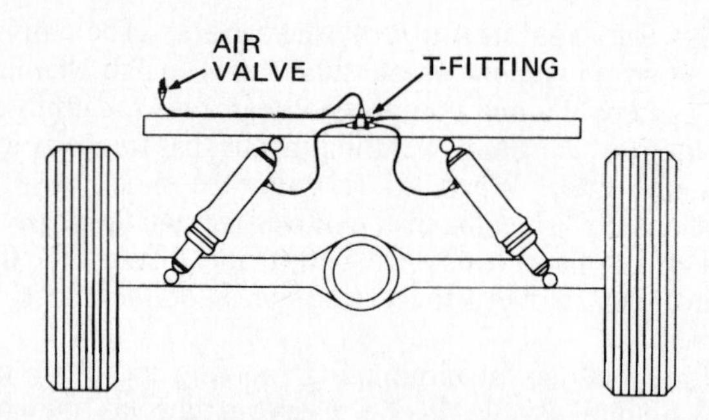

Fig. 9–14 Typical installation of air shock absorbers, as viewed from the rear of the car. (*Goerlich*)

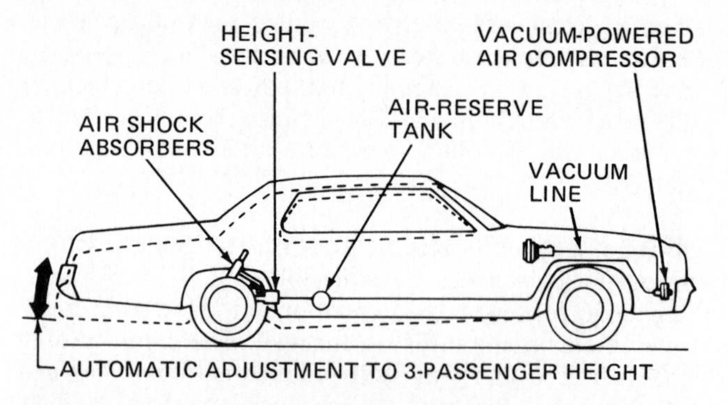

Fig. 9–15 Action of an automatic-level-control system. The dotted lines show the lower height of the car before the automatic level control restores the correct height. (*Chrysler Corporation*)

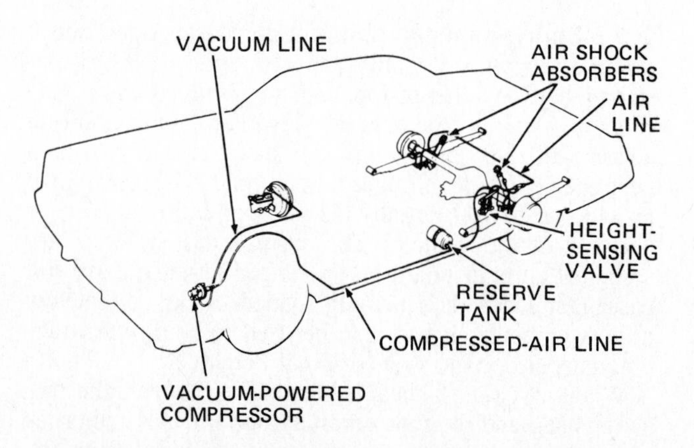

Fig. 9–16 Automatic-level-control system using a vacuum-powered compressor. (*Chrysler Corporation*)

Fig. 9–17 Installation of an electric air compressor for the automatic-level-control system. (*Buick Motor Division of General Motors Corporation*)

The compressor builds up air pressure in the reserve tank. Then, when a load is added to the rear of the car, additional air is passed through the height-control valve to the two rear air shock absorbers. The air entering the air chamber (Fig. 9-13) raises the air dome of the shock absorber to bring the rear of the car back up to its normal ride height.

The height-control valve (Fig. 9-16) has a linkage to the rear suspension. When this linkage is raised by the addition of a load, it opens the intake valve, allowing compressed air to flow to the shock absorbers. When the load is removed, the rear of the car rises. Then the linkage operates the exhaust valve, allowing air to exit from the shock absorber until the correct level is achieved. The height-control valve has a time-delay mechanism that allows the valve to respond only after several seconds. This eliminates fast valve action, which could cause the system to function every time a wheel encountered a bump in the road. Therefore, the system functions on load changes only, and not on road shocks.

The vacuum-operated compressor (Fig. 9-16), mounted in the engine compartment, works on the difference in pressure between atmospheric pressure and intake-manifold vacuum. Atmospheric pressure is applied to one side of a diaphragm in the compressor. Intake-manifold vacuum is applied to the other side of the diaphragm. This difference in pressure moves the diaphragm along with a piston attached to it.

A pulsating action of the diaphragm is set up by a small valve which opens each time the diaphragm moves to the vacuum side. Now this valve admits air to the vacuum side and vacuum to the other side, so the diaphragm moves to the opposite side. These diaphragm pulsations move the piston so that it pumps air into the reservoir, building up a pressure that may go as high as 275 psi [1896 kPa]. A relief valve prevents excessive pressure.

Figure 9-18 is a schematic diagram of the automatic-level-control system with an air compressor powered by an electric motor, and with an attached air dryer. This system has a height sensor that operates electronically

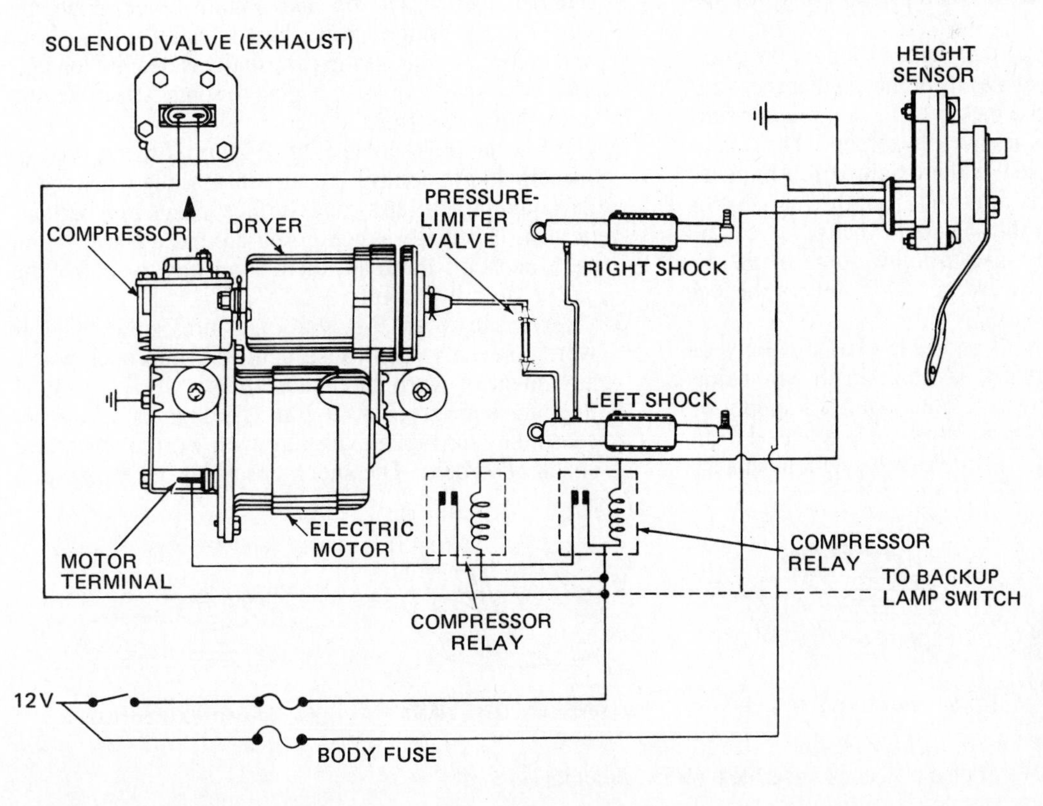

Fig. 9–18 Schematic of an automatic-level-control system using an electric air compressor and a mechanical height sensor.

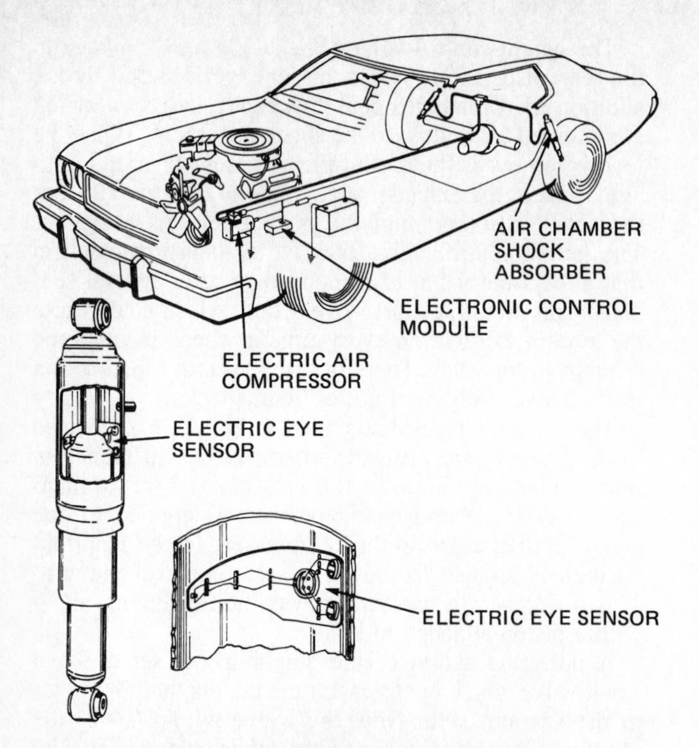

Fig. 9–19 An electronic automatic-level-control system. It uses an electronic eye to switch the electric air compressor on and off. (*Monroe Auto Equipment Company*)

rather than mechanically. The electronic height sensor has a shutter connected to the control arm. A suspension arm is connected to the control arm so that it moves as the height of the car rear-end changes. The shutter interrupts a beam of light inside the height sensor if the height is not correct. If it is either two low or too high, the sensor triggers either the compressor relay or the solenoid relay.

If the height is too low, the shutter triggers the compressor relay. The relay closes its points so that the compressor is connected to the battery. The compressor runs and supplies air to the rear shock absorbers. This raises the rear of the car to level. If the height is too high, as it would be after a load is removed from the rear seat or trunk, the sensor triggers the solenoid exhaust valve. The solenoid then opens the valve to allow some of the air in the shock absorbers to escape. This lowers the car rear-end to level.

The air dryer (Figs. 9-17 and 9-18) has a supply of chemical which absorbs any moisture in the air being pumped in by the compressor. This assures a supply of dry air for the shock absorbers. When air is released from the shock absorbers by the solenoid-valve action, this air passes back through the air dryer, where the air picks up and carries away the moisture from the chemical. The air dryer also has a valve that maintains some air pressure in the shock absorbers to improve riding characteristics.

Figure 9-19 shows an electronic automatic-level-control system that is similar to the one shown in Fig. 9-18. The major difference is that the height control is installed inside one of the rear shock absorbers. A photo-optic sensor (or *electric eye*) is built into the shock absorber. The sensor ''tells'' the electronic control module when any change in height has occurred. Then the electronic module either sends air to the shock absorbers or releases air from them to adjust the height to the correct level.

The sensor circuitry provides an 8- to 14-second time delay before it triggers either the compressor or the exhaust-valve solenoid. This keeps normal riding motions from operating the system. In addition, the operating time is limited to a maximum of 3½ minutes. This time limit prevents continued operation in case of trouble in the system, such as leakage or a malfunction of the solenoid. Turning the ignition off and on resets the system.

Shock-absorber service

❂ 9-9 Shock-absorber troubles Most shock-absorber troubles cause some problem in car handling or ride. Here are various troubles that can result from worn or defective shock absorbers.

1. Topping or bottoming out (Fig. 9-20*a*). If the shock absorbers do not control the suspension, the springs will compress and expand to their extremes. With no restraint on the springs, the suspension hits the rubber bumpers (❂ 8-21), top and bottom, every time the wheel meets rough spots. This is called *topping out* and *bottoming out*. If the car usually is heavily loaded, correction may require changing to some type of heavy-duty shock absorbers.
2. Hard-to-handle vehicle (Fig. 9-20*b*). If the shock absorbers do not control the suspension, the wheels will be bouncing off the road surface. This can be dangerous, especially when going around a curve. With the wheels off the road, there is no way to control the car or steer it properly.
3. Uneven tire wear (Fig. 9-20*c*). If tires are cupped or worn unevenly, the cause could be improper wheel alignment or worn shock absorbers.
4. Sagging springs (Fig. 9-20*d*). Springs may sag because they have been working with worn or defective shock absorbers. The shock absorbers are not able to

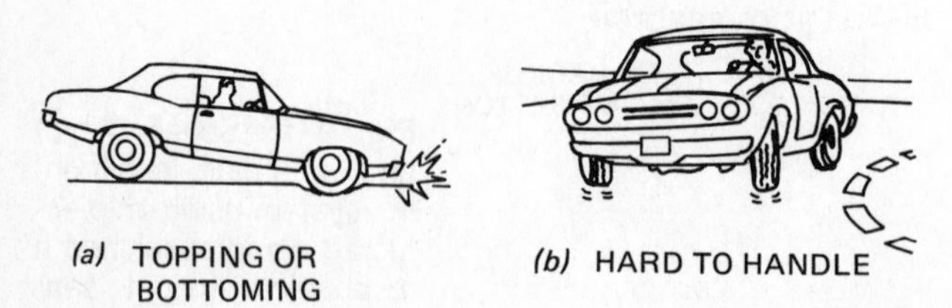

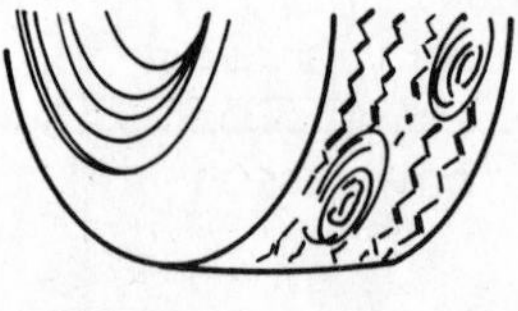

Fig. 9–20 Troubles caused by defective shock absorbers. (*Moog Automotive, Inc.*)

protect the springs from abnormal movements, and they can take a permanent set. New shocks can improve handling of the car, but they will not restore the original effectiveness of the springs. The torsion-bar spring can be adjusted, and in some cases coil spring action can be improved by adding shims in the spring seats. However, for other installations, the only remedy for sagged springs is replacement.

5. Leaking or damaged shocks (Fig. 9-21). If you see fluid on the shock lower cylinder, it may indicate fluid leakage. It is normal for some fluid to appear on the lower cylinder. This provides lubrication. There is a sufficient reserve of fluid to take care of this. Wipe the fluid away and recheck it later. If there is a heavy accumulation of fluid, the shock is leaking and should be replaced. If the shock or the shock mounting has been damaged in any way, replace it.
6. High mileage. Shock absorbers do wear. The amount of wear they get depends on the type of operation the car has. If the car is driven in rough areas where the shocks are constantly working hard, the shocks will wear out faster. Under severe usage, some shock-absorbers may require replacement at around 25,000 miles [40,234 km].

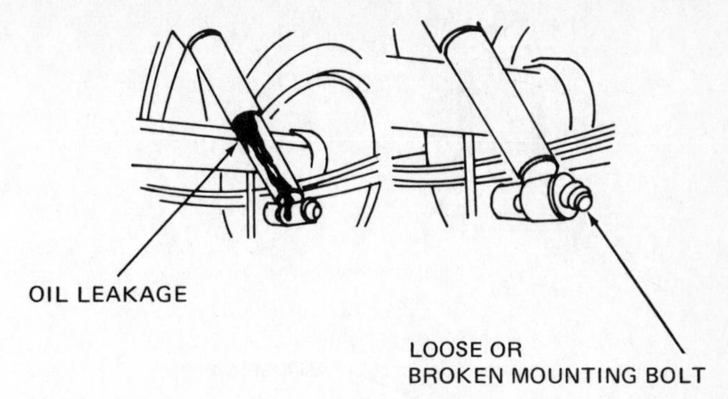

Fig. 9–21 Left, check the shock absorber for oil leakage. Right, check for loose or broken shock-absorber mounting bolts or brackets.

Fig. 9–22 Making a bounce test to check the action of the shock absorber.

❁ 9-10 Checking shock absorbers

Shock absorbers should be checked for proper action, loose attachments, and leakage. Two tests can be made for proper shock-absorber action. The *bounce test* is made with the shock absorbers on the car. The *bench test* is made with the shock absorbers removed from the car.

1. Bounce test (Fig. 9-22). You can make a rough test of the shocks by trying to bounce one corner of the car up and down. With the car on a level surface, push down on one corner of the car and then release it. The car should come back to its original height and stay there. If the car continues to bounce up and down more than two times after you release it, the shock absorber is probably defective and should be replaced.
2. Bench test (Fig. 9-23). The bounce test of a shock absorber on the car does not always clearly indicate whether the shock is good or bad. When there is any question about the condition of a shock absorber, remove it from the car.

CAUTION: **Always use safety stands to support the car before going under it. Never go under a car when it is supported only by a hydraulic jack. The jack could slip, and the car could fall. Always pin or lock a shop lift before going under a raised car.**

Hold the shock in its normally installed position, and pull it to its full extension. Then turn it over and compress it to its shortest length. Repeat this procedure three times to expel any air trapped in the shock absorber.

Clamp the lower end of the shock in a vise in a vertical position. Extend the shock to its full length and then compress it to its shortest length. You should feel a constant drag throughout the complete stroke. Any sudden loss of drag indicates that there is air in the shock absorber, or that the valves in it are faulty. If the piston travel is not smooth and even, or if a scraping sound is heard, the shock absorber is defective. A new shock absorber should be installed.

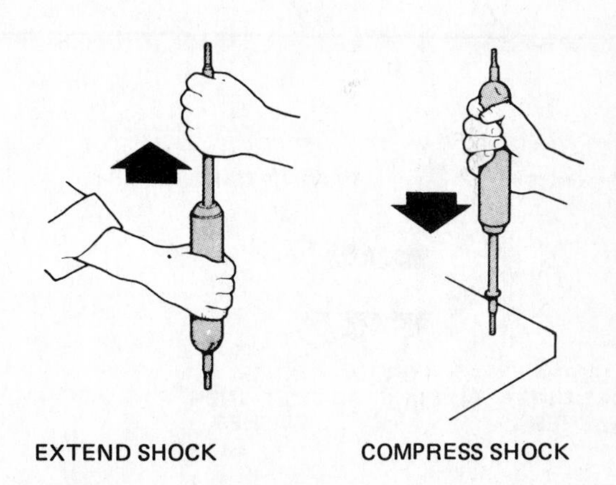

Fig. 9–23 Bench testing a shock absorber. (*Ford Motor Company*)

❁ 9-11 Shock-absorber replacement

Some early-model shock absorbers could be disassembled and reassembled in the shop. However, the shock absorbers now in general use are serviced by complete replacement. Actually, comparing the cost of a new shock absorber with the labor cost of servicing a shock absorber by disassembling and reassembling it, installing a new shock absorber is less expensive.

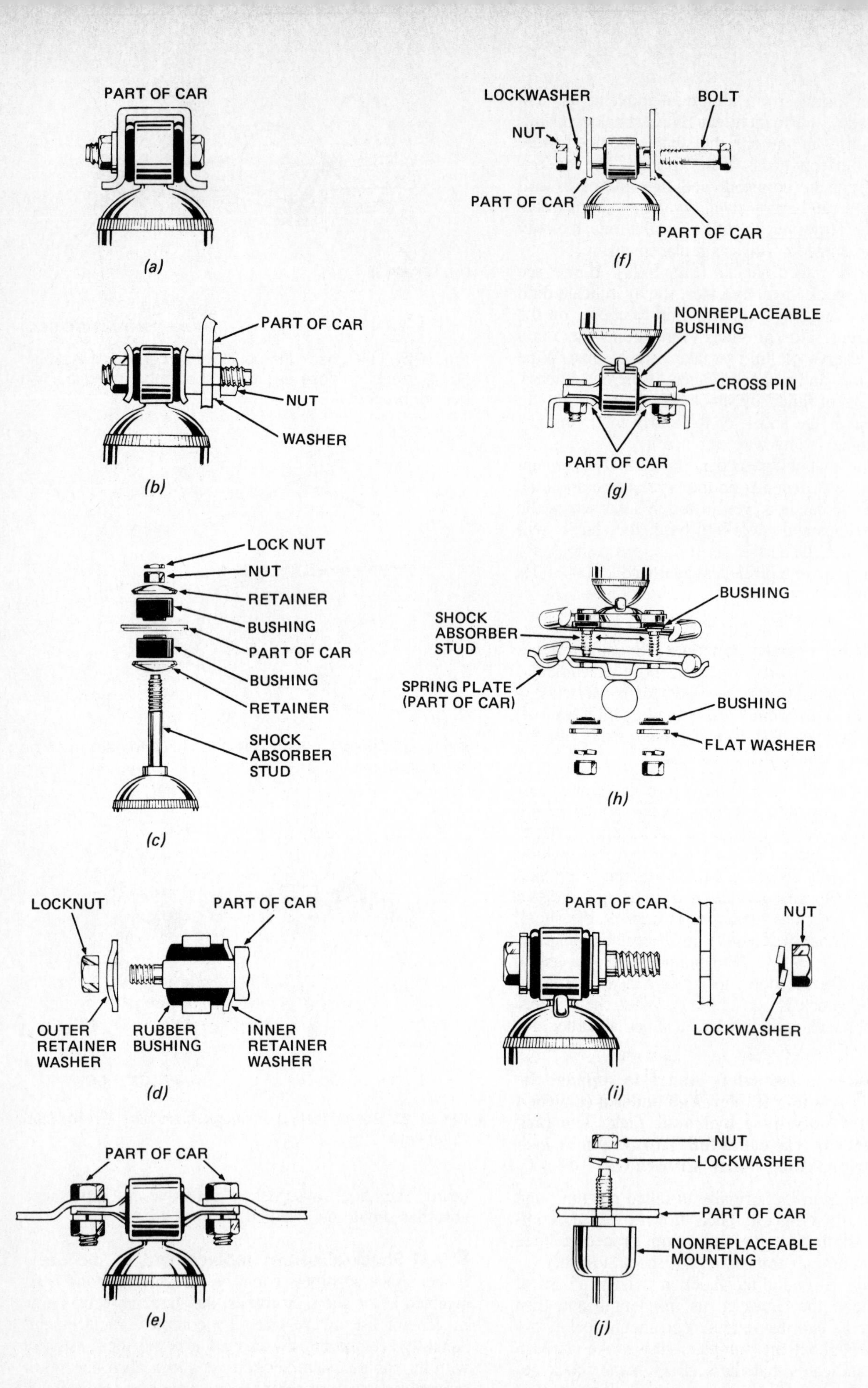

Fig. 9–24 Various types of shock-absorber mountings. (*Goerlich*)

The shock absorbers that are originally installed on a car during manufacture usually are called *standard shocks*. These provide adequate service for normal on-road driving. However, shock absorbers do not last forever, and they will eventually fail to perform properly. Sometimes they may get damaged or leak.

When the suspension system is not worn, and the problem is a leaking or damaged shock absorber or a loose or floating suspension, then new standard shock absorbers may be installed. Also, standard shocks should be installed if replacement becomes necessary at low mileage. This is due to the stiffness of the suspension system on a new car.

At high mileage, replacement or *after-market* shock absorbers should be installed. These often have different valving to provide more shock control and a firmer ride. This compensates for changes in suspension stiffness as the car ages and accumulates mileage. Ford recommends that these shocks be installed in pairs to maintain balanced shock control. For conditions other than those listed above, either heavy-duty or extra-heavy-duty shock absorbers should be installed.

Careful! On a car with coil-spring rear suspension, never lift the car and allow the suspension to hang while removing the shock absorber. The spring could become dislodged and fly out. Always support the wheel before removing the shock absorber.

Figure 9-24 shows various types of shock-absorber mountings. Most of the mounting illustrations are self-explanatory. Always read the instructions that come with the new shock absorber and follow the procedures outlined. Use the mounting parts that are included with the shock. You use some of the original mounting parts when installing some shocks. On others, the old parts are discarded and the new parts included in the shock-absorber package are used.

When installing the stud-type shock absorber (Fig. 9-24*c*), do not overtighten the nut. Overtightening squeezes the rubber bushings too much. Figure 9-25 shows the correct and incorrect tightening of the bushings. Tighten the stem nut only enough so that the rubber bushings are expanded to the diameter of the retaining washers.

Before installing any shock absorber, be sure to purge it of air. The procedure is the same as shown in Fig. 9-23. With the shock absorber in its normal upright position, extend or open it as far as possible. Then turn the shock absorber upside down and push it together.

⚙ 9-12 Overload shock-absorber installation If the vehicle carries heavier than normal loads, some type of overload shock absorbers should be installed to handle the extra loads. Spring-assisted shock absorbers (⚙ 9-6) and air shock absorbers (⚙ 9-7) are often used on these vehicles.

Figure 9-26 shows the installation of spring-assisted shock absorbers. They are installed the same as standard shock absorbers. On some vehicles, you may have to slightly enlarge the hole in the lower control arm. File or grind the hole to permit the shock to slide up into place. Always check for interference between the shock absorber and the suspension parts.

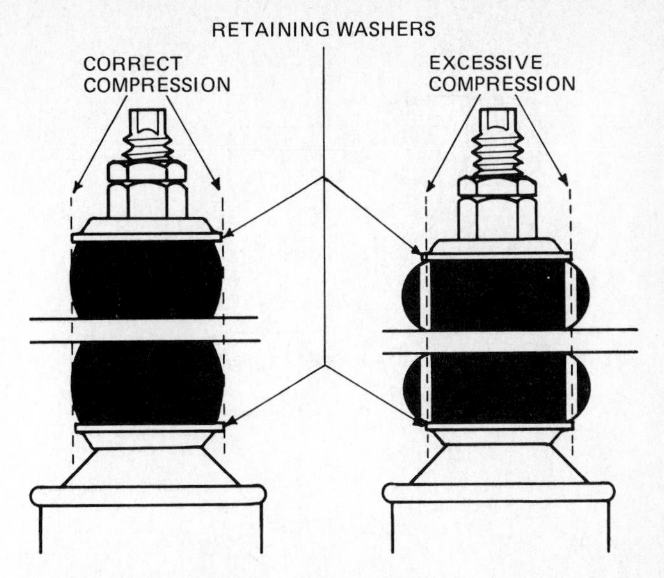

Fig. 9–25 Correct and excessive tightening of the stud nut. (*Moog Automotive, Inc.*)

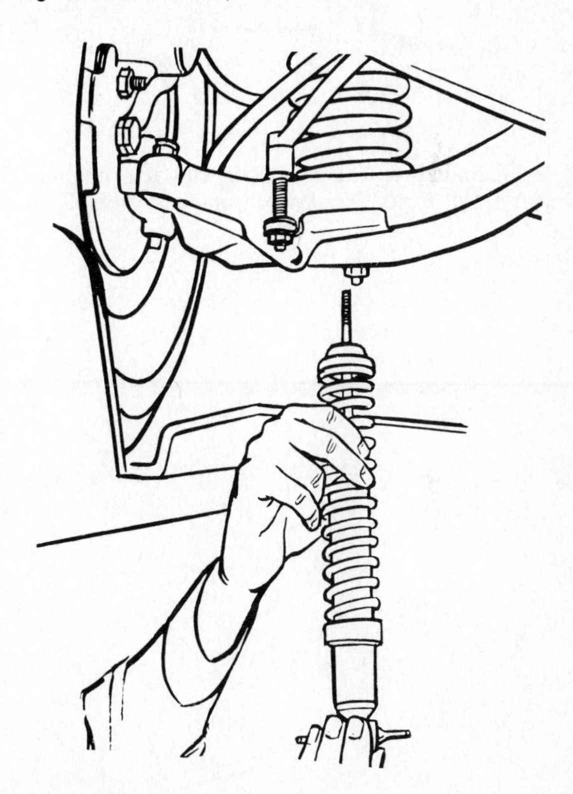

Fig. 9–26 Installation of an overload shock absorber, which is spring-assisted. (*Goerlich*)

⚙ 9-13 Replacing air shock absorbers Air shock absorbers are removed and installed in the same way as the shocks described in ⚙ 9-11 and 9-12. However, there is one added item to watch, and that is the air line connections. These are made with special clips as shown in Fig. 9-27.

⚙ 9-14 MacPherson-strut service Figure 9-28 shows a front suspension system, with drive train and MacPherson struts, for a front-wheel-drive car. Figure 9-29

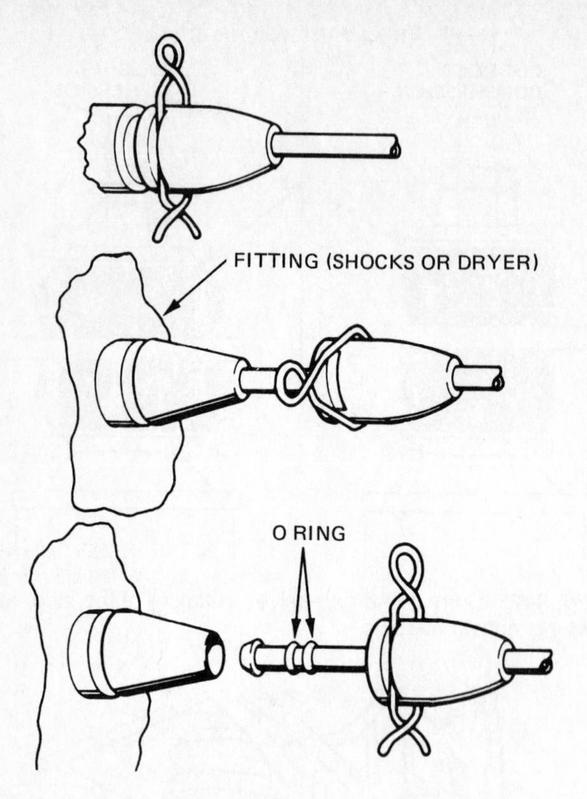

Fig. 9–27 Special air-line clips used with air shock absorbers. (*Chevrolet Motor Division of General Motors Corporation*)

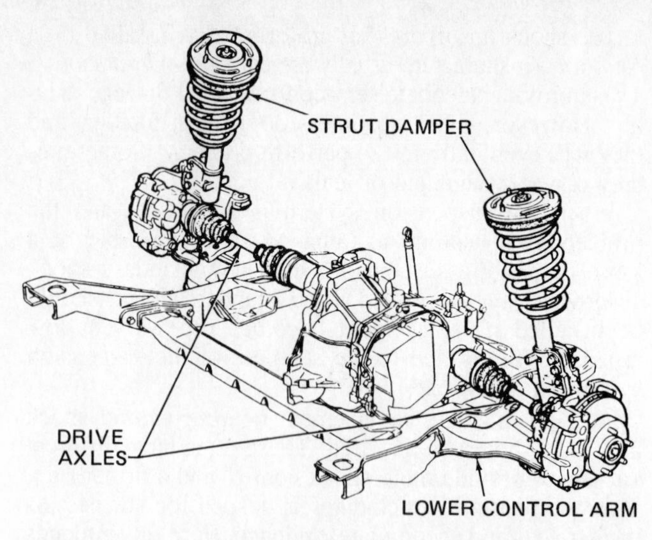

Fig. 9–28 MacPherson struts in a front-suspension system, with power train for a front-wheel drive car. (*Chevrolet Motor Division of General Motors Corporation*)

shows the mounting arrangement for one strut. Note how to mark the camber-adjusting cam before removing the bolts. After installation, the camber and toe must be checked and adjusted (Chap. 13). The strut assembly, complete with spring, must be removed. After it is removed, a pair of spring compressors must be used as

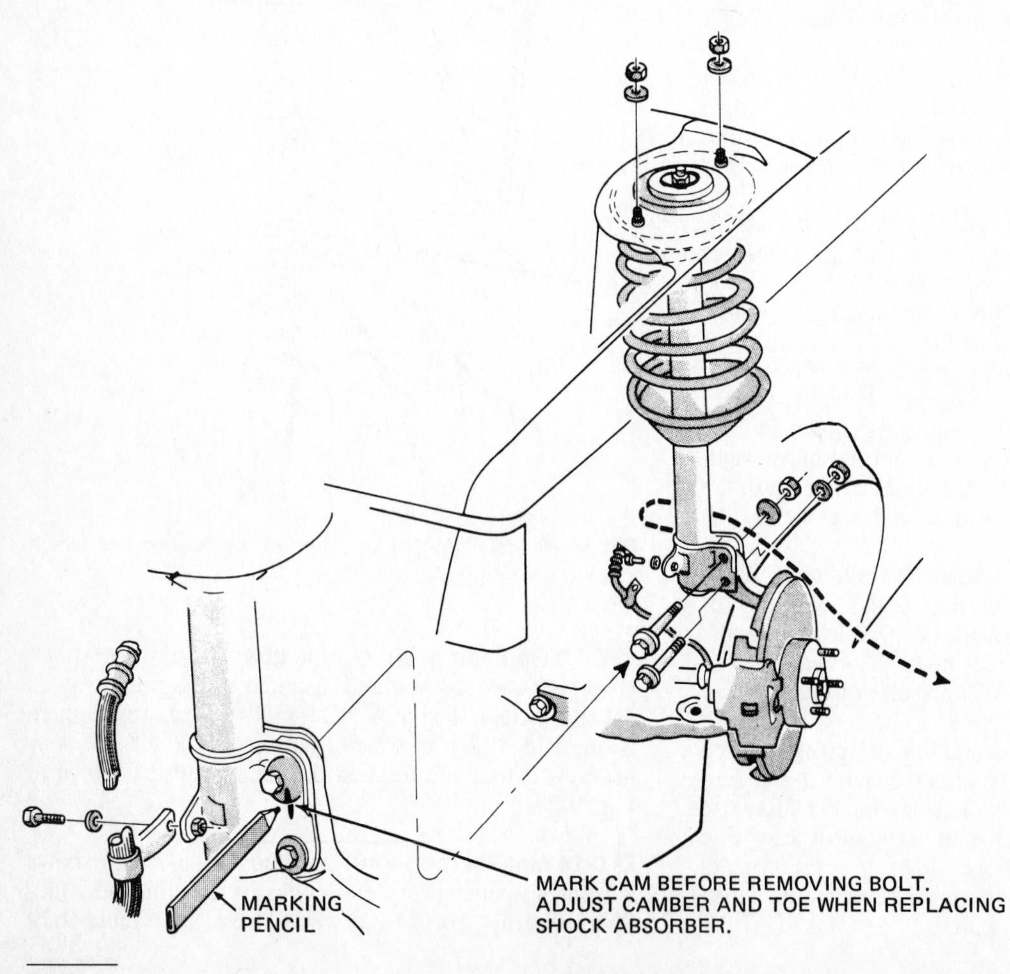

Fig. 9–29 Strut mounting arrangement. (*Chrysler Corporation*)

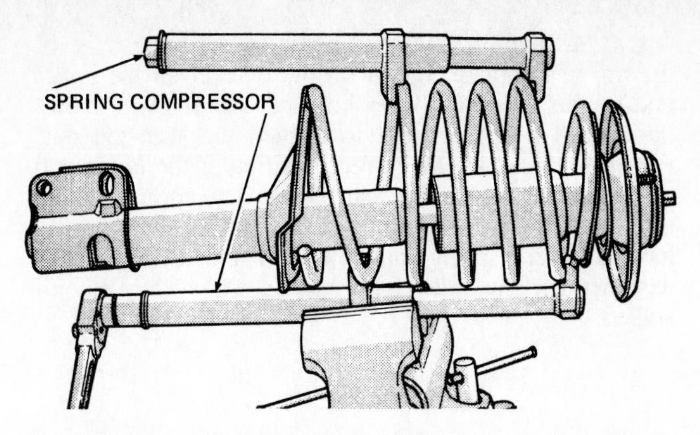

Fig. 9–30 Using a strut-spring compressor to compress the spring. (*Chrysler Corporation*)

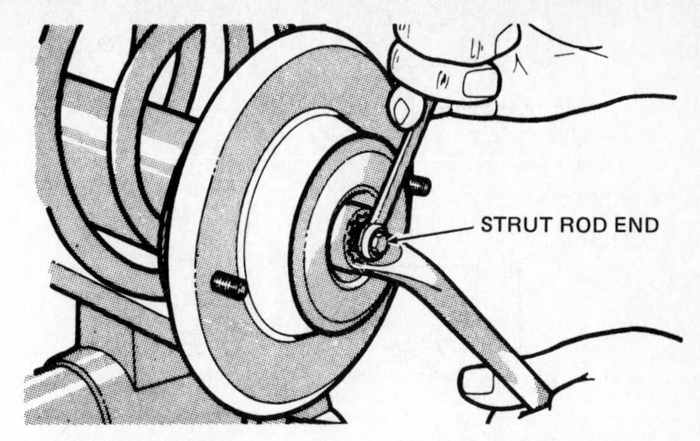

Fig. 9–31 Loosening the strut-rod nut. (*Chrysler Corporation*)

shown in Fig. 9-30 to compress the spring. Then, hold the strut rod while loosening the nut (Fig. 9-31). Remove the nut.

CAUTION: **Use care in releasing the spring tension. If the compressor should happen to slip off while the spring is still in compression, it could jump out and hurt you or anyone nearby.**

Figure 9-32 shows the top of the strut disassembled. You must keep track of all these parts and replace them in the proper positions when reassembling the strut.

With the spring and other parts off, check the shock absorber and purge it of air as shown in Fig. 9-33. Extend the shock absorber to its limit. Look for leakage on the rod and rod seal. Turn the shock upside down and close it, as shown to the right in Fig. 9-33.

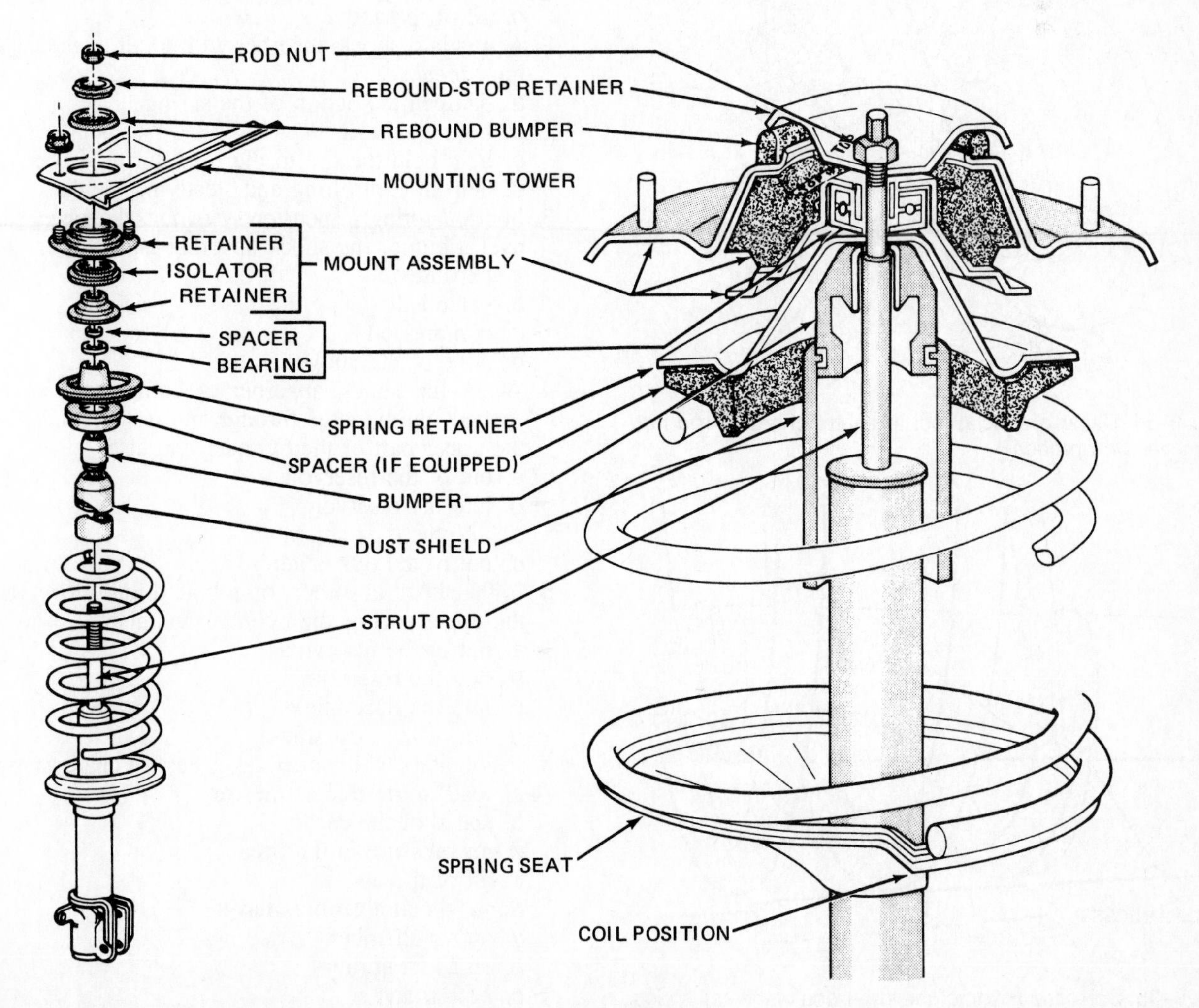

Fig. 9–32 Disassembled upper end of the strut. (*Chrysler Corporation*)

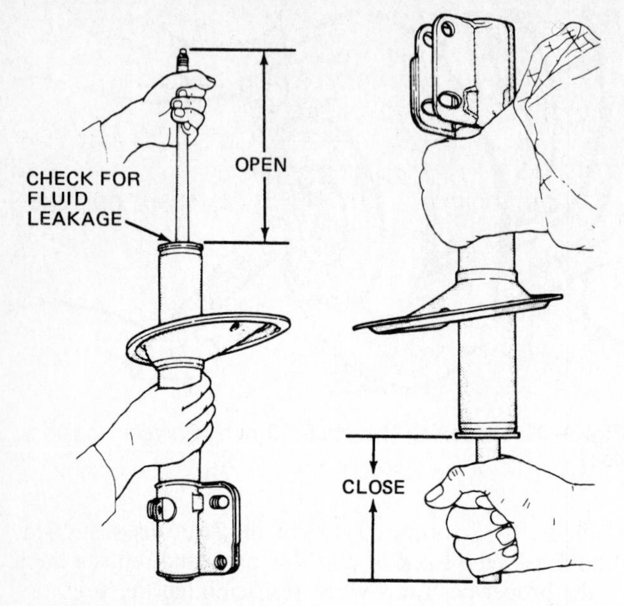

Fig. 9–33 Testing and expelling air from the shock absorber. (*Chrysler Corporation*)

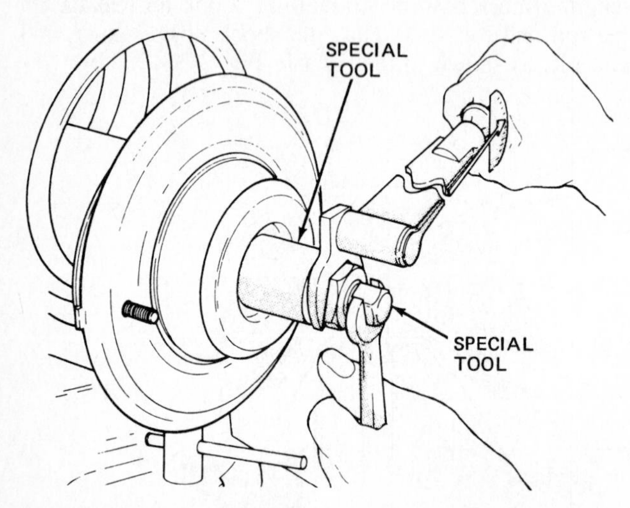

Fig. 9–34 Using a special tool to tighten the strut-rod nut. (*Chrysler Corporation*)

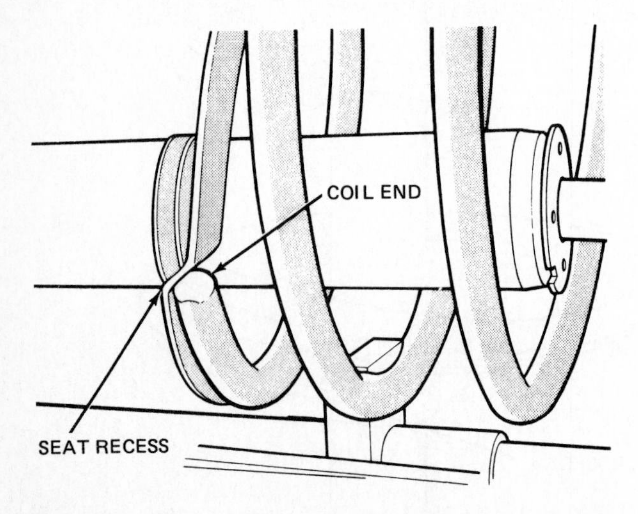

Fig. 9–35 Correct position of the lower end of the coil spring in the seat recess. (*Chrysler Corporation*)

On reassembly, align the notch in the outer edge of the upper spring retainer with the bracket on the lower end of the shock-absorber bracket. Figure 9-34 shows the use of a special tool to tighten the strut-rod nut. Figure 9-35 shows the correct position of the lower end of the coil spring in the seat recess. The springs on the two sides of the car are not interchangable. They usually have different spring rates (✪ 8-7) and are especially designed for the side of the car on which they are installed.

Chapter 9 review questions

Select the *one* correct, best, or most probable answer to each question. Then check your answers against the correct answers given at the end of the book.

1. A rear shock absorber should never be removed without first supporting the wheel on a car with which type of suspension?
   a. torsion-bar
   b. leaf-spring
   c. coil-spring
   d. all of the above
2. A standard shock absorber will do all of the following *except:*
   a. dampen the action of the spring
   b. hold up the car
   c. help hold the tire to the road during driving
   d. help in controlling and steadying the car
3. In a coil-spring suspension system, as the wheel passes over a bump, the shock absorber is:
   a. expanded
   b. extended
   c. compressed
   d. none of the above
4. When the shock absorber is compressed or telescoped, fluid passes through the piston orifices into the upper part of the cylinder and also:
   a. out of the reservoir
   b. into the reservoir
   c. into the dust shield
   d. out of the dust shield
5. In the shock absorber, on rebound fluid flows out of the upper part of the cylinder and also:
   a. out of the reservoir
   b. into the reservoir
   c. into the dust shield
   d. out of the dust shield
6. Automatic level control takes care of changes in the:
   a. load in the rear of the car
   b. speed of the car
   c. air pressure in the tires
   d. all of the above
7. Some shock absorbers have:
   a. an air chamber
   b. an assist spring
   c. both *a* and *b*
   d. neither *a* nor *b*

8. A shock absorber has a very slight trace of fluid on the outside of the lower cylinder. Mechanic A says the shock absorber is defective and should be replaced. Mechanic B says the condition is normal and the shock absorber can stay in service. Who is right?
   a. A only
   b. B only
   c. both A and B
   d. neither A nor B
9. When installing a stud-type shock absorber, the nut is properly torqued when:
   a. the rubber bushing begins to split
   b. you cannot turn it further
   c. the retaining washers have flattened
   d. the rubber bushings are expanded to the diameter of the retaining washers
10. Never attempt to remove the spring from a MacPherson strut unless you:
   a. cut the spring with a torch or hacksaw
   b. first remove the strut-rod nut
   c. have a spring-compressor tool in place to relieve the tension on the spring
   d. have drained the fluid from the shock cartridge

CHAPTER 10

STEERING SYSTEMS

After studying this chapter, you should be able to:

1. Explain the purpose, construction, and operation of manual steering systems.
2. Discuss front-end geometry and the various angles involved.
3. Describe the construction and operation of a rack-and-pinion manual-steering gear.
4. Define *steering ratio.*
5. Explain the purpose, construction, and operation of power-steering systems.
6. Describe the difference between the integral and linkage types of power-steering systems.
7. Discuss the construction and operation of tilt and telescoping steering wheels.
8. Explain the purpose and construction of three types of collapsible steering columns.

10-1 Function of the steering system A simplified drawing of a steering system is shown in Fig. 1-20. Various methods of supporting the front-wheel spindles so that the wheels can be swung to the left or right for steering were described in Chap. 8. This movement is produced by gearing and linkage between the steering wheel and the steering knuckle. The complete arrangement is called the *steering system.*

The steering system is composed of two elements: a steering gear at the lower end of the steering column, and the linkage between the gear and the wheel steering knuckle. Following sections describe the steering system from the standpoint of geometry, or the angles involved.

Wheel-alignment angles

10-2 Front-end geometry Front-end geometry (also called *wheel-alignment angularity*) is the relationship of the angles among the front wheels, the front-wheel attaching parts, and the ground. The various factors that enter into front-end geometry are classified under the following six terms: front-suspension height, camber, steering-axis inclination, caster, toe, and turning radius (Fig. 10-1).

Wheel-alignment angles affect the steering ease, steering stability, and riding qualities of the car. They also may affect tire wear. The angles vary as the suspension does its job, and the body moves up and down in relation to the wheels. The load in a car and its speed may also cause the angles to change. In addition, they are affected by changes in vehicle attitude caused by acceleration and braking, the type of road surface, and the cornering forces. How each angle affects wheel alignment is described in following sections.

10-3 Camber Camber is the amount that the front wheels tilt in or out at the top when viewed from the front of the car (Fig. 10-2). When the top of the wheel leans out, the camber is positive (Fig. 10-2*a*). When the top of the wheel leans in (Fig. 10-2*b*), the camber is negative. The amount of the tilt is measured in degrees from true vertical. This measurement is called the *camber angle*. It is checked and adjusted during the wheel-alignment procedure.

On many cars, the wheels are given a slight outward tilt as the specified setting. When the car is loaded and rolling along the road, the wheels should run straight up and down in a true vertical position. This ''zero camber'' position puts the full width of the tire tread on the road surface. The load and wear are evenly distributed across the tread.

However, an average ''running'' camber of zero seldom ocurs during driving. This is because camber changes as the body and wheels move up and down. Figure 10-3 shows how the camber goes negative when the tire hits a bump. When the tire drops into a hole in the road, the camber changes from zero to slightly positive (Fig. 10-4). In wheel-alignment and suspension work, the upward movement of the wheel is called *jounce* and the downward movement of the wheel is called *rebound*.

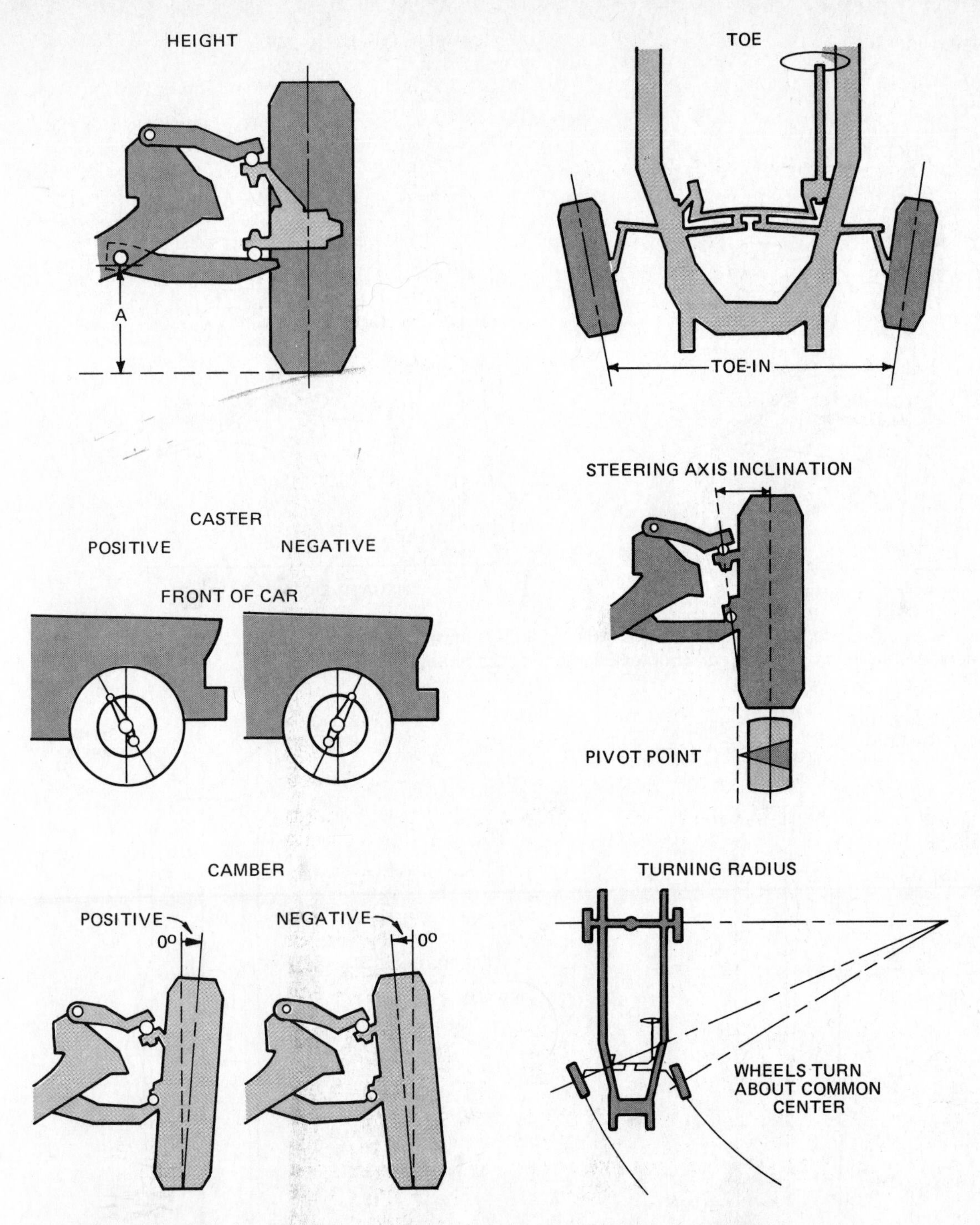

Fig. 10–1 The six factors included in front-wheel alignment. (*Chrysler Corporation*)

Any amount of camber, either positive (+) or negative (−), tends to cause uneven and more rapid tire wear. Tilting the wheel puts more load on one side of the tire tread than on the other side. This is why camber is called a *tire-wear angle*. Excessive positive camber causes the outside of the tire tread to wear. Excessive negative camber causes the inside of the tire tread to wear. The camber angle specified by the manufacturer is the setting that will normally provide maximum tire life.

If the vehicle is rolling on a perfectly level road, the ideal camber would be the same for both front wheels. This ideal camber would be just sufficient to bring the front wheels to the vertical position when the vehicle is normally loaded and moving. However, roads are seldom perfectly level.

Most roads are crowned, slightly higher at the center than on the two sides. When a car is moving along one side of a crowned road, the car tends to lean out slightly and drift to the right. This could cause the outside of the

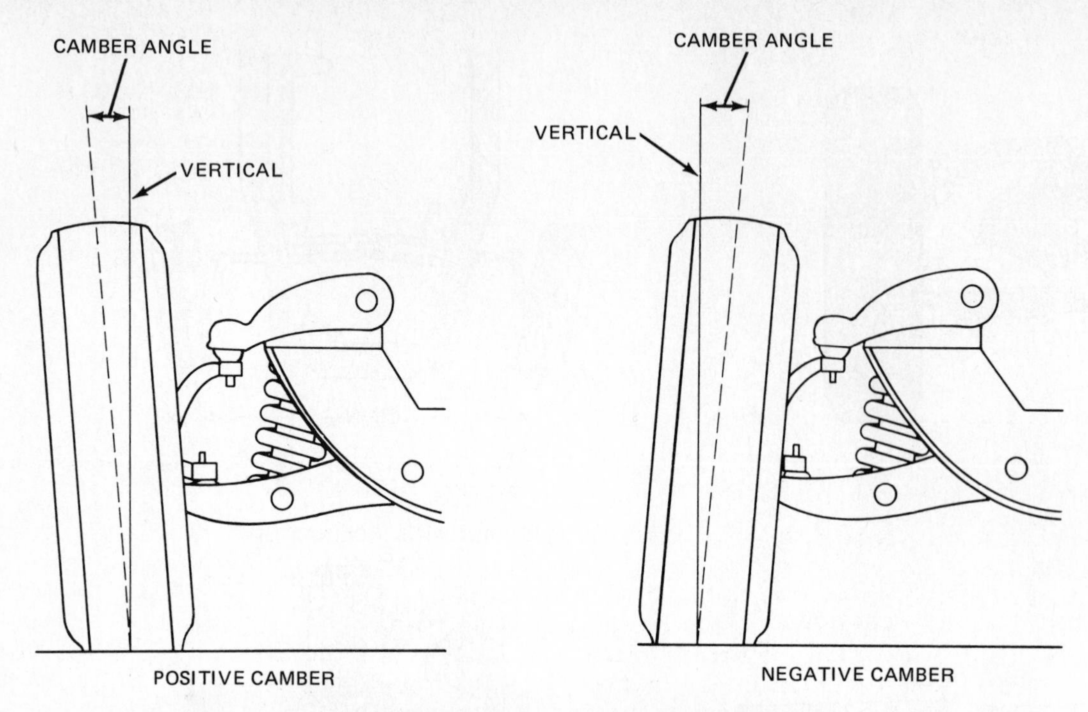

Fig. 10–2 Camber is the amount in degrees that a wheel tilts outward (left) or inward (right) from vertical when viewed from the front of the car. (*Hunter Engineering Company*)

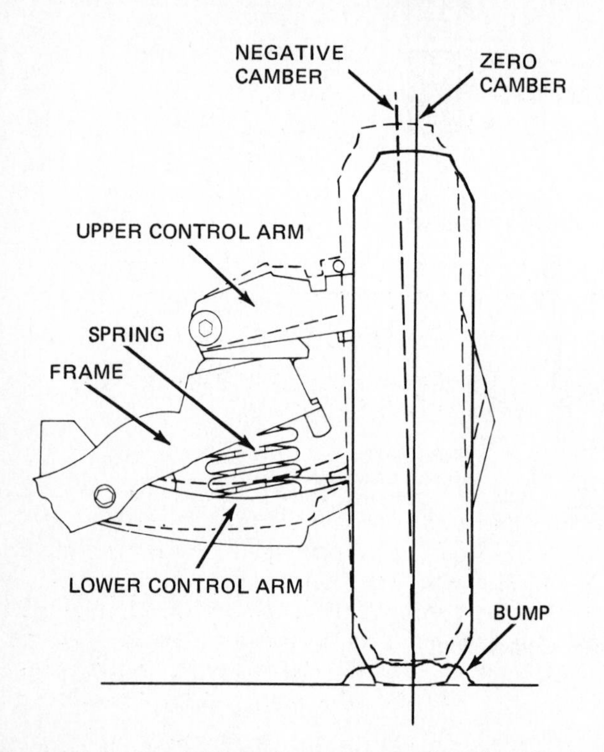

Fig. 10–3 Front-wheel camber goes negative when the tire hits a bump on the road.

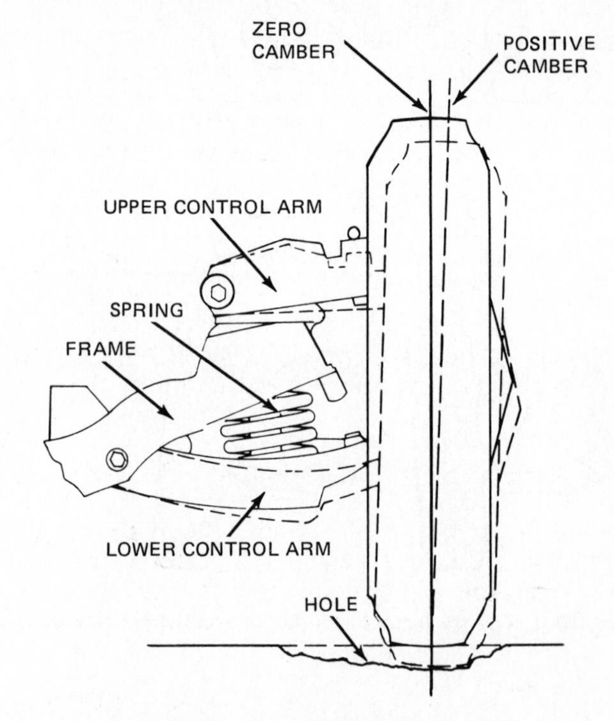

Fig. 10–4 Camber goes slightly positive when the tire drops into a hole in the road.

tread on the right front tire to wear excessively because it carries more than its share of car weight. To overcome this, some vehicles have camber settings that give the left front wheel ¼° more positive camber than the right. Since the car tends to pull toward the wheel with the most positive camber, the effects of the average crowned road are overcome. The car travels forward in a straight line.

Another reason for the increased positive camber on the left front wheel is that most cars are operated with only the driver in them. When the driver gets in the car, the driver's weight tends to slightly lower the left side

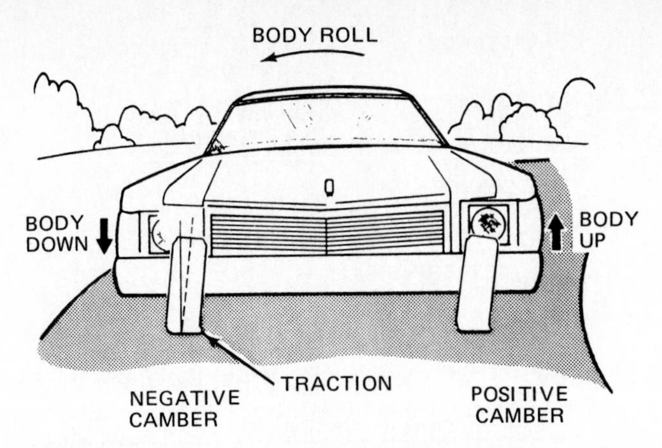

Fig. 10–5 Changes in camber during a turn. (*Ford Motor Company*)

of the car. This, in turn, reduces the positive camber of the left front wheel.

Camber changes have little effect on directional stability when the car is traveling straight ahead. However, in cornering, centrifugal force causes the body to lean toward the outside of the turn (Fig. 10-5). Tire traction resists the resulting tendency to skid. As a result, the side forces against the bottom of the tires causes their tops to tilt toward the inside of the turn. In relation to the car frame, the outer wheel tilts in at the top, producing negative camber. The inner wheel tilts out at the top, producing positive camber.

Improper camber on both wheels can cause hard steering, unstable steering, and wander. Excessive, unequal camber between front wheels contributes to low-speed shimmy.

Incorrect chassis height, caused by sagging front or rear springs, can also affect camber. When a rear spring is weak, it can cause a camber change at the diagonally opposite front wheel. In some cars, this camber change may be as much as ¾° for each inch [25 mm] of rear-spring sag. This is the reason that front-suspension height is checked and corrected before the front wheels are aligned.

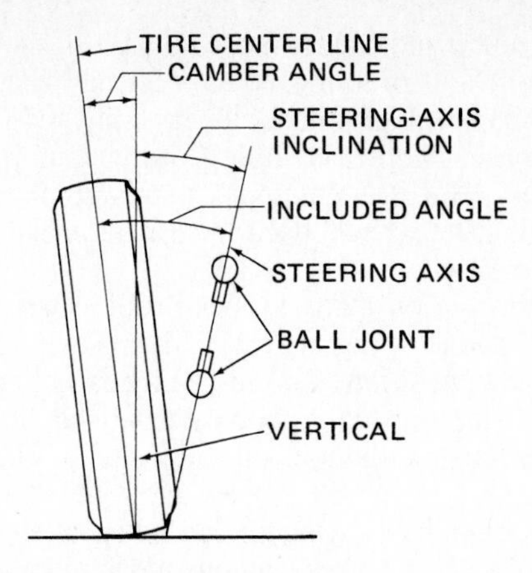

Fig. 10–6 Camber angle and steering-axis inclination. Positive camber is shown.

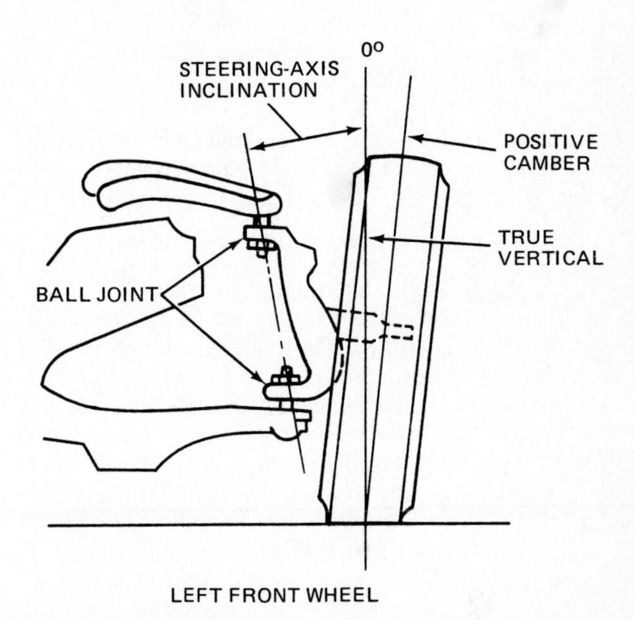

Fig. 10–7 Steering-axis inclination. (*Chevrolet Motor Division of General Motors Corporation*)

❁ 10-4 Steering-axis inclination In older cars, all steering systems had a kingpin that attached the steering knuckle to a support (Fig. 8-50). Then ball joints (❁ 8-24) were adopted, allowing the steering knuckle and steering-knuckle support to be combined into a single part. With ball joints, no kingpin is used. Instead, the steering knuckle is supported through ball joints by upper and lower control arms.

A line drawn through the centers of the ball joints is called the *steering axis* (Fig. 10-6). This is the line around which the steering knuckle pivots as the wheel swings to the right or left. Therefore, *steering axis* is defined as *the centerline around which the front wheel swings for steering*.

Steering-axis inclination (SAI) is also called *ball-joint inclination* (or *kingpin inclination* [KPI] on vehicles that have kingpins). It is the angle, measured in degrees, between the vertical and a line drawn through the centers of the ball joints, when viewed from the front of the vehicle (Fig. 10-7). Another definition is the inward tilt of the ball-joint centerline from the vertical (Fig. 10-6).

Inward tilt of the ball-joint centerline is desirable for several reasons. First, it helps provide steering stability by returning the wheels to a straight-ahead position after the car has turned. This is called *returnability*. Second, steering-axis inclination reduces steering effort, particularly when the car is stationary. Third, it reduces tire wear.

Also, steering-axis inclination tends to keep the front wheels rolling straight ahead. This is because the inward tilt of the steering axis causes the car to be raised slightly as the front wheels are swung away from the straight-ahead position.

When the front wheel is in the straight-ahead position, the wheel spindle is at its highest point. As the spindle pivots forward or backward, it begins to drop. This is because the spindle pivots around the steering axis, which is tilted inward. However, the tire is already in contact with the ground and cannot move down. So, as the wheel is swung away from the straight-ahead position, the steering knuckle, ball joints, and car body and frame are lifted upward. The lift is slight—only about 1 inch [25 mm] or less. But it is enough for the weight of the car to help bring the wheels back to straight ahead after the turn is completed.

This same action tends to make rolling wheels resist any small force trying to move them away from the straight-ahead position. The resistance is not enough to cause hard steering, only good steering stability.

Steering-axis inclination is not adjustable. The amount of the inclination, or tilt, is designed into the steering knuckle. Generally, if camber can be adjusted to specifications, the steering-axis inclination is correct. However, a change in camber will cause a similar change in SAI. This is discussed further in the following section on included angle (❁ 10-5). When the SAI is not within specifications, the spindle, ball joints, or other parts are bent and should be replaced.

❁ 10-5 Included angle In front-end geometry, the *included angle* is the camber angle plus the steering-axis inclination angle (Fig. 10-7). This can be written as

Included angle = camber + SAI

The included angle determines the point of intersection of the tire centerline with the steering axis or ball-joint centerline (Fig. 10-8). This point determines whether the rolling wheel tends to toe in or toe out.

Toe-in (❁ 10-7) is the amount that the front wheels point inward (Fig. 10-9). It is measured in inches, millimeters, or degrees. Toe-out is just the opposite. A wheel that has toe-out points outward as it rolls. The tire on a wheel that rolls with toe-in or toe-out will wear rapidly. The tire has to travel in the direction that the car is moving. But if the tire has toe-in or toe-out, it is dragged sideways as it rolls forward. The more toe-in or toe-out the tire has, the more it is dragged sideways and the faster it wears. Ideally, the wheel rolls straight ahead, with neither toe-in nor toe-out.

Figure 10-8 shows two opposing forces acting on the wheel. One is the forward push through the ball joints. The other is the road resistance to the tire. If these two forces are exactly in line, the wheel will have no tendency toward toe-in or toe-out. However, the two forces are in line with each other only when the point of intersection of the tire centerline with the steering axis is at the road surface.

When the point of intersection is below the road surface (Fig. 10-8*a*), the wheel tends to toe out. Since the forward push through the centerline of the ball joints is inside the tire centerline at the road surface, the wheel attempts to swing outward. When the point of intersection is above road level (Fig. 10-8*b*), the wheel attempts to swing inward, or toe in.

❁ 10-6 Caster Caster is the angle (measured in degrees) formed by the forward or rearward tilt of the steering axis from vertical, when viewed from the side of the wheel (Fig 10-10). The angle is positive (+) when the steering axis tilts backward (Fig. 10-11, left). With positive caster, the upper ball joint is behind the lower ball joint. Caster is negative (−) when the steering axis tilts forward (Fig. 10-11, right). Then the upper ball joint is ahead of the lower ball joint. When the upper ball joint is directly above the lower one, there is zero caster.

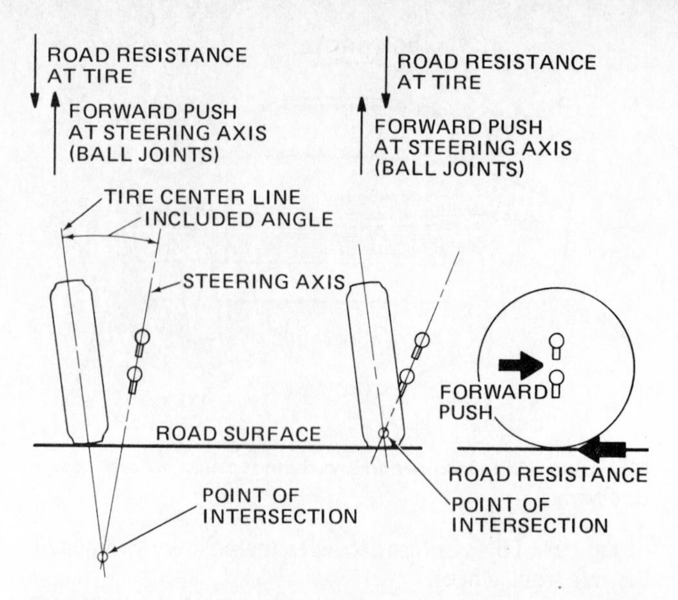

Fig. 10–8 Effect when the point of intersection is (left) below the road surface and (center) above the road surface. The left front wheel as viewed from the driver's seat is shown. To the right is a side view of the wheel to show the two forces acting on the tire and ball joints.

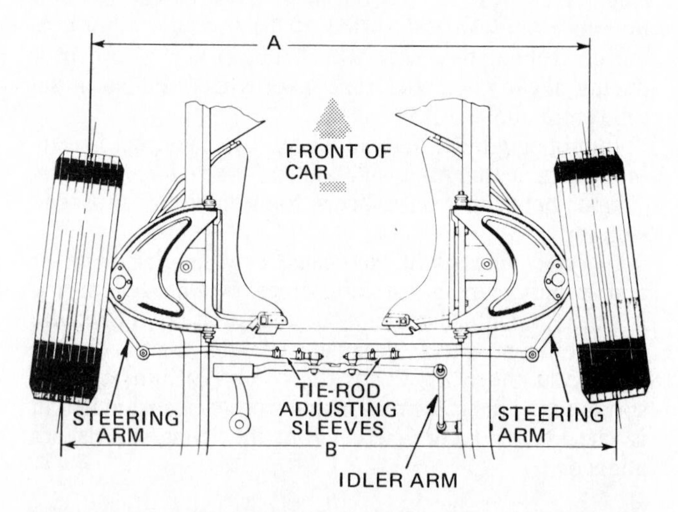

Fig. 10–9 Toe-in. The wheels are viewed from above, with the front of the car at the top of the illustration. Distance *A* is less than distance *B*. Toe-in angles are shown greatly exaggerated. (*Bear Manufacturing Company*)

Although the caster angle is adjustable, it has little effect on tire wear. There are three reasons why caster is used. (Note that tire wear is *not* one of them.) These are:

1. To maintain directional stability and control
2. To increase steering returnability
3. To reduce steering effort

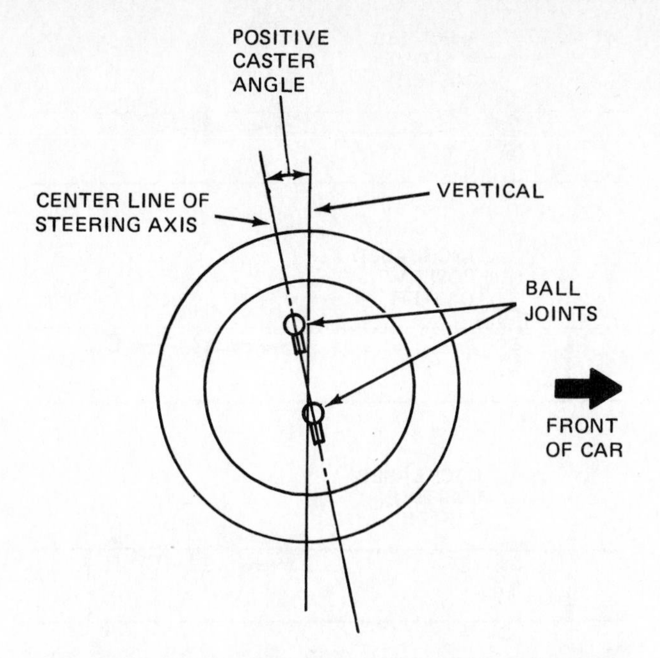

Fig. 10–10 Caster of the left front wheel as viewed from the driver's seat. The view is from the inside so that the backward tilt of the steering axis from the vertical can be seen. This backward tilt is called *positive caster.*

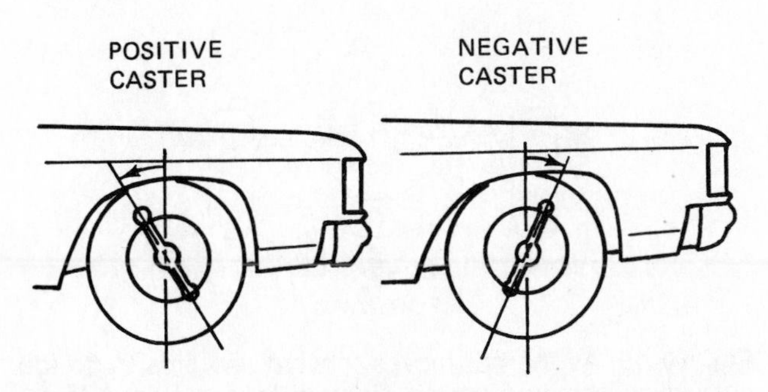

Fig. 10–11 Left, positive caster angle. Right, negative caster angle. (*Ford Motor Company*)

Positive caster aids directional stability. The centerline of the ball joints passes through the road surface ahead of the centerline of the wheel (Fig. 10-10). Therefore, the push on the ball joints is ahead of the road resistance to the tire (Fig.10-8). The tire is trailing behind, just as the caster on a table leg trails behind when the table is pushed (Fig. 10-12). Like the table-leg caster, a car wheel that is pulled has greater directional and steering stability than a wheel that is pushed. In addition, positive caster tends to keep the wheels pointed straight ahead. It helps overcome any tendency for the car to wander or steer away from the straight-ahead position.

Negative caster does not aid directional stability. Steering-axis inclination contributes more than caster to directional stability of the tire. This is why cars with manual steering, which often have a slight negative caster, still retain good directional stability.

Caster also affects the tendency of the steering wheel to return to its center position after making a turn. This is called *returnability*. Positive caster increases steering-

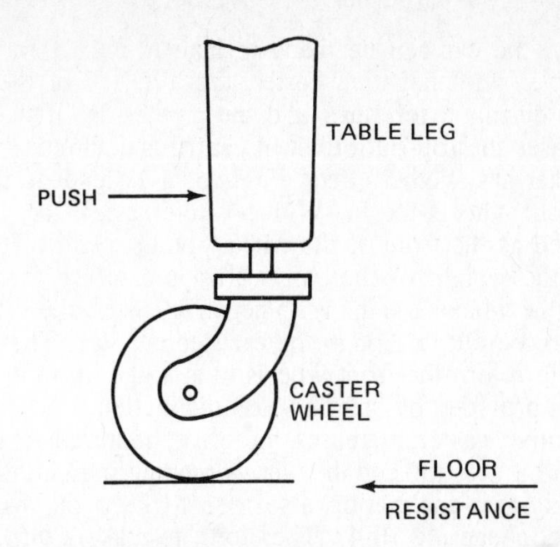

Fig. 10–12 The wheel of the caster trails behind and follows in the direction of the push when the table leg is moved.

wheel returnability. Therefore, cars with power steering often have slightly more positive caster than manual-steering cars. The additional positive caster helps overcome the tendency of the power steering to hold the front wheels in a turn. Although the additional positive caster requires greater turning effort, the driver does not notice it because of the power steering.

Positive and negative caster have different effects on the actions of the wheels and car body. When both front wheels have positive caster, the body leans toward the outside of the turn (Fig. 10-5). But if the front wheels have negative caster, the car tends to lean into the turn (Fig. 10-13).

When the left front wheel has positive caster, during a left turn the left front wheel tries to move down against the road surface. This causes the ball joints to be lifted, which lifts the left side of the car body. At the same time, positive caster causes the right side of the car to be lowered slightly during a left turn.

When the right side of the car is lowered and the left side is lifted as a left turn is made, the car leans out on the turn (Fig. 10-5). This is just the opposite of what is desirable, since it adds to the effect of centrifugal force on the turn. By using negative caster, which tilts the centerline of the ball joints forward as in Fig. 10-11

Fig. 10–13 Negative caster counteracts the roll-out effects of centrifugal force during a high-speed turn. (*McQuay-Norris, Inc.*)

(right), the car can be made to lean in on a turn. For example, with negative caster, the left side of the car drops during a left turn and the right side lifts. This decreases the roll-out effect of centrifugal force.

Caster has another effect. Positive caster tends to make the front wheels toe in. With positive caster, the car is lowered as the front of the wheel pivots inward. Therefore, the weight of the car is always exerting force to make the wheels toe in. With negative caster, the wheels tend to toe out, although the car steers easier. Then the force to return the front wheels to straight ahead after a turn is provided by steering-axis inclination.

Positive caster increases the effort required to steer because it tries to keep the wheels running straight ahead. Steering-axis inclination also tries to keep the wheels straight ahead (❁ 10-4). Therefore, to make a turn, the steering effort must overcome the effects of both positive caster and steering-axis inclination.

Excessive positive caster can cause problems. These include too much steering effort, snap-back of the steering wheel after a turn, wander, low-speed shimmy, and an increased amount of road shock. Improper caster may occur as a result of spring sag. For example, front-spring sag decreases positive caster (Fig. 10-14). For this reason, front-suspension height is checked and corrected before the front end is aligned.

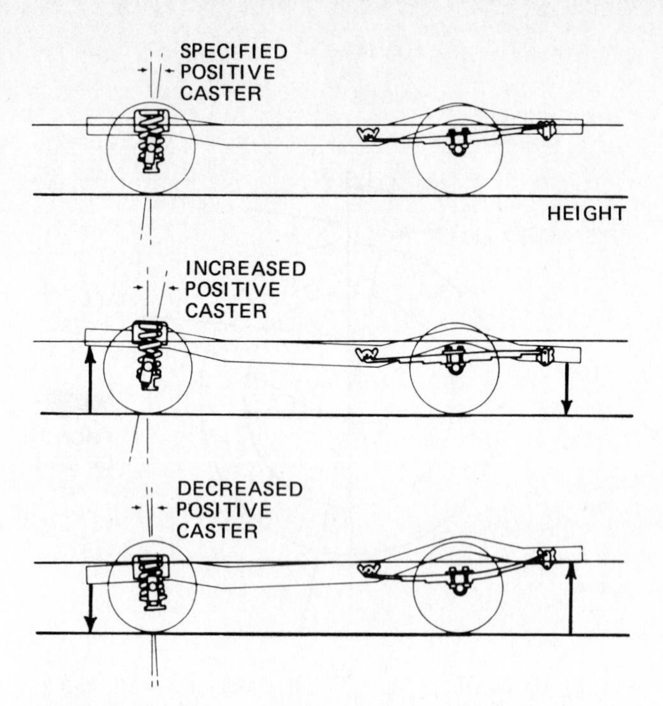

Fig. 10–14 Effects of sagging springs on caster. (*Ford Motor Company*)

❁ 10-7 Toe Toe is the amount in inches, millimeters, or degrees by which the front wheels point inward or outward. Its purpose is to ensure parallel rolling of the front wheels, to stabilize steering, and to prevent side-slipping and excessive wear of the tires. With toe-in, the front wheels attempt to roll inward instead of straight ahead. This is because the tires are slightly closer together at the front than at the rear (Fig. 10-4). A typical toe-in setting is about ⅛ inch [3.2 mm].

Front-wheel toe-in is set with the car standing still. It offsets the play in the steering linkage, which is eliminated when the car is moving forward (Fig. 10-15). This change in toe is due to the rolling resistance of the tires against the road. The amount of toe-in should be sufficient to prevent any toe-out when the car is moving forward. This is called *running toe* and may be measured by some types of dynamic wheel analyzers (Chap.13). Ideally, the toe setting will bring the actual running toe to very nearly zero when the car is moving.

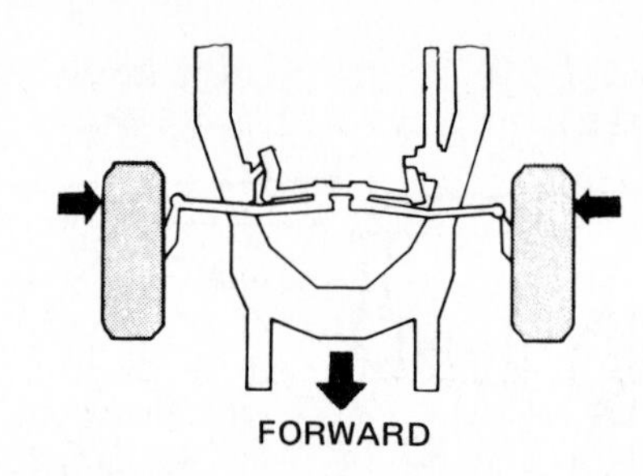

Fig. 10–15 As the car moves forward, the tires try to toe out, which compresses the steering linkage. (*Ford Motor Company*)

As a car with toe-in begins to move forward, the backward push of the ground againt the tires (shown in Fig. 10-15) tries to force the wheels to toe out. Since the tie rods are behind the wheels, this force compresses the steering linkage and takes up any play. As a result, the tires become parallel and roll straight ahead.

Toe greatly affects tire wear. Therefore toe can be adjusted. Excessive toe-in or toe-out causes the tire tread to scrub the road. This happens when the car is moving straight, with the tire turned slightly in or out. Whatever the toe setting, the wheels must not run with toe-out. This causes wander.

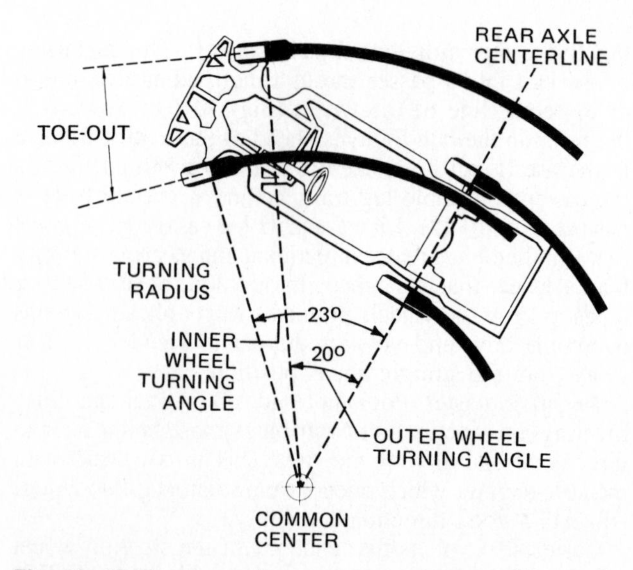

Fig. 10–16 Turning radius, or toe-out-on-turns. (*Hunter Engineering Company*)

❁ 10-8 Turning radius Turning radius is sometimes called *toe-out during turns* and *turning angle*. It is the difference between the two angles formed by the two front wheels and the car frame during turns. Then the inner wheel is following the radius of a smaller circle than the outer wheel (Fig. 10-16). Therefore the inner wheel must toe out more to prevent tire sideslip and excessive wear.

When the front wheels are steered to make the turn shown in Fig. 10-16, the inner wheel turns at an angle of 23°, while the outer wheel turns only 20°. This permits the inner wheel to follow a shorter radius than the outer wheel. The two front wheels turn on concentric circles, which have a common center (Fig. 10-16).

Toe-out during turns is achieved by the proper relationship between the steering arms, tie rods, and steering gear (Fig. 10-17). For this reason, the steering arms are angled inward (Fig. 10-9). If the steering arms were parallel to the wheels, then the wheels would remain parallel during a turn. This would scuff the tires. The relationship of the parts ensures that the inner wheel on a curve always has more toe-out than the outer wheel.

As the dotted line in Fig. 10-17 shows, when the tie rod is moved to the left during a right turn, it pushes at almost a right angle against the left steering arm. At the same time, the right end of the tie rod moves to the left and swings forward. This turns the right wheel an additional amount. When a left turn is made, the left wheel is turned more than the right wheel.

Figure 10-17 shows a parallelogram type of steering linkage. Other types of steering linkage provide a similar toe-out of the inner wheel during turns. Turning radius cannot be adjusted. If it is not correct, one or both of the steering arms are bent and should be replaced.

10-9 Steering linkages Many types of steering linkages have been made to connect the front-wheel steering knuckles to the steering gear. The pitman arm swings from one side to the other—or forward and backward, on some cars—as the steering wheel is turned. This movement is carried to the wheel spindle by the steering linkage. The linkage must be designed so that the driver can easily and accurately control the position of the wheels. However, the position of the wheels, and any up-and-down motion that they may go through, should not affect the driver's control of the car.

To some extent, the type of front-suspension system determines the design of the steering linkage. When a car has independent front suspension, there are two tie rods and usually an idler arm (Fig. 10-9). However, when a solid front axle is used, there is only one long tie rod (Fig. 10-17). No idler arm is required.

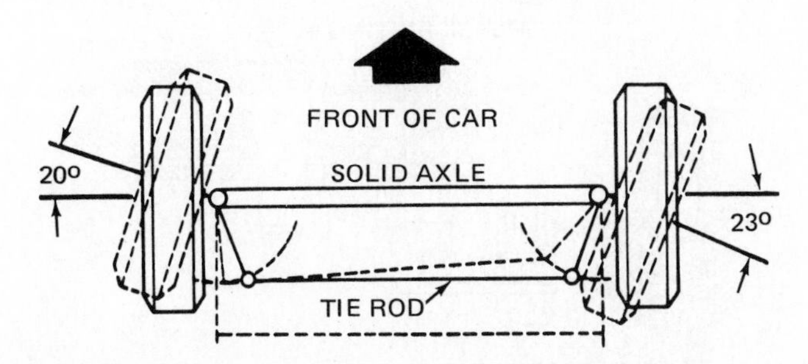

Fig. 10–17 How toe-out-on-turns is obtained. (*Chevrolet Motor Division of General Motors Corporation*)

Figure 10-18 shows several types of steering linkage. All of these have some means of adjusting the lengths of the tie rods or links so that proper alignment of the front wheels can be established. This alignment gives the front wheels a slight toe-in when the car is at rest. When the car begins to move forward, the toe-in practically disappears as all looseness in the steering system is taken up.

The most common type of steering linkage is some form of the parallelogram system (Fig. 10-19). Points of attachment between the metal parts of the tie rods, pitman and idler arms, and spindles are insulated by bushings. Tie-rod ends, which include a tapered ball stud and socket similar to a ball joint, attach the tie rods to the steering arms on the spindles. Adjusting sleeves that connect the tie rods to the tie-rod ends are threaded, as are the mating ends of the tie rods. This permits adjustment of the front-wheel toe. To change the effective length of the tie rods, the sleeves can be turned one way or the other. This alters how much the front wheels point it.

Figure 10-20 is a schematic layout of a similar steering linkage. This system is the same for both manual and power steering.

10-10 Ball sockets and tie-rod ends The various parts of the steering linkage are connected by ball sockets or tie-rod ends of several kinds (Fig. 10-21). Some have a grease fitting for lubrication. Others are prelubricated for life at the time of manufacture.

On many cars, the idler arm is connected through rubber bushings. These bushings twist as the idler arm swings to one side or the other. They then supply some force to help return the wheels to center after a turn is completed. Ball sockets are smaller, but are similar in construction to the front-suspension ball joints (8-24).

Manual steering

10-11 Steering gears The steering gear converts the rotary motion of the steering wheel into straight-line motion of the linkage. There are two basic types of steering gears, the pitman-arm type and the rack-and-pinion type. Either of these can be straight manual or powered (power steering).

The pitman-arm type has a gear box at the lower end of the steering shaft (Fig. 10-20). The rack-and-pinion type has a small gear (a *pinion*) at the lower end of the steering shaft (Fig. 10-22). The action is the same in either system. When the steering wheel and shaft are turned by the driver, the rotary motion is changed into straight-line motion. This causes the front wheels to pivot or swing from one side to the other to steer the car.

10-12 Pitman-arm steering gears Steering linkages using pitman arms are shown in Figs. 10-18 to 10-20. The steering gear at the lower end of the steering shaft consists essentially of two parts: a worm on the end of the steering shaft, and a pitman-arm shaft (or cross-

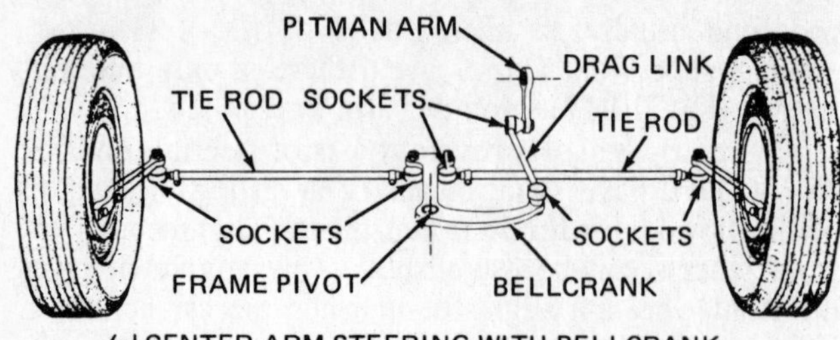

(a) CENTER-ARM STEERING WITH BELLCRANK INTERMEDIATE ARM

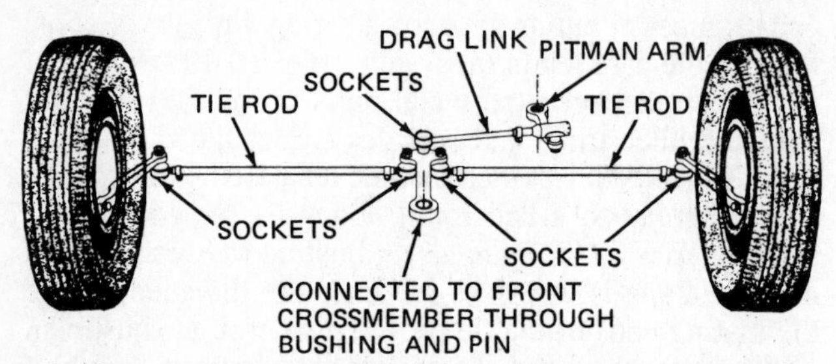

(b) CENTER-ARM STEERING WITH TRANSVERSE DRAG LINK

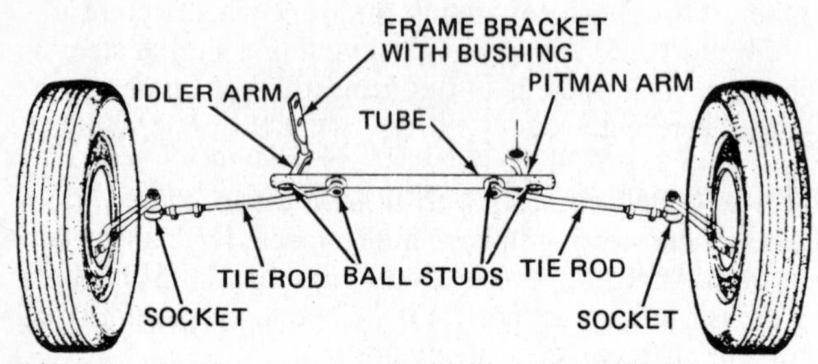

(c) PARALLELOGRAM LINKAGE WITH TUBULAR CENTER LINK

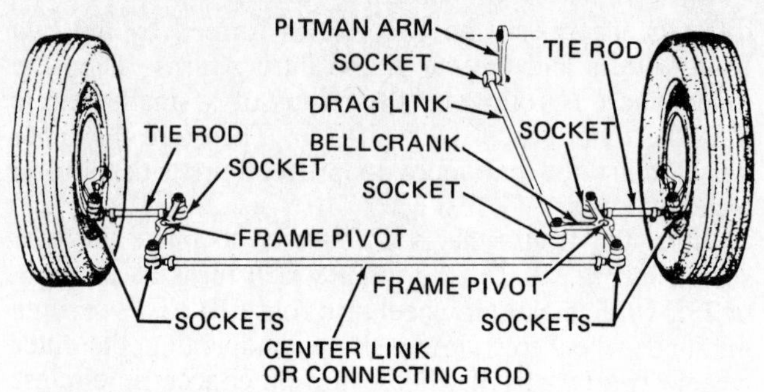

(d) PARALLELOGRAM LINKAGE WITH CENTER LINK AHEAD OF AXLE

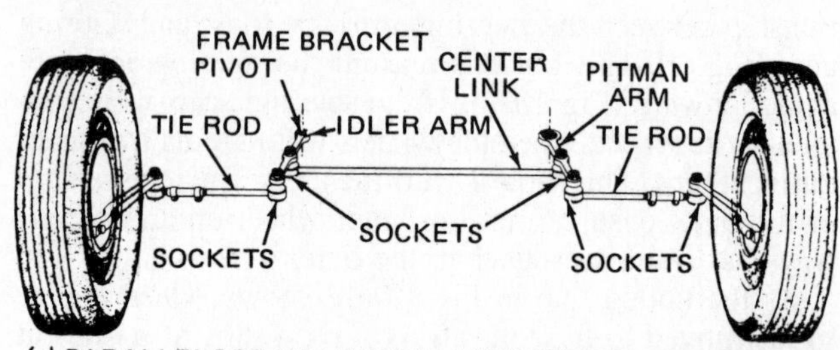

(e) PARALLELOGRAM LINKAGE WITH CENTER LINK BEHIND AXLE

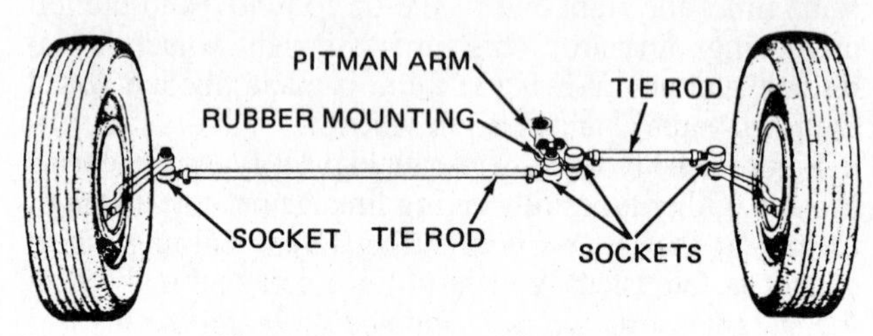

(f) LONG-ARM, SHORT-ARM LINKAGE

Fig. 10–18 Various types of steering linkages. (*TRW, Inc.*)

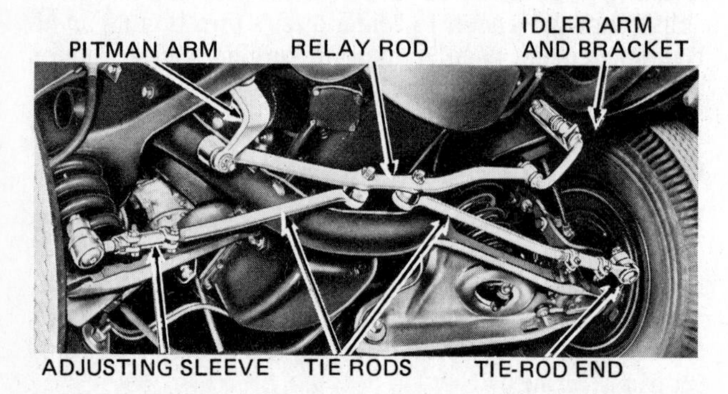

Fig. 10–19 Steering linkage with transverse drag link, which usually is called a relay rod. (*Ford Motor Company*)

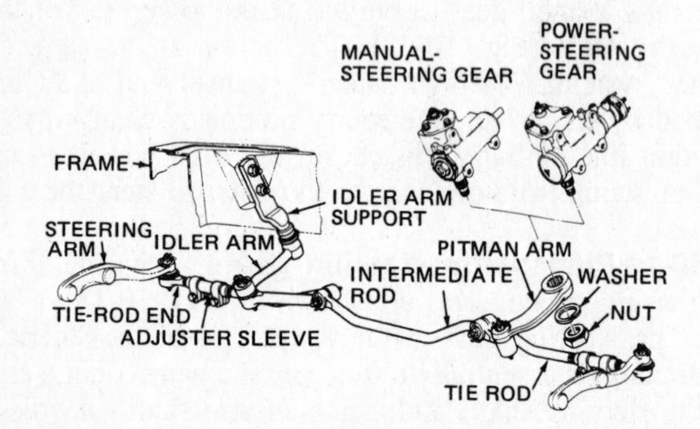

Fig. 10–20 Basic parts of a steering linkage.

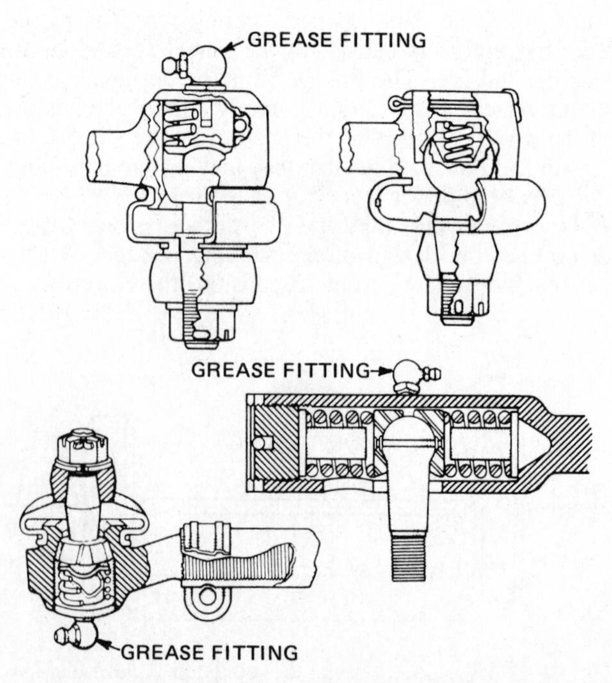

Fig. 10–21 Ball sockets and tie-rod ends. (*Ford Motor Company*)

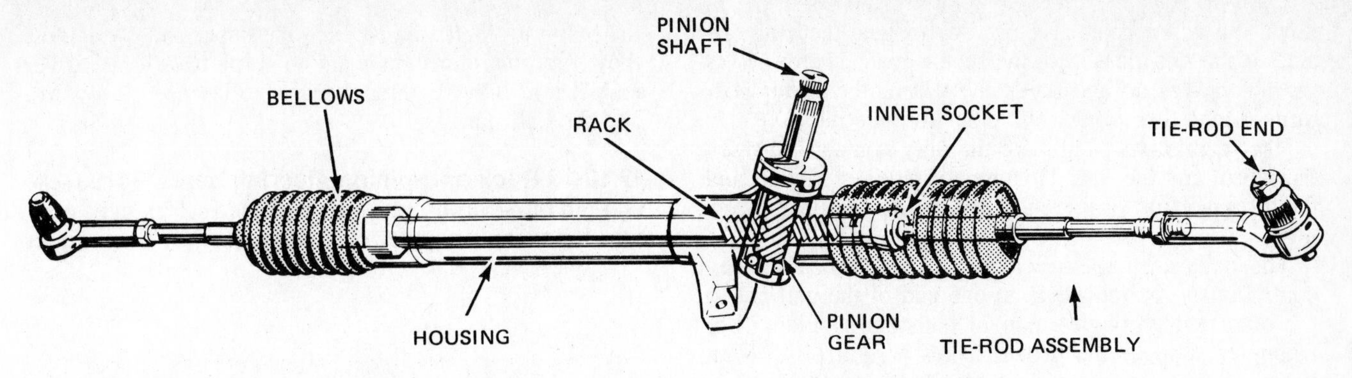

Fig. 10–22 Phantom view of a manual rack-and-pinion steering gear. (*Chrysler Corporation*)

shaft) on which there is a gear sector, a toothed roller, or a stud.

The gear sector, toothed roller, or stud meshes with the worm (Fig. 10-23). In Fig. 10-23, the steering gear uses a toothed roller. The roller and worm teeth mesh. When the worm is rotated (by rotation of the steering wheel), the roller teeth must follow along. This action causes the pitman-arm shaft to rotate. The other end of the pitman-arm shaft carries the pitman arm. Rotation of the pitman-arm shaft causes the arm to swing in one direction or the other. This motion is then carried through the linkage to the steering knuckles at the wheels.

NOTE: The pitman-arm shaft is also called the *cross-shaft, pitman shaft, roller shaft, steering-arm shaft,* and *sector shaft.*

Two versions of the recirculating-ball steering gear are shown in Figs. 10-24 and 10-25. In these units, the worm gear on the end of the steering shaft has a special ''nut'' running on it. The nut rides on rows of small recirculating balls. The recirculating balls move freely through grooves in the worm and inside the nut. As the steering shaft is rotated, the balls force the nut to move up and

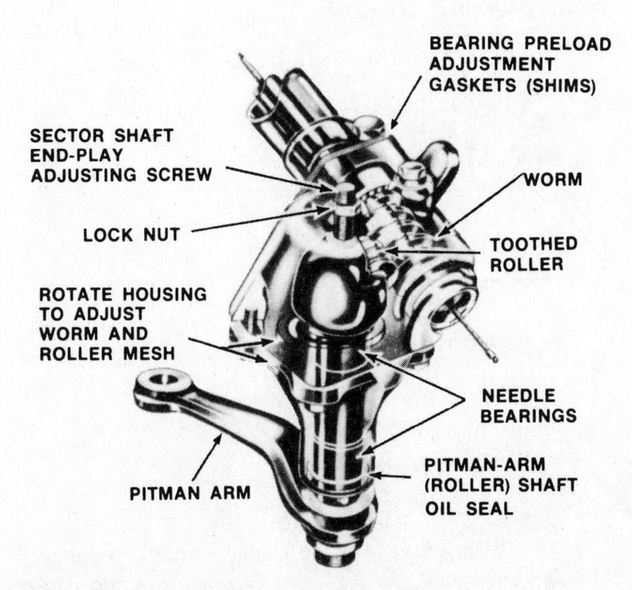

Fig. 10–23 Phantom view of a steering gear using a toothed roller attached to the pitman-arm shaft. The teeth of the worm mesh with the teeth of the roller. (*Ford Motor Company*)

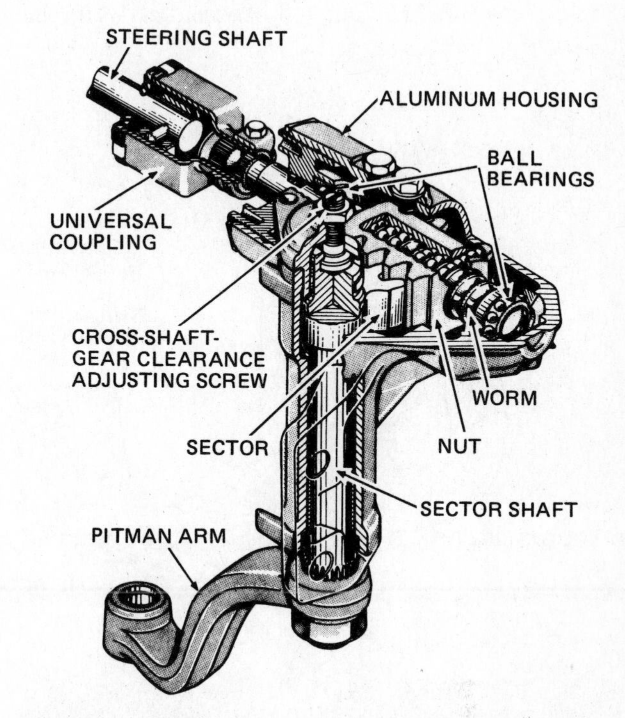

Fig. 10–24 Cutaway view of a recirculating-ball steering gear. (*Chrysler Corporation*)

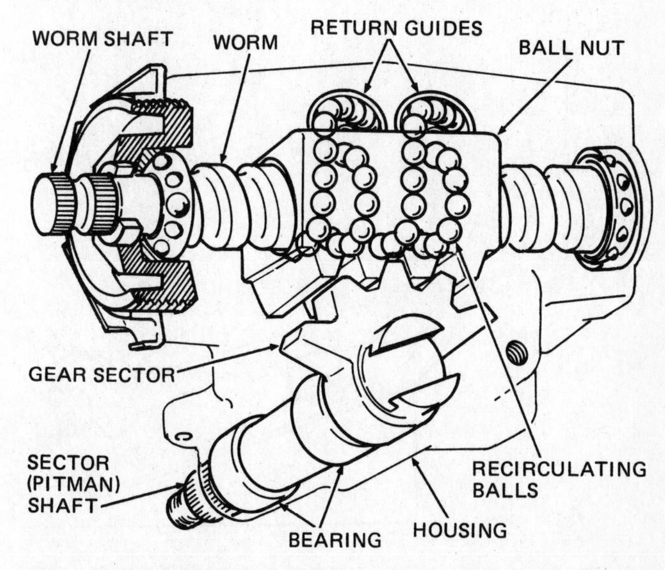

Fig. 10–25 Phantom view of a recirculating-ball steering gear. (*Ford Motor Company*)

down the worm gear. A short rack of gear teeth on one side of the nut mesh with the sector gear. Therefore, as the nut moves up and down the worm, the sector gear turns in one direction or the other for steering.

The recirculating balls are the only contacts between the worm and the nut. This greatly reduces friction and the turning effort or force applied by the driver for steering.

The balls are called recirculating balls because they continuously recirculate from one end of the ball nut to the other end through a pair of ball-return guides. For example, suppose the driver makes a right turn. Then the worm gear is rotated in a clockwise direction when viewed from the driver's seat. This causes the ball nut to move upward. The balls roll between the worm and the ball nut. As the balls reach the upper end of the nut, they enter the return guide and then roll back to the lower end. There they re-enter the groove between the worm and the ball nut.

❁ 10-13 Rack-and-pinion steering gears The rack-and-pinion steering gear has become increasingly popular for today's smaller cars. It is simpler, more direct

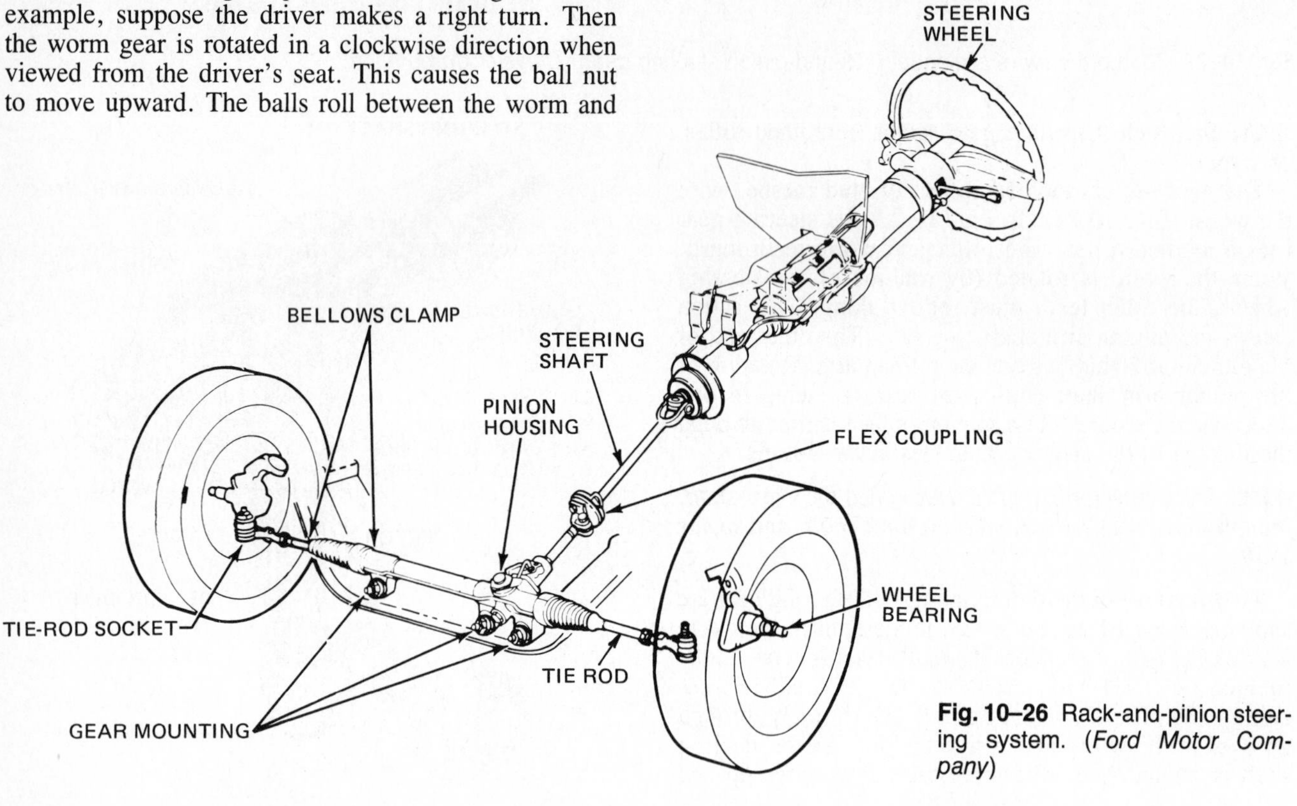

Fig. 10–26 Rack-and-pinion steering system. (*Ford Motor Company*)

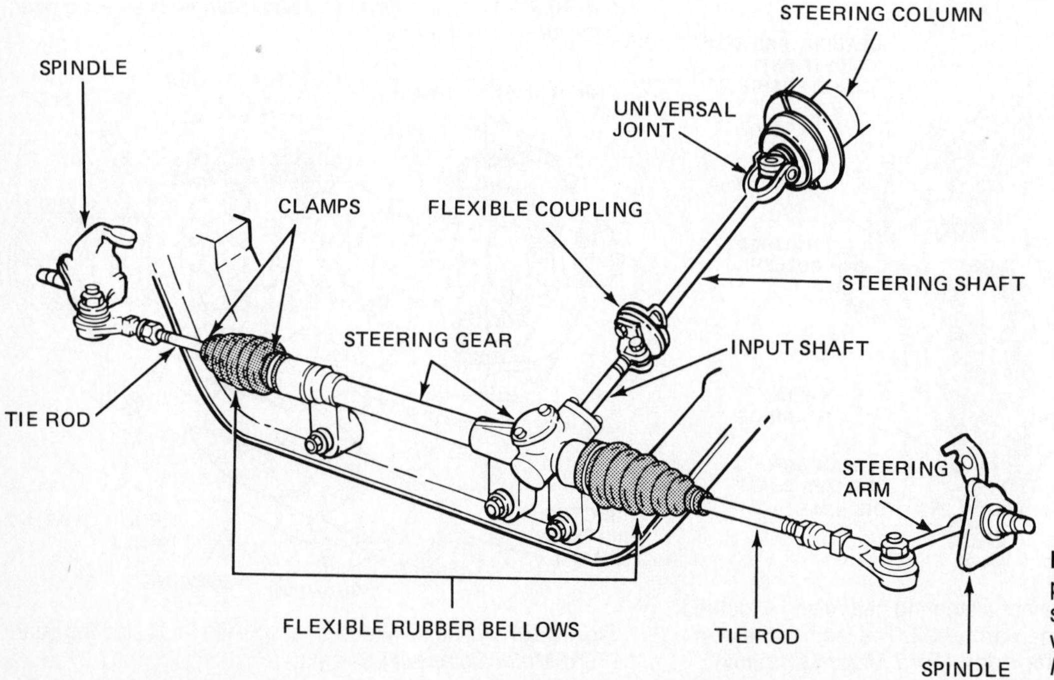

Fig. 10–27 Rack-and-pinion steering gear showing linkage to the wheel spindles. (*Ford Motor Company*)

acting, and may be straight mechanical or power-assisted in operation. Figure 10-26 shows a complete rack-and-pinion steering system, set apart from the rest of the car. As the steering wheel and shaft are turned, the rack moves from one side to the other. This pushes or pulls on the tie rods, forcing the wheel spindles to pivot on their ball joints. This turns the wheels to one side or the other so that the car is steered.

Figure 10-27 shows the steering gear and tie rods. The universal joint at the upper end of the steering shaft and the flexible coupling at the lower end help eliminate road shock. They also prevent road shocks and noise from passing up through the steering column to the driving compartment.

Figure 10-28 shows a completely disassembled manual rack-and-pinion steering gear. The pinion gear is integral with the input shaft. The inner ends of the tie rods have balls which fit into the ball sockets on the two ends of the rack. The ball-and-socket arrangement allows the spindle ends of the tie rods to move up and down with the spindles and wheels.

Rack-and-pinion steering is basically a very simple type of steering gear. It can be used in small cars where the steering forces are light. However, the gear ratio is limited by the diameter of the steering wheel and the pinion gear. On larger, heavier vehicles, this can be a disadvantage. Therefore, other types of steering gears, such as the recirculating ball (❁ 10-12), are usually found

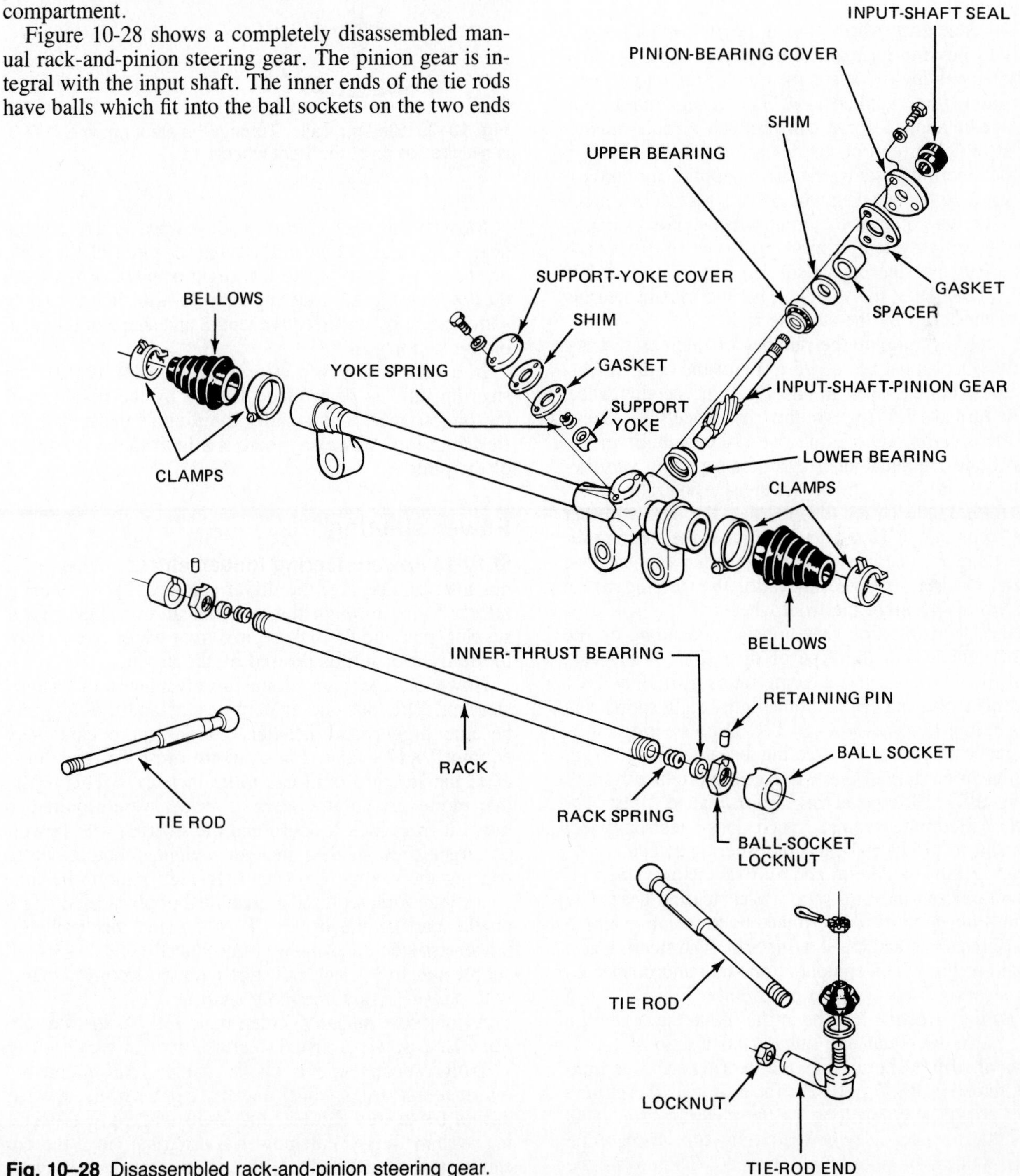

Fig. 10–28 Disassembled rack-and-pinion steering gear.

on larger cars and trucks. The greater mechanical advantage possible with the other types reduces the effort required to steer a larger vehicle.

In a small car, rack-and-pinion steering is quick and easy. It provides the maximum amount of road feel as the tires meet irregularities in the road. There is no damping out of road shocks and vibration. Other types of steering systems usually provide some damping action. They can also be adjusted to eliminate almost all play in the steering wheel. Mechanical advantage and the meaning of "quick" steering are discussed in ❁ 10-14.

❁ 10-14 Steering ratio One of the jobs of the steering gear is to provide mechanical advantage. In a machine or mechanical device, this is the ratio of the output force to the input force applied to it. This means that a relatively small applied force can produce a much greater force at the other end of the device.

In the steering system, the driver applies a relatively small force to the steering wheel. This results in a much larger steering force at the front wheels. For example, in one steering system, applying a force of 10 pounds [44.5 N] to the steering wheel can produce up to 270 pounds [1201 N] at the wheels. This increase in steering force is produced by the steering ratio.

The steering ratio is the number of degrees that the steering wheel must be turned to pivot the front wheels one degree. For example, in Fig. 10-29 the steering wheel must be turned 17.5° to pivot the front wheels 1°. Therefore, the steering ratio is 17.5:1. In cars built in the United States, manual-steering ratios typically vary between 15:1 and 33:1. The higher the steering ratio (33:1, for example), the easier it is to steer the car, all other things being equal. However, the higher the steering ratio, the more the steering wheel has to be turned to achieve steering. With a 30:1 steering ratio, the steering wheel must turn 30° to pivot the front wheels 1°.

Actual steering ratios vary greatly, depending on the type of vehicle and the type of operation. Many cars with manual steering use steering ratios as high as 28:1 to minimize steering effort. Some lightweight sports cars have steering ratios as low as 10:1. High steering ratios are often called "slow" steering because the steering wheel has to be turned many degrees to produce a small steering effect. Low steering ratios, called "fast" or "quick" steering, require much less steering-wheel movement to produce the desired steering effect.

Steering ratio is determined by two factors: steering-linkage ratio and the gear ratio in the steering gear. The steering-linkage ratio is determined by the relative length of the pitman arm and the steering arm. The steering arm is bolted to the wheel spindle at one end and connected to the steering linkage at the other end.

When the effective lengths of the pitman arm and the steering arm are equal, the linkage has a ratio of 1:1. If the pitman arm is shorter than the steering arm, the ratio is less than 1:1. In Fig. 10-29, for example, the pitman arm is about twice as long as the steering arm. This means that for every degree the pitman arm swings, the wheels will pivot about two degrees. Therefore, the steering-linkage ratio is about 1:2.

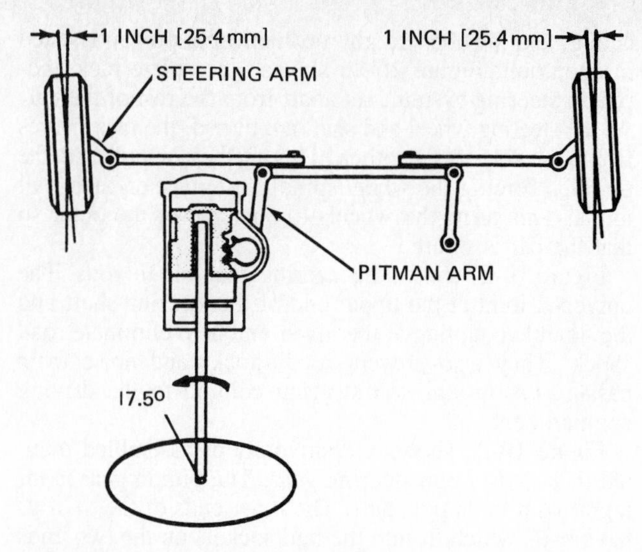

Fig. 10–29 Steering ratio. Turning the steering wheel 17.5° is required to pivot the front wheels 1°.

Most of the steering ratio is developed in the steering gear. The ratio is due to the angle or pitch of the teeth on the worm gear and to the angle or pitch of the teeth on the sector gear. Steering ratio is also determined to some extent by the effective length and shape of the teeth on the sector gear.

In a rack-and-pinion steering system (❁ 10-13), the steering ratio is determined largely by the diameter of the pinion gear. The smaller the pinion, the higher the steering ratio. However, there is a limit to the smallness of the pinion.

Power steering

❁ 10-15 Power-steering fundamentals When a car has manual steering, the driver supplies all the steering effort. Then, through the mechanical advantage of the steering gear and the linkage, the front wheels are pointed to the right or left as desired by the driver.

However, there are some disadvantages to manual steering. This became evident as cars and their engines became bigger and heavier, and as more cars were equipped with wider, low-pressure tires. To steer these cars, the steering ratio had to be increased. This meant that more turns of the steering wheel were required to move it from lock to lock, and the steering was slower.

Larger tires, with a heavier weight on them, made parking the car more difficult. In fact, parking became a task that often required a great deal of physical strength on the part of the driver. To overcome this problem, power-assisted steering was introduced (Fig. 10-30). It supplements the steering effort required from the driver with power from some other source.

Automobiles usually do not have full power steering. They have power-assisted steering, which we call *power steering*. When there is no mechanical connection between the steering wheel and the front wheels, the car cannot be steered if the engine stalls or the power-steering system fails. With power-assisted steering, the car can be steered manually if the power assist is not available.

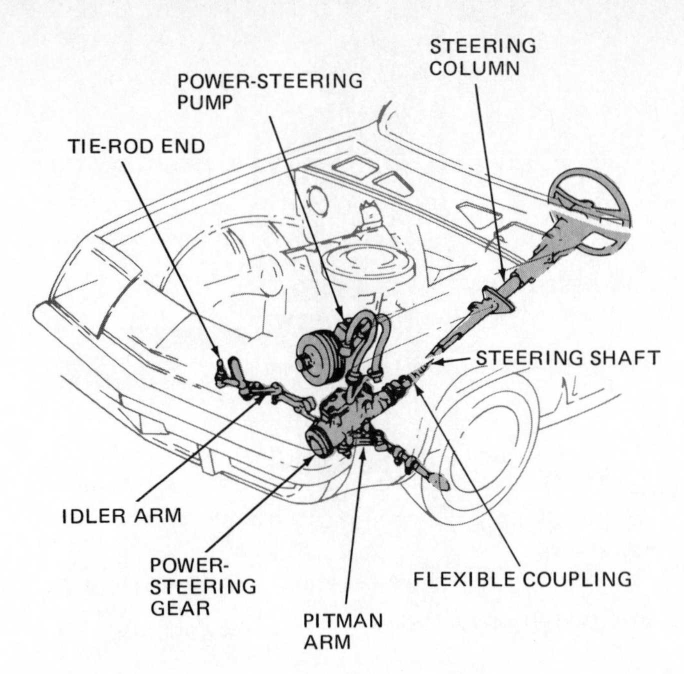

Fig. 10–30 Power steering for a car with an in-line engine mounted in the front.

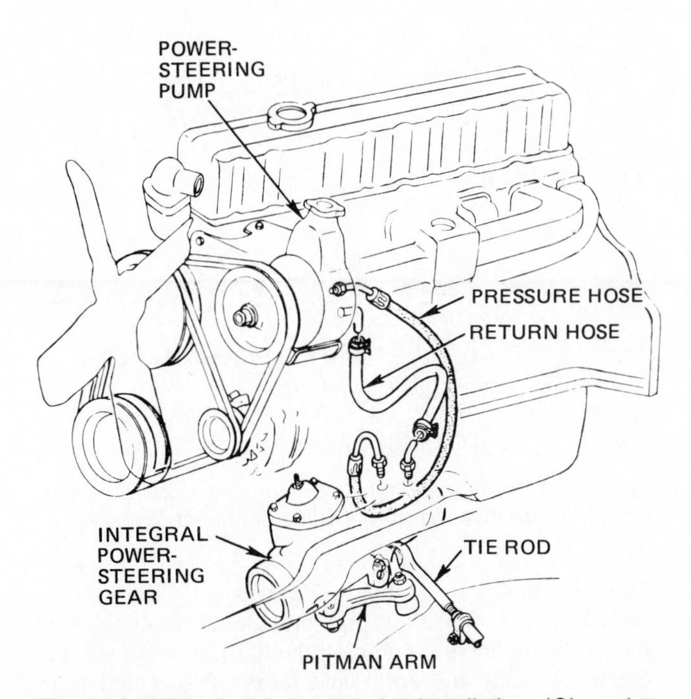

Fig. 10–31 A typical power-steering installation. (*Chevrolet Motor Division of General Motors Corporation*)

Power-steering systems have used compressed air, electrical devices, and hydraulic pressure. Some trucks still have air-operated power steering. However, today hydraulic (oil) pressure is used in all car and most truck power-steering systems. The principles of hydraulics are covered in Chap. 2.

⚙ 10-16 Power-steering operation The power-steering system used in automobiles is basically a modified manual-steering system (Fig. 10-30). The steering column, steering gear, and steering linkage are little changed from the manual-steering systems discussed earlier in this chapter. The main difference is that a power booster has been added to assist the driver.

In the basic system of power-assisted steering, the booster is set into operation when the steering shaft is turned. Then, after the steering effort exceeds a certain force, the booster takes over and provides most of the force required for steering. For example, some cars have power assistance when the force applied at the steering wheel is more than about 3 pounds [13 N]. The force varies with the make and model of the car, but for most cars it ranges from 1.5 to 7 pounds [7 to 31 N].

In the power-steering system, a continuously operating pump provides hydraulic pressure when needed (Fig. 10-31). As the steering wheel is turned, valves are operated to admit this hydraulic pressure to a cylinder that contains the power piston. Then the pressure causes the piston to move, and it provides most of the steering force.

Two general types of power-steering systems are found in automobiles (Fig 10-32). In the integral type, the power piston is built into the steering gear. In the linkage type, the power piston is attached between the frame of the vehicle and the steering linkage. The integral type is the more widely used. Usually it must be installed at the factory as the vehicle is built. The linkage type is widely used on trucks. It also can be installed after manufacture on some vehicles.

⚙ 10-17 Power-steering-system components In addition to the steering gear, the power-steering system has an oil pump, a control valve, a power cylinder, an oil tank or reservoir, and connecting hoses or lines. These are shown in the illustration of the linkage-type power-steering system shown in Fig. 10-32. In the illustration of the integral type in Fig. 10-32, notice that the control valve and power piston are built into, or ''integral'' with, the steering gear. That is where the name ''integral power steering'' comes from. Now, let's briefly describe each part of the power-steering system.

Oil pump The oil pump, or power-steering pump (Fig. 10-33), is driven by the engine. The purpose of the oil pump is to deliver oil under pressure to the control valve. The pump has a fairly low flow rate. It may be less than two gallons [7.6 L] per minute. However, the typical power-steering pump can produce pressures of more than 1000 psi [6895 kPa]. The power-steering pump is discussed further in ⚙ 10-19.

Control valve The control valve directs oil to the power piston to produce steering assistance. The valve is connected to the steering system in such a way that it senses the amount of steering effort being applied. At the right time the control valve opens and closes ports to direct the oil flow from the pump to the power cylinder.

Power cylinder The power cylinder is a hydraulic cylinder that is linked mechanically to the steering system. The cylinder contains a piston that has a pressure chamber at each end. Oil enters one chamber to aid in right turns and the other chamber for left turns.

Figure 10-34 shows a power-steering system for a front-wheel-drive car. Instead of the power-steering gear being located at the lower end of the steering shaft, it is set forward and connected to the steering shaft by a second-

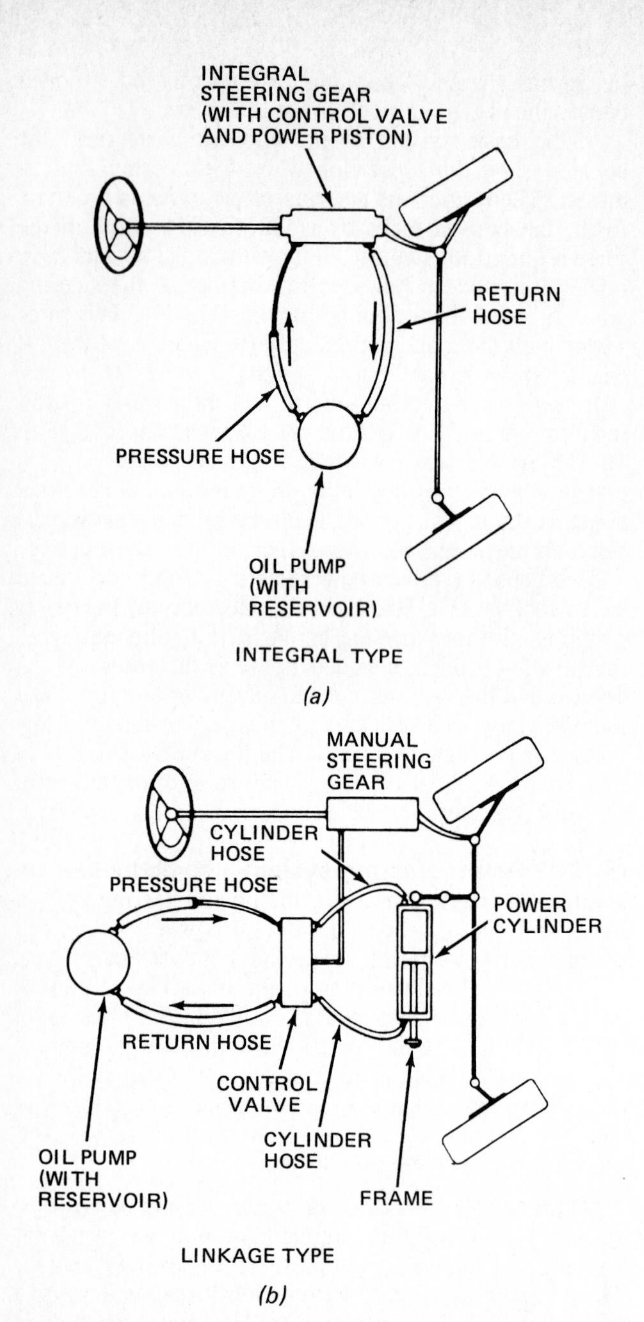

Fig. 10–32 The two basic types of power-steering systems. (*Lempco Industries, Inc.*)

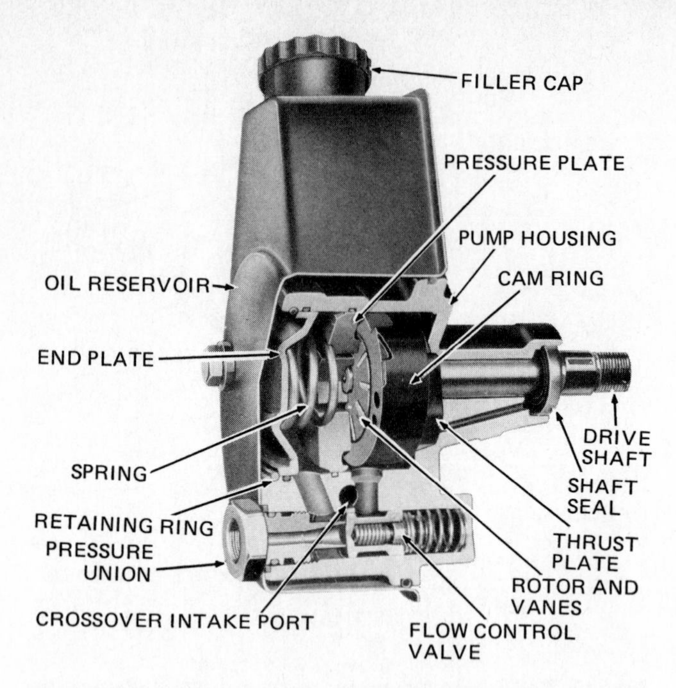

Fig. 10–33 Cutaway view of a vane-type power-steering pump. (*Cadillac Motor Car Division of General Motors Corporation*)

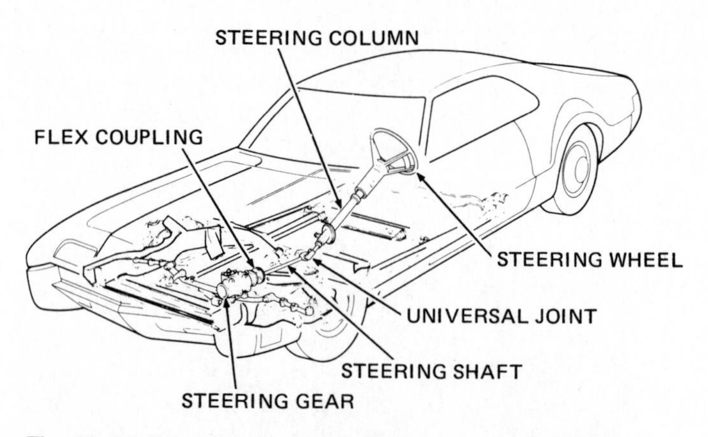

Fig. 10–34 Power-steering system for a car with front-wheel drive. (*Oldsmobile Division of General Motors Corporation*)

ary steering shaft. The two shafts meet at an angle and are connected by a universal joint. This arrangement is required by the front-drive design so that there is room above the steering shaft and linkage for the transmission and other power-train components.

Reservoir The reservoir (Fig. 10-33) is a supply tank for the power-steering fluid, or oil. The tank forms part of the power-steering pump, and is usually mounted on top or behind it. Figure 10-31 shows a typical installation.

Hoses and fittings Special high-pressure hoses, metal lines, and fittings are used to carry hydraulic fluid between the parts of the system (Figs. 10-31 and 10-32). The pressures can go up to 2000 psi [13,790 kPa] in some systems. When this high pressure is combined with the heat generated in pumping the fluid, it becomes necessary to make the hoses and tubing strong and heat-resistant. Notice in Figs. 10-31 and 10-32 that integral power steering has only two hoses. However, linkage-type power steering (Fig. 10-32) may have four hoses.

Fluid cooler Some vehicles with air conditioning have a small oil cooler for the power-steering fluid. This is required because vehicles with air conditioning usually have a higher under-hood temperature than vehicles without air conditioning. The small, finned oil cooler lowers the temperature of the fluid, preventing heat damage to the hoses and seals.

Power-steering fluid The fluid used in power-steering systems is a special type of oil. It must withstand wide variations in temperature and pressure, and resist foam-

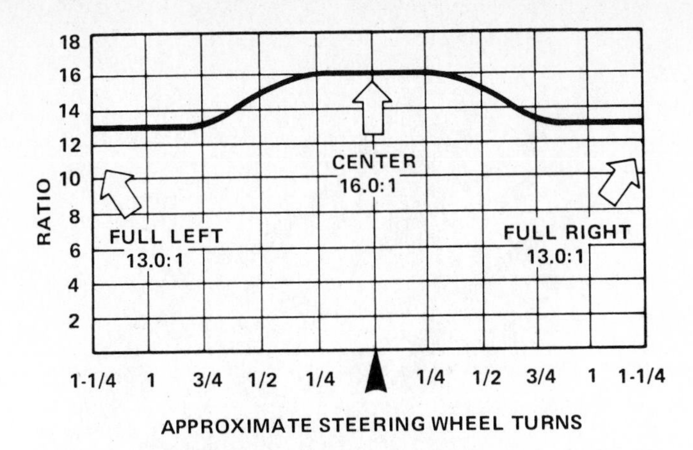

Fig. 10–35 Graph showing the relationship, or ratio, between steering-wheel turns and movement of the front wheels in right and left turns. (*American Motors Corporation*)

ing. A detergent additive helps keep the hydraulic system clean. A stabilizer additive helps to reduce noise and wear of the bearings, power piston, and control valve. In addition, the fluid must be noncorrosive. It must also prevent hardening and shrinking of the seals. These are the major causes of fluid leaks and seepage from the power-steering system.

In past years, some manufacturers allowed the use of automatic-transmission fluid in the power steering. However, this was only when the system needed the addition of fluid to the reservoir. Now it is recommended that only an approved power-steering fluid be used. This is especially important after flushing the power-steering system, after a major overhaul of the pump, or after performing work on the system that requires draining the system.

⚙ 10-18 Variable-ratio steering The steering ratio is the number of degrees that the steering wheel must be turned to pivot the front wheels one degree (⚙ 10-14). In power-steering systems, a variable steering ratio may be used. The steering ratio may vary from 16:1 for straight-ahead driving to as low as 13:1 during full turns (Fig. 10-35). How this variable ratio is achieved is described below.

From the straight-ahead position, the steering ratio stays constant for the first 40° of steering-wheel movement in either direction. Then the steering ratio decreases gradually with further steering-wheel movement. This provides good steering control for highway driving, such as maneuvering and passing. Highway driving seldom requires more than a quarter turn of the steering wheel.

In city driving, when turning corners or parking, the steering wheel is turned much more than 40°. As the steering wheel is turned well past 40°, the steering ratio drops, as shown in Fig. 10-35. Typically, it drops to as low as 13:1. Notice that by the time the steering wheel has been turned a full turn, the steering ratio is 13:1. The advantage of this is that when a short turn must be made (when rounding a corner, backing, and parking), the steering-wheel movement has a greater effect. Therefore, with a variable steering ratio, steering response increases as the need increases.

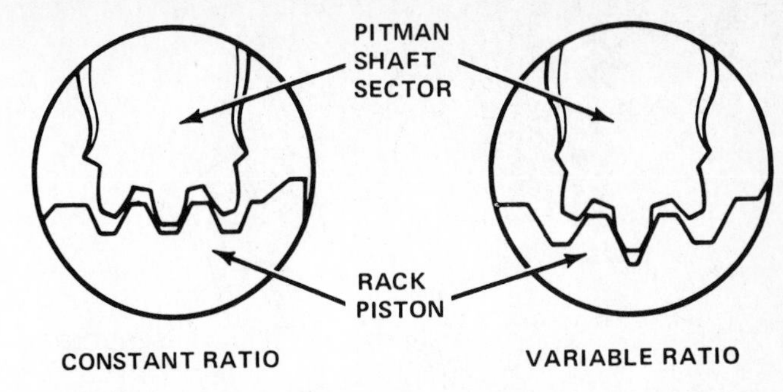

Fig. 10–36 Left, shape of teeth on piston rack and pitman-shaft sector for constant-ratio steering. Right, shape of teeth for variable-ratio steering. (*American Motors Corporation*)

The variable ratio is achieved by the shape of the teeth in the piston rack and the sector gear (Fig. 10-36). In a constant-ratio steering gear, the teeth are all the same size. But with a variable ratio, the center teeth are larger than the outer teeth. Therefore, when the piston rack has moved enough to bring one of the outer teeth into contact, the effective leverage is changed. Then further movement of the piston rack produces a greater movement of the sector gear.

⚙ 10-19 Power-steering pump All power-steering systems require a pump that will produce high pressures. The pressures may range from 800 to 2000 psi [5516 to 13,789 kPa]. A variety of designs, and mounting and driving arrangements, have been used. A typical arrangement has the pump mounted on a bracket at the front of the engine and driven by a belt from the crankshaft pulley (Fig. 10-31).

Three types of power-steering pumps are shown in Figs. 10-37 through 10-39. All operate on a similar principle. A rotor turns in an oval body. The rotor has a

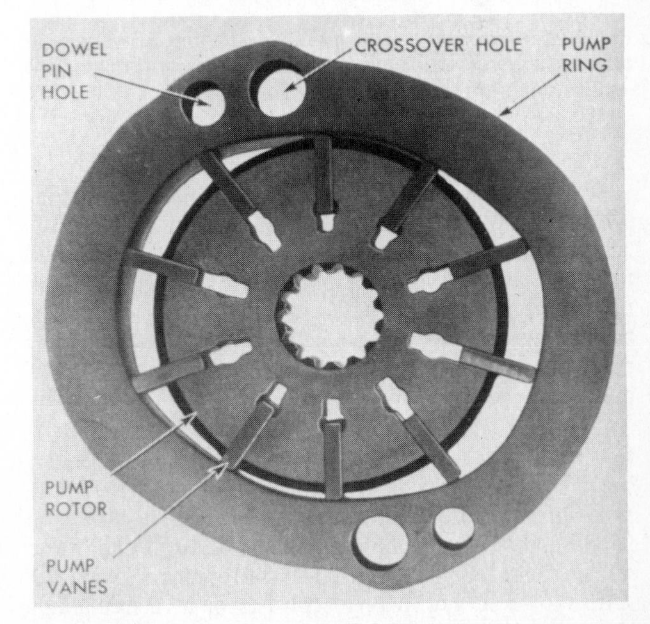

Fig. 10–37 Rings, rotor, and vanes of a vane-type power-steering pump. (*Chevrolet Motor Division of General Motors Corporation*)

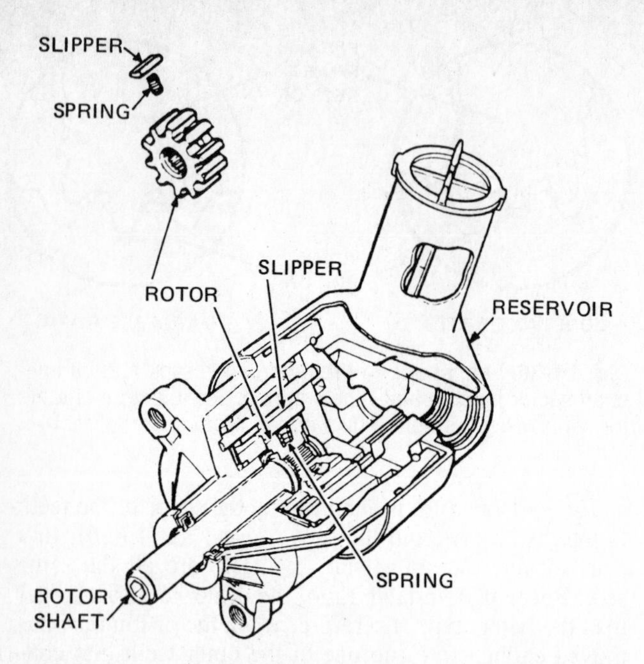

Fig. 10–38 A power-steering pump that uses spring-loaded slippers in the rotor. (*Ford Motor Company*)

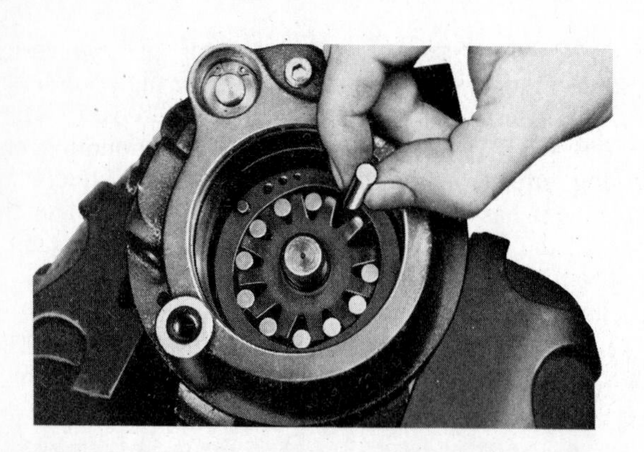

Fig. 10–39 A roller-type power-steering pump. The pressure plate and cover are removed to show how the rollers are installed into the rotor. (*Chrysler Corporation*)

series of vanes (Figs. 10-33 and 10-37), slippers (Fig. 10-38), or rollers (Fig. 10-39) set in slots in the rotor. As the rotor turns, the pockets formed between the rotor, the body, and the vanes, slippers, or rollers increase and then decrease in size. As the pockets increase in size, they draw oil into the pockets. Then, as the pockets decrease in size, the oil is forced out through the exit ports.

10-20 General Motors power steering General Motors cars use a Saginaw in-line, rotary-valve (or *torsion-bar*) type of power-steering gear (Fig. 10-40). This is an integral unit which mounts on the lower end of the steering shaft (Fig. 10-34). The power-steering gear itself, shown in Fig. 10-41, consists of a recirculating-ball steering gear to which has been added a rack piston that is actuated by the pressure from the hydraulic oil pump. In operation, movement of the steering wheel by

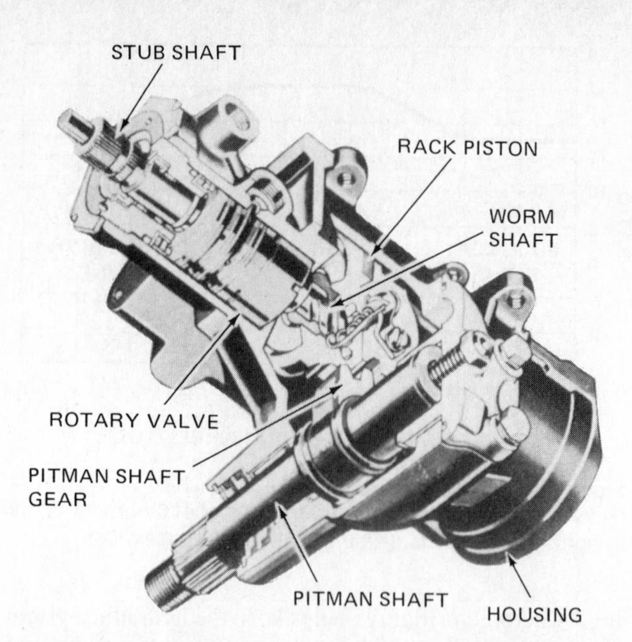

Fig. 10–40 Cutaway view of a Saginaw rotary-valve power-steering gear. (*Cadillac Motor Car Division of General Motors Corporation*)

the driver, and the resulting twisting action on the steering shaft, actuate a valve that directs oil pressure to the piston. The piston then supplies most of the steering force. This greatly reduces the force the driver must apply to the steering wheel.

The recirculating-ball power-steering gear is very similar to the manual type shown earlier in Figs. 10-24 and 10-25. In both types, manual and power, there is a ball nut surrounding the worm gear. Balls circulate between grooves in the worm gear and in the ball nut when the steering wheel is turned. This moves the ball nut up or down on the worm gear. The nut movement causes the pitman-shaft gear to turn so that the pitman arm swings to one side or the other. Linkages from the pitman arm to the front wheels then cause the wheels to pivot to one side or the other for steering.

The ball nut in the power-steering gear has a different shape and an additional job. It also serves as a piston to provide an assisting force that does most of the work of steering. Whenever more than about 3 pounds [13.3 N] of force is required to turn the steering wheel, oil from the pump is forced into one end of the piston. This helps the piston move so that most of the steering effort is provided by hydraulic fluid pressure. As a result, steering is easier.

The hydraulic fluid is controlled by a rotary valve that is connected through a torsion bar to the steering shaft (Fig. 10-41). When the steering wheel is turned, steering resistance at the front wheels causes the torsion bar to twist. In turn, the twisting of the torsion bar causes the rotary valve to turn slightly, so that high-pressure oil is directed to one end of the piston to produce steering assistance.

In the straight-ahead position (Fig. 10-42), the rotary-valve spool is positioned in the neutral position. Oil from the oil pump flows equally through the various ports in the valve. Therefore, no pressure builds up on the piston.

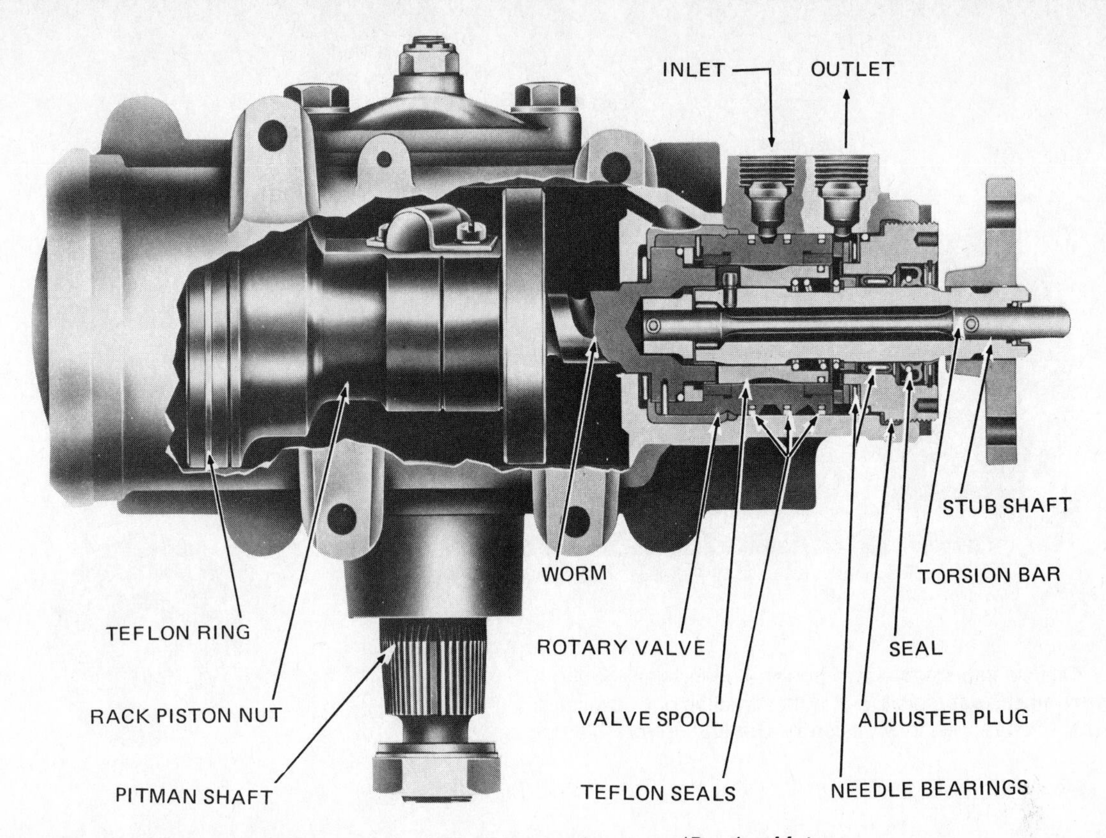

Fig. 10–41 Sectional view of a Saginaw rotary-valve power-steering gear. (*Pontiac Motor Division of General Motors Corporation*)

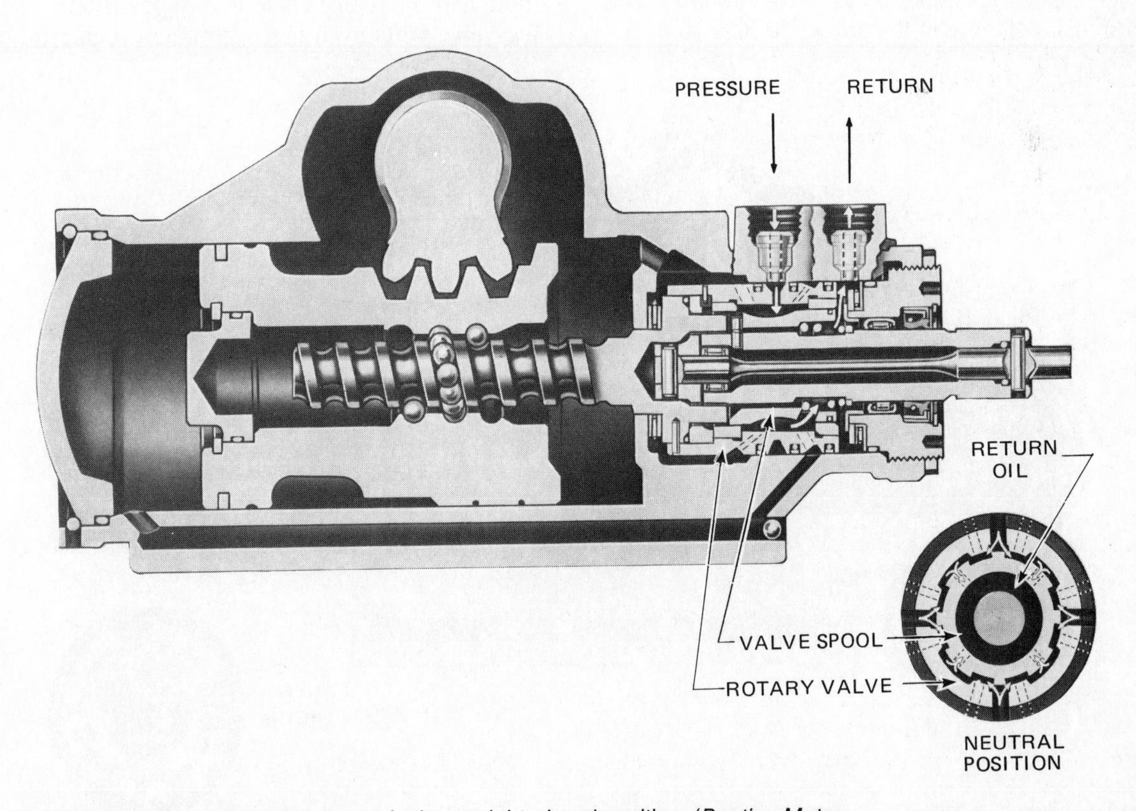

Fig. 10–42 Rotary-valve power-steering gear in the straight-ahead position. (*Pontiac Motor Division of General Motors Corporation*)

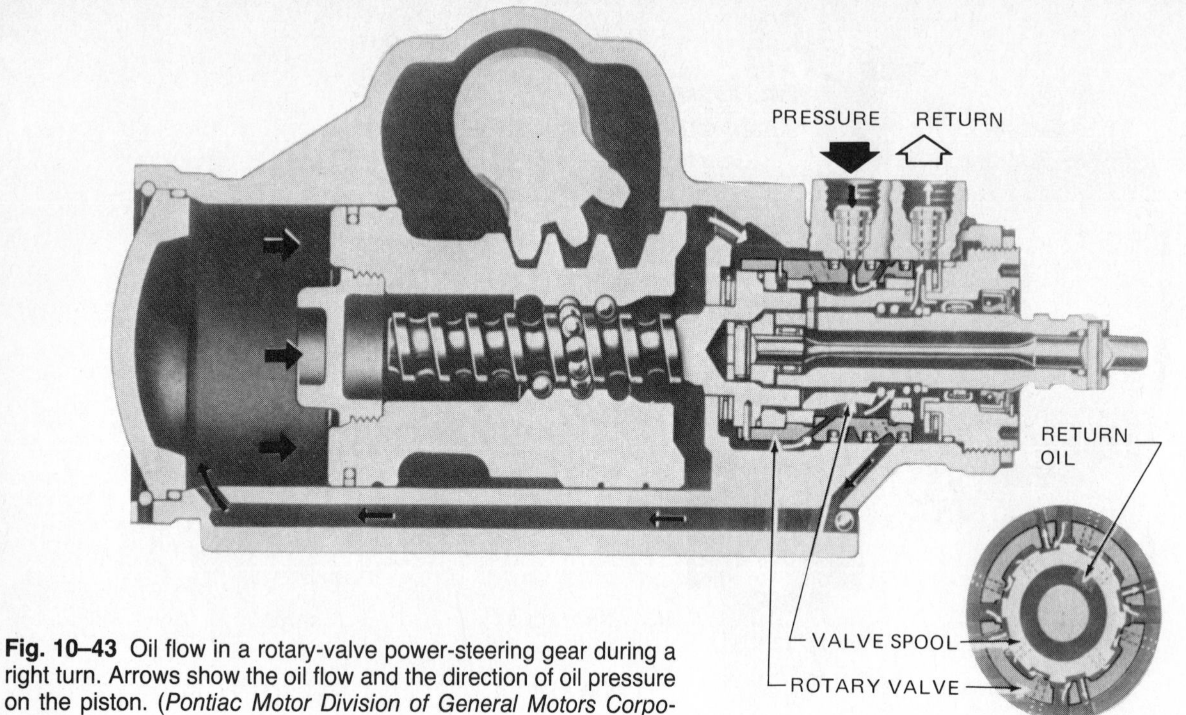

Fig. 10–43 Oil flow in a rotary-valve power-steering gear during a right turn. Arrows show the oil flow and the direction of oil pressure on the piston. (*Pontiac Motor Division of General Motors Corporation*)

When the steering wheel is rotated for a right turn (Fig. 10-43), the car wheels resist the movement. However, the turning effort applied to the steering wheel twists the torsion bar, causing the valve spool to rotate slightly in its housing. This partly closes the ports so that oil pressure builds up. Now oil flows to the lower end of the piston, as shown by the arrows in Fig. 10-43.

Oil pressure applied to the end of the piston helps to move the piston rack. This action takes over most of the steering effort. The more steering resistance the front wheels of the car offer, the harder the driver must twist the steering wheel. As a result, the torsion bar is twisted more and the rotary valve is further rotated. This closes the valve ports even more, so that a higher oil pressure

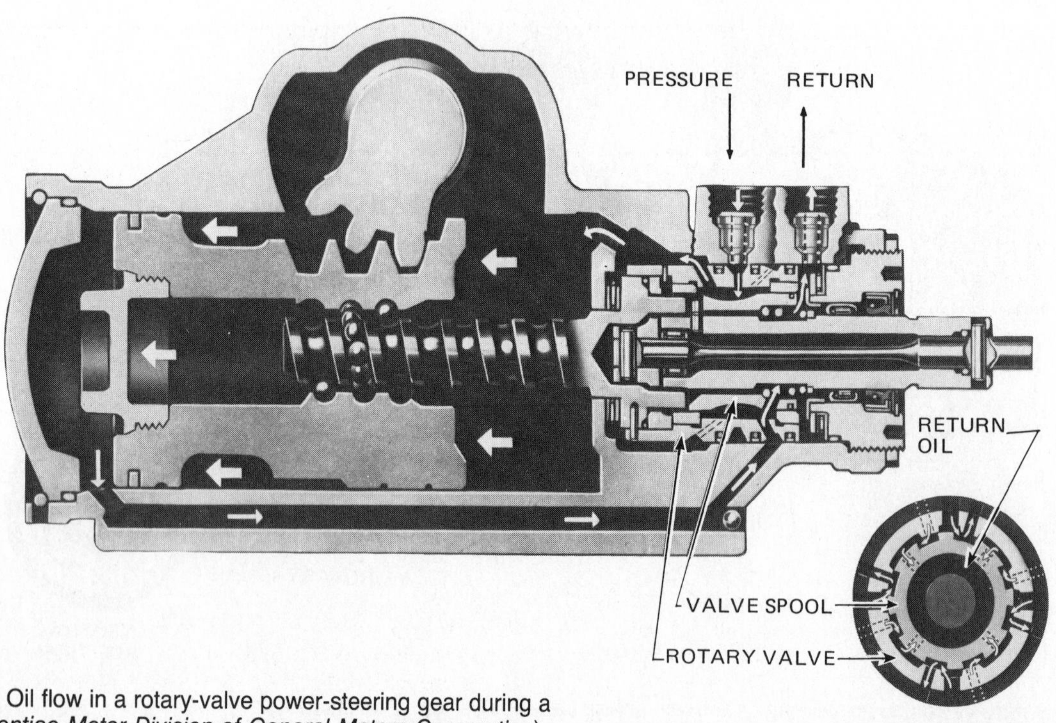

Fig. 10–44 Oil flow in a rotary-valve power-steering gear during a left turn. (*Pontiac Motor Division of General Motors Corporation*)

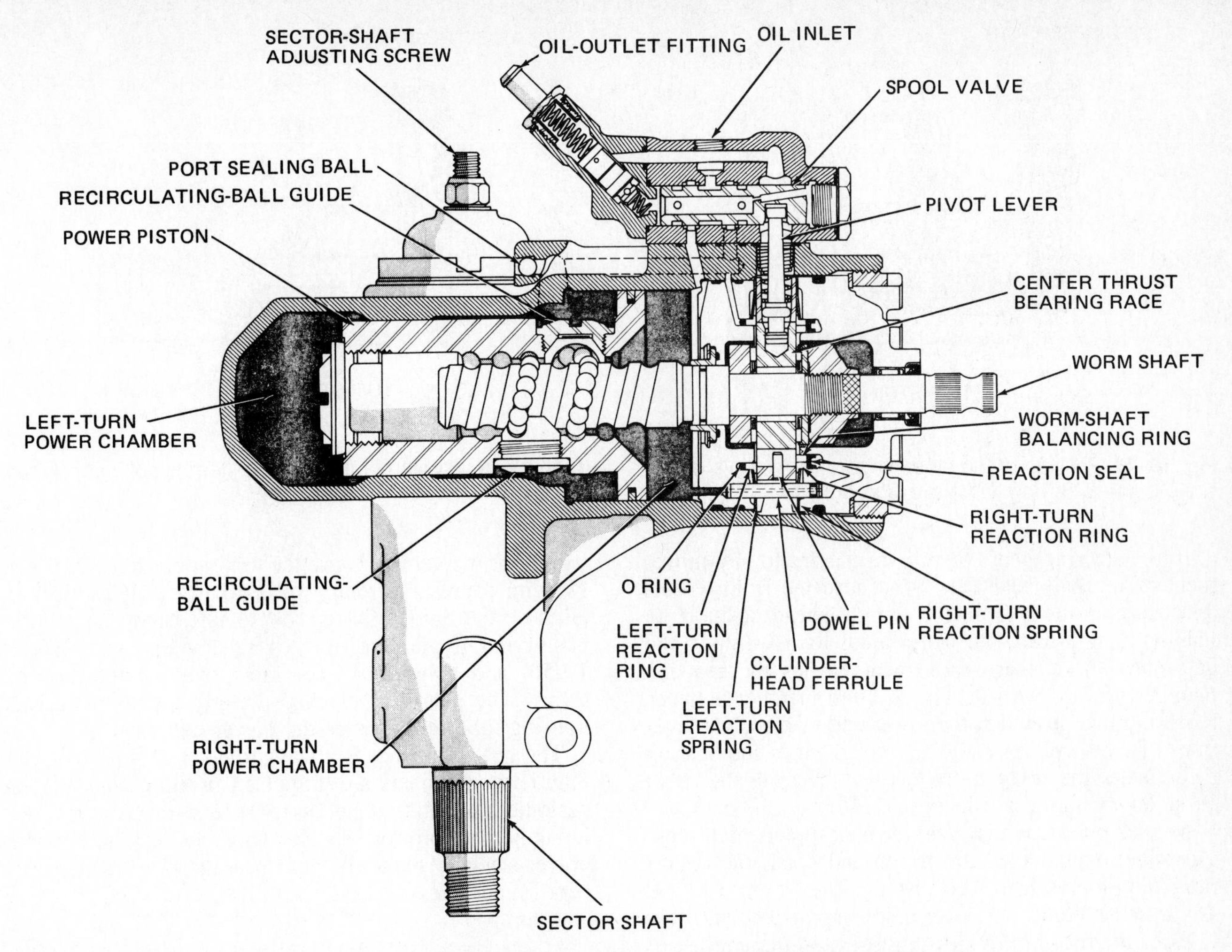

Fig. 10–45 Sectional view of a constant-control power-steering gear. (*Chrysler Corporation*)

is built up and directed to the piston. This provides a proportional steering effort, with most of the force provided by the oil pressure.

When the driver returns the steering wheel to the straight-ahead position, the rotary valve centers itself as the torsion bar resumes the neutral position. Now the front wheels return to the straight-ahead position. Figure 10-44 shows the oil flow through the steering unit when a left turn is made. Notice that the rotary valve is turned in the opposite direction from that shown in Fig. 10-43. This directs oil pressure to the opposite end of the rack piston.

❁ 10-21 Chrysler power steering

Chrysler uses a constant-control power-steering gear (Fig. 10-45). This unit is similar to the Saginaw unit (❁ 10-20). However, instead of a rotary valve, the Chrysler steering gear uses a spool valve that is actuated by a pivot lever.

A simplified spool valve is shown in Fig. 10-46. It has holes and grooves in it through which oil can circulate. In addition, it has a hole in which the end of the pivot lever (shown in Fig. 10-45) rides. The lower end of the pivot lever is held in a reaction member that is positioned between two flat springs. The springs, shaped like washers, are slightly bowed to provide tension.

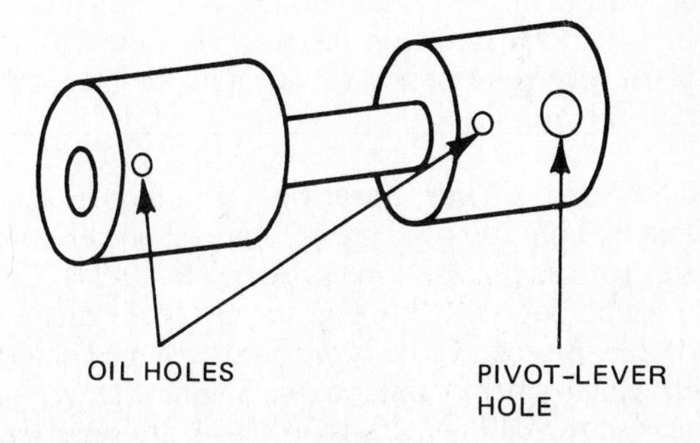

Fig. 10–46 Simplified drawing of a spool valve.

During straight-ahead driving, the springs keep the reaction member centered so that the spool valve is centered. In this position, the oil circulates as shown by the dashed arrows in Fig. 10-47. The same pressure is applied to both sides of the piston. Notice that the piston is part of the ball-nut assembly, just as in the Saginaw power-steering gear shown in Fig. 10-44.

When a turn is made that requires more than a slight force on the steering wheel, a strong end thrust develops on the worm shaft (Fig. 10-45). This is the same effect as backing a screw out of a nut while the nut is held stationary. The screw rises up out of the nut.

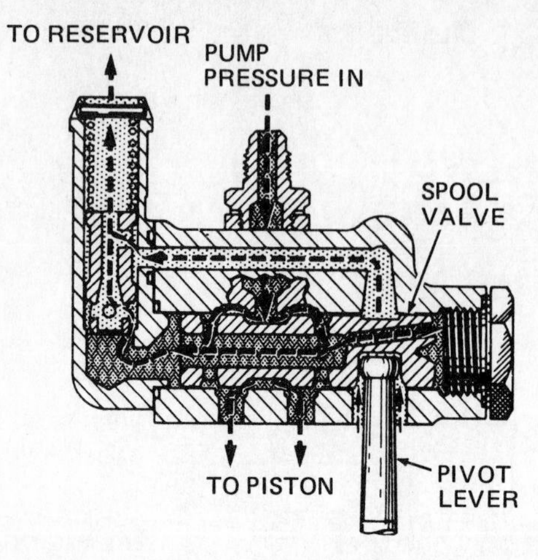

Fig. 10–47 Position of the steering-gear spool valve during straight-ahead driving. (*Chrysler Corporation*)

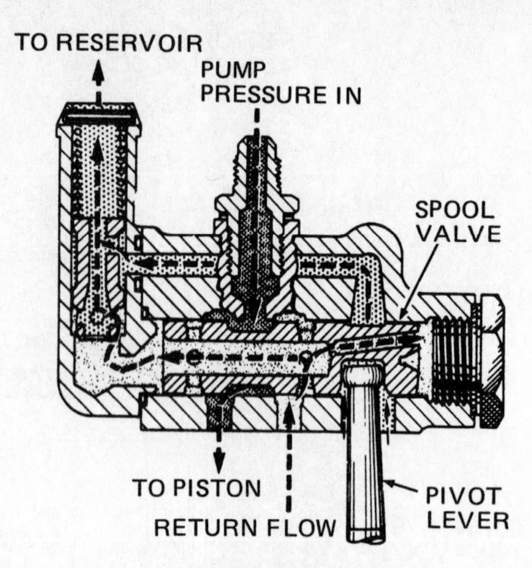

Fig. 10–48 Position of the steering-gear spool valve during a left turn. (*Chrysler Corporation*)

In the steering gear, there is resistance to any movement of the ball nut. This is because the front wheels resist turning away from the straight-ahead position. In making a right turn, the worm tends to move up in the ball nut. The upward movement carries the reaction member upward with it. The reaction member includes a bearing race in which the lower end of the pivot lever rides. Therefore, the endwise movement of the bearing race causes the pivot lever to pivot, moving the spool valve down (to the left in Fig. 10-48).

As the spool valve moves down, it directs high-pressure oil to one side of the piston and opens the circuit from the other side of the piston to the reservoir. The resulting difference in oil pressure applied to the two sides of the piston provides most of the force needed to turn the front wheels from the straight-ahead position.

When the turning force applied to the steering wheel is stopped and the wheel is returned to the straight-ahead position, the end thrust on the worm is relieved. Then, the pivot lever and spool valve return to the centered position (Fig 10-47).

✪ 10-22 Ford power steering Automobiles manufactured by Ford use two types of integral power-steering gears as well as linkage-type systems (✪ 10-24). One of the integral power-steering gears is the Saginaw unit (✪ 10-20). A second type is the Ford-designed unit (Fig. 10-49) which is very similar to the Saginaw power-steering gear shown in Figs. 10-41 to 10-44. It uses a torsion bar to control the movement of a spool valve. When the spool valve moves off center, it directs fluid under pressure to one or the other end of the piston. Then the fluid pressure provides most of the steering force.

✪ 10-23 Rack-and-pinion power steering A power rack-and-pinion steering gear is another design of integral power steering (Fig. 10-50). The rack functions as the power piston. The control valve is connected to the pinion gear.

Operation of the control valve is similar to that for the Saginaw power-steering gear (✪ 10-20). When the steering wheel is turned, the resistance of the wheels and the weight of the vehicle cause the torsion bar to twist. This twisting causes the rotary valve to move in its sleeve, aligning the fluid passages for the left, right, or neutral position. Oil pressure exerts force on the piston (Fig. 10-50) and helps move the rack to assist the turning effort. The piston is attached directly to the rack. The housing tube functions as the power cylinder.

The gear assembly is always filled with fluid, and all internal components are immersed in fluid. This makes periodic lubrication unnecessary, and also acts as a cushion to help absorb road shocks. On some rack-and-pinion power-steering gears, all fluid passages are internal except for the pressure and return hoses between the gear and pump.

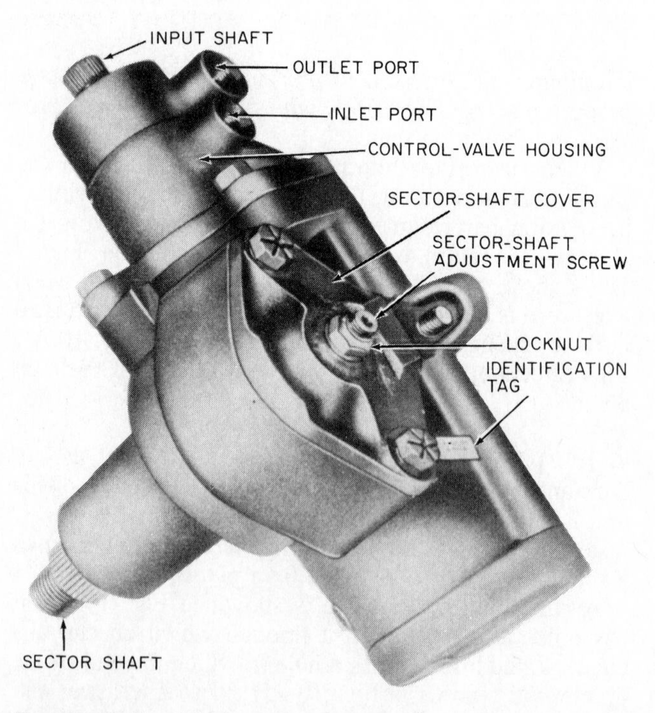

Fig. 10–49 Ford-designed integral power-steering gear. (*Ford Motor Company*)

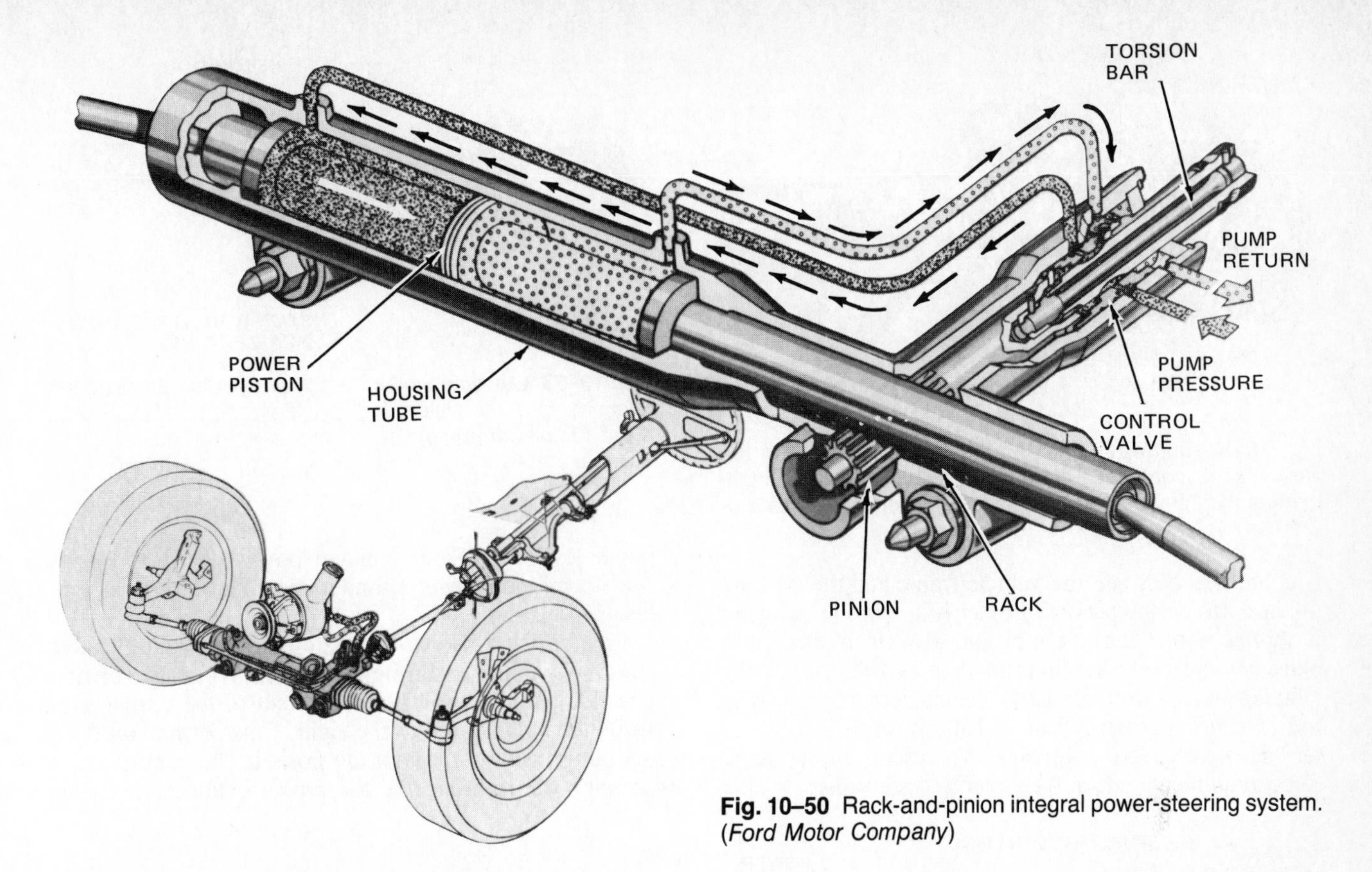

Fig. 10–50 Rack-and-pinion integral power-steering system. (*Ford Motor Company*)

10-24 Linkage-type power steering In the linkage-type power-steering system, the power cylinder is not part of the steering gear (Fig. 10-32). Instead, the power cylinder is attached to the steering linkage. In addition, the control valve is also attached to the steering linkage, either separately or combined with the power cylinder.

Figure 10-51 shows a linkage power-steering system in which the power cylinder and control valve are separate units. In operation, the steering gear works the same way as the manual-steering systems described earlier. However, in this power-steering system, the swinging end of the pitman arm is not directly connected to the steering linkage. Instead, the pitman arm is connected to the control valve.

As the end of the pitman arm swings when a turn is made, it actuates the control valve (Fig. 10-52). The control valve then sends oil to the power cylinder, which

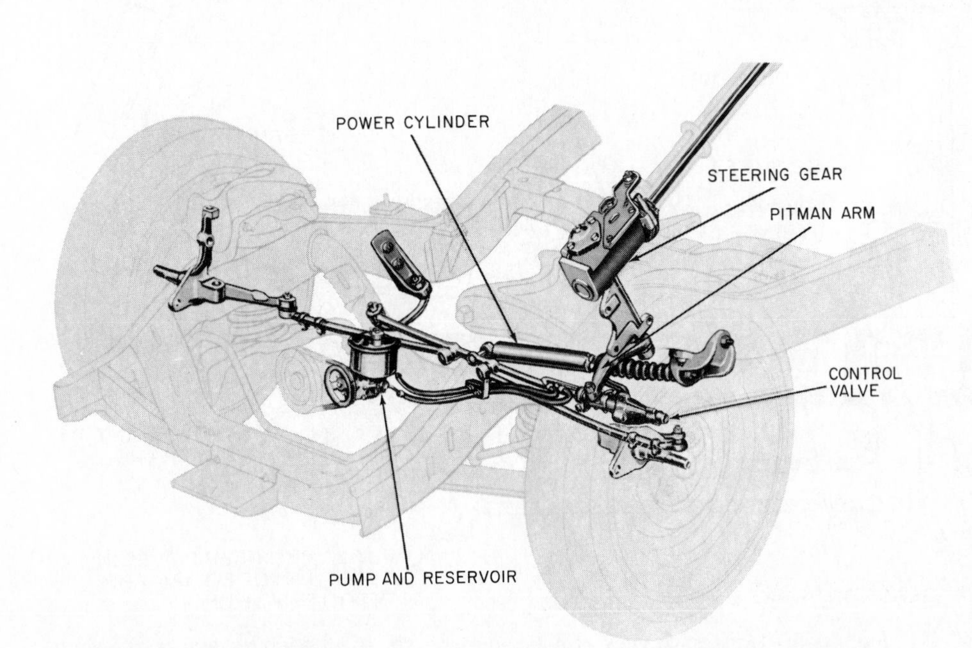

Fig. 10–51 Linkage-type power-steering system. (*Ford Motor Company*)

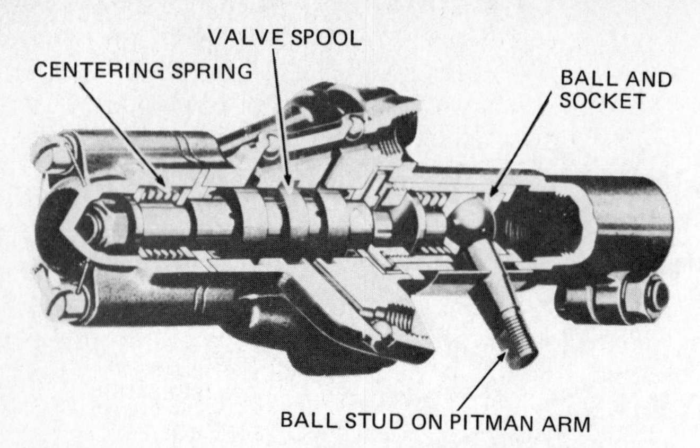

Fig. 10–52 Cutaway view of a control valve showing the valve spool, centering spring, and ball-and-socket attachment of the pitman arm to the valve-spool stem. (*Ford Motor Company*)

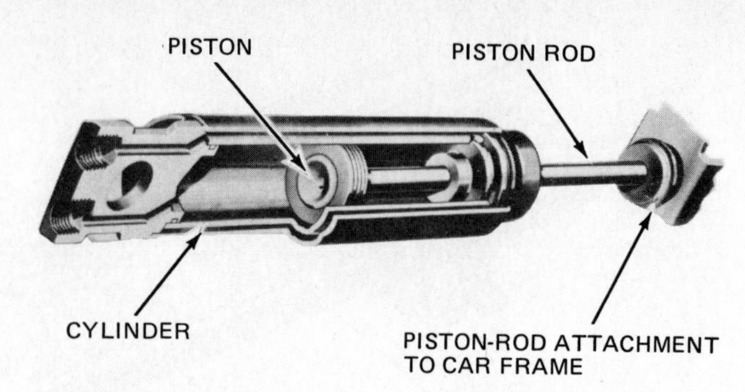

Fig. 10–53 Cutaway view of a power cylinder, showing the relationship of the piston, rod, cylinder, and rod attachment. (*Ford Motor Company*)

is connected between the vehicle frame and the steering linkage. Inside the power cylinder (Fig. 10-53), pressure is applied to one end of the piston. Movement then takes place. In the power cylinder shown in Fig. 10-53, the cylinder moves instead of the piston, because the piston rod is attached to the frame. This movement is transferred to the steering linkage. Therefore, most of the force required to steer the vehicle is furnished by the power cylinder. Some linkage power-steering systems use a combined power-cylinder and control-valve assembly (Fig. 10-54).

Figure 10-55 shows the actions in the control valve and power cylinder during a left turn. When the steering wheel is turned, the ball on the end of the pitman arm shifts the valve spool to the right. This permits oil from the pump to flow through the ports in the control valve and into the right side of the power cylinder. Then the

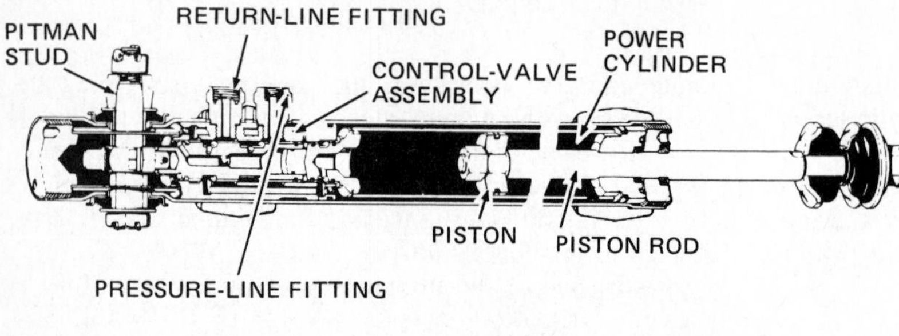

Fig. 10–54 Cutaway view of a combined power cylinder and control valve. (*Monroe Auto Equipment Company*)

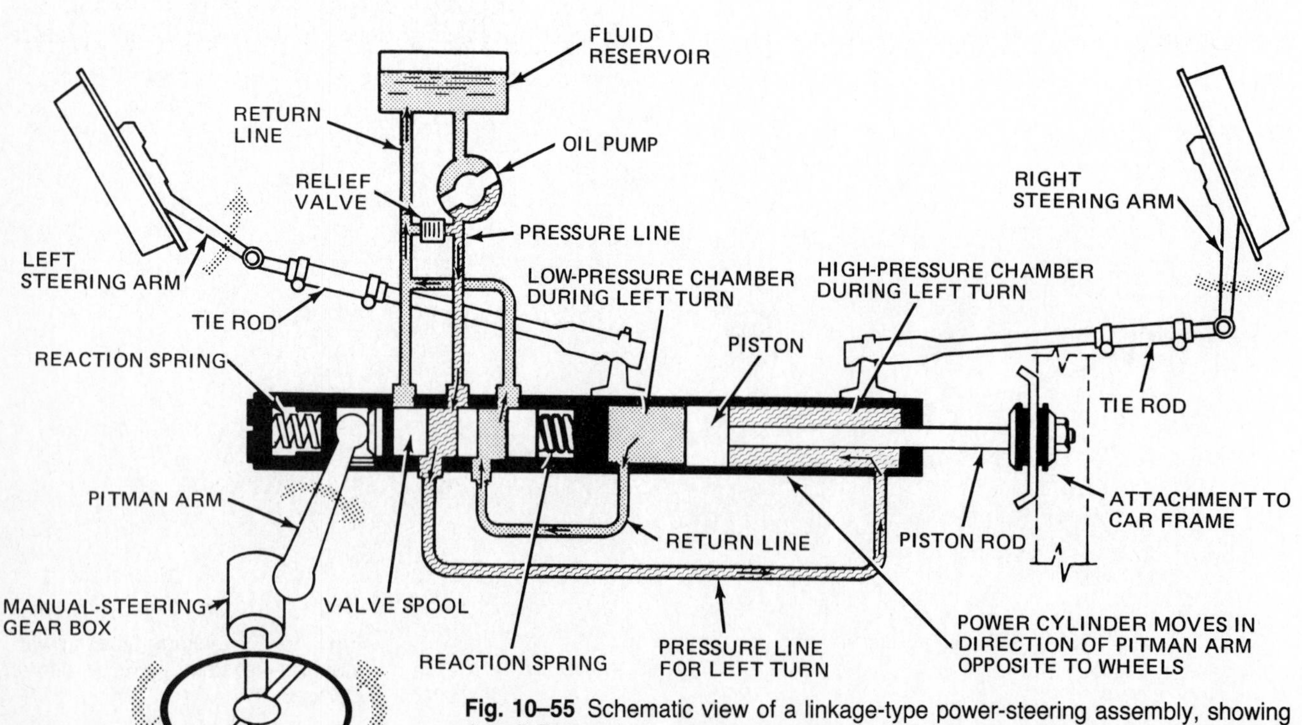

Fig. 10–55 Schematic view of a linkage-type power-steering assembly, showing the oil flow and movement of parts during a left turn. (*Monroe Auto Equipment Company*)

high-pressure oil forces the cylinder to move to the right. This movement of the cylinder provides most of the required steering force.

⚙ 10-25 Air power steering Some heavy-duty trucks and buses have air-powered steering, using compressed air supplied by the engine-driven air-brake compressor (Fig. 10-56). The system shown in Fig. 10-56 is the linkage type, with the control valve actuated by the pitman arm. The control valve is moved one way or the other by the pitman arm as the steering wheel is turned. This causes compressed air to be admitted to one end of the power cylinder. The compressed air then forces the piston to move in the cylinder. This movement is carried by the piston rod to the steering linkage.

⚙ 10-26 Four-wheel steering On some vehicles, all four wheels are steered as the steering wheel is turned. This provides greater maneuverability of the vehicle. Both front and rear axles have universal joints that permit the wheels to turn at various angles to the frame of the vehicle. The driving axles continue to deliver power to the wheels through the universal joints, regardless of the angle at which the wheels are turned.

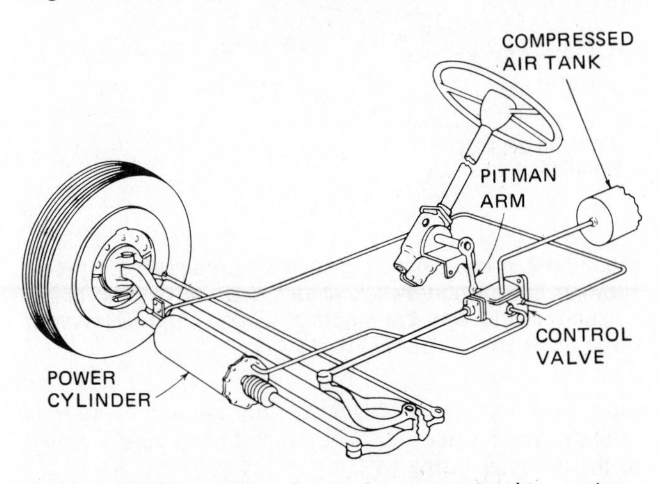

Fig. 10–56 Schematic of an air power-steering system. *(Bendix-Westinghouse Automotive Air Brake Company)*

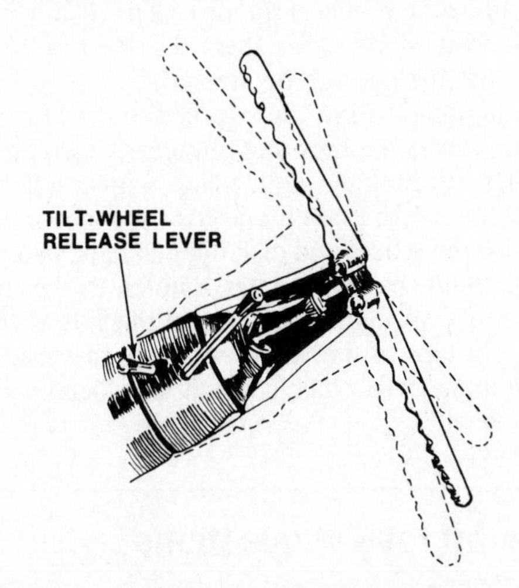

Fig. 10–57 Tilt steering wheel. Lifting the release lever permits the steering wheel to be tilted to various positions. *(Buick Motor Division of General Motors Corporation)*

The steering arms on the rear wheels are linked by tie rods to the same pitman arm to which the front-wheel steering arms are linked. Often, a vehicle with four-wheel steering also has provision for four-wheel drive. In this arrangement, all four wheels are linked by drive shafts to the engine.

Steering wheels and columns

⚙ 10-27 Tilt and telescoping steering wheels Many cars have steering wheels that tilt up or down and also can be moved out of or into the steering column (Figs. 10-57 and 10-58). This makes it easier for the driver to get into or out of the car. The driver also can vary the position of the wheel to suit his or her build and can change the position during a long drive to vary driving posture.

Some automobiles have a steering column that can be moved inward toward the center of the car to make it easier for the driver to get in and out of the car. The pivot point on this arrangement is just above the steering gear, at the lower end of the steering column. At that point there is a flex joint that connects between the upper steering shaft and the worm shaft in the steering gear. This permits the steering shaft and steering column to be pivoted toward the center of the car.

A locking mechanism is connected to the transmission selector lever. The steering column is locked in the DRIVE position and in all selector-lever positions except PARK. To unlock the steering column, the selector must be moved to PARK. Also, if the steering column is moved out of the driving position, the selector lever is locked in PARK. This interlocking is a safety feature which prevents the steering column from being accidentally moved while the car is in operation.

⚙ 10-28 Collapsible steering column The collapsible steering column (Figs. 10-59 to 10-61), used on modern cars as a protective device, will collapse on impact. If the car should become involved in a front-end collision that throws the driver forward, the steering column will absorb the energy of this forward movement

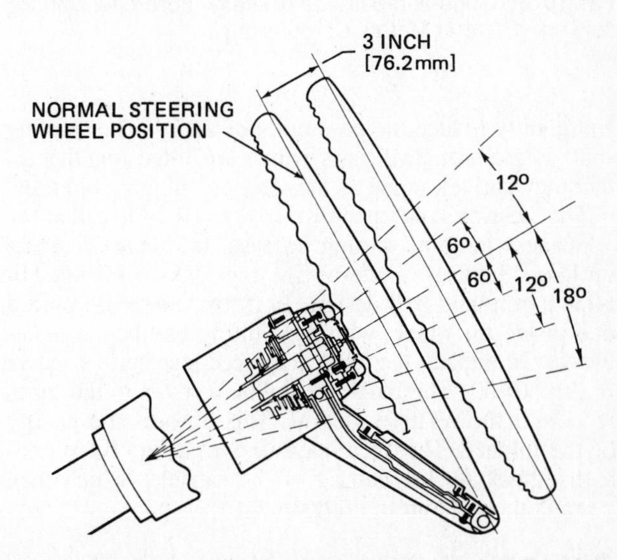

Fig. 10–58 Tilting and telescoping steering wheel. *(Cadillac Motor Car Division of General Motors Corporation)*

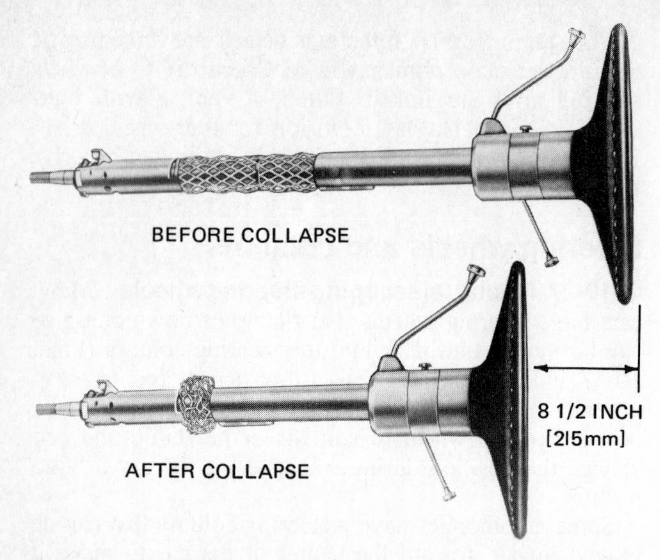

Fig. 10–59 "Japanese lantern" type of energy-absorbing steering column which can collapse during impact. (*Cadillac Motor Car Division of General Motors Corporation*)

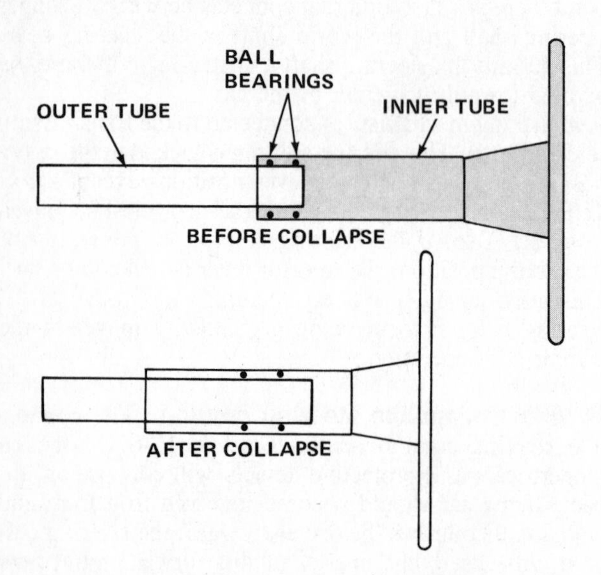

Fig. 10–60 Tube-and-ball type of energy-absorbing steering column. (*General Motors Corporation*)

and greatly reduce the possibility of injury. The steering shaft is made in two parts which are fitted together so that they can telescope as the steering column collapses.

The steering column shown in Fig. 10-59 is called the "Japanese lantern" design because, on impact, it folds up like a Japanese lantern. The type shown in Fig. 10-60 is a tube-and-ball design. In it, two tubes are placed one inside the other, with tight-fitting ball bearings between. On impact, the tubes are forced together, as shown in Fig. 10-60. The balls must plow furrows in the tubes to permit the relative motion. This absorbs the energy of the impact. The shear-capsule type (Fig. 10-61) absorbs shock by the cutting of the capsule, which then permits the column to collapse.

❁ 10-29 Steering and ignition lock Automobiles are equipped with a combination ignition switch and steering-wheel lock (Fig. 10-62). The ignition switch is mounted on the steering column, and has a gear attached to the cylinder in the lock.

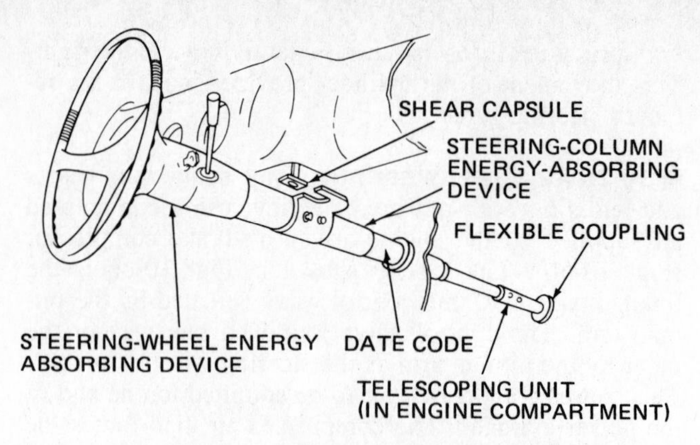

Fig. 10–61 Sheer-capsule type of collapsible steering column.

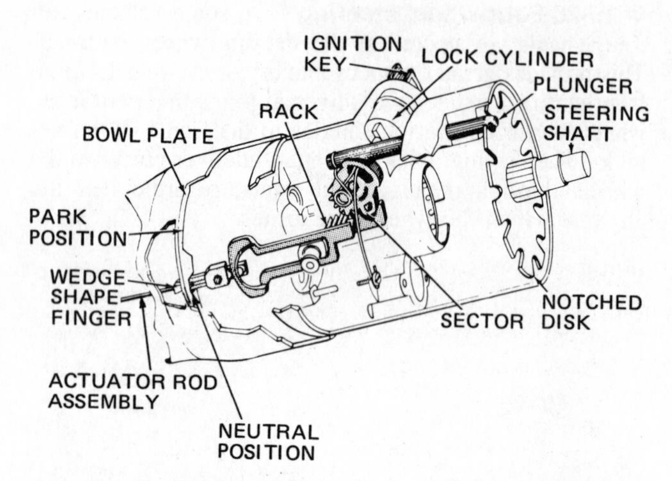

Fig. 10–62 A combination ignition switch and steering lock, showing the ignition switch at OFF and the plunger in the notched disk locking the steering. (*Buick Motor Division of General Motors Corporation*)

When the ignition key is inserted and the ignition switch is turned to ON, the gear rotates and pulls the rack and plunger out of the notch in the disk. The disk is mounted on the steering shaft. This frees the steering shaft and wheel so that the car can be steered.

When the ignition switch is turned to OFF, the rotation of the gear moves the rack and plunger toward the locked position. If the plunger is lined up with a notch in the disk, it will enter the notch and lock the steering wheel. However, if the wheel and disk happen to be in a position where the plunger cannot enter a notch in the disk, the plunger will be spring loaded against the side of the disk. Now a slight turn of the wheel will turn the disk enough that the plunger will enter a notch and thereby lock the steering wheel.

Chapter 10 review questions

Select the *one* correct, best, or most probable answer to each question. Then check your answers against the correct answers given at the end of the book.

1. The purpose of the caster angle on an automobile is to:
   a. prevent tire wear
   b. bring the road contact of the tire under the point of load
   c. compensate for wear in the steering linkage
   d. maintain directional control
2. When turning a corner:
   a. the front wheels are toeing out
   b. the front wheels are turning on different angles
   c. the inside front wheel has a greater angle than the outside wheel
   d. all of the above
3. The tilting of the front wheels away from the vertical is called:
   a. camber
   b. caster
   c. toe-in
   d. toe-out
4. The inward tilt of the centerline of the ball joints is called:
   a. caster
   b. camber
   c. steering-axis inclination
   d. included angle
5. Camber angle plus steering-axis-inclination angle is called the:
   a. caster
   b. included angle
   c. point of intersection
   d. toe-out
6. The point at which the centerline of the wheel and the centerline of the ball joints cross is called the:
   a. included angle
   b. point of departure
   c. point of intersection
   d. point of included angle
7. When the point of intersection is below the road surface, the front wheel will tend to:
   a. toe out
   b. toe in
   c. roll straight
   d. none of the above
8. The backward tilt of the centerline of the ball joints from the vertical is called:
   a. positive caster
   b. negative caster
   c. positive camber
   d. negative camber
9. Positive caster tends to make front wheels:
   a. toe in
   b. toe out
   c. have neutral camber
   d. none of the above
10. In the steering gear, a gear sector or toothed roller is meshed with:
    a. a worm
    b. a ball bearing
    c. a roller bearing
    d. a steering wheel
11. Two basic types of power-steering systems are the:
    a. integral and valve
    b. booster and power
    c. integral and linkage
    d. rack and pinion
12. The valve spool in the Saginaw in-line power-steering unit is centered, in neutral, by:
    a. the torsion bar
    b. oil pressure
    c. two disk clutches
    d. the rotary valve
13. In the Saginaw rotary-valve unit, rotation of the valve spool from the neutral position results in application of hydraulic oil pressure from the oil pump to one or the other end of the:
    a. pitman arm
    b. connecting rod
    c. piston
    d. disk clutch
14. Three types of oil pumps used in power-steering systems are the vane, roller, and
    a. shoe
    b. slipper
    c. gear
    d. all of the above
15. In the Ford-design power-steering gear, the valve spool is moved by:
    a. coil springs
    b. a torsion bar
    c. the ball nut
    d. the rotary valve
16. In the constant-control power-steering unit, the spool valve is moved by:
    a. a pivot lever
    b. a valve rotor
    c. high oil pressure
    d. none of the above
17. In the linkage type of power-steering system, the swinging end of the pitman arm actuates:
    a. a spool valve
    b. a tie rod
    c. an oil pump
    d. an idler lever
18. In the linkage type of power-steering system, the piston rod is attached at one end to:
    a. a tie rod
    b. a connecting rod
    c. the vehicle frame
    d. none of the above
19. In a rack-and-pinion steering gear, the tie rods are attached to the
    a. rack
    b. spindle steering arm
    c. both *a* and *b*
    d. neither *a* nor *b*
20. In a rack-and-pinion power-steering gear, the rack functions as the
    a. control valve
    b. power piston
    c. pinion gear
    d. torsion bar

LEFT 5/16 Forward & out

BOTH LEFT

CHAPTER 11

DIAGNOSING STEERING AND SUSPENSION TROUBLES

After studying this chapter, and with proper instruction and equipment, you should be able to:

1. Diagnose troubles in manual-steering systems.
2. Diagnose troubles in power-steering systems.
3. Diagnose troubles in the suspension system.
4. Inspect ball joints.
5. Inspect wheel bearings.

11-1 Steering and suspension trouble-diagnosis charts If you can relate various complaints to the conditions that cause them, you will know which items to check and correct to eliminate a trouble. Trouble-diagnosis charts help pinpoint trouble causes.

Figure 11-1 is a trouble-diagnosis chart for manual-steering systems. Many problems noticed by the driver in the manual-steering system may be caused by suspension troubles.

Figure 11-2 is a trouble-diagnosis chart for power-steering systems. Note how many more possible causes

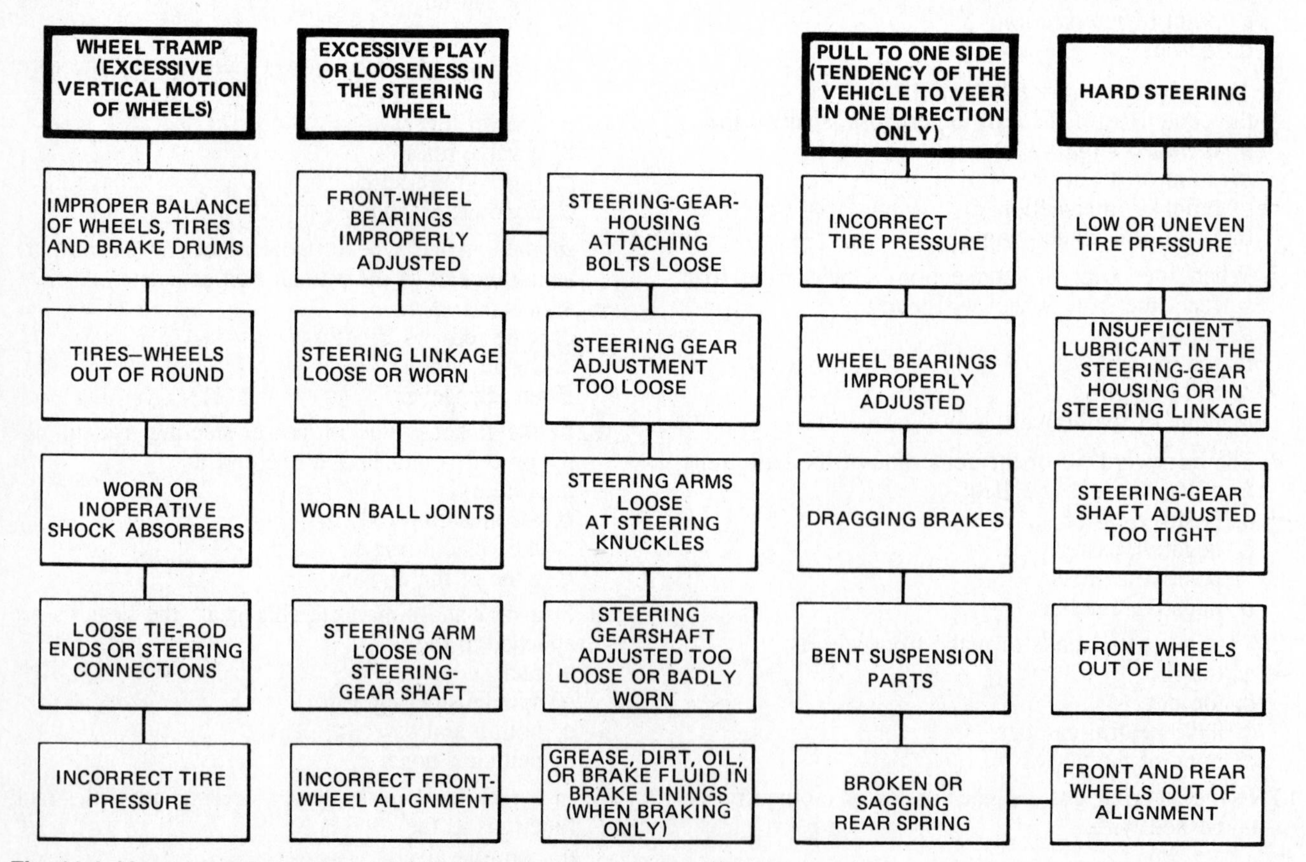

Fig. 11-1 Manual-steering-system trouble-diagnosis chart. (*Chrysler Corporation*)

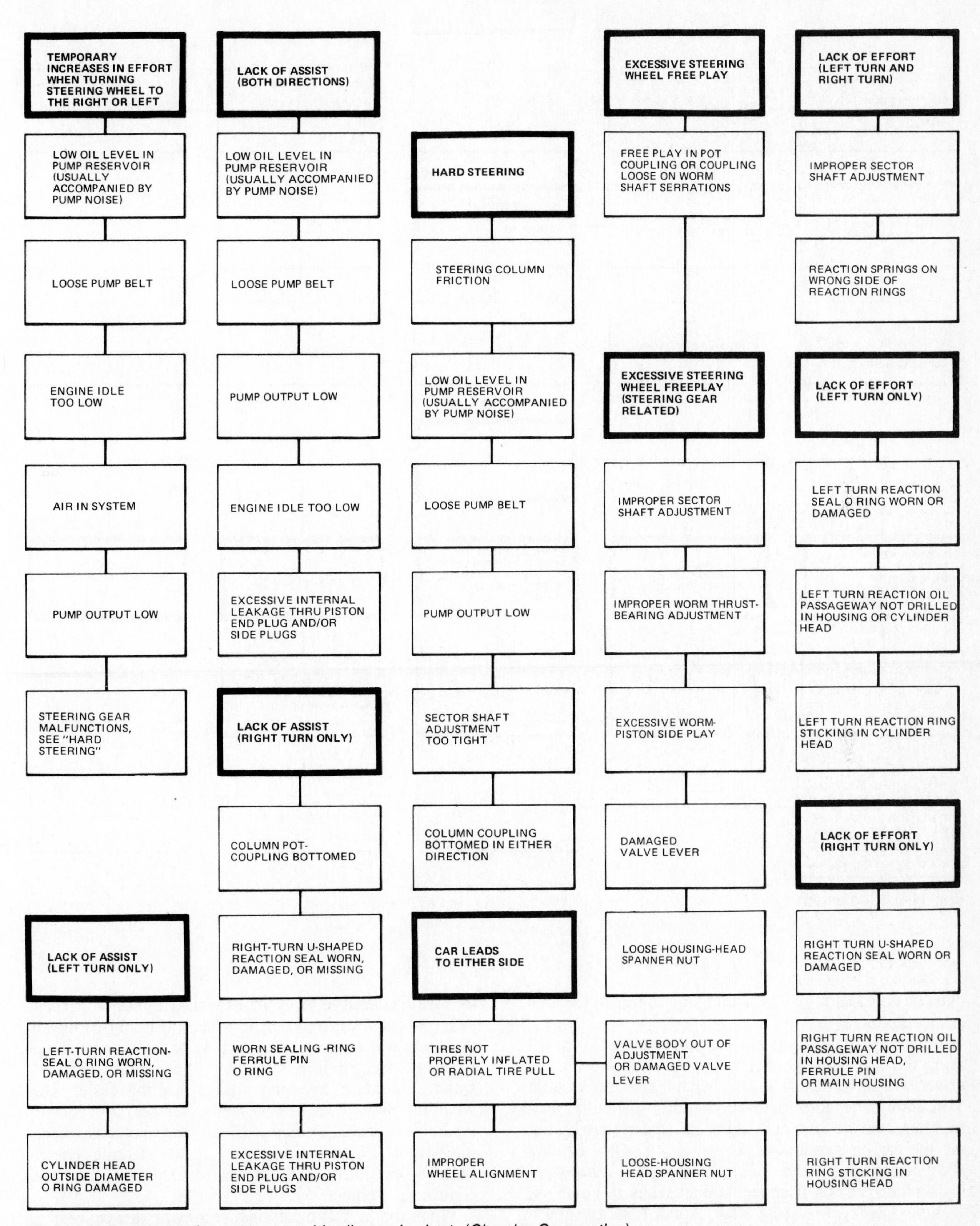

Fig. 11-2 Power-steering-system trouble-diagnosis chart. (*Chrysler Corporation*)

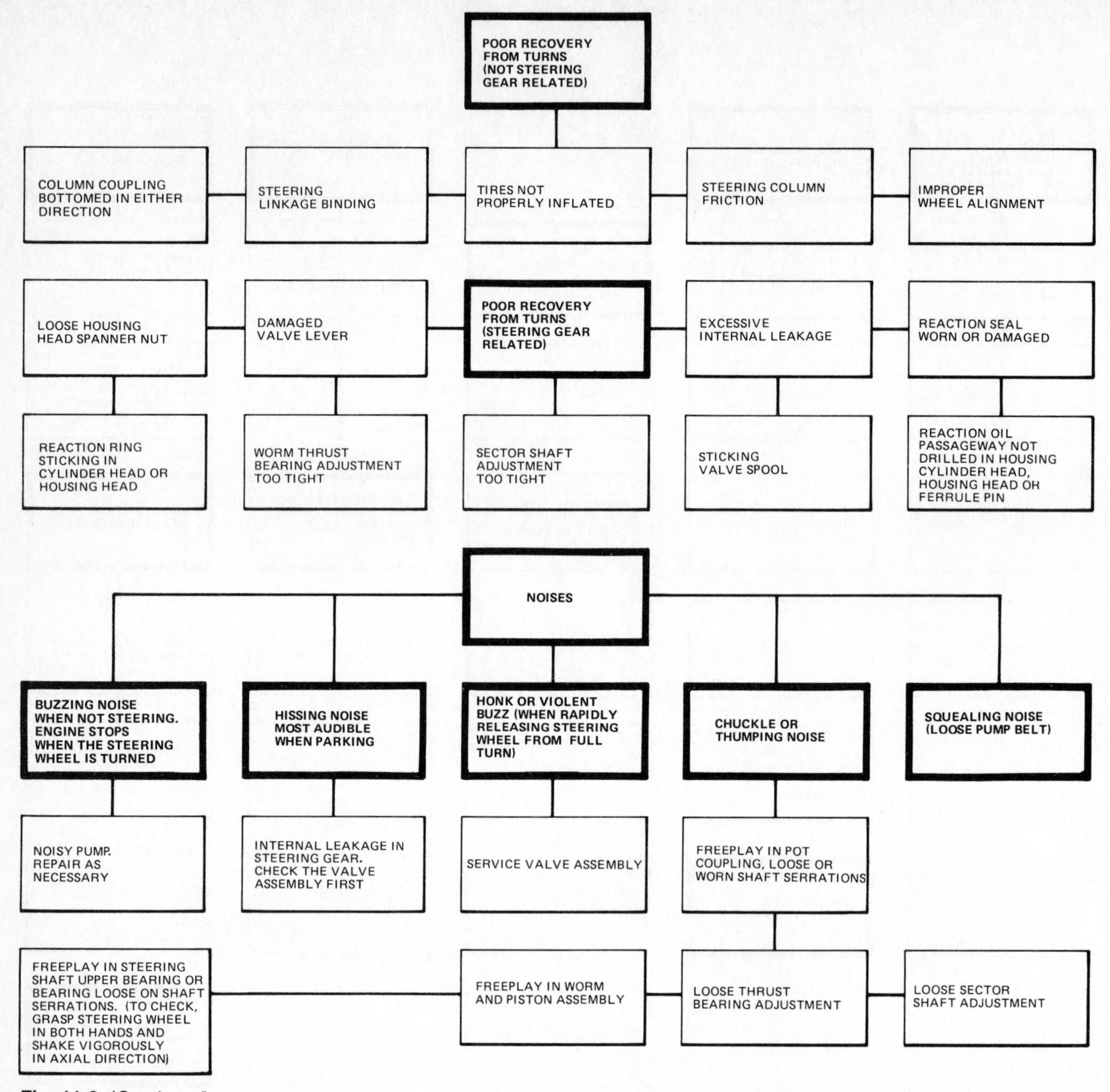

Fig. 11-2 *(Continued)*

must be considered in determining the source of the driver's complaints.

A variety of steering and suspension problems will bring the driver to the mechanic, but it is rare that the driver will have a clear idea of what causes the trouble. You should be able to detect steering difficulty, hard steering, or excessive play in the steering system by taking the car on a road test. The chart on pages 179 and 180 lists possible causes of these, as well as other steering and suspension troubles. It then refers to numbered sections later in the chapter for explanations of how to find and correct the troubles.

❁ 11-2 Excessive play in steering system Excessive looseness in the steering system (Fig. 11-3) means that there will be excessive free play of the steering wheel without corresponding movement of the front wheels. A small amount of free play makes steering easier. But when the play, or lash, becomes excessive, it is considered objectionable by many drivers and may make steering harder. Excessive free play in the steering system reduces the ability of the driver to accurately steer and control the vehicle.

When the play is excessive, it may be due to wear or improper adjustment of the steering gear, worn parts or

Steering and Suspension Trouble-Diagnosis Chart

(See ❁ 11-2 to 11-16 for detailed explanations of trouble causes and corrections listed below.)

COMPLAINT	POSSIBLE CAUSE	CHECK OR CORRECTION
1. Excessive play in steering system (❁ 11-2)	a. Looseness in steering gear	Readjust, replace worn parts
	b. Looseness in linkage	Readjust, replace worn parts
	c. Worn ball joints or steering-knuckle parts	Replace worn parts
	d. Loose wheel bearing	Readjust
2. Hard steering (❁ 11-3)	a. Power steering inoperative	See Chaps. 15 to 18
	b. Low or uneven tire pressure	Inflate to correct pressure
	c. Friction in steering gear	Lubricate, readjust, replace worn parts
	d. Friction in linkage	Lubricate, readjust, replace worn parts
	e. Friction in ball joints	Lubricate, replace worn parts
	f. Alignment off (caster, camber, toe, steering-axis inclination)	Check alignment and readjust as necessary
	g. Frame misaligned	Straighten
	h. Front spring sagging	Replace or adjust
3. Car wander (❁ 11-4)	a. Low or uneven tire pressure	Inflate to correct pressure
	b. Linkage binding	Readjust, lubricate, replace worn parts
	c. Steering gear binding	Readjust, lubricate, replace worn parts
	d. Front alignment off (caster, camber, toe, steering-axis inclination)	Check alignment and readjust as necessary
	e. Looseness in linkage	Readjust, replace worn parts
	f. Looseness in steering gear	Readjust, replace worn parts
	g. Looseness in ball joints	Replace worn parts
	h. Loose rear springs	Tighten
	i. Unequal load in car	Readjust load
	j. Stabilizer bar ineffective	Tighten attachment, replace if damaged
4. Car pulls to one side during normal driving (❁ 11-5)	a. Uneven tire pressure	Inflate to correct pressure
	b. Uneven caster or camber	Check alignment, adjust as necessary
	c. Tight wheel bearing	Readjust, replace parts if damaged
	d. Uneven springs (sagging, broken, loose attachment)	Tighten, replace defective parts
	e. Wheels not tracking	Check tracking, straighten frame, tighten loose parts, replace defective parts
	f. Uneven torsion-bar adjustment	Adjust
	g. Brakes dragging	Repair
5. Car pulls to one side when braking (❁ 11-6)	a. Brakes grab	Readjust, replace brake lining, etc. (see Chaps. 5 and 6)
	b. Uneven tire inflation	Inflate to correct pressure
	c. Incorrect or uneven caster	Readjust
	d. Causes listed under item 4	
6. Front-wheel shimmy at low speeds (❁ 11-7)	a. Uneven or low tire pressure	Inflate to correct pressure
	b. Loose linkage	Readjust, replace worn parts
	c. Loose ball joints	Replace worn parts
	d. Looseness in steering gear	Readjust, replace worn parts
	e. Front springs too flexible	Replace, tighten attachment
	f. Incorrect or unequal camber	Readjust
	g. Irregular tire tread	Replace worn tires, match treads
	h. Dynamic imbalance	Balance wheels
7. Front-wheel tramp (high-speed shimmy) (❁ 11-8)	a. Wheels out of balance	Rebalance
	b. Too much wheel runout	Balance, remount tire, straighten or replace wheel
	c. Defective shock absorbers	Repair or replace
	d. Causes listed under item 6	

Steering and Suspension Trouble-Diagnosis Chart (*Continued*)

(See ❁ 11-2 to 11-16 for detailed explanations of trouble causes and corrections listed below.)

COMPLAINT	POSSIBLE CAUSE	CHECK OR CORRECTION
8. Steering kickback (❁ 11-9)	a. Tire pressure low or uneven	Inflate to correct pressure
	b. Springs sagging	Tighten attachment, replace
	c. Shock absorbers defective	Repair or replace
	d. Looseness in linkage	Readjust, replace worn parts
	e. Looseness in steering gear	Readjust, replace worn parts
9. Tires squeal on turns (❁ 11-10)	a. Excessive speed	Take curves at slower speed
	b. Low or uneven tire pressure	Inflate to correct pressure
	c. Front alignment incorrect	Check and adjust
	d. Worn tires	Replace
10. Improper tire wear (❁ 11-11)	a. Wear at tread sides from underinflation	Inflate to correct pressure
	b. Wear at tread center from overinflation	Inflate to correct pressure
	c. Wear at one tread side from excessive camber	Adjust camber
	d. Featheredge wear from excessive toe-in or toe-out on turns	Correct toe-in or toe-out in turns
	e. Cornering wear from excessive speeds on turns	Take turns at slower speeds
	f. Uneven or spotty wear from mechanical causes	Adjust brakes, align wheels, balance wheels, adjust linkage, etc.
	g. Rapid wear from speed	Drive more slowly for longer tire life
11. Hard or rough ride (❁ 11-12)	a. Excessive tire pressure	Reduce to correct pressure
	b. Defective shock absorbers	Repair or replace
	c. Excessive friction in spring suspension	Lubricate, realign parts
12. Sway on turns (❁ 11-13)	a. Loose stabilizer bar	Tighten
	b. Weak or sagging springs	Repair or replace
	c. Caster incorrect	Adjust
	d. Defective shock absorbers	Replace
13. Spring breakage (❁ 11-14)	a. Overloading	Avoid overloading
	b. Loose center or U bolts	Keep bolts tight
	c. Defective shock absorber	Repair or replace
	d. Tight spring shackle	Loosen, replace
14. Sagging springs (❁ 11-15)	a. Broken leaf	Replace
	b. Spring weak	Replace
	c. Coil spring short	Install shim
	d. Defective shock absorber	Repair or replace
15. Noises (❁ 11-16)	Could come from any loose, worn, or unlubricated part in the suspension or steering system.	

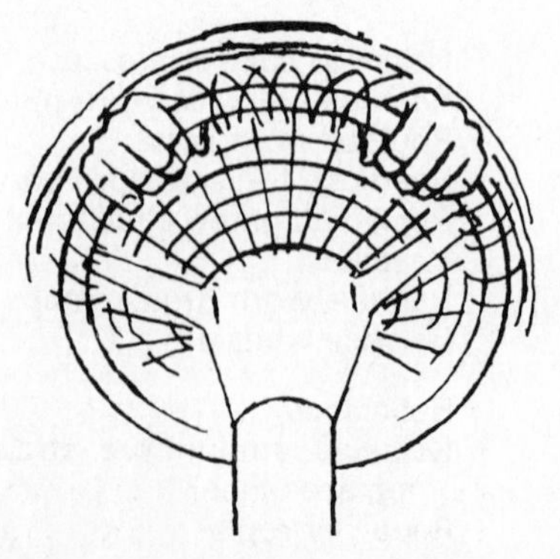

Fig. 11-3 A driver may detect excess play in the steering but will usually not know what causes it. (*ATW*)

improper adjustments in the steering linkage, worn ball joints or steering-knuckle parts, or loose wheel bearings. In most cars with power steering, the steering-wheel rim should move 2 inches [51 mm] or less before the front wheels begin to move. On cars with manual steering, the maximum allowable free play is 3 inches [76 mm]. Figure 11-4 shows this measurement.

To check the amount of play in the steering system on vehicles with power steering, check the condition and tension of the drive belt for the power-steering pump. Then check the fluid level in the pump reservoir. Start the engine. Next, with the front wheels in the straight-ahead position, turn the steering wheel until the front wheels begin to move. Align a reference mark on the

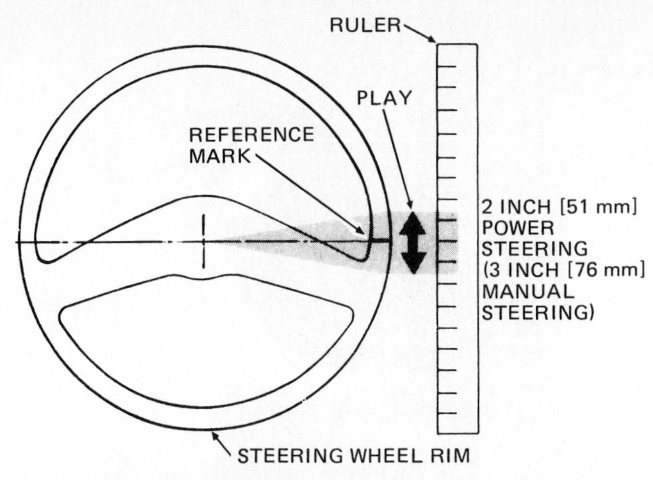

Fig. 11-4 Checking for play in the steering wheel. (*Motor Vehicle Manufacturers Association*)

Fig. 11-5 Checking looseness in steering linkage and tie rods. (*ATW*)

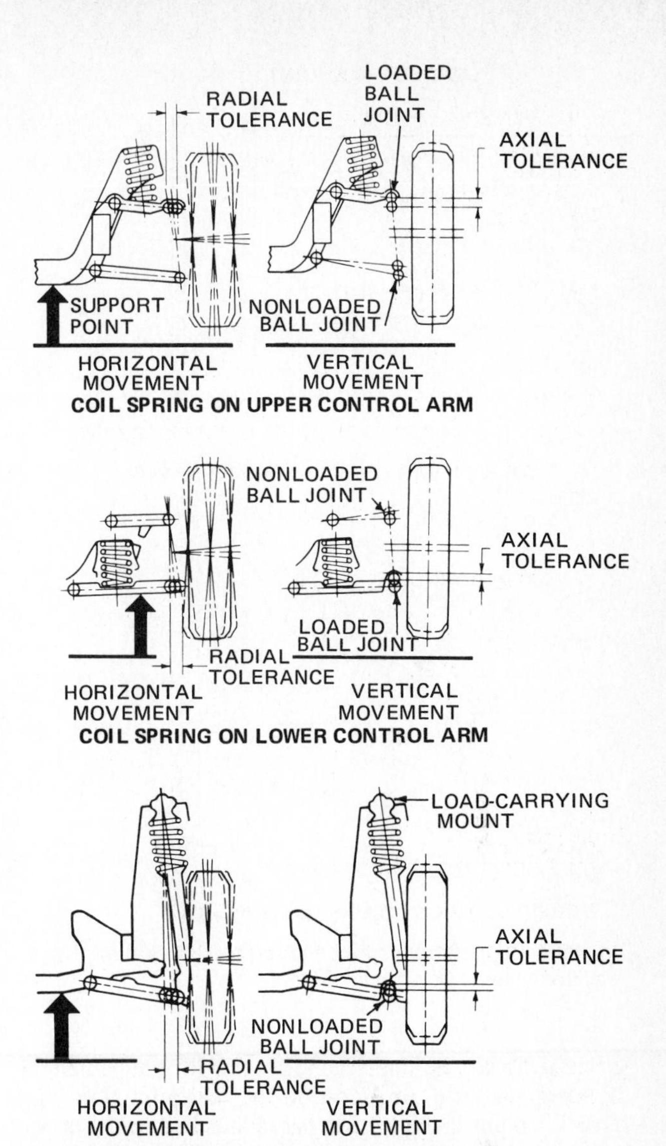

Fig. 11-6 Support points for checking ball joints in various front-suspension systems using coil springs. (*Motor Vehicle Manufacturers Association*)

steering wheel with a mark on a ruler or scale (Fig. 11-4).

Now slowly turn the steering wheel in the opposite direction until the front wheels start to move again. The distance that the steering-wheel reference mark has moved along the ruler is the amount of free play in the steering system. If the steering-wheel rim moves too much before the front wheels begin to move, there is excessive play.

1. Steering-linkage check Steering linkage, including tie rods, can be checked for looseness. Raise the front of the car until the bottoms of the tires are slightly off the floor. Then grasp both front tires and push out on both at the same time (Fig. 11-5). Next pull in on both tires at the same time. Excessive movement means worn linkage parts.

Excessive movement in the steering linkage can cause wheel shimmy, vehicle wander, uneven braking, steering-control problems, and excessive tire wear. On vehicles with 16-inch [406-mm] diameter (or smaller) wheels, the maximum movement should be ¼ inch [6.35 mm] or less. Other steering and suspension check points for each car are shown in the manufacturer's service manual.

NOTE: If the movement is excessive and the wheel bearings have not been adjusted, they may be loose. This will give a false reading by allowing excessive movement. To eliminate the effect of the wheel bearings, apply the brakes during the check. An assistant can do this, or you can use a portable brake depressor.

2. Inspecting front-wheel bearings Loose wheel bearings can cause poor steering control, car wander, uneven front-brake action, and rapid tire wear. To check the front-wheel bearings, raise the car on a lift or use floor jacks, properly placed. The lift points differ according to the type of front end. Lift points for cars with coil springs are shown in Fig. 11-6. Figure 11-7 shows the lift points for cars with torsion bars. If the spring is between the frame and the lower control arm, the car should be lifted at the frame crossmember. Use this same lift point for torsion-bar suspension systems that have the torsion bar attached to the lower control arm. If the

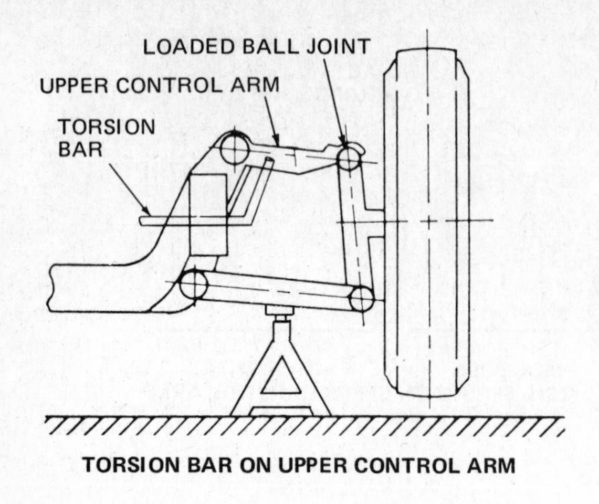

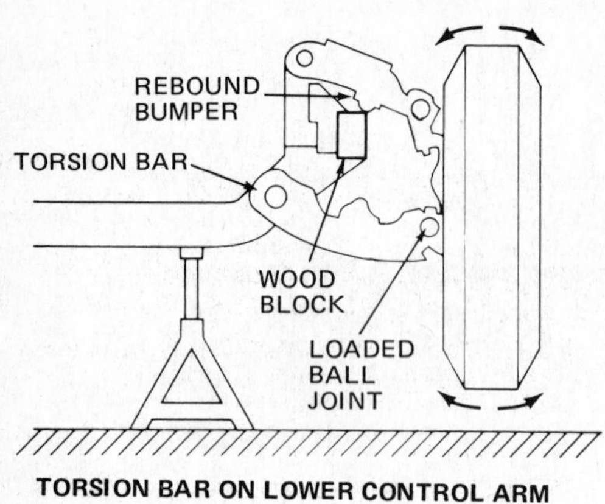

Fig. 11-7 Support points for checking ball joints in front-suspension systems using torsion bars.

spring is above the upper control arm, lift the vehicle at the lower control arm, close to the ball joint. Use this same lift point if the torsion bar is attached to the upper control arm. In either case, the weight of the wheel takes up any play in the ball joints.

Now grasp the tire at the top and bottom (Fig. 11-8), and rock it in and out. Any movement is the result of looseness in the wheel bearing or in the ball joints. Look at the brake drum or disk and the backing plate or shield as you rock the wheel. If you see movement between the drum or disk and the plate or shield, the looseness is in the wheel bearing. Have an assistant apply the brakes as you try rocking the wheel. If applying the brakes eliminates the free play, a loose wheel bearing is permitting the rocking motion.

In some inspection programs, the vehicle should be rejected if the wheel can be rocked more than ⅛ inch [3.2 mm], measured at the outer circumference of the tire. This amount of wheel wobble can make the vehicle unstable and hard to steer. The wheel bearing should be adjusted.

3. Inspecting ball joints Worn ball joints and loose wheel bearings can be detected by raising the front end of the car, grasping the top and bottom of the tire, and checking for play (Fig. 11-8). Try to rock the wheel in and out at the top and bottom. Excessive play indicates loose wheel bearings or worn ball joints. Have someone

Fig. 11-8 Checking for wear in the ball joints and wheel bearings.

apply the brakes as you try to rock the wheel again. If applying the brakes eliminates the free play, the wheel bearing is loose and should be adjusted. If the play is still there, it probably is in the ball joints.

Ball joints can be checked for wear while the wheel is supported as shown in Figs. 11-6 and 11-7. Axial play or tolerance is also called "vertical movement." It is checked by moving the wheel straight up and down. Radial play or tolerance also is called "horizontal movement." It is measured by rocking the wheel in and out at the top and bottom. Figure 11-8 shows this measurement.

The actual amount of play in a ball joint is measured with a dial indicator. In Fig. 11-9, the dial indicator is clamped to the lower control arm. The plunger tip rests against the steering-knuckle leg. With a pry bar, try to raise and lower the steering knuckle. As you do this, the play in the ball joint will show on the dial indicator. On the front-wheel-drive car shown in Fig. 11-9, vertical movement of the ball joint must not exceed 0.050 inch [1.2 mm]. The axial (up-and-down) play in a typical ball joint should not exceed ¹⁄₁₆ inch [1.6 mm].

The ball joints on vehicles manufactured prior to 1973 must be inspected with the ball joints unloaded. This means that the weight of the car must be removed from them. Beginning with some 1973 vehicles, manufacturers used wear-indicating ball joints. It is much easier to check this type of ball joint for wear. A quick visual check is all that is necessary. The check is made with the ball joints loaded, or carrying the weight of the car.

Figures 11-6 and 11-7 identify the loaded and nonloaded ball joints for various suspension systems. The load is carried by the ball joint for the control arm on which the spring rests. For example, when the spring is mounted on the upper control arm, the upper ball joint carries the load. When the spring is mounted on the lower control arm, the lower ball joint carries the load.

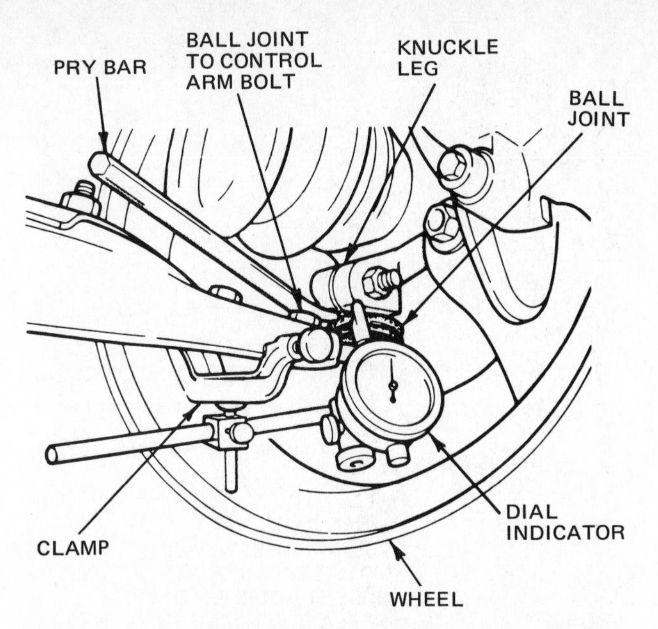

Fig. 11-9 A dial indicator is used to measure the actual amount of play in a ball joint. (*Motor Vehicle Manufacturers Association*)

In general, most wear occurs in the load-carrying ball joints. Nonloaded ball joints usually wear relatively little.

Figure 11-10 shows the use of a pry bar to check the ball joints. Pry under the front tire to see how much vertical movement the ball joints allow. Use only enough force to overcome the weight of the wheel assembly. If you use too much force, the ball joint may give you a false reading. You want to measure the movement of the wheel and ball joint as the joint is moved up to the "load" position. Note the movement as indicated on the dial indicator.

Fig. 11-10 Pry upward on the tire with a bar to check for loose ball joints. (*ATW*)

Next, grasp the tire at top and bottom (Fig. 11-8), and try to wobble it. This is the test described earlier for inspecting front-wheel bearings. However, here we are assuming that the wheel bearings have been checked and either adjusted or found to be properly tightened. Therefore, we are now checking the horizontal movement of the ball joints. However, some manufacturers do not accept horizontal movement as an indicator of ball-joint wear.

The actual specifications for the allowable wear limits of the ball joints are listed in the manufacturer's service manual. Refer to the specifications for the car you are checking.

NOTE: Some ball joints are preloaded with rubber or springs under compression. They should have very little movement in a vertical direction. These ball joints are marked as "preloaded" in specification tables.

Any ball joint should be replaced if there is excessive play in it. On certain vehicles with the spring on the lower control arm, some manufacturers specify replacement of the lower ball joint whenever the upper ball joint is replaced. When checking ball joints on 1955 through 1970 Chevrolet cars, the lower ball joint should be replaced if the radial play exceeds 0.250 inch [6.35 mm]. Today this tolerance is less on most ball joints. The lower ball joint should also be replaced if axial play between the lower control arm and the spindle exceeds the tolerance specified in the manufacturer's service manual. This specification may vary from 0 to 0.150 inch [3.81 mm], so always refer to the manufacturer's service manual.

When the spring is on the upper control arm, the specifications for some cars do not allow any play in the lower ball joint. The lower ball joint should be replaced if it has any noticeable looseness. The upper ball joint should be replaced if the radial play exceeds 0.250 inch [6.35 mm].

The upper ball joint should also be replaced if the axial play between the upper control arm and the spindle exceeds the tolerance specified by the vehicle manufacturer. This specification may vary from 0 to 0.095 inch [2.41 mm] or more. For example, Ford specifies that no check for vertical movement, or axial play, is necessary when the spring is on the upper control arm. They recommend only that the radial play should be checked, using a dial indicator.

4. Inspecting wear-indicating ball joints Many cars have ball joints with wear indicators (⚙ 8-24). The wear of this joint can be checked by visual inspection alone (Fig. 11-11). The amount of wear is indicated by the recession of the grease-fitting nipple into the ball-joint socket.

On a new ball joint, the nipple protrudes from the socket 0.050 inch [1.27 mm]. As the ball joint wears, the nipple recedes into the socket. When the wear has caused the nipple to recede 0.050 inch [1.27 mm] or more, the nipple will be level with or below the socket. Then replace the ball joint.

To check a wear-indicating ball joint, the vehicle should be supported on its wheels so that the ball joints are

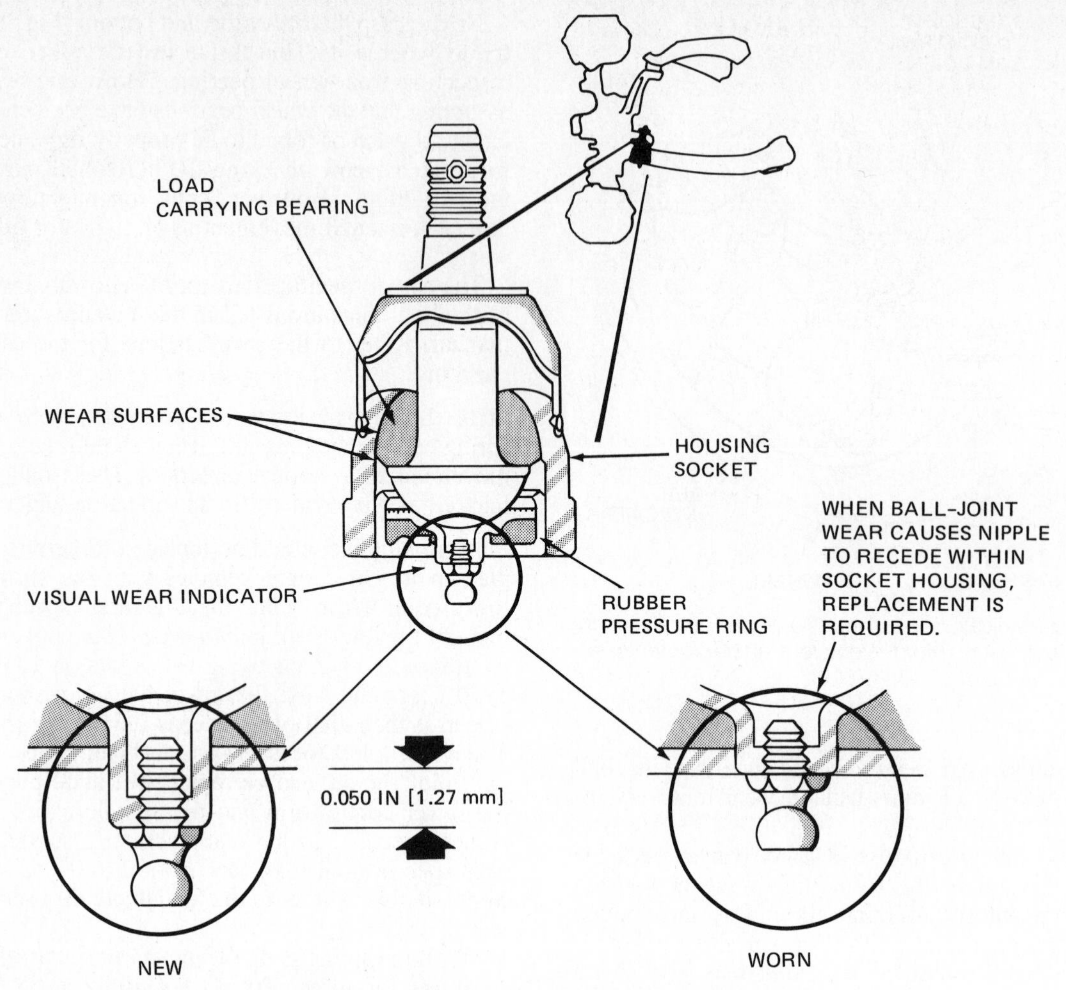

Fig. 11-11 How wear-indicating ball joints show that ball-joint replacement is necessary. In a worn ball joint, the grease-fitting nipple recedes into the socket, as shown to the right.

loaded. You may find it convenient to raise the car on a drive-on type of lift so that you can get to the ball joints easily.

Wipe the grease fitting to remove all dirt and grease. Then note the position of the grease-fitting nipple and compare it with Fig. 11-11. Use a steel scale to check the position of the nipple. However, the wear can also be checked with a screwdriver or even your fingernail. If the scale, screwdriver, or your fingernail passes over the nipple because it has recessed into the socket, then the ball joint is worn and should be replaced. Any ball joint that has cut, torn, or damaged seals should also be replaced.

5. Steering-gear check A quick check for looseness in the steering gear can be made by watching the pitman arm while an assistant turns the steering wheel one way and then the other, with the front wheels on the floor. If, after reversal of steering-wheel rotation, excessive movement of the steering wheel is required to move the pitman arm, then the steering gear is worn or in need of adjustment. Steering-gear service is covered in later chapters.

11-3 Hard steering If hard steering occurs, it is probably due to excessively tight adjustments in the steering gear or linkages. Hard steering can also be caused by low or uneven tire pressure; abnormal friction in the steering gear, in the linkage, or at the ball joints; or improper wheel or frame alignment.

If the car has power steering, its failure causes the steering system to revert to straight mechanical operation. A much greater steering effort is then required from the driver. When this happens, the power-steering gear and the pump should be checked as outlined in a later chapter.

The steering system may be checked for excessive friction by raising the front end of the car, turning the steering wheel, and checking the steering-system components to locate the source of excessive friction. Disconnect the linkage at the pitman arm. If this eliminates the frictional drag that makes the steering wheel hard to turn, then the friction is either in the linkage itself or at the steering knuckles. If the friction is not eliminated when the linkage is disconnected at the pitman arm, the steering gear is probably at fault. Steering-gear service is discussed in a later chapter.

If hard steering does not seem to be due to excessive friction in the steering system, the cause is probably incorrect front-wheel alignment, a misaligned frame, or sagging springs. Excessive caster, especially, causes hard steering. Chapter 13 describes wheel alignment.

⚙ 11-4 Car wander Wander is the tendency of a car to veer away from a straight path without driver control. Frequent steering-wheel movements are necessary to prevent the car from weaving from one side of the road to the other (Fig. 11-12). An example is when the driver must continually move the steering wheel back and forth to keep the car on the right side of the road or in the proper lane of traffic.

A variety of conditions can cause car wander. Low or uneven tire pressure, binding or excessive play in the linkage or steering gear, or improper front-wheel alignment will cause car wander. Any condition that causes tightness in the steering system will keep the wheels from automatically seeking the straight-ahead position. The driver has to correct the wheels constantly. This condition would probably cause hard steering, which is described in the previous section. Looseness or excessive play in the steering system might also cause car wander. These conditions tend to allow the wheels to waver slightly from their normal running position.

Several improper wheel-alignment angles may cause car wander. Excessively negative caster or uneven caster on the front wheels will tend to cause the wheels to swing away from the straight-ahead direction so that the driver must steer continually. The wrong camber angle will do the same thing. Excessive toe-in may also cause the same condition. The wheel-alignment angles mentioned previously are discussed in Chap. 13.

⚙ 11-5 Car pulls to one side during normal driving Sometimes a car pulls to one side so that force must constantly be applied to the steering wheel to maintain straight-ahead travel. The cause could be uneven tire pressure, uneven caster or camber, a tight wheel bearing, uneven springs, uneven torsion-bar adjustment, or wheels not tracking. A lack of tracking means that the rear wheels are not following a path that is parallel to the path of the front wheels. Figure 11-13 shows this problem.

Anything that makes one wheel drag or toe in or toe out more than the other causes the car to pull to that side. The methods used to check tracking and front-wheel alignment are covered in Chap. 13.

⚙ 11-6 Car pulls to one side when braking The most likely cause of pulling to one side when braking is grabbing brakes. This happens when the brake lining on the shoes or pads becomes soaked with oil or brake fluid, when brake shoes are unevenly or improperly adjusted, or when a stuck wheel cylinder or caliper piston causes the shoes at one wheel to apply less braking force than the shoes at the wheel on the other side of the axle. Chapter 6 covers brake service. The conditions listed in ⚙ 11-4 on car wander could also cause the car to pull to one side when braking. A pulling condition, from whatever cause, tends to become more noticeable when the car is braked to a stop.

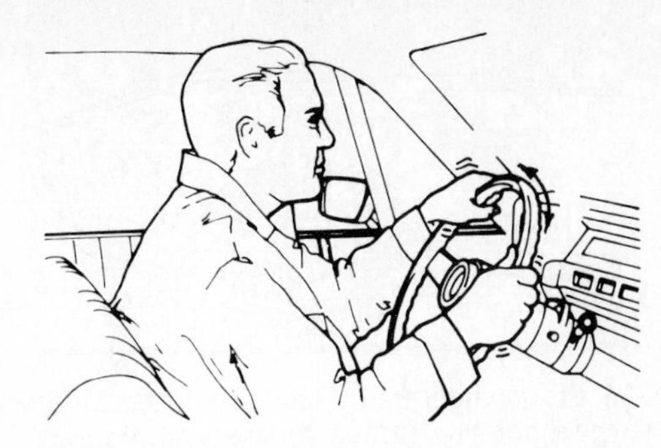

Fig. 11-12 Frequent steering-wheel movements are needed to overcome the effects of wander. (*Ford Motor Company*)

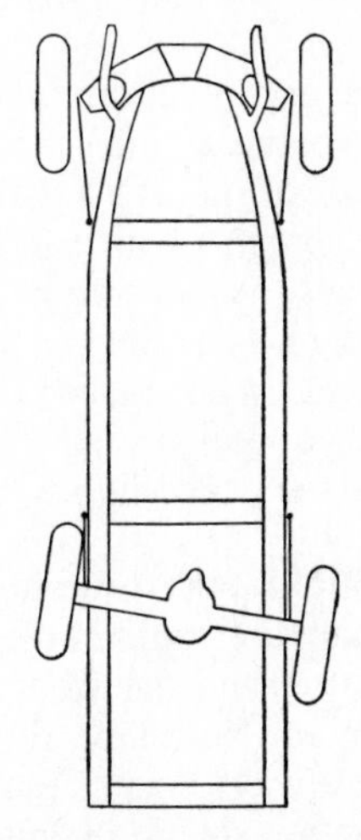

Fig. 11-13 If the frame is bent or the rear axle is swung back on one side, the car will not track properly. (*Applied Power, Inc.*)

⚙ 11-7 Low-speed front-wheel shimmy Front-wheel shimmy and front-wheel tramp are sometimes confused. Low-speed shimmy is the rapid oscillation of the wheel on the steering-knuckle support. The wheel tries to turn in and out alternately. This action causes the front end of the car to shake from side to side. Front-wheel tramp, or high-speed shimmy, is the tendency for the wheel-and-tire assembly to hop up and down and try to leave the pavement. Even when the tire does not leave the pavement, tramp can be observed as a rapid flexing-unflexing action of that part of the tire in contact with the pavement. The bottom of the tire first appears deflated as the wheel moves down, then appears inflated as the wheel moves up.

Low-speed shimmy can result from low or uneven tire pressure, excessive lateral runout, looseness in linkage, excessively soft springs, incorrect or unequal wheel camber, dynamic imbalance of the wheels, or irregularities in the tire treads.

⚙ 11-8 Front-wheel tramp Front-wheel tramp is often called high-speed shimmy. This condition causes the front wheels to move up and down alternately. One of the most common causes of front-wheel tramp is unbalanced wheels, or wheels that have too much radial runout. An unbalanced wheel is heavy in one spot, like the wheel

Fig. 11-14 Front-wheel tramp can be caused by an out-of-balance wheel and tire. (*Moog Automotive, Inc.*)

in Fig. 11-14. As it rotates, the heavy part sets up a circulating outward thrust that tends to make the wheel hop up and down.

A similar action occurs if the wheel has too much radial runout. This is the amount that the wheel rotates out-of-round, instead of making a true circle as it turns. Defective shock absorbers, which fail to control spring oscillations, also cause wheel tramp. Any of the causes described in the previous section on front-wheel shimmy may also cause wheel tramp. Later sections describe the servicing of the wheel and tire so that they can be restored to proper balance and alignment.

⚙ 11-9 Steering kickback Steering shock, or kickback, consists of sharp and rapid movements of the steering wheel that occur when the front wheels encounter obstructions in the road. Normally, some kickback to the steering wheel will always occur. When it becomes excessive, an investigation should be made. This condition could be the result of incorrect or uneven tire inflation, sagging springs, defective shock absorbers, or looseness in the linkage or steering gear. Any of these defects could permit road shock to carry excessively to the steering wheel.

⚙ 11-10 Tires squeal on turns If the tires skid or squeal on turns, the cause may be excessive speed on the turns. If this is not the cause, it is probably low or uneven tire pressure, worn tires, or misalignment of the front wheels. Improper camber and toe settings may tend to cause tire squeal.

⚙ 11-11 Improper tire wear Various types of abnormal wear occur on tires. The type of tire wear is often a good indication of a particular defect in the suspension or steering system, or improper operation or abuse. For example, if the tire is underinflated, the sides will bulge over, and the center of the tread will be lifted clear of the road (Fig. 11-15). The sides of the tread will take all the wear. The center will be barely worn.

Uneven tread wear shortens tire life. But, even more damaging is the excessive flexing of the tire sidewalls that takes place as the underinflated tire rolls on the pavement. The repeated flexing causes excessive heat, the fabric in the sidewalls to crack or break, and the plies to separate. This cracking and separation seriously weakens the sidewalls and may soon lead to complete tire failure. In addition, the underinflated tire is unprotected against rim bruises. For example, when the tire hits a rut or stone on the road, or bumps a curb too hard, the tire will flex so much under the blow that it will actually be pinched on the rim. Pinching causes plies to break and leads to early tire failure.

NOTE: The radial tire applies a much larger area of tread to the pavement, as explained in Chap. 20. Therefore, radial tires may appear to be running underinflated, when compared with bias-ply tires.

Continuous high-speed driving on curves, both right and left, can produce tread wear that looks almost like underinflation wear. The side thrust on the tires as they round the curves causes the sides of the tread to wear. The only remedy is to reduce car speed on turns.

Overinflation causes the tire to ride on the center of its tread (Fig. 11-15), so that only the center of the tread wears. Uneven tread wear shortens tire life. But, equally damaging is the fact that the overinflated tire does not have normal "give" or flex when it meets a rut or bump in the road. Instead of flexing normally, the tire fabric takes the major shock. As a result, the fabric may crack or break so that the tire quickly fails.

Excessive toe-in or toe-out on turns causes the tire to be dragged sideways while it is moving forward. The tire on a front wheel that toes in 1 inch [25.4 mm] from straight ahead will be dragged sideways about 150 feet every mile [28.6 m every km]. This sideward drag scrapes off rubber. Characteristic of this type of wear are featheredges of rubber that appear on one side of the tread design (Fig. 11-16). If both front tires show this type of wear, the front end has improper toe. But if only one tire shows this type of wear when both tires have been

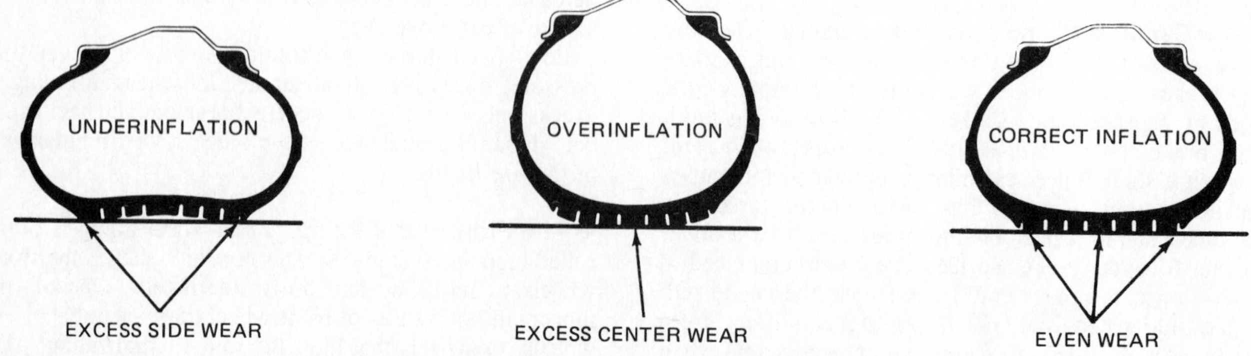

Fig. 11-15 Effects of overinflation and underinflation on tires. (*Moog Automotive, Inc.*)

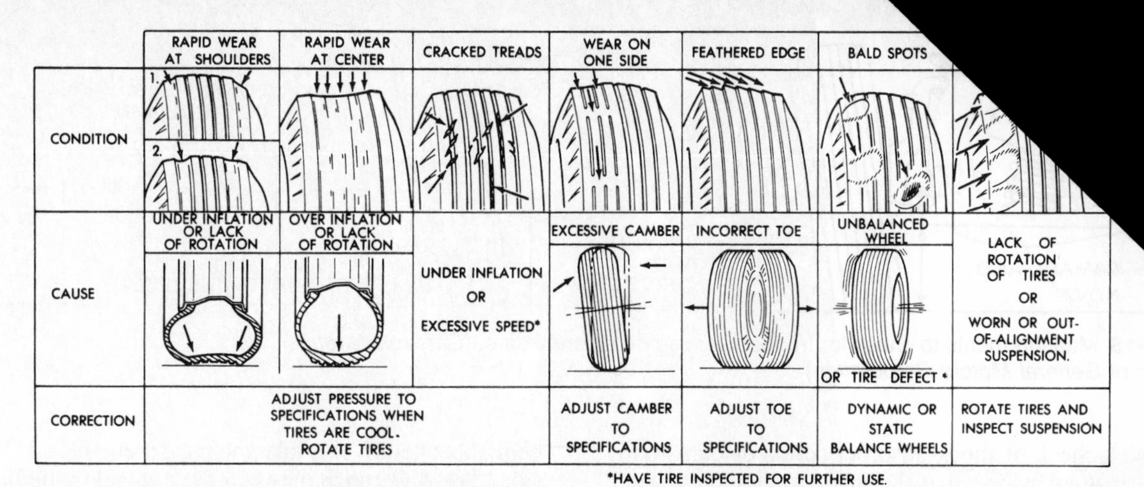

Fig. 11-16 Types of tire wear, and their causes and corrections. (*Chrysler Corporation*)

in the same poisition on the car for some time, a bent steering arm is indicated. This causes one wheel to toe-in more than the other.

Excessive camber of a wheel causes one side of the tire tread to wear more quickly than the other. This kind of wear is also shown in Fig. 11-16. If the camber is positive, the tire will tilt outward. If the camber is negative, the tire will tilt inward. In this case, heavy tread wear will appear on the inside.

Cornering wear, caused by taking curves at excessively high speeds, may be mistaken for camber wear or toe-in or toe-out wear. Cornering wear is due to centrifugal force acting on the car and causing the tires to roll and skid on the road. This produces a diagonal type of wear that rounds the outside shoulder of the tire and roughens the tread surface near the outside shoulder. In severe cornering wear, fins or sharp edges will be found along the inner edges of the tire treads. There is no adjustment that can be made to correct the steering system for this type of wear. The only solution is for the driver to slow down on curves.

Uneven tire wear, such as bald spots and scalloped wear, causes the tread to be unevenly or spottily worn (Fig. 11-16). This type of wear can result from a number of mechanical conditions. These conditions include misaligned wheels, unequal or improperly adjusted brakes, unbalanced wheels, overinflated tires, out-of-round brake drums, and incorrect linkage adjustments.

High-speed operation causes much more rapid tire wear because of the higher temperature and greater amount of scuffing and rapid flexing to which the tires are subjected. The chart in Fig. 11-17 shows just how much tire wear increases with car speed. According to the chart, tires wear more than three times as fast at 70 mph [113 km/h] as they do at 30 mph [48 km/h]. Careful, slow driving with correct tire inflation pressures will greatly increase tire life.

⚙ 11-12 Hard or rough ride A hard or rough ride could be due to excessive tire pressure, improperly operating shock absorbers, or excessive friction in the spring suspension. The spring suspension can be checked for excessive friction in leaf-spring suspension systems. With a yardstick, measure from the floor to the lower edges of the car body, front and back. Then lift the front end of the car as high as possible by hand, and very slowly let it down. Carefully measure again and write down the distance.

Push down on the car bumper at the front end, and again slowly release the car. Remeasure the distance from the floor to the body. Note the difference in measurements. Repeat this action several times to obtain accurate measurements. The difference is caused by the friction in the suspension system and is called friction lag. After determining friction lag at the front end, check it at the rear of the car.

Excessive friction lag is corrected by lubricating the springs, shackles, and bushings (on types where lubrication is specified) and by loosening the shock-absorber mounts, shackle bolts, and U bolts. Then retighten the U bolts, shackle bolts, and shock-absorber mounts, in that order. This procedure permits realignment of parts that might have become misaligned and caused excessive friction.

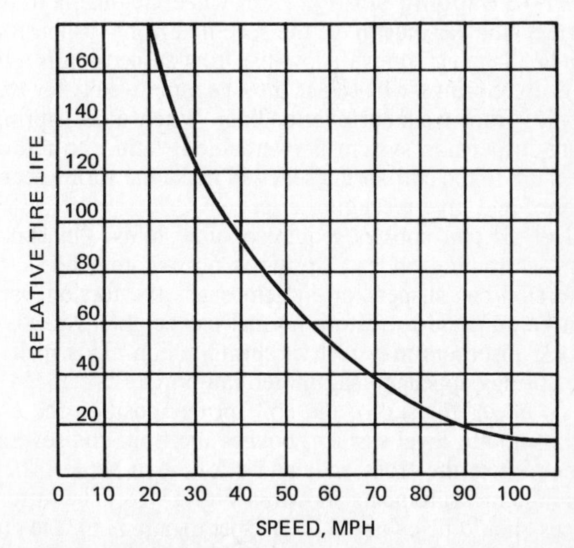

Fig. 11-17 Graph showing how tire wear increases with speed.

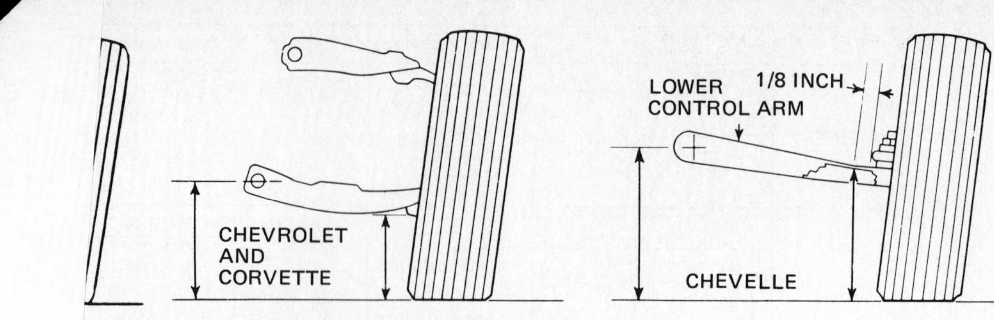

ɔr front-spring sag on different cars. (*Chevrolet Motor* ... *on*)

A quick check ... r action on cars giving a hard or uneven ride may be made by bouncing each corner of the car in turn (Fig. 9-22). Shock-absorber inspection, bounce testing, and bench testing are described in ✪ 9-10.

✪ 11-13 Sway on turns Sway of the car body on turns or on rough roads may be due to a loose stabilizer bar or sway bar. Stabilizer-bar attachments to the frame, axle housing, or suspension arms should be checked. Weak or sagging springs could also cause excessive sway. If the shock absorbers are ineffective, they may permit excessive spring movement. This could cause strong body pitching and sway, particularly on rough roads. If the caster is excessively positive, it will cause the car to roll out, or lean out, on turns. Front-wheel alignment should be corrected.

✪ 11-14 Spring breakage Breakage of leaf springs can result from excessive overloading; loose U bolts that cause breakage near the center bolt, a loose center bolt that causes breakage at the center-bolt holes, an improperly operating shock absorber that causes breakage of the master leaf, or a tight spring shackle that causes breakage of the master leaf near or at the spring eye. Determining the point at which breakage has occurred usually indicates the cause.

✪ 11-15 Sagging springs For wheel alignment to be correct, the car must have the specified front-suspension height. Sagging springs allow the front-suspension height to drop. Springs will sag if they become weak, for example, from frequent overloading. When a coil-spring front-suspension system is overhauled, failure to return the shim to the coil-spring seat will make the spring seem shorter and give a sagging effect.

Not all coil springs require or use shims. On many cars, shimming under a spring is not recommended. If a torsion-bar suspension system sags, the torsion bars can be adjusted to restore normal car height. Defective shock absorbers may restrict spring action. This makes the springs appear to sag more than normal.

To check for sag of the coil spring, position the car on a smooth, level surface, bounce the front end several times, raise the front end, and allow it to settle. Then take the measurements shown in Fig. 11-18. The differences should be as noted in the specifications for the car being checked. Next, take the measurements on the other side. The difference between the two sides should be no greater than ½ inch [12.7 mm]. To make a correction, the springs must be replaced, or the torsion bar adjusted.

✪ 11-16 Noises The noises produced by defects in springs or shock absorbers will usually be either rattles or squeaks. Rattling noises can be produced by looseness of leaf-spring U bolts, metal spring covers, rebound clips, spring shackles, or shock-absorber mounts. These noises can usually be located by a careful examination of the various parts in the suspension system. Spring squeaks can result from a lack of lubrication in the spring shackles, at spring bushings, or in a leaf spring itself.

Squeaks in the shock absorber can result from tight or dry bushings. Rattles in the steering linkage may develop if linkage parts become loose. Sometimes squeaks during turns can develop because of a lack of lubrication in the joints or bearings of the steering linkage. This condition also produces hard steering.

Some of the connections between steering-linkage parts are made with ball sockets that can be lubricated. Others are permanently lubricated on original assembly. If these develop squeaks or excessive friction, they must be replaced. Lubricating and replacing steering-linkage ball sockets are covered later.

Chapter 11 review questions

Select the *one* correct, best, or most probable answer to each question. Then check your answers against the correct answers given at the end of the book.

1. A driver says that the front end of the car vibrates up and down while traveling at most road speeds. Mechanic A says that too much radial runout of the front tires could be the cause. Mechanic B says that static out-of-balance of the front tires could be the cause. Who is right?
   a. A only
   b. B only
   — *c.* either A or B
   d. neither A nor B

2. A car wanders on a level road. Mechanic A says that too much negative camber could be the cause. Mechanic B says that too much positive caster could be the cause. Who is right?
   a. A only
   b. B only
   c. either A or B
   d. neither A nor B
3. A car with power steering has no assist to the right. All of these could cause this *except:*
   a. a binding spool valve
   b. a leaking seal inside the steering gear
   c. a sticking relief valve in the pump
   d. an incorrectly installed torsion (steering-shaft) bar
4. A car with power steering is running at fast idle. A buzzing noise is heard when the wheels are straight ahead. The noise stops as the wheels are turned. Which of these could be the cause?
   a. a blocked high-pressure line
   b. a sticking pressure-control valve
   c. a misadjusted steering gear
   d. a loose pump belt
5. When a vehicle is road-tested, the steering wheel shakes from side to side at higher speeds. Mechanic A says that this could be caused by front wheels not statically balanced. Mechanic B says that this could be caused by front wheels not dynamically balanced. Who is right?
   a. A only
   b. B only
   c. either A or B
   d. neither A nor B
6. With power steering, temporary increases in effort when turning the steering wheel in either direction may be caused by:
   I. Engine idle too slow
   II Air in the system
   a. I only
   b. II only
   c. either I or II
   d. neither I nor II
7. As a general rule, axial (up-and-down) play in a ball joint should not exceed:
   a. ¼ inch [6.35 mm]
   b. 1/16 inch [1.6 mm]
   c. ⅛ inch [3.2 mm]
   d. ½ inch [12.7 mm]
8. Mechanic A says that tire cupping is caused by wheel tramp. Mechanic B says that the cause is loose tie-rod ends. Who is right?
   a. A only
   b. B only
   c. either A or B
   d. neither A nor B
9. The tie rods move forward and backward but do not move up or down. Mechanic A says that they are good. Mechanic B says that the steering knuckles should be replaced. Who is right?
   a. A only
   b. B only
   c. both A and B
   d. neither A nor B
10. Load-carrying ball joints with load indicators:
   a. should be unloaded for checking
   b. should be checked with ball joints loaded
   c. should be checked on a frame-contact hoist
   d. should be checked with a pry bar and ball-joint checking gauge

CHAPTER 12

SERVICING STEERING LINKAGE AND SUSPENSION SYSTEMS

After studying this chapter, and with proper instruction and equipment, you should be able to:

1. Lubricate the steering linkage.
2. Replace defective components in the steering linkage.
3. Replace and adjust front-wheel bearings.
4. Replace defective components in the front-suspension system.
5. Check and replace ball joints.
6. Replace control-arm bushings.
7. Replace coil springs and torsion bars in front-suspension systems.
8. Replace coil springs and leaf springs in rear-suspension systems.

Steering linkage

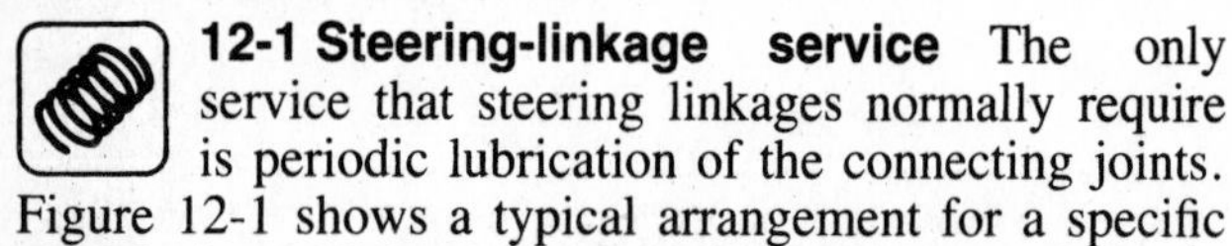

12-1 Steering-linkage service The only service that steering linkages normally require is periodic lubrication of the connecting joints. Figure 12-1 shows a typical arrangement for a specific model of car. Other cars have different linkage arrangements.

NOTE: Inspect the steering linkage and suspension systems whenever you are under a car. It takes only a moment to check seals and note the condition of the ball joints, ball sockets, and links. The steering linkage must be in good condition for the safe operation of the car. Check that the steering linkage of every car you service is in safe operating condition.

If ball joints or other parts are worn, or have been bent or damaged, they should be replaced. Never straighten and reuse bent steering-linkage parts. This could lead to failure later.

If the linkage is disassembled, check the front-wheel alignment after all parts have been put back together again. Chapter 13 covers wheel alignment.

12-2 Steering-linkage lubrication The connecting joints between the links normally require periodic lubrication. The connecting joints are ball joints and ball sockets. In Fig. 12-1, the arrows indicate the joints and sockets that can be lubricated. Notice that arrows are not pointed at the sockets at the pitman arm and the idler arm. These are permanently lubricated on assembly and do not require further lubrication.

The lubricating procedure for the steering linkage varies slightly on different cars. It depends on whether the ball joints and sockets have plugs to be removed, as in Fig. 12-2, or regular grease fittings, as in Fig. 12-3.

The frequency of lubrication for the steering linkage varies from one model of car to another. Ford recommends that the front suspension and steering linkage on new cars be lubricated every 30 months or 30,000 miles [48,000 km], whichever occurs first. On the Omni and Horizon, Chrysler recommends lubricating the grease fittings and checking for damaged seals (Fig. 12-3) every 36 months or 30,000 miles [48,000 km]. Chevrolet recommends a chassis lubrication every 12 months or 7500 miles [12,000 km] for the Chevrolet Citation.

1. Lubricating ball joint with plug First, wipe the plug and the area around it so that no dirt will get into the lubrication hole. Remove the plug. Then use a rubber-tipped, hand-operated grease gun filled with the proper lubricant. Chevrolet recommends the use of a water-resistant chasis lubricant rated ''EP'' for extreme pressure. Apply the tip to the plug hole and operate the grease gun at low pressure. This forces the lubricant into the joint. When the seal or dust cover begins to swell, stop. Do not overlubricate or you will destroy the weathertight seal.

2. Lubricating ball joint with grease fitting First, wipe the grease fitting so that no dirt will get into the ball joint. Then take a hand-operated grease gun with the proper connector for the grease fitting, and fill it with the proper lubricant. Operate the grease gun at low pres-

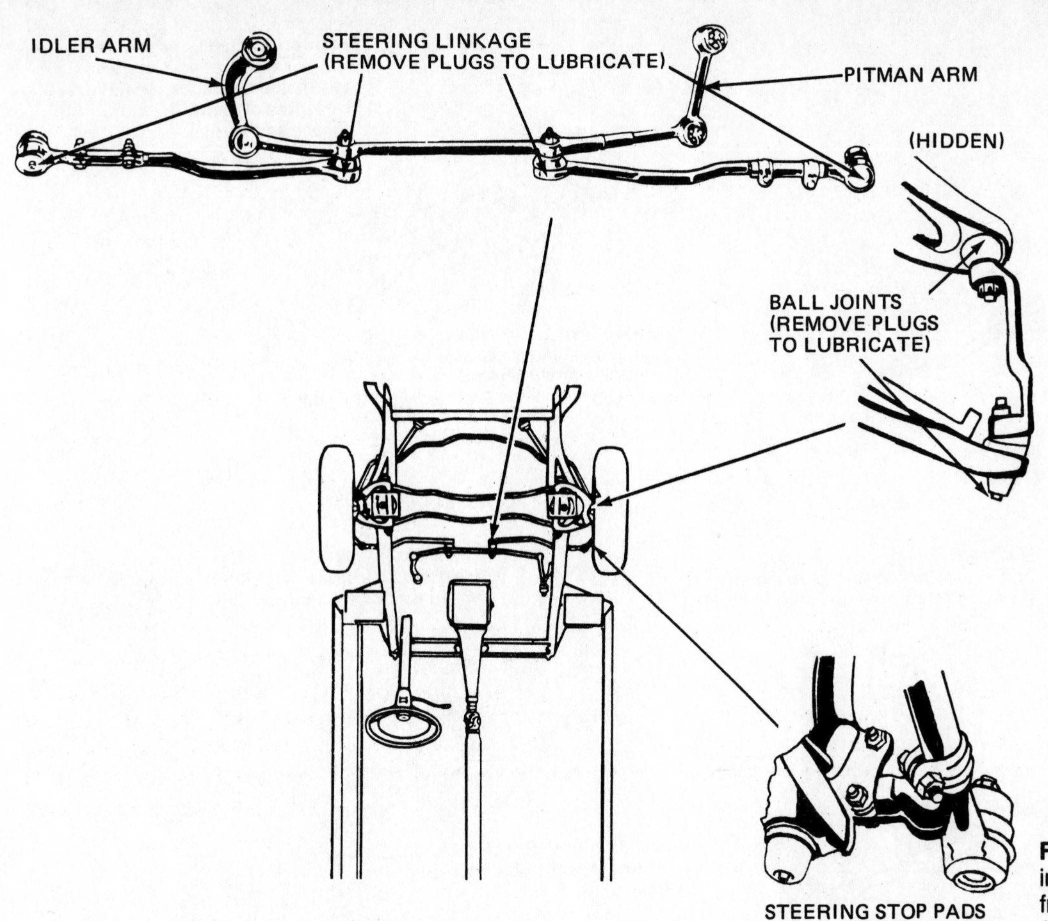

Fig. 12-1 Lubrication points in a steering linkage and front-suspension system. (*Ford Motor Company*)

sure. Chrysler recommends to "full and flush" the joint with lubricant. Stop filling when the grease begins to flow freely from the base of the seal or if the seal begins to balloon.

Careful! When assembling automotive components, always use the proper fasteners such as bolts, nuts, and cotter pins. This means using parts from the original manufacturer, or their equivalent. Never use a replacement part of lesser quality or different design in servicing front suspension and steering systems unless you *know*

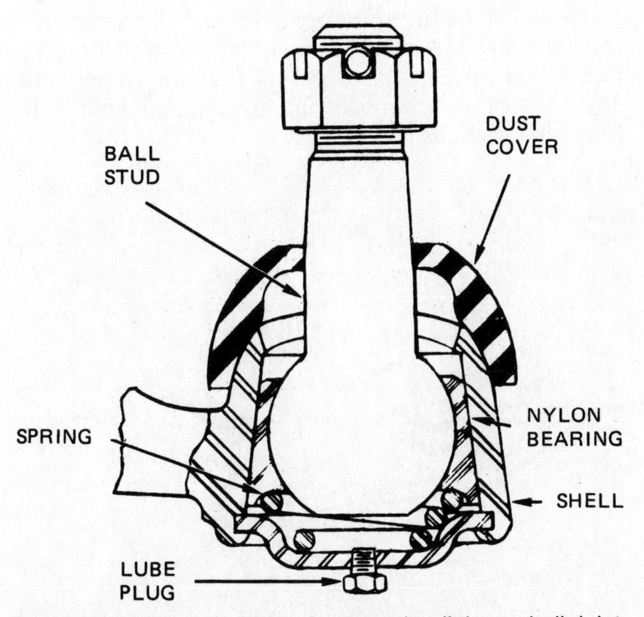

Fig. 12-2 Sectional view of a steering-linkage ball joint. (*American Motors Corporation*)

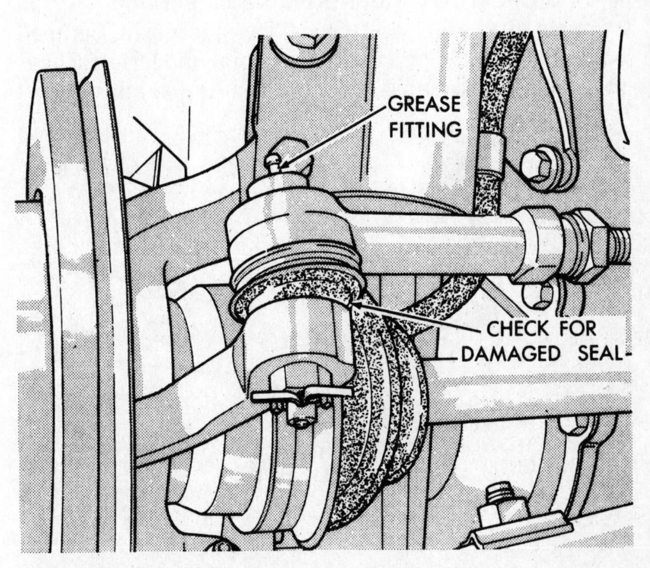

Fig. 12-3 Grease fitting in a tie-rod-end ball socket. (*Chrysler Corporation*)

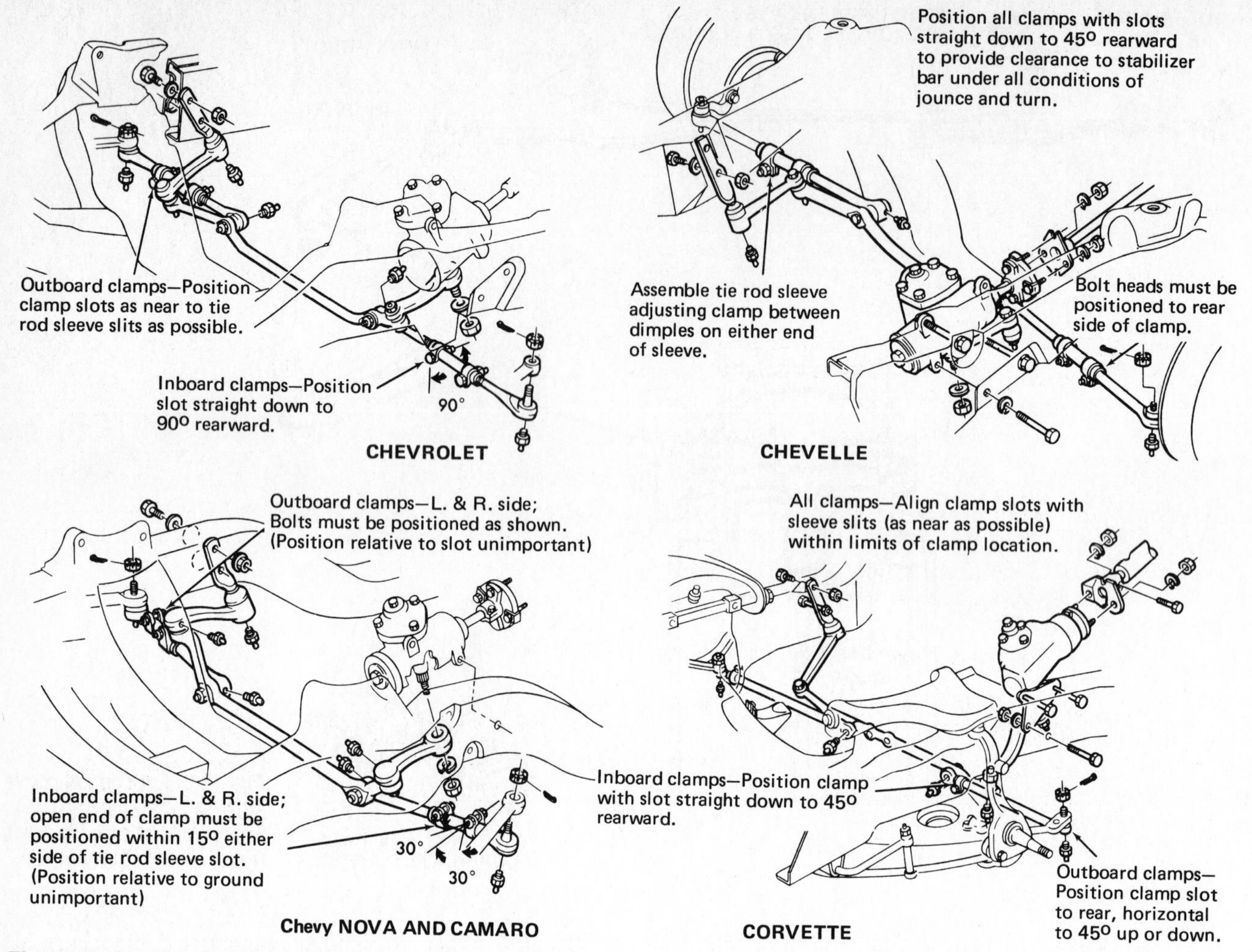

Fig. 12-4 Steering-linkage arrangements. (*Chevrolet Motor Division of General Motors Corporation*)

that it is safe to do so. After installing parts and fasteners, always torque all nuts and bolts to specifications.

⚙ 12-3 Chevrolet steering-linkage service As an example of steering-linkage service procedures, outlined below are typical procedures for some models of Chevrolet cars. Procedures for other cars are similar. However, always check the service manual for the car you are servicing before you begin the job so that you will know exactly how to proceed. Figures 12-4 and 12-5 show the steering linkages described in this section. Removal and installation of the tie rods, relay rod, idler arm, pitman arm, and steering arms are covered in the following sections.

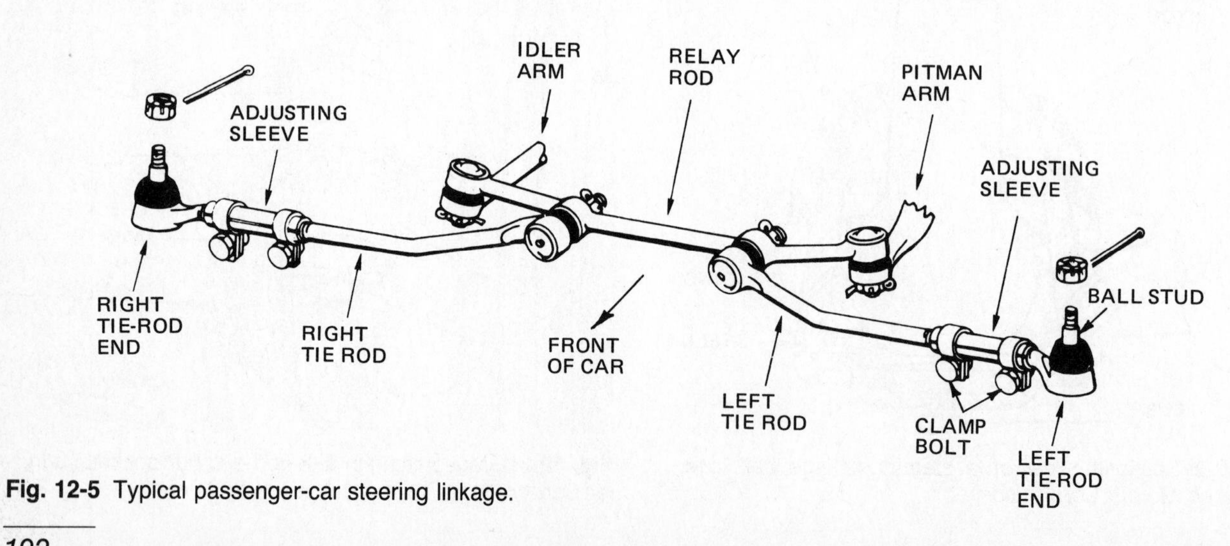

Fig. 12-5 Typical passenger-car steering linkage.

1. Tie-rod removal Remove the cotter pins from the ball studs and take off the nuts. To disconnect the outer ball stud, tap on the steering arm at the tie-rod end with a hammer. Use a heavy hammer as a backing (Fig. 12-6). Pull down on the tie rod, if necessary, to free it from the steering arm. Disconnect the other end of the tie rod from the relay rod in the same manner. If necessary, take the tie-rod end off by loosening the clamp bolts and screwing it off.

NOTE: Other manufacturers recommend the use of a puller (Fig. 12-7) to separate a ball stud from the tapered hole in the linkage.

2. Tie-rod installation When installing the tie-rod ends, lubricate the threads with EP chassis lubricant. Make sure both ends are threaded an equal distance on the tie rod. The threads on the ball-stud nuts must be clean and smooth. If they bind, they may turn the ball studs in the tie-rod ends when the nuts are tightened. Install the seals, and place the ball studs in the steering arm and relay rod. Put the nut on, torque it to specifications, and install the cotter pin. Lubricate the tie-rod ends and adjust the toe-in.

Before locking the clamp bolts, be sure the tie-rod ends are aligned with their ball studs. Each ball stud should be in the center of its travel. Otherwise, binding will result. The bolts must be installed facing forward, with the centerlines of the bolts within the angles shown in Fig. 12-8. The slot of the adjusting sleeve may be in any position, but not closer than 0.010 inch [0.254 mm] to the edge of the clamp jaw, or between the clamp jaws. These angles are specified for the Chevrolet models listed in Fig. 12-8. Other cars require different angles and arrangements. Refer to the service manual for the car you are servicing.

3. Relay-rod removal Remove the inner ends of the tie rods from the relay rod as discussed and shown in Figs. 12-5 and 12-6. Remove the damper, where present with manual steering, from the relay rod. Remove the cotter pin and nut from the relay-rod ball-stud attachment at the pitman arm. Figure 12-5 shows these parts. Detach the relay rod from the pitman arm. Move the steering linkage as necessary to free the pitman arm from the relay rod. Remove the cotter pin and nut from the idler arm and take the relay rod from the idler arm.

4. Relay-rod installation Install the relay rod on the idler arm, making sure the stud seal is in place. Then install and tighten the nut. Advance the nut just enough beyond the specified torque to align the castellated nut with the cotter-pin hole. Install the pin. Raise the end of the rod and install it on the pitman arm. Secure the rod with the nut and cotter pin. Install the tie-rod ends on the relay rod as described earlier for tie-rod installation. Install the damper, if used. Adjust the toe and center the steering wheel. These steps are explained in Chap. 13 on wheel alignment.

5. Idler-arm removal The idler-arm assembly should be replaced if an up-and-down push of 25 pounds [111 N] applied at the relay-rod end produces a vertical movement of more than 0.125 inch [3.18 mm]. Movement

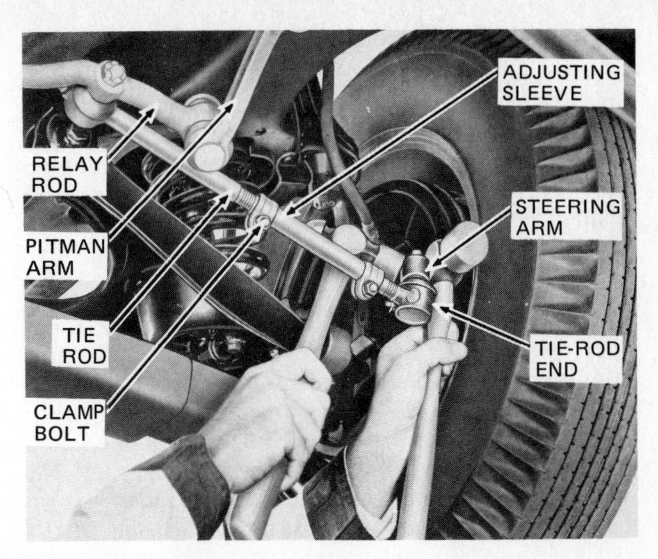

Fig. 12-6 Freeing the tie-rod-end ball stud from the steering arm. (*Chevrolet Motor Division of General Motors Corporation*)

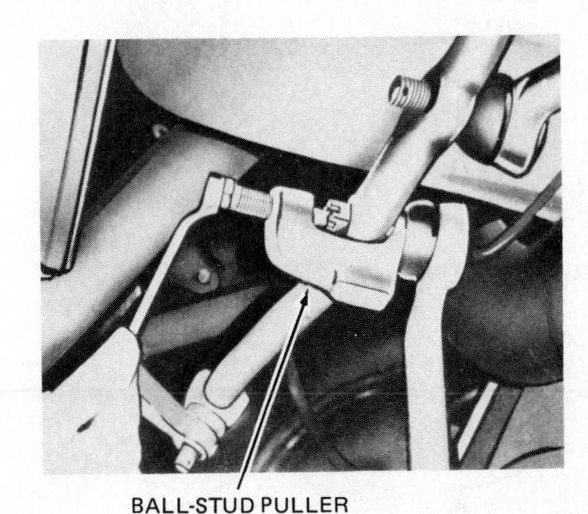

Fig. 12-7 Removing the ball stud from the rod. (*Ford Motor Company*)

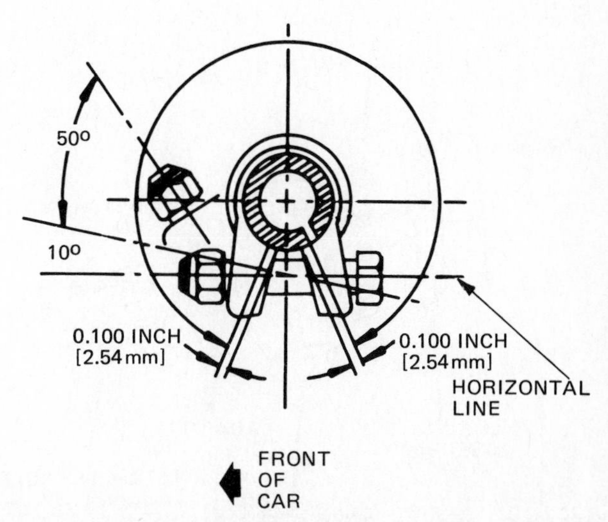

Fig. 12-8 Proper installation of bolts and tie-rod clamps for Chevrolet, Chevelle, and Monte Carlo models. (*Chevrolet Motor Division of General Motors Corporation*)

means a worn ball stud. To remove the idler-arm assembly, raise the car. Remove the nuts, washers, and bolts attaching the idler arm to the frame. These parts are shown in Figs. 12-4 and 12-5. Remove the cotter pin and nut from the ball stud attaching the idler arm to the relay rod. Detach the relay rod from the idler arm by tapping the relay rod with a hammer (Fig. 12-6). Remove the idler arm.

6. **Idler-arm installation** The installation procedure differs from the various models shown in Fig. 12-4. On Chevrolet cars, put the seal in position on the idler-arm stud, position the stud up through the frame, and secure the stud with a lock washer and nut. On the Monte Carlo, Nova, and Corvette, position the idler arm on the frame and install the mounting bolts, washers, and nuts. No washer is used on the Corvette. Install the relay rod on the idler arm, making sure the seal is on the stud. Install

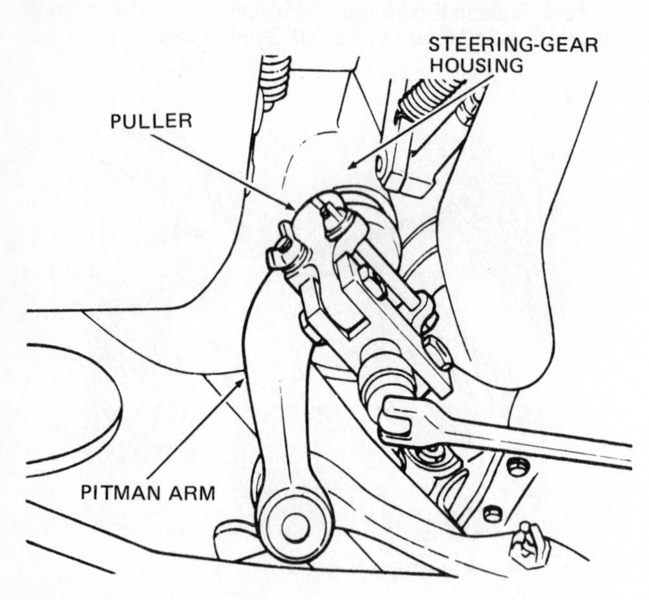

Fig. 12-9 Removing the pitman arm from the sector shaft in the steering gear. (*Ford Motor Company*)

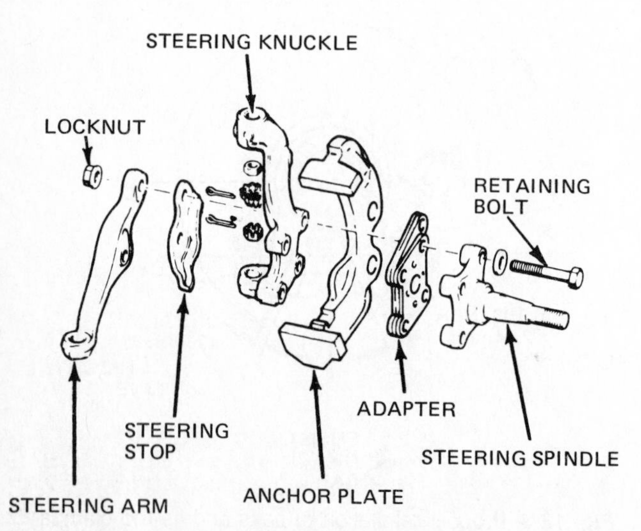

Fig. 12-10 To replace a steering arm, unbolt it from the steering knuckle. (*American Motors Corporation*)

and tighten the nut to the specified torque and install the cotter pin.

7. **Pitman-arm removal** Remove the cotter pin and nut from the pitman-arm ball stud (Fig. 12-5). Remove the relay rod from the pitman arm by tapping with a hammer on the side of the rod or arm in which the stud mounts, in the same manner as shown in Fig. 12-6. Pull down on the relay rod to remove it from the stud. Remove the pitman-arm nut. Mark the relationship of the pitman arm to the sector shaft in the steering gear. Then use a puller to pull the pitman arm off the shaft (Fig. 12-9).

8. **Pitman-arm installation** Install the pitman arm on the sector shaft, aligning the marks. Install the sector-shaft nut. Position the relay rod on the pitman arm. Then install the nut and cotter pin.

9. **Steering-arm removal** The only time a steering arm requires replacement is if it is bent. To remove an arm, first detach the tie-rod as explained in item 1 above. Then remove the front wheel, hub, and brake drum as a unit by taking off the hubcap, dust cap, cotter pin, and spindle nut. Pull the assembly outward. If the assembly is hard to remove, back off the brake adjustment to increase the shoe-to-drum clearance. On cars with disk brakes, remove the caliper and disk. Now, the steering-arm retainer nuts and bolts can be removed so the steering arm can be taken off (Fig. 12-10).

10. **Steering-arm installation** Put the steering arm in place and install the retaining bolts (Fig. 12-10). Secure the bolts with the special locknuts. Pack the wheel bearings with the specified lubricant. Install the bearings and wheel-hub-brake assembly. On disk-brake cars, install the disk, and caliper. Install the keyed washer and spindle nut and adjust the front-wheel bearing as outlined later. Install the tie-rod ball stud in the steering arm. Make sure the dust cover is in place on the ball stud. Install the castellated nut and cotter pin. Check the toe and steering-wheel centering as explained in Chap. 13 on wheel alignment.

⚙ 12-4 Ford steering-linkage service Different models require different service procedures. Figures 12-11 and 12-12 show the different linkage arrangements used on various Ford models.

1. **Connecting-rod assembly** The ball studs in the connecting-rod ends are nonadjustable and cannot be lubricated. The rod-end assembly must be replaced if the ball studs become loose.

a. Removal Remove the cotter pin and nut from the ball stud. Disconnect the end from the spindle arm or center link. Loosen the connecting-rod-sleeve clamp bolt and count the number of turns needed to remove the rod end from the sleeve. Discard all rod-end parts that have been removed. Always use all new parts when replacing a connecting rod.

b. Installation Thread the new rod end into the sleeve, but do not tighten the clamp bolt. Insert the stud into the spindle arm or center link and install the stud nut, tightening it to the specified torque. Check and adjust the toe

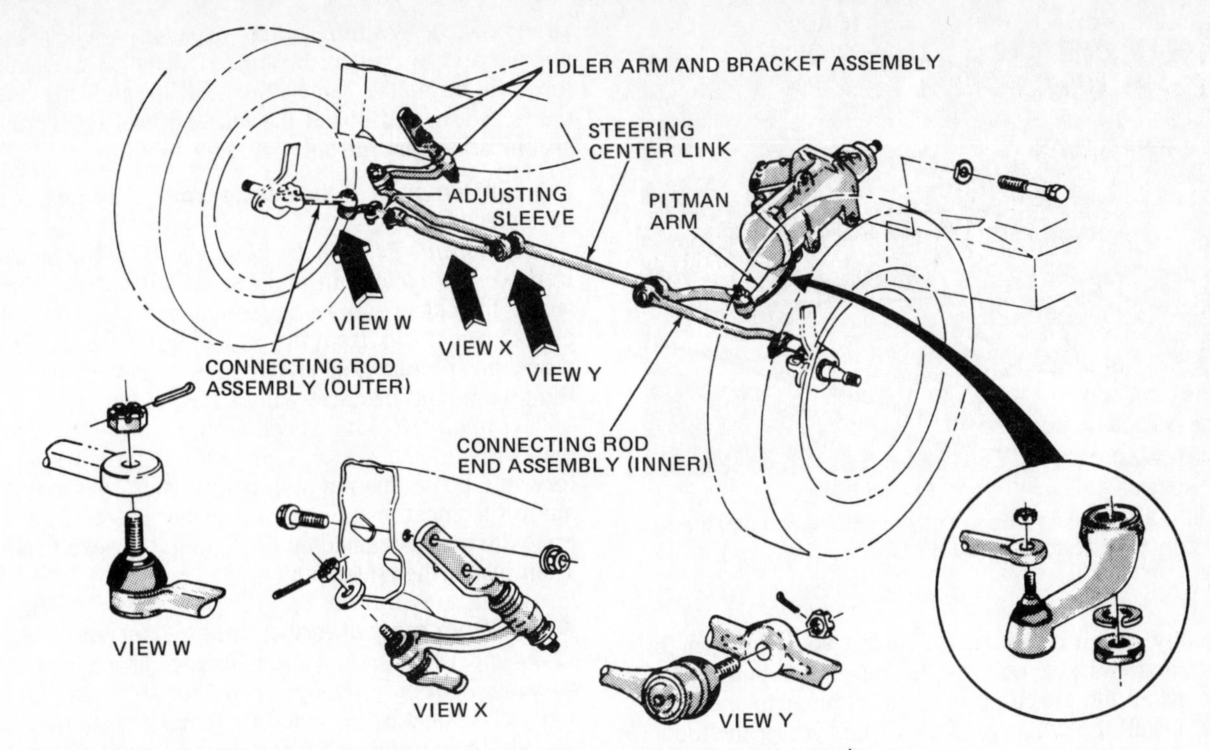

Fig. 12-11 Details of the steering linkage for either manual or integral power steering. (*Ford Motor Company*)

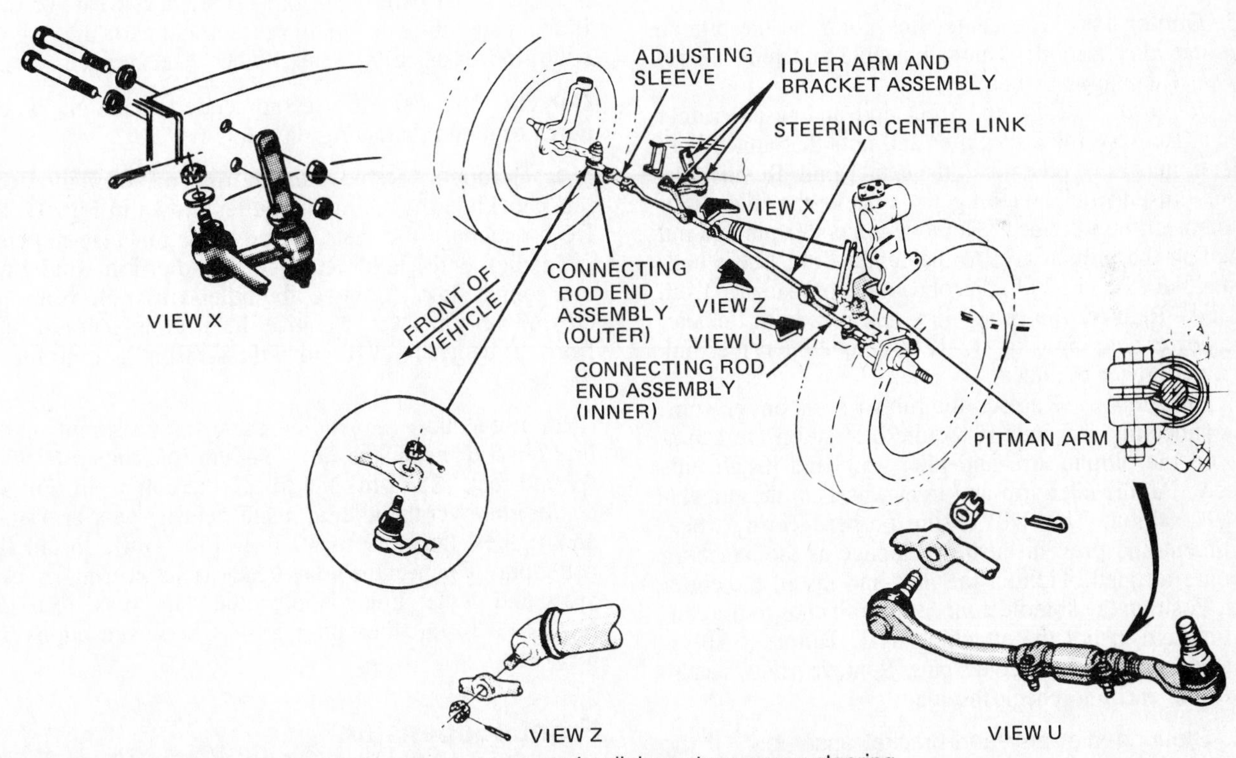

Fig. 12-12 Details of the steering linkage for a system using linkage-type power steering. (*Ford Motor Company*)

(Chap. 13). After this, loosen the adjusting-sleeve clamps; oil the sleeve, clamps, and bolts; tighten the nuts to specified torque. Make sure the sleeve clamps are installed as shown in Fig. 12-13 to avoid interference with the side rail.

2. Replacement of adjusting sleeve A damaged sleeve should be replaced. To remove it, first remove the spindle connecting-rod assembly. Screw the spindle connecting-rod end assembly into the new sleeve the same number of turns as required to remove the old

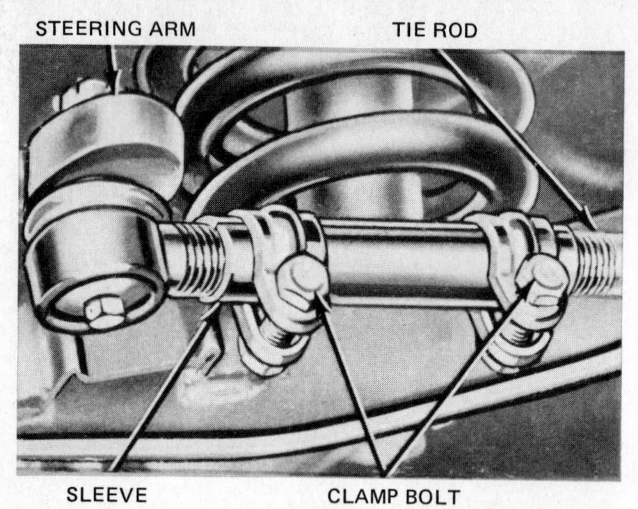

Fig. 12-13 Adjusting sleeve between the tie rod and the tie-rod end. (*Ford Motor Company*)

assembly. Do not tighten the clamp bolt. Position the sleeve and end assembly on the center link and spindle arm. Install the attaching nut and torque it to specifications; install the cotter pin. Check and adjust the toe as necessary. Then oil the sleeve clamp bolts and torque them to specifications. The clamps must be in the position shown in Fig. 12-13.

3. Center link The center link connects the pitman arm and idler arm. It is nonadjustable and must be replaced if damaged or bent.

a. Removal Raise the front end and install safety stands. Remove the cotter pins and nuts that attach both inner connecting-rod ends to the center link. Remove the cotter pin and nut attaching the link to the idler arm. Disconnect the idler arm. Remove the cotter pin and nut attaching the pitman arm to the center link. Use a ball-stud puller (Fig. 12-7) to disconnect the pitman arm from the link. Remove the center link. On cars with linkage-type power steering (Fig. 12-12), remove the center link from the power cylinder.

b. Installation Replace the rubber seals on the spindle connecting-rod ends if required. Position the center link on the pitman arm and idler arm, and install nuts loosely. Put the idler arm and front wheels in the straight-ahead position. This will ensure proper steering-wheel alignment and prevent bushing damage as the attaching nuts are torqued. Tighten the nuts and install the cotter pins. Position the spindle connecting-rod ends to the center link and install the attaching nuts. Torque to specifications and install the cotter pins. Remove safety stands, lower the car, and check the toe.

4. Steering-idler-arm-and-bracket assembly If the idler-arm bushings are worn, the complete assembly must be replaced.

a. Removal Remove the cotter pin and nut attaching the center link. Disconnect the center link from the idler arm. Remove the two bolts that attach the idler-arm-and-bracket assembly to the frame and remove the assembly.

b. Installation Secure the new idler-arm-and-bracket assembly to the frame with two attaching bolts, nuts, and washers. Place the idler arm and front wheels in the straight-ahead position so that the steering wheel will be aligned and to prevent bushing damage as the nuts are torqued. Insert the center-link stud through the hole in the idler arm and install the nut and washer. Torque to specifications and install the cotter pin.

5. Pitman arm If the pitman arm is damaged, it must be replaced.

a. Removal Position the front wheels to straight ahead. Remove the cotter pin and the castellated nut that attaches the center link to the pitman arm. Disconnect the pitman center link from the pitman arm (Fig. 12-7). Remove the pitman-arm attaching nut and lock washer. Remove the pitman arm with a special puller.

b. Installation With wheels straight ahead, put the pitman arm on the sector shaft with it pointing straight forward. Install the nut and lock washer, and torque the nut to specifications. Secure the center link to the pitman arm with the castellated nut, torquing it to specifications. Then install the cotter pin.

❁ 12-5 Plymouth steering-linkage service Figures 12-14 and 12-15 show various linkage arragements used on some Plymouth, Dodge, and Chrysler cars. Tie-rod-end seals should be inspected for damage at all oil-change periods. About the only time steering-linkage service is required is when seals require replacement or parts are damaged or worn. If seals are damaged, remove them and check the parts for loss of lubricant, wear, or rust. If adequate lubricant is still present and parts are in good condition, replace the seals. Otherwise, replace parts.

Careful! Use only the recommended tools and procedures to avoid damaging the seals.

1. Removal Remove the tie-rod ends from the steering-knuckle arms, using the puller shown in Fig. 12-16. Do not damage the seals. Remove the inner tie-rod ends from the center link. Remove the idler-arm stud from the center link. Remove the idler-arm bolt from the crossmember bracket. Remove the steering-gear-arm stud from the center link. Remove the steering-gear arm from the gear.

2. Installation Position the idler-arm assembly in the bracket and install the bolt. Tighten the nut to 65 lb-ft (pound-feet) [88 N-m] and install the cotter pin. Put the center link over the idler arm and steering-gear-arm studs and tighten the nuts to 40 lb-ft [54 N-m]. Install the cotter pins. Connect the tie-rod ends to the steering-knuckle arms and center link. Tighten the nuts to 40 lb-ft [54 N-m] and install the cotter pins. Check and adjust the toe.

Front suspension

❁ 12-6 Front-suspension service In general, front-suspension service is similar for various types of cars. This service includes checking and adjusting front-wheel bearings, replacing bearings if they are worn or damaged, lubricating bearings periodically (those that need it—some bearings are prelubricated for the life of the car), and replacing defective or worn parts in the suspension system. These include ball joints, bushings, springs, control arms, and shock absorbers.

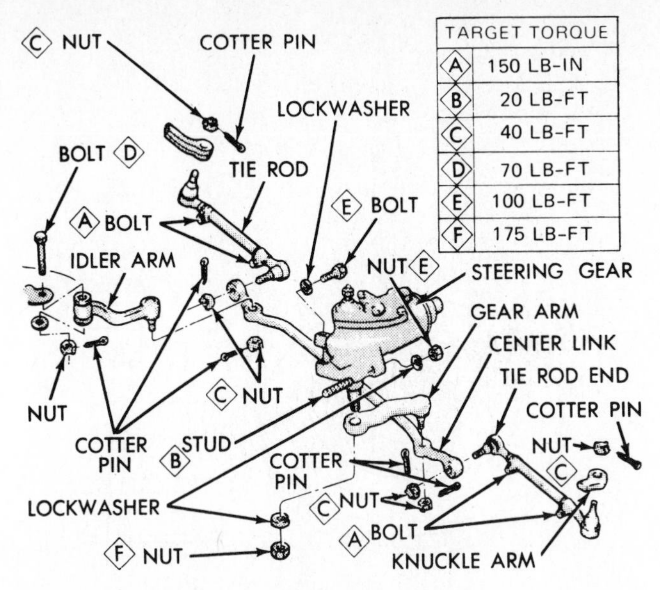

Fig. 12-14 Steering linkage for some models of Chrysler-built cars. (*Chrysler Corporation*)

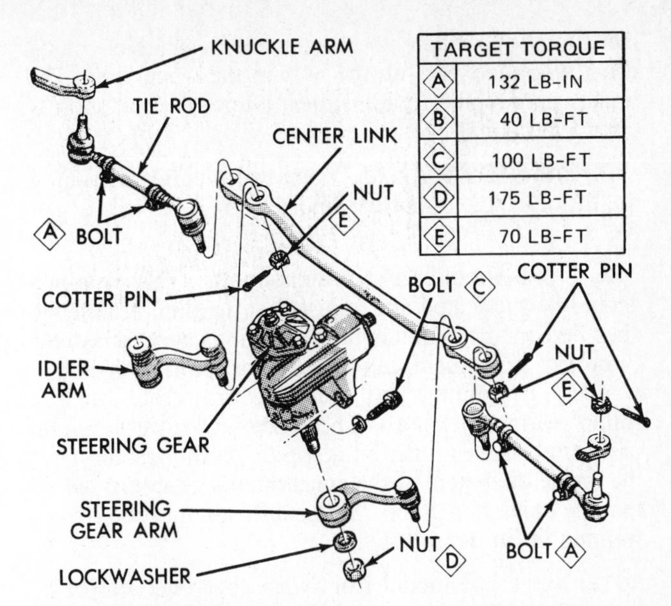

Fig. 12-15 Steering linkage for other models of Chrysler-built cars. (*Chrysler Corporation*)

Included in following sections are typical front-suspension servicing procedures on some models of Chevrolet, Ford, and Plymouth cars. Procedures for other cars are similar. However, before starting any job, check the service manual for the car you are servicing for the specifications and procedures.

⚙ 12-7 Chevrolet front-suspension service Figure 12-17 shows a front suspension system used in several models of front-engine rear-drive cars built by Chevrolet. Servicing procedures are outlined below for front-wheel bearings, hubs, shock absorbers, coil springs, and upper and lower control arms.

1. Brake-drum removal To remove the brake drum, raise the car, take off the wheel, and pull off the brake drum. It may be necessary to back off the brake adjustment. Check and service the drum (⚙ 6-9). On rein-

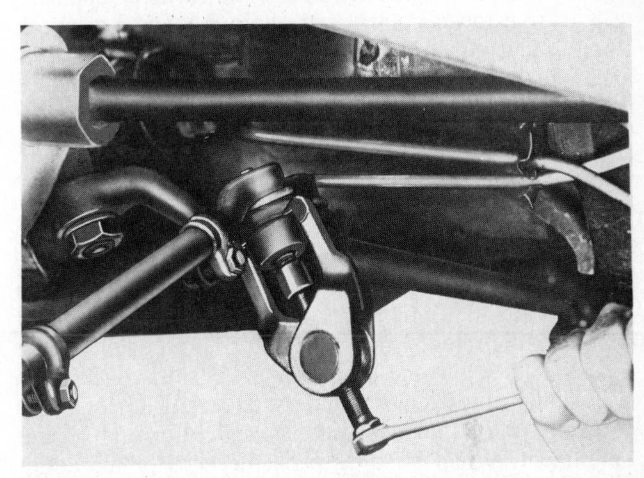

Fig. 12-16 Using the special puller to remove the ball stud in the steering linkage. (*Chrysler Corporation*)

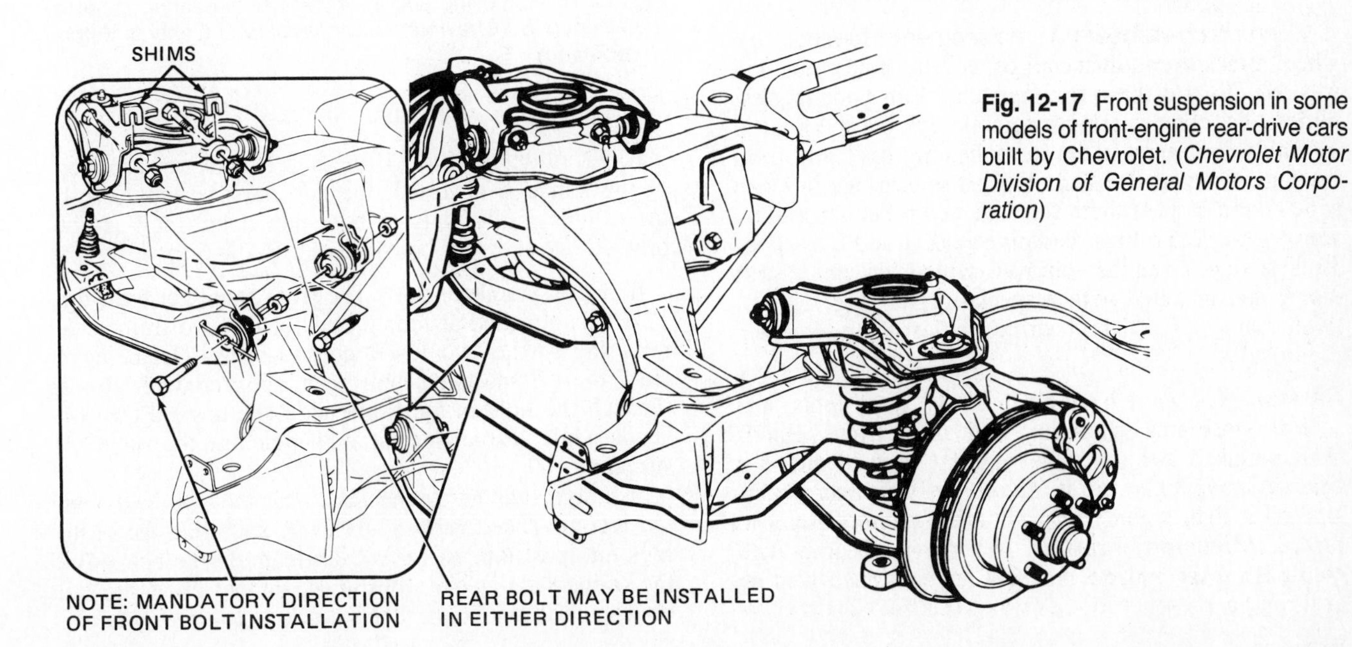

Fig. 12-17 Front suspension in some models of front-engine rear-drive cars built by Chevrolet. (*Chevrolet Motor Division of General Motors Corporation*)

stallation, make sure the alignment dowel pin on the drum web indexes with the hole in the wheel hub. The brakes may require readjustment after the installation is completed (Chap. 6).

2. Brake-disk removal Removal requires detaching the disk caliper so that the rotor can be removed (Chap. 6).

3. Front-wheel-bearing adjustment The bearings should be lubricated and adjusted at the time or mileage specified in the manufacturer's maintenance schedule. Improper adjustment can cause poor steering stability, wander or shimmy, and excessive tire wear. Tapered roller bearings are used. These bearings must *never* be preloaded. Cones must be a slip fit on the spindle, and the inside diameters of the cones should be lubricated so that the cones can creep. The spindle nut must be a free-running fit on the threads.

NOTE: Some late-model front-wheel bearings are prelubricated at the factory and do not require periodic lubrication.

To check the adjustment, raise the car and support it at the front lower control arm. Spin the wheel to check for unusual noise or roughness. If bearings are noisy, tight, or too loose, they should be cleaned, inspected, and, if okay, relubricated. To check for looseness, grip the tire at top and bottom. Push and pull with both hands to see if you can move the hub on the spindle. If you cannot move it 0.001 inch [0.025 mm], or if it moves more than 0.005 inch [0.127 mm], adjust the bearings.

To adjust, raise the front end and remove the hubcap or wheel disk and the dust cap. Take out the cotter pin and discard it. Tighten the spindle nut to the specified torque while rotating the wheel. Back off the adjustment nut one flat or to the "just loose" position. Insert the new cotter pin. If the slot and the cotter pin do not align, back off the adjustment nut just barely enough to get alignment. When bending back the ends of the cotter pin, bend them inboard or cut the ends off to avoid the possibility of their damaging the static collector in the dust cap. Reinstall the dust cap and the hubcap.

4. Front-wheel-bearing replacement Remove the wheel, brake drum, dust cap, cotter pin, spindle nut, and washer. Discard the old cotter pin. With your fingers, remove the outer bearing assembly. Pry out the lip-seal assembly and remove the inner bearing assembly from the hub. Discard the seal. Wash the bearings in clean solvent and inspect them for damaged roller separators, worn or cracked rollers, and pitted or cracked races (Fig. 12-37). Races can be removed with a driver or drift punch and installed with a special driver (Fig. 12-18). After reassembly of the bearings, adjust them (see item 3, above).

Careful! You *must* have clean hands, clean tools, and a clean work area when you work on bearings. Hands must be clean *and* dry. The slightest trace of dirt in a bearing can quickly ruin it. As soon as you wash a bearing, oil it lightly and wrap it in clean oilproof paper to protect it from dirt or rusting. Never spin a bearing with compressed air and do not spin an uncleaned bearing even by hand.

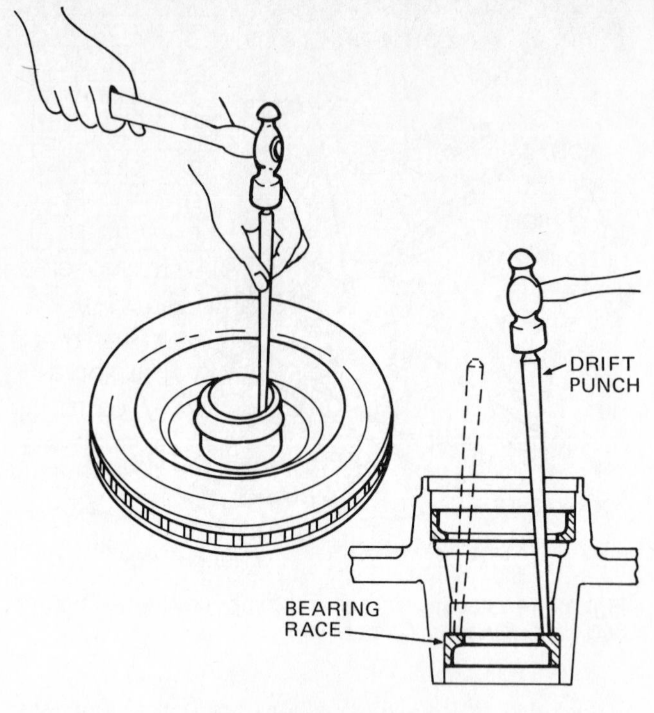

(a) REMOVAL

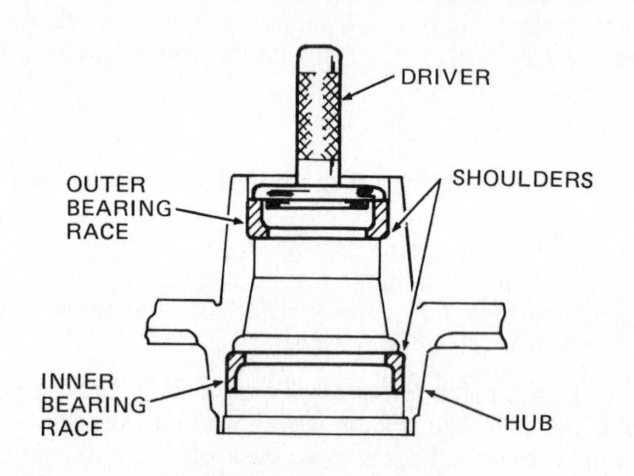

(b) INSTALLATION

Fig. 12-18 Removing and installing wheel-bearing races in the wheel hub. (*Chevrolet Motor Division of General Motors Corporation*)

5. Front-wheel hub If the hub bore is out of round or the flange is distorted, the hub must be replaced. If the trouble is due only to bent hub bolts, they can be pressed out and new bolts pressed in (✲ 6-9).

6. Front stabilizer bar The front stabilizer bar (Fig. 12-19) is attached at four places: at the two frame side rails and to the two lower control arms. Detaching it from these four points will permit its removal. Reach through the hole in the frame side rail to hold the bolt heads while unscrewing the nuts attaching the stabilizer support.

If new insulators are needed, coat the stabilizer with the recommended rubber lubricant, and slide the bushings into position. Never get lubricant on the outside of the frame-stabilizer-bar bushings, or they may slip out of the brackets. Connect the brackets to the frame and

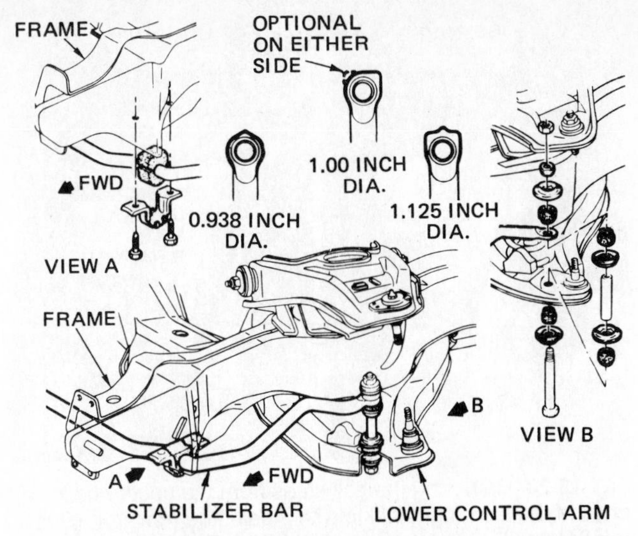

Fig. 12-19 Details of stabilizer-bar attachments. To identify the bar, some have a tape with the part number stamped on it. (*Chevrolet Motor Division of General Motors Corporation*)

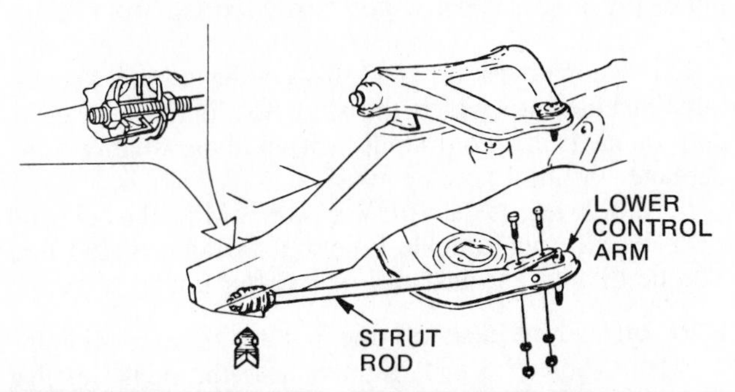

Fig. 12-20 Location of the strut rod. (*Chevrolet Motor Division of General Motors Corporation*)

attach the stabilizer ends to the lower control arms. Tighten the bracket bolts and link nuts to the specified torque.

7. Strut rod Raise the car on a lift and remove the nut, retainer, and rubber bushing from the front end of the strut rod (Fig. 12-20). Remove the two bolts attaching the rear end of the strut rod to the lower control arm. Pull the strut rod from the bracket.

If the rear nut on the front end of the strut rod has been removed, turn it on until it is about ¾ inch [19 mm] from the end of the threads. Install the rear retainer, sleeve, and bushing on the rod so that the pilot diameter faces forward. Insert the rod into the bracket and install the front bushing on the sleeve so that the raised pilot diameter faces the rear to enter the hole in the bracket and rear bushing. Install the forward retainer and nut on the rod. Attach the strut rod to the top of the lower control arm with two bolts, washers, and nuts. Torque to specifications. Check the caster, camber, and toe (Chap. 13).

8. Front-coil-spring check To check for spring sag, position the car on a smooth, level surface, bounce the front end several times, raise up on the front end, and allow it to settle. Then take the measurements as shown in Fig. 11-18. The differences should be as noted in the specifications for the car being checked. Next, take the

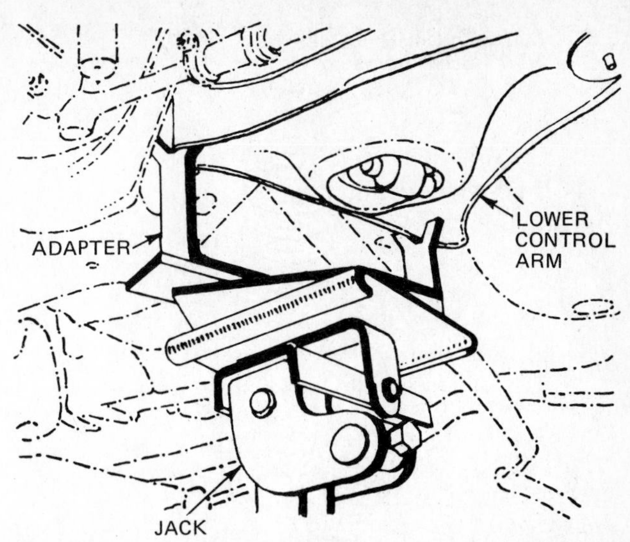

Fig. 12-21 Using a jack adapter placed between the inner bushings to lower and raise the lower control arm. (*ATW*)

measurements on the other side. The difference between the two sides should be no greater than ½ inch [12.7 mm]. To make a correction, the springs must be replaced. Placing a shim under the spring is *not recommended*.

a. Removal Raise the car on a lift. Remove the two shock-absorber attaching screws and push the shock up through the control arm and into the spring. The car should be supported so that the control arms hang free. Put the jack adapter shown in Fig. 12-21 into position so that it cradles the inner bushings of the lower control arm.

Detach the stabilizer bar from the lower control arm. Raise the jack to remove the tension on the lower-control-arm pivot bolts. Install a spring compressor or chain so that the spring cannot jump out as the tension is released. Remove the rear bolt and nut, and then the front bolt and nut. Lower the jack to lower the control arm. When all spring compression is removed, remove the spring.

Careful! Do not apply force on the lower control arm and ball joint as you remove the spring. Maneuver the spring so it will come out easily.

b. Installation Position the spring properly on the control arm (Fig. 12-22) and lift the control arm with the adapter and jack (Fig. 12-21). Position the control arm into the frame and install the pivot bolts (front bolt first) and nuts. Figure 12-17 shows the mandatory bolt directions. Torque to specifications and lower the jack. Reattach the stabilizer bar and shock absorber.

9. Upper-control-arm ball-joint check Figures 11-6 and 11-10 show methods of checking ball joints for wear. A second method which requires partial disassembly follows. Raise the car and place a safety stand under each lower control arm, as near as possible to each lower ball joint. Remove the tire and wheel assembly. Remove the upper-ball-stud cotter pin and loosen the nut one turn. Install the ball-stud remover between the ball studs (Fig. 12-23). Turn the threaded end of the tool until the stud is free of the steering knuckle.

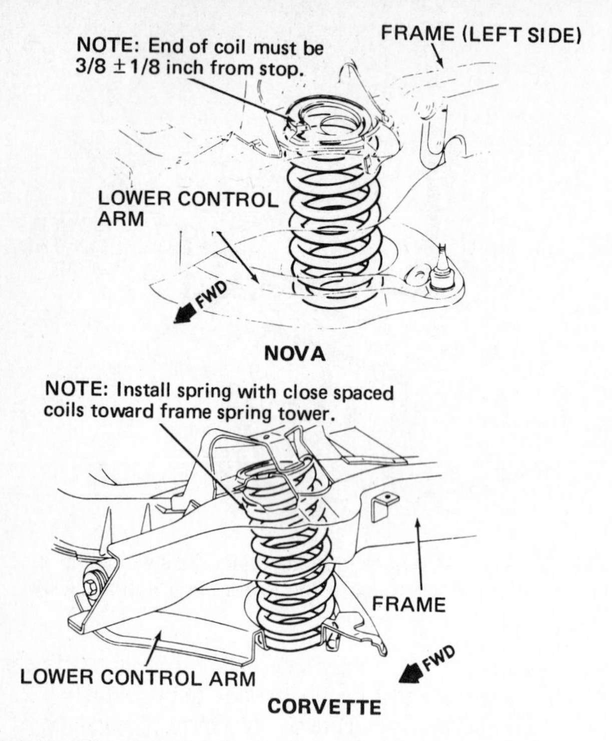

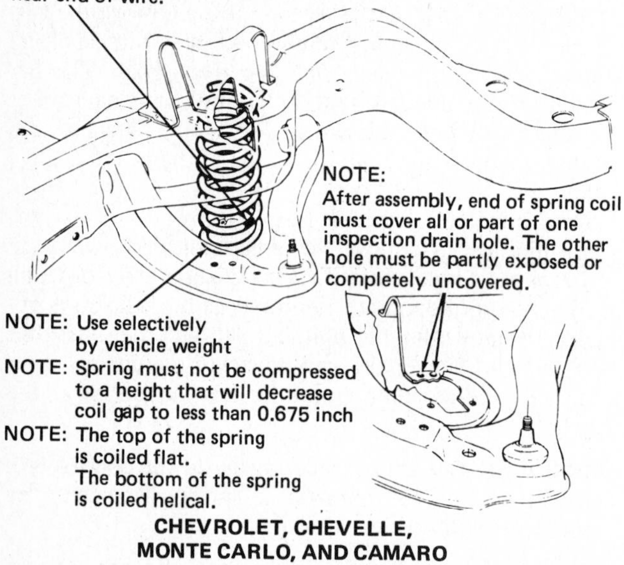

Fig. 12-22 Positioning of the coil spring on different models. (*Chevrolet Motor Division of General Motors Corporation*)

Remove the upper-ball-joint-stud nut and allow the steering knuckle to swing out of the way. Lift the upper arm and put a block of wood between the frame and arm to act as a support.

Check the ball joint for wear and looseness. If the stud has any looseness, or if it can be twisted in its socket with the fingers, replace the ball joint.

a. Replacement Use drills and a punch to remove the rivet heads (Fig. 12-24). Do not damage the control arm or ball-joint seat. Install the new ball joint in the upper control arm, using the bolts and nuts supplied with

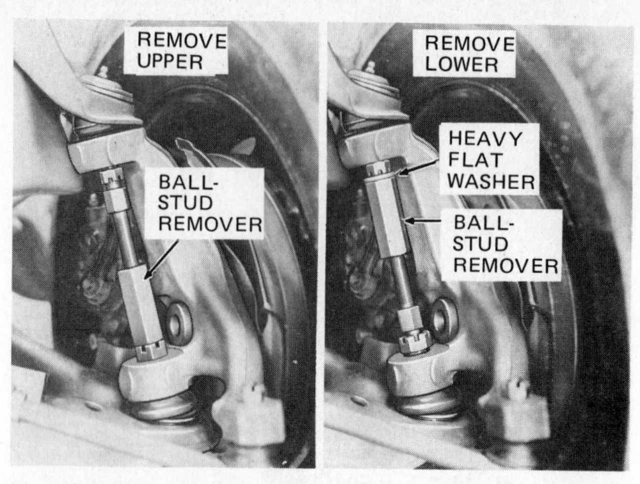

Fig. 12-23 Removing the ball studs from the upper and lower ends of the steering knuckle with a ball-stud remover. (*Chevrolet Motor Division of General Motors Corporation*)

the replacement ball joint (Fig. 12-25). Install the bolts pointing up. Torque to specifications. Turn the ball-stud cotter-pin hole so that it points toward the front of the car.

Remove the block of wood. Inspect the steering-knuckle tapered hole into which the stud fits. It must be clean and round. If it is not round, or if you note wear or other damage, install a new knuckle.

b. Attachment to steering knuckle Mate the ball stud with the steering-knuckle hole and install the stud nut. Torque the nut and install a new cotter pin.

Careful! Never back off the nut to align it with the cotter-pin hole. Instead, tighten it to the next slot that lines up with the hole. Lubricate the ball joint. Then install the tire and wheel.

10. Lower-control-arm ball-joint check Late-model cars are equipped with lower-control-arm ball joints with wear indicators (Fig. 11-11). On earlier models, the lower ball joint can be checked for wear by measuring from the top of the lubrication fitting to the bottom of the ball stud with a micrometer. Take the measurement with the car supported on its wheels. Then support the car on the outer ends of the lower control arms and remeasure. If the difference is greater than $\frac{1}{16}$ inch [1.6 mm], the ball joint is worn and must be replaced.

Ball-stud tightness in the knuckle can be checked by shaking the wheel and noting if there is any movement of the stud or nut in the knuckle. It can also be checked by removing the cotter pin and checking the torque on the ball-stud nut.

a. Removal Raise the car and place safety stands under the frame. Remove the wheel-and-tire assembly. Then place a floor jack under the spring seat in the lower control arm. Remove the lower-ball-stud cotter pin. Loosen the stud nut one turn. Install the ball-joint remover between the ball studs and turn the threaded end of the tool until the stud is free of the steering knuckle (Fig. 12–23). Remove the stud nut and free the knuckle from the ball stud. Guide the lower control arm out of the opening in the splash shield with a putty knife or similar tool.

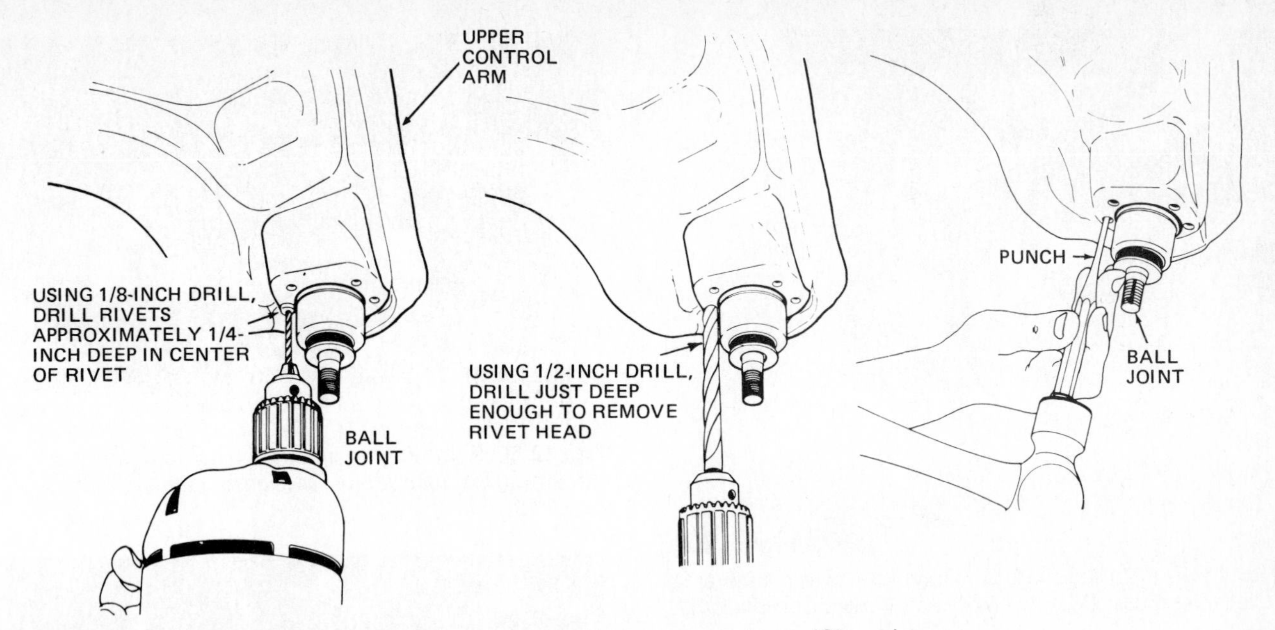

Fig. 12-24 Steps in removing a ball joint that is riveted in the upper control arm. (*Chevrolet Motor Division of General Motors Corporation*)

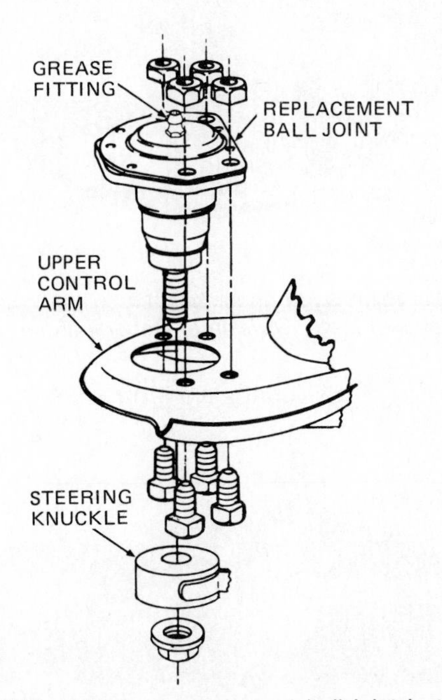

Fig. 12-25 Installing a replacement ball joint in the upper control arm. (*Chevrolet Motor Division of General Motors Corporation*)

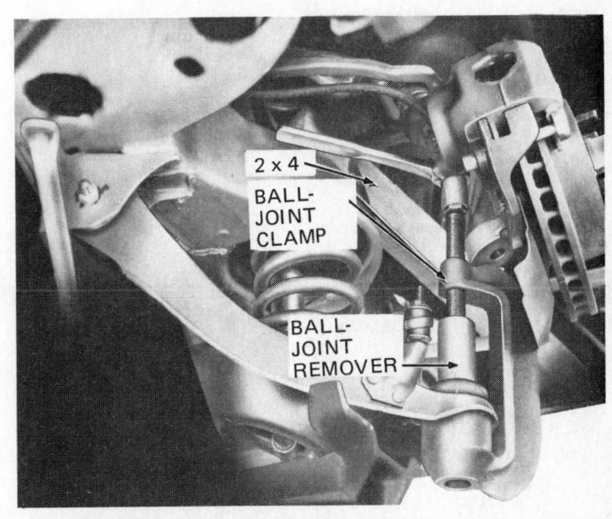

Fig. 12-26 Pushing the ball joint out of the lower control arm. (*Chevrolet Motor Division of General Motors Corporation*)

Lift the upper control arm, with the knuckle-and-hub assembly attached. Place a block of wood between the frame and the upper control arm.

Careful! Do not pull on the brake hose when lifting the assembly.

If the tie-rod end of the steering knuckle is in the way, detach it. Then position the ball-joint clamp and ball-joint remover (Fig. 12–26). Turn the bolt until the ball joint is pushed out of the lower control arm.

b. Installation Position the ball joint in the lower control arm, using the ball-joint clamp and the installing tool (Fig. 12–27). Position the bleed vent of the new ball joint facing inward. Turn down the bolt until the new ball joint is seated in the control arm. Remove the tools. Turn the stud cotter-pin hole fore and aft. Remove the block of wood holding the upper control arm out of the way.

Inspect the tapered hole in the steering knuckle. If it is out of round or otherwise damaged, replace the steering knuckle. Mate the stud with the hole and install the stud nut. Torque the nut and install the cotter pin.

Careful! Never back off the nut to align it with the cotter-pin hole. Instead, tighten the nut to the next slot that lines up with the stud hole.

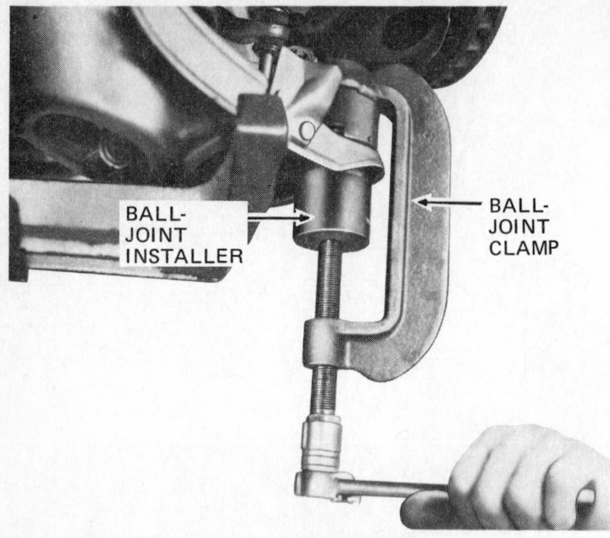

Fig. 12-27 Installing the ball joint in the lower control arm. (*Chevrolet Motor Division of General Motors Corporation*)

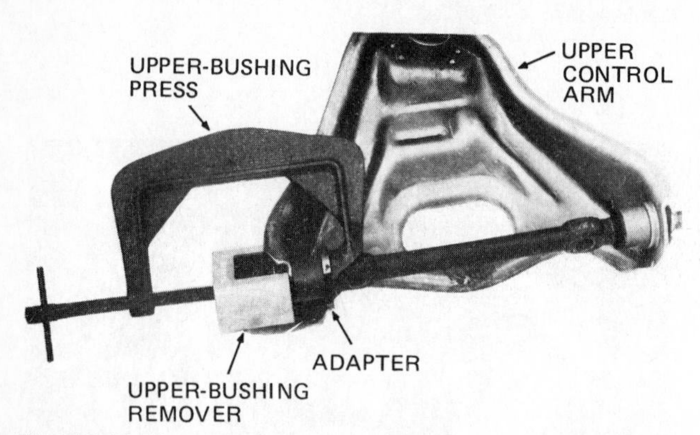

Fig. 12-28 Removing an inner bushing from the upper control arm. (*Chevrolet Motor Division of General Motors Corporation*)

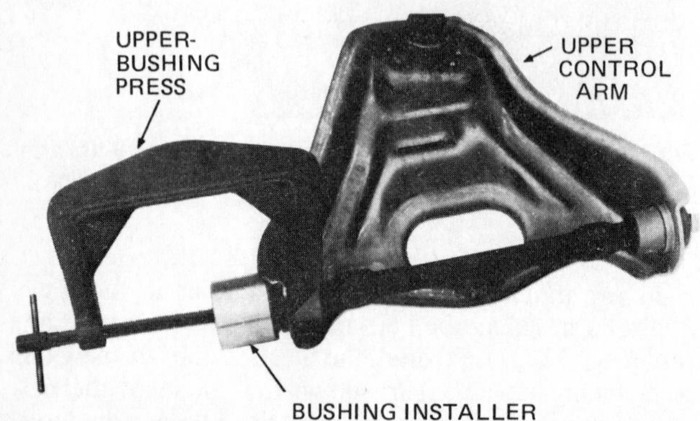

Fig. 12-29 Installing an inner bushing in the upper control arm. (*Chevrolet Motor Division of General Motors Corporation*)

Install a lubrication fitting and lubricate the joint. Then install the wheel and tire, if they have been removed.

11. Control-arm bushings The upper-control-arm bushings are removed and replaced with an upper-bushing press and adapters as shown in Figs. 12–28 and 12–29. Note that when the second bushing is installed, the cross shaft must be in place.

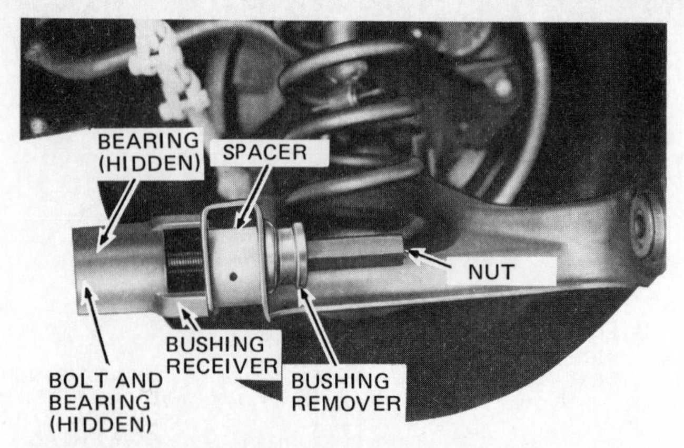

Fig. 12-30 Removing an inner bushing from the lower control arm. (*Chevrolet Motor Division of General Motors Corporation*)

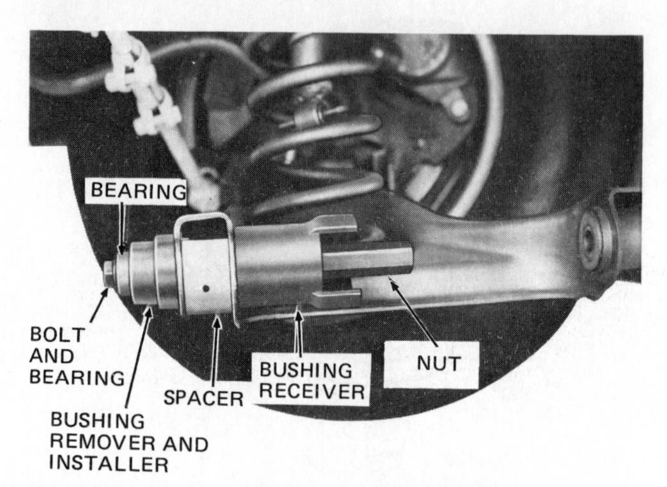

Fig. 12-31 Installing an inner bushing in the lower control arm. (*Chevrolet Motor Division of General Motors Corporation*)

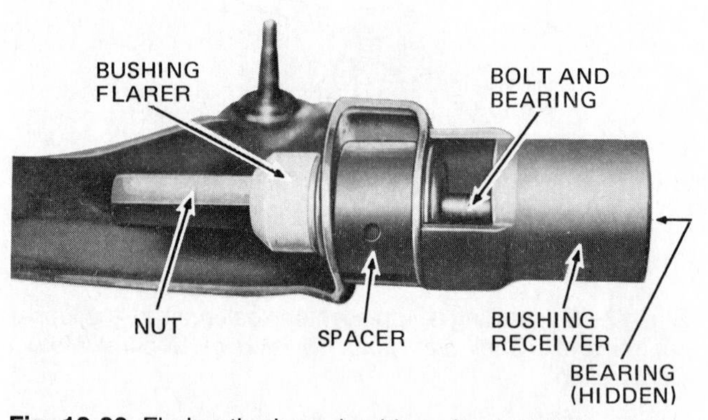

Fig. 12-32 Flaring the inner bushing after installation in the lower control arm. (*Chevrolet Motor Division of General Motors Corporation*)

The lower-control-arm bushings are removed and replaced with the tools shown in Figs. 12–30 and 12–31. The metal collars on the bushings are flared on the inner ends after installation (Figs. 12–32 and 12–33). This flare must be removed by tapping on the edge with a hammer before the bushing is pressed out. Then, when the new bushing is in place, flare it with the special tools shown in Fig. 12–32.

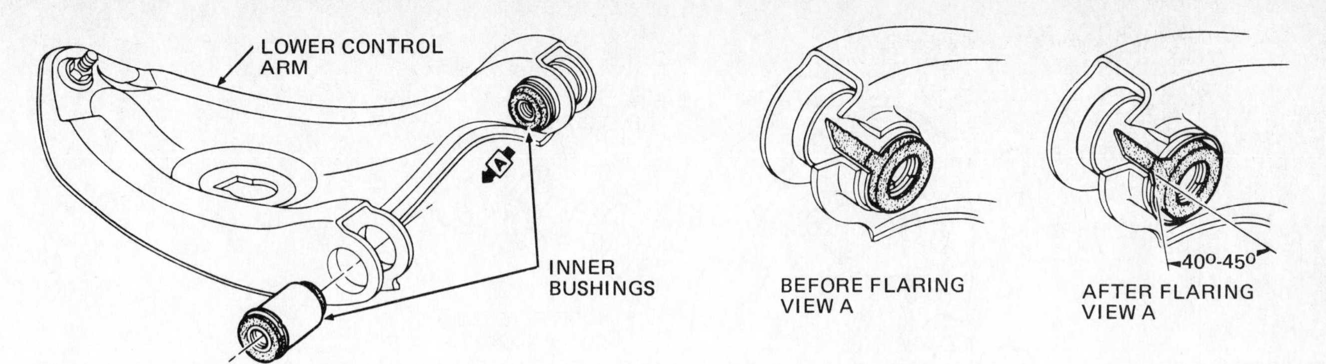

Fig. 12-33 How the inner bushings are installed and flared in the lower control arm. (*Chevrolet Motor Division of General Motors Corporation*)

NOTE: When reattaching the lower control arm to the frame, observe the mandatory bolt directions (Fig. 12–17).

12-8 Ford front-suspension service In servicing Ford front-suspension systems, the upper or lower control arm must be serviced as a unit. Ford does not recommend installing new ball joints or other components in used Ford control arms. On other makes of cars, new ball joints can be installed on the original control arms. Figure 12–34 shows various views of a front-suspension system used in Ford cars. The following sections describe servicing procedures for front-suspension systems with the spring between the frame and the lower control arm, and the procedures for systems with the spring above the upper control arm.

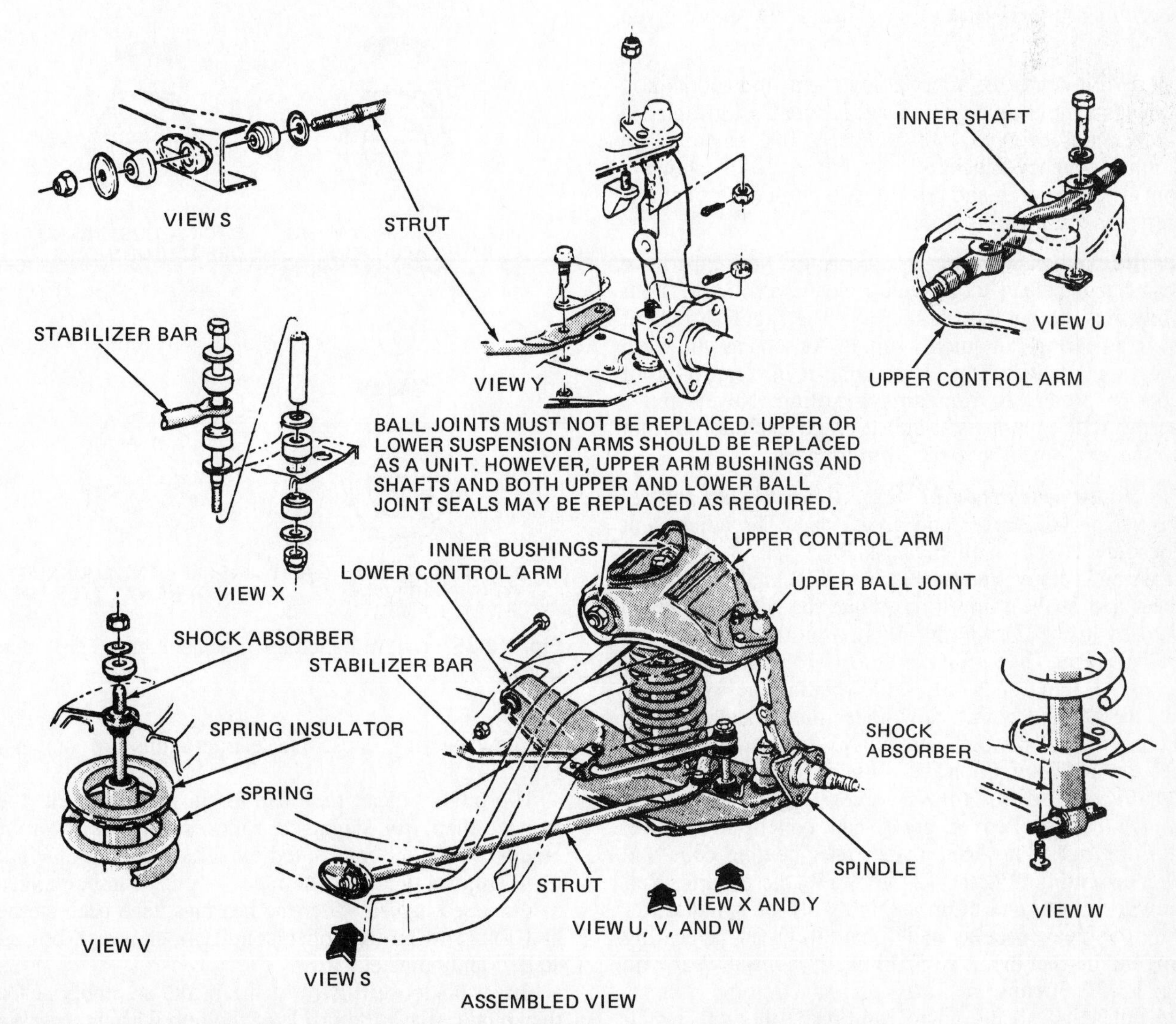

Fig. 12-34 Details of a front-suspension system used in Ford cars. (*Ford Motor Company*)

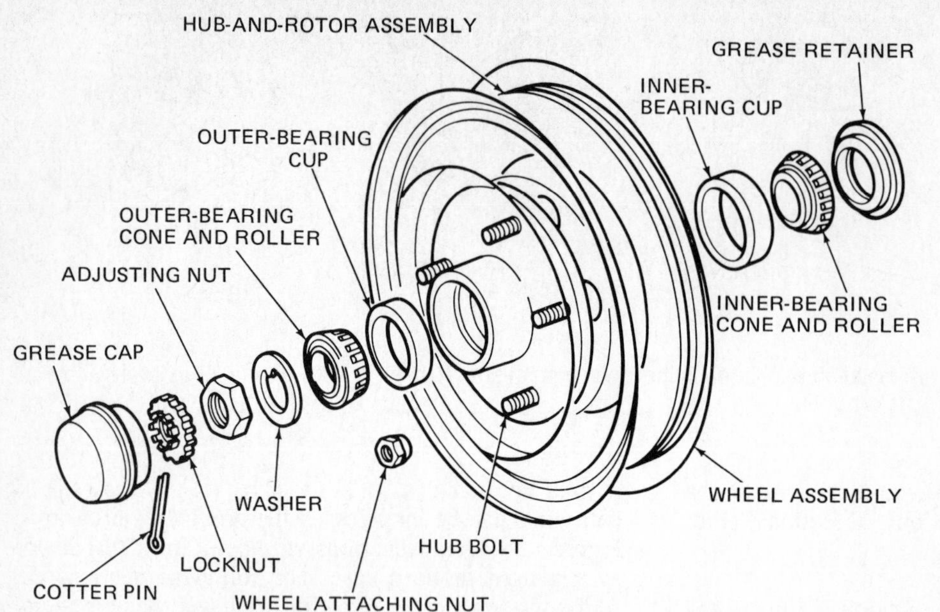

Fig. 12-35 A disassembled front-wheel bearing. (*Ford Motor Company*)

12-9 Ford—spring on lower control arm Eight procedures for this type of suspension system are given below.

1. Front-wheel bearing adjustment and lubrication Bearings should be adjusted and lubricated, or replaced if necessary, every 37,500 miles [60,000 km] or 37½ months, whichever occurs first. Figure 12–35 shows a front-wheel bearing and related parts. Procedures are different for drum and disk brakes.

Careful! When working with bearings, you must have clean hands, clean tools, and a clean work area. Your hands must be both clean and dry. The slightest trace of dirt in a bearing can quickly ruin it. As soon as you wash a bearing, oil it lightly. Then wrap it in clean oilproof paper to protect it from dirt or rusting. Never spin a bearing with compressed air. It could explode. Do not spin an uncleaned bearing, even by hand.

a. Adjustment (drum brakes) Raise the car and remove the wheel cover and grease cap. The adjustment procedure is shown in Fig. 12–36. Wipe away excess grease and remove the cotter pin and locknut. Rotate the wheel and at the same time torque the adjusting nut to 17 to 25 lb-ft [23 to 34 N-m]. Loosen the adjusting nut one-half turn. Then retighten it to 10 to 15 lb-in (pound-inches) [1.1 to 1.7 N-m], while rotating the wheel. Install the locknut with a new cotter pin. Recheck wheel rotation. If it is rough and noisy, disassemble for inspection and lubrication.

b. Bearing service (drum brakes) Raise the car and remove the wheel cover, grease cap, cotter pin, locknut, adjusting nut, flat washer, and outer bearing cone and roller assembly. Figure 12–35 shows these parts. Pull the wheel, hub, and drum assembly off the spindle. Remove the grease retainer and discard it. Clean the bearing cups and inspect them, comparing any signs of wear with Fig. 12–37. If cups are worn or damaged, remove them. After cleaning all lubricant from inside the hub, install

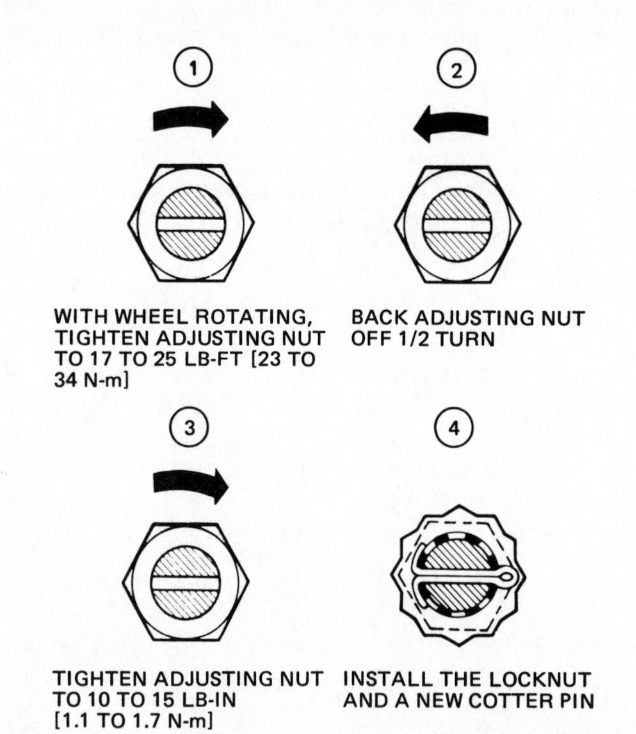

Fig. 12-36 Front-wheel-bearing adjustment. (*Ford Motor Company*)

the new bearing cups. These operations are shown in Fig. 12–18.

Thoroughly clean the bearings in clean solvent, then inspect them for damaged roller separators, worn or cracked rollers, and pitted or cracked races. Figure 12–37 shows defects. The races or cups can be replaced as discussed earlier. After the bearings have been cleaned and found to be in proper condition, they can be reinstalled and adjusted.

Brush all loose dirt from the brake assembly. Clean the spindle. Pack the inside of the hub with the specified

ABRASIVE ROLLER WEAR

PATTERN ON RACES AND ROLLERS CAUSED BY FINE ABRASIVES.

CLEAN ALL PARTS AND HOUSINGS. CHECK SEALS AND BEARINGS AND REPLACE IF LEAKING, ROUGH OR NOISY.

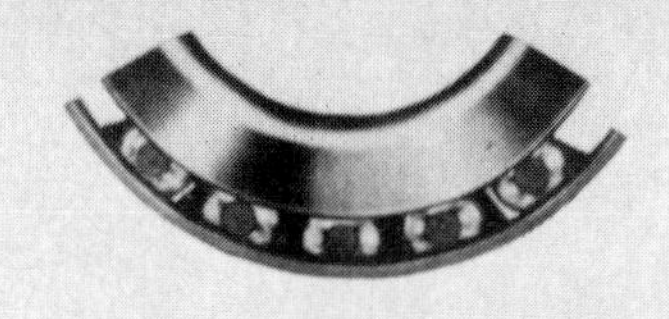

GALLING

METAL SMEARS ON ROLLER ENDS DUE TO OVERHEAT, LUBRICANT FAILURE, OR OVERLOAD.

REPLACE BEARING. CHECK SEALS AND CHECK FOR PROPER LUBRICATION.

ETCHING

BEARING SURFACES APPEAR GRAY OR GRAYISH BLACK IN COLOR WITH RELATED ETCHING AWAY OF MATERIAL USUALLY AT ROLLER SPACING.

REPLACE BEARINGS. CHECK SEALS AND CHECK FOR PROPER LUBRICATION.

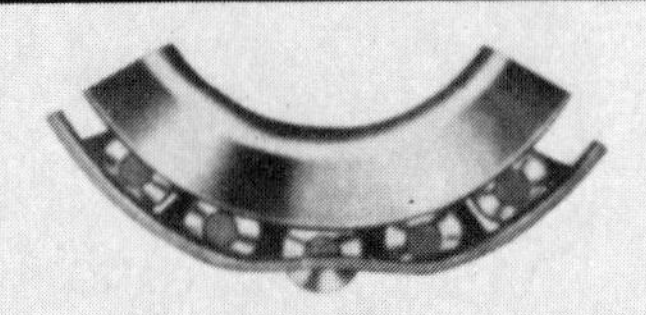

BENT CAGE

CAGE DAMAGE DUE TO IMPROPER HANDLING OR TOOL USAGE.

REPLACE BEARING.

INDENTATIONS

SURFACE DEPRESSIONS ON RACE AND ROLLERS CAUSED BY HARD PARTICLES OF FOREIGN MATERIAL.

CLEAN ALL PARTS AND HOUSINGS. CHECK SEALS AND REPLACE BEARINGS IF ROUGH OR NOISY.

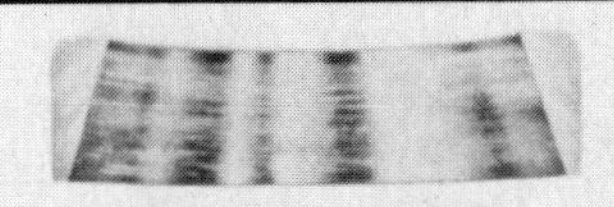

CAGE WEAR

WEAR AROUND OUTSIDE DIAMETER OF CAGE AND ROLLER POCKETS CAUSED BY ABRASIVE MATERIAL AND INEFFICIENT LUBRICATION.

CLEAN RELATED PARTS AND HOUSINGS. CHECK SEALS AND REPLACE BEARINGS.

MISALIGNMENT

OUTER RACE MISALIGNMENT.

CLEAN RELATED PARTS AND REPLACE BEARING. MAKE SURE RACES ARE PROPERLY SEATED.

CRACKED INNER RACE

RACE CRACKED DUE TO IMPROPER FIT, COCKING, OR POOR BEARING SEATS.

REPLACE BEARING AND CORRECT BEARING SEATS.

FATIGUE SPALLING

FLAKING OF SURFACE METAL RESULTING FROM FATIGUE.

REPLACE BEARING. CLEAN ALL RELATED PARTS.

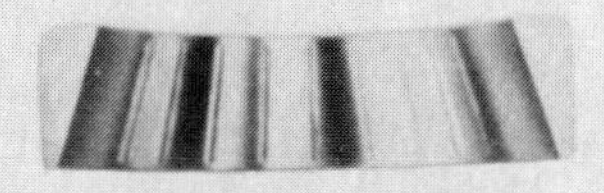

BRINELLING

SURFACE INDENTATIONS IN RACEWAY CAUSED BY ROLLERS EITHER UNDER IMPACT LOADING OR VIBRATION WHILE THE BEARING IS NOT ROTATING.

REPLACE BEARING IF ROUGH OR NOISY.

FRETTING

CORROSION SET UP BY SMALL RELATIVE MOVEMENT OF PARTS WITH NO LUBRICATION.

REPLACE BEARING. CLEAN RELATED PARTS. CHECK SEALS AND CHECK FOR PROPER LUBRICATION.

HEAT DISCOLORATION

HEAT DISCOLORATION CAN RANGE FROM FAINT YELLOW TO DARK BLUE RESULTING FROM OVER LOAD OR INCORRECT LUBRICANT.

EXCESSIVE HEAT CAN CAUSE SOFTENING OF RACES OR ROLLERS.

TO CHECK FOR LOSS OF TEMPER ON RACES OR ROLLERS A SIMPLE FILE TEST MAY BE MADE. A FILE DRAWN OVER A TEMPERED PART WILL GRAB AND CUT METAL, WHEREAS A FILE DRAWN OVER A HARD PART WILL GLIDE READILY WITH NO METAL CUTTING.

REPLACE BEARINGS IF OVER HEATING DAMAGE IS INDICATED. CHECK SEALS AND OTHER PARTS.

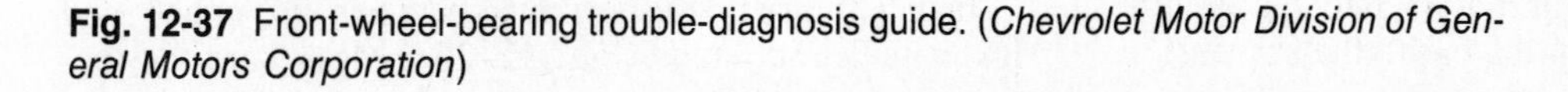

Fig. 12-37 Front-wheel-bearing trouble-diagnosis guide. (*Chevrolet Motor Division of General Motors Corporation*)

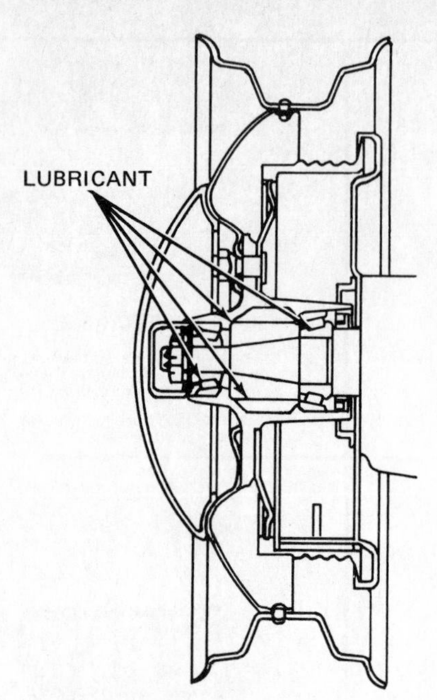

Fig. 12-38 Lubrication points for front-wheel bearings. (*Ford Motor Company*)

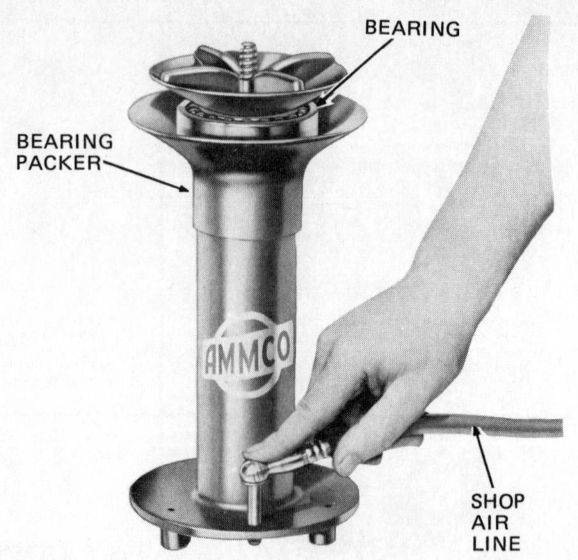

Fig. 12-39 Using a bearing packer to pack a bearing with grease. (*Ammco Tools, Inc.*)

wheel-bearing grease. Do not overlubricate. Add grease at the points shown in Fig. 12–38 until it is flush with the inside diameters of the bearing cups.

Pack the bearings with the specified grease, using a bearing packer (Fig 12–39). Grease the cone surfaces. Install a new grease retainer. Then install other parts and adjust the bearings as discussed above.

c. Adjustment (disk brakes) Adjusting wheel bearings on a car with disk brakes is similar to working on a car with drum brakes, with the following exceptions. Loosen the adjusting nut three turns instead of one-half turn. Rock the wheel-and-rotor assembly several times to push the shoes and linings away from the rotor. Then tighten the adjusting nut while rotating the wheel (Fig. 12–36). Finish the adjustment as previously explained. Before driving the car, pump the brake pedal several times to restore braking.

d. Bearing service (disk brakes) On a car with disk brakes, wheel-bearing service is similar to that on a car with drum brakes. However, the caliper must be detached and wired up out of the way. On reassembly, after lubrication and bearing adjustment, install the caliper to the anchor plate. Finally, pump the brake pedal several times before driving the car to restore braking.

2. Stabilizer-bar attachments The stabilizer bar is attached at its two ends to the lower suspension arms through bushings shown in view 12–34. It is attached at two points to the frame through insulators. Bushings can be replaced by removing the attaching nut and bolt. Observe the relationship of the washers, insulators, and spacers so that all parts can be returned to their correct positions. Use a new nut and bolt.

To replace insulators, remove the stabilizer bar from the car. Coat the ends of the bar with Ruglyde or similar lubricant and slide the new insulators into place. Use new bolts to attach the ends of the bar to the suspension arms. Attach the insulators to the frame.

3. Lower-arm-strut bushing The strut, shown in view S of Fig. 12–34, must be removed from the car to replace the bushing. After reinstalling the strut, check caster, camber, and toe (Chap. 13).

4. Front-spring removal and installation With the car raised and set on safety stands, disconnect the lower end of the shock absorber from the suspension arm. A pry bar may be needed to free the shock absorber from the arm. On some cars, the shock absorber must be removed. Then put a jack under the lower arm. Remove the bolts attaching the strut and stabilizer bar to the control arm. Disconnect the inner end of the lower arm from the frame. Then slowly and carefully lower the jack to relieve the spring force (Fig. 12–40). You may need to use a pry bar to free the spring.

CAUTION: **Use a stable jack and relieve spring force carefully. If you do not work carefully, the spring may fly out and cause injury.**

To install the spring, tape the insulator to the spring and position the spring on the lower arm. The end of the spring must be no more than ½ inch [12.7 mm] from the end of the depression in the arm. Raise the lower arm carefully to compress the spring and attach the inner end to the frame with a nut and bolt. Reattach the shock absorber, strut, and stabilizer bar to the lower control arm. Remove the safety stands and lower the car.

5. Lower-control-arm removal and installation Raise the front of the car and support it with safety stands under both sides of the frame just behind the lower arms. Remove the wheel-and-tire assembly, caliper and brake hose, and hub-and-rotor assembly. Disconnect the lower end of the shock absorber and push it up out of the way. Disconnect the stabilizer bar and strut. Remove the cotter pins from the ball joints. Loosen the lower-ball-joint stud nut one or two turns.

Install the ball-stud remover between the upper- and lower-ball joint studs (Fig. 12–23). Make sure the tool is seated on the studs and not on the nuts. Turn the nut

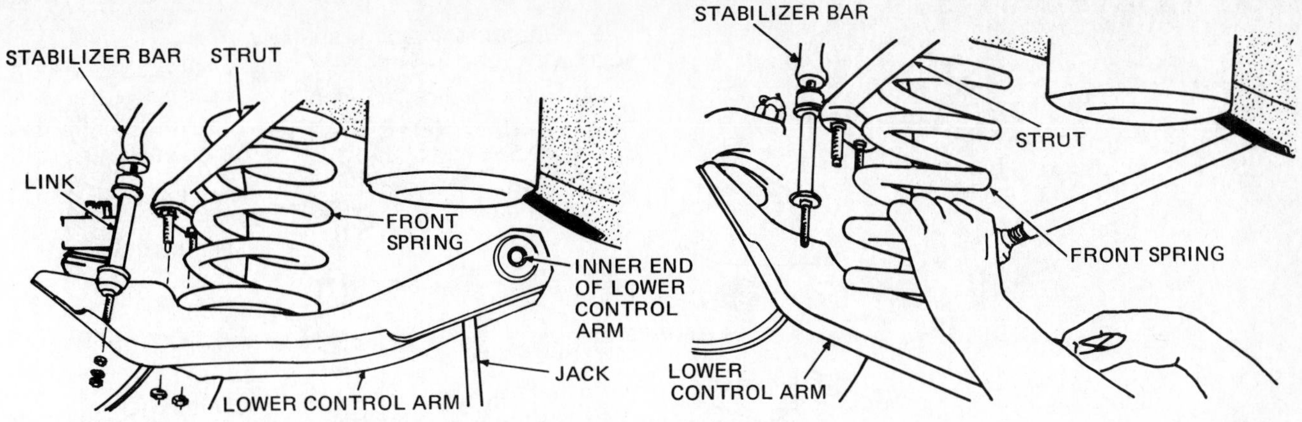

Fig. 12-40 Removing and installing a coil front spring. (*Ford Motor Company*)

on the ball-stud remover to apply force to the stud. Tap the spindle near the lower stud with a hammer to loosen the stud in the spindle. Do not loosen the stud with tool force alone. Position a jack under the lower control arm (Fig. 12–41). Remove the stud nut, then lower the arm and detach it from the frame.

To install the arm, loosely attach it to the spindle. Do not tighten the stud nut. Position the spring and insulator to the upper spring pad and lower arm. Use a floor jack and raise the control arm to align with the frame connection for the inner end of the arm. Attach the arm with a through-bolt and nut. Remove the jack. Tighten the ball-joint attaching nut to specifications. Install the cotter pin. Attach the shock absorber, strut, and stabilizer bar. Install the hub and rotor, caliper, and wheel and tire. Then adjust the wheel bearing. Install the grease cap and wheel cover. Check caster, camber, and toe (Chap. 13).

6. Upper-control-arm removal and installation Raise the car and support it with safety stands under both sides of the frame just behind the lower arms. Remove the wheel cover, wheel and tire, and cotter pin from the upper-ball-joint stud nut. Loosen the nut one or two turns. Use the ball-stud remover (Fig. 12–23) to loosen the upper-ball-joint stud from the spindle. Do not loosen the stud by tool force alone. Tap the spindle near the stud with a hammer while the tool is applying force to the stud. Put a floor jack under the lower control arm and raise it to relieve the force from the upper ball joint. Remove the nut, then remove the attaching bolts of the upper-arm inner shaft. Now take off the arm as an assembly.

To install the upper control arm, attach the inner end to the frame bracket and the ball-joint stud at the outer end to the spindle. Tighten the nut to the specified torque, then tighten further to align the cotter-pin hole in the stud with the nut bolts. Install the new cotter pin. Then install the wheel and tire and adjust the wheel bearings. Install the wheel cover. Remove the safety stands and lower the car. Adjust caster, camber, and toe (Chap. 13).

7. Upper-control-arm bushing service Upper-control-arm inner bushings are shown in Fig. 12–34. If the bushings require replacement, remove the nuts and washers from both ends of the upper arm shaft. Then press the bushings out and instsall the new bushings with the special tools required. The operation is similar to that shown in Figs. 12–28 and 12–29.

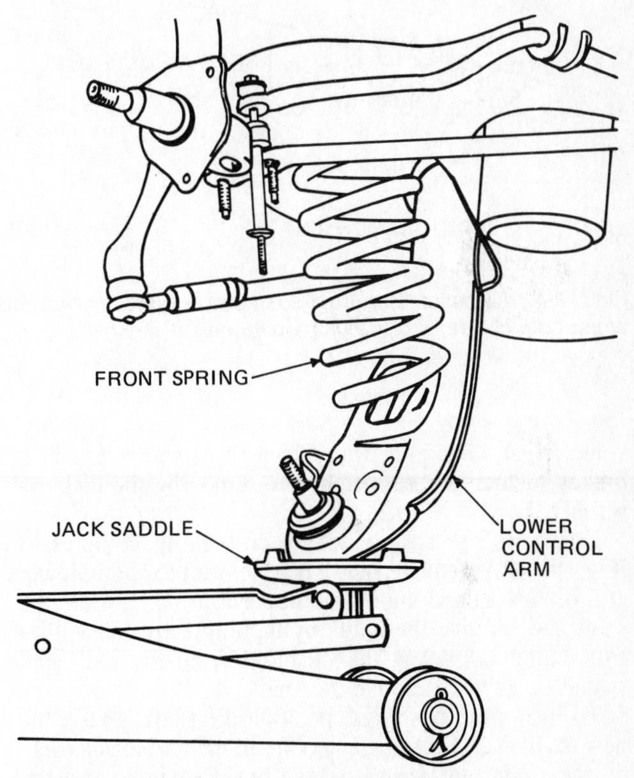

Fig. 12-41 Removing the coil spring so that the lower control arm can be removed. (*Ford Motor Company*)

❁ 12-10 Ford—spring on upper control arm-

Figure 12–42 shows this kind of front suspension. When the spring is mounted on the upper control arm, the upper or lower control arm must be installed as a unit. Replace the stabilizer-bar bushings and insulators, and the strut-rod bushings, as explained in ❁ 12–9 for the suspension system with the spring between the arms.

1. Upper-arm-bushing service Remove the shock absorber. Raise and support the car on safety stands. Remove the wheel cover, grease cap, cotter pin, locknut, adjusting nut, and outer bearing from the hub. Pull the

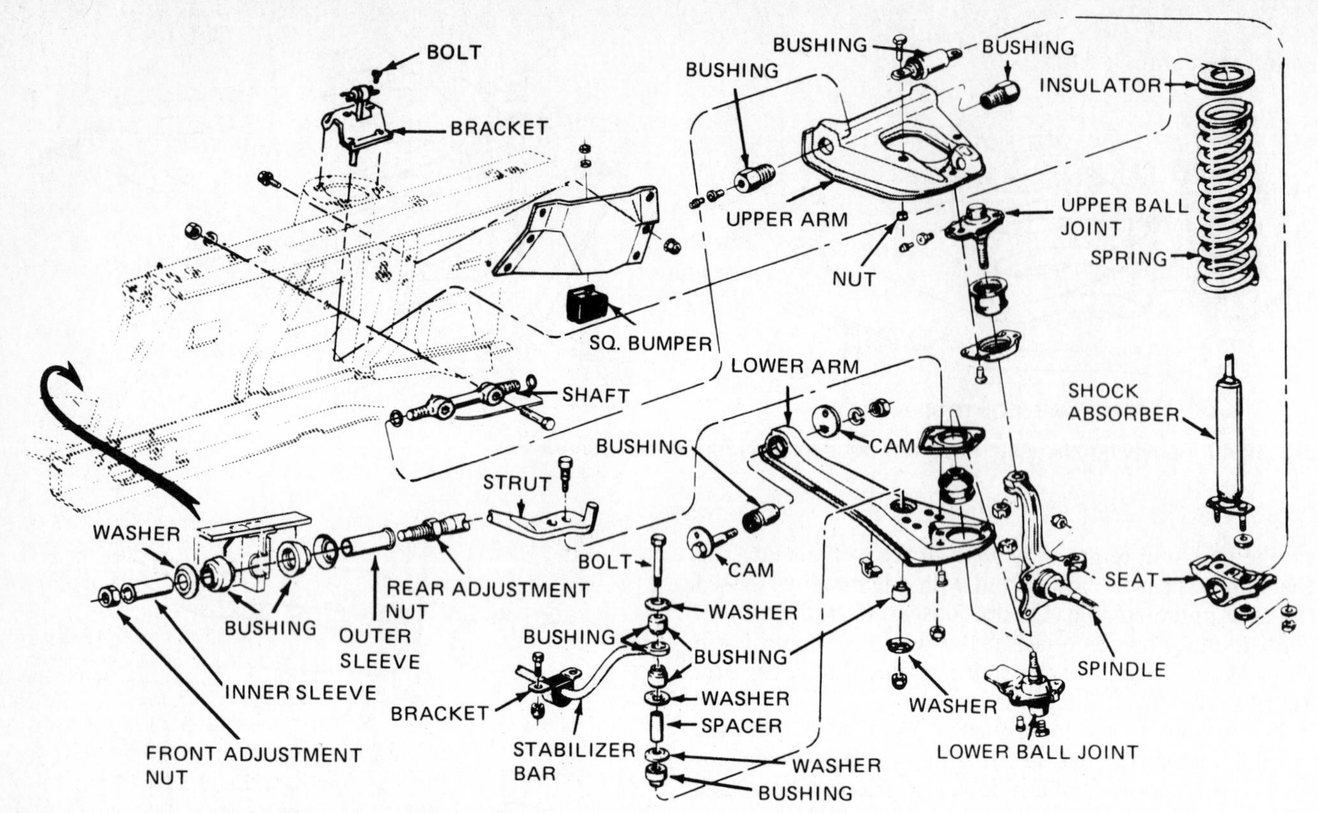

Fig. 12-42 Disassembled front-suspension system which has the spring mounted on the upper control arm. (*Ford Motor Company*)

wheel, tire, hub, and drum from the spindle. On disk brakes, detach the caliper before removing the disk assembly.

Compress the spring with a coil-spring compressor (Fig. 12–43). Remove the two upper-arm-to-spring-tower attaching nuts and swing the upper arm out. Rotate the shaft and remove the studs by tapping them out with a soft mallet. Unscrew the bushings from the shaft and suspension arm. Remove the shaft.

To install the bushings, position the shaft, grease the new bushings and O rings, and install the bushings loosely on the shaft and arm. Turn the bushings in so that the shaft is exactly centered, as in Fig. 12–44. Using ¾-inch [19-mm] pipe, make a spacer 8 1/16 inches [205 mm] long, and position it parallel to the shaft, as shown in Fig. 12–45. If it will not fit, the arm is distorted and must be discarded.

With the spacer positioned as shown in Fig. 12–45, torque the bushings to specifications. Make sure the arm can move on the shaft. Then remove the spacer. Attach the upper suspension arms to the underbody. Release the spring. Install the parts removed and adjust the wheel bearing. Lower the vehicle, install the shock absorber, and adjust caster, camber, and toe (Chap. 13).

2. Front-spring removal and installation Proceed as for the replacement of upper arm bushings as discussed earlier to get the upper arm out of the way. Then use the spring compressor and remove the spring (Fig. 12–43). Installation is the reverse of removal.

3. Lower-arm removal and installation With the car raised and supported on safety stands, remove the wheel and tire. Disconnect the stabilizer bar and strut. Figure 12–42 shows how they are connected. Remove the cotter pin from the nut on the lower-ball-joint stud and loosen the nut one or two turns. Use the ball-stud remover to apply force on the lower stud (Fig. 12–23). Tap the spindle to loosen the stud from the spindle. Remove the nut from the stud and lower the arm. Detach the arm from the underbody by removing the cam bolt, nut, and washer. To install the arm, reattach it to the underbody and spindle.

4. Upper-arm removal and installation With the car raised and supported on safety stands, remove the wheel and tire. Remove the shock-absorber attaching nuts and lift out the shock absorber (Fig. 12–42). On eight-cylinder cars, remove the air cleaner. Install the spring-compressor tool and compress the spring (Fig. 12–43). Disconnect the upper-ball-joint stud from the spindle using the ball-stud remover (Fig. 12–23) and a hammer. Now detach the upper-arm-shaft nuts and remove the arm. Installation is the reverse of removal. Be sure to use the specified keystone type of lock washers to attach the shaft-bolt nuts.

5. Front-wheel-spindle removal and installation Removal and installation procedures vary slightly because of the different steps required for drum and disk brakes. Basically, the procedure is to detach the wheel-

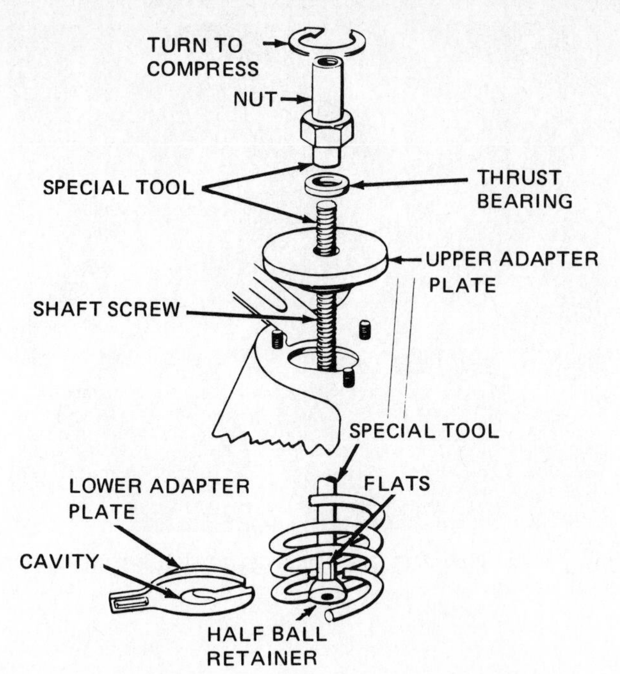

Fig. 12-43 Installing one type of coil-spring compressor. (*Ford Motor Company*)

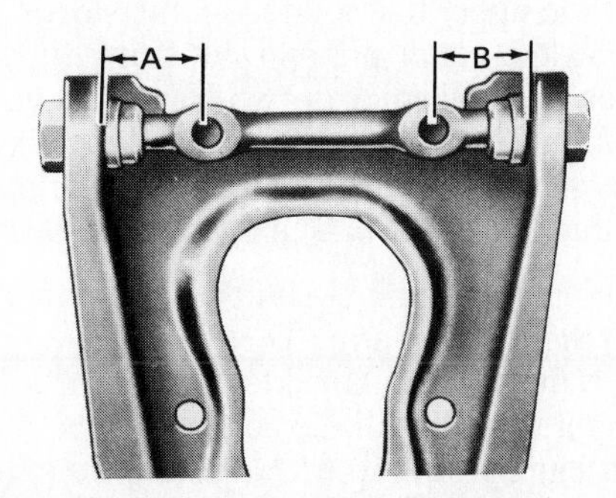

Fig. 12-44 When new bushings are installed, they must be turned in so that the shaft is exactly centered in the control arm. (*Ford Motor Company*)

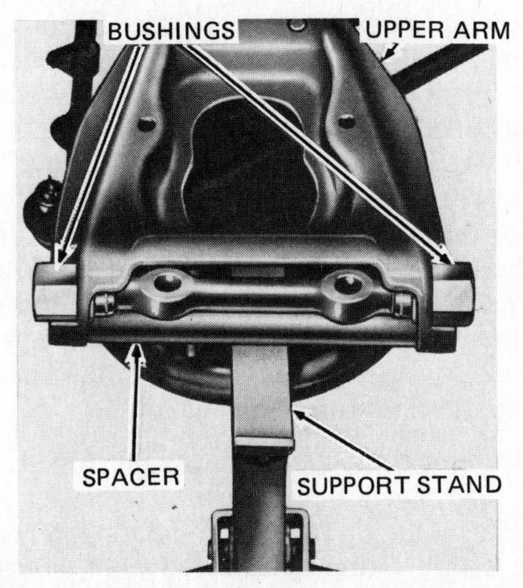

Fig. 12-45 With the spacer between the inner sides of the control arm, tighten the bushings to the specified torque. (*Ford Motor Company*)

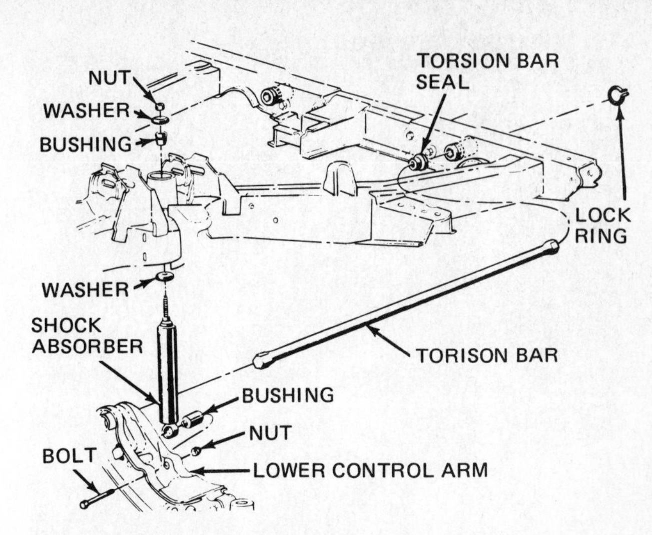

Fig. 12-46 Location and attachment of a longitudinal torsion bar. (*Chrysler Corporation*)

and-tire assembly and the brake assembly from the spindle and move them out of the way. Figure 12–42 shows the spindle with these assemblies removed. Then loosen the upper- and lower-ball joint studs with the ball-stud remover (Fig. 12-23) and a hammer. With the studs detached from the spindle, the spindle is free for removal. Installation is the reverse of removal.

12-11 Plymouth front-suspension service The front-suspension system on some Plymouth cars uses longitudinal torsion bars (Fig. 12–46) instead of coil springs. The procedures are similar for other Chrysler-built cars with this type of suspension system. Service operations include height adjustment, torsion-bar replacement, upper- and lower-control-arm replacement, and ball-joint and sway-bar replacement.

1. Height adjustment With the vehicle on a level floor, the tires at the proper pressure, a full tank of gas, and no passengers, jounce the car a few times to settle the suspension. Release the car on the downward motion.

Measure the distance from the adjustment blade to the floor and the distance from the lowest point of the steering knuckle to the floor. These are labeled A and B in Fig. 12–47. Find the difference between the two measurements. The difference varies with different models, but on one car the specification is 1⅝ to 1⅞ inches [41.3 to 47.6 mm]. Also, the difference between the two sides of the car should be no more than ⅛ inch [3.2 mm]. To correct, turn the torsion-bar adjustment bolt. Figure 12–48 shows the position of this bolt. After each adjustment, jounce the car before rechecking the measurement.

2. Torsion-bar replacement The torsion bars are not interchangeable between left and right. They are marked either right or left by an R or L stamped on one end of the bar. They may also be identified by the part number stamped on the torsion bar. An odd number indicates a left-side torsion bar; even indicates right. To remove a torsion bar, raise the front of the car. If you use a lift,

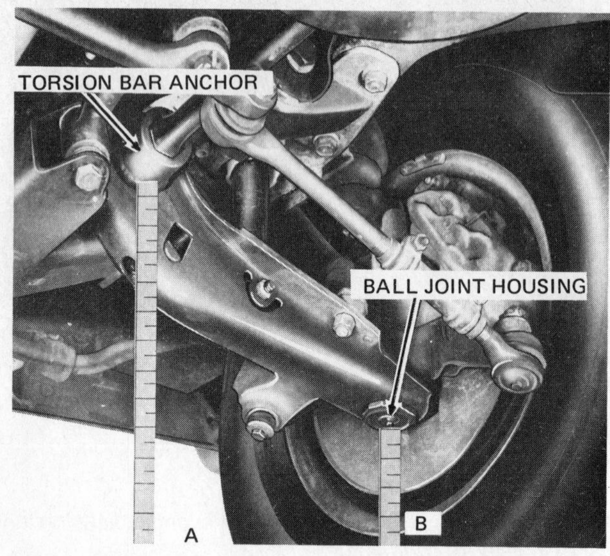

Fig. 12-47 Measuring front-suspension height. (*Chrysler Corporation*)

Fig. 12-48 Removing a longitudinal torsion bar. (*Chrysler Corporation*)

the pads should be on the body so that the front suspension is under no load. If you use jacks, first place a support under the frame crossmember to avoid damaging it.

Release the load from the torsion bar by backing off the adjusting bolt. Remove the lock ring from the rear end of the torsion bar. Attach the special striking tool to the torsion bar and knock the bar loose. (Fig. 12–48). Then remove the tool, slide the rear-anchor balloon seal off the anchor, and slide the bar out through the rear of the anchor. Try not to damage the balloon seal.

Check the torsion bar for scratches or nicks. Dress them down and paint the repaired area with rustproof paint. Check the bar attachments or anchors. Replace any damaged parts. Clean all parts.

Install the torsion bar by sliding it forward through the rear anchor (Fig. 12–46) Slip the balloon seal over the front of the torsion bar with the cupped end toward the rear. Coat both ends of the bar with the specified lubricant. Slide the bar forward so that the hex head enters the opening in the lower control arm. Install the lock ring at the rear. Pack the annular opening in the rear anchor completely full of multipurpose grease. Position the balloon seal on the rear anchor so that the lip engages in the groove of the anchor. Tighten the adjusting bolt, shown in Fig 12–48 to place a load on the torsion bar. Lower the vehicle to the floor and adjust its height as described earlier. Install the upper bumper.

3. Steering-knuckle removal Figure 12–49 shows various steering-knuckle arrangements used. To replace a steering knuckle, turn the ignition switch to OFF or UNLOCKED. Remove the rebound bumper. Raise the vehicle to remove all load from the front suspension. Place safety stands under the frame. Remove the wheel cover, wheel, and tire assembly. On cars with disk brakes, remove the brake caliper. Support it with a piece of wire so that it does not hang from the brake hose. Remove the hub and disk or drum assembly and brake splash shield. Remove all load from the torsion bar by backing off the adjusting bolt.

Remove the upper ball joint from the steering knuckle by removing the cotter pin and nut from the upper ball joint. Force the ball joint out with a ball-stud remover (Fig. 12–23). Remove the bolts attaching the steering arm to the steering knuckle. Take the steering knuckle off. Note that the lower ball joint is in the steering arm.

4. Steering-knuckle installation Attach the steering knuckle to the steering arm. Then install the upper-ball-joint stud in the steering knuckle and secure with the nut properly tightened. Install the cotter pin. Put the load on the torsion bar by turning the adjusting bolt. Install the parts removed, including the splash shield, hub and disk or drum, and caliper, if so equipped. Adjust wheel bearings. Install the wheel and cover. Then lower the car to the floor. Install the rebound bumper. Adjust front-suspension height and wheel alignment.

5. Steering-knuckle-arm removal Turn the ignition switch to OFF or UNLOCKED. Remove the rebound bumper (Fig. 12–49). Raise the vehicle so that the front suspension is unloaded. Put safety stands under the frame. Remove the wheel cover, wheel, and tire. Remove the brake caliper, where present, and hang it by a wire to prevent damage to the brake hose. Remove the hub and brake disk or drum assembly. Remove the brake splash shield. Unload the torsion bar. Disconnect the tie rod from the steering-knuckle arm by removing the cotter pin and nut. Remove the lower-ball-joint stud from the knuckle arm and detach the arm from the knuckle.

6. Steering-knuckle-arm installation Install the ball-joint stud and attach it with a nut and cotter pin (Fig. 12–49). Attach the tie rod. Load the torsion bar. Install the brake disk or drum assembly. Install the caliper, if present. Adjust wheel bearings, install the wheel and cover, then lower the car. Install the rebound bumper. Adjust the front-suspension height and wheel alignment.

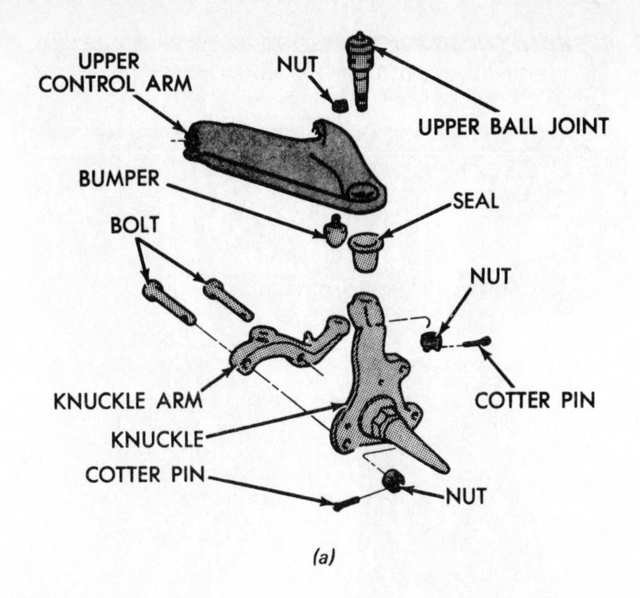

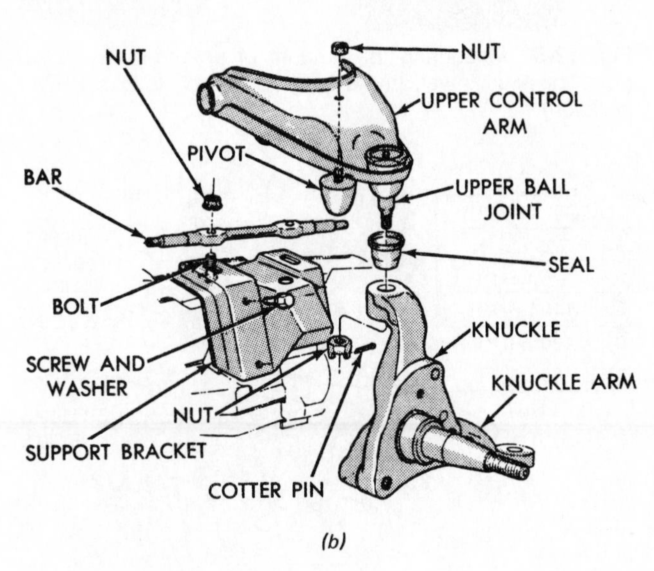

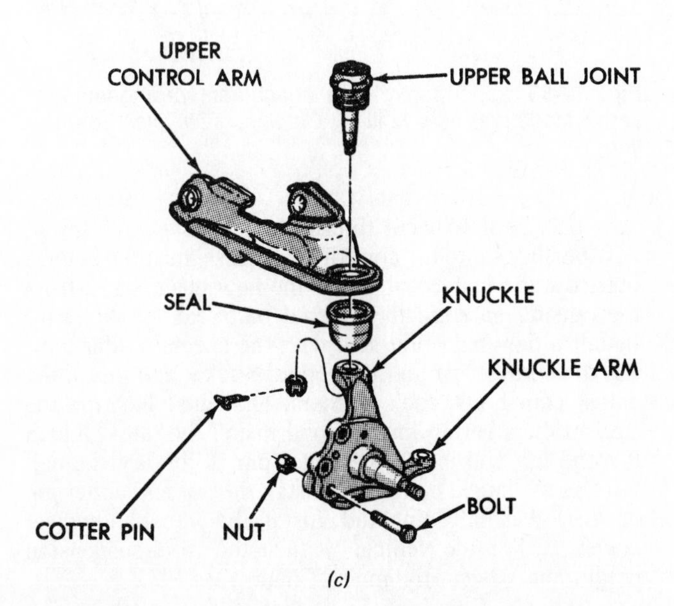

Fig. 12-49 Various steering-knuckle arrangements used on cars built by Chrysler. (*Chrysler Corporation*)

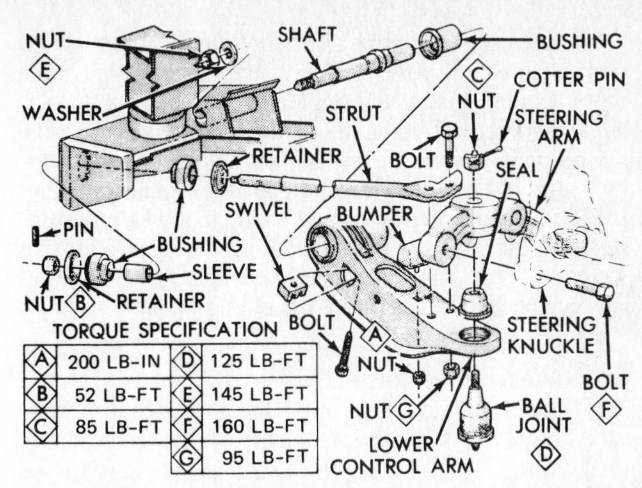

Fig. 12-50 Lower-control-arm attachment arrangement on some models of cars built by Chrysler. (*Chrysler Corporation*)

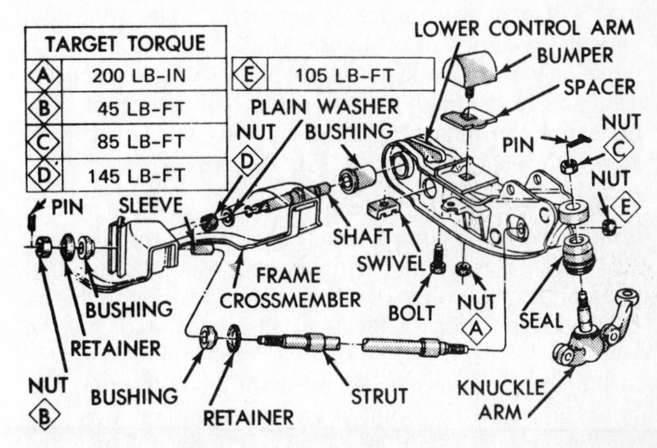

Fig. 12-51 Lower-control-arm attachment. In this type, the arm is removed from the car with the strut still attached. (*Chrysler Corporation*)

7. Lower-control-arm and shaft removal Figure 12–50 shows the lower-control-arm assembly. With the ignition switch at the OFF or UNLOCKED position and the car supported as explained earlier, remove the rebound bumper, wheel cover, wheel, brake caliper, hub-and-rotor assembly or brake drum, and splash shield.

Disconnect the lower end of the shock absorber. Disconnect the strut and sway or stabilizer bar. On some cars, the strut is removed with the control arm as an assembly (Fig. 12–51). Also, on some cars, the automatic-transmission/gearshift/torque-shaft assembly must be removed. Measure the depth of the torsion-bar anchor bolt in the lower control arm, then relieve the tension on the torsion bar. Remove the torsion bar as explained earlier.

Separate the lower ball joint from the knuckle arm, using the ball-stud remover (Fig. 12–23). Remove the nut from the lower-control-arm shaft and push the shaft out of the frame crossmember. If necessary, tap the threaded end of the shaft with a soft hammer to loosen it. Remove the lower control arm and shaft as an assembly. If the shaft bushing is worn, replace it.

8. Lower-control-arm and shaft installation Position the lower control arm with the shaft in the frame crossmember. Install and tighten the nut finger-tight. Attach the lower-ball-joint stud to the knuckle arm with the nut properly tightened. Install the torsion bar and load it by returning the adjusting bolt to its original position. Install the transmission torque shaft, if it was removed. Reattach the strut and sway bar. Reinstall other parts that were removed. Adjust the wheel bearing. Check the front-suspension height and front-wheel alignment.

9. Lower-ball-joint removal Figure 12–52 shows the procedure for measuring the lower-ball-joint axial travel. With the weight of the car on the lower control arm, raise and lower the wheel with a pry bar under the center of the tire.

Removal of the ball joint requires all the preliminary steps outlined above. Remove the cotter pins and nuts from both the upper- and lower-ball-joint studs and install the ball-stud remover (Fig. 12–23). The removal tool will now rest on the lower stud. Tighten the tool enough to apply force to the stud, but do not try to remove the ball joint by tool force alone. Strike the knuckle arm with a hammer to loosen the ball-joint stud. The ball joint can then be pressed out of the lower control arm with a ball-joint press (Fig. 12–26).

10. Lower-ball-joint installation Press the new ball joint in the lower control arm (Fig. 12–27). Install a new seal over the ball joint, if necessary, using the tool that is essentially a collar of the proper size. Figure 12–50 shows the position of the seal. Insert the stud into the hole in the knuckle arm. Install the retainer nut and tighten to the specified torque. Secure with the cotter pin. Lubricate the new ball joint as explained earlier. Load the torsion bar and install the parts you removed. Lower the vehicle to the floor. Adjust the front-suspension height and front alignment as needed.

11. Upper-control-arm and ball-joint removal Figure 12–53 shows one arrangement of how the upper control arm is attached. To remove it, position the ignition switch to OFF or UNLOCKED. Raise the front of the car with a jack and remove the wheel cover and wheel. Position a short safety stand under the lower control arm near the splash shield, and lower the jack. Make sure that the safety stand does not touch the shield and that the rebound bumpers are under no load. Remove the cotter pin and nut from both ball joints. Slide the ball-stud remover into place with the lower end resting on the steering-knuckle arm and the upper end on the upper ball-joint stud (Fig. 12–23). Then strike the steering knuckle with a hammer to loosen the stud. Do not loosen the stud with tool force alone.

Remove the tool and disengage the ball joint from the steering knuckle. Remove the rubber engine splash shield and the pivot-shaft-bolt nuts or bolts. Lift the upper control arm along with the ball joint and pivot-bar nuts, retainers, and bushings. Install new bushings if the old ones are worn. Remove the ball joint from the upper arm with the ball-joint press.

12. Upper-control-arm and ball-joint installation Install the new ball joint with the ball-joint press. The new ball joint will cut threads into a new arm. Install new bushings into the control arm. Press the old bushings out from the inside out. Press the new bushings in from the outside in, until the tapered part seats on the arm. Install a new ball-joint seal with the special collar tool. Put the control arm in the support bracket and install the cams, cam bolts, lock washers, and nuts. Position the stud in the steering knuckle and install the nut. Tighten the end nut and install the cotter pin. Lubricate the ball joint as explained earlier. Reinstall the nut and cotter pin on the lower-ball-joint stud. Install the wheel and wheel cover. Lower the vehicle. Adjust the front-suspension height and wheel alignment (Chap. 13).

14. Front-wheel bearings The front-wheel bearings are similar to those in Ford cars (Fig. 12–38). Front-

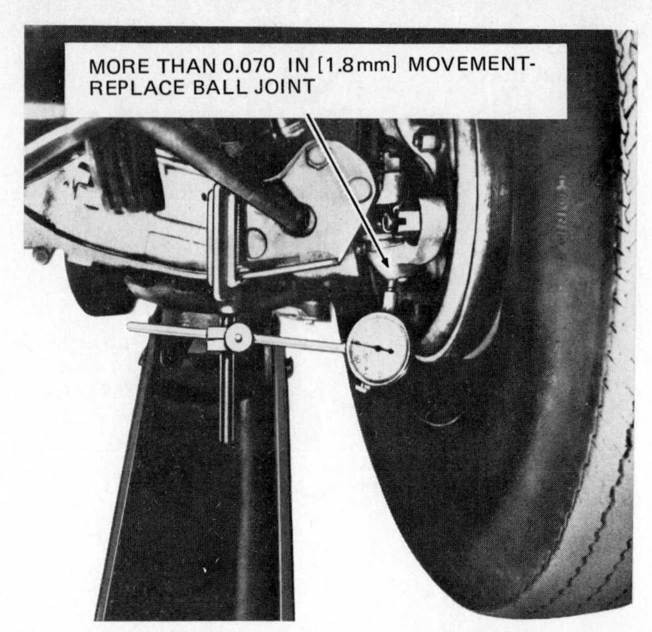

Fig. 12-52 Measuring the amount of up-and-down movement, or axial travel, in the lower ball joint. (*Chrysler Corporation*)

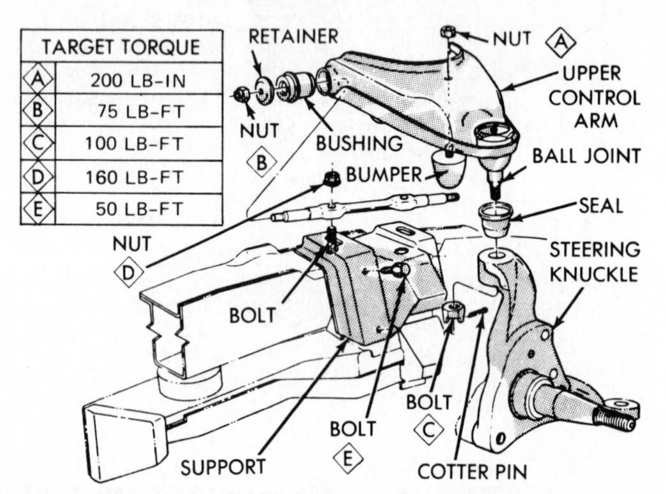

Fig. 12-53 Upper-control-arm attachment arrangement in some models of cars built by Chrysler. (*Chrysler Corporation*)

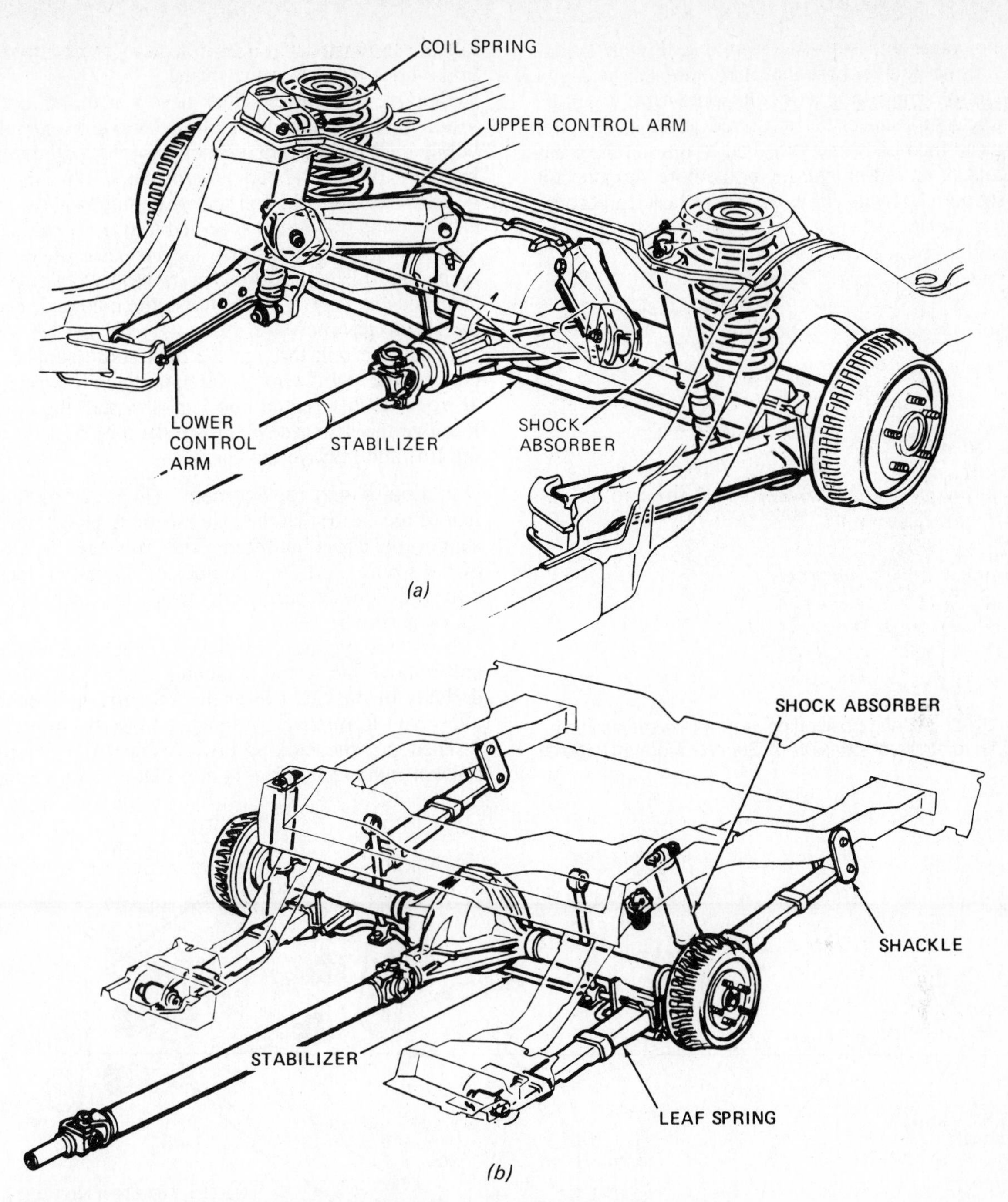

Fig. 12-54 Typical coil-spring and leaf-spring rear-suspension systems used by Chevrolet. (*Chevrolet Motor Division of General Motors Corporation*)

wheel bearings in Chrysler products are inspected, removed, cleaned, replaced, and adjusted as discussed in ⚙ 12-9.

Rear suspension

⚙ 12-12 Rear-suspension service Normally, rear-suspension systems require no special service. However, when parts are worn or broken, they must be replaced. The sections that follow describe specific replacement procedures for certain models of Chevrolet, Ford, and Plymouth cars. These procedures are typical for many cars. The parts that most often require replacement are the rubber grommets in spring eyes, control arms, and track bars. In addition, cars using independent rear suspension, such as the Corvette shown in Fig. 8–24, require periodic rear-wheel alignment (Chap. 13).

⚙ 12-13 Chevrolet rear-suspension service Chevrolet cars use two general types of rear-suspension system: coil spring and leaf spring (Fig. 12–54).

1. Coil-spring replacement (Fig. 12-54) When replacing coil springs, raise the rear of the car with a lift

under the frame and support the rear axle with an adjustable lifting device. Disconnect both the left and right upper control arms at the axle. On some cars, it will be necessary to disconnect the track rod at the differential housing and then disconnect the end of the stabilizer bar at the control arm. Remove the brake-hose support bolt at the support. Do not allow the brake hose to become kinked or twisted. You do not need to disconnect any brake lines to replace the spring.

Remove the lower shock-absorber mounting. Then lower the lifting device to allow the axle to extend down below the fully extended shock absorber. When the axle has been lowered far enough, you can remove the spring.

When replacing the coil spring, be sure to always tighten each nut and bolt to the specified torque as it is installed. To install the spring, place the insulator on top of the spring and place the spring in its proper position on the axle housing (Fig. 12-55). Raise the axle and reconnect the shock absorber.

Install the bolt through the brake-hose support. Then connect the stabilizer bar. Reconnect both upper control arms at the differential housing, or install the track rod. Remove the lifting device from the axle housing, lower the lift, and remove the car.

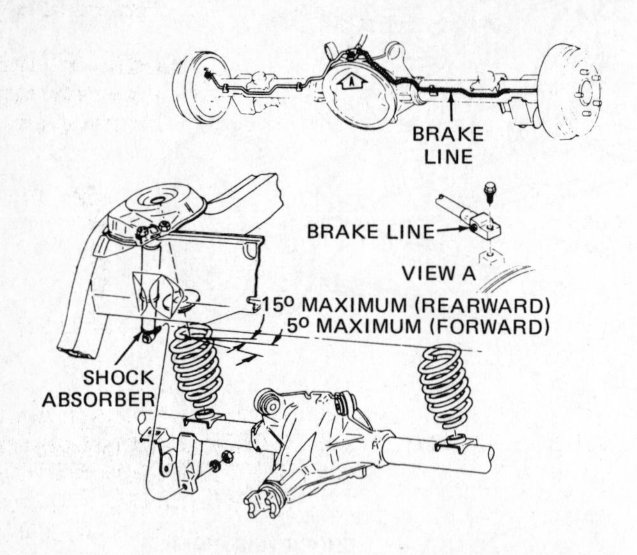

Fig. 12-55 Coil-spring positioning in a rear-suspension system. (*Chevrolet Motor Division of General Motors Corporation*)

2. Leaf-spring replacement (Fig. 12-54) Raise the rear of the car so that the axle assembly hangs free. Then support the car at both frame side rails near the front eye of the spring. Lift the axle housing so that all tension is removed from the spring and detach the lower end of the shock absorber.

Next, loosen the spring-eye-to-bracket retaining bolt and remove the screws attaching the bracket to the underbody of the car. Lower the axle assembly enough to allow you to remove the bracket from the spring.

Then pry the parking-brake cable from the retainer bracket mounted on the spring plate. Remove the nuts

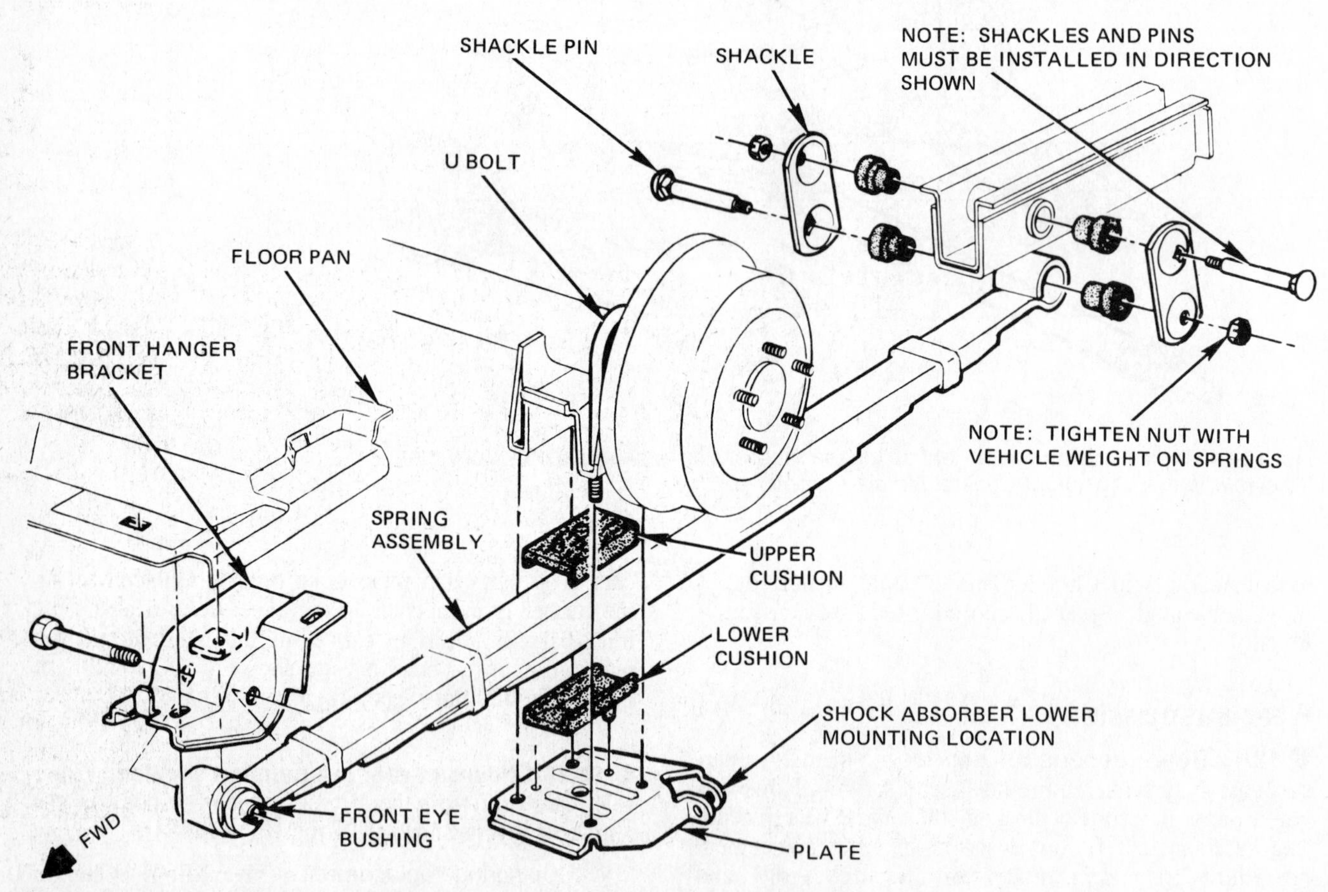

Fig. 12-56 Disassembled left rear-suspension system using a leaf spring. (*Chevrolet Motor Division of General Motors Corporation*)

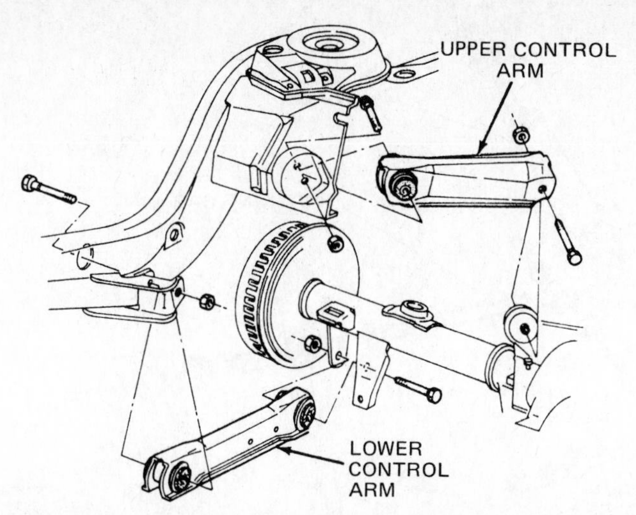

Fig. 12-57 Arrangement of the control arms in a coil-spring rear-suspension system. (*Chevrolet Motor Division of General Motors Corporation*)

from the U bolts that attach the spring to the axle housing (Fig. 12-56). Support the spring and remove the lower bolt from the spring rear shackle. Now the spring can be removed from the car.

If a spring leaf requires replacement, both it and any damaged spring-leaf insert can be replaced by removing the center-bolt nut.

To install the leaf spring, loosely attach the bracket to the spring eye, put the spring in place, and attach the rear of the spring to the shackle. Then loosely attach both the front bracket to the underbody and the spring to the mounting pad on the axle housing. Be sure that all insulators and cushions are in place.

Attach the parking-brake cable and lower the car to the floor. Then tighten the shackle nuts and all other parts to the specified torque.

3. Control-arm replacement Figure 12-57 shows the attaching points for the upper and lower control arms. When replacing the control arm, the rear axle must be supported in such a way that the housing will be prevented from rotating when the control arm is detached.

4. Stabilizer-bar replacement Figure 12-58 shows the stabilizer-bar attaching points for typical coil-spring and leaf-spring rear-suspension systems. Not all cars use a rear stabilizer bar. To replace this bar, raise the car on a lift, support the axle housing, and then remove and replace the stabilizer bar.

5. Corvette rear suspension The Corvette has independently sprung rear wheels (Fig. 8-24). Only one leaf spring, which is transversely mounted, is used. Most parts in this rear suspension can be replaced easily by following the manufacturer's procedures. After service or repair, the rear wheels must be checked for alignment (Chap. 13). On this suspension system, camber and toe can be adjusted.

To align the rear wheels, back the car onto the wheel aligner. Check that the strut rods are straight. If they are

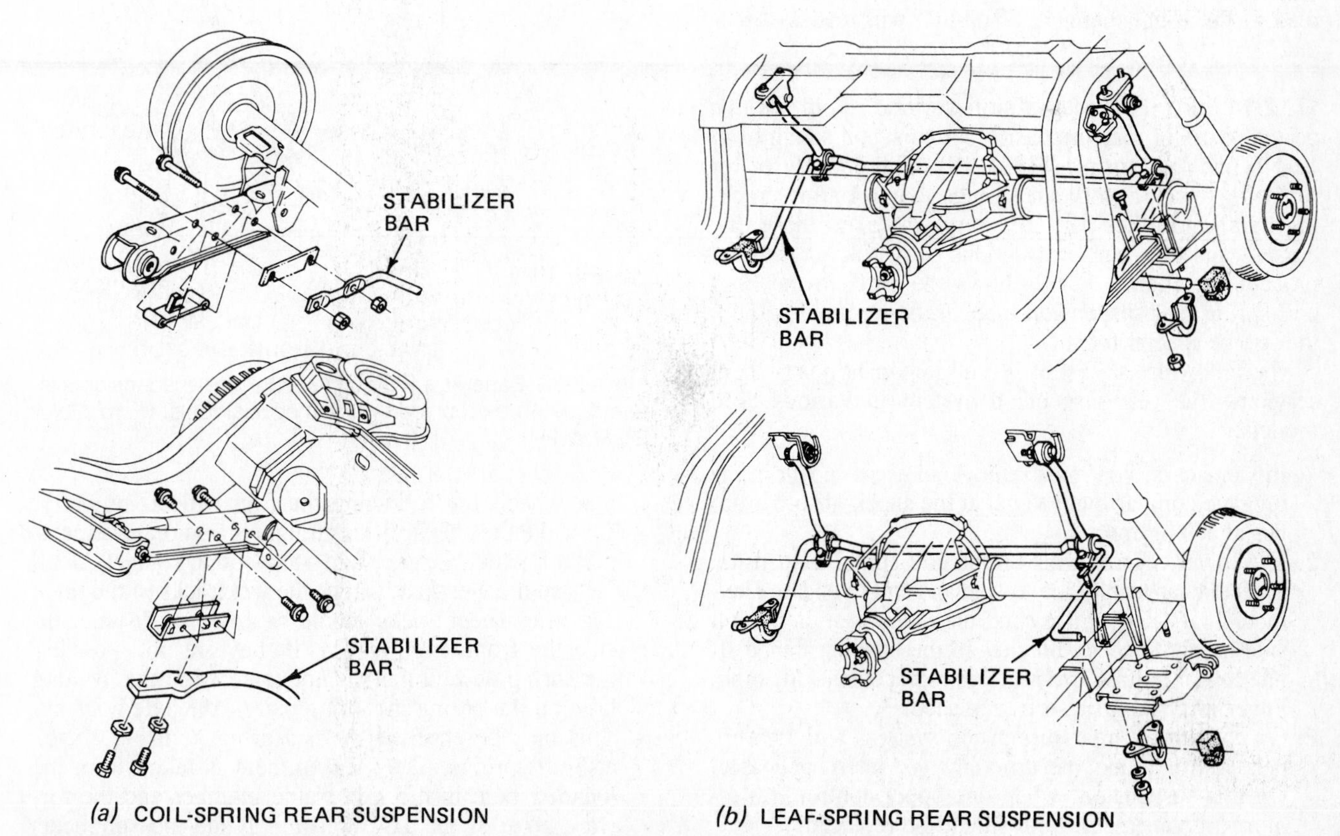

Fig. 12-58 Arrangement of the stabilizer bar in (left) a coil-spring rear suspension, and (right) a leaf-spring suspension system. (*Chevrolet Motor Division of General Motors Corporation*)

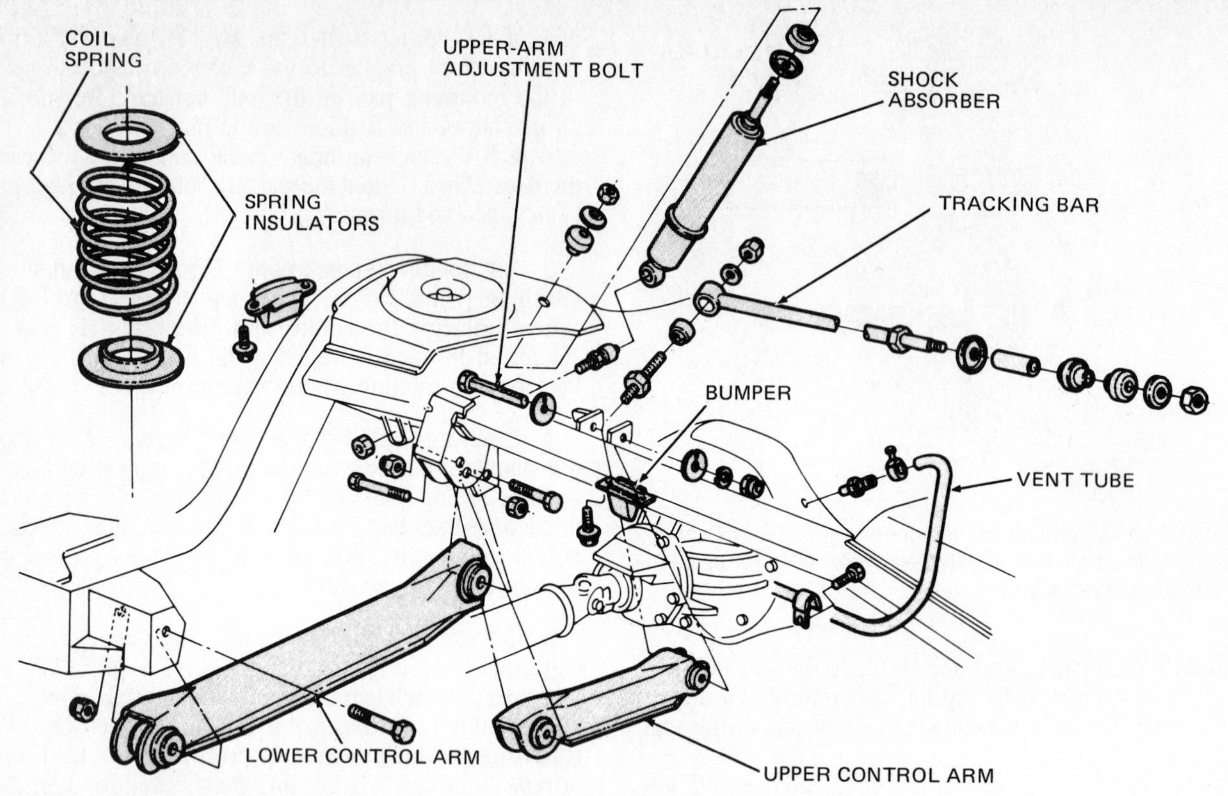

Fig. 12-59 Typical Ford coil-spring rear-suspension system that uses only one upper control arm. (*Ford Motor Company*)

bent, they should be straightened. Camber can now be read in the usual manner. "Toe-in" will read as "toe-out."

12-14 Ford rear-suspension service Ford has two general types of rear-suspension systems: coil spring (Fig. 12-59) and leaf spring (Fig. 8-19). The procedures for replacing shock absorbers, springs, control arms, tracking bars, and other parts are very similar to those discussed in the earlier sections covering Chevrolet procedures. Figure 12-60 shows the parts in the Ford coil-spring rear suspension that can be replaced without the use of special tools.

The following is a list of several important points about servicing the rear-suspension system in various Ford models:

1. In some cars, you must remove an access cover in the luggage compartment to get at the shock-absorber upper attaching nut.
2. The lower control arms on the coil-spring suspension are interchangeable side-to-side and end-to-end on later-model cars. On earlier cars, the lower arm on the left side is identified by notches in the bushing flange. If one lower control arm requires replacement, the other lower arm should also be replaced.
3. On coil-spring, rear-suspension systems with two upper control arms, the arms are *not* interchangeable. On this suspension, when one upper control arm requires replacement, both should be replaced.
4. One check that can be made on the leaf-spring suspension is for tracking. This determines whether the rear wheels are following the front wheels properly. To make the check, drive straight ahead on pavement, part of which is wet, and stop about 10 feet [3 m] beyond the wet area. Check the wet tracks of the tires. The rear-wheel tracks should be an equal distance inside the front (Fig. 12-61). If they are not, possibly the spring tie-bolt head is not centered in the locating hole on the spring mounting pad of the axle housing. This may be checked by measuring *X* in Fig. 12-62 at both springs. The measurement is taken from the locating hole in the side frame member and the forward edge of the axle housing. If the measurements differ by more than ⅛ inch [3.2 mm], you'll need to reposition the U bolts as follows:

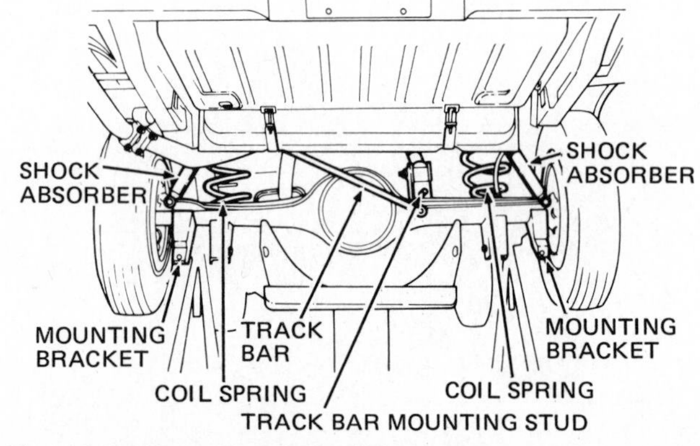

Fig. 12-60 Parts in a Ford coil-spring rear-suspension system that can be replaced without special tools. (*Ford Motor Company*)

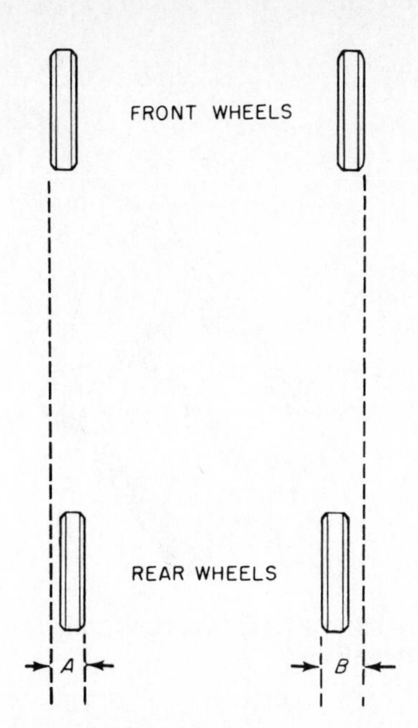

Fig. 12-61 Alignment check of the rear suspension. *A* should equal *B*. (*Ford Motor Company*)

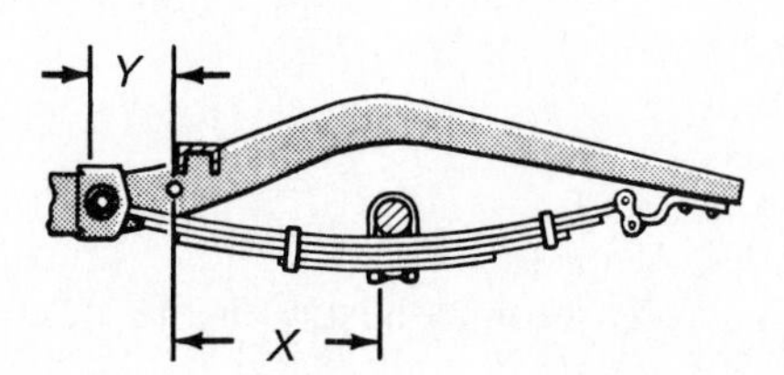

Fig. 12-62 Alignment check for cars with leaf-spring rear suspension. *X* should be the same (or within ⅛ inch [3.2 mm]) at both springs. (*Ford Motor Company*)

Loosen the four U-bolt nuts and use a jack to push the axle housing into position. Then move the U bolts into line and tighten the nuts to the specified torque. If this does not correct the tracking, the frame may be out of line because of a collision.

The dimension *Y* in Fig. 12-62 is a measurement to be taken to determine whether the front hanger should be replaced. When this dimension is not correct, replace the front hanger. If the hanger is welded to the frame, cut off the old hanger with an acetylene torch, and then weld a new hanger to the frame.

12-15 Plymouth rear-suspension service Figure 12-63 shows disassembled views of some Plymouth rear-suspension systems. Various service operations on this suspension system are described below.

1. Measuring spring height Jounce the car several times, first at the front and then at the back, releasing the bumpers at the same point in each cycle. Next, locate the highest point on the underside of the rear-axle bumper strap at the rear of the bumper, and measure the distance from here to the top of the axle housing. Take measurements on both sides. If the difference is more than ¾ inch [19 mm], one of the rear springs needs replacement.

2. Rear-spring removal Disconnect the rear shock

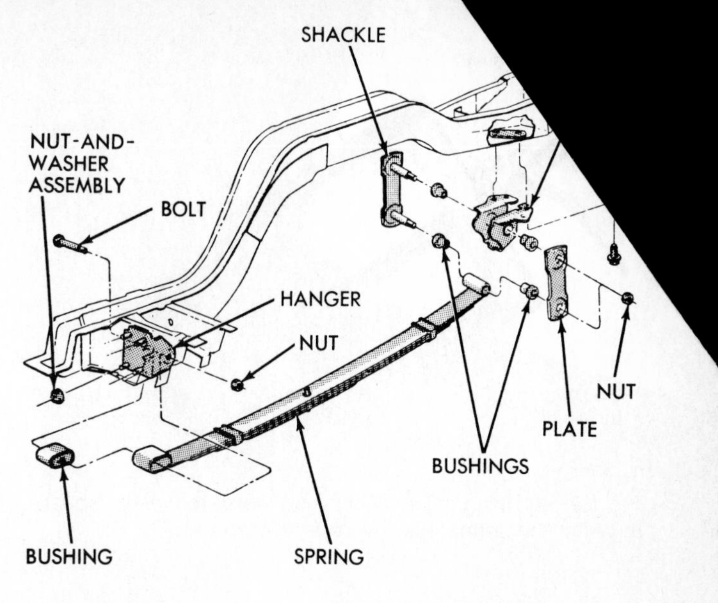

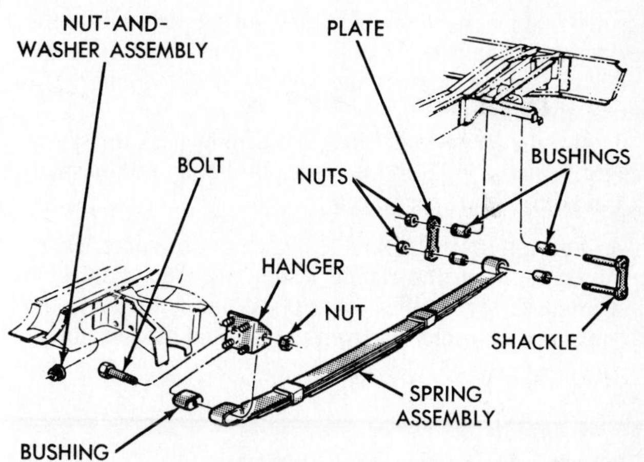

Fig. 12-63 Disassembled leaf-spring rear-suspension systems. (*Chrysler Corporation*)

absorbers at the lower mounting studs. Raise the vehicle at the lifting points so that the rear springs hang free, and use a jack to support the axle housing in this position. Then remove the nuts that attach the spring front hanger to the frame, remove the U-bolt nuts and the spring plate, remove the rear hanger bolts, and take the spring off the car. Take out the front pivot bolt to remove the front hanger from the spring, and remove the rear shackle and bushings from the spring.

3. Rear-spring installation If the front pivot-bolt bushing needs replacement in the spring, use the pivot-bushing tool to remove the old bushing (Fig. 12-64). Then the same tool is used to install the new bushing (Fig. 12-65).

Attach the front hanger with the front pivot bolt and run on the nut, but do not tighten it. Assemble the rear shackle and bushing as shown in Fig. 12-63. Do not lubricate the rubber bushings. Start the shackle-bolt nuts but do not tighten them.

Next, put the spring in place on the car and attach the hangers to the car frame. Tighten the bolts and nuts to the proper torque. Remove the axle support, and put the center hole of the axle spring seat over the head of the

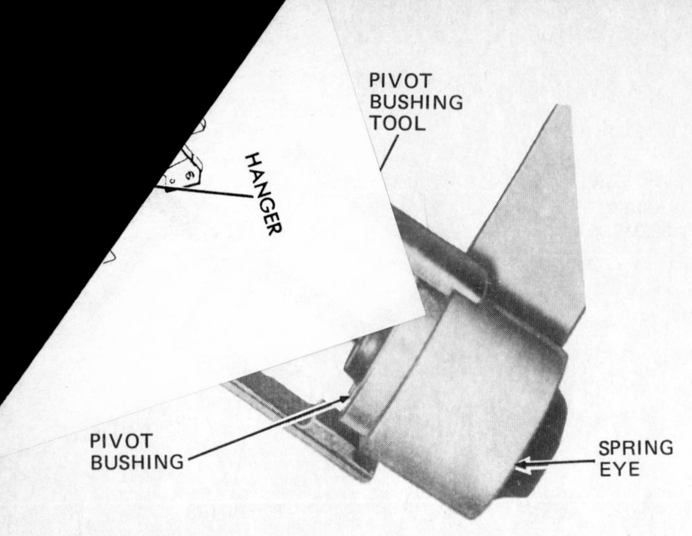

Fig. 12-64 Using the pivot-bushing tool to remove the bushing from the spring eye. (*Chrysler Corporation*)

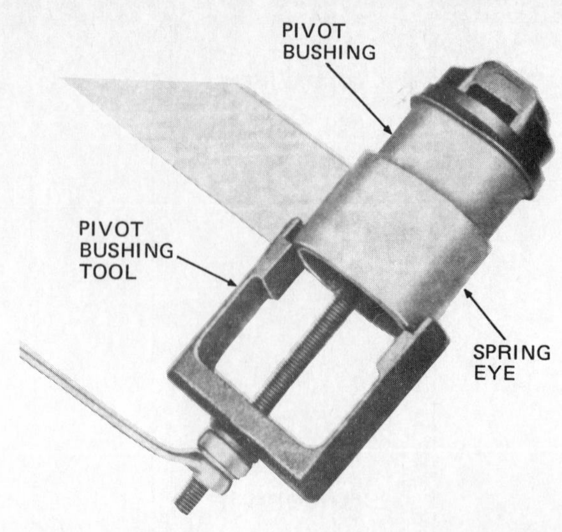

Fig. 12-65 Installing a new pivot bushing in the spring eye. (*Chrysler Corporation*)

spring center bolt. Then place the spring plate under the spring and install the U bolts, tightening the nuts to the specified torque. Lower the vehicle and reconnect the shock absorbers.

Jounce the car several times and then recheck the spring height. Finally, tighten the front pivot bolt and the shackle nuts to the proper torque.

4. Interleaves In some leaf springs, zinc interleaves are used between the spring leaves to reduce corrosion and improve spring life. To replace the interleaves, the spring must be removed from the car and disassembled.

Chapter 12 review questions

Select the *one* correct, best, or most probable answer to each question. Then check your answers against the correct answers given at the end of the book.

1. The only service that the steering linkage normally requires is periodic:
   a. tie-rod adjustment
   b. lubrication
   c. ball-joint replacement
   d. none of the above
2. If a steering-linkage part is bent, it should be:
   a. heated and straightened
   b. cold-straightened
   c. discarded
   d. welded
3. The only time the steering arm would require replacement would be after:
   a. wheel-bearing replacement
   b. a collision
   c. wheel replacement
   d. all of the above
4. If the ball stud in a tie-rod end has excessive looseness:
   a. it should be adjusted
   b. the ball stud should be replaced
   c. the tie-rod end should be replaced
   d. the tie rod should be replaced
5. The front-wheel bearings of a car are to be adjusted. Mechanic A says that a torque wrench can be used to set the intial preload. Mechanic B says that a dial indicator can be used to set the initial preload. Who is right?
   a. A only
   b. B only
   c. both A and B
   d. neither A nor B
6. The Chevrolet recommendation is that if a front coil spring is sagging:
   a. shims should be installed
   b. the shock absorber should be replaced
   c. the spring should be replaced
   d. all of the above
7. On the typical front suspension, the shims between the frame and the upper-control-arm shaft provide adjustments of the front-wheel:
   a. caster and camber
   b. toe-in and toe-out
   c. included angle and toe-in
   d. all of the above
8. On the Plymouth front suspension, if there is spring sag, it can be corrected by:
   a. adding shims
   b. replacing springs
   c. adjusting torsion bars
   d. all of the above
9. When removing the ball stud from the steering knuckle:
   a. do not loosen by tool force alone
   b. tighten the tool until the ball joints pull loose
   c. drive the ball joints out with a soft hammer
   d. first drill out the retaining rivets
10. When adjusting front-wheel bearings, always:
    a. tighten adjustment or spindle nut with wheel stationary
    b. tighten adjustment or spindle nut with wheel rotating
    c. tighten adjustment or spindle nut until wheel stops rotating
    d. none of the above

CHAPTER 13

WHEEL ALIGNMENT

After studying this chapter, and with proper instruction and equipment, you should be able to:

1. Perform all required prealignment checks.
2. Demonstrate how each wheel-alignment angle is checked using the gauges and equipment in your shop.
3. Align the front wheels on a car with coil-spring front suspension.
4. Align the front wheels on a car with MacPherson-strut front suspension.
5. Align the front wheels on a car with torsion-bar front suspension.
6. Align the front wheels on a vehicle with an I-beam front axle.
7. Align the front wheels on a vehicle with twin I-beam front suspension.
8. Align the rear wheels on a car with independent rear suspension.

Prealignment checks

13-1 Prealignment checks Many conditions, in addition to wheel alignment, influence how a vehicle steers. Before caster, camber, toe, turning radius, and steering-axis inclination are checked, these other factors should be checked and corrected, if necessary. These include tire pressure and condition, wheel-bearing condition and adjustment, wheel and tire balance and runout, ball-joint and steering-linkage looseness, rear-leaf-spring condition, and front-suspension height. If any of these factors is not correct, you cannot accurately align the wheels. It is even possible that a ''wheel alignment'' could make the abnormal conditions worse. Checking ball-joint and steering-linkage looseness, rear-leaf-spring condition, and front-suspension height are covered in previous chapters.

13-2 Tires Check that all tires are inflated to the specified air pressure. The tires should be the same size, in good condition, and have about the same amount of wear. Look at the tread wear. Check that bias-ply and radial tires are not on the same axle. Then run your hand across the tread to feel for uneven wear and sharp edges on the tread ribs. Compare the tread wear with the examples shown in Fig. 11-16. The condition of the tread always indicates steering and suspension problems that cause abnormal wear. Chapter 21 describes tire service.

13-3 Wheel bearings Raise the front wheels and check the condition and adjustment of the front wheel bearings (Chap. 12). If the wheel bearing is worn or defective, a slight vibration or grinding may be felt as the wheel is spun.

13-4 Tire and wheel runout Each front tire should be checked to determine if it wobbles sideways or is out-of-round. Sideways wobble is called *lateral runout*. Out-of-round is called *radial runout*.

Radial runout can be checked by spinning the tire slowly and bringing the pointer of a dial indicator or runout gauge toward the center of the tread until it touches (Fig. 13-1). If the tire touches uniformly all around, the tire tread is not out-of-round. However, there usually is some radial runout in a tire. If the tread has excessive radial runout, the dial indicator will show it. Typically, if the radial runout of a tire exceeds 0.070 inch [1.8 mm], then the radial runout of the wheel should be checked at the point shown in Fig. 13-1. A typical specification for wheel runout is 0.045 inch [1.1 mm]. If the radial runout of the wheel is excessive, replace the wheel.

NOTE: Never straighten a bent wheel. Straightening could weaken the wheel and cause it to fail later.

If the radial runout of the tire is excessive, and the wheel is within specifications, the problem is in how the tire is mounted, or in the tire itself. To correct excessive tire runout, it may be necessary only to deflate the tire and work it around to another position on the wheel. Align the highest radial point on the tire with the lowest radial point on the wheel. On some tires, you may have to remove some of the rubber from the tread by use of a tire truing machine that trims off the excess. But if these ''fixes'' don't work, replace the tire.

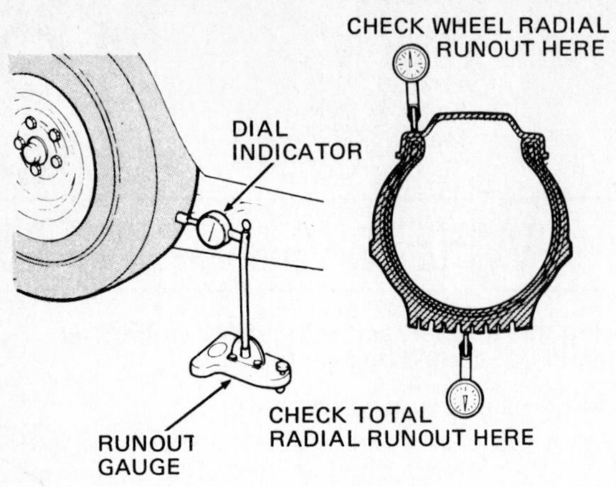

Fig. 13-1 Checking radial runout of a wheel and tire. (*Ford Motor Company*)

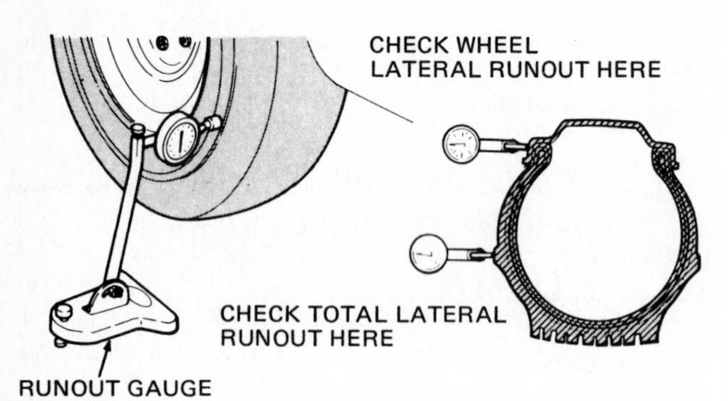

Fig. 13-2 Checking lateral runout with a dial indicator. (*Ford Motor Company*)

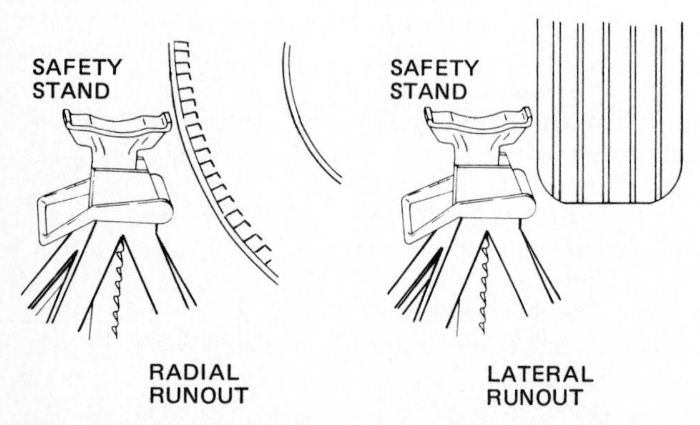

Fig. 13-3 Making a quick check of radial and lateral runouts with a safety stand. (*Ford Motor Company*)

To check lateral runout of a tire, place the pointer of a dial indicator or runout gauge against the scrub rib on the sidewall of the tire (Fig. 13-2). If the reading is excessive, check the lateral runout of the wheel. Typically, the specifications for the lateral runout of the tire and the wheel are about the same as those specified for radial runout.

NOTE: A safety stand can be used as shown in Fig. 13-3 to make a quick check of radial and lateral runout. However, the use of a dial indicator is more accurate.

Fig. 13-4 Balancing an assembled tire and wheel on a static, or bubble, balancer. (*Ammco Tools, Inc.*)

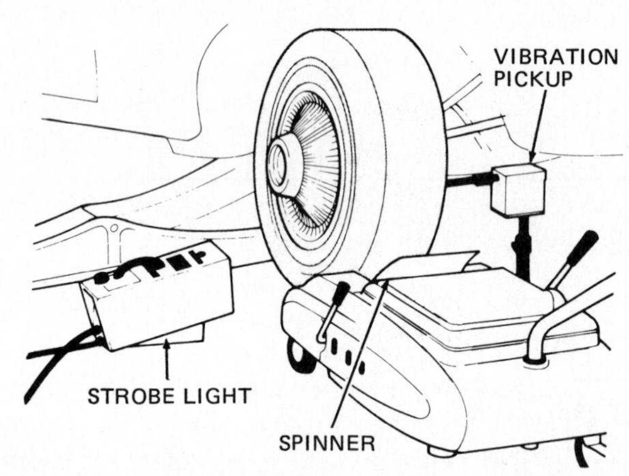

Fig. 13-5 An electronic wheel balancer. A magnet is attached to the brake backing plate. Through a short arm, any movement of the magnet is sensed by a vibration pickup. This causes the strobe light to flash, indicating where to attach a wheel weight. (*Ford Motor Company*)

13-5 Wheel balance If a wheel-and-tire assembly is out of balance, the car will be hard to steer, the ride will be rough, and the tire wear will be rapid. The wheel may be checked for balance on or off the car. This job is done by two methods: static balancing and dynamic balancing. Static balancing usually will cure any problems with low-speed shake. However, dynamic balancing may be required to eliminate high-speed shake and dynamic imbalance.

To static-balance or "bubble-balance" a wheel, it must be taken off the car. The wheel is then placed on a static balancer to detect any imbalance (Fig. 13-4). A wheel that is statically out of balance is heavier in one section. This will cause the bubble in the center of the balancer to move off center. To balance the wheel, weights are added until the bubble returns to the center.

To dynamic-balance or "spin-balance" a wheel, it is run at high speeds either on or off the car. Figure 13-5 shows an electronic wheel balancer being used to balance

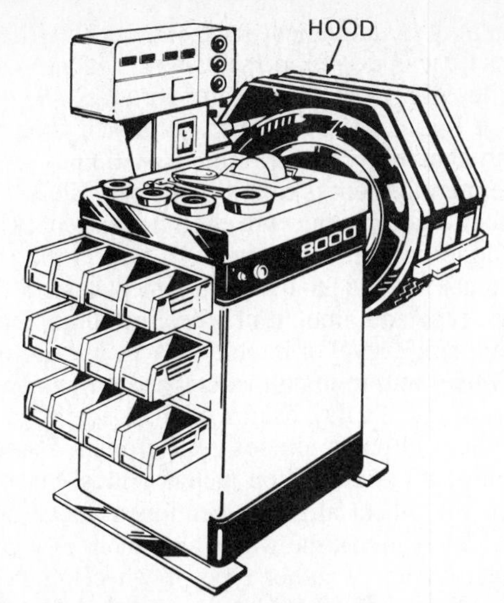

Fig. 13-6 Wheel balancer with a safety hood installed over the tire. (*Stewart-Warner Alemite*)

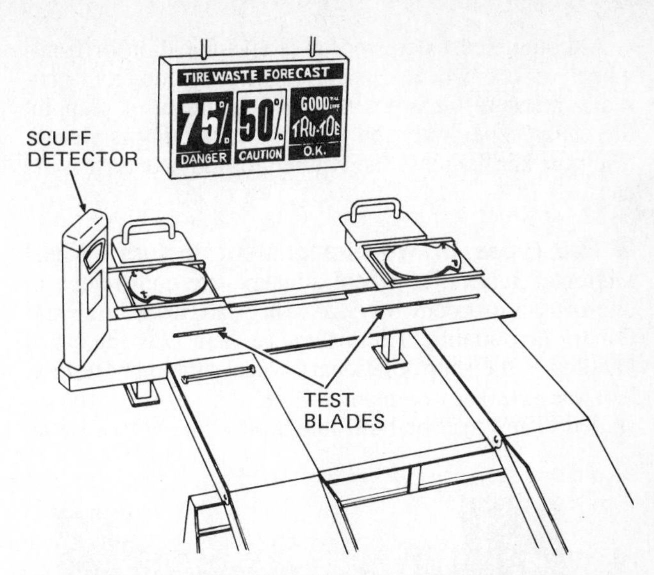

Fig. 13-7 A scuff detector with metal blades, mounted on an alignment rack. (*Bear Division of Applied Power, Inc.*)

a wheel on the car. Lack of balance shows up as a tendency of the wheel to move off center or out of line. If a wheel is out of balance, one or more weights are installed on the wheel rim.

CAUTION: **To prevent injury caused by stones thrown out of the spinning tire, off-the-car wheel balancers should have a safety hood (Fig. 13-6). This hood fits around or over the tire while it is spinning to catch any stones that fly from the tire tread.**

Wheel-alignment testers

13-6 Wheel-alignment testers Numerous devices have been used to check wheel alignment. These have included rolling the car along lines marked on the floor and over strips of toilet tissue. Other methods have used tape measures and yardsticks. However, today, these methods are not accurate enough. To get the accuracy needed, some type of wheel-alignment tester must be used that includes meters or gauges.

Various types of wheel aligners available include mechanical, light-beam, optical, and electronic aligners, and various combinations of these types. However, regardless of how the wheel aligners vary in complexity and construction, they all check the same basic wheel-alignment angles.

13-7 Scuff-detectors Special scuff detectors give evidence of an improper toe setting when the car is driven over them. A scuff tester usually indicates how much the tires are being dragged sideways, in feet per mile of car travel. This is an indication of whether the car has excessive toe-in or toe-out. Several types of scuff detectors are available.

The scuff tester shown in Fig. 13-7 has a pair of test blades over which the front wheels of the car are driven. If the wheels toe in or out excessively, they will exert a sideways push on the blades. The amount of sideways push on the blades is then registered on the tester.

On some scuff testers a series of indicator lights show whether alignment is needed. A green light indicates that the alignment is okay. Yellow, or caution, indicates that alignment is needed to reduce excessive tire wear. Red indicates that the misalignment is so bad that the wheels should be aligned immediately. On some scuff detectors, a record of the tester reading can be punched on a card by pressing a button. Then the card can be given to the customer.

To use a scuff detector, drive the front wheels in the straight-ahead position over the tester blades or plates. Then record the misalignment reading. Back up the car so that the front wheels roll back over the plates or blades. Then drive the car forward again and record the reading. The first reading indicates the amount of misalignment. The difference between the first and second readings indicates the amount of play or looseness in the steering system. If the difference is excessive, then a complete front-end check is needed to find the cause.

Rear wheels can also be checked for misalignment with a scuff detector (Fig. 13-8). Excessive mialignment indicates a bent rear axle or housing, a misaligned spring, or worn or bent attachment parts or control arms.

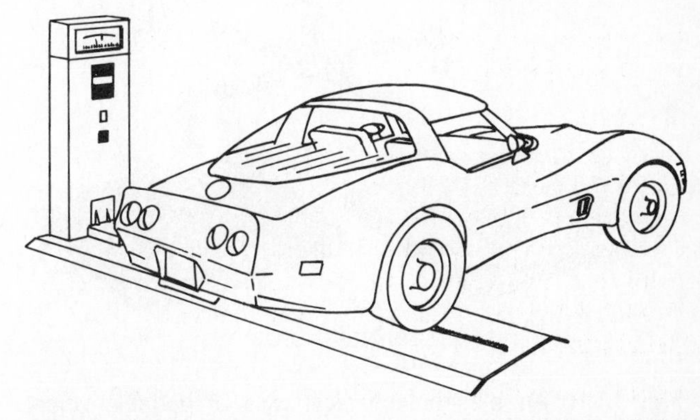

Fig. 13-8 Rear-wheel toe can be checked with a scuff detector. (*ATW*)

Although scuff detectors give an indication of front-wheel or rear-wheel misalignment, they do not accurately measure the amount of misalignment or pinpoint the cause. They only tell you that something is wrong. Then further diagnosis is required to find and correct the cause.

❁ 13-8 Types of wheel-alignment testers Wheel alignment testers, or wheel aligners, are constructed in one of two different ways. A wheel aligner can be stationary or portable. The difference is in how the car is handled in the shop. Stationary wheel-alignment testers, like the one shown in Fig. 13-9, are permanently installed. They may be built into a pit along with a frame straightening or alignment rack. To check wheel alignment, the car is driven to the tester.

Fig. 13-9 A light-beam type of wheel aligner, installed as a stationary pit unit. (*Hunter Engineering Company*)

Mobile alignment testers are portable, like the one shown in Fig. 13-10. They can be moved from place to place so that the car can remain stationary while the alignment equipment is rolled to the car. However, even with mobile equipment, wheel-alignment checking usually is done only in a certain work area in the shop.

There are several kinds of alignment testers that will show the type and amount of wheel misalignment. They vary from spirit-level or bubble devices that are mounted on the wheel hub to instruments that are mounted on the wheel rim and display beams of light against a screen. Other wheel aligners are fully electronic. These report the alignment conditions on meters almost instantly.

All of the wheel aligners mentioned above are static aligners. They check the wheel alignment of a car while it is not moving. Another type of wheel aligner is the dynamic aligner (❁ 13-16). It checks the alignment while the wheels are being spun by motors in the tester. Meters in the tester show the actual running alignment. On some dynamic aligners, adjustments can be made to the alignment while the car wheels are spinning. This gives you the opportunity to see instantly what effects the changes you are making have on the alignment angles.

❁ 13-9 Measuring devices for wheel toe Various types of measuring devices have been used to check wheel toe setting. Many toe testers are basically very accurate tape measures, as can be seen in Fig. 13-11. These are called mechanical toe testers. Other types of toe testers

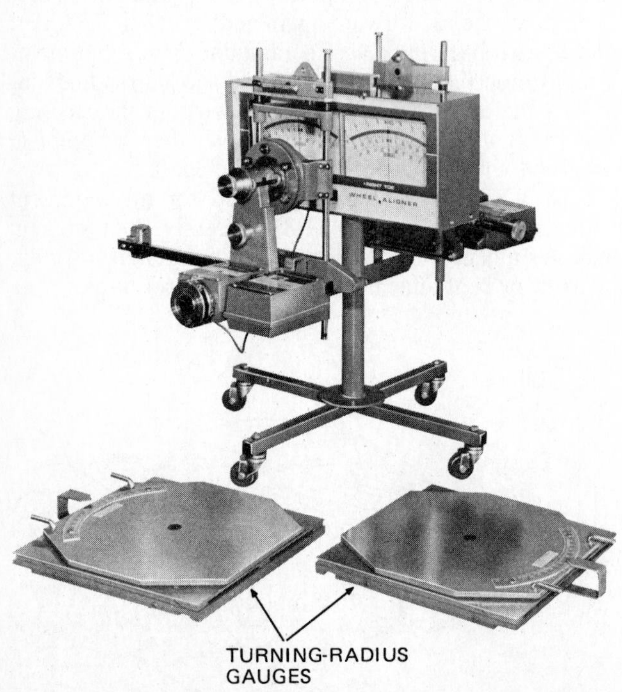

Fig. 13-10 An electromechanical type of portable wheel aligner. The toe reading is taken electronically and displayed on the meters. All other readings are taken mechanically. (*Hunter Engineering Company*)

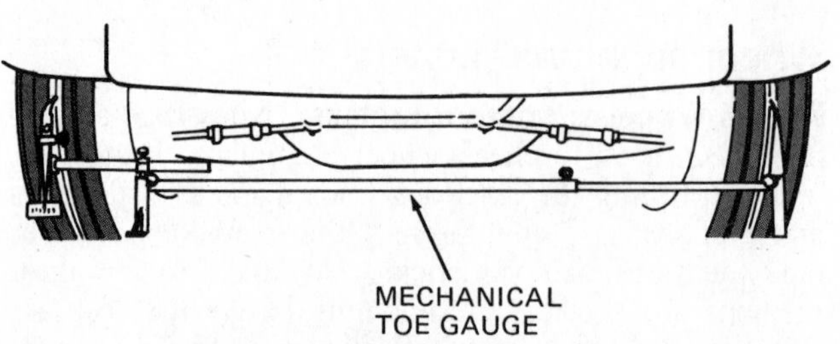

Fig. 13-11 Toe-in gauge for measuring wheel toe. (*Bear Division of Applied Power, Inc.*)

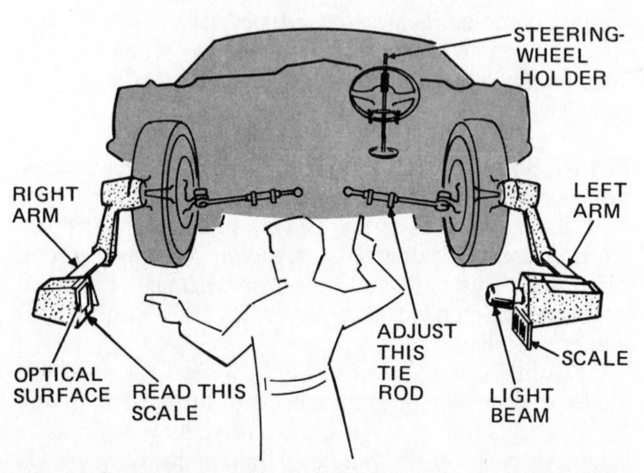

Fig. 13-12 Checking the toe with an optical toe gauge. (*Snap-on Tools Corporation*)

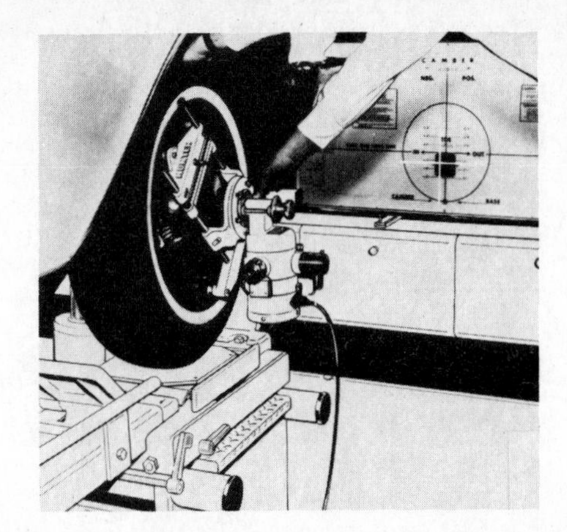

Fig. 13-13 Adjusting the projector-head adapter to compensate for lateral runout in the wheel, prior to checking alignment angles. Note the horizontal and vertical lines of light shining on the screen. (*Hunter Engineering Company*)

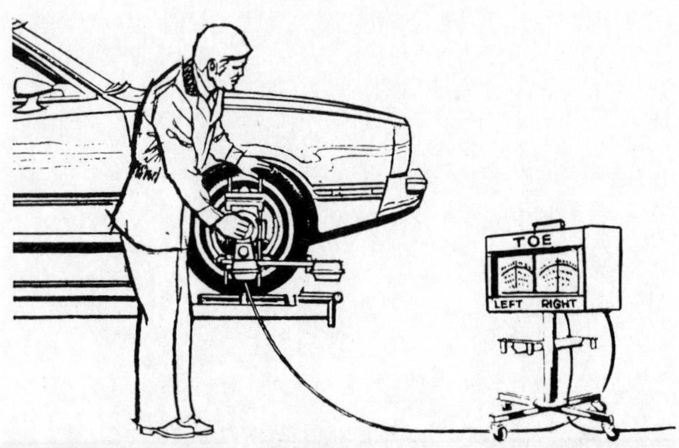

Fig. 13-14 Electronic toe meters. (*Hunter Engineering Company*)

are light-beam testers (Fig. 13-9) and electronic testers (Fig. 13-10).

Figure 13-12 shows the use of an optical toe gauge. This type of gauge has a light beam that is projected onto a mirror and then onto a small screen that has a scale on it. When the wheels are parallel, there is zero toe. The light beam shines against the zero marking on the toe scale. If the wheels toe in or out, the light beam indicates the amount on the scale.

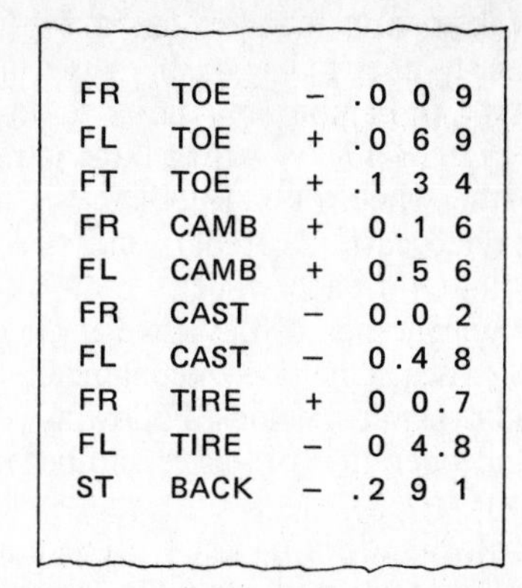

FR	TOE	−	.0 0 9
FL	TOE	+	.0 6 9
FT	TOE	+	.1 3 4
FR	CAMB	+	0.1 6
FL	CAMB	+	0.5 6
FR	CAST	−	0.0 2
FL	CAST	−	0.4 8
FR	TIRE	+	0 0.7
FL	TIRE	−	0 4.8
ST	BACK	−	.2 9 1

Fig. 13-16 A printed tape showing the readings taken while running on a dynamic wheel aligner. (*Hunter Engineering Company*)

With light-beam testers, a projector head, which contains a light that shines a beam straight ahead, is attached to an adapter mounted on the wheel rim. The light beam is projected against a sliding wall chart or screen (Fig. 13-13). The position of the beam indicates the alignment angles.

Electronic toe gauges are also available. An adapter is fitted to each wheel rim (Fig. 13-14). Then a sensor is mounted on the adapter. The toe reading is displayed electronically on toe meters, one for each wheel. This type of gauge also measures steering setback and allows for this measurement in the toe reading (❁ 13-24).

A toe reading can be taken almost instantly by the dynamic wheel aligner shown in Fig. 13-15. This type of aligner has two sets of motorized rollers that spin the wheels. Adjustments can be made while the wheels are turning. The readings shown on the meters are actual running toe, caster, and camber. A printed tape of the readings may be obtained by pressing a button (Fig. 13-16).

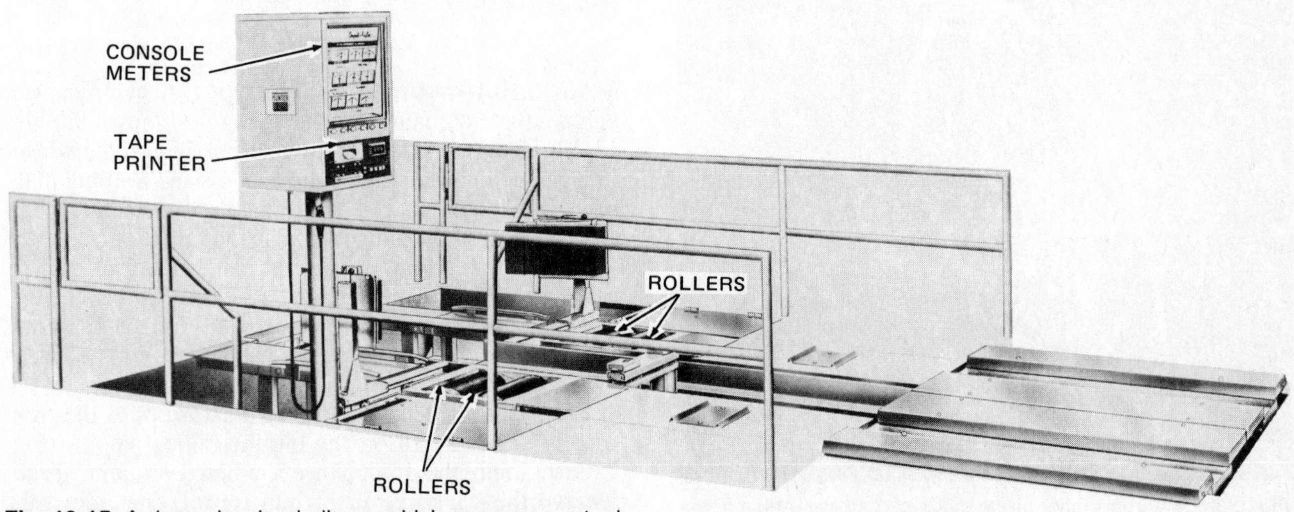

Fig. 13-15 A dynamic wheel aligner which measures actual running toe, camber, and caster. (*Hunter Engineering Company*)

⚙ 13-10 Camber and caster testers Camber and caster are usually checked with the same instrument or gauge. One type of camber and caster tester is the *magnetic gauge* (Fig. 13-17). A strong built-in magnet holds the gauge to the wheel hub. Bubble-level indicators in the gauge provide caster, camber, and steering-axis inclination readings for each wheel.

There are many types of devices for checking caster and camber. These include mechanical, light-beam, electronic, and dynamic methods. Following sections cover installation and operation of caster-camber testers.

⚙ 13-11 Turning-radius gauges Toe-out on turns or turning radius is measured with turntables or *turning-radius gauges* (Fig. 13-10). The turning-radius gauges are placed under the front wheels. Then each wheel is turned 20° and the angle of the other wheel is measured. If the toe-out on turns is not within specifications, one of the steering arms or tie rods is bent.

⚙ 13-12 Using camber-caster testers Camber-caster testers mount either to the wheel hub (Fig. 13-17) or to the wheel rim (Fig. 13-14). They may be of the mechanical type, with built-in bubble levels, or they may use a light beam or electronic sensors which display the readings directly on meters. The following sections describe how to mount and read various types of camber-caster testers.

Before camber and caster can be checked, all of the prealignment checks and adjustments to the car must be made (⚙ 13-1 to 13-5). These include checks on the tires, wheel bearings, wheels, ball joints, springs, and steering linkage. Any out-of-specification condition must be corrected. On some cars, front-suspension height must also be checked and corrected before caster and camber are checked.

When using any type of camber-caster tester, follow the operating instructions for the tester. In addition, follow any special instructions or procedures that may apply to the car you are checking.

Fig. 13-17 A magnetic-gauge type of camber and caster tester. This gauge also measures steering-axis or kingpin inclination. (*Snap-on Tools Corporation*)

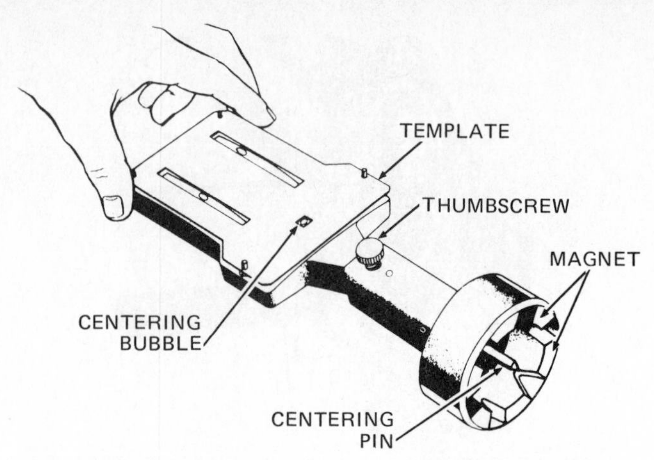

Fig. 13-18 Magnetic gauge using a template. The tapered pin centers the gauge on the end of the spindle, and the series of strong magnets holds the gauge to the hub. (*Bear Division of Applied Power, Inc.*)

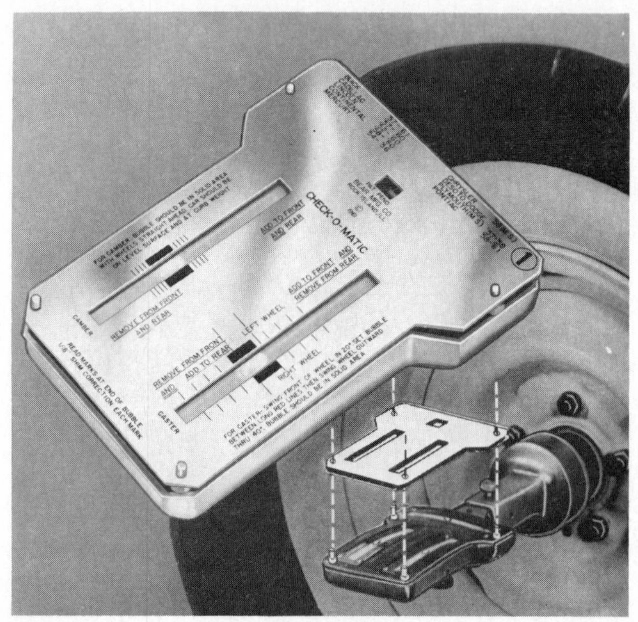

Fig. 13-19 A template being placed on a bubble type of magnetic camber-caster gauge. (*Bear Division of Applied Power, Inc.*)

⚙ 13-13 Hub-mounted bubble type A hub-mounted, bubble type of camber-caster gauge is shown in Fig. 13-18. Figure 13-17 shows it being placed on the wheel. The gauge contains three bubble levels and a strong magnet to hold it in place. With this gauge, a series of templates is part of the gauge set. Each template is calibrated to apply to a certain year, make, and model of car. As new car models come out, and as specifications change, new templates are made available. Figure 13-19 shows a template in position above the gauge, ready to be placed over the locating pins.

To use the gauge, roll the car forward until the front wheels are centered on the turning-radius gauges (Fig. 13-10). With the front wheels pointed straight ahead, remove the wheel cover or hub cap. Then remove the dust cap from the hub and wipe off any excess grease from the end of the spindle. Clean the face of the wheel

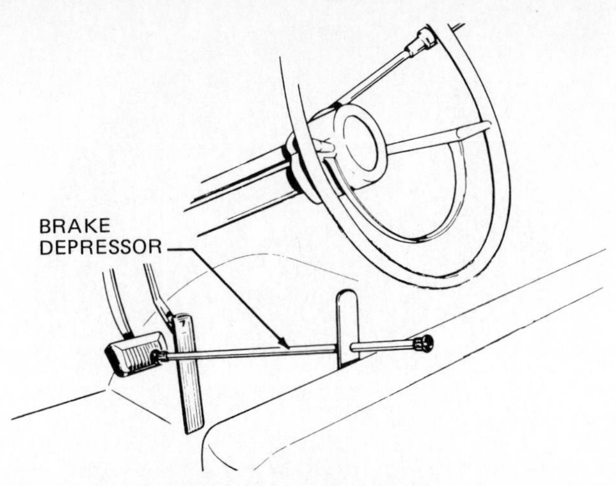

Fig. 13-20 Lock all four wheels by holding the brake pedal down with a brake depressor. (*Hunter Engineering Company*)

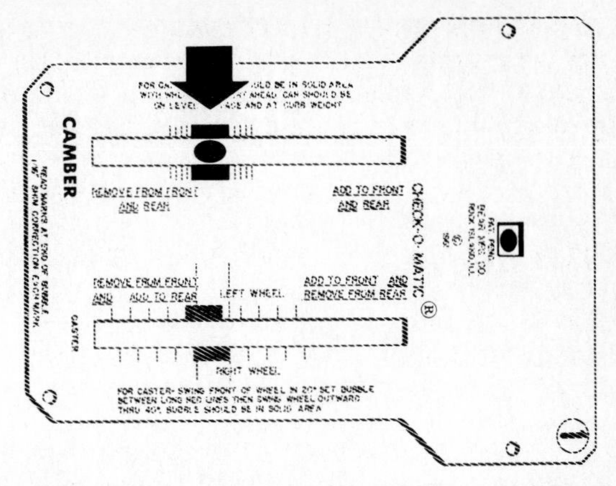

Fig. 13-21 If camber is correct, the bubble will lie entirely within the solid area of the camber scale. (*Bear Division of Applied Power, Inc.*)

hub. Install the brake-pedal depressor, so that the brake pedal is held down to lock all four wheels (Fig. 13-20).

Select the template for the year, make, and model of the car being checked. Place the template on the gauge as shown in Fig. 13-19. Then, install the gauge on the wheel hub as in Fig. 13-17. Center the gauge on the spindle with the centering pin and turn the gauge until it is horizontal. This is shown when the centering bubble is centered in the small window closest to the wheel hub, shown in Fig. 13-18.

1. **Checking camber** Camber can now be checked by noting the location of the camber bubble in relation to the markings on the template. Figure 13-21 gives an example of the template. The bubble should lie entirely within the solid area of the camber scale on the template. If the bubble is not entirely within the solid area, the amount that the bubble is outside determines the amount and direction of adjustment required.

For example, assume you are checking a car in which the camber adjustment is made by shims. If the end of the bubble lies two graduations or marks outside the solid area, the camber adjustment must be changed by two $\frac{1}{16}$-inch [1.6-mm] shims. If the bubble is on one side, the shims must be added. If it is on the other side, the shims must be removed. Later sections describe how to make camber adjustments on various cars.

2. **Checking caster** With the gauge in place on the wheel and the wheel pointed straight ahead (Fig. 13-17), set the turning-radius gauge under the wheel to zero. Then turn the front of the wheel in until the turning-radius gauge reads 20°. With the thumbscrew on the gauge (Fig. 13-18), adjust the caster level until the bubble lies squarely between the two long red crosslines on the caster scale.

Turn the wheel back in the opposite direction, stopping it when the turning-radius gauge reads 20°. This is a total wheel swing of 40°. Now the entire bubble should lie in the solid area on the left side of the caster scale when the left wheel is being checked (Fig. 13-22). The bubble should lie entirely in the solid area at the right side of the caster scale when the right wheel is being checked (Fig. 13-23).

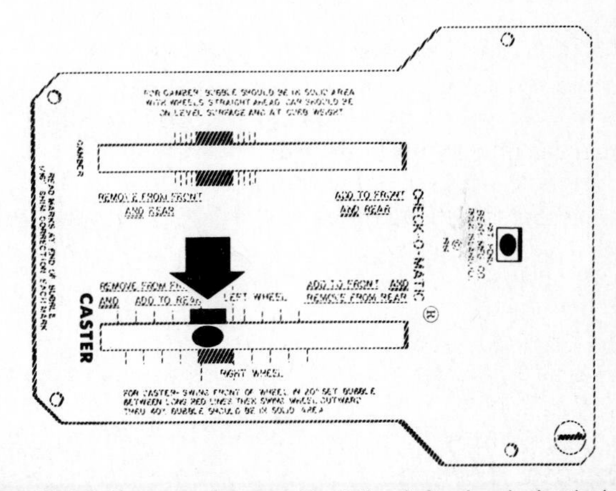

Fig. 13-22 If caster is correct at the left wheel, the bubble will lie entirely within the solid area of the left caster scale. (*Bear Division of Applied Power, Inc.*)

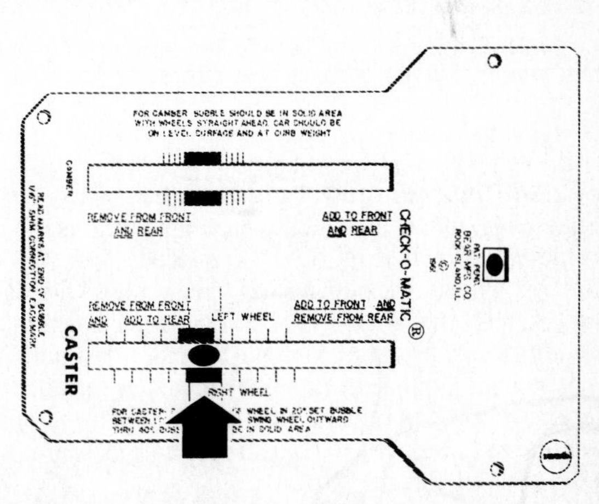

Fig. 13-23 If caster is correct at the right wheel, the bubble will lie entirely within the solid area of the right caster scale. (*Bear Division of Applied Power, Inc.*)

If the bubble location is not correct, the caster is out of specification. The amount and direction that it is off indicates the amount and direction of the adjustment that needs to be made. Caster adjustment is described in later sections.

Fig. 13-24 A rim-mounted type of camber-caster gauge. (*Hunter Engineering Company*)

Fig. 13-25 Compensating the adapter for a bent rim and for wheel runout. (*Hunter Engineering Company*)

Fig. 13-26 A light-beam projector mounted on the wheel rim. (*FMC Corporation*)

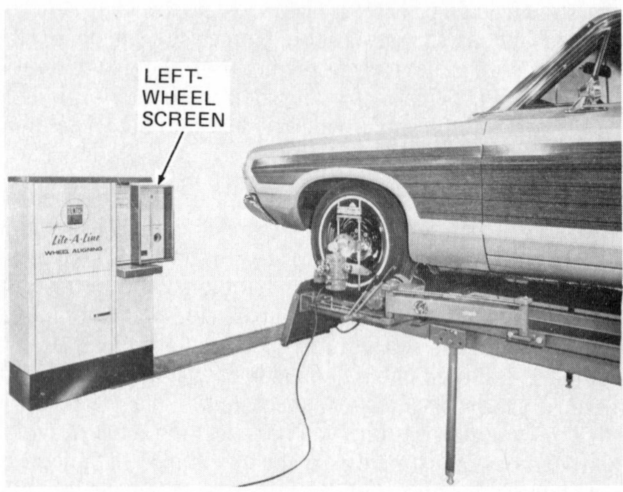

Fig. 13-27 A light-beam wheel aligner includes two light projectors, one attached to each wheel, and a screen in front of each projector. (*Hunter Engineering Company*)

⚙ 13-14 Rim-mounted bubble type This type of alignment gauge is attached to an adapter that is mounted on the wheel rim (Fig. 13-24), instead of on the hub. For most wheels, the adapter has three screws with threads that catch on the rim lip. Then the adapter is lengthened slightly to securely lock it to the rim. For cast aluminum wheels that have no rim lip, the adapter is installed with the screws on the outside of the rim. Then the adapter is shortened slightly, so that the locking action is from the outside of the rim inward.

Most rim adapters have a runout compensator. By turning the screws on the compensator (Fig. 13-25), corrections can be made for bent rims, wheel runout, and a bent adapter. When the adapter is properly compensated, the adapter spindle has the same centerline as the center of the wheel. The bubble levels are then mounted to an arm that is parallel to the wheel.

After being attached to the adapter, and with the wheels pointed straight ahead, the camber knob is turned until the camber level is centered (Fig. 13-25). Camber is then read from a scale around the camber knob at the end of the gauge. The procedure for checking caster is similar to the steps outlined for using the hub-mounted gauge (⚙ 13-13).

⚙ 13-15 Light-beam type There are several variations of the light-beam type of alignment tester, one of which is shown in Fig. 13-26. Basically, a light-beam wheel aligner includes two light projectors, one attached to each wheel. To display the lights, there must be screens straight ahead of them (Figs. 13-9 and 13-27). These sometimes are in the back of cabinets or in shadow boxes for easier viewing.

Figure 13-13 shows a projector attached to the right front wheel. A light inside the projector is directed to a

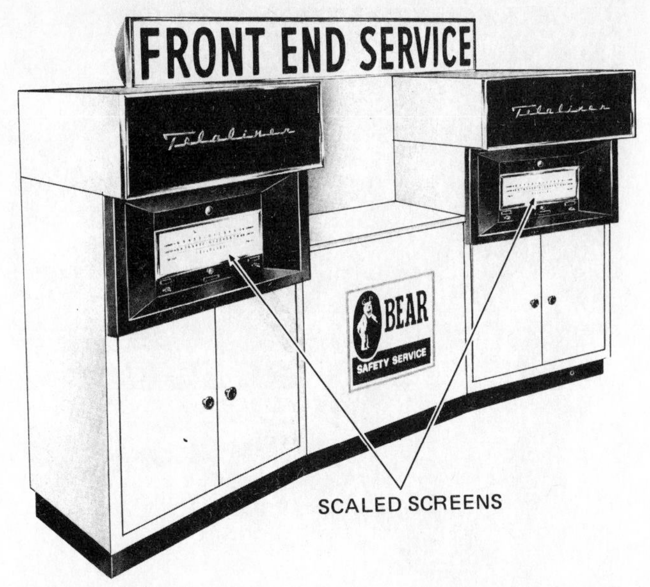

Fig. 13-28 As the wheels move, lines of light move across the screens, showing the alignment angles. (*Bear Division of Applied Power, Inc.*)

mirror on the wheel-mounted gauge. The light passes through aiming lenses and then strikes the screen in front of the car. The position of the spot or lines of light indicates the camber and caster angles. Checking procedures are similar to those for bubble-type gauges (⚙ 13-13 and 13-14).

The projector shown in Fig. 13-13 throws the lines onto a screen that has a scale on it. However, the projector shown in Fig. 13-26 projects the entire chart onto a blank screen. Another variation of the light-beam type of wheel aligner has sensors mounted at each wheel. An electric cable connects the sensors to scaled screens in a cabinet in front of the car (Fig. 13-28). As the wheel is moved, a line of light moves across the screens to indicate the alignment angles.

⚙ 13-16 Dynamic-type wheel-alignment testers The dynamic type of floor-roller alignment tester is the most advanced type of electronic wheel aligner (Fig. 13-15). Figure 13-29 shows the rollers on which the tires ride. There is a pair of rollers for each tire. Each roller pair is free to tilt sideways and also swing through the horizontal plane.

In operation, the rollers are driven by motors so the wheel spins at a speed of about 10 mph [16 km/h]. This is fast enough for the tire to assume its normal running position. The rollers align with the tire tread, and their position is sensed and displayed on the console meters. The meters report camber and caster of each wheel as well as toe and other angles. Pushing the control buttons and moving the steering wheel are the only operations required.

The console for one type of dynamic aligner is shown in Fig. 13-15. It includes the meters and printer. This console can print out the meter readings on a paper tape whenever the "printout" button is pushed (Fig. 13-16). This gives a permanent record of the readings.

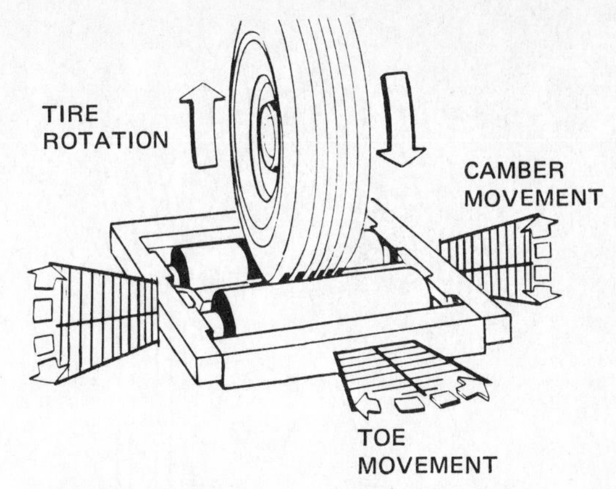

Fig. 13-29 Movements of the rollers in a dynamic wheel aligner. (*Hunter Engineering Company*)

A dynamic analyzer has the advantage of taking the alignment angles directly from the tire-tread contact with the road surface. All other aligners, whether hub-mounted or wheel-mounted, take the readings from the spindle.

Checking alignment angles

⚙ 13-17 Methods for adjusting camber and caster Several different ways to adjust camber and caster have been provided for in different models of cars. The methods include adjustment by removing or installing shims, by turning a cam, by shifting the inner shaft, and by shortening or lengthening the strut rod. Each method is discussed below.

1. Installing or removing shims The adjustment shims are located at the upper-control-arm shafts. They are placed either inside or outside the frame bracket. Figure 13-30 shows the location of the shims in many General Motors cars. In this suspension, the shims are inside the frame bracket. Figure 13-31 shows the location of the shims in many Ford cars, outside the frame bracket. When the

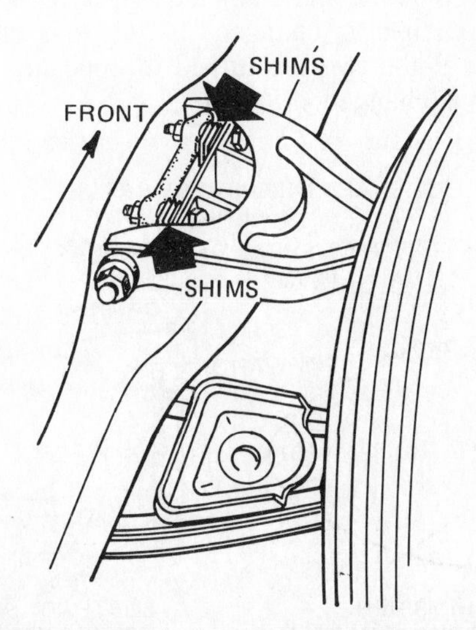

Fig. 13-30 Location of caster and camber adjusting shims (indicated by heavy arrows) on many General Motors cars. Note that the shims and upper-control-arm shaft are inside the frame bracket. (*Snap-on Tools Corporation*)

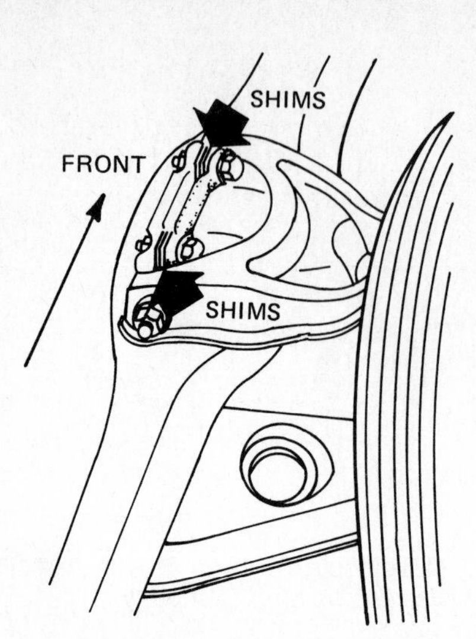

Fig. 13-31 Location of caster and camber adjusting shims (indicated by heavy arrows) on many Ford and other cars. Note that the shims and upper-control-arm shaft are outside the frame bracket. (*Snap-on Tools Corporation*)

shims are inside the frame bracket (Fig. 13-30), adding shims moves the upper control arm inward. This reduces positive camber. When the shims and shaft are outside the frame bracket (Fig. 13-31), adding shims moves the upper control arm outward. This increases positive camber. If shims are added at one of the attachment bolts and removed from the other, the outer end of the upper control arm shifts one way or the other. This increases or decreases caster. Figure 13-32 shows these adjustments.

2. Turning a cam Different cars have had several variations of this adjustment method. Figure 13-33 shows an arrangement used on some GM big cars with front-wheel drive. The two bushings at the inner end of the upper control arm are attached to the frame brackets by two attachment bolts and cam assemblies. When the cam bolts are turned, the camber and caster are changed. If both are turned the same amount and in the same direction, the camber is changed. If only one cam bolt is turned, or if the two are turned in opposite directions, the caster is changed.

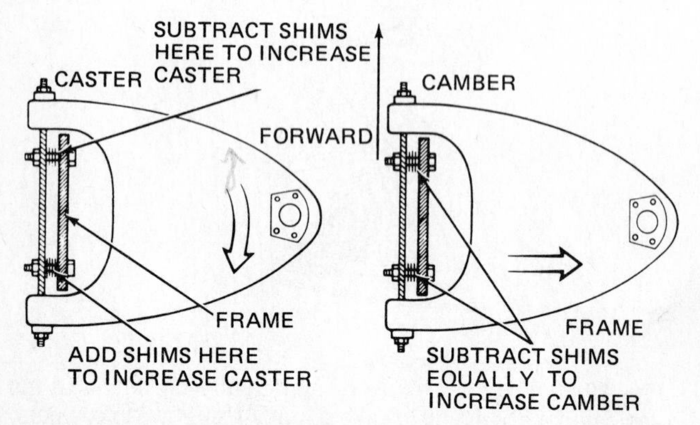

Fig. 13-32 Caster and camber adjustments on some cars using shims. (*Chevrolet Motor Division of General Motors Corporation*)

Fig. 13-33 Turning the cam bolts moves the upper control arm toward or away from the frame to adjust caster and camber. (*Cadillac Motor Car Division of General Motors Corporation*)

Fig. 13-34 Adjusting caster and camber by shifting the position of the inner shaft using slots in the frame. (*Chrysler Corporation*)

One complete rotation of a cam bolt represents the full amount of possible adjustment. Always recheck the final caster and camber settings after tightening the cam-bolt nuts.

3. Shifting inner shaft This system uses slots in the frame at the two points where the inner shaft is attached (Fig. 13-34). When the attaching bolts are loosened, the inner shaft can be shifted in or out to change camber. To change caster, only one end is shifted.

4. Changing length of strut rod Figure 13-35 shows the adjustment points for this system. This arrangement allows changing the caster by changing the length of the

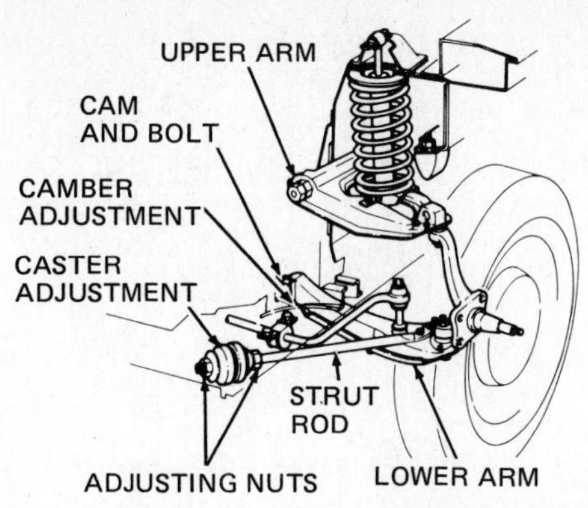

Fig. 13-35 Adjusting caster by changing the length of the strut rod. (*Ford Motor Company*)

strut rod. Camber is changed by turning the cam on which the inner end of the lower control arm is mounted.

13-18 MacPherson-strut camber and caster adjustments Some cars with MacPherson struts do not have any adjustment for camber or caster. Other cars with this type of suspension have a camber adjustment only. On these cars, turn the cam bolt at the lower end of the strut to move the top of the wheel in or out, adjusting the camber (Fig. 13-36).

Wrong alignment settings can cause excessive tire wear. Therefore, a caster-camber adjustment kit is available for installation on some cars with nonadjustable MacPherson struts (Fig. 13-37). After the kit is installed, caster and camber can be adjusted by moving the top of the strut. The kit includes a slotted plate that is installed between the strut and the inner fender. Unless a kit of this type is installed, the top of the strut is fixed in position because it is bolted through the holes in the inner fender.

13-19 I-beam front-axle camber and caster adjustments On the straight, solid, or I-beam front axle, camber and steering-axis inclination can be adjusted by bending the axle on a frame or alignment rack. However, air or hydraulic jacks and a variety of adapters are needed. Correction can be made by bending only when both camber and steering-axis inclination are off by the same amount. When one is off more than the other, the wheel spindle has been bent and should be replaced. Then the axle may require additional bending to bring the two angles within specifications.

Caster adjustment on a straight axle can be made by inserting tapered caster shims between the spring seat on the axle and the spring (Fig. 13-38), or by bending the axle. In general, when caster variation between wheels is 1° or less, a caster shim may be installed at only one wheel to correct it. If caster variation exceeds 1°, the

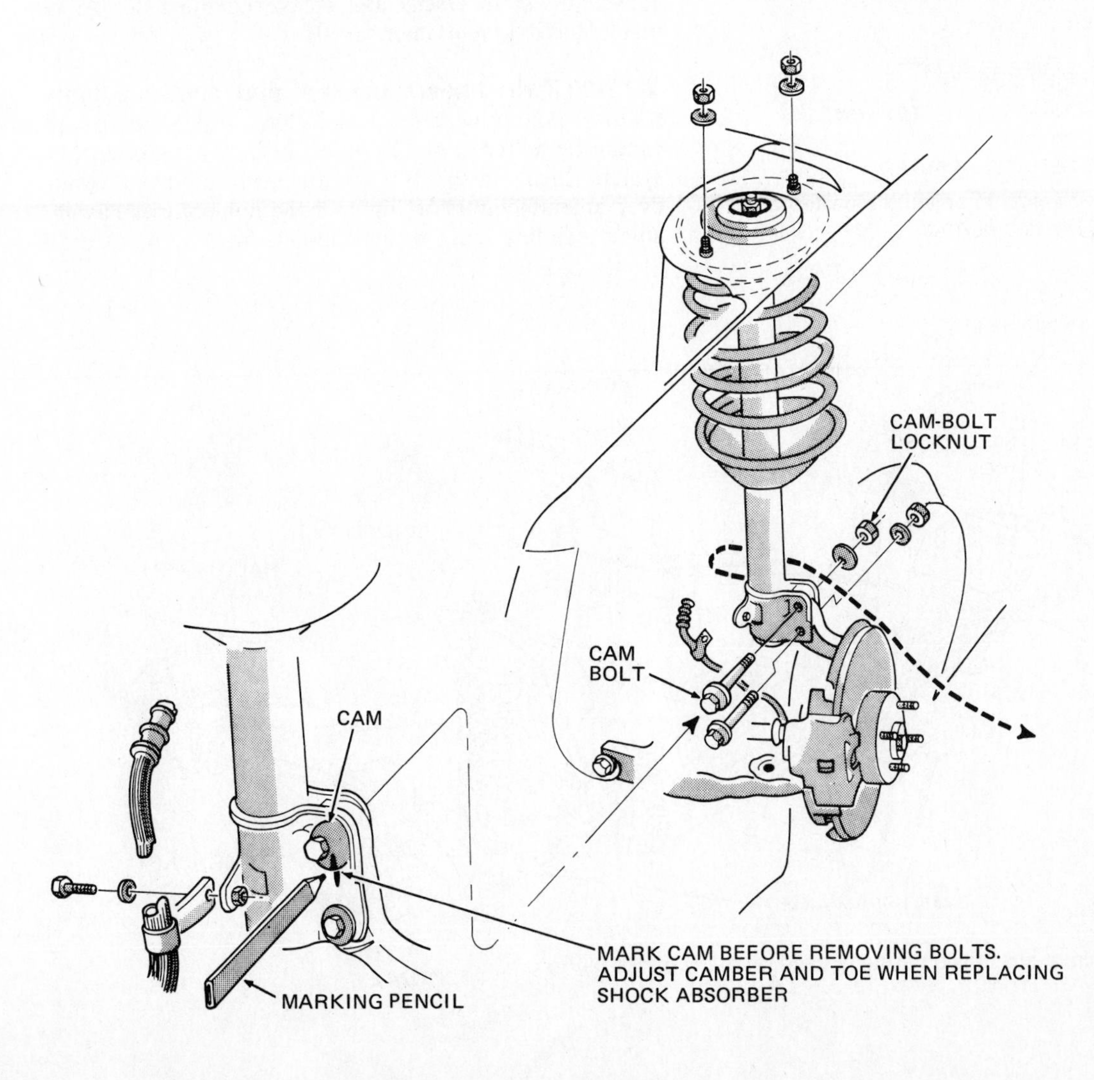

Fig. 13-36 On some cars with MacPherson struts, camber is adjusted by turning a cam bolt. (*Chrysler Corporation*)

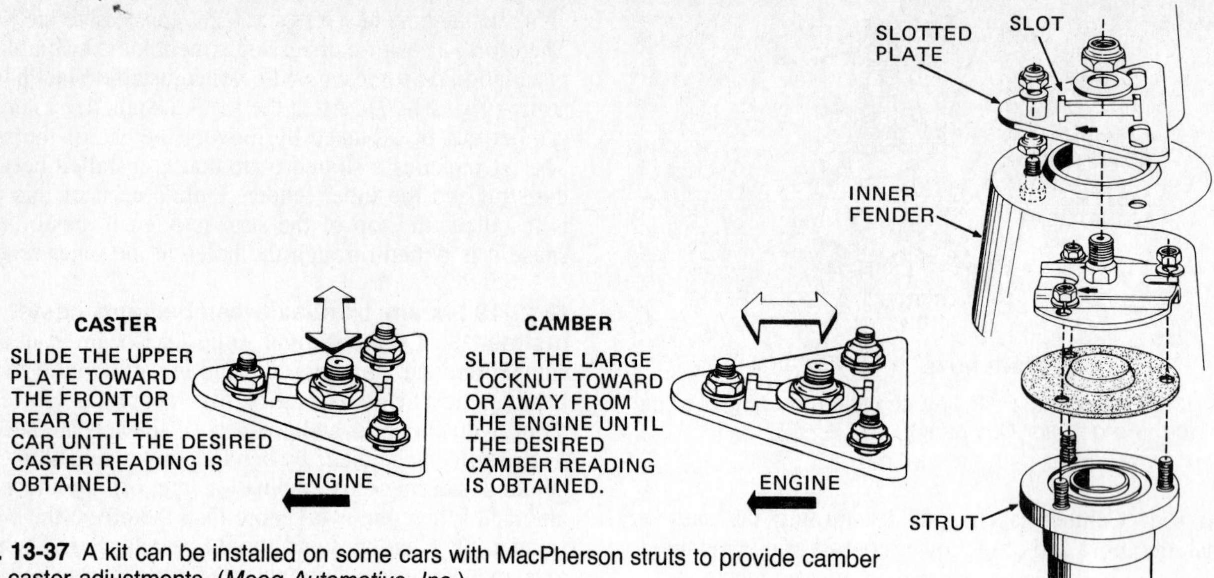

Fig. 13-37 A kit can be installed on some cars with MacPherson struts to provide camber and caster adjustments. (*Moog Automotive, Inc.*)

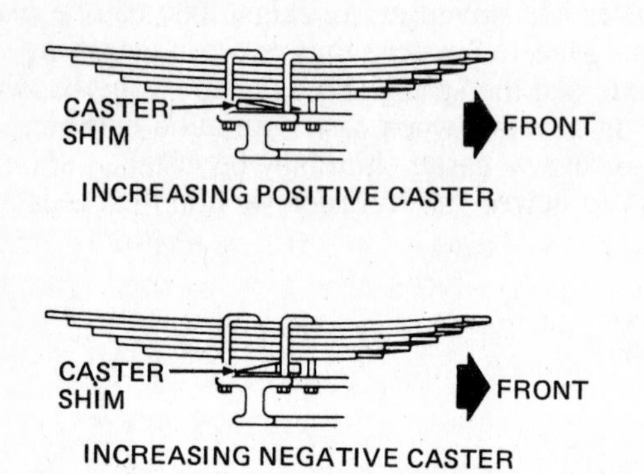

Fig. 13-38 Using tapered shims to adjust the caster on an I-beam front axle. (*Bear Division of Applied Power, Inc.*)

caster should be adjusted by bending the axle or by replacing it.

NOTE: The I-beam axle should be replaced if excessive bending is required. A bad bend in an axle may have caused invisible cracks and weakened the axle. As a result, the axle may later fail.

13-20 Twin I-beam camber and caster adjustments According to the manufacturer, camber and caster cannot be adjusted on the twin I-beam front-suspension system (Fig. 13-39). Only the toe can be adjusted. However, manufacturers of alignment equipment make available a bending tool and attachments that can be used to

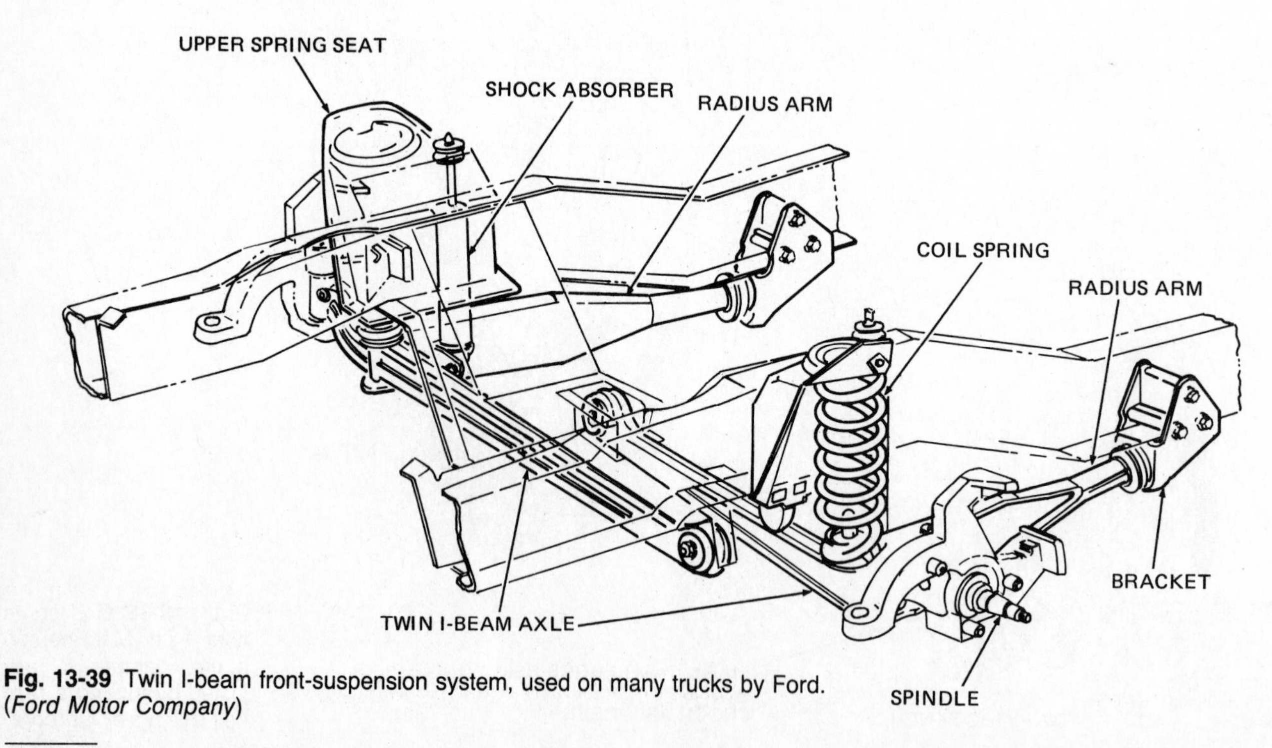

Fig. 13-39 Twin I-beam front-suspension system, used on many trucks by Ford. (*Ford Motor Company*)

Fig. 13-40 Adjusting caster on a truck with twin I-beam front suspension. (*ATW*)

bend the axles or the radius arms. Bending the axle for each wheel will correct improper camber and steering-axis inclination. The caster can be corrected by bending the radius arm (Fig. 13-40).

13-21 Toe adjustments An incorrect toe setting can cause rapid tire wear. Therefore, manufacturers have designed most automobile suspension systems so that the toe can be adjusted. However, steering-axis inclination and turning radius are not adjustable angles. If caster and camber are correct, then steering-axis inclination and turning radius should also be correct. If not, then one or more suspension parts are bent. To get steering-axis inclination and turning radius back within specifications, replace the defective parts.

After camber and caster have been adjusted, toe can be measured and adjusted. Measurement of toe is shown in Fig. 10-9. Some of the various testers that can be used to measure toe are shown in Figs. 13-9 to 13-14. The adjustment procedure is the same on most cars.

Resetting toe may affect the position of the steering wheel. To align the steering wheel, place the front wheels in the straight-ahead position. Then check the position of the spokes in the steering wheel. If they are not aligned, Fig. 13-41 shows how they can be properly positioned while setting toe. Toe is adjusted by turning the adjusting sleeves in the steering linkage.

Figure 13-42 shows a typical steering linkage. To adjust the toe-in, change the effective lengths of the two tie rods. This can be done by first loosening the clamp bolts on the adjusting sleeves, then turning the sleeves as required. The sleeves and the matching ends of the rod and tie-rod end have right-hand and left-hand threads. Turning the sleeve in one direction increases the effective length of a tie-rod, and turning the sleeve in the other direction shortens the length.

Adjustment of the sleeves can also affect the steering-wheel position in straight-ahead driving. For example, if both sleeves are turned in the same direction and by the same amount, the toe-in will remain unchanged but the steering-wheel position will be changed. This is the procedure to use for aligning the spokes in the steering wheel (Fig. 13-41).

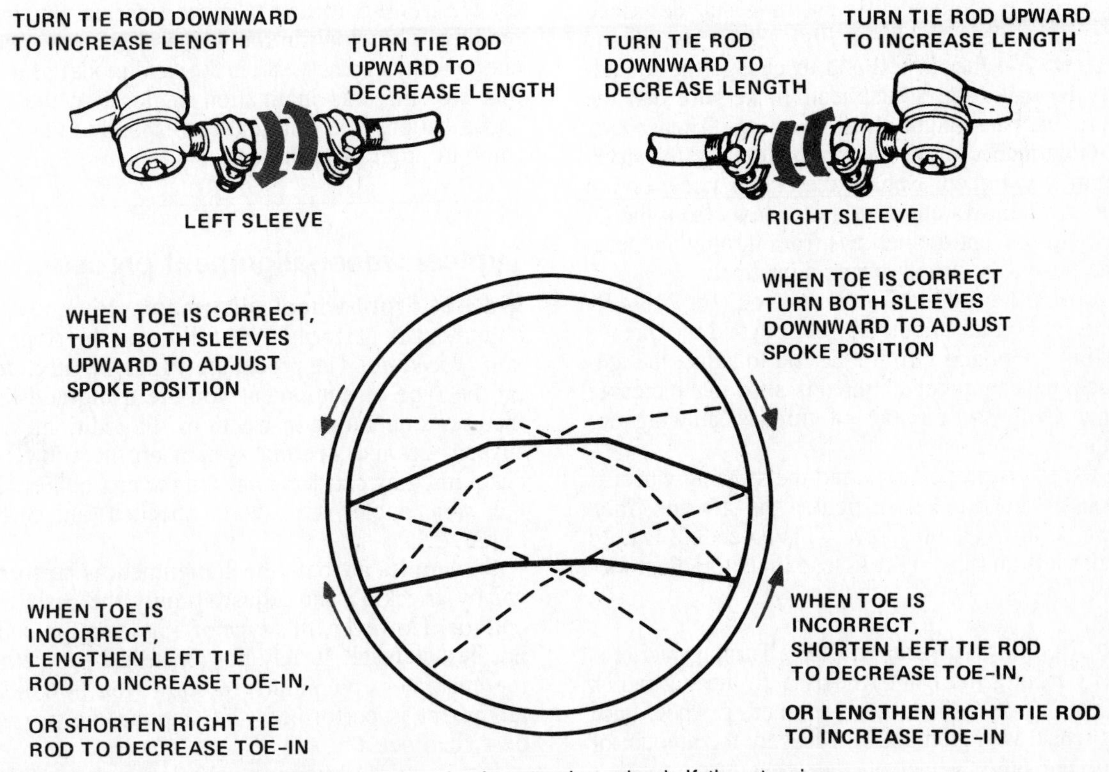

Fig. 13-41 Adjusting toe and aligning the spokes in the steering wheel. If the steering wheel is in its proper position, adjust both tie rods equally to maintain the position of the spokes. (*Ford Motor Company*)

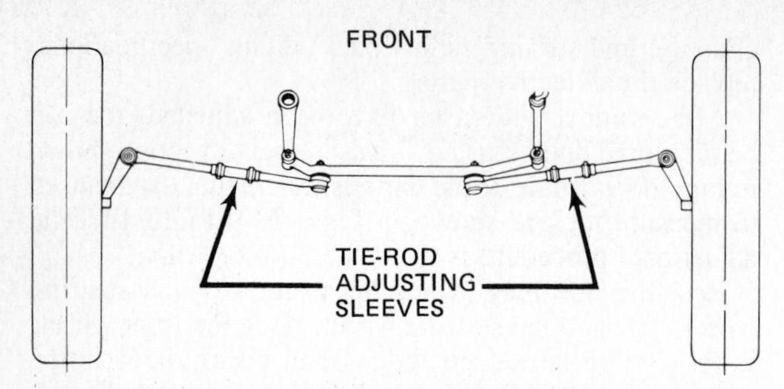

Fig. 13-42 Turn the tie-rod adjusting sleeves to set the toe. (*Chevrolet Motor Division of General Motors Corporation*)

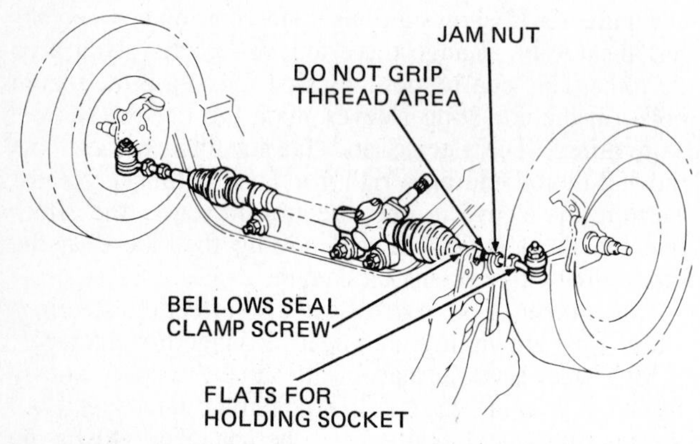

Fig. 13-43 Adjusting toe on a car with rack-and-pinion steering. (*Ford Motor Company*)

The procedure for adjusting toe on cars with rack-and-pinion steering is slightly different than that described above. In the system shown in Fig. 13-43, the tie rod itself is rotated in the tie-rod end to change the toe adjustment. To make the adjustment, make sure that the wheels are in the straight-ahead position. Then check that the index marks on the steering wheel and the steering column are aligned. Loosen the clamp screw on the rubber tie-rod bellows and free the bellows from the tie rod. This will prevent the bellows from turning and tearing as the tie-rod is turned during adjustment.

Then loosen the jam nut on the tie rod (Fig. 13-43). Grip the tie rod on its smooth surface beyond the threads with locking pliers and turn the tie rod to adjust the toe-in. Reducing the number of threads showing increases toe-in Increasing the number of threads showing decreases toe-in.

When the toe is properly set and the steering wheel is in the straight-ahead position, tighten the jam nut. Then tighten the bellows clamp screw. This procedure is used on both the left and the right tie-rod ends to adjust toe-in.

13-22 Checking turning radius Turning radius is checked by rolling the front wheels onto turning-radius gauges (Fig. 13-10). Then turn the steering wheel until one front wheel is angled at 20°. Measure the amount of turning of the other wheel on the other turning-radius gauge. A typical specification for a car with power steering is that when the inside wheel is turned 20°, the outside wheel should have turned 17°30′. For a car with manual steering, the outside wheel should have turned 17°21′.

NOTE: This may be written as 17 degrees, 21 minutes. In wheel-alignment angles, the term *minutes* is often used. When one degree is divided into 60 parts, each part is called one minute, abbreviated 1′.

If the turning radius is not correct, the car has a bent steering arm that should be replaced. Then wheel alignment must be rechecked. Never straighten a bent steering or suspension part. This could weaken the part and cause it to fail.

13-23 Checking steering-axis inclination Steering-axis inclination is checked with the same equipment used to check caster and camber. With some gauges, the steering-axis inclination can be checked at the same time the caster is being checked. Lock the front brakes. Then adjust the steering-axis inclination scale on the gauge so that the bubble is at zero when the wheel is turned in 20°. Then when the wheel is turned out 20°, for a total swing of 40°, the steering-axis inclination scale will indicate the SAI angle.

If the steering-axis inclination is wrong but other angles are correct, the spindle probably is bent and should be replaced. When both steering-axis inclination and camber are wrong, check for a bent lower control arm and excessive wear in the lower ball joint. Also, the inner bushings in the lower control arm could be worn.

For recreational vehicles, such as a motor home built on a new cab-and-chassis, there often are no published alignment specifications. One way to find specifications for the truck is to first take the steering-axis inclination reading. Then look in the wheel-alignment specification charts for the same year, make, and model of truck using that steering-axis inclination angle. Use the alignment specifications you find that correspond to the SAI on the truck to align the front end.

Typical wheel-alignment procedures

13-24 Front-wheel alignment Basically, a wheel alignment is performed by following the steps in a certain procedure. The procedure varies slightly, depending on the type of equipment you are using and the manufacturer's operating instructions. In addition, the type of suspension and steering system on the vehicle you are checking may cause changes in the procedure. However, the same angles are always checked and corrected as required.

Sometimes a front-wheel alignment is performed on a car by checking and adjusting only the caster, camber, and toe. However, this type of alignment procedure may not be adequate. It may not locate the problem causing the customer complaint. To show you how a complete alignment is performed, the steps in the procedure are described below.

Figure 13-44 shows a wheel-alignment check sheet that is available from Hunter Engineering Company, a manufacturer of many types of wheel-alignment and bal-

WHEEL ALIGNMENT CHECK SHEET

NAME ____________________

ADDRESS ____________________

CITY __________ STATE __________ ZIP __________

PHONE __________ OFFICE PHONE __________

MAKE __________ YEAR __________

MODEL __________ COLOR __________

LICENSE ____________________

MILEAGE __________ DATE __________

PRE-ALIGNMENT INSPECTION

- ☐ NORMAL VEHICLE LOAD
- ☐ TIRE PRESSURES
- ☐ UNEVEN & WORN TIRES
- ☐ WHEEL BEARINGS
- ☐ DRAGGING BRAKES
- ☐ SHOCK ABSORBERS
- ☐ BALL-JOINTS (KING-PINS)
- ☐ IDLER-ARM
- ☐ STABLIZER
- ☐ TIE-RODS & ENDS
- ☐ STEERING ARMS
- ☐ PITMAN ARM
- ☐ SPINDLE
- ☐ BUSHINGS
- ☐ REAR-AXLE
- ☐ STEERING WHEEL

• CHECK ALL WHEELS & TIRES . . .

CONDITION	LF	RF	LR	RR
PRESSURE				
RUN-OUT				
OUT-OF-ROUND				
EXCESSIVE WEAR				
BALANCE				

• TORSION-BAR VEHICLES ONLY . . .

SUSPENSION HEIGHT	LEFT	RIGHT
INITIAL READINGS		
FACTORY SPECS.		
FINAL SETTINGS		

• TRUCKS ONLY . . .

FRAME ANGLE LEFT	FRAME ANGLE RIGHT

SUSPENSION HEIGHT LEFT	SUSPENSION HEIGHT RIGHT

RIGHT SIDE	CAMBER ANGLE	CASTER ANGLE	TOE
• INITIAL READING			
• FACTORY SPEC.			
• FINAL SETTING			
TOTAL TOE			(FINAL SETTING)

STEERING AXIS INCLIN.	INCLUDED ANGLE	TURNING ANGLE	WHEEL SET BACK

LEFT SIDE	CAMBER ANGLE	CASTER ANGLE	TOE
• INITIAL READING			
• FACTORY SPEC.			
• FINAL SETTING			
			PLUS

STEERING AXIS INCLIN.	INCLUDED ANGLE	TURNING ANGLE	WHEEL SET BACK

REMARKS ____________________

HUNTER®

Fig. 13-44 A wheel-alignment check sheet. (*Hunter Engineering Company*)

ancing equipment. When the check sheet is followed, no step can be bypassed, and few problem-causing conditions can be overlooked.

All of the prealignment checks discussed in ✪ 13-1 to 13-5 must be made before the wheel-alignment procedure begins. The items to be checked during the prealignment inspection are listed in the box on the Wheel Alignment Check Sheet in Fig. 13-44. Figure 13-45 shows where to check on typical coil-spring and torsion-bar front-suspension systems.

As you perform the prealignment inspection, you are continually making a series of go/no-go decisions. Each item you inspect must be either okay or not okay. If it is not okay, then it must be corrected by adjustment or replacement before the check of the wheel-alignment angles begins. A vehicle with loose, worn, or broken parts cannot be aligned properly.

When you check the wheels and tires, fill in the information in the box labeled "Check All Wheels and Tires" in Fig. 13-44. Next, if the vehicle is equipped

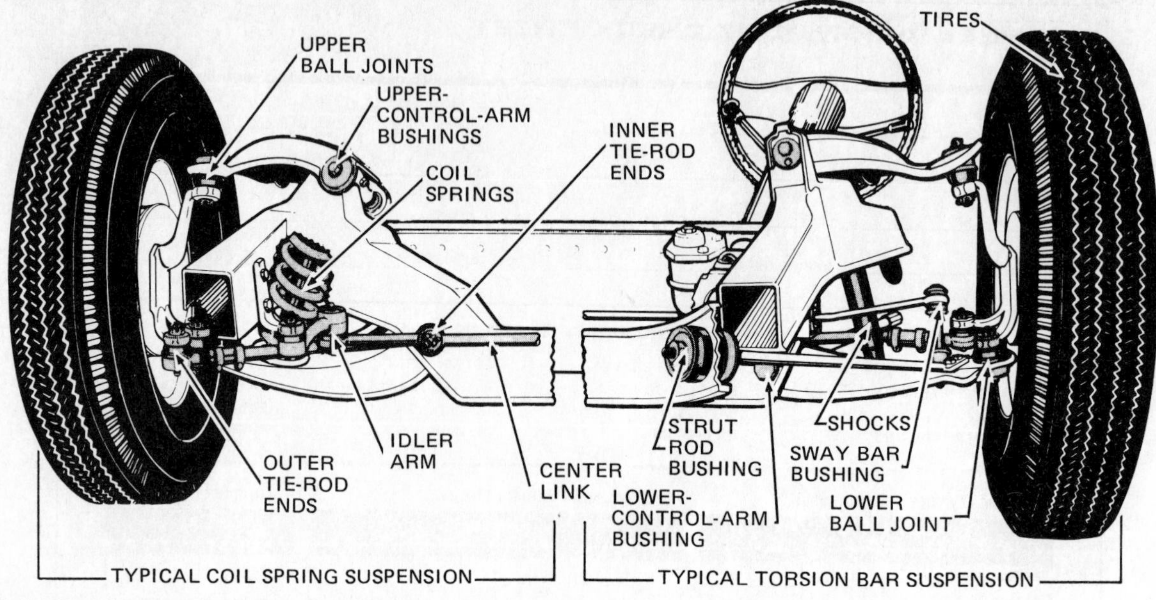

Fig. 13-45 Checks to make during a prealignment inspection. (*Moog Automotive, Inc.*)

with torsion bars, fill in the center box for "Torsion-Bar Vehicles Only" as you check front-suspension height.

On trucks, the frame angle must be checked and the angle entered in the box marked "Trucks Only." Frame angle is a required check on most trucks. It affects caster only. When the truck is built, the frame is horizontal with the ground. It has zero frame angle. Since frame angle affects the caster, to find the actual caster, use this equation:

Actual caster =
frame angle + caster angle of wheel

To check the frame angle, place a bubble protractor on the frame in the areas shown in Fig. 13-46. If the front is higher than the rear, *subtract* the frame angle from the caster reading. If the rear is higher than the front, *add* the frame angle to the caster reading.

Drive the car forward enough to establish the straight-ahead position of the front wheels. Then mark the steering-wheel hub and column (Fig. 13-47). This establishes the position that the steering wheel actually takes when the front wheels point straight ahead. If the steering wheel spokes are not horizontally aligned, adjustment should be made when the toe is adjusted.

Check and adjust the camber, caster, and toe of each wheel. After completing each adjustment, tighten the nuts to the proper torque.

Check and compare steering-axis inclination, included angle, and turning angle with the manufacturer's specifications. If any of these measurements are not within specifications, check for bent parts.

The last item on the Check Sheet in Fig. 13-44 is "Wheel Setback." Setback is a measure of how much one wheel is ahead or behind the other (Fig. 13-48). The cause of setback is the manufacturing tolerances allowed during vehicle assembly. Production tolerance for each side of the frame is ¼ inch [6.35 mm] each way. Excessive setback is any amount greater than ¾ inch [19.05 mm]. This indicates bent parts. It has not been proven

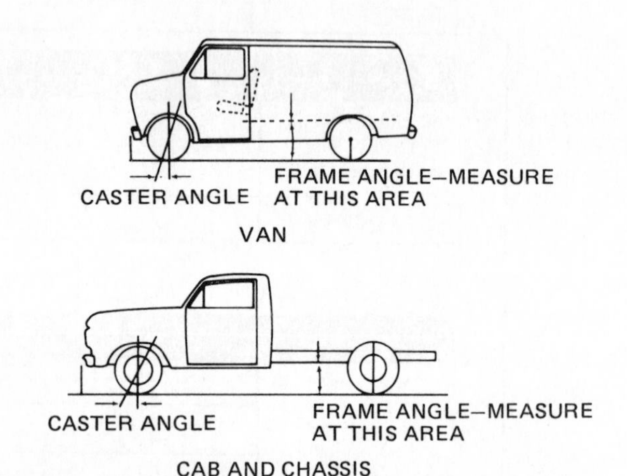

Fig. 13-46 Frame angle is a required check on most trucks. (*Ford Motor Company*)

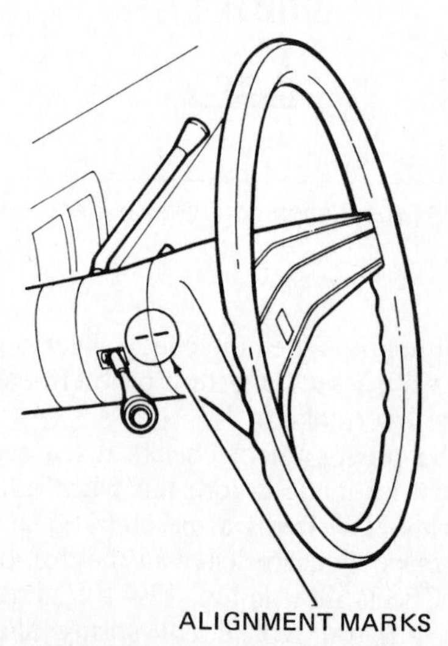

Fig. 13-47 Straight-ahead position marks. (*Ford Motor Company*)

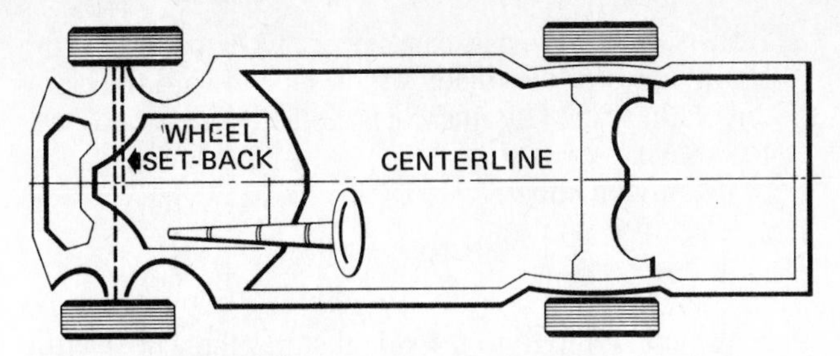

Fig. 13-48 Setback is a measure of how far one wheel is ahead of or behind the other. (*Hunter Engineering Company*)

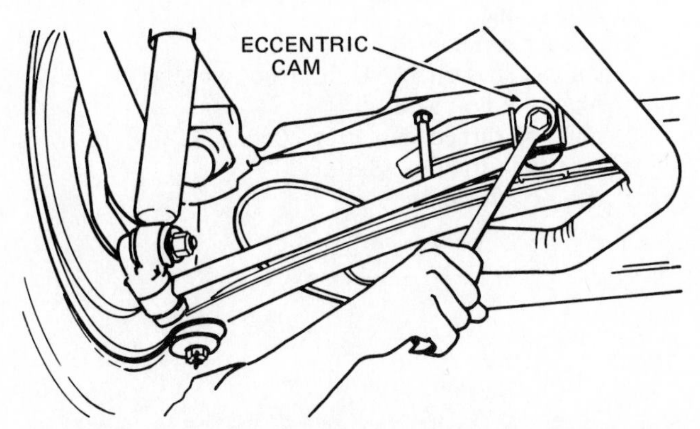

Fig. 13-49 Adjusting rear-wheel camber on a Chevrolet Corvette, which has independent rear suspension. (*Chevrolet Motor Division of General Motors Corporation*)

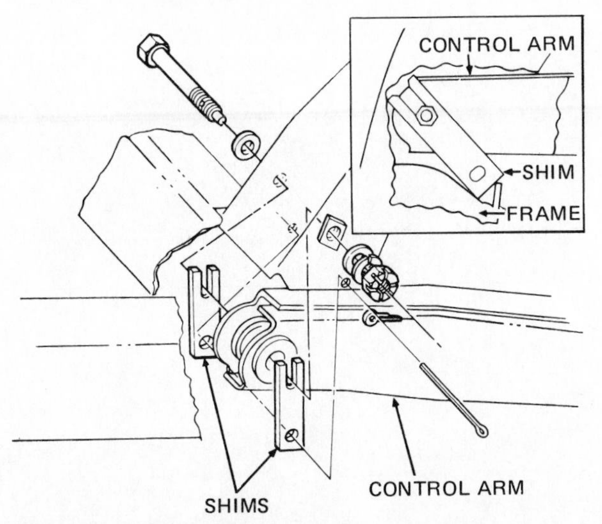

Fig. 13-50 Shim location to adjust rear-wheel toe on the Chevrolet Corvette. (*Chevrolet Motor Division of General Motors Corporation*)

that setback affects wheel alignment or tire wear. However, it will cause the center point of the steering gear to be off.

13-25 Rear-wheel alignment On cars with independent rear suspension, such as the Chevrolet Corvette, the rear-wheel alignment can also be checked. To do this, back the car onto the gauges used to align the front end. Camber will read in the normal manner. However, toe-in will now read as toe-out, and toe-out will read as toe-in. Caster is usually set at zero during manufacturing and needs no further adjustment. Figure 8-24 shows the rear suspension of a Corvette. Check that the strut rods are straight. If they are bent, they should be replaced.

1. **Camber adjustment** Adjust camber by turning the eccentric cam and bolt (Fig. 13-49). Loosen the cam-bolt nut and turn the cam-and-bolt assembly to get the correct camber. The location of the camber cam is shown to the lower right of the differential in Fig. 8-24. Tighten the locknut after making the adjustment.

2. **Toe-in adjustment** Adjust toe-in by inserting shims inside the frame side member on both sides of the torque-control-arm pivot bushing (Fig. 13-50).

Chapter 13 review questions

Select the *one* correct, best, or most probable answer to each question. Then check your answers against the correct answers given at the end of the book.

1. The alignment settings on a car are shown below.

	READINGS		SPECIFICATIONS
	LEFT	RIGHT	LEFT OR RIGHT
Camber	+¾° or +45′	−1½° or −1° 30′	0° to +½° or 0° to +30′
Caster	0°	0°	0° to +1°
Toe-in	1/16 inch or 0.16 mm		1/16 to 3/16 inch or 0.16 to 0.48 mm

Which of the following conditions would result from these alignment settings?
a. Left tire wear on inside; car does not pull to either side.
b. Right tire wear on inside; car pulls to left.
c. Right tire wear on outside; car pulls to left.
d. Right tire wear on outside; left tire wear on inside; car pulls to left.

2. Which of these statements is true about the strut rod in a coil-spring front suspension system?
a. Decreasing the rod length changes steering-axis inclination.
b. The length of the rod will not affect caster or camber.
c. The rod acts as the stabilizer bar.
d. Increasing the rod length changes caster.

3. While aligning the front end, the mechanic finds the toe-out-on-turns to be incorrect. Which, if any, of these could be the cause?
a. bent pitman arm
b. bent tie rod
c. bend idler arm
d. none of the above

4. Too much negative caster on the left front wheel will cause the
a. car to pull to the left
b. left tire to wear on the outside edge
c. car to pull to the right
d. left tire to wear on the inside edge

5. A front-suspension system which has the coil spring mounted on the lower control arm is being checked for ball-joint wear. Mechanic A says that the check can be made with the front end jacked up under the front crossmember. Mechanic B says that the check can be made with the front end jacked up under the lower control arm. Who is right?
   a. A only
   b. B only
   c. both A and B
   d. neither A nor B
6. The steering wheel of a car is not centered when traveling straight down the road. Mechanic A says that the steering column and its position changed. Mechanic B says that the steering wheel can be turned to center and the toe readjusted. Who is right?
   a. A only
   b. B only
   c. both A and B
   d. neither A nor B
7. The front wheels of a car with torsion-bar front suspension are being aligned. Mechanic A says that the suspension height should be adjusted before caster and camber are adjusted. Mechanic B says that the suspension height should be adjusted before toe is adjusted. Who is right?
   a. A only
   b. B only
   c. both A and B
   d. neither A nor B
8. Both front tires on a car have a wear pattern with sharp edges of the treads which face toward the center of the car. This indicates that the front end has too much:
   a. positive camber
   b. negative camber
   c. toe-in
   d. toe-out
9. A vehicle will pull to the side that has the wheel with the:
   a. most positive camber
   b. least positive caster
   c. both *a* and *b*
   d. neither *a* nor *b*
10. When caster is checked, the front wheels:
    a. must not be moved
    b. must be turned a total of 20°
    c. must be turned a total of 40°
    d. none of the above

CHAPTER 14

MANUAL STEERING-GEAR SERVICE

After studying this chapter, and with proper instruction and equipment, you should be able to:

1. Adjust the Saginaw steering gear.
2. Overhaul the Saginaw steering gear.
3. Adjust the Ford steering gear.
4. Overhaul the Ford steering gear.
5. Adjust the Plymouth steering gear.
6. Overhaul the Plymouth steering gear.

14-1 Steering-gear adjustments Many manual-steering gears are of the recirculating ball-and-nut type (Fig. 14-1). They provide easy steering and precise steering control. These units seldom require service beyond adjustments. This chapter covers the adjustment, disassembly, repair, and reassembly procedures for manual-steering gears used in several makes of cars.

Manual-steering gears have two basic adjustments (Fig. 14-2). One adjustment is for taking up the steering-shaft or worm-gear end play. The other adjustment is for removing backlash between the worm gear and the sector—or the roller or lever studs in other steering-gear designs. In addition, some designs have a means of adjusting the sector-shaft or pitman-arm-shaft end play.

Steering-shaft endplay or worm-shaft end play is also called *bearing preload* or *thrust-bearing preload*. In many steering gears, steering-shaft end play is adjusted by moving the thrust bearing in or out against the end of the shaft.

Lash or backlash between the worm gear and the sector is caused by the clearance between these parts. It is also called pitman-shaft lash or over-center preload. The adjustment may be called the *over-center adjustment*. Basically, the adjuster moves the worm gear and the sector closer together to reduce lash, or farther apart if they are binding.

Before attempting to adjust the steering gear to take up the excessive end play or to relieve binding, make sure that the condition is not a result of faulty front-wheel

WORMSHAFT
WORM BEARING ADJUSTER
BALLS AND GUIDES
LOCK NUT
SEAL
WORM BEARING
WORM BEARING
BALL NUT
SECTOR

Fig. 14-1 Sectional view of the Saginaw (General Motors) recirculating-ball-and-nut manual-steering gear. (*Chevrolet Motor Division of General Motors Corporation*)

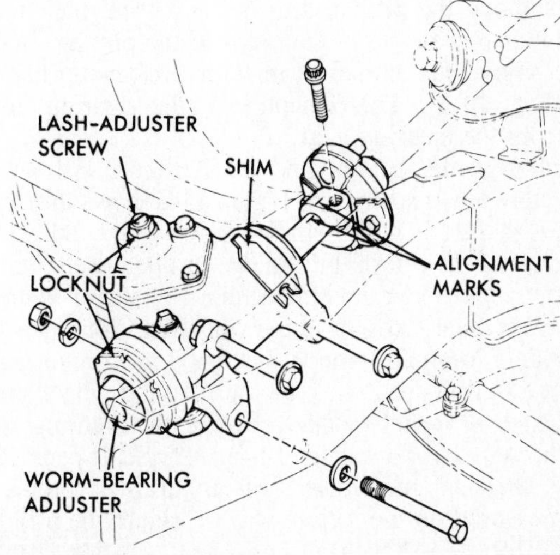

Fig. 14-2 Adjustment points on a recirculating-ball-and-nut manual-steering gear. (*Chevrolet Motor Division of General Motors Corporation*)

alignment or wear in some other part of the linkage or front-suspension system.

A steering and suspension trouble-diagnosis chart is given in ❁ 11-1.

❁ 14-2 Saginaw (General Motors) steering-gear service The Saginaw manual-steering gear (Figs. 14-1 and 14-2) has been used on General Motors cars since 1961. This steering gear is filled at the factory with steering-gear lubricant. No lubrication is required for the life of the steering gear. It should not be drained and refilled with lubricant if it continues to operate normally. However, the gear should be inspected every 30,000 miles [48,000 km] for seal leakage. Look for seepage past the seals and deposits of actual solid grease, not just a light oil film.

If a seal is replaced or the steering gear is overhauled, then the housing should be filled with the steering-gear lubricant specified by General Motors, or its equivalent. Chassis lubricant should not be used in the steering gear. Be careful never to overfill the steering-gear housing, because this could damage the seals.

Adjustment of the steering gear is required if the steering is too loose or too tight. If it is too loose, it will have too much lash. If it is too tight, excessive turning effort will be required. However, before adjusting the steering gear, make sure that all other front-end factors are within specifications. Be sure that no other front-end condition is causing the problem. Check the tires, wheel balance, wheel alignment, steering linkage, and shock absorbers. Make any corrections that are required. If the condition still exists, steering-gear adjustments should be made.

❁ 14-3 Saginaw steering-gear adjustments Two adjustments can be made to the Saginaw manual-steering gear. These are (1) worm-bearing end play or bearing preload, and (2) pitman-shaft lash or over-center preload. Adjustment points are shown in Fig. 14-2. To make these adjustments, proceed as follows:

1. Disconnect the battery ground cable. Raise the car. Remove the pitman-arm nut and then mark the relationship of the pitman arm to the pitman shaft.
2. Remove the pitman arm with a pitman-arm puller (Fig. 14-3). This disconnects the steering linkage from the steering gear.
3. Loosen the locknut and back off the lash-adjuster screw one-quarter turn. Figure 14-2 shows this screw.
4. Measure the worm-shaft bearing drag. To do this, remove the horn button or shroud, then turn the steering wheel gently in one direction until it is stopped by the gear. Now turn the wheel back one-half turn.
5. Apply a torque wrench that has a maximum reading of 50 lb-in [6 N-m] to the steering-wheel nut by using a ¾-inch socket. Rotate the steering wheel through a 90° arc (Fig. 14-4). A typical specification is that the maximum bearing drag or preload as measured on the torque wrench should be 5 to 8 lb-in [0.6 to 1.0 N-m].

Careful! Never turn the steering wheel hard against the stops with the steering linkage disconnected. This could damage the ends of the ball guides in the steering gear.

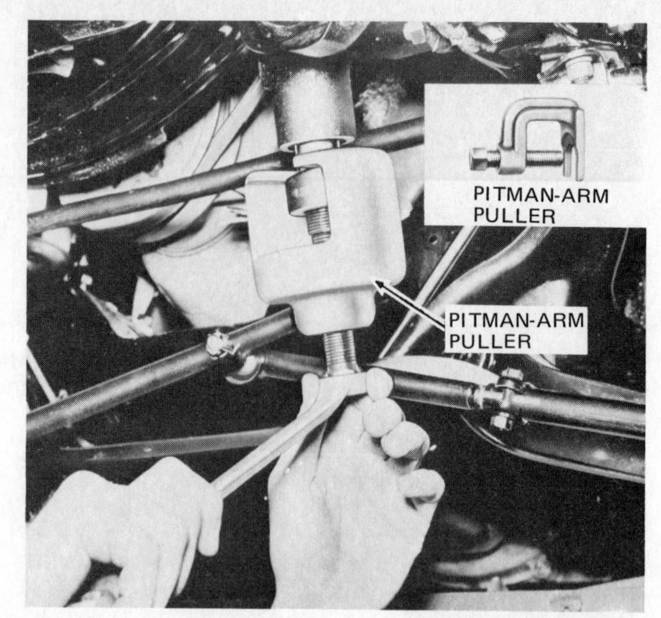

Fig. 14-3 Removing the pitman arm from the sector shaft. (*Chevrolet Motor Division of General Motors Corporation*)

Fig. 14-4 Using a torque wrench to measure the worm-shaft bearing preload, or drag. (*Chevrolet Motor Division of General Motors Corporation*)

6. If the torque or bearing preload is not within specifications, adjust the worm bearings. Loosen the worm-bearing-adjuster locknut and turn the worm-bearing adjuster to obtain the proper bearing preload. The adjuster is shown in Fig. 14-2. Tighten the locknut and recheck the bearing drag with the torque wrench (Fig. 14-4).

NOTE: If the gear feels "lumpy" or rough after adjustment, the bearings probably have been damaged by severe impact or improper adjustment. The steering gear must be disassembled for replacement of the defective parts.

7. Check the pitman-shaft lash or over-center preload with the steering wheel centered. To center the steering wheel, turn it gently from stop to stop, while counting the total number of turns. Turn the wheel back exactly halfway.

8. Turn in the lash-adjuster screw to take out all lash between the ball nut and the teeth on the pitman-shaft sector gear. Tighten the locknut.
9. Now check the steering-wheel torque (Fig. 14-4) while the wheel is rotated through the center position. If necessary, adjust the over-center preload by loosening the locknut and readjusting the lash-adjuster screw (Fig. 14-2). A typical specification is that the torque reading should be 4 to 10 lb-in [0.5 to 1.2 N-m] in excess of the worm-bearing preload.

NOTE: If, while you make an over-center preload adjustment the maximum specification is exceeded, back off the lash-adjuster screw. Turn the screw in to obtain the specified adjustment. Hold the screw and torque the locknut to specifications.

10. Reassemble the pitman arm to the pitman shaft, aligning the marks made during disassembly. Tighten the pitman-shaft nut to specifications. Reinstall the horn button cap or shroud to the steering wheel.

⚙ 14-4 Saginaw manual-steering-gear overhaul

Figure 14-1 is a sectional view of the Saginaw steering gear discussed in this section. Figure 14-5 is a view of the disassembled steering gear. To overhaul the Sainaw manual-steering gear, proceed as follows:

1. Removal Raise the car so that you can get under it conveniently. Disconnect the pitman arm from the shaft as described in ⚙ 14-3 and shown in Fig. 14-3. Remove the splash-pan attachment bolts and the pan. Remove the nuts and lock washers from the steering-gear attachment bolts and take out the bolts and any shims. Now the steering gear is loose from the frame and can be taken off the car as soon as the gear is detached from the intermediate shaft. This is done by loosening the bolt on the lower universal-joint clamp.

2. Disassembly Clamp one of the steering-gear mounting tabs in a vise, with the worm shaft horizontal. Rotate the worm shaft from one extreme to the other, then turn it back exactly halfway so that the gear is centered. Now, loosen the locknut and back off the lash-adjuster screw several turns. Then loosen the locknut and back off the worm-bearing adjuster a few turns.

Put a pan under the unit to catch the lubricant. Remove the three bolts and washers attaching the side cover to the housing (Fig. 14-6). Pull the side cover with the pitman-shaft assembly from the housing. If the sector gear does not clear the housing opening, turn the worm shaft by hand until it does.

Remove the worm-bearing adjuster, locknut, and lower ball bearing from the housing. Now draw the worm shaft and nut from the housing (Fig. 14-7). Remove the upper ball bearing.

NOTE: Do not allow the ball nut to run down to either end of the worm. This could damage the ends of the ball guides.

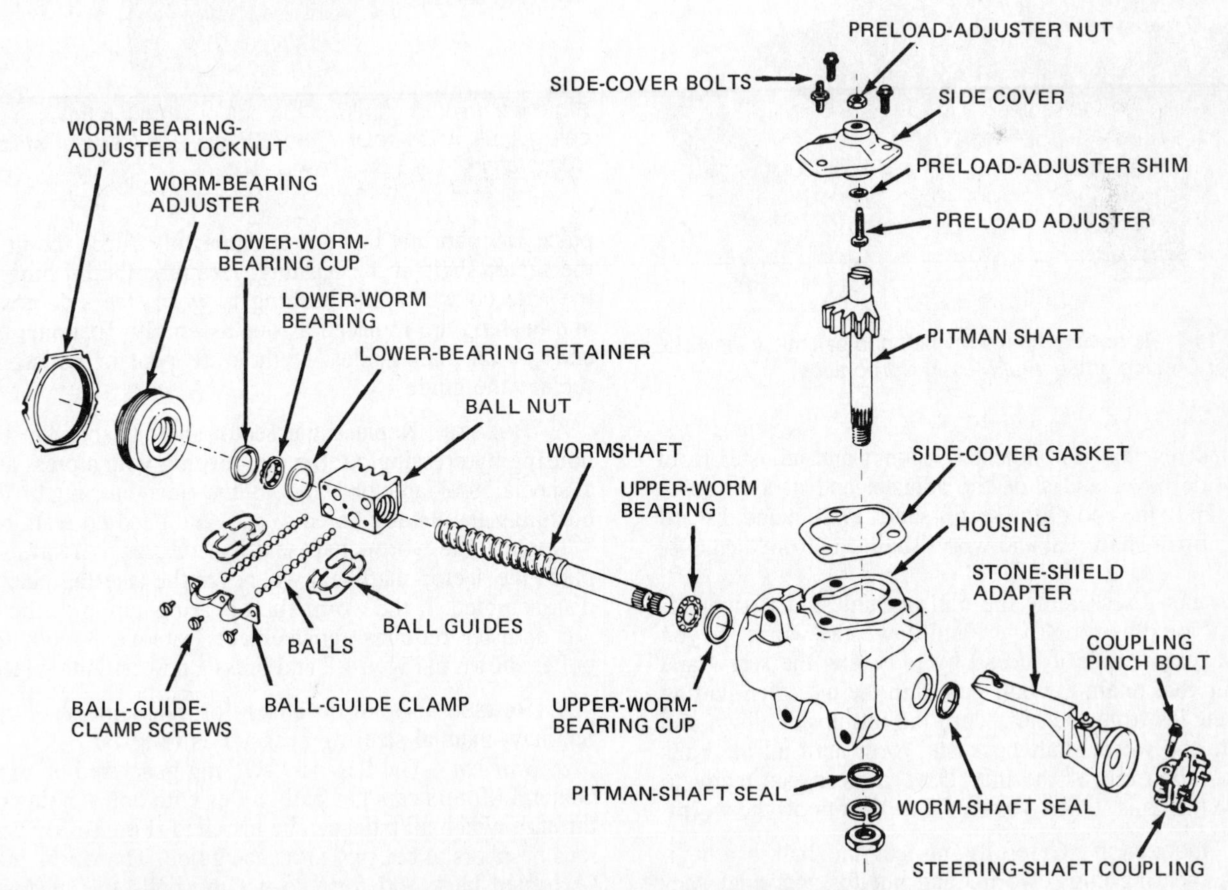

Fig. 14-5 Exploded view of a Saginaw manual-steering gear. (*Oldsmobile Division of General Motors Corporation*)

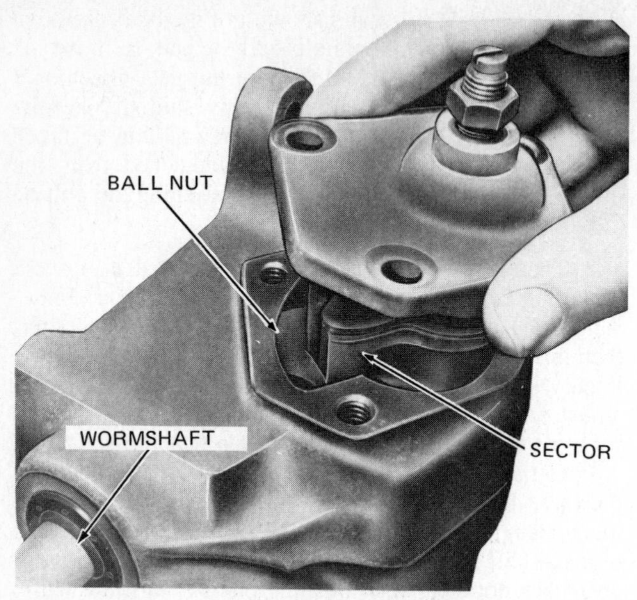

Fig. 14-6 Removing the pitman-shaft assembly. (*Chevrolet Motor Division of General Motors Corporation*)

Fig. 14-7 Removing the worm shaft and ball nut. (*Chevrolet Motor Division of General Motors Corporation*)

Unscrew the lash-adjuster locknut and adjuster from the side cover and slide the adjuster and its shim from the slot in the end of the sector shaft. Pry out and discard the pitman-shaft seal and worm-shaft seal from the housing.

Do not disassemble the ball nut unless it is tight, or binds, or runs roughly up and down the worm. If you must disassemble it, do so by removing the screw and clamp that retain the ball guides in the nut, then pulling the guides from the nut. Turn the nut upside down and rotate the worm shaft back and forth until all the balls have fallen out of the nut. Have a clean pan ready to catch the balls. Now, you can slide the nut off the worm.

3. Inspection Carefully inspect the ball bearings, balls, bearing cups, worm, and nut for abnormal conditions such as wear, dents, cracks, and chipping. Re-

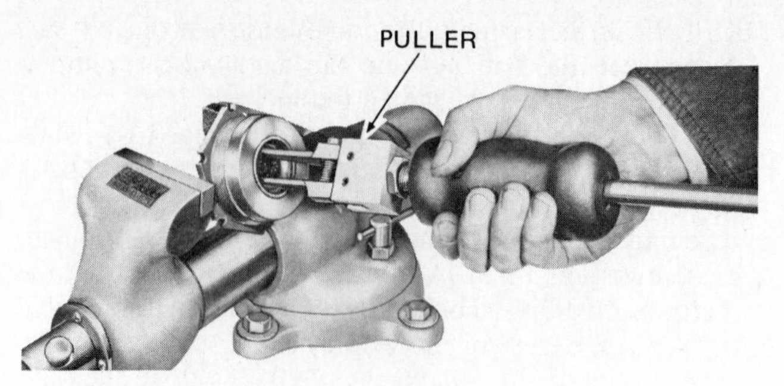

Fig. 14-8 Using a puller to remove the bearing cup from the adjuster. (*Chevrolet Motor Division of General Motors Corporation*)

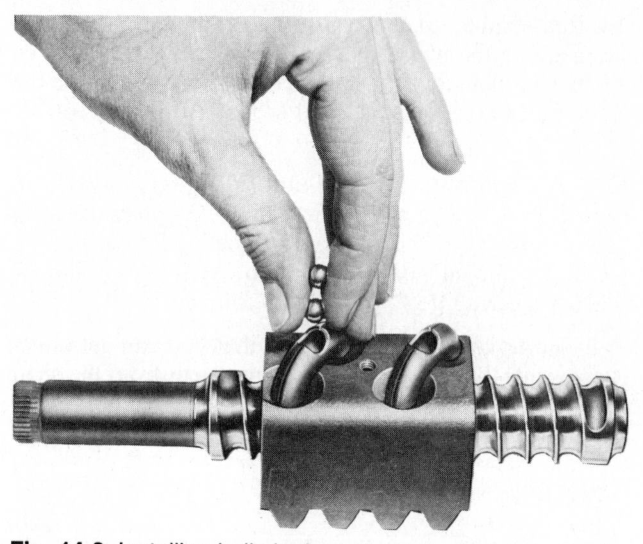

Fig. 14-9 Installing balls in the ball nut through holes in the ball guides. (*Chevrolet Motor Division of General Motors Corporation*)

place any part that is worn or damaged. Check the fit of the sector shaft in the bushing. Examine the bushing in the side cover. If this bushing is worn, the side cover and bushing are replaced as one assembly. Examine the ends of the ball guides. If they are bent or damaged, replace the guides.

4. Repairs Replace the sector-shaft bushing in the housing by pressing out the old bushing with a press and a special tool and pressing in the new bushing. New bushings are already bored to size and need no reaming.

Replace the worm-shaft seal if necessary. Always replace the sector-shaft seal whenever the steering gear is disassembled. If the worm-shaft bearing cup in the bearing adjuster requires replacement, remove it with the puller shown in Fig. 14-8 and press a new cup into place.

5. Reassembly Procedures for reassembly of the Saginaw manual-steering gear are as follows:

a. Ball nut The Saginaw steering gear used on most General Motors cars has ball guides with holes in the top through which all balls can be installed (Fig. 14-9). Various numbers of balls go into each guide. The worm must be turned back and forth to get the balls to run down into the guides and fill them. Before assembly, the worm and the inside of the nut should be coated with steering-gear lubricant. When the balls are all in place, the guide clamp should be attached with screws.

On some Saginaw steering gears, there are no holes in the ball guides. Therefore, a different method of installing the balls is required. The procedure is covered in the manufacturer's service manuals.

b. Steering gear Cleaning the parts will remove the sealing compound from the screw threads. Therefore, the threads on the adjuster, side-cover bolts, and lash adjuster should be coated with a sealing compound, such as Permatex No. 2. Do not get sealer on the worm-shaft bearing in the adjuster. Apply steering-gear lubricant to the worm bearings, sector-shaft bushings, and ball-nut teeth.

With the worm-shaft seal, bushings, and bearing cups installed, slip the upper ball bearing over the worm shaft and insert the worm-shaft-and-nut assembly into the housing. Put the ball bearing in the adjuster cup, press the retainer into place, and install the adjuster and locknut in the lower end of the housing. Figure 14-5 shows these parts.

Assemble the lash or preload adjuster with a shim in the slot in the end of the sector shaft (Fig. 14-5). End clearance between the bottom of the slot and the head of the lash adjuster should be no greater than 0.002 inch [0.051 mm]. If it is greater, install another shim.

Start the end of the sector shaft into the side-cover bushing, and pull the shaft into place by turning the lash adjuster. Rotate the worm shaft to put the nut in the center of its travel. This is to make sure that the rack and sector will mesh properly, with the center tooth of the sector entering the center-tooth space in the rack on the nut (Fig. 14-1).

Put a new gasket on the side cover and push the cover-and-sector-shaft assembly into the housing, making sure that the teeth mesh properly. Check that there is some lash between the sector and the rack, then secure the side cover with screws.

c. Adjustment on the bench The steering gear can be adjusted on the bench. Tighten the worm-bearing adjuster to remove all shaft end play, tighten the locknut, and then install the steering wheel on the worm shaft. The adjustment procedure is then the same as described in ⚙ 14-3.

Careful! Do not force the steering wheel on the shaft, but tap the steering wheel into place lightly. Forcing the wheel on could damage the bearings.

⚙ 14-5 Ford steering-gear adjustments Figure 14-10 is a phantom view of the recirculating-ball-and-nut steering gear described in this section. Figure 14-11 is an exterior view. Two adjustments are required: one to provide minimum worm-shaft end play or steering-shaft end play, and the other to provide minimum backlash between the sector and the ball nut. Adjustments are to be made in the following order.

1. Disconnect the pitman arm from the relay rod. Figure 10-19 shows the steering linkage on one Ford model. On cars with power steering, disconnect the arm from the control-valve ball stud.

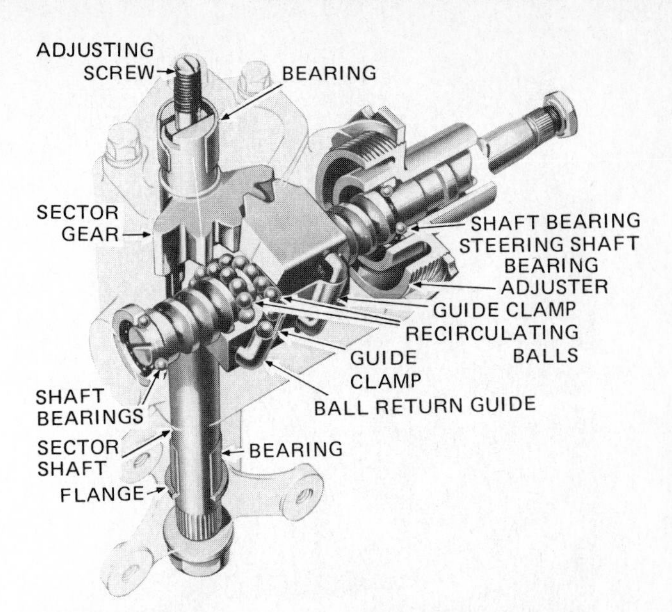

Fig. 14-10 Phantom view of the Ford recirculating-ball steering gear. (*Ford Motor Company*)

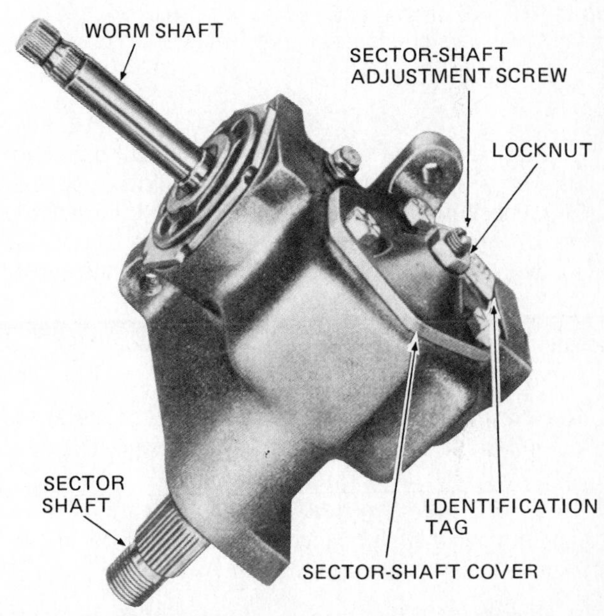

Fig. 14-11 Ford recirculating-ball manual-steering gear. (*Ford Motor Company*)

2. Loosen the locknut on the sector-shaft adjustment screw (Fig. 14-12). Turn the adjustment screw counterclockwise. This relieves the load on the teeth.
3. Use a torque wrench on the steering-wheel nut to measure the worm-bearing preload (Fig. 14-4). With the steering wheel off center, read the torque required to move the shaft at points about 1½ turns to either side of center. If the required torque is not within specifications, adjust it by loosening the bearing-adjuster locknut, then tighten or back off the adjuster. Tighten the locknut and recheck the torque.
4. Adjust the backlash next. Turn the steering wheel slowly to either stop. Do not bump the steering wheel against the stop because this could damage the ball

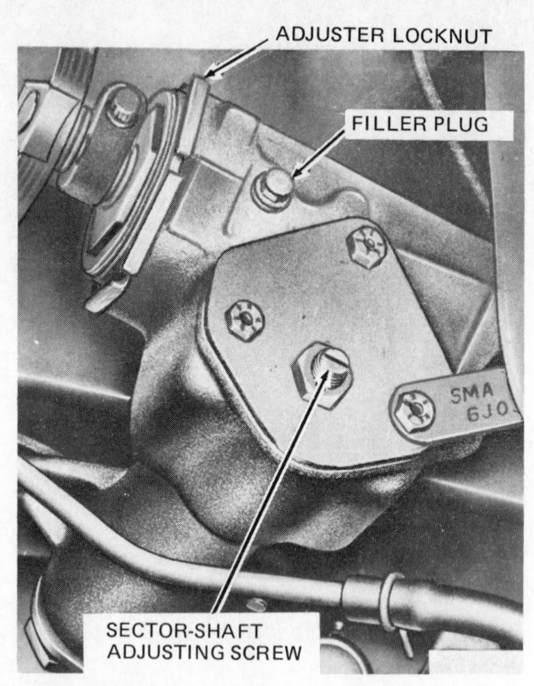

Fig. 14-12 Adjustment points on a Ford recirculating-ball-and-nut manual steering gear. (*Ford Motor Company*)

guides. Rotate the wheel to center the ball nut. Turn the sector-shaft adjustment screw clockwise, repeatedly checking the torque required to pull the steering wheel through center (Fig. 14-4). When the correct torque is attained, hold the adjustment screw and tighten the locknut. Recheck the torque.

NOTE: At 30° on either side of center you should not feel any backlash.

5. Reconnect the pitman arm to the relay rod, or to the control-valve ball stud on cars with power steering.

⚙ 14-6 Ford steering-gear service Figure 14-10 is a phantom view of the steering gear described in this section.

1. Removal If the steering column is of the movable type, remove the steering-column pivot-plate bolts. On all types, remove the bolt that attaches the flex coupling to the worm shaft of the steering gear. Raise the front end of the car and disconnect the pitman arm from the sector shaft (Fig. 14-3). Remove the bolts that attach the steering gear to the frame. On some cars, it may be necessary to disconnect the muffler inlet pipe and the clutch linkage.

2. Disassembly Rotate the worm shaft to the center position. Remove the locknut from the sector-shaft adjustment screw and remove the three cover bolts. Now the cover and sector shaft can be taken from the housing (Fig. 14-13).

Loosen the locknut and back off the worm-shaft-bearing adjuster nut so that the shaft and ball nut can be withdrawn from the housing. Figure 14-14 shows these parts.

Careful! Do not let the nut run down to the end of the worm. This could damage the ball guides.

Remove the ball-guide clamp, turn the nut upside down over a clean pan, and rotate the worm shaft back and forth until all the balls fall out. Now the nut will slide off the shaft.

Remove the needle bearings only if they require replacement. They can be pressed out with a press and a special remover. The bearing cups can be removed from the housing and adjuster by tapping the housing or adjuster on a wooden block to jar them loose.

3. Reassembly Install new needle bearings and a new oil seal if the old ones have been removed. Apply steering-gear lubricant to the bearings. Install sector-shaft bearing cups in the housing and adjuster. Install a new seal in the adjuster.

Insert the ball guides into the holes in the ball nut, tapping them lightly with the wood handle of a screwdriver, if necessary, to seat them. Put the ball nut into position on the steering shaft and drop the specified number of balls into the hole in the top of each ball guide. Rotate the shaft slightly back and forth to distribute the balls.

Install ball-guide clamps. Check the steering shaft to make sure it is free to turn in the ball nut.

Coat the threads of the steering-shaft bearing adjuster, the housing-cover bolts, and the sector adjustment screw with oil-resistant sealing compound. Do not get sealer on the internal threads or on the bearings. Coat the worm bearings, sector-shaft bearings, and gear teeth with steering-gear lubricant.

Clamp the housing in a vise, with the sector-shaft axis horizontal, and position the worm-shaft lower bearing in its cup. Install the worm shaft, with its nut, in the housing. Put the upper bearing in place on the worm, then with the cup in place, run the bearing adjuster down. Adjust the worm-bearing preload as described earlier.

Put the sector-shaft adjustment screw, with a shim, in the slot in the end of the sector shaft. Clearance should be less than 0.002 inch [0.051 mm] between the end of the screw and the bottom of the slot in the shaft. If the clearance is greater, add shims.

Install a new gasket on the housing cover. Start the adjustment screw into the cover. Apply enough lubricant to fill the pocket in the housing between the sector-shaft bearings. This fills the housing about 30 percent full. Rotate the worm shaft so that the ball nut will mesh properly with the sector teeth.

Put the sector shaft and cover into place, then turn the cover out of the way and pack about 0.7 pound [0.32 kg] of lubricant into the gear. Push the sector shaft and cover into place and install the top two cover bolts. Do not tighten them until you are sure that there is some lash between the teeth. Then tighten the top two cover bolts and adjust the lash as explained in ⚙ 14-5. The lower bolt goes in after the final filling of the steering gear with lubricant.

4. Installation With the steering wheel and the sector shaft in their center, or straight-ahead, positions, attach the steering gear to the frame. Tighten the attaching

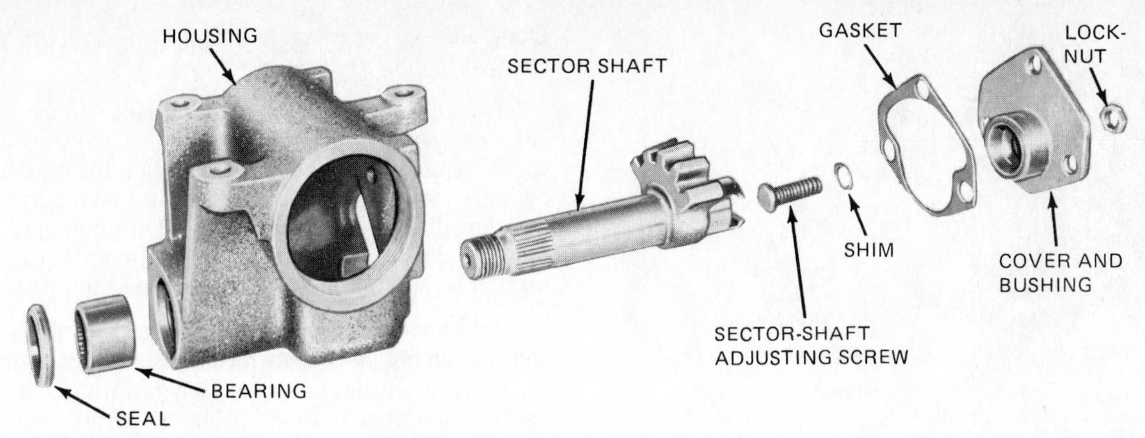

Fig. 14-13 Sector shaft and housing disassembled. (*Ford Motor Company*)

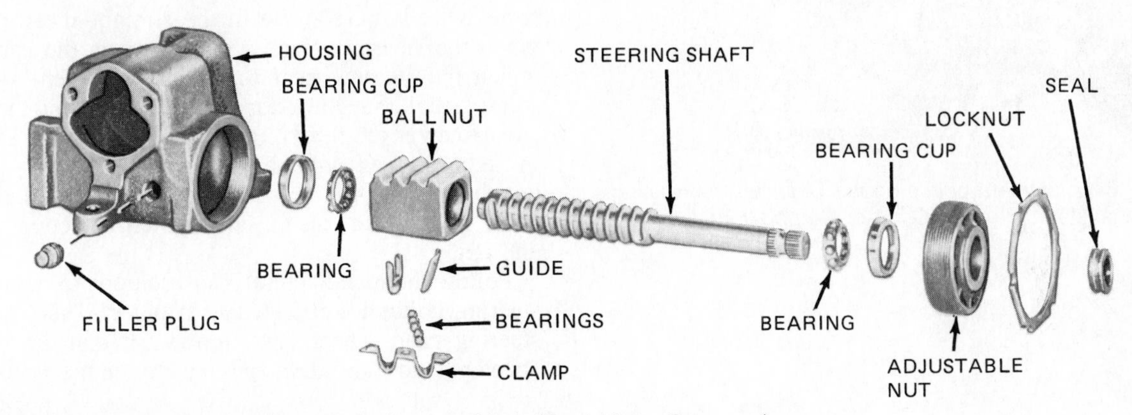

Fig. 14-14 Steering shaft and related parts disassembled. (*Ford Motor Company*)

bolts to the specified torque. With the front wheels straight ahead, connect the pitman arm to the sector shaft. Reconnect the muffler inlet pipe if it has been disconnected. Reinstall the flex-coupling bolt, and reinstall the pivot-plate bolts on the movable-column type.

Turn the steering wheel to the left to move the ball nut away from the filler hole. Fill the steering gear with lubricant until it comes out the lower cover-bolt hole. Then install the lower bolt and tighten it to the specified torque.

❁ 14-7 Plymouth steering-gear adjustments Figure 14-15 shows the steering gear described in this section. There are two adjustments, one for worm-shaft-bearing preload, and the other for the sector-shaft gear clearance. These are shown in Fig. 14-16. They are the same as those on the Ford and General Motors recirculating-ball steering gears. The adjustments are checked and corrected, if necessary, as described earlier.

❁ 14-8 Plymouth steering-gear service Figure 14-15 illustrates the steering gear described in this section. Figure 14-17 shows the relationship of the steering gear to the steering column.

1. Removal To remove the steering gear, first remove the pitman-arm nut and pull the arm from the sector shaft with a puller. Then remove the coupling-clamp bolt from the upper end of the worm shaft. To get enough

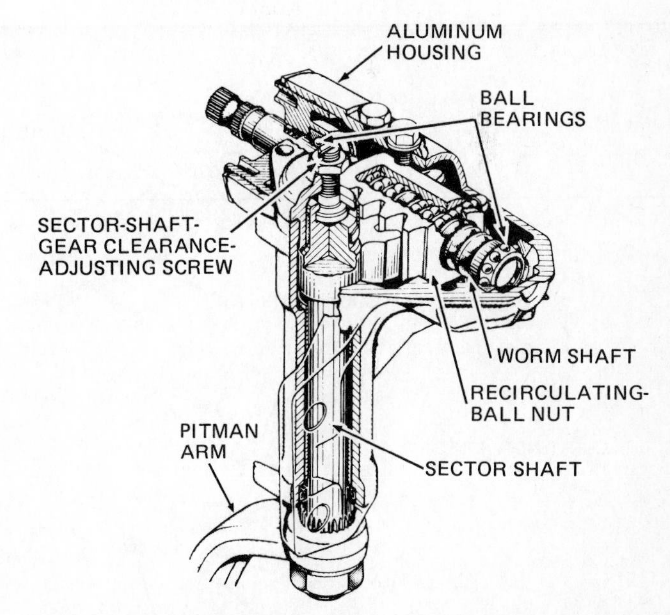

Fig. 14-15 Cutaway view of the Chrysler recirculating-ball manual-steering gear. (*Chrysler Corporation*)

room to remove the steering gear, loosen the column-jacket clamp bolts. Next, slide the column assembly up far enough to disengage the coupling from the worm shaft (Fig. 14-17).

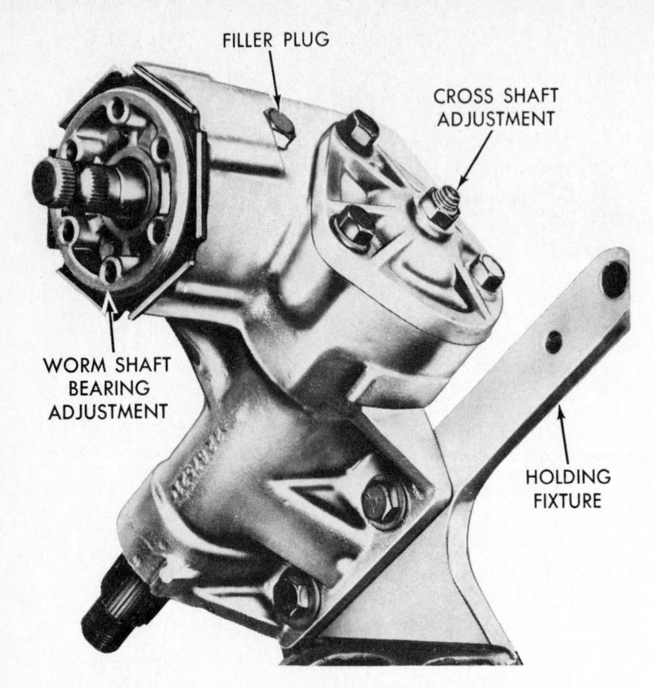

Fig. 14-16 Adjustment points on the Chrysler recirculating-ball manual-steering gear. (*Chrysler Corporation*)

Careful! Do not scratch the steering column on the clamps!

Remove the mounting bolts and take off the steering gear. On cars with six-cylinder engines the steering gear can be removed from the engine compartment. On eight-cylinder models the steering gear must be removed from underneath. On some cars, the left front engine mount must be detached and the engine raised 1½ inches [38 mm]. The starting motor may also need to be removed.

2. Disassembly and reassembly The disassembly and reassembly procedures for this steering gear are about the same as for the Ford and General Motors units previously described. Some special tools are required to remove and replace oil seals, bearings, and bearing cups.

3. Installation The steering wheel, steering gear, and front wheels must all be in the straight-ahead position when the steering gear is reinstalled on the car. First, attach the steering gear to the frame. Then, slide the steering column down far enough to permit the worm shaft to enter the flexible coupling. The master serration must be aligned with the notch mark on the coupling housing. When the grooves in the coupling and worm shaft are aligned, install and tighten the coupling bolt and nut.

Follow the manufacturer's instructions to reinstall the steering column and fasten it in place. Then, with the steering gear, wheel, and front wheels in the straight-ahead position, install the pitman arm on the sector shaft.

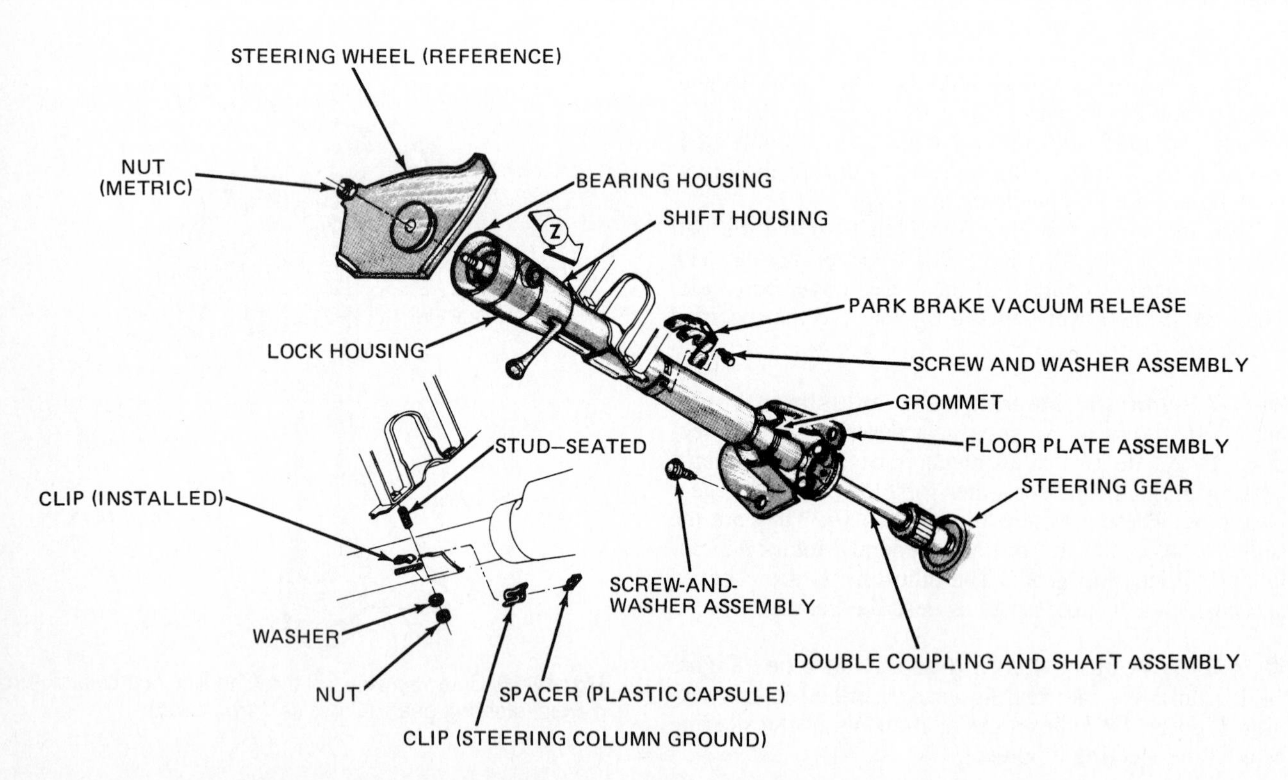

Fig. 14-17 Relationship and attachment points of the Chrysler steering column and steering gear. (*Chrysler Corporation*)

Chapter 14 review questions

Select the *one* correct, best, or most probable answer to each question. Then check your answers against the correct answers given at the end of the book.

1. After the linkage is disconnected, which of these is the correct order for adjusting a manual-steering gear?
   a. Center the steering wheel, adjust worm-shaft end play, reconnect linkage.
   b. Adjust sector shaft, adjust worm-shaft bearing preload, reconnect linkage.
   c. Center steering wheel, adjust sector shaft to remove free play, reconnect linkage.
   d. Adjust worm-shaft bearing preload, adjust sector shaft, reconnect linkage.
2. There are two basic adjustments on steering gears, one for taking up worm-gear and steering-shaft end play, the other for removing steering-gear:
   a. frontlash
   b. side play
   c. backlash
   d. end play
3. When turning the steering wheel with the linkage disconnected from the pitman arm, avoid:
   a. bumping at the end of turns
   b. bumping at the center of turns
   c. spinning the wheel slowly
   d. all of the above
4. On some cars, the worm-bearing or preload adjustment is made by turning a bearing adjuster. On other cars, the adjustment is made by means of:
   a. adjustment nuts
   b. roller bearings
   c. shims
   d. bushings
5. Steering-gear backlash is usually adjusted by moving the sector in relation to the:
   a. pitman shaft
   b. worm
   c. steering wheel
   d. pitman arm

CHAPTER 15

SAGINAW ROTARY-VALVE POWER-STEERING SERVICE

After studying this chapter, and with proper instruction and equipment, you should be able to:

1. Diagnose troubles in the Saginaw power-steering gear.
2. Bleed the hydraulic system.
3. Adjust the steering gear.
4. Overhaul the Saginaw power-steering gear.
5. Overhaul the Saginaw power-steering pump.

Trouble diagnosis

15-1 Rotary-valve power-steering trouble-diagnosis chart This power-steering gear was described in ❁ 10-20. The chart on the following page lists possible trouble symptoms or complaints, causes, and checks or corrections. The chart gives you a means of logically analyzing trouble and quickly locating the cause. The adjustment or repair can be made following the procedures described in later sections.

❁ 15-2 Trouble-diagnosis checks When a driver complains of trouble in the Saginaw power-steering system (Fig. 15-1), the following checks should be made to pinpoint the cause. Some troubles, such as a loose belt, may be easy to correct. Other troubles may require the use of a pressure gauge.

1. Check belt tension To have power-steering action, the power-steering-pump pulley must be turning. If the drive belt has low tension and slips, it will cause low oil pressure in the power-steering system and hard steering. A quick check of the belt tension can be made with the engine running and oil pump warm. Turn the steering wheel from lock to lock with the front wheels on a dry surface. As the steering wheel is turned, maximum pressure builds up. This imposes a full load on the drive belt. If the belt slips, the tension is too low or there is mechanical damage in the pump.

Another way to check drive-belt tension is to turn the engine off and push in on the belt midway between the pulleys. The specifications for one car state that the belt should deflect ½ to ¾ inch [13 to 19 mm] with a push of 15 pounds [67 N] halfway between the fan and pump pulleys. To increase tension, loosen the mounting bolts and move the pump out. Then tighten the bolts.

Most belt-adjustment procedures specify the use of a belt-tension gauge (Fig. 15-2). The gauge fits on the belt

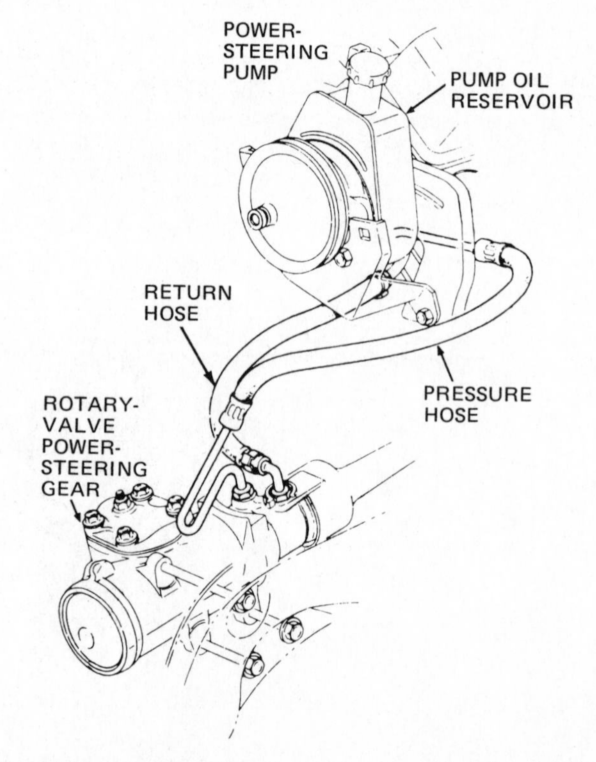

Fig. 15-1 Power-steering system using a Saginaw rotary-valve power-steering gear and a Saginaw vane-type power-steering pump. (*Buick Motor Division of General Motors Corporation*)

and measures the force required to deflect the belt a standard distance. Typical belt tension specifications for various sizes of belts are given in Fig. 15-3.

2. Check oil level If the power-steering oil level is low, add oil to bring the level up to the marking on the side of the reservoir. Use only the recommended type of

Rotary-Valve Power-Steering Trouble-Diagnosis Chart

(See ⚙ 15-2 to 15-11 for details of checks and corrections listed.)

COMPLAINT	POSSIBLE CAUSE	CHECK OR CORRECTION
1. Hard steering	a. Tight steering-gear adjustment	Readjust
	b. Pump drive belt loose	Tighten
	c. Low oil pressure	Check (see item 3, below)
	d. Air in hydraulic system	Bleed system
	e. Low oil level	Add oil
	f. Lower coupling flange rubbing on adjuster plug	Loosen flange bolt and adjust to 1/16-in [1.6 mm] clearance
	g. Internal leakage	Check pump pressure (see item 3, below)
	h. Tire pressure low	Inflate to correct pressure
	i. Frame bent	Repair
	j. Front springs weak	Check suspension height; replace springs
2. Poor centering (or recovery from turns)	a. Valve sticky	Free
	b. Steering shaft binding	Align, replace bushings
	c. Incorrect steering-gear adjustments	Readjust
	d. Lower coupling flange rubbing on adjuster plug	Loosen flange bolt, adjust to 1/16-in [1.6 mm] clearance
	e. Front end needs alignment	Align
	f. Steering gear out of adjustment	Adjust
	g. Steering linkage or ball joints binding	Replace defective parts
3. Low oil pressure	a. Loose pump belt	Tighten
	b. Low oil level	Add oil
	c. Mechanical trouble in pump	Check relief valve, rotor parts
	d. Oil leaks, external	Check hose connections, O sealing rings at cover, etc.
	e. Oil leaks, internal	Replace cylinder adapter, valve cover, or upper housing seal
	f. Engine idling too slowly	Set idle speed to specifications, check idle solenoid
4. Excessive wheel kickback or loose steering	a. Steering-linkage ball joints loose	Replace
	b. Steering-gear adjustments loose	Adjust
	c. Front-wheel bearings out of adjustment or worn	Adjust or replace
	d. Air in system	Fill, bleed
	e. Steering gear loose on frame	Tighten attaching bolts
	f. Flexible coupling loose	Tighten pinch bolts
5. Pump noise	a. Oil cold	Oil will warm up in a few minutes
	b. Air in system	Bleed
	c. Oil level low	Add oil
	d. Air vent plugged	Open
	e. Dirt in pump	Clean
	f. Mechanical damage	Disassemble pump, replace defective parts
6. Gear noise	a. Loose over-center adjustment	Adjust
	b. Loose thrust-bearing adjustment	Adjust
	c. Air in system	Bleed
	d. Oil level low	Add oil
	e. Hose rubbing body or chassis part	Relocate hose

power-steering oil. If the oil level is low, there may be a leak. Check all hose and power-steering connections. Leakage may occur at various points in the power-steering gear if the seals are defective. Check around the piston and valve housings for leakage signs. Replace the seals or tighten the connections to eliminate leaks.

3. Check turning force A check of the power-steering action can be made with a spring scale hooked to the wheel rim (Fig. 15-4). Oil in the system must be warm before the test is made, and the front wheels must be resting on a level, dry floor. If the oil is cold, set the parking brake, start the engine, and allow it to idle for 3 minutes. While the engine is idling, turn the steering wheel left and right to warm the power-steering fluid.

With the oil warm, hook the spring scale on the steering-wheel rim. Check the pull required to turn the steering wheel one complete revolution in each direction. The

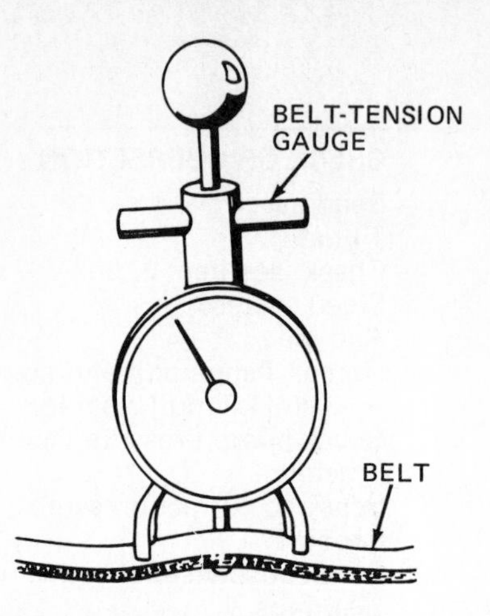

Fig. 15-2 Checking the tension of the power-steering-pump drive belt with a belt-tension gauge. (*Oldsmobile Division of General Motors Corporation*)

Belt Width

	5/16 INCH WIDE	3/8 INCH WIDE	15/32 INCH WIDE
New belt	350 N max. 80 lb max.	620 N max. 140 lb max.	750 N max. 165 lb max.
Used belt	200 N min. 50 lb min.	300 N min. 70 lb min.	400 N min. 90 lb min.
Used cogged belt		250 N min. 60 lb min.	

Fig. 15-3 Chart showing typical belt tension for various sizes of power-steering-pump drive belts. (*Oldsmobile Division of General Motors Corporation*)

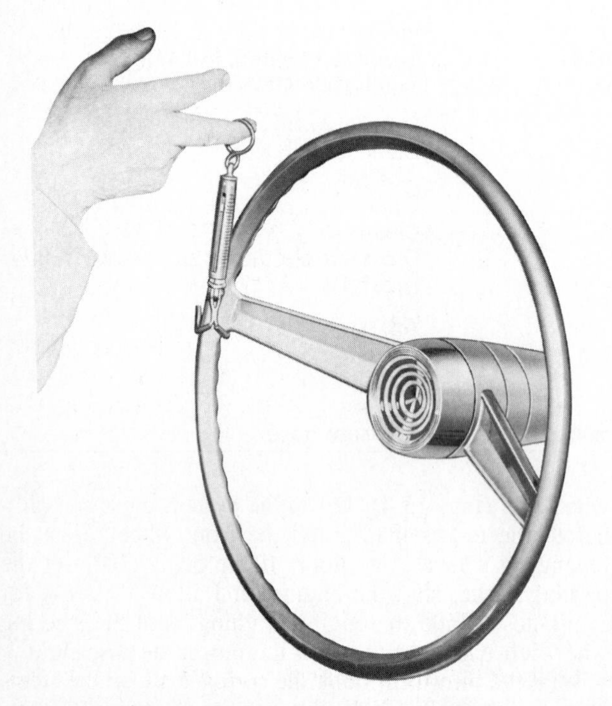

Fig. 15-4 Using a spring scale to measure the force required to turn the steering wheel. (*Chevrolet Motor Division of General Motors Corporation*)

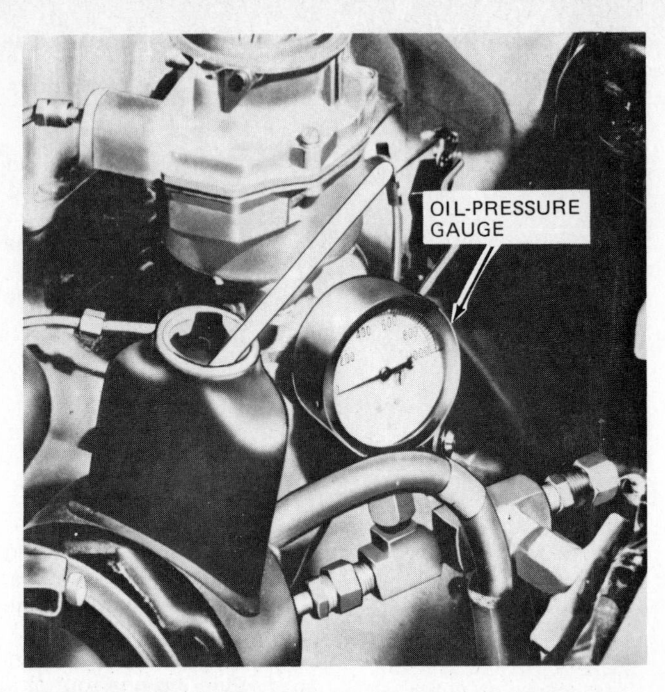

Fig. 15-5 The pressure gauge is connected into the line from the pump to check maximum pump pressure. (*Chevrolet Motor Division of General Motors Corporation*)

amount of pull specified varies with different cars because of conditions such as tire and wheel size and steering-linkage design. In general, if the force or pull required exceeds about 10 pounds [44.5 N], then the power steering is not working properly. The oil pressure should be checked.

NOTE: Be sure the engine is idling at the specified speed. If it is idling too slowly, it may not be turning the pump fast enough to build up normal pressure. Also, the tire pressure must be correct and the front wheels must be properly aligned.

4. Check oil pressure A power-steering analyzer or oil-pressure gauge is required to check oil pressure from the pump (Fig. 15-5). To use the gauge, disconnect the pressure hose from the pump and connect the gauge between the pressure fitting on the pump and the pressure hose.

To make the test, the pump reservoir must be filled, the tires inflated to the correct pressure, and the engine idling at the specified speed. Hold the steering wheel against one stop and check the connections at the gauge for leaks. Bleed the system as explained in the next section. Insert a thermometer in the pump reservoir. Move the steering wheel from one stop to the other several times until the thermometer registers 150 to 170°F [65.6 to 76.7°C].

NOTE: To prevent scrubbing wear and flat spots on the tires, move the car if you have to turn the steering wheel more than five times.

Hold the wheel against the stop and read the pressure on the gauge. If the pressure is below specifications, additional checks must be made. A typical pump should provide a minimum of 1100 psi [7590 kPa] pressure.

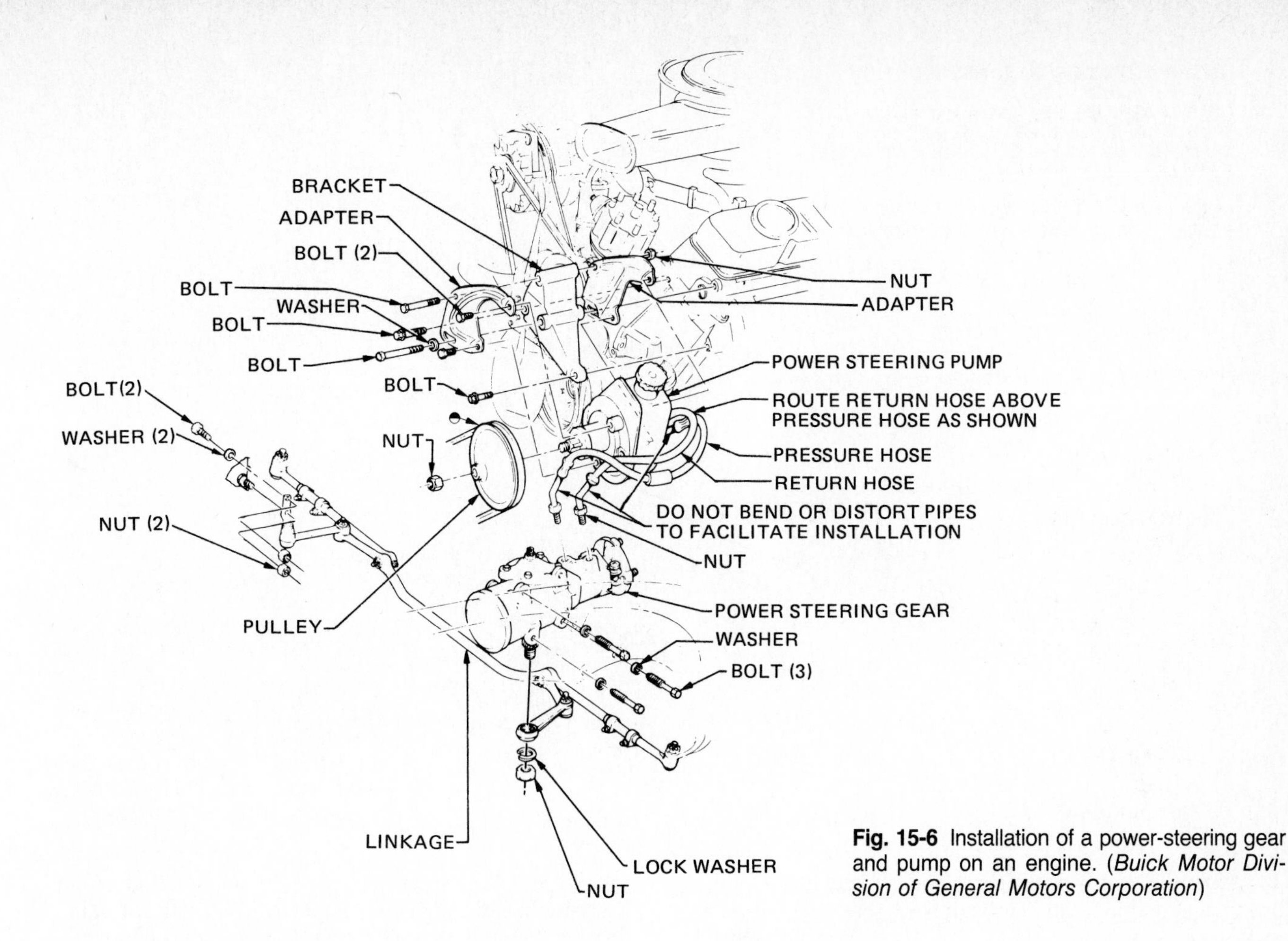

Fig. 15-6 Installation of a power-steering gear and pump on an engine. (*Buick Motor Division of General Motors Corporation*)

Slowly turn the shutoff valve to the closed position and read the pressure. Quickly reopen the valve to prevent damage to the pump. If the pressure is low, the pump is at fault. If the pressure comes up to normal, the trouble is in the hoses, connections, valve, or steering gear.

15-3 Bleeding the hydraulic sysem Air will enter the hydraulic system while oil lines are being disconnected and reconnected, after the pump or steering gear has been removed and replaced, or possibly because of a low oil level in the reservoir. The air must be removed, or the power-steering unit will operate noisily and unsatisfactorily.

To bleed the power-steering hydraulic system, fill the oil reservoir to the proper level and allow the car to sit for at least 2 minutes. Start the engine and let it run for a few seconds, then add oil if necessary. Repeat until the oil level remains constant.

Raise the car so that the front wheels are off the ground. Start the engine and run it at about 1500 rpm. Turn the steering wheel right and left. Turn to the stops, but don't bang them hard. Add more oil if needed. Lower the car to the ground and turn the wheel right and left. Add more oil if necessary.

If the oil is foamy, wait for 3 minutes and repeat the procedure. Then check the pulley for wobble. Also check the hoses and connections to make sure they are tight and not leaking. Foamy oil may indicate air leakage into the system, so check all points at which this could occur.

Adjustments

15-4 Adjusting steering gear on car At one time, an over-center adjustment could be made on the car. However, now many manufacturers discourage adjustment of the steering gear in the car. Getting to the steering gear is difficult, and the hydraulic fluid tends to confuse the adjustment. Therefore, the steering gear should be removed from the vehicle before any adjustments are attempted.

15-5 Removing steering gear from car The removal procedure varies from car to car because of differences in the installations. Figure 15-6 illustrates one installation. A typical removal procedure follows.

Disconnect the pressure and the return hoses from the steering gear. Elevate both hose ends so that the oil will not run out. Cap both the hoses and the steering-gear outlets to keep dirt from getting in. Remove the two nuts that attach the coupling lower flange to the steering-shaft coupling. On other cars, remove the pinch bolt that fastens the coupling to the stub end of the worm shaft (Fig. 15-7). Raise the car and remove the pitman arm from the pitman-arm shaft. Loosen the bolts that attach the steering gear to the frame, and remove the steering gear.

15-6 Adjusting steering gear off car Take the steering gear off the car as explained above. Drain the fluid and clamp the steering gear in a vise (Fig. 15-8). Make the adjustments as follows:

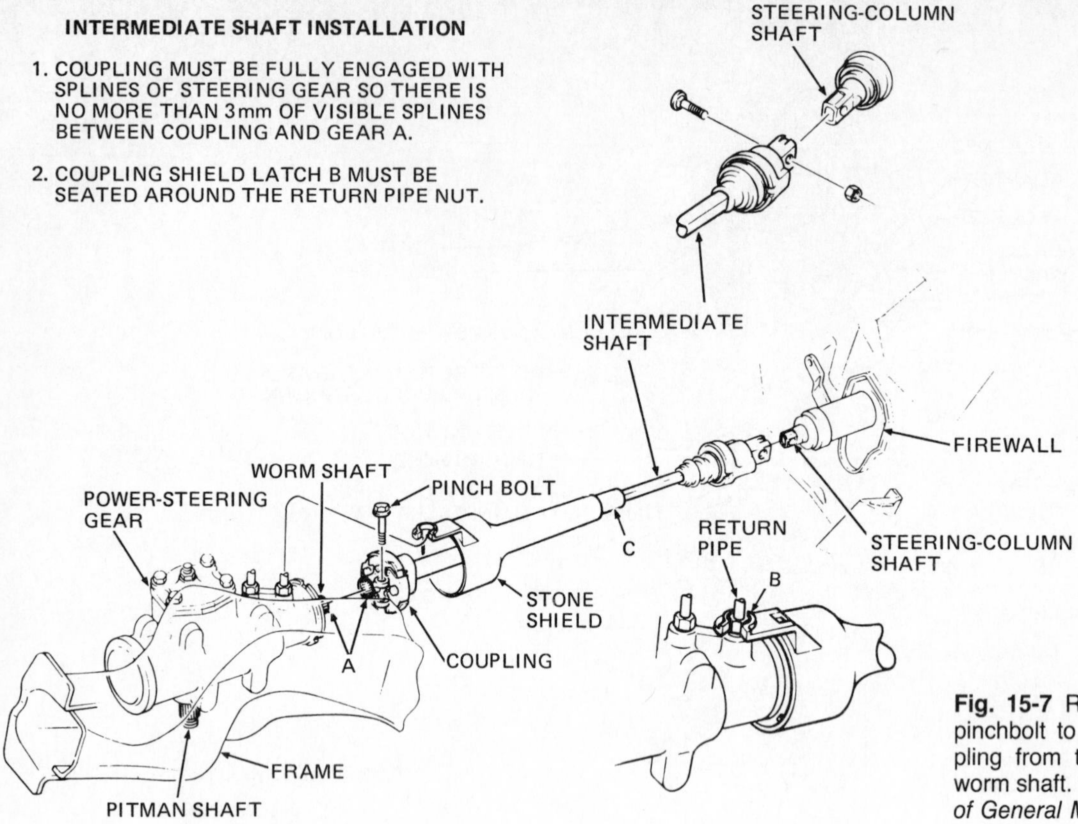

INTERMEDIATE SHAFT INSTALLATION
1. COUPLING MUST BE FULLY ENGAGED WITH SPLINES OF STEERING GEAR SO THERE IS NO MORE THAN 3mm OF VISIBLE SPLINES BETWEEN COUPLING AND GEAR A.
2. COUPLING SHIELD LATCH B MUST BE SEATED AROUND THE RETURN PIPE NUT.
STEERING-COLUMN SHAFT
INTERMEDIATE SHAFT
FIREWALL
WORM SHAFT
PINCH BOLT
POWER-STEERING GEAR
RETURN PIPE
C
STEERING-COLUMN SHAFT
B
STONE SHIELD
A
COUPLING
FRAME
PITMAN SHAFT

Fig. 15-7 Remove the coupling pinchbolt to disengage the coupling from the stub end of the worm shaft. (*Oldsmobile Division of General Motors Corporation*)

1. Adjust the worm-bearing preload by backing off the locknut and pitman-shaft adjuster screw 1½ turns. Loosen the adjuster-plug locknut. Use a spanner wrench to bottom the adjuster plug by turning it clockwise. Avoid excessive torque!

   Back off the adjuster plug 5 to 10°, or about 3⁄16 inch [5 mm] at the outside diameter of the adjuster plug. Use the torque wrench to make sure the torque required to turn the worm shaft is within specifications. Next, hold the adjuster plug with the spanner and tighten the locknut.
2. Over-center preload is adjusted by turning the pitman-shaft adjusting screw (Fig. 15-9) while turning the worm shaft through its center position with a torque wrench. First, check the torque with the adjusting screw backed out all the way. Then turn the adjusting screw in slowly, continuing to swing the torque wrench through center while turning the screw. Typically, the adjustment is correct when the torque wrench reads 4

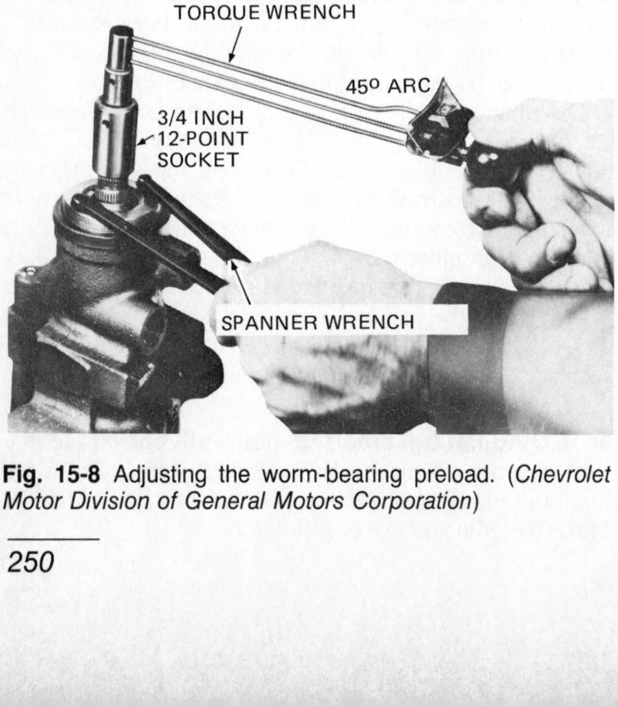

TORQUE WRENCH
45° ARC
3/4 INCH
12-POINT
SOCKET
SPANNER WRENCH

Fig. 15-8 Adjusting the worm-bearing preload. (*Chevrolet Motor Division of General Motors Corporation*)

120° ARC
3/4-INCH
12-POINT SOCKET
TORQUE WRENCH
PITMAN-SHAFT
END-PLAY
ADJUSTING SCREW
ALLEN WRENCH

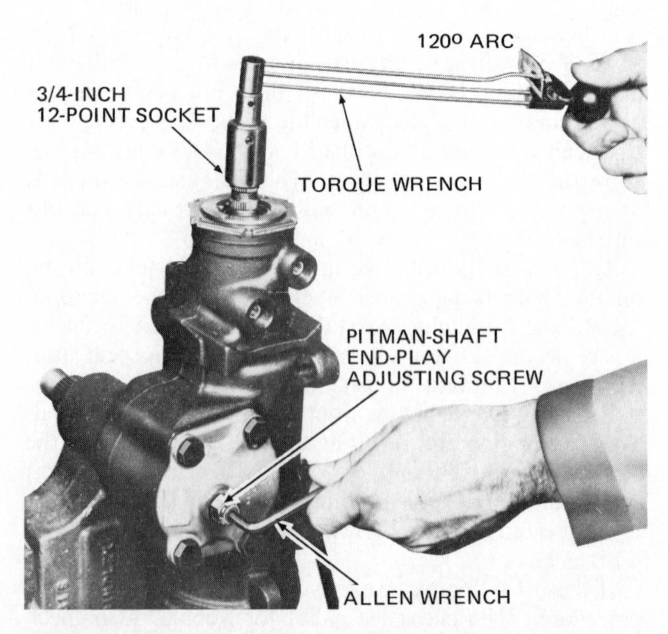

Fig. 15-9 Adjusting the over-center preload. (*Chevrolet Motor Division of General Motors Corporation*)

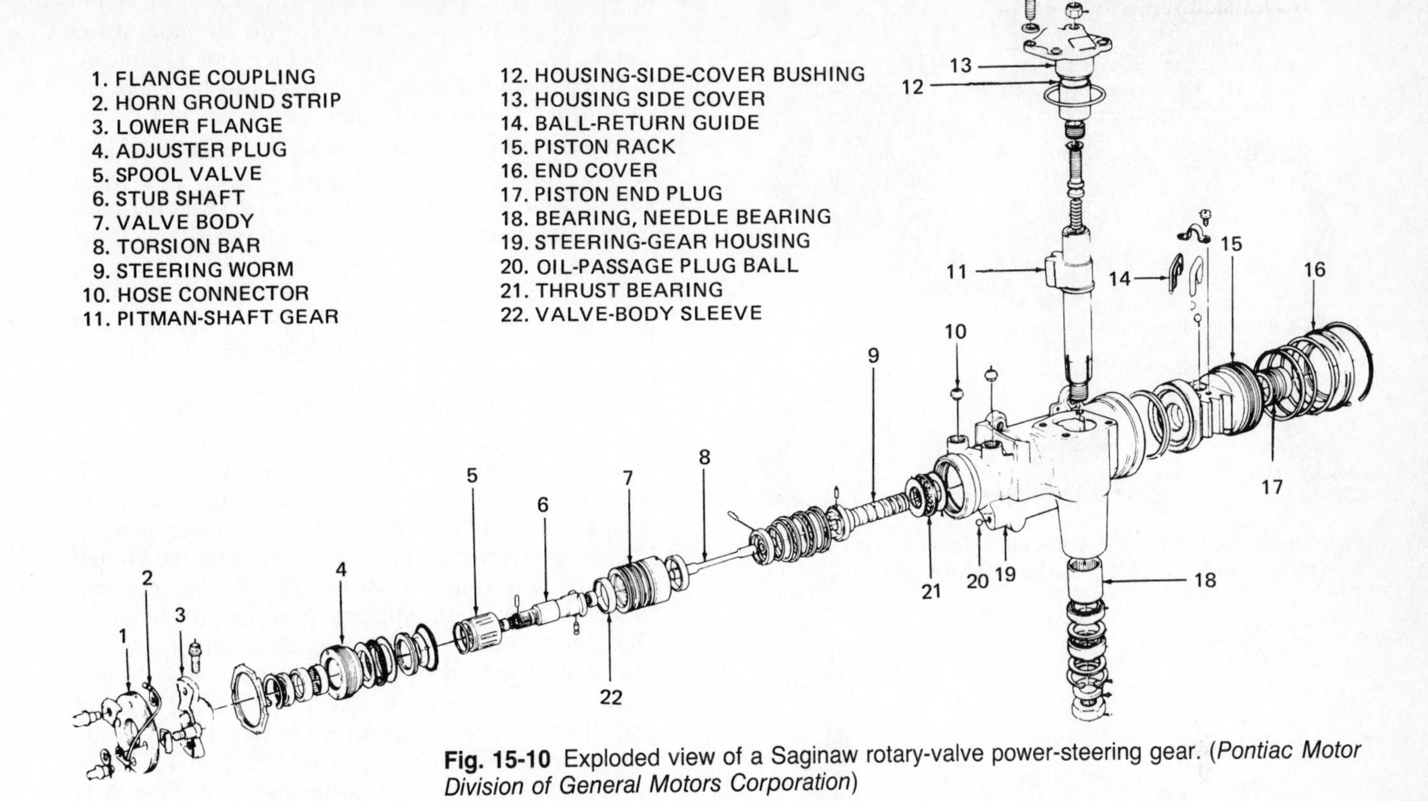

Fig. 15-10 Exploded view of a Saginaw rotary-valve power-steering gear. *(Pontiac Motor Division of General Motors Corporation)*

to 8 lb-in [0.452 to 0.904 N-m] more than the torque with the adjusting screw backed out. When adjustment is correct, tighten the locknut.

NOTE: The permissable additional torque for a steering gear that has been in service 400 miles [644 km] or more is 4 to 5 lb-in [0.452 to 0.565 N-m]. But the total torque should not exceed 14 lb-in [1.58 N-m].

Overhaul

15-7 Disassembly of Saginaw power-steering gear Figure 15-10 is an exploded view of the Saginaw rotary-valve power-steering gear. After removing the steering gear from the car, mount the steering gear ina holding fixture which can be clamped in a vise (Fig. 15-11). Never clamp the housing in a vise. This could distort the housing. Clean the steering gear and then drain out all lubricant. Turn the worm shaft through its entire range several times to assist drainage.

NOTE: It is not always necessary to completely disassemble the steering gear to find and fix a problem. Most of the components can be removed from the housing without complete disassembly. However, the complete procedure is described below.

Careful! When disassembling a steering gear, your work area, your tools, and the parts must be kept clean. A trace of dirt can cause malfunctioning of the steering gear. If a broken component or dirt is found in either the steering gear or pump, the entire hydraulic system must be disassembled and cleaned. The system must be flushed out and fresh power-steering fluid added after reassembly.

Fig. 15-11 Steering gear mounted on a holding fixture. *(Pontiac Motor Division of General Motors Corporation)*

1. Use a punch and a screwdriver to remove the end-plug retaining ring (Fig. 15-12). Turn the worm-shaft to push the end plug out. Discard the O ring. Do not turn the shaft farther than necessary. This could allow the balls to drop out of the ball nut. Remove the rack piston end plug with a ½-inch (12.7-mm) socket extension (Fig. 15-13).

Fig. 15-12 Removing the end-plug retaining ring. (*Pontiac Motor Division of General Motors Corporation*)

Fig. 15-13 Removing the end plug. (*Chevrolet Motor Division of General Motors Corporation*)

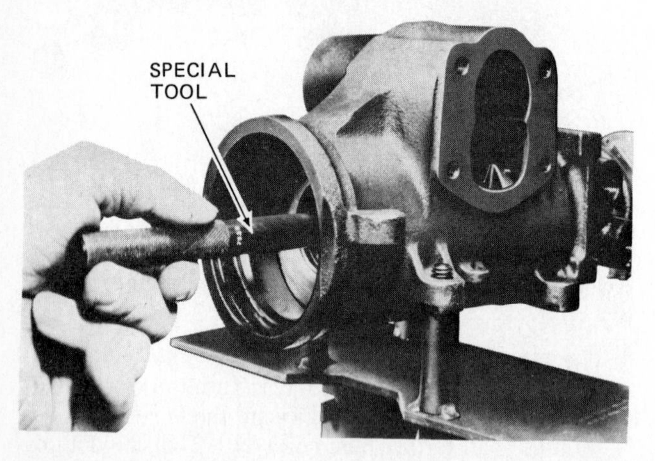

Fig. 15-14 Holding the special tool against the end of the rack piston to keep the balls from falling out. (*Pontiac Motor Division of General Motors Corporation*)

2. Remove the side-cover screws and washers. Move the cover around so that you can see the location of the sector. Turn the worm shaft until the sector is centered so that the pitman shaft and cover can be removed. Discard the side-cover O ring.
3. Remove the rack piston by using a special tool (Fig. 15-14). Hold the tool in the end of the piston while you turn the worm shaft. The tool prevents the balls from falling out while the worm is threaded out of the rack nut.
4. Take the coupling flange off the worm shaft by removing the locking bolt. Remove the adjuster-plug locknut with a punch or a spanner wrench and then remove the adjuster plug. Push on the end of the worm with a hammer handle while turning the shaft to slip the assembly out of the housing. Figure 15-15 shows these parts. Pull the adjuster plug off the shaft. Pull the worm shaft from the rotary valve and discard the O ring. Discard the O ring from the adjuster plug.
5. If necessary, separate the side cover from the pitman shaft by removing and discarding the locknut and backing out the lash adjuster from the pitman shaft. These parts are serviced as a single assembly.
6. The rotary valve should be disassembled only if necessary. If a "squawk" has developed in the steering gear, the valve-spool-dampener O ring probably needs replacement. This ring is shown in Fig. 15-16. Work the spool spring onto the bearing diameter of the shaft so that the spring can be removed. Tap the end of the shaft gently down against the workbench so that the spool comes off.

Careful! The clearance is small and the spool could cock and jam in the valve body.

Remove and discard the O ring. Put a new O ring in the spool groove, lubricate it with power-steering fluid, and install the spool in the valve body. Do not allow the O ring to twist in the groove. *Extreme* care is required to prevent damage to the O ring! The notch in the spool must align with the pin in the stub shaft.

7. On the housing, the pitman-shaft lower seal and bearing can be replaced, if necessary. Special drivers are used to drive out the old bearing and drive in the new bearing.

⚙ 15-8 Inspection of Saginaw power-steering gear parts Figure 15-10 shows the power-steering gear disassembled and ready for inspection. If the pitman-shaft bearing in the side cover is worn, replace the side-cover assembly. Replace the pitman-shaft-and-lash-adjuster assembly if the sector teeth or bearing surfaces are worn or damaged or if the lash adjuster has end play in the shaft.

The worm groove and rack-piston interior grooves and balls should be checked for wear. If replacement is required, both the worm and the rack piston must be replaced as a matched assembly.

Check the ball-return guides, making sure the ends, where the balls enter and leave, are not damaged. Replace the lower thrust bearing and races if they are worn or otherwise damaged.

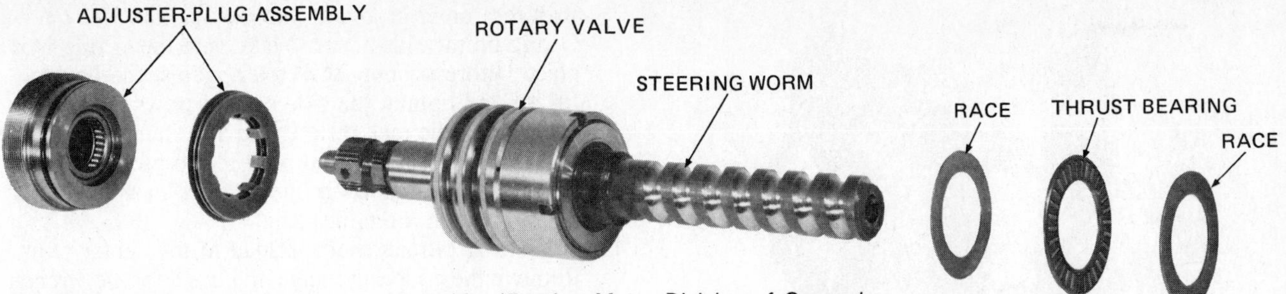

Fig. 15-15 Rotary valve and worm assembly. (*Pontiac Motor Division of General Motors Corporation*)

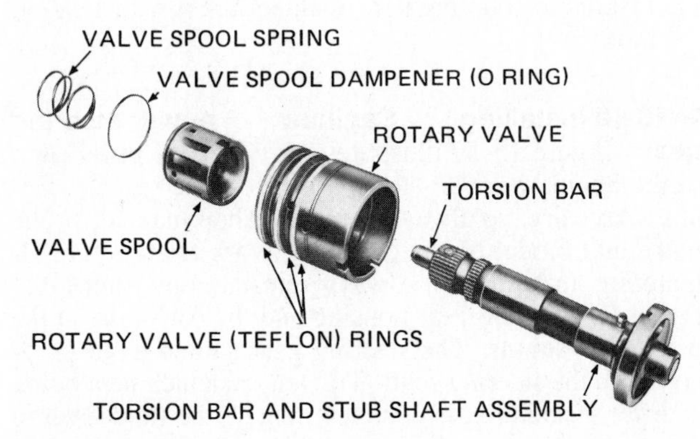

Fig. 15-16 Disassembled rotary valve. (*Pontiac Motor Division of General Motors Corporation*)

If the rotary valve is damaged, the valve must be replaced as an assembly. The valve parts are matched and are not serviced separately.

The housing will require replacement if there are defects in the piston bore or rotary-valve bore. A slight polishing of the bores is normal.

15-9 Assembly of Saginaw power-steering gear

Lubricate all parts as they are assembled.

1. Screw the lash adjuster through the side cover until the cover bottoms on the shaft. Install but do not tighten the locknut.
2. Use great care when assembling the rotary valve, and make sure the valve spool does not cock and jam in the valve body. Figure 15-16 shows this valve. The notch in the spool must line up with the pin in the shaft.
3. Figure 15-17 shows the proper relationship of the pitman-shaft seals and washers. If the old shaft seals and washers have been removed, new ones should be installed by using special drivers.
4. Working on the rack piston, lubricate and install the O rings and piston rings. Insert the worm all the way into the rack piston. Align the ball-return guide holes with the worm groove. Load the proper number of balls into the guide hole nearest the piston ring while slowly rotating the worm to the left. Alternate black and silver balls. Fill one half of a ball-return guide with seven balls. Place the other guide over the balls and plug the ends with grease. Insert the guide into the guide holes. Secure with a guide clamp and screws.
5. Check the worm preload. The worm has a high point at the center, which should cause a small increase in torque when the rack piston passes over this point. If this torque is not as specified in the manufacturer's instructions, it may be necessary to fit smaller or larger balls. However, this is not necessary unless the driver has complained of loose steering. Normally, a thrust and over-center adjustment, as explained earlier, will correct the problem. Put a special arbor into the end of the rack piston and turn the worm out of the rack piston. Keep the arbor in contact with the worm so that the balls do not fall out.
6. Install the lower thrust bearing and races on the worm. Assemble the valve assembly to the worm, making sure the slot aligns with the pin on the worm head. Make sure the O ring is between the valve body and the worm head. Install a new O ring on the adjuster plug and put the plug on the shaft.
7. Put the worm-valve assembly into the housing and turn the adjuster plug into the housing until it is snug. Back it off about one-eighth turn. Use a torque wrench to check the torque required to turn the worm shaft. Tighten the adjuster plug to obtain the correct reading. Install and tighten the locknut and recheck the torque.
8. Put the coupling flange on the worm shaft. Use a special piston-ring compressor so that the ring will

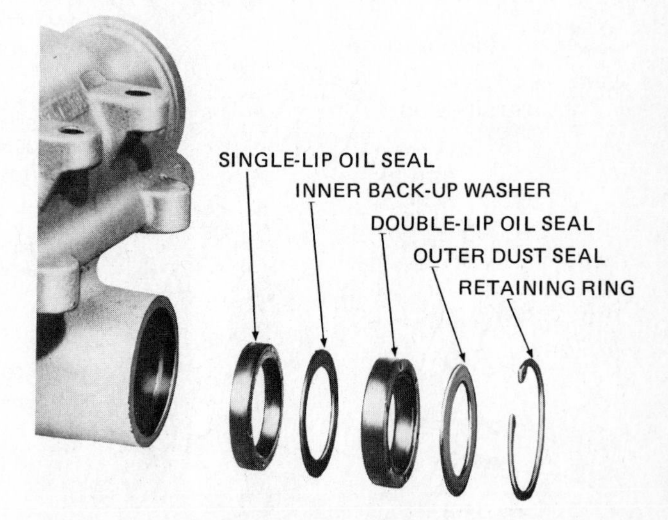

Fig. 15-17 Pitman-shaft seals and washers. (*Pontiac Motor Division of General Motors Corporation*)

Fig. 15-18 Installing the rack-piston assembly with a ring compressor and special arbor. (*Pontiac Motor Division of General Motors Corporation*)

go into the housing. Push the rack piston into the housing until the piston-rack arbor contacts the end of the worm (Fig. 15-18). Hold the arbor tightly against the end of the worm and turn the coupling flange to draw the rack piston onto the worm and into the housing. Do not drop the balls out of the rack piston!

9. Install the pitman shaft and the side cover. Make sure that the center tooth of the shaft sector aligns with the center groove of the rack piston. Use a new O ring on the side cover. Make sure the O ring is in place before pushing the cover against the housing. Install and tighten the side-cover screws. Install the end plug in the rack piston with a ½-inch (12.7-mm) socket drive. Tighten the plug to the proper torque.
10. Install the housing lower end plug with a new O ring. Secure it with a retainer ring.
11. Adjust the pitman-shaft preload at the center point. Remove the coupling flange and use a torque wrench to check the preload at the center point. Adjust the lash adjuster to increase the torque within specifications. Tighten the locknut and recheck the preload.
12. Install the coupling flange and secure it with a clamp bolt.

⚙ 15-10 Installing Saginaw power-steering gear Figure 15-19 illustrates the details of installation of the Saginaw power-steering gear on one vehicle. Details may vary, so always check the shop manual for the make and model of the car being worked on before attempting installation. Always reinstall any shims between the steering-gear housing and the car frame in the original positions. The steering gear should align properly with the steering shaft. Tighten the attachment bolts.

After installation is complete, check the fluid level in the pump reservoir. Add fluid as necessary to bring it to the full mark. With the car wheels off the floor, start the engine. Turn the steering wheel right and left to the stops several times to bleed out all air. Make a final check of the through-center pull after the installation is complete to make sure there is no misalignment.

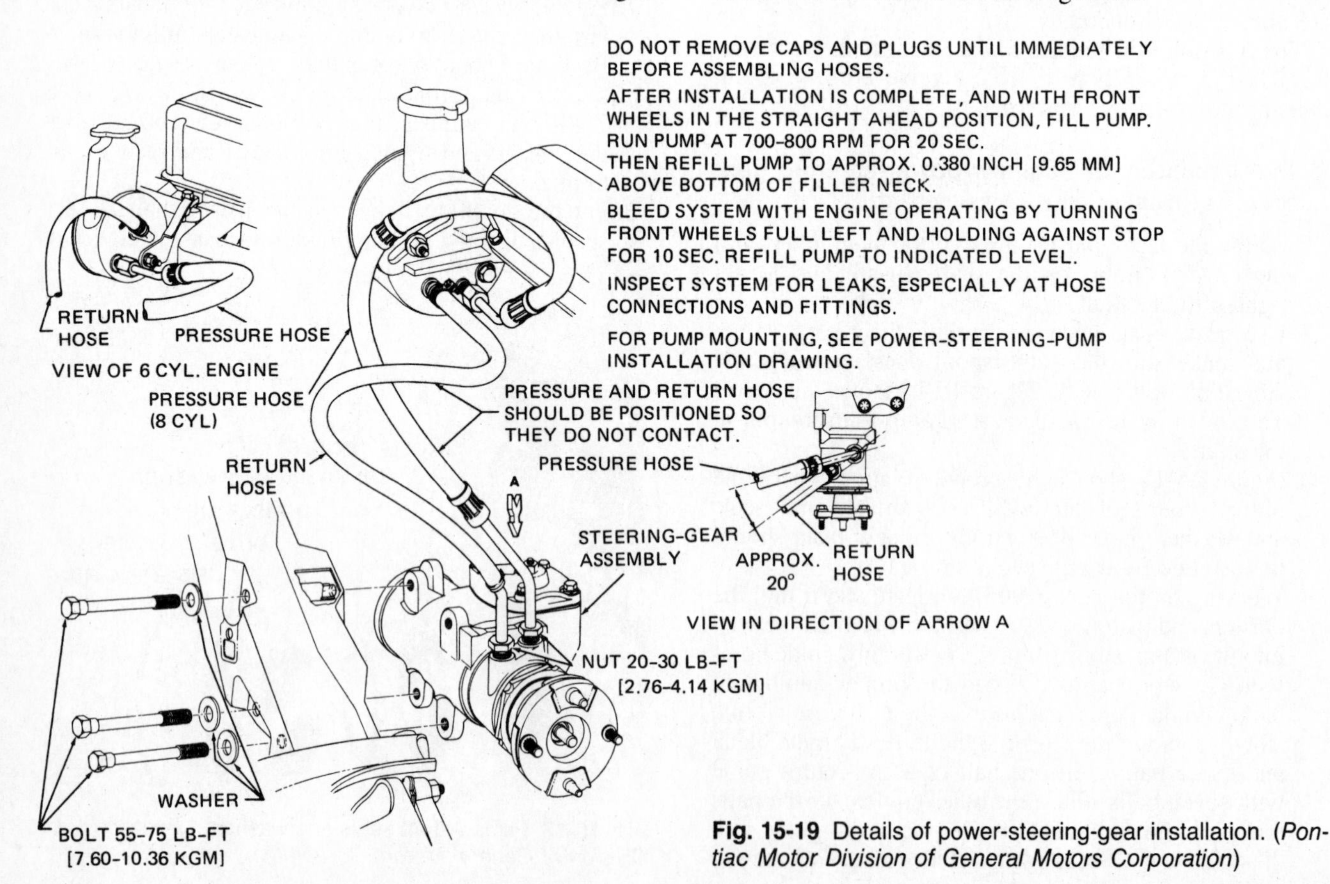

Fig. 15-19 Details of power-steering-gear installation. (*Pontiac Motor Division of General Motors Corporation*)

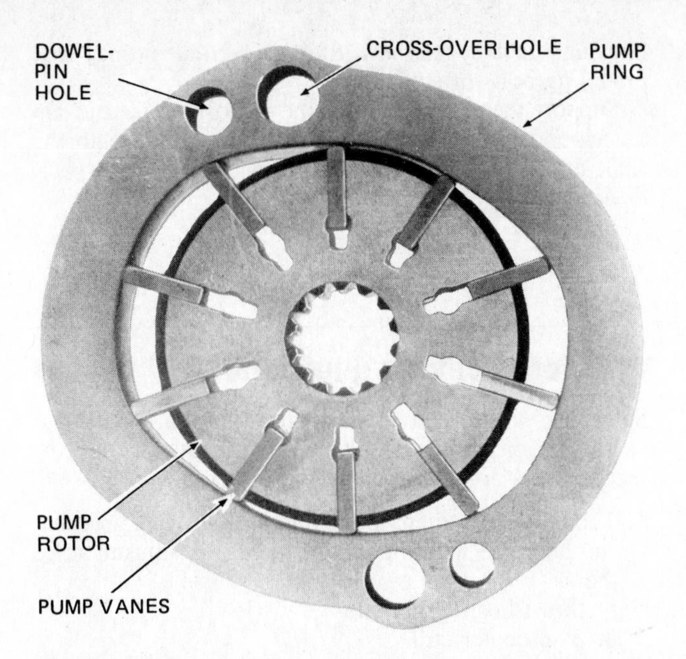

Fig. 15-20 Ring, rotor, and vanes of a vane-type-power-steering pump. Notice how the vanes fit into the pump rotor. (*Chevrolet Motor Division of General Motors Corporation*)

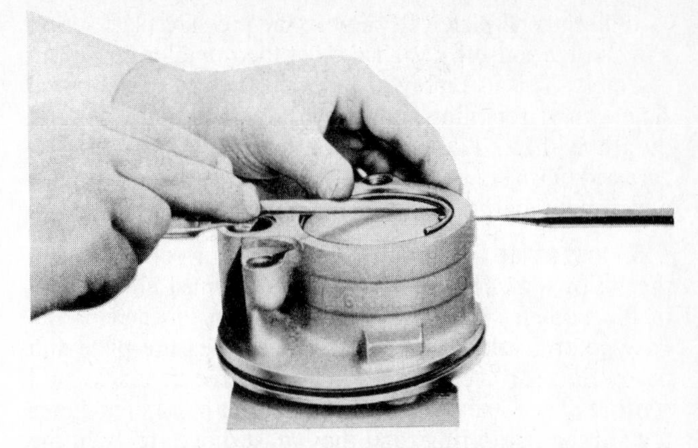

Fig. 15-22 Removing the end-plate ring with a punch and screwdriver. (*Chevrolet Motor Division of General Motors Corporation*)

Pump service

15-11 Saginaw power-steering pump service The vane pump shown in Figs. 15-19 and 15-20 is widely used in power-steering systems. It has a series of vanes assembled in slots in a rotor. A typical overhaul procedure follows.

1. Pump removal To remove the pump from the engine, disconnect the hoses from the pump and fasten them in an elevated position so that they will not drain. Put caps on all pump connections. Take off the pump-mounting bolts. Remove the pump. Drain all oil from the reservoir after removing the cover.

2. Disassembly Clean the outside of the pump with solvent. Remove the pulley retaining nut and pulley. Lightly clamp the pump in the soft jaws of a vise. Remove the fitting, O ring seal, and mounting bolts or studs from the reservoir. Figure 15-21 shows these parts. Tap lightly on the outer edge of the reservoir to break it loose. Remove the reservoir and O ring seal from the housing. Discard the O ring seal. Remove and discard the O ring seals from the mounting bolts or studs. On the Corvette, remove and discard the filter assembly from the end plate.

Remove the end-plate retaining ring, using a small punch to compress the ring and a screwdriver to pry it out (Fig. 15-22). Remove the end plate. The end plate is spring-loaded and will usually rise above the housing when the ring is removed. If it sticks, tap it lightly with a soft hammer or rock it.

Fig. 15-21 Exploded view of a vane-type power-steering pump. (*Pontiac Motor Division of General Motors Corporation*)

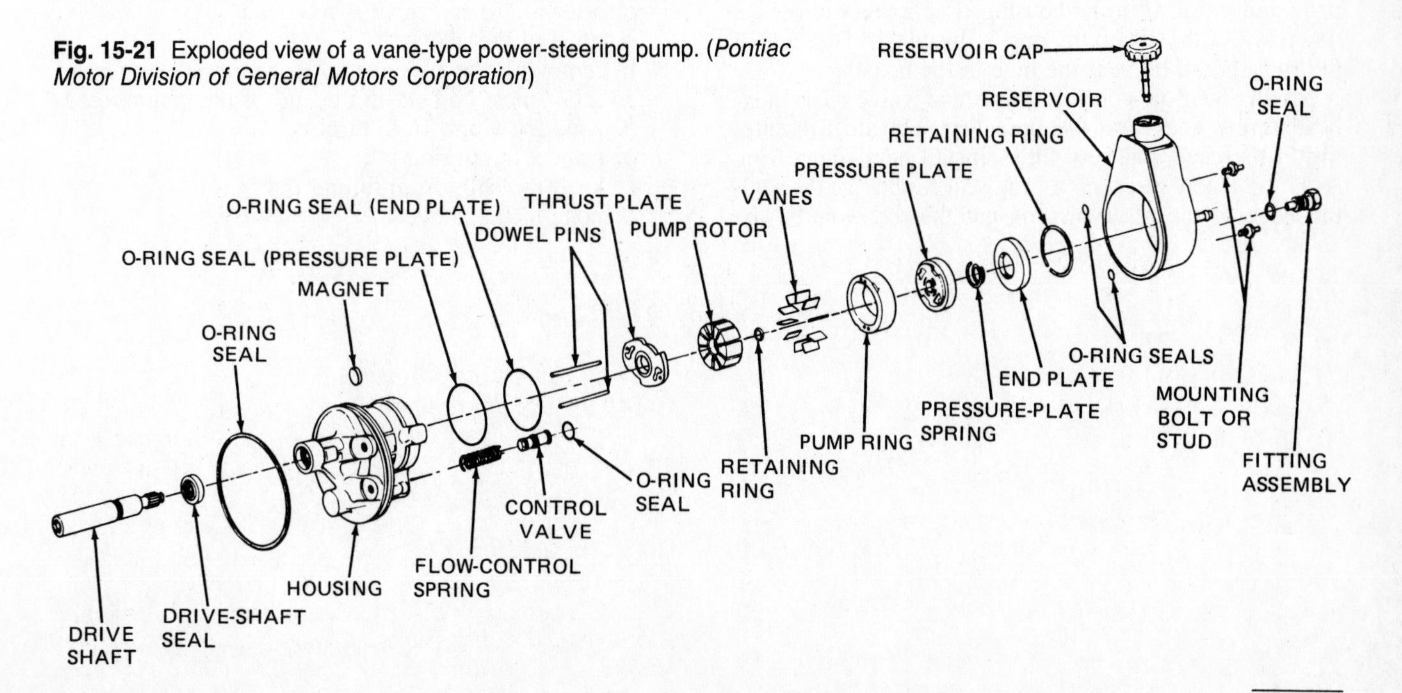

With the end plate off, remove the pressure-plate spring and shaft woodruff key. Take out the impeller unit as an assembly. This assembly includes the pressure plate, pump ring, vanes, retaining ring, pump rotor, thrust plate, and the drive shaft. These parts can then be separated. Discard old O rings and seals, including the drive-shaft seal, which must be pried out of the housing.

3. Inspection After cleaning all parts in solvent, check the following. The flow-control valve must slide freely in the housing bore. The cap screw in the end of the flow-control valve must be tight. The pressure-plate and pump-ring surfaces must be flat and free of cracks and scoring. The vanes must be installed with rounded edges toward the pump ring, and they must move freely in the rotor slots. Check the drive shaft for worn splines and cracks.

4. Reassembly Make sure all parts are clean during reassembly. Install a new shaft seal in the pump housing. Put dowel pins in the pump housing and install a new pressure-plate O ring lubricated with power-steering fluid. Install the thrust plate on the shaft with ports facing toward the splined end of the shaft.

Install the rotor on the shaft with the countersunk side toward the thrust plate. Install the new shaft retaining ring, tapping it onto the shaft with a drift punch. Tap the retaining ring into place with a ⅜-inch socket. Put the housing in the vise and install the shaft, thrust plate, and rotor assembly. Align the holes in the thrust plate with the dowels in the housing. Install the pump ring on the dowels with the arrow that shows the direction of rotation positioned to the rear of the housing. Install the vanes in the rotor slots with rounded edges out. Lubricate the pressure plate with power-steering fluid and install it with ports toward the pump ring. Seat the plate by pressing down with a large socket on top of the plate. Put the pressure-plate spring in place.

Lubricate a new end-plate O ring and install it in the housing groove. Lubricate the outer diameter of the end plate and install it in the housing. Use a press to hold it down while the retaining ring is installed. The end of the ring should be near the hole in the housing.

Install the flow-control spring and valve. The hex-head screw goes into the bore first. On the Corvette, install the cage and new filter. Install new square ring seals and a new reservoir O ring seal. Lubricate the sealing edge of the reservoir and put the reservoir on the housing. Install the new fitting O ring, fitting, and mounting bolts or studs.

Support the drive shaft on the opposite side and tap the woodruff key into place. Slide the pulley onto the shaft and install the pulley nut, then torque it to specifications.

Chapter 15 review questions

Select the *one* correct, best, or most probable answer to each question. Then check your answers against the correct answers given at the end of the book.

1. The only adjustment to the Saginaw steering gear that can be made with the unit in the car is adjustment of the:
   a. thrust-bearing preload
   b. over-center pull
   c. oil pressure
   d. none of the above
2. Turning the adjustment screw to make the over-center adjustment causes:
   a. endwise movement of the ball nut
   b. endwise movement of the pitman shaft
   c. endwise movement of the worm shaft
   d. endwise movement of the steering shaft
3. In the Saginaw steering gear, the thrust-bearing preload is adjusted by:
   a. turning an adjustment screw
   b. adding or removing shims
   c. turning an adjustment plug
   d. all of the above
4. In disassembling the steering gear, the pitman-shaft assembly must be removed:
   a. after the rack piston is removed
   b. before the rack piston is removed
   c. after the rotary valve is removed
   d. none of the above
5. In removing the rack piston from the steering gear, an arbor must be held in the end of the piston to:
   a. keep the worm from turning
   b. protect the O ring
   c. keep the balls from falling out
   d. none of the above

CHAPTER 16

CHRYSLER POWER-STEERING SERVICE

After studying this chapter, and with proper instruction and equipment, you should be able to:

1. Diagnose troubles in the Chrysler power-steering gear.
2. Adjust the steering gear.
3. Overhaul the steering gear.
4. Overhaul the power-steering pump.

16-1 Chrysler power-steering trouble diagnosis The Chrysler power-steering gear is an integral unit which combines the control valve for the hydraulic fluid and the power piston with the steering gear (Fig. 10-45). On older cars a roller-type power-steering pump was used. Now the system uses a vane-type pump similar to the General Motors Saginaw pump that was discussed earlier. In the Chrysler power-steering system there is always some pressure on both sides of the power piston. It is the difference in pressure that provides the power assist. This power-steering gear is described in ⚙ 10-21.

The trouble-diagnosis chart on the following page lists trouble-symptoms, causes, and checks or corrections that will help you to diagnose troubles and locate their causes in the Chrysler power-steering system. Following sections describe the procedures required to eliminate the troubles when servicing or repairing the system.

⚙ 16-2 In-vehicle service The sector shaft can be adjusted, the valve body reconditioned, and the sector-shaft oil seal replaced without removing the steering gear from the vehicle.

1. Sector-shaft adjustment Disconnect the center link from the steering-gear arm. Start the engine and run at idle speed. Turn the steering wheel gently from one stop to the other, counting the number of turns. Then turn the wheel back exactly halfway, to the center position. Loosen the sector shaft adjusting screw until backlash is felt in the steering-gear arm. Figure 10-45 shows this adjusting screw.

You can feel the backlash by holding the end of the arm with a very light grip and attempting to move it. Tighten the adjusting screw until backlash just disappears. Continue to tighten beyond this point three-eighths to one-half turn. Then tighten the locknut to specifications to maintain the setting.

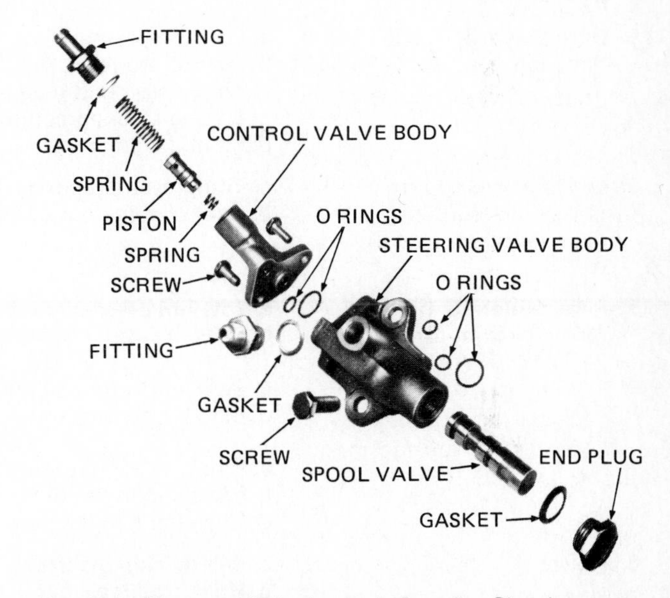

Fig. 16-1 Disassembled valve body from the Chrysler power-steering gear. (*Chrysler Corporation*)

2. Valve-body reconditioning Figure 16-1 shows the disassembled valve body. To recondition it on the vehicle, disconnect the hoses and tie the ends up to avoid loss of fluid. Remove the screws and lift the valve body up and off. Separate the control-valve body from the steering-valve body by removing the screws. Then, from the control-valve body, remove the fitting, gasket, spring, piston, and spring. Carefully shake out the spool valve from the steering-valve body. Do not remove the end plug unless the gasket is leaking so that it must be replaced.

Clean the parts with solvent and blow out the passages with compressed air. Examine the spool valve for burrs and nicks. These may be removed with fine crocus cloth, provided you do not round off the sharp edges. If the

Chrysler Power-Steering Trouble-Diagnosis Chart

COMPLAINT	POSSIBLE CAUSE	CHECK OR CORRECTION
1. Hard steering in both directions	a. Leak in hydraulic system	Correct leak, refill
	b. Fluid level low	Check for and correct leak, refill reservoir
	c. Pump belt slipping or broken	Tighten, install new belt
	d. Linkage not lubricated	Lubricate
	e. Tire pressure low	Inflate tires properly
	f. Oil pressure low	See item 2, below
	g. Bind in steering column or gear	Align or adjust
	h. Front alignment off	Align front end
2. Low oil pressure	a. Belt loose	Tighten, replace if necessary
	b. Pump valve stuck	Free, clean if necessary
	c. Mechanical trouble in pump	Repair or replace pump (⚙ 16-8)
	d. Pressure loss in steering gear	Repair, adjust steering gear
3. Oil leaks	a. Hose adapter	Tighten or replace adapters or gaskets
	b. At pump	Repair or replace pump (⚙ 16-8)
	c. Between gear and worm housings	Tighten attaching screws, replace O ring
	d. Gear-shaft oil seal	Replace oil seal
4. Smaller turning radius in one direction	Wheel stops out of adjustment	Readjust wheel stops
5. Hard steering in one direction	a. Low tire pressure	Inflate tires properly
	b. Internal troubles, oil leak past seal ring, piston end plug loose, etc.	Recondition steering gear
	c. Bind in steering column or gear	Align or readjust
	d. Front alignment off	Realign front end
6. Car attempts to turn unless pressure is maintained on steering wheel	a. Tire pressure uneven	Inflate tires properly
	b. Control valve out of adjustment	Readjust
7. Poor centering, or recovery from turns	a. Low tire pressure	Inflate tires properly
	b. Balls in worm connector are binding	Disassemble unit to clear or replace balls and connector
	c. Bind in steering column or gear	Align or readjust
	d. Bind in steering knuckles	Check ball joints and bushings, replace as necessary
	e. Worm-bearing adjustment too tight	Readjust
	f. Front alignment off	Align front end
	g. Gear-shaft adjustment too tight	Adjust
8. Noises	a. Belt tension incorrect	Adjust
	b. Fluid level low	Check for leaks, fill oil reservoir
	c. Worn pump bearings	Repair or replace pump (⚙ 16-8)
	d. Dirt or sludge in pump	Disassemble and clean pump, drain system, refill, change filter element
	e. Noise in power unit	Check for air in system, hose clearance, gear-shaft adjustment

valve or body is damaged, replace the valve-and-body assembly.

Lubricate parts with power-steering fluid. Install the spool valve in the body so that the valve-lever hole aligns with the lever opening in the valve body. Use new O rings during the reassembly.

To reassemble the valve body, put the cushion spring or inner spring in the counterbore in the bottom of the control-valve body. Lubricate the piston and insert the nose end into the bore. Make sure that the cushion spring is not cocked and that the piston slides easily in the bore. Install the other spring, washer, and fitting. Tighten to specifications.

Using new O rings, attach the control-valve body to the steering-valve body. Tighten the attaching screw to specifications. Install the assembly on the steering gear with new O rings, making sure that the valve lever enters the hole in the valve spool. Be sure that the key section on the bottom of the valve body nests with the keyway in the gear housing. Secure the assembly with attaching

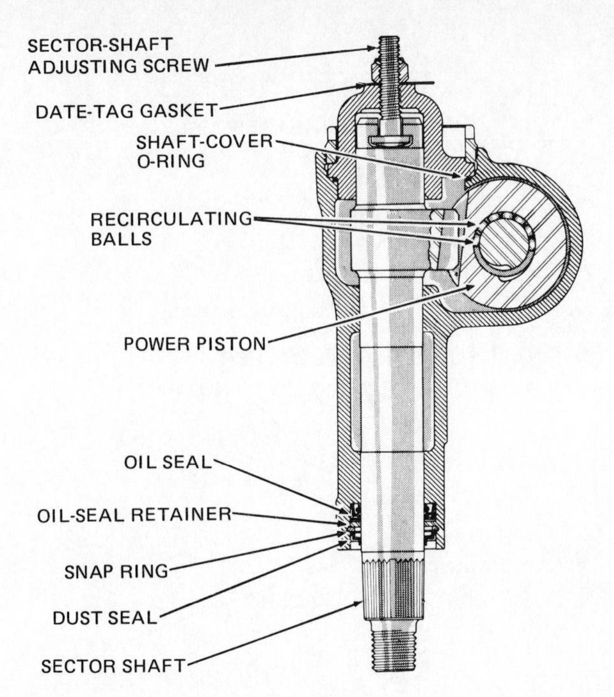

Fig. 16-2 Cutaway view of the sector shaft and oil seal. (*Chrysler Corporation*)

screws tightened to specifications. This should be just enough to hold the assembly in place during the adjusting procedure.

Connect the hoses to the valve body and start the engine. If the assembly tries to steer itself, the valve body is not in the correct position. Tap it up or down to move the valve body and stop the steering effect. When tapping down, tap on the end plug. When tapping up, tap on the head of the screw attaching the control-valve body to the steering-valve body. Do not hit the control-valve body. Turn the steering wheel gently from one stop to the other to eliminate air from the system. Refill the reservoir as necessary.

With the steering wheel straight ahead, start and stop the engine several times. Tap the valve body up or down as necessary to eliminate any steering-wheel movement when the engine is started or stopped. Tighten the attaching screws to specifications.

3. **Sector-shaft oil-seal replacement** This seal is located at the outer end of the sector shaft (Fig. 16-2). It can be replaced by removing the nut from the shaft and pulling off the steering arm. Carefully pry the dust seal out of the housing. Remove the oil-seal snap ring with snap-ring pliers, and remove the seal backup washer. Place a drain pan under the steering gear and start the engine. Turn the steering wheel all the way to the left. The oil pressure against the seal will force it out of the housing.

Using a seal protector, lubricate and install the new seal. Secure it in place with a new seal backup washer, snap ring, and a new dust seal. Install the steering arm in its original position on the sector shaft, and tighten the retaining nut to specifications.

❁ 16-3 Steering-gear removal To avoid damaging the energy-absorbing steering column, the column should be completely detached from the floor and instrument panel. Figure 16-3 shows the attachment points of a

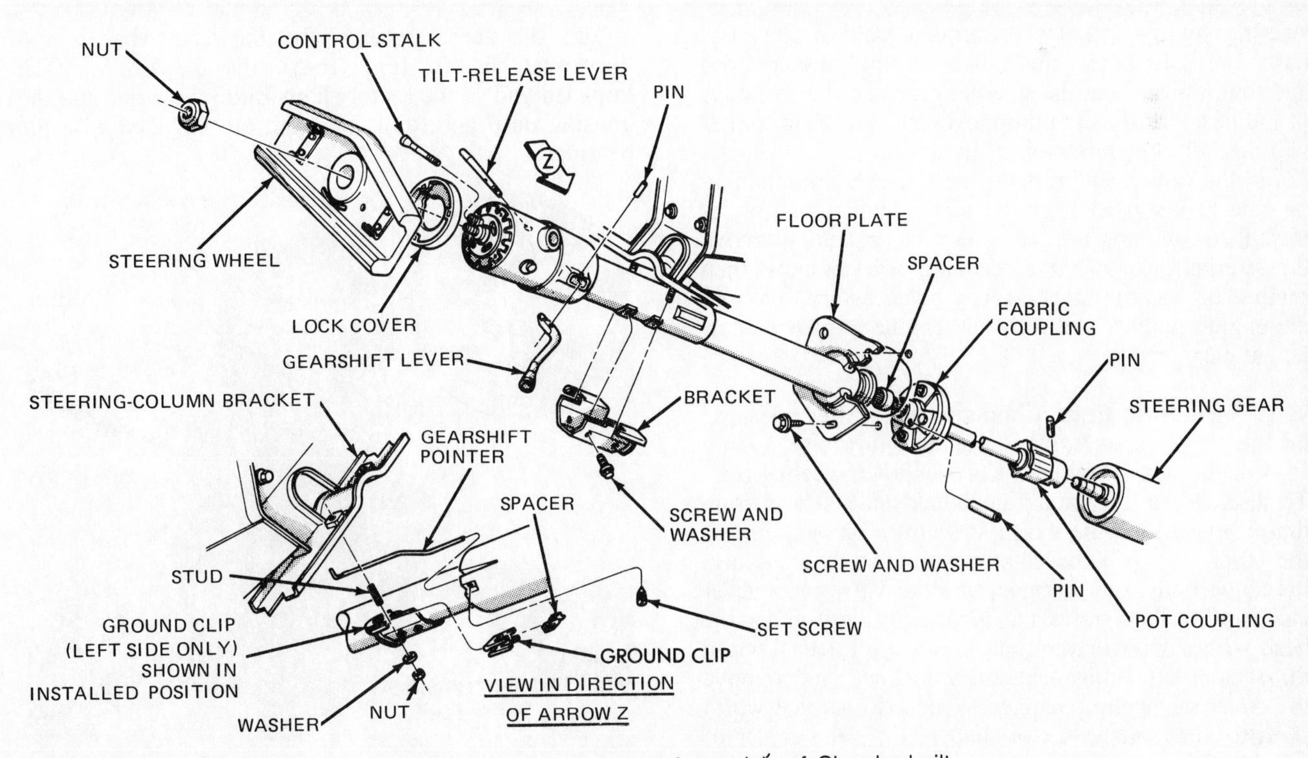

Fig. 16-3 Attachment points of the steering column on certain models of Chrysler-built cars. (*Chrysler Corporation*)

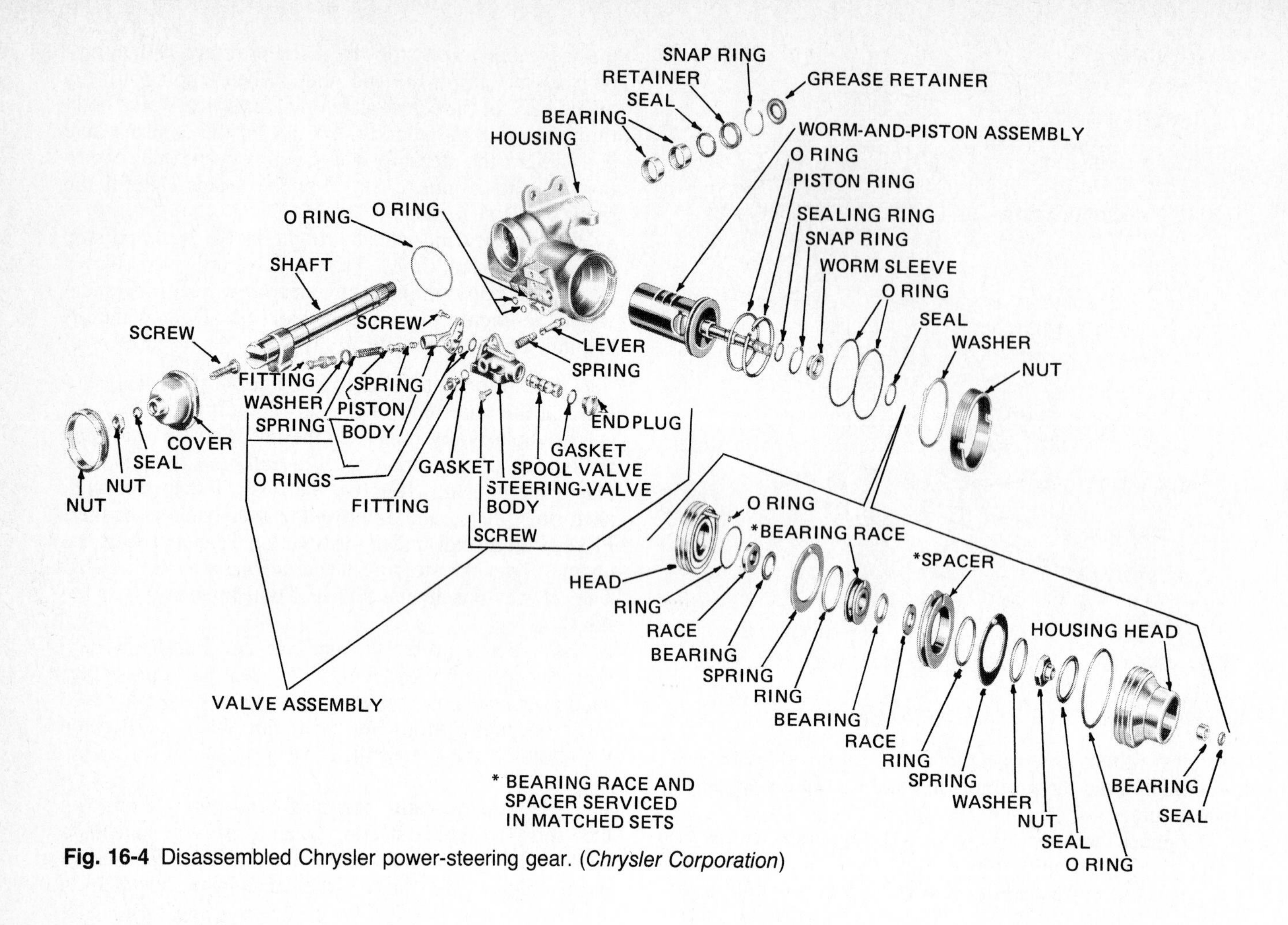

Fig. 16-4 Disassembled Chrysler power-steering gear. (*Chrysler Corporation*)

steering column found on several models of Chrysler-built cars. Disconnect the power-steering pressure hose and return hose from the steering gear. Tie the free ends of the hoses above the pump level and cap them to prevent loss or contamination of fluid.

Lift the vehicle and remove the steering-arm retaining nut and lock washer from the sector shaft. Use a puller to pull the steering arm from the sector shaft. Remove the steering-gear-to-frame retaining bolts or nuts, then remove the steering gear. It may be necessary to loosen the engine mounts and raise the engine to provide sufficient clearance.

⚙ 16-4 Steering-gear disassembly After cleaning the steering gear in solvent, clamp it in a vise. Figure 16-4 shows the disassembled Chrysler power-steering gear. To disassemble it, place a container under it and drain the oil by turning the worm shaft from one extreme to the other. Remove the coupling pin and the coupling, the valve body and O rings, and the valve-pivot lever and spring. To remove the pivot lever, pry under the head with a screwdriver, but do *not* use pliers. Loosen the sector-shaft adjusment-screw locknut, and remove the cover nut with a spanner wrench. Rotate the worm shaft to position the sector-shaft teeth at the center of piston travel. Loosen the steering-gear power-train retaining nut with a spanner wrench.

Turn the steering gear so that the sector shaft is in a horizontal position (Fig. 16-5). Put the arbor on the threaded end of the sector shaft. Slide the arbor into the housing until both tool and shaft are engaged with the bearings.

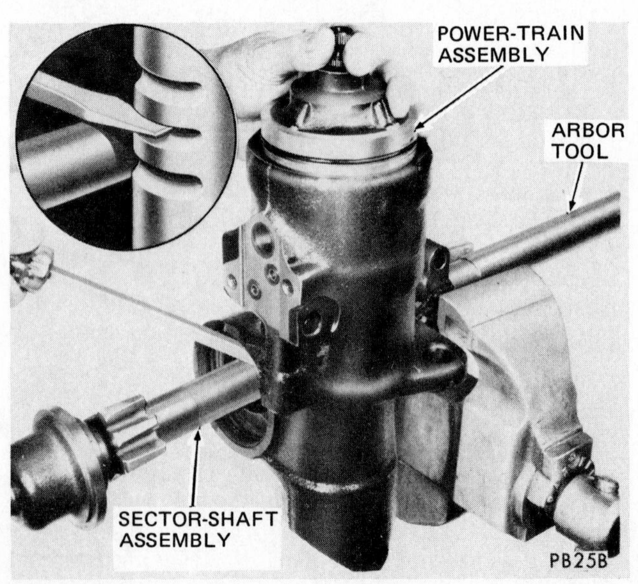

Fig. 16-5 Removing the power-train assembly. (*Chrysler Corporation*)

Rotate the worm shaft by hand to full left position to compress the power-train parts. Now, remove the power-train retaining nut, using a spanner wrench (Fig. 16-6). Hold the power-train parts fully compressed and pry on the piston teeth with a screwdriver, using the sector shaft as a fulcrum. The action is shown in the circled inset to the upper left in Fig. 16-5. This will push the power train up so that it can be removed.

Careful! Be sure to keep the power train compressed. Otherwise the reaction rings may slip out of their grooves, or the center spacer may become cocked in the housing. This makes it impossible to remove the power train without damaging the spacer, the housing, or both.

1. Power-train disassembly Put the power train in the soft jaws of a vise. Never turn the worm shaft more than one-half turn during disassembly. Remove the housing head from the worm shaft. Also remove the large O ring from the groove in the housing head. Use air pressure to remove the reaction seal from the groove in the face of the housing head, or worm-shaft support (Fig. 16-7). Now the reaction spring, reaction ring, balancing ring and spacer, O ring, and center-bearing spacer can be removed.

Hold the worm, turn the nut with sufficient force to free it from the knurled section, and remove the nut. Wire-brush the knurled section to remove any chips. Blow out the nut and worm.

Remove the upper and lower bearings and races, the reaction ring and spring, and the cylinder-head assembly. Remove the O rings from the cylinder head. Use air pressure in the oil hole between the two O ring grooves to blow loose the reaction O ring in the cylinder face.

Remove the retainer, backup ring, and oil seal from the cylinder-head counterbore. Test the operation of the worm shaft to measure the torque required to turn the shaft in the piston. If it is not correct, or if there is other damage, discard the assembly. The worm and piston are serviced as a unit and must not be disassembled.

2. Control-valve disassembly Figure 16-1 is a disassembled view of the control valve. Follow this illustration when disassembling the valve. The seals in the steering-gear housing can be replaced.

⚙ 16-5 Steering-gear reassembly

On reassembly, all new O rings and seals must be used. All parts should be lubricated with power-steering fluid during reassembly.

To reassemble the power train, first put the piston assembly on the bench, worm shaft up. Put the cylinder head, ferrule up, on the worm shaft and against the piston flange. Make sure the gap on the worm-shaft ring is closed to avoid breaking the shaft-seal ring. On some units, the ends of the ring lock together. A special tool is required to compress and lock the ring in the piston groove.

Install the thick lower-bearing race, thrust bearing, reaction spring and ring (flange up), and center-bearing race. Install the outer spacer, thin upper thrust-bearing race, and a new worm-shaft thrust-bearing nut.

Fig. 16-6 Removing the power-train retaining nut with a spanner wrench. (*Chrysler Corporation*)

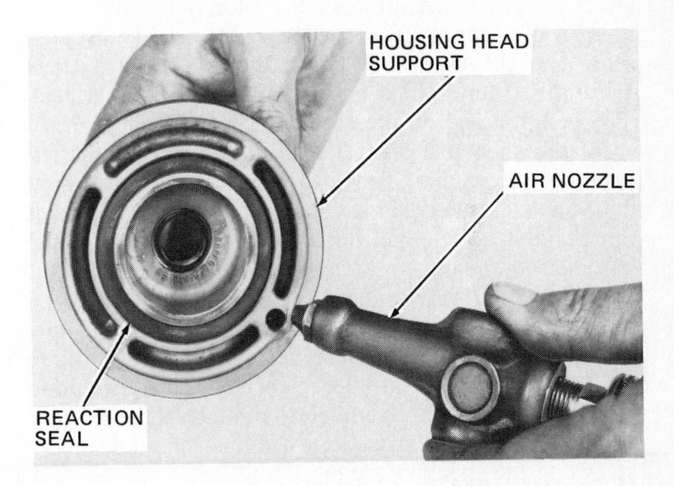

Fig. 16-7 Removing the reaction seal from the worm-shaft support with air pressure. (*Chrysler Corporation*)

Tighten the nut by turning the worm shaft counterclockwise one-half turn, hold the shaft, and tighten the nut to the specified torque. (Do not turn the worm shaft more than one-half turn!) Rotate the worm center-bearing race several turns to position the parts, loosen the adjustment nut, and retighten it to give the specified bearing torque. Check this by wraping a cord around the center-bearing race and measuring the pull required to turn the race with a spring scale. Stake the nut into the slot in the worm shaft.

Install the center-bearing spacer, engaging the dowel pin with the slot. Put the reaction rings over the center spacer and install the upper reaction spring with the cylinder-head ferrule through the hole in the spring. Install a new O ring in the ferrule groove. Install the jacket support, engaging the cylinder-head ferrule and O rings, making sure that the reaction rings enter the groove in the jacket support. Align the parts so that the valve-lever hole in the center-bearing spacer is 90° from the piston-rack teeth. Lock the parts to the worm shaft with a drill rod through the jacket support and the worm-shaft holes.

Install the power train in the housing with the center-bearing-spacer valve-lever hole up. Align the hole with the clearance hole in the housing with an alignment tool. Install the column-support spanner nut and tighten to the specified torque. Set the piston at its center of travel and install the gear shaft and cover. Install the cover spanner nut and tighten it to the specified torque. Install the valve lever and valve body. Be sure the three O rings are in place.

❁ 16-6 Adjustments Back off the gear-shaft adjustment screw. With the steering unit in the test fixture, connect the test hoses to the hydraulic pump on the car with a pressure gauge to check pressures. With pressure applied, position the steering valve by tapping lightly on one of the pressure-control valve screws or on the valve end plug. This should give equal gear-shaft torque (within 5 lb-ft [6.8 N-m] and not to exceed 20 lb-ft [27 N-m] in either direction) when the shaft is slowly turned.

With the gear shaft on center, tighten the adjustment screw until the backlash just disappears. Tighten the screw 1¼ turns. Then hold it and tighten the locknut. Check the operation and oil pressure.

Readjust the gear-shaft backlash by loosening the adjustment screw until backlash is evident. Retighten it until the backlash disappears, and then tighten it further three-eighths to one-half turn. Tighten the locknut to the specified torque.

The steering valve must be positioned to give equal torque. Test the torque by turning the worm shaft one turn to either side of center and checking the torque required to turn the shaft through center. If it varies more than 5 lb-ft [6.8 N-m] from left to right, shift the steering valve to get equal torque. Then tighten the attachment screws to the specified torque.

Although the valve body can be centered as explained above, the latest recommendation from Chrysler is to make the final centering adjustment after the steering gear is reinstalled on the car. This procedure is explained in ❁ 16-2.

❁ 16-7 Installation Installation of the power-steering gear is essentially the reverse of removal. After installation, fill the oil reservoir in the pump, start the engine, and let it idle until the steering gear is up to operating temperature. Turn the steering wheel to the right and left to expel air. Refill the reservoir.

❁ 16-8 Pump service Chrysler-built cars use a vane-type power-steering pump, similar to the Saginaw pump used on General Motors cars. The service procedure for the Saginaw pump is covered in ❁ 15-11.

Chapter 16 review questions

Select the *one* correct, best, or most probable answer to each question. Then check your answers against the correct answers given at the end of the book.

1. During disassembly of the steering gear, the old O rings and seals should be:
   a. cleaned and reused
   b. greased and reused
   c. thrown away
   d. none of the above
2. If the steering gear tries to steer itself:
   a. the valve body is incorrectly positioned
   b. the idle speed is too high
   c. the steering linkage needs adjustment
   d. all of the above
3. To remove the reaction seal from the groove in the face of the worm-shaft support use:
   a. a screwdriver
   b. a special puller
   c. air pressure
   d. an arbor press
4. The steering valve must be positioned to give:
   a. equal torque in each direction
   b. quick recovery
   c. both *a* and *b*
   d. neither *a* nor *b*
5. During reassembly, all parts must be lubricated with:
   a. engine oil
   b. power-steering fluid
   c. petroleum jelly
   d. locking sealer

CHAPTER 17

FORD POWER-STEERING SERVICE

After studying this chapter, and with proper instruction and equipment, you should be able to:

1. Perform the valve-spool centering check.
2. Adjust the steering gear in the car.
3. Remove and install the steering gear.
4. Overhaul the power-steering gear.
5. Overhaul a roller-type power-steering pump.

17-1 In-vehicle service The Ford power-steering gear is shown in Fig. 10-49. The control valve is operated from the worm shaft, and the power piston is integral with the ball-nut assembly. This steering gear is very similar to the Saginaw power-steering gear (Chap. 15).

One check and one adjustment can be made with the power-steering gear in the car.

1. Valve-spool centering check Install a 2000 psi [13,800 kPa] pressure gauge between the power-steering pump outlet port and the steering-gear inlet port. If a power-steering analyzer is used (Fig. 17-1), then the flow-control valve in the analyzer must be fully open. Check the fluid level in the reservoir and fill it to the proper level. Start the engine, then turn the steering wheel gently from stop to stop to bring the power-steering fluid up to normal operating temperature. Recheck the reservoir and add more fluid if it is necessary. Remove the cover from the center of the steering wheel.

Run the engine at 1000 rpm and, with the steering wheel centered, use a torque wrench to apply enough torque to the steering-wheel nut to get a reading of 250 psi [1724 kPa] on the pressure gauge. Apply the torque first in one direction and then in the other. If the difference in readings in the two directions is more than 4 lb-in [0.45 N-m], a correction must be made.

To correct, remove the steering gear. Then remove the valve housing and substitute either a thicker or a thinner valve-centering shim (⚙ 17-3). Use only one shim. If steering effort is heavy, substitute a thicker shim.

2. Steering-gear adjustment Only the mesh load can be adjusted on the car. To adjust the mesh load, disconnect the pitman arm from the shaft. Also disconnect the fluid-return line at the reservoir. Put the end of the return line in a clean container. Start the engine and turn the steering wheel from stop to stop to discharge the fluid from the steering gear. Remove the cover from the steering wheel and turn the steering wheel to 45° from the left stop. Place a torque wrench that reads in pound-inches on the steering-wheel nut. Now check the torque required to turn the steering wheel through one-eighth turn from the 45° position.

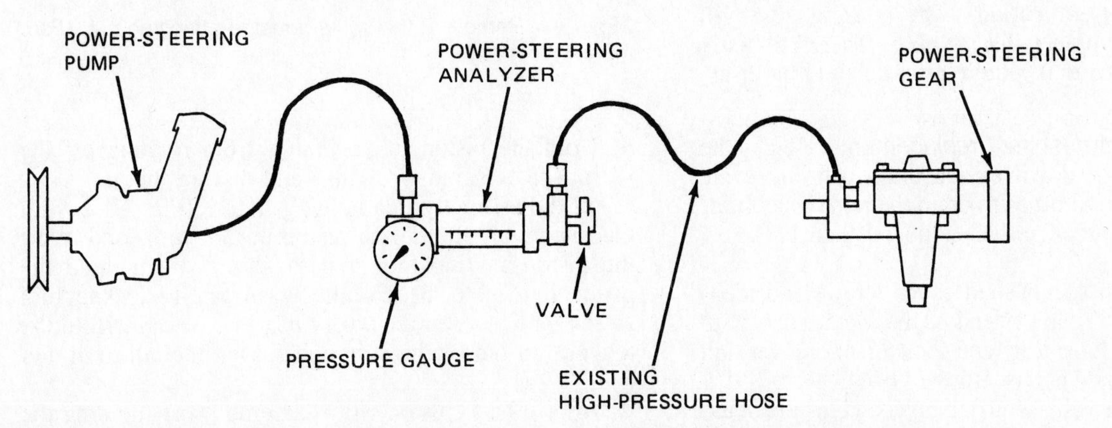

Fig. 17-1 Installation of an oil-pressure gauge, or a power-steering analyzer, in the power-steering hydraulic system. (*Ford Motor Company*)

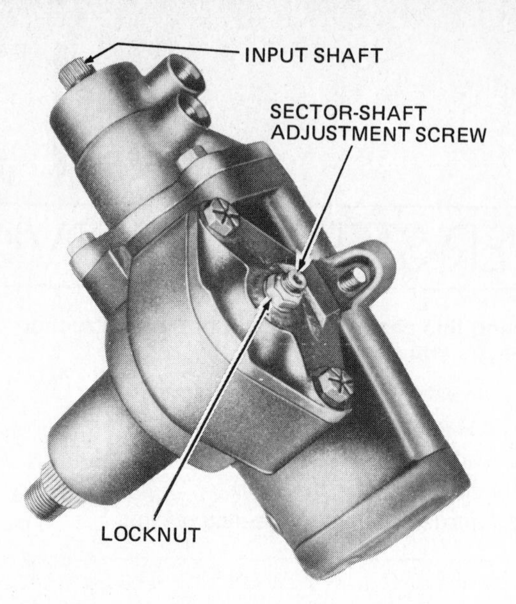

Fig. 17-2 Sector-shaft adjustment screw to adjust mesh load. *(Ford Motor Company)*

Center the steering wheel and check the torque required to turn the wheel back and forth across the center position. Loosen the locknut and turn in the sector-shaft adjustment screw (Fig. 17-2) until the reading is 14 to 18 lb-in [1.6 to 2 N-m] greater at the center position than at the 45° position. Tighten the locknut while holding the adjustment screw stationary. Recheck the readings. Install the steering-wheel hub cover and pitman arm. Refill the reservoir after reconnecting the return line.

17-2 Steering-gear removal Disconnect the fluid lines from the steering gear, and plug the lines and ports to prevent the entrance of dirt. Remove the two bolts connecting the flex coupling to the steering gear.

Raise the vehicle and remove the pitman-arm attaching nut and the pitman arm, using a puller. If the car has a standard transmission, remove the retracting spring from the clutch release lever to provide clearance. Support the steering gear and remove the attaching bolts. Work the steering gear free of the flex coupling and remove the steering gear from the car. If the flex coupling stays on the steering gear, remove it.

17-3 Steering-gear repair The steering gear will seldom require complete disassembly. Therefore, only those subassemblies that require repair should be disassembled.

1. Valve-centering-shim replacement Hold the steering gear upside-down over a drain pan and cycle the input shaft several times to drain the remaining fluid. Turn the input shaft to either stop and then back 1¾ turns to center it.

Remove the two sector-shaft-cover screws and identification tag. Tap the lower end of the sector shaft with a soft hammer to loosen it, and then lift the cover and shaft from the housing. Discard the O ring.

Remove the four valve-housing bolts and lift the housing off. The worm shaft and piston will come off with it. Hold the piston to prevent it from rotating off the worm shaft. Remove O rings and discard them.

Put the assembly in a holding fixture (Fig. 17-3) with the piston up. Rotate the piston up 3½ turns and insert the piston-holding tool into the bolt hole to hold the piston up (Fig. 17-3). Use the worm-bearing locknut tool as shown to loosen the worm-bearing locknut. Hold the locknut up out of the way and loosen the attaching nut (Fig. 17-4).

Now, lift the piston-worm assembly off, holding the piston to prevent it from spinning off the worm.

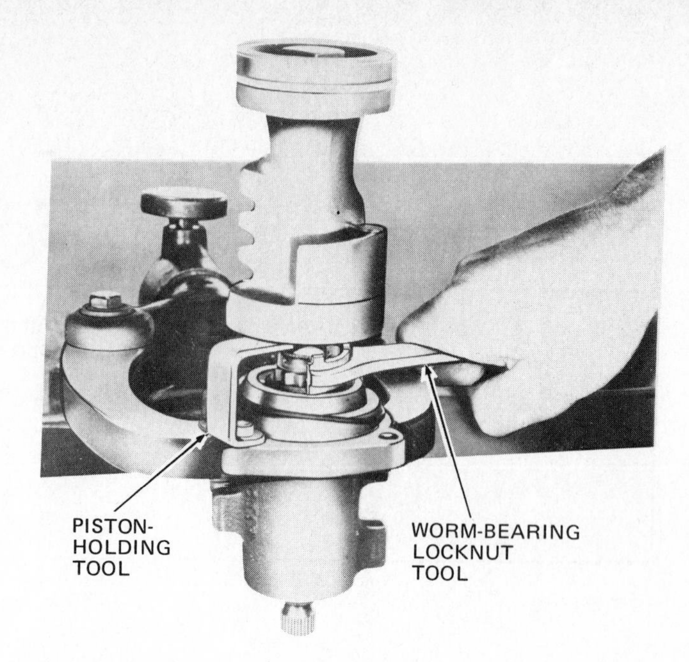

Fig. 17-3 Removing the worm-bearing locknut. *(Ford Motor Company)*

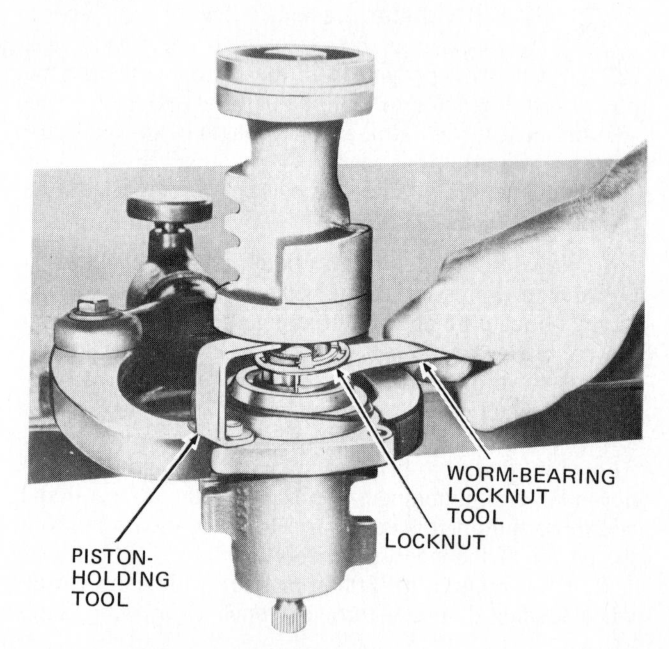

Fig. 17-4 Removing the valve-housing attaching nut. *(Ford Motor Company)*

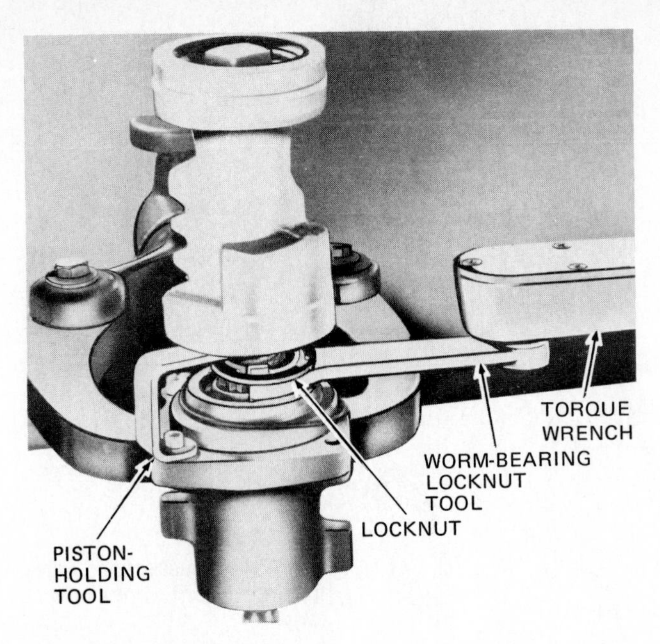

Fig. 17-5 Installing the valve-housing attaching nut. (*Ford Motor Company*)

Change the centering shim, as described in ⚙ 17-1.

Put the piston-worm assembly on the valve housing, holding the piston to prevent it from spinning off the worm. Install the attaching nut, and torque to specifications (Fig. 17-5). Install the locknut, and torque to specifications. Rotate the piston up one-half turn and remove the piston-holding tool (Fig. 17-3). Take the assembly off the holding fixture. Position a new O ring in the counterbore of the gear housing. Apply petroleum jelly to the piston seal. Put a new O ring on the valve housing. Slide the piston and valve into the gear housing carefully to avoid damaging the piston seal.

Align the fluid passages in the valve and gear housings, and install but do not tighten the attaching bolts.

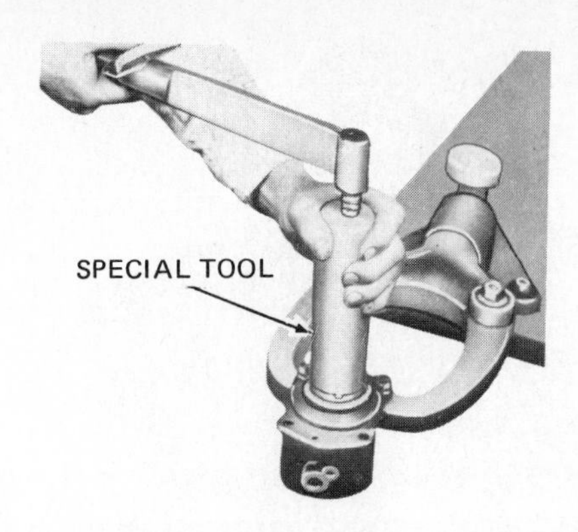

Fig. 17-7 Removing the locknut. (*Ford Motor Company*)

Rotate the piston so that the teeth align with the sector-shaft teeth. Tighten the four valve-housing attaching bolts to specifications.

Put the sector-shaft O ring in the steering-gear housing. Turn the input shaft as required to center the piston. Slide the sector-shaft-and-cover assembly into position and secure with two bolts. Put the identification tag under one bolt head.

Adjust the mesh load to about 4 lb-in (0.45 N-m). Then torque the cover bolts to specifications. Finally, readjust the mesh load to specifications (⚙ 17-1).

2. Steering-gear disassembly (Fig. 17-6) If the steering gear requires repair, remove the worm-shaft-and-piston assembly, as previously described. Stand the assembly on end, piston down, and rotate the input shaft counterclockwise out of the piston, allowing the balls to drop into the piston. Put a cloth over the open end of the piston and turn it upside down to catch the balls.

To remove the worm-and-valve assembly from the housing, install the housing in a holding fixture and use a special tool to remove the lock and attaching nuts (Fig. 17-7). Then slide the assembly from the housing, using extreme care not to cock it. This could cause it to jam and ruin the housing and valve.

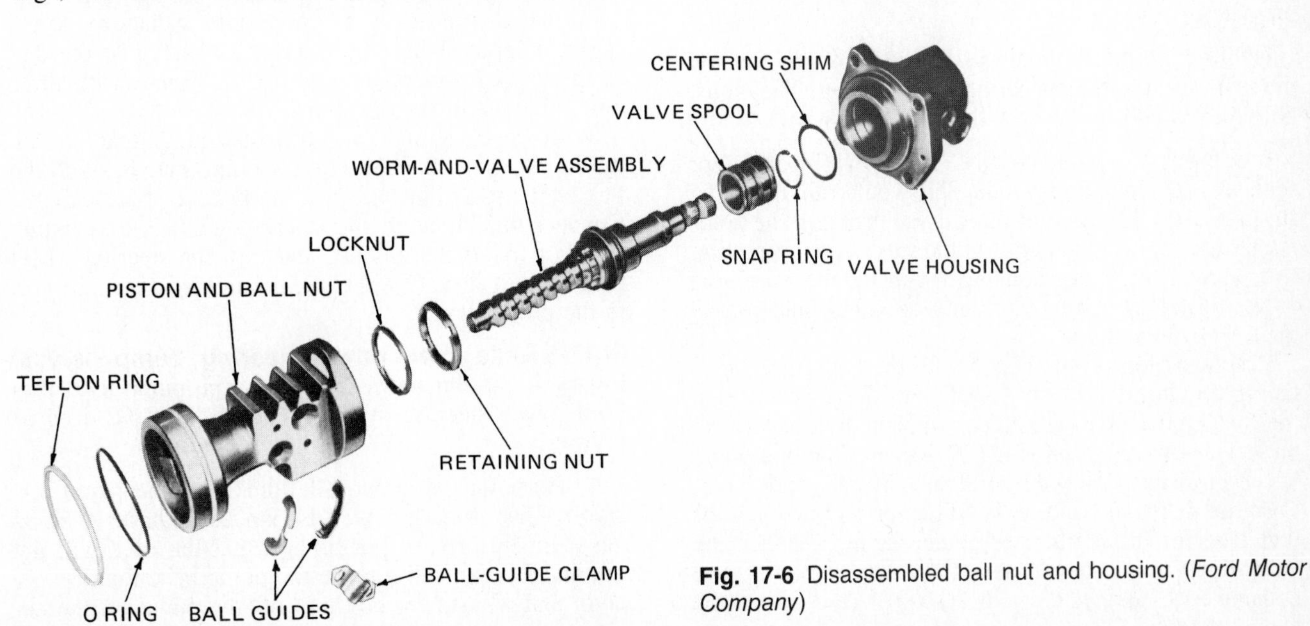

Fig. 17-6 Disassembled ball nut and housing. (*Ford Motor Company*)

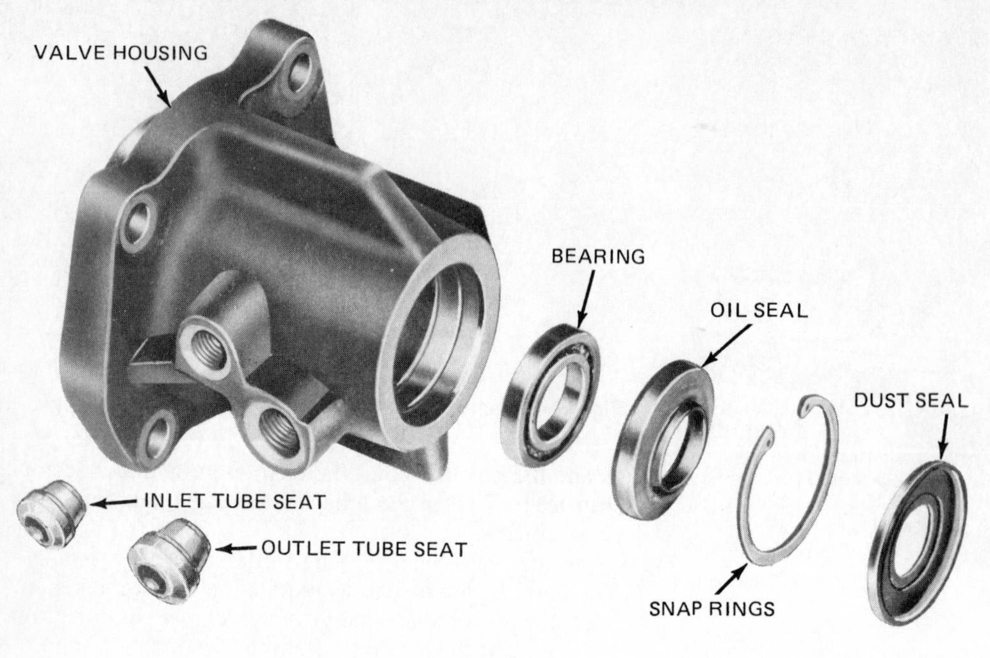

Fig. 17-8 Disassembled valve housing. (*Ford Motor Company*)

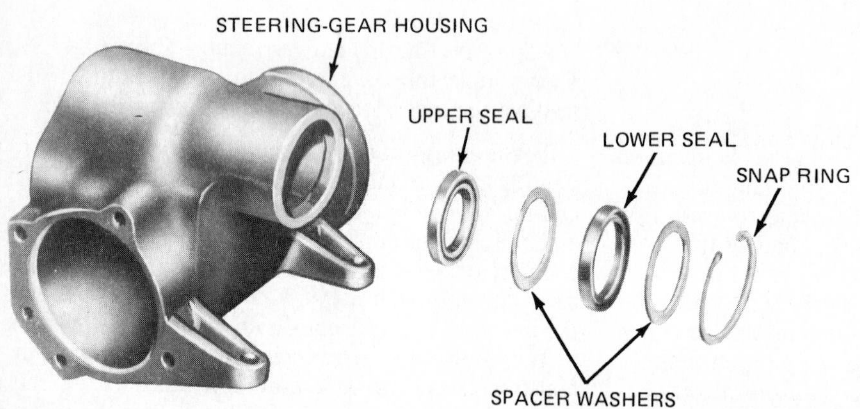

Fig. 17-9 Disassembled steering-gear housing. (*Ford Motor Company*)

3. Valve housing Figure 17-8 shows the housing parts. Special tools are required to remove and replace the bearing and oil seal.

4. Steering-gear housing Figure 17-9 shows how the seals line up in the steering-gear housing. They must be removed and replaced with special tools.

5. Steering-gear assembly Never wash, clean, or soak seals in cleaning solvent. This would ruin them. To start the assembly, mount the valve housing in the holding fixture, flanged end up. Put the valve-spool centering shim (Fig. 17-6) in the housing and install the worm and valve. Install attaching nuts and locknuts, and torque them to specifications.

Put the piston on the bench with the ball-guide holes facing up. Insert the worm shaft into the piston so that the first groove is in alignment with the hole nearest to the center of the piston (Fig. 17-10). Put the ball guide in the piston and drop 27 balls into the ball-guide hole. Turn the worm in a clockwise direction so that the balls will feed in. Install the ball-guide clamp. Reattach the valve-housing-and-piston assembly to the steering-gear housing as explained in item 1 above. Then adjust the mesh load (⚙ 17-1).

⚙ 17-4 Steering-gear installation On installation, slide the flex coupling into place on the steering shaft. Turn the steering wheel to the middle. Center the steering-gear input shaft. Slide the input shaft into the flex coupling and attach the gear to the frame with three bolts. Torque to specifications.

With wheels straight ahead, install the pitman arm on the sector shaft. Secure with attaching nut. Position the flex coupling and install attaching bolts to specifications. Connect fluid lines to the steering gear. Fill the pump reservoir, start the engine, and turn the steering wheel from stop to stop. Check for leaks and refill the reservoir to the proper level.

⚙ 17-5 Roller-type power-steering pump service Figure 17-11 illustrates the roller-type pump used in a Ford power-steering system. The pump is serviced as follows.

1. Removal Remove all fluid from the pump reservoir with a suction gun. Disconnect the hoses and fasten them in a raised position so that the fluid will not run out. Loosen and remove the pump belt. Remove the pivot and adjustment bolt and lift the pump off the engine.

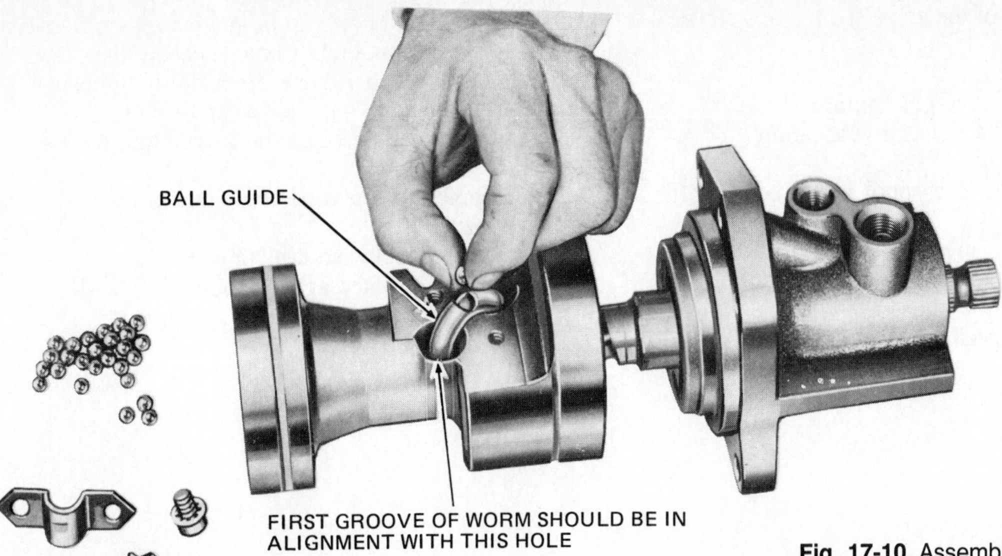

Fig. 17-10 Assembling the piston on the worm shaft. (*Ford Motor Company*)

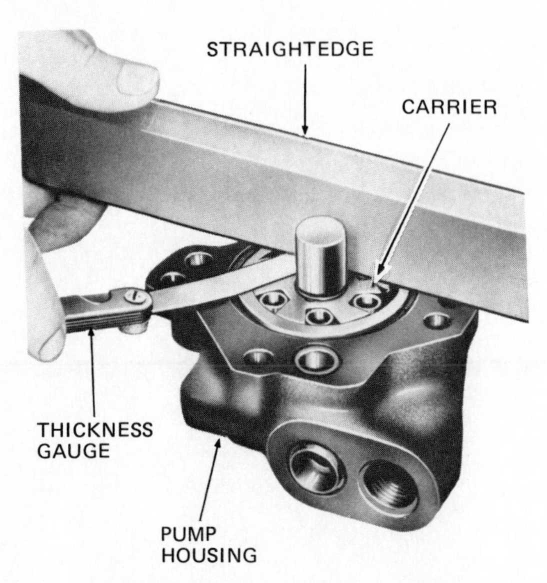

Fig. 17-11 Checking end clearance of the carrier and rollers in the pump. (*Ford Motor Company*)

2. Disassembly Handle all parts carefully to avoid nicks or scratches which would ruin them. Detach the reservoir by removing the retaining nut from inside the reservoir. Remove the two orifice O rings from the pump. Take off the pulley and the key. Remove all the bolts so that the bracket, pump housing, and cover can be separated. Lift the cover vertically so that the rollers will not fall out. Tap the cover loose if it sticks.

Use a straightedge and a thickness gauge to check the end clearance of the carrier and rollers (Fig. 17-11). If the clearance is excessive, or if parts are damaged, replace all rollers and the carrier. These parts come in a service kit.

Check the valve in the cover by removing the plug retainer and O ring. It should be free of nicks or scratches and slide easily in the valve bore.

3. Assembly Coat all parts with power-steering fluid on assembly. The carrier teeth should point in a counterclockwise direction when the carrier is installed in the housing. Do not damage the seal when installing the shaft in the housing. After installing the valve in the cover, tighten the plug retainer and install the O rings. Put the cover in place and attach it with bolts. Attach the adjustment bracket to the housing. Put the pulley and its key on and secure them with a bolt. Put new O rings on the grooves on top of the housing and install the housing. Cement a new gasket around the inside of the reservoir cover.

4. Installation On installation, make the attachment bolts finger-tight. Check the alignment of the crankshaft and pump pulleys, install the belt, and adjust tension to the proper specification. Tighten the attachment bolts. Connect the two hoses. Position the filter seat in the reservoir, put the filter on, and secure with a washer and a spring. Fill the reservoir with fluid. Start the engine and allow it to run while cycling the steering gear to eliminate air from the system. Check all connections for fluid leaks. Check the fluid level in the reservoir and add fluid if necessary.

Chapter 17 review questions

Select the *one* correct, best, or most probable answer to each question. Then check your answers against the correct answers given at the end of the book.

1. On reassembly, old O rings and seals should be:
   a. thrown away
   b. relubricated and reused
   c. washed carefully and, if not cut, reused
   d. none of the above

2. To adjust the centering of the valve spool, the steering gear must be:
   a. installed in the car
   b. connected to the steering-gear pump
   c. disconnected from the steering-gear pump
   d. removed from the car
3. If the centering of the valve spool is not correct, it can be adjusted by:
   a. changing the oil pressure
   b. adding or removing shims
   c. turning the valve-spool adjuster
   d. grinding the valve spool
4. One of the main checks to be made on the roller-type pump during disassembly is to measure the
   a. clearance between the carrier and the housing
   b. end clearance of the carrier and rollers
   c. side clearance between the carrier and rollers
   d. all of the above
5. On reassembly of the roller-type pump, make sure the carrier teeth:
   a. point in a clockwise direction
   b. are firmly in place against the shaft
   c. point in a counterclockwise direction
   d. none of the above

CHAPTER 18

LINKAGE-TYPE POWER-STEERING SERVICE

After studying this chapter, and with proper instruction and equipment, you should be able to:

1. Diagnose trouble in the linkage-type power-steering system.
2. Replace piston-rod seals.
3. Service the control valve.
4. Service the power-steering pump.

18-1 Trouble-diagnosis chart The construction and operation of the linkage-type power-steering system was described in ❁ 10-24 and illustrated in Fig. 10-51. The trouble-diagnosis chart for this system is on the following page. The list of trouble symptoms, causes, and checks or corrections in the chart will give you a means of logically analyzing troubles so that their causes can be quickly located and eliminated. Not all the troubles, causes, and corrections that are listed apply to all models of linkage power-steering units. There are some variations in design and operation.

❁ 18-2 Trouble-diagnosis checks When a driver complains about a linkage-type power-steering system, make the following checks. Before checking steering effort and oil pressure, make sure that the wheel alignment, tire pressure, suspension, and shock absorbers are in normal condition.

1. Check belt tension A quick check of the tension of the drive belt for the power-steering pump can be made while the engine is running. Turn the steering wheel to the extreme position while watching the belt. If the belt slips, the tension is too low. The pump-mounting bracket has slotted holes that permit the pump to be shifted outward to increase the tension as necessary.

2. Check oil level If oil level in the pump reservoir is low, add oil to bring it up to the proper level. Also, check the connections at the pump, power cylinder, and valve for signs of leakage.

3. Check steering effort After checking the condition of the tires, the front suspension, and the wheel alignment, use a spring scale to check the force required to pull the steering wheel through center, first in one direction, then in the other (Fig. 15-4). Specifications vary, so refer to the shop manual on the car being checked.

Careful! Do not turn and hold the wheel at the extreme left or right position.

4. Check oil pressure Install an oil-pressure gauge in the pressure line from the pump (Fig. 17-1). With the engine idling and the oil in the system at normal operating temperature, turn the steering wheel a full turn either to the left or the right. Check the reading on the gauge. It should be within specifications.

If the pressure is low, close the shutoff valve, quickly read the pump pressure, and open the valve again. Do not leave the valve closed more than a few seconds, since this might damage the pump. If the pump pressure goes up to specifications as the valve is closed, the pump is okay. The loss of pressure is due to leakage at the power cylinder or valves. If the pump pressure stays low as the valve is closed, the loss of pressure is due to a defect in the pump. This could be a stuck or damaged relief or flow-control valve, rotors worn or broken, or other causes.

❁ 18-3 Bleed hydraulic system If necessary, bleed the system by adding oil to the reservoir, then turning the steering wheel to the left and right several times. Do not bump the wheel at the extreme positions. Add more oil if the level in the reservoir falls too low.

❁ 18-4 Power-cylinder service The power cylinder is constructed so that it usually cannot be repaired in the shop if internal damage has occurred. However, the piston-rod seals can be replaced.

To remove the power cylinder, first raise the car on a lift. Wipe the hose fittings clean at the power cylinder, detach the hoses, and let the system drain. For complete drainage, move the wheels back and forth several times. Do not reuse the old fluid.

Several methods of attaching the piston rod to the frame bracket are used. Figure 18-1 shows one method

Linkage-Type Power-Steering Trouble-Diagnosis Chart

(See ⚙ 18-2 to 18-6 for details of checks and corrections listed.)

COMPLAINT	POSSIBLE CAUSE	CHECK OR CORRECTION
1. Hard steering	a. Low oil or leaks	Check for leaks, add oil
	b. Low oil pressure	See item 2, below
	c. Binding in steering linkage	Check, adjust linkage as needed
	d. Power-cylinder piston rod bent	Replace rod or cylinder
	e. Low tire pressure	Inflate to proper pressure
	f. Incorrect front alignment	Align front end
	g. Binding in steering column	Align steering column
	h. Valve stuck or out of adjustment	Adjust. Remove valve and check for cause. Replace valve if it is damaged.
2. Low oil pressure	a. Pump belt slipping	Tighten to proper tension
	b. Relief and flow-control valve stuck	Remove and clean
	c. Valve spring weak or broken	Replace
	d. Drive coupling broken	Replace
	e. Pump rotors, body, or cover worn or broken	Replace rotors or pump
	f. Leaks past piston in power cylinder	Replace piston rings or cylinder
	g. Leaks past valve	Replace seal rings
3. Hard steering—one direction only	This could be caused by d, e, f, or g listed under item 1, or by a misadjusted or sticky valve in the power cylinder or valve body.	
4. Poor centering or recovery from turns	a. Pitman arm and stud binding on power cylinder	Adjust to proper tension
	b. Bind in steering column or gear	Align column, adjust gear
	c. Bind in steering linkage	Check, adjust linkage as needed
	d. Bent power-cylinder piston rod	Replace piston rod or cylinder
	e. Incorrect front alignment	Align front end
	f. Valve misadjusted or stuck	Free up, adjust
	g. Antiroll pin and bracket binding (on some models)	Align pin with bracket
5. Car wander	a. Uneven tire pressure	Inflate correctly
	b. Pitman arm and stud binding on power cylinder	Adjust to proper torque
	c. Valve stuck or misadjusted	Clean, replace
	d. Incorrect front alignment	Align front end
6. Noises	a. Low oil level	Replace oil, check for leaks
	b. Pump bushing worn	Replace bushing
	c. Dirty pump	Clean after disassembly, drain system, and refill with clean oil
	d. Looseness in steering linkage	Check, adjust, and tighten
	e. Tie rods improperly attached	Loosen clamps and retighten with bolts in proper position

of attachment. On this car, the cotter pins and nuts at the piston-rod end and the ball-stud end are removed so that the power cylinder can be taken off. Figure 18-2 shows the method of attaching the ball stud to the piston body.

Figure 18-3 illustrates a typical arrangement for sealing the piston-rod end. The seals can be replaced after removing the snap ring. New seals should be lubricated and pressed into place over the piston rod with a deep socket.

To install the power cylinder, reverse the removal procedure. If the grommet at the piston-rod end is worn, replace it. Grease the ball joint after the attachment is complete. Fill the pump reservoir, start the engine, and turn the steering wheel right and left to bleed the system. Refill the reservoir.

⚙ 18-5 Control-valve service The following steps cover how to remove, disassemble, and service the control valve.

1. Valve removal Attachments on different cars may require slightly different removal procedures. A typical procedure follows. Disconnect the fluid lines from the control valve. Drain the fluid, turning the steering wheel to the right and left several times to force the fluid out. Loosen the clamping nut and bolt at the right end of the sleeve. Locate the roll pin in the steering-arm-to-idler-arm rod and remove it through the slot in the sleeve. Remove the control-valve ball-stud nut, then remove the ball stud from the control valve with the special tool shown in Fig. 18-4. Turn the front wheels to the left and unthread the control valve from the rod.

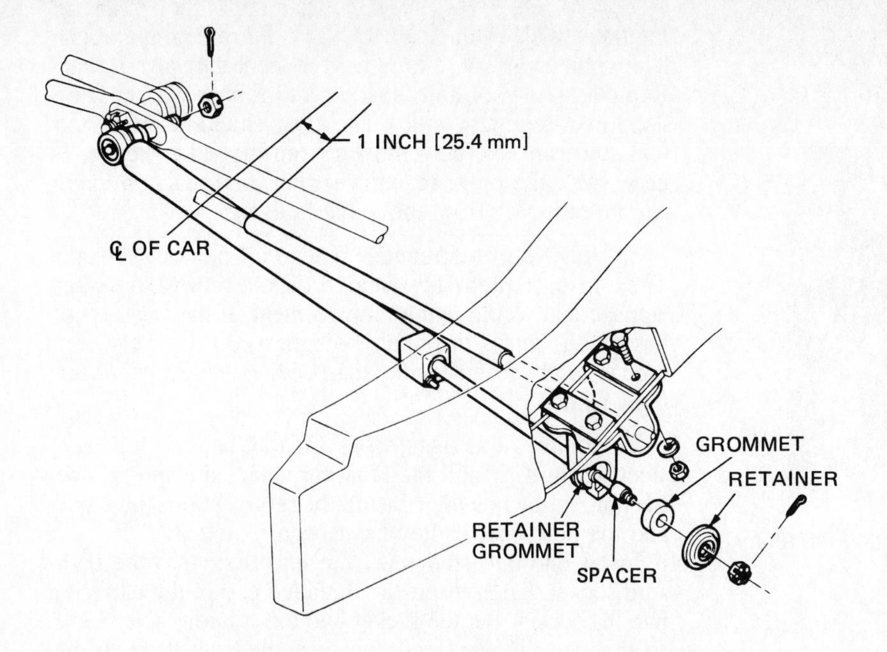

Fig. 18-1 Power-cylinder attachment. (*Chevrolet Motor Division of General Motors Corporation*)

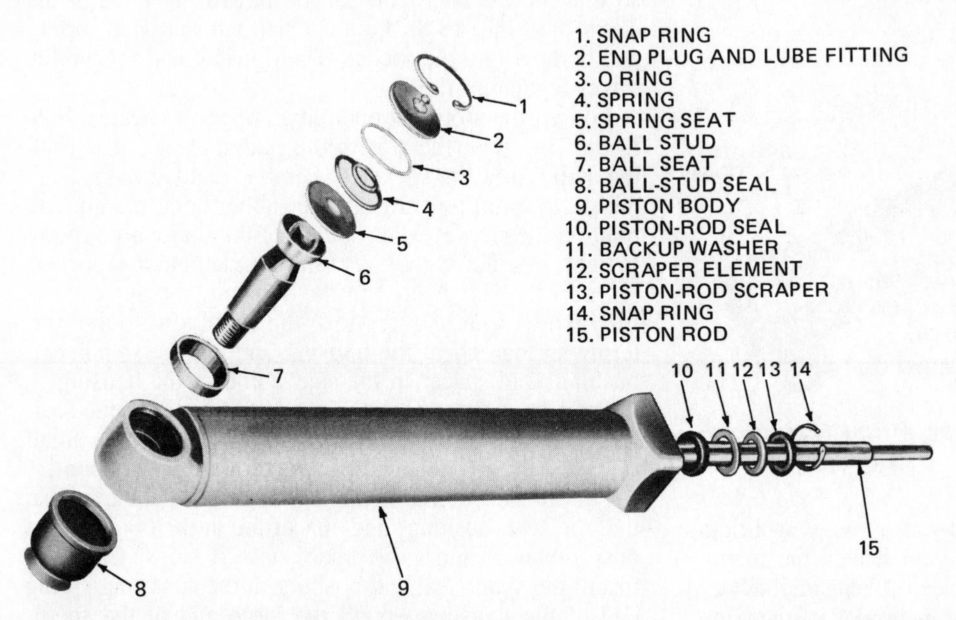

Fig. 18-2 Power-cylinder piston-rod seal and ball-stud attachment. (*Chevrolet Motor Division of General Motors Corporation*)

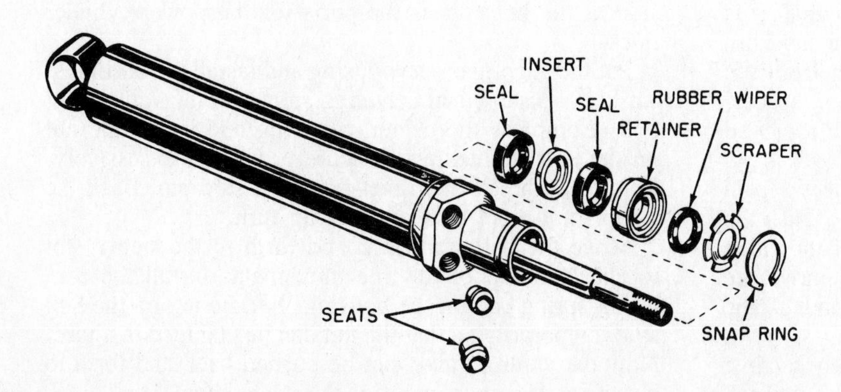

Fig. 18-3 Piston-rod seals in the power cylinder. (*Ford Motor Company*)

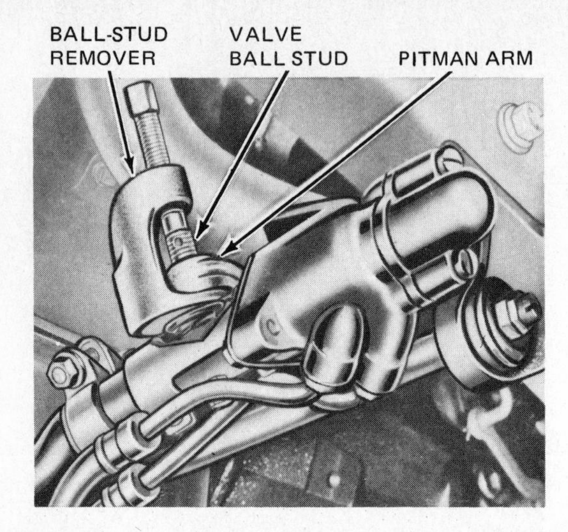

Fig. 18-4 Removing the control-valve ball stud. (*Ford Motor Company*)

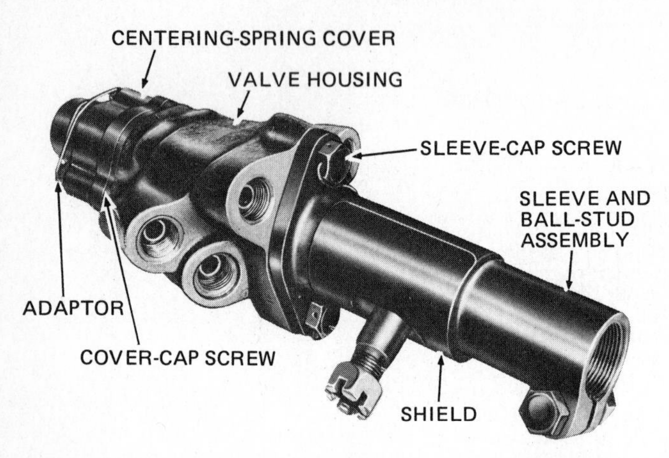

Fig. 18-5 Valve-and-sleeve assembly. (*Ford Motor Company*)

2. Valve disassembly Be very clean when working on the control valve. Make sure your hands, the workbench, and the tools you use are clean. Even small traces of dirt may cause damage or malfunctioning of the control valve.

Put the valve assembly (Fig. 18-5) in the soft jaws of the vise. Do not tighten the vise more than necessary, since this could distort the valve. Clamp the vise around the sleeve flange.

Figure 18-6 shows a disassembled control valve. It is disassembled by removing the centering-spring cap from the valve housing and then removing the nut from the end of the valve-spool bolt. Now, the washers, spacer, centering spring, adapter, and bushing can be removed from the bolt and the valve housing.

Remove the bolts and separate the valve sleeve from the housing. Remove the plug from the sleeve. Push the valve spool out of the centering-spring end of the valve housing and take the seal from the spool. Remove the spacer, bushing, and seal from the sleeve end of the housing.

Pull the head of the valve-spool bolt tightly against the travel-regulator stop and drive the stop pin out of the regulator with a punch (Fig. 18-7). Turn the travel regulator counterclockwise to remove it from the valve sleeve. Remove the spool bolt, spacer, and rubber washer from the travel-regulator stop. The dust shield, clamp, and ball stud can now be removed from the valve sleeve. If necessary, the plug, reaction spring, and reaction valve can be removed from the valve housing.

3. Valve-part inspection Clean all parts in solvent. Then inspect them for wear, cracks, scores, or other damage that would require replacement. If the valve spool has small burrs, they can be removed with very fine crocus cloth provided the sharp edges of the spool are not rounded off.

4. Control-valve assembly Coat all parts with power-steering fluid. Install the reaction valve, the spring, and the plug in the housing. Install the return-port relief valve and the hose seat if they have been removed.

Insert one ball-stud seat, flat end first, into the ball-stud socket. Then insert the threaded end of the ball stud into the socket. Put the socket into the control-valve sleeve so that the threaded end can be pulled out through the sleeve slot (Fig. 18-8). Put the other ball-stud seat, spring, and bumper into the socket. Then, install and tighten the travel-regulator stop.

Loosen the stop just enough to align the nearest hole in the stop with the slot in the ball-stud socket. Install the stop pin in the socket, the travel-regulator stop, and the valve-spool bolt. Install the rubber boot, clamp, and plug on the sleeve. Make sure that the lubrication fitting is tight, but that it does not bind on the ball-stud socket.

Insert the valve spool in the housing, rotating it carefully while installing it. Move the spool toward the centering-spring end of the housing and put the small seal, bushing, and spacer in the sleeve end of the housing.

Press the valve spool against the inner lip of the seal, and guide the lip of the seal over the spool with a small screwdriver. Do not nick or scratch the seal or spool!

Put the sleeve end of the housing on a flat surface so that the seal, bushing, and spacer are at the bottom, and push down on the valve spool until it stops. Carefully install the spool seal and bushing in the centering-spring end of the housing around the large end of the spool. Guide the seal over the spool with a small screwdriver. Be careful not to nick or scratch the seal or the spool!

Pick up the housing and move the spool back and forth to make sure it moves freely. Attach the sleeve assembly to the housing, making sure the ball stud is on the same side of the housing as the ports for the power-cylinder lines.

Put the adapter on the housing and install the bushings, washers, spacers, and centering spring on the valve-spool bolt. Compress the centering spring and install the nut on the bolt, tightening it securely, but not excessively. This could break the travel-regulator stop pin. Back the nut off not more than one-quarter turn.

Move the ball stud back and forth in the sleeve slot to check the spool for free movement. Install the centering-spring cap on the housing. Put the nut on the ball stud temporarily so that the nut can be clamped in a vise. Now the control valve can be pushed back and forth to check for free movement of the valve spool.

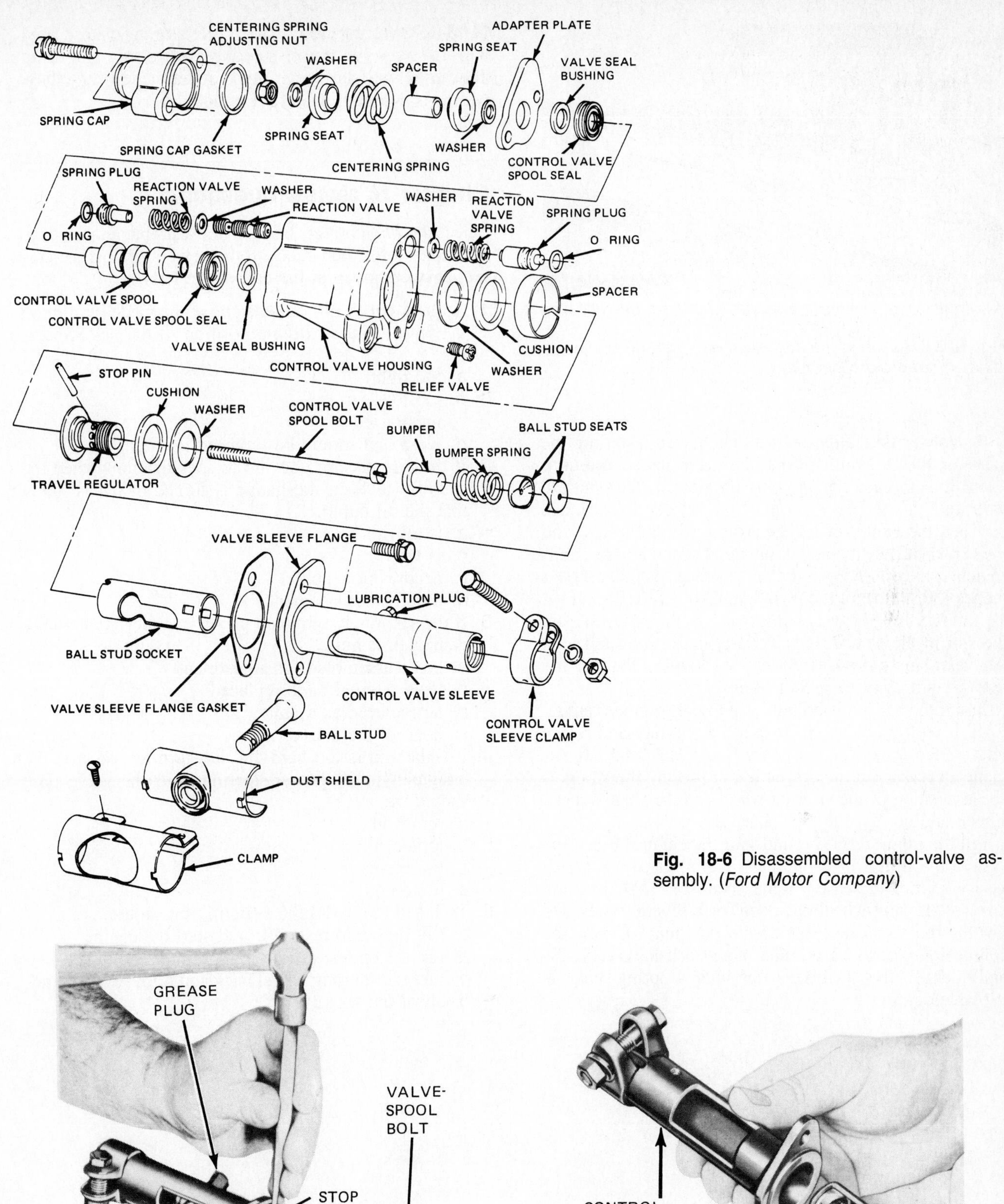

Fig. 18-6 Disassembled control-valve assembly. (*Ford Motor Company*)

Fig. 18-7 Driving the stop pin out of the travel regulator. (*Ford Motor Company*)

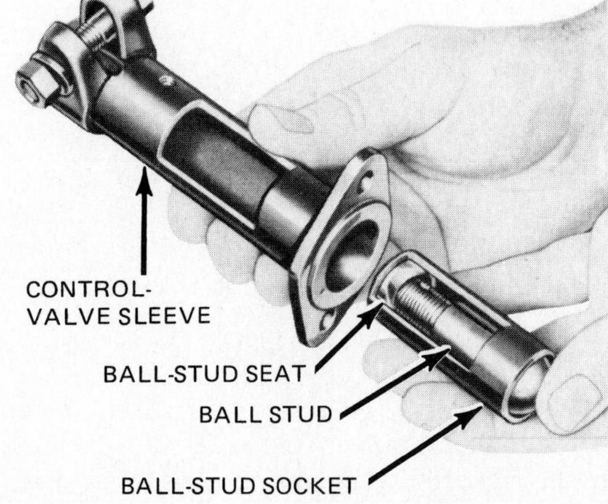

Fig. 18-8 Installing the ball-stud seal and socket. (*Ford Motor Company*)

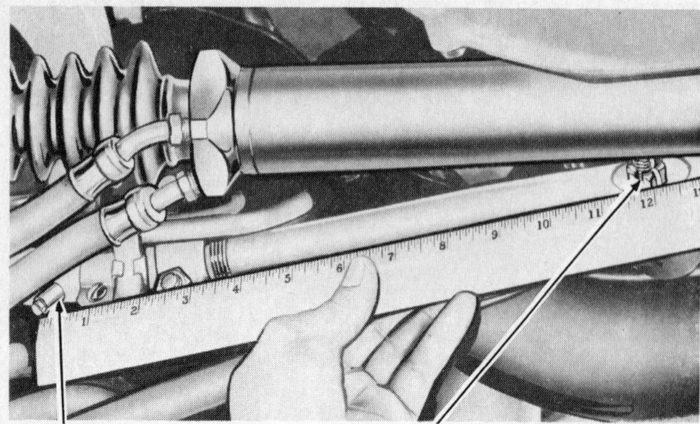

Fig. 18-9 Measuring the control-valve length on the installation. (*Ford Motor Company*)

5. Valve installation As a first step in installing the valve on the car, screw it onto the steering-arm rod until about four threads are showing on the rod, then put the ball stud in the pitman arm. Now check the distance between the center of the grease plug in the sleeve and the center of the stud at the inner end of the left-steering-spindle-arm connecting rod. This distance should be 11 15/16 inches [303 mm] on the installation shown in Fig. 18-9. To adjust, turn the valve in or out on the rod. Line up the slot in the sleeve with the hole in the rod, and lock the valve in this position with a roll pin. Tighten the sleeve-clamp bolt to specifications.

Install the ball stud and tighten it to the specified torque. Lock it with a cotter pin. Install the pressure and return lines. Then lower the car to the floor. Make any adjustments necessary at the tie-rod ends to secure the straight-ahead position of the steering wheel and the front wheels. Check and correct the toe, if necessary.

Fill the pump reservoir and start the engine. Bleed the system by idling for 2 minutes, adding more oil if necessary. Now, increase engine speed to 1000 rpm and turn the steering wheel back and forth several times. Do *not* hit the stops at the extreme positions! Check the connections for leaks and add more oil if necessary. Finally, check the steering effort with a spring-scale as explained earlier.

⚙ 18-6 Pump service The power-steering pumps used with linkage-type power-steering systems are similar to those used with the integral-type systems. Servicing these pumps is covered in ⚙ 15-11 and 17-5.

Chapter 18 review questions

Select the *one* correct, best, or most probable answer to each question. Then check your answers against the correct answers given at the end of the book.

1. A quick check of belt tension can be made by turning the wheel to the extreme position. If the belt slips, it is:
   a. too tight
   b. too loose
   c. undersized
   d. worn and should be replaced
2. If the oil pressure goes up to specifications when the shutoff valve on the gauge is turned off, it is likely that the oil pump:
   a. is all right
   b. is worn
   c. needs a new belt
   d. requires replacement
3. If the separately mounted power cylinder is internally damaged, it must:
   a. be disassembled for adjustment
   b. have internal parts replaced
   c. be replaced as a unit
   d. none of the above
4. If right turn is too hard and left turn too easy on the integral type of power-steering unit, the valve plug should be:
   a. removed
   b. replaced
   c. turned in
   d. turned out
5. To bleed the hydraulic system, you should:
   a. fill the pump reservoir and start the engine
   b. run the engine at idle for 2 minutes
   c. turn the steering wheel back and forth several times
   d. all of the above

CHAPTER 19

RACK-AND-PINION STEERING-GEAR SERVICE

After studying this chapter, and with proper instruction and equipment, you should be able to:

1. Diagnose trouble in rack-and-pinion manual-steering gears.
2. Diagnose trouble in rack-and-pinion power-steering gears.
3. Service rack-and-pinion manual-steering gears.
4. Service rack-and-pinion power-steering gears.
5. Service the power-steering pump.

19-1 Servicing rack-and-pinion manual-steering gears Rack-and-pinion steering is widely used on small cars. It combines the steering gear and the steering linkage into a single compact assembly (Fig. 10-26). The construction and operation of this type of steering gear was covered in ⚙ 10-13. Following sections describe the trouble diagnosis and service of typical rack-and-pinion steering gears.

In the rack-and-pinion manual-steering gear, if the steering rack pinion shaft, housing-rack bushing, or tube-and-housing assembly is damaged, the complete steering-gear assembly must be replaced. However, the linkage may be reused, if it is not damaged.

⚙ 19-2 Servicing rack-and-pinion power-steering gears The rack-and-pinion power-steering system consists of the same four major components as other power-steering systems. These are the power-steering pump, the power-steering gear, the pressure hose, and the return hose. The pump and hose are similar to those of other power-steering systems. However, the rack-and-pinion power-steering gear is different in design from other power-steering gears. Therefore, different diagnosis and service procedures are required.

⚙ 19-3 Rack-and-pinion steering-gear trouble diagnosis When diagnosing trouble in the rack-and-pinion steering gear, always drive the car to verify the driver's complaint. Check for obvious malfunctions such as uneven tire wear, incorrect tire pressure, or loose steering components. Then visually check the flexible coupling and intermediate-shaft universal joint in the steering shaft for being loose, worn, or broken. Tighten or replace any obviously defective part before road testing the car. Raise the car on a lift and inspect under the car for loose or damaged parts. Rotate the front wheels and check for out-of-round tires or bent wheel rims. Check the wheel bearings for looseness or a rough feel when the wheel is rotated. If the preliminary checks do not reveal the problem cause, then follow the steps listed in the appropriate diagnosis and repair chart in ⚙ 19-4. The diagnosis and repair charts provide typical procedures to follow when test driving and an under-the-car inspection fail to locate the cause of the problem.

⚙ 19-4 Diagnosis and repair charts These charts provide a graphic method of diagnosis and troubleshooting through the use of pictures and symbols. They are not go/no-go decision trees, and they are not the same as the tables that you have frequently used before. Instead, this type of chart uses pictures plus a few key words to help you solve a problem. Symbols and words are used to help guide you through each step.

The charts are divided into three sections. They are *step, sequence,* and *result*. Select the chart with the heading for the problem you are trying to solve. Then start at the first step and go through the complete sequence from left to right. If the problem is solved, the symbol (OK) will send you to (STOP). If the problem is not solved, the symbol (~~OK~~) will send you through another sequence of checks that end with a result and tell you the next step to go to. Work through each step of the chart until the system is repaired.

PROBLEM: HARD STEERING—EXCESSIVE EFFORT REQUIRED AT STEERING WHEEL OR POOR RETURNABILITY OF STEERING WHEEL AFTER MAKING TURN

Chart 1

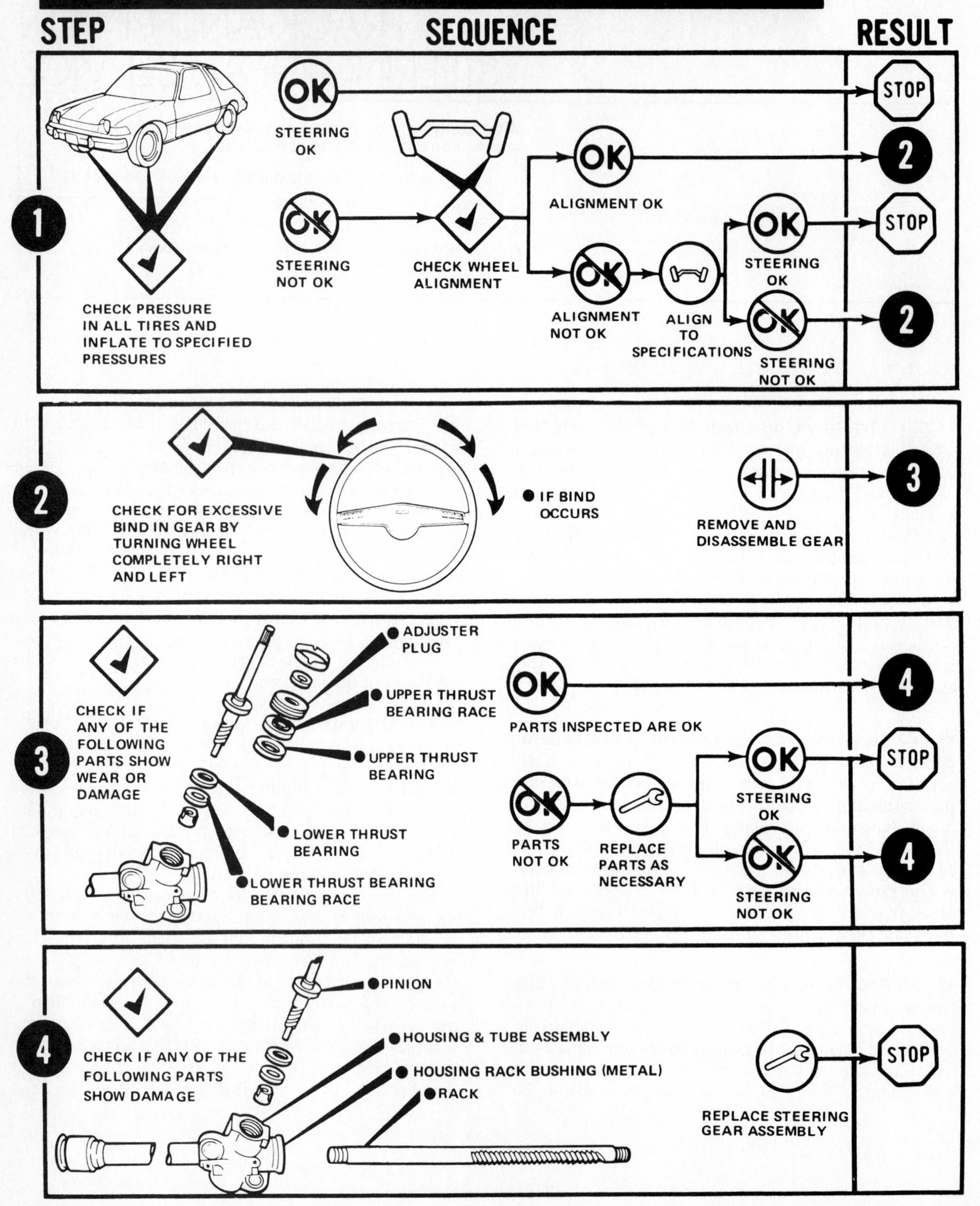

SOURCE: *American Motors Corporation.*

PROBLEM: EXCESSIVE PLAY OR LOOSENESS IN STEERING SYSTEM

Chart 2

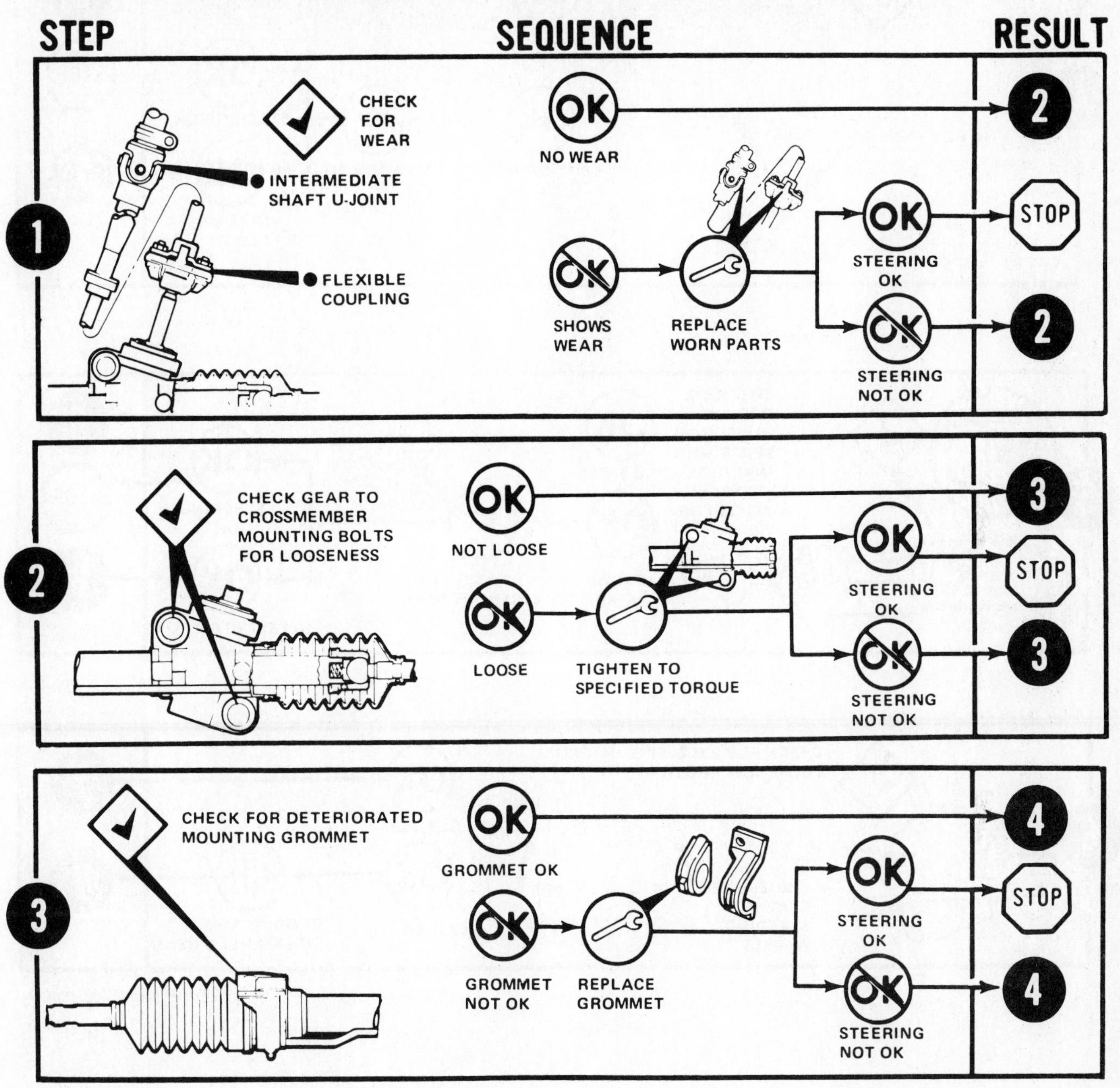

SOURCE: American Motors Corporation.

STEP SEQUENCE RESULT

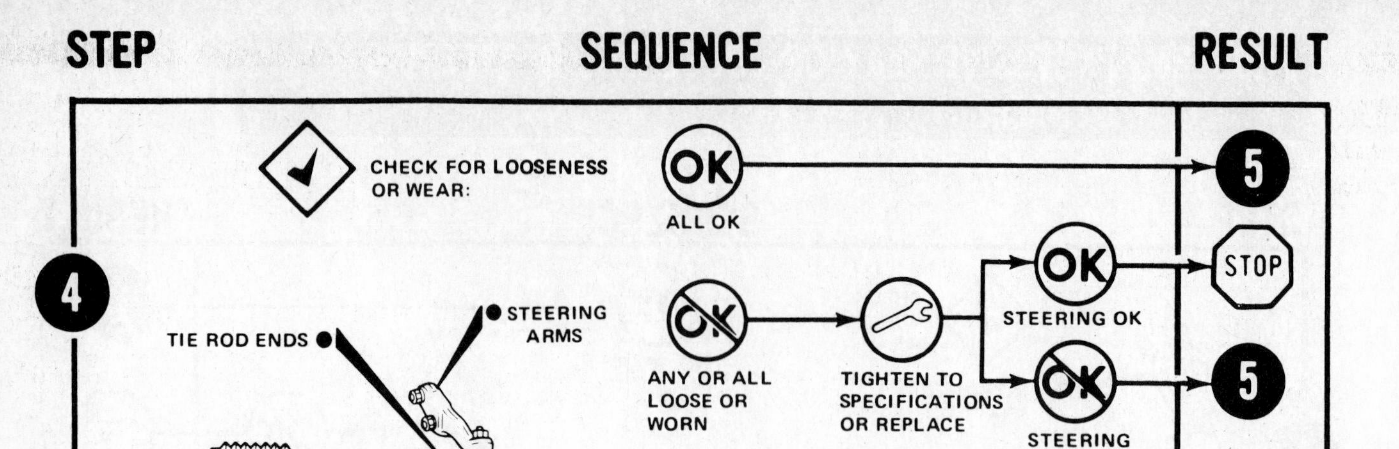

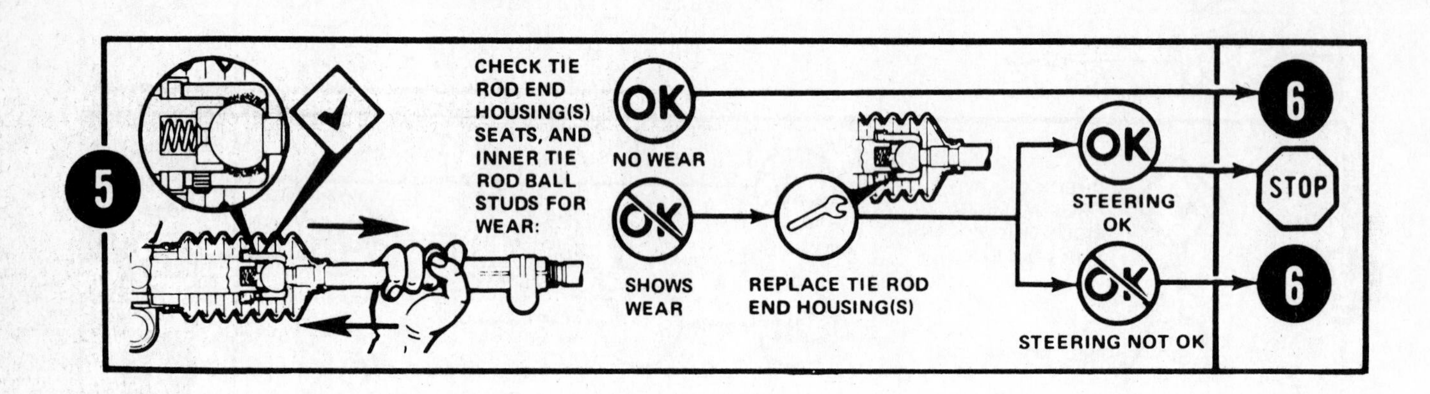

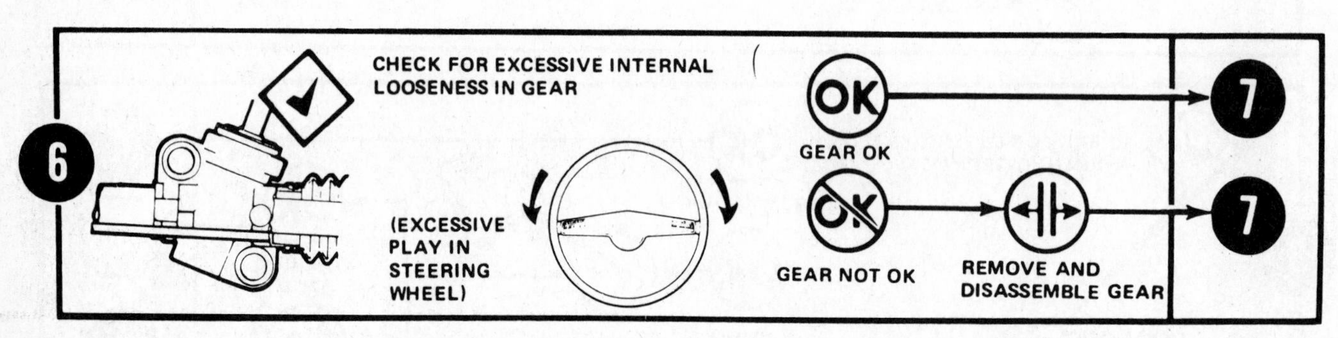

SOURCE: American Motors Corporation.

STEP SEQUENCE RESULT

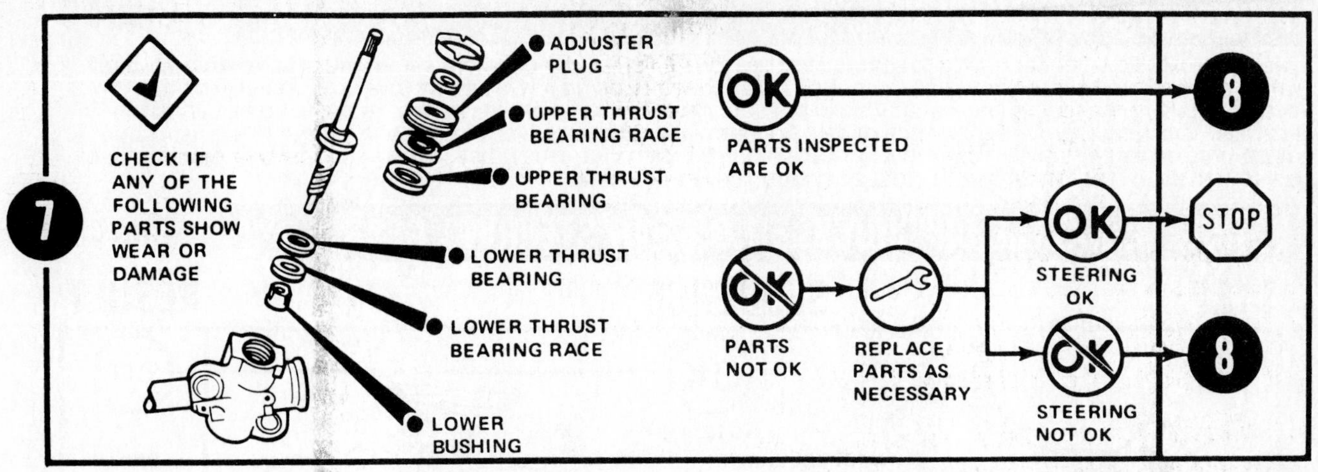

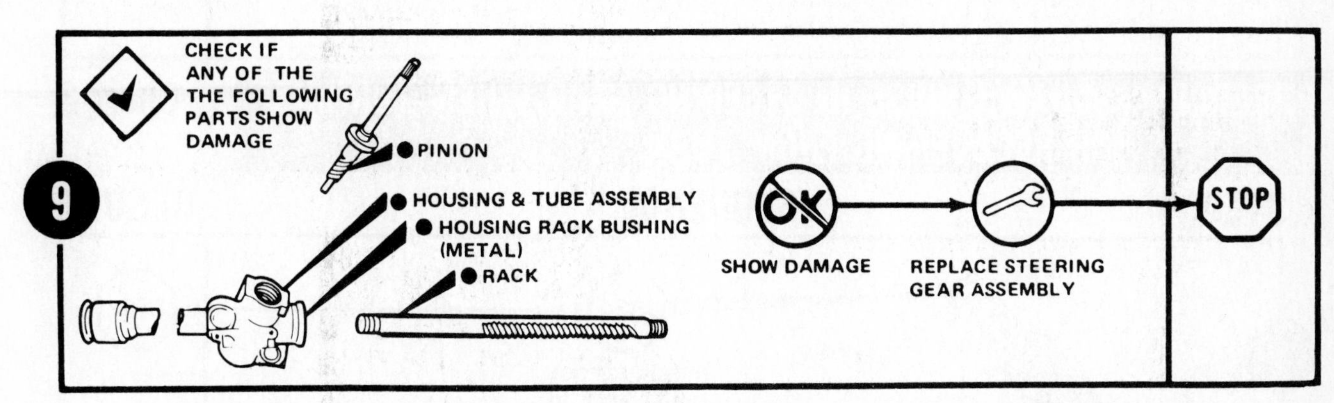

SOURCE: American Motors Corporation.

PROBLEM: HISSING NOISE — Chart 1

THERE IS SOME NOISE IN ALL POWER STEERING SYSTEMS. ONE OF THE MOST COMMON IS A HISSING SOUND MOST EVIDENT WHILE AT PARKING. HISS IS A NOISE THAT SOUNDS LIKE SLOWLY CLOSING A WATER TAP. THE NOISE IS PRESENT IN EVERY VALVE AND RESULTS FROM HIGH VELOCITY FLUID PASSING VALVE ORFICE EDGES. THERE IS NO RELATIONSHIP BETWEEN THIS NOISE AND PERFORMANCE OF THE STEERING. HISS MAY BE EXPECTED WHEN STEERING WHEEL IS AT END OF TRAVEL OR WHEN SLOWLY TURNING AT STANDSTILL. TRANSMITTING THIS NOISE INTO THE PASSENGER COMPARTMENT IS PREVENTED BY THE USE OF THE FLEXIBLE STEERING SHAFT COUPLING.

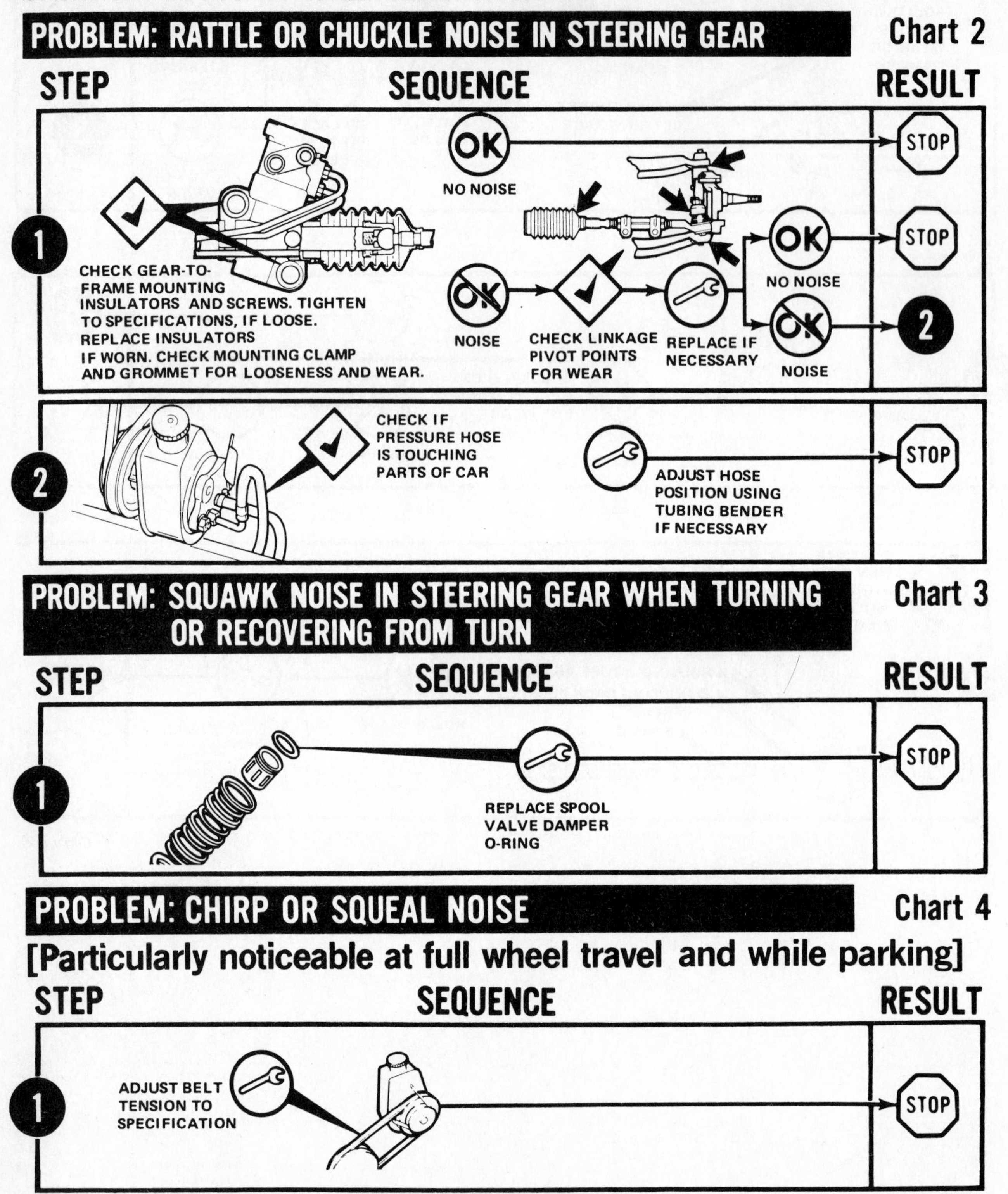

SOURCE: American Motors Corporation.

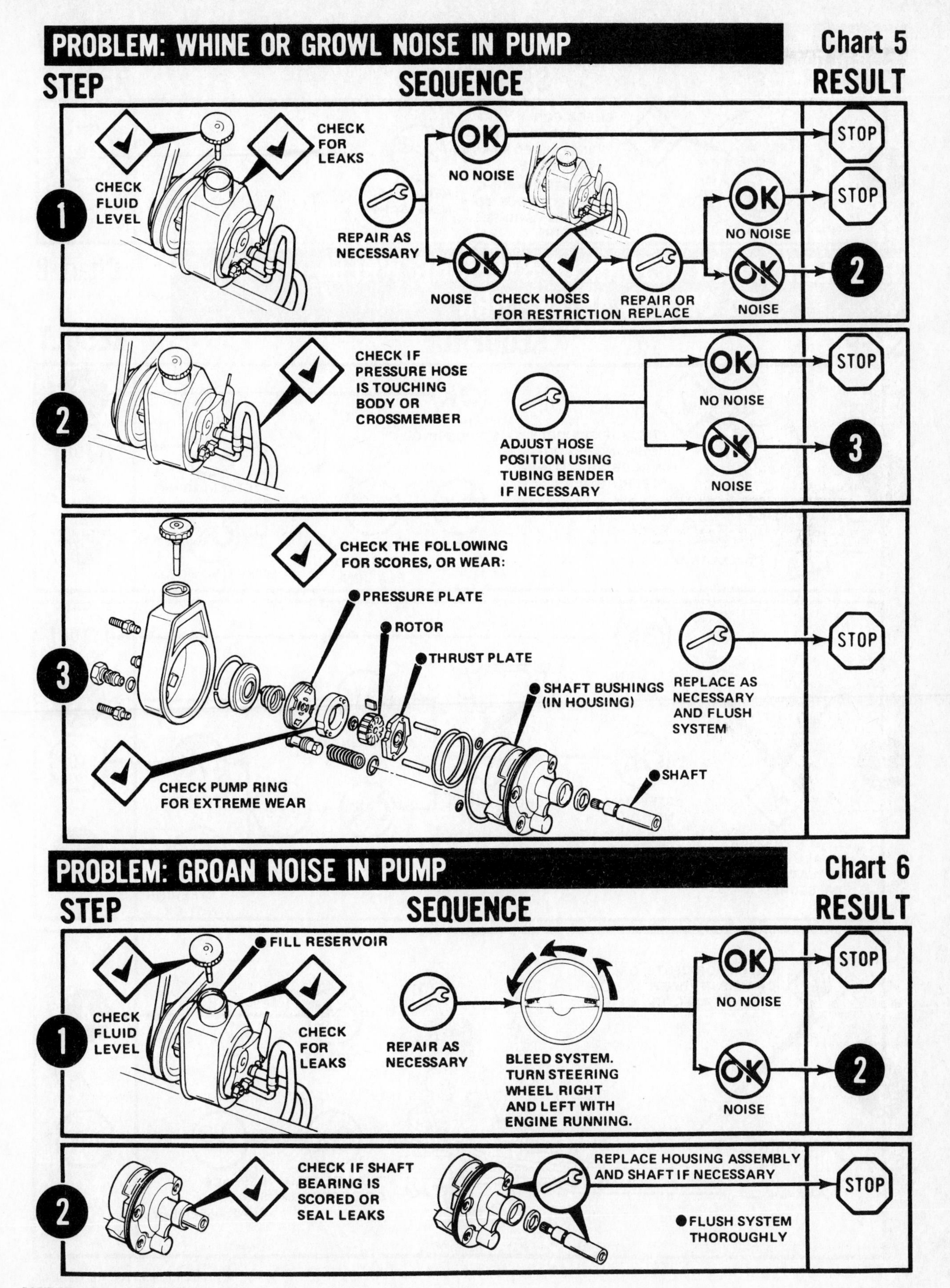

SOURCE: *American Motors Corporation.*

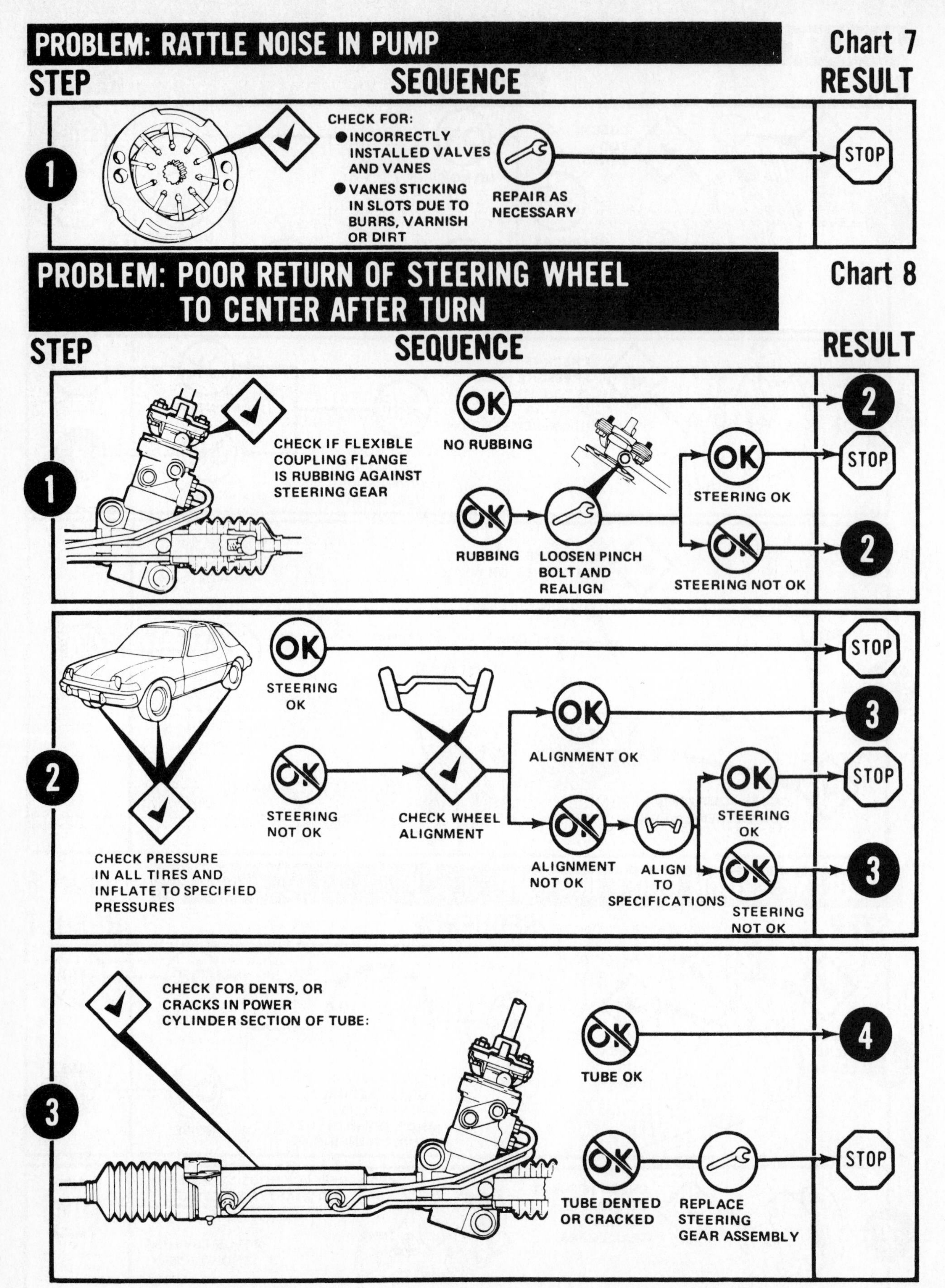

SOURCE: American Motors Corporation.

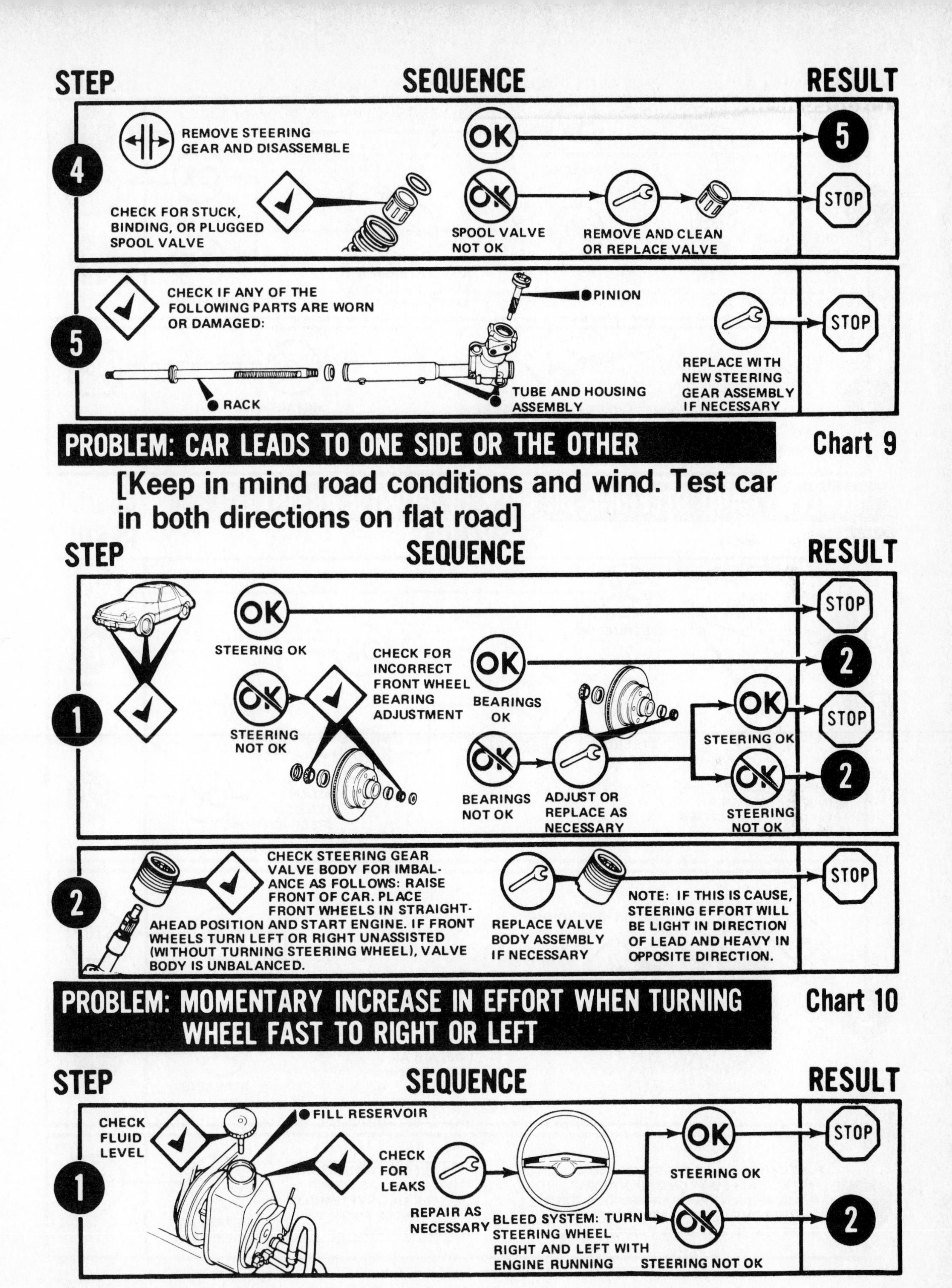

SOURCE: *American Motors Corporation.*

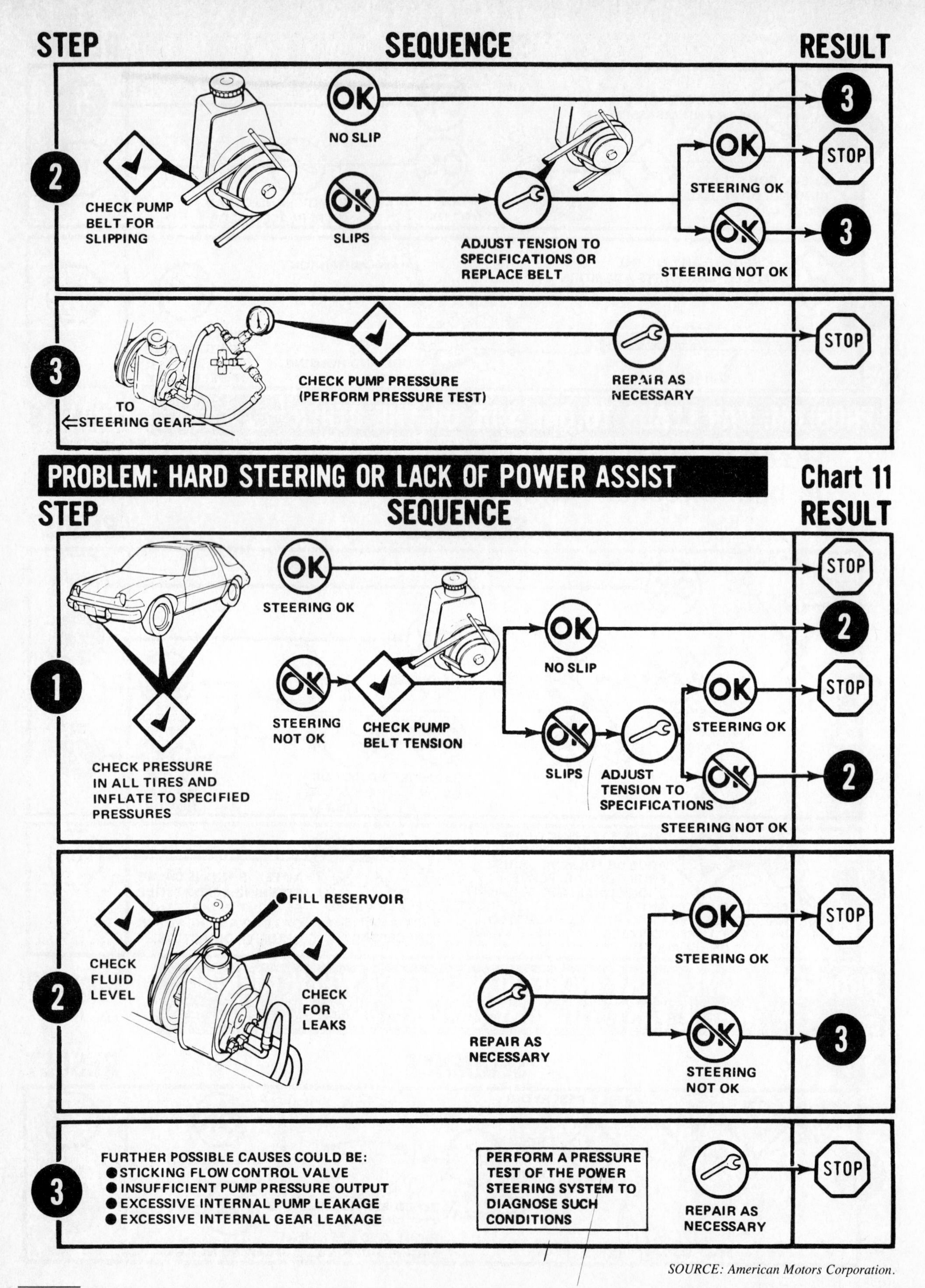

SOURCE: American Motors Corporation.

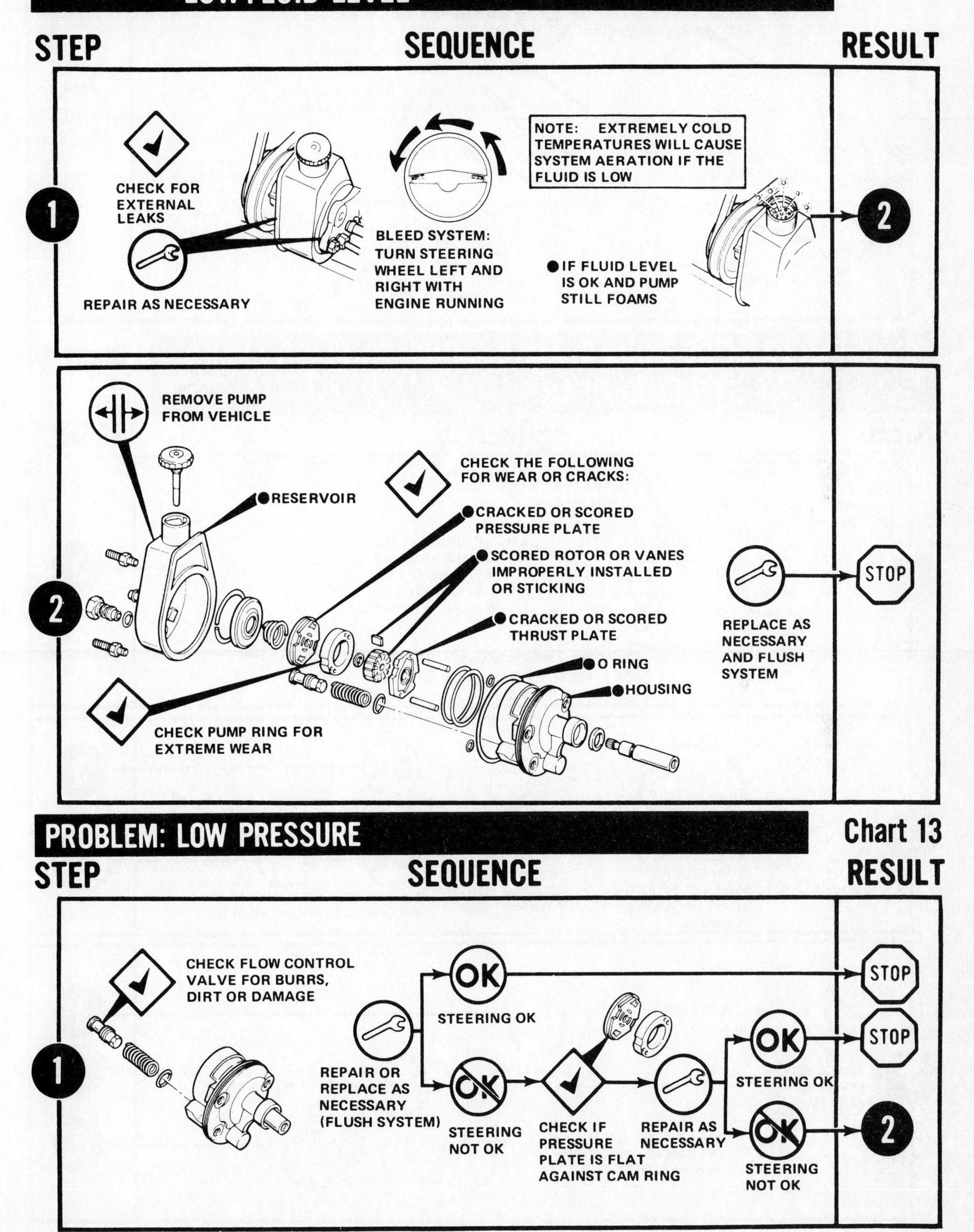

SOURCE: *American Motors Corporation.*

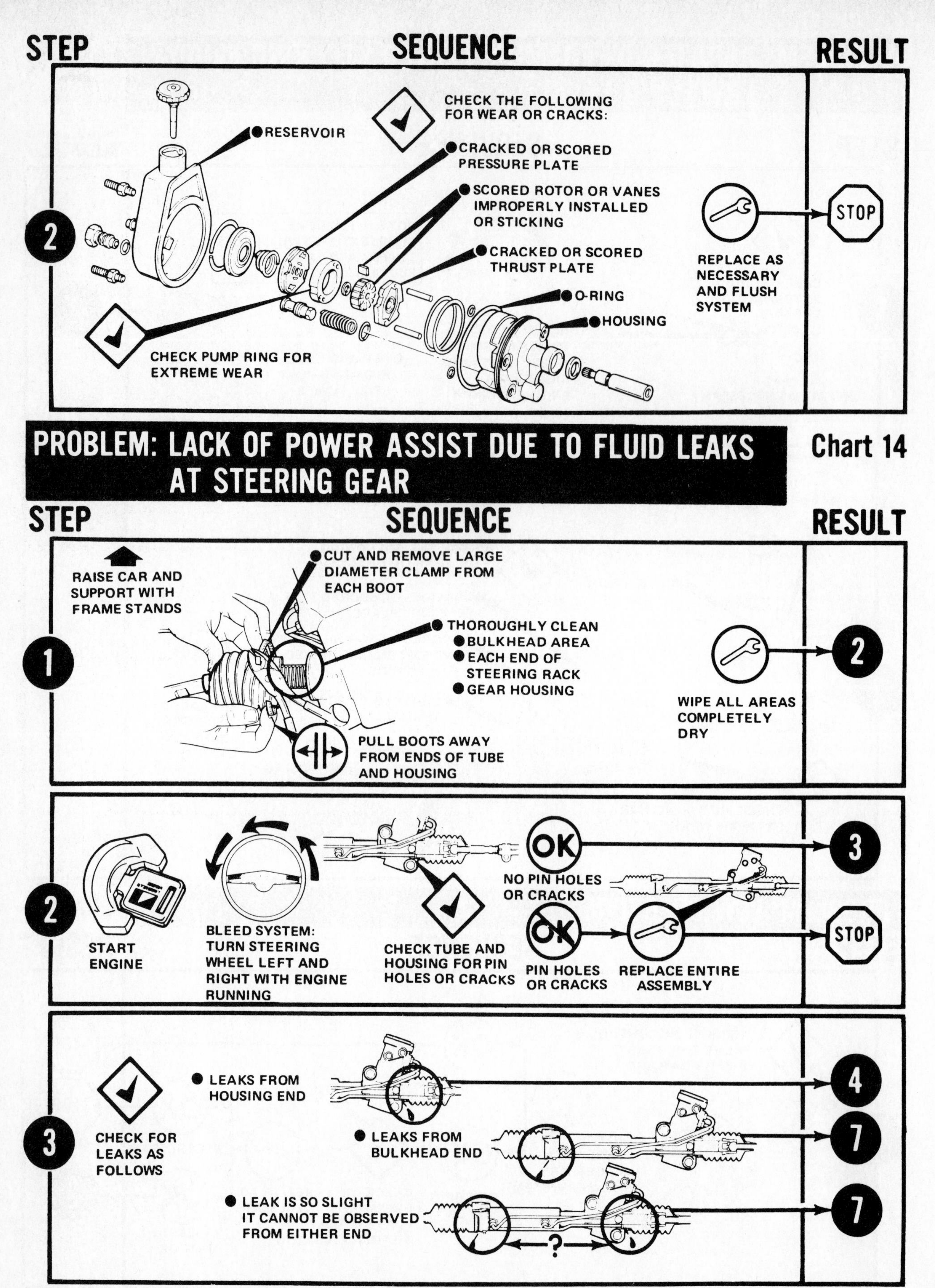

SOURCE: *American Motors Corporation.*

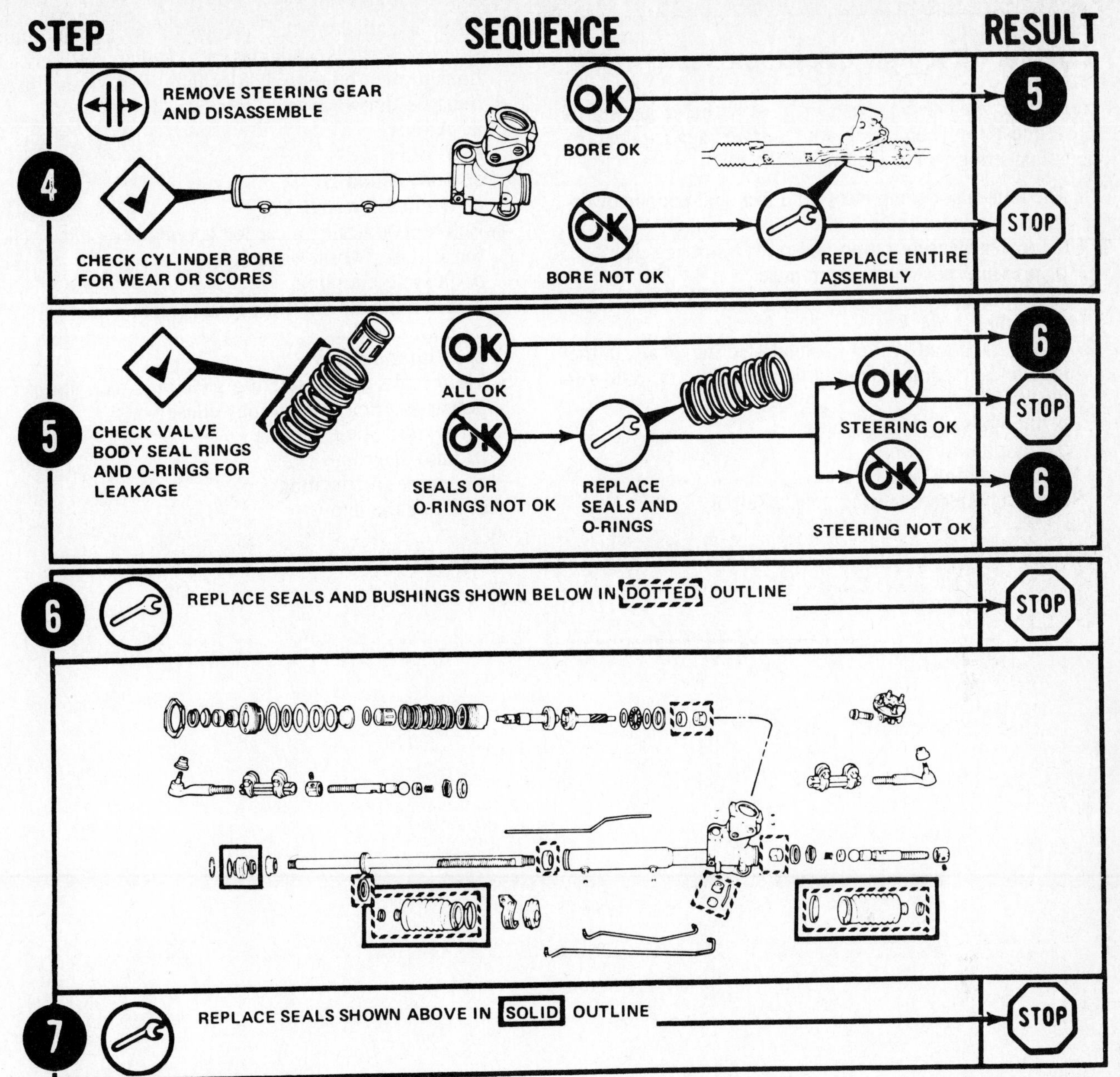

SOURCE: American Motors Corporation.

Chapter 19 review questions

Select the *one* correct, best, or most probable answer to each question. Then check your answers against the correct answers given at the end of the book.

1. The four major components of a rack-and-pinion power-steering system include the:
   a. power-steering pump and power-steering gear
   b. pressure hose and return hose
   c. both *a* and *b*
   d. neither *a* nor *b*
2. In the rack-and-pinion manual-steering gear, if the tube-and-housing assembly is damaged, you must replace
   a. the housing-rack bushing
   b. the pinion shaft
   c. the steering rack
   d. the complete steering-gear assembly
3. After a collision, the linkage on a rack-and-pinion steering gear is undamaged. Mechanic A says the linkage may be reused. Mechanic B says the linkage must be thrown away. Who is right?
   a. A only
   b. B only
   c. both A and B
   d. neither A nor B
4. Before road testing a car for steering complaints, check for a loose, worn, or broken:
   a. flexible coupling
   b. intermediate-shaft universal joint
   c. both *a* and *b*
   d. neither *a* nor *b*
5. Fluid leaks at the steering gear in a rack-and-pinion power-steering system may cause
   a. oversensitive steering
   b. hard steering
   c. engine overheating
   d. all of the above

CHAPTER 20

WHEELS AND TIRES

After studying this chapter, you should be able to:

1. Explain the difference in construction between radial and bias-ply tires.
2. Interpret the markings on the side of a tire by explaining the meaning of each.
3. Discuss the Uniform Tire Quality Grading System and explain how it grades tires.
4. Describe the differences between a conventional inner tube and an inner tube for a radial tire.
5. Explain how a puncture-sealing tire prevents air leakage after a puncture.
6. Describe the differences between a collapsible spare tire and a compact spare tire.

20-1 Purpose of tires Tires have two functions. First, they are air-filled cushions that absorb most of the shocks caused by road irregularities. The tires flex, or give, as they meet these irregularities. Therefore they reduce the effect of the shocks on the passengers in the car. Second, the tires grip the road to provide good traction. Good traction enables the car to accelerate, brake, and make turns without skidding.

20-2 Tire construction There are two general types of tires: those with inner tubes and those without tubes, called *tubeless* tires. On the inner-tube type, both the tube and the tire casing are mounted on the wheel rim. The tube is a hollow rubber doughnut. It is inflated with air after it is installed inside the tire and the tire is put on the wheel rim (Fig. 20-1). This inflation causes the tire to resist any change of shape.

Tubes are used in some truck tires and in motorcycle tires. Tubes are seldom used in passenger-car tires today. Cars use tubeless tires. The tubeless tire does not use an inner tube. Instead, the tubeless tire is mounted on the wheel rim so that the air is retained between the rim and the tire (Fig. 20-2).

The amount of air pressure used in the tire depends on the type of tire and the operation. Passenger tires are inflated to about 22 to 36 psi [155 to 248 kPa]. Heavy-duty tires on trucks or buses may be inflated to 100 psi [690 kPa].

Tire casings and tubeless tires are made in about the same way. Layers of cord, called *plies,* are shaped on a form and impregnated with rubber. The rubber sidewalls and treads are then applied (Fig. 20-3). They are vulcanized into place to form the completed tire. The term

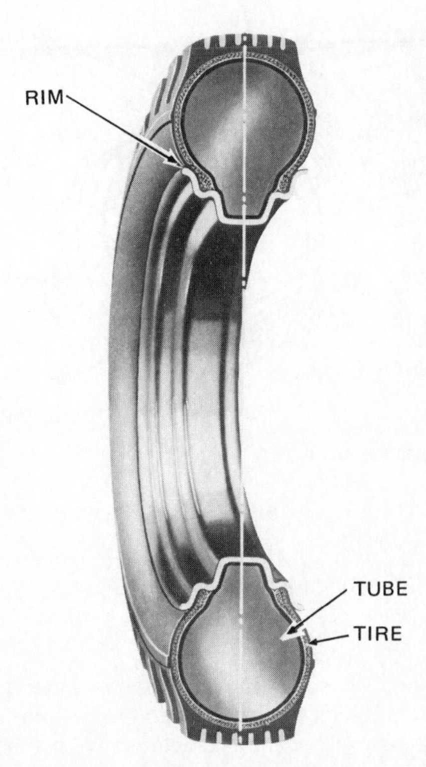

Fig. 20-1 Tire and tire rim cut away so that the tube can be seen.

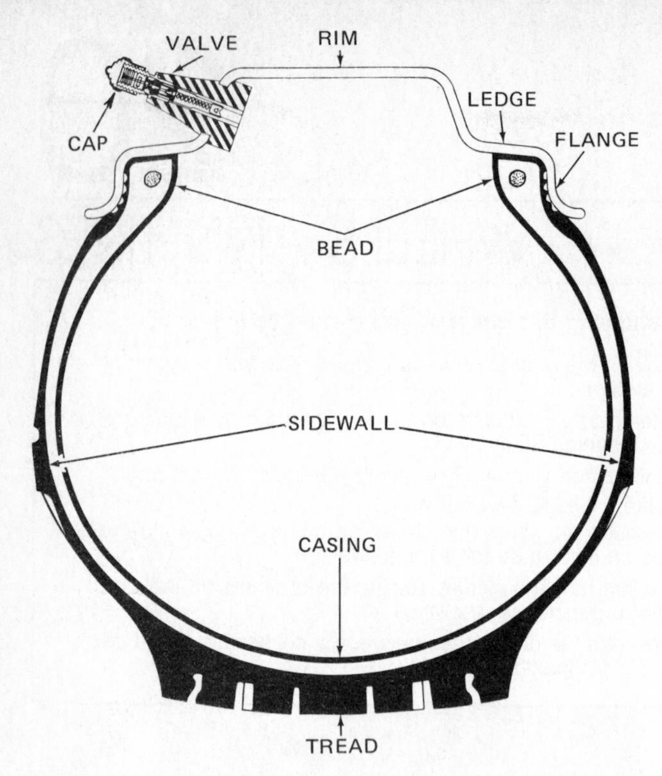

Fig. 20-2 Sectional view of a tubeless tire, showing how the tire bead rests between the ledges and flanges of the rim to produce a good seal. (*Pontiac Motor Division of General Motors Corporation*)

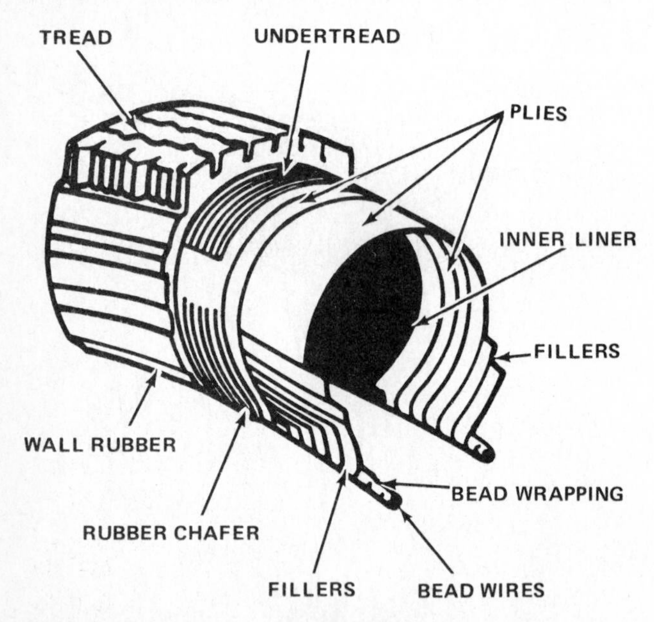

Fig. 20-3 Cutaway view of a tubeless tire, showing tire construction. (*Chevrolet Motor Division of General Motors Corporation*)

vulcanizing means heating the rubber under pressure. This process molds the rubber into the desired form and gives it the proper wear characteristics and flexibility. The number of layers of cord, or plies, varies according to the intended use of the tire. Passenger-car tires have 2, 4, or 6 plies. Heavy-duty truck and bus tires may have up to 14 plies. Tires for heavy-duty service, such as earthmoving machinery, may have up to 32 plies.

⚙ 20-3 Bias vs. radial plies There are two ways to apply the plies, on the bias and radially. For many years most tires were of the bias type, as shown to the left in Fig. 20-4. These tires have the plies criss-crossed. One layer runs diagonally one way, and the other layer runs diagonally the other way. This arrangement makes a carcass that is strong in all directions because of the overlapping plies. However, the plies tend to move against each other in bias tires. This movement generates heat, especially at high speed. Also, the tread tends to "squirm," or close up, as it meets the road. This increases tire wear.

Tires with radial plies, as shown to the right in Fig. 20-4 and in Fig. 20-5, were introduced to remedy these problems. In a radial tire, all plies run parallel to each other and vertical to the tire bead. Belts are applied on top of the plies to provide added strength parallel to the bead. Then the tread is vulcanized on top of the belts, which are made of rayon, nylon, glass fiber, or steel mesh. All radial tires work in the same way, regardless of the belt material.

Radial tires are installed on about 95 percent of all cars built in the United States. The radial is more flexible than a bias-ply tire, so more of the tread stays on the pavement (Fig. 20-6). On a radial, the tread has less tendency to heel up when the car goes around a curve (Fig. 20-7). This keeps more rubber on the road and reduces the tendency of the tire to skid.

Also, better fuel economy is claimed for cars with belted tires. The tires roll on the road with less resistance and use up less power. Radial tires wear more slowly than bias-ply tires. This is because the radial-tire tread does not squirm as the tire meets the pavement. The bias-ply tire tread tends to squirm (Fig. 20-8). As the treads pinch together, they slide sideways. This causes tread wear. There is less heat buildup on the highway in the radial tire. This also slows radial-tire wear.

Bias-ply tires may also be belted (Fig. 20-9). However, even some manufacturers who make the belted-bias tire recommend the belted radial as the superior tire.

Careful! Never mix belted-radial and bias-ply tires, either belted or unbelted, on a car. Mixing the two types can cause poor car handling and increase the possibility of skidding. This is especially important with snow tires. Bias-ply snow tires on the rear and belted radials on the front can result in oversteer and spin-out on wet or icy roads.

⚙ 20-4 Tire tread The tread is the part of the tire that rests on the road. There are many different tread designs. Figure 20-10 shows a few. Snow tires have large rubber cleats molded into the tread. The cleats cut through snow to improve traction.

Some tires have steel studs that stick out through the tread. Studs help the tire get better traction on ice and

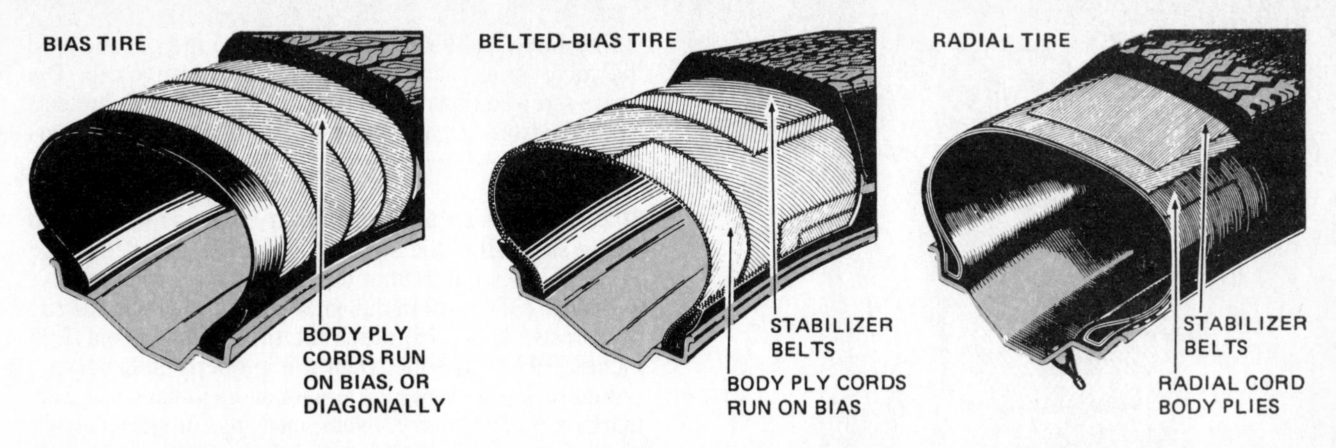

Fig. 20-4 Cutaway views of the three basic tire constructions. (*Firestone Rubber Company*)

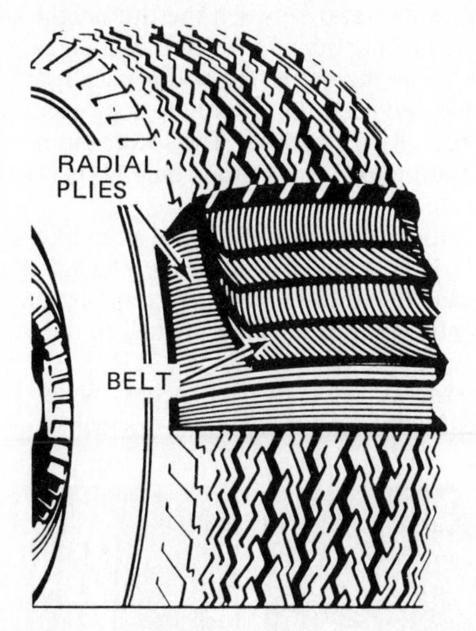

Fig. 20-5 Belted radial tire, partly cut away to show radial plies and belt. (*B. F. Goodrich Company*)

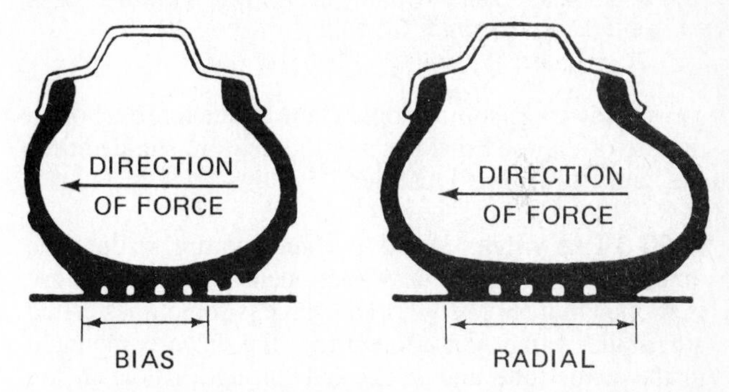

Fig. 20-7 Difference in the amount of tread a nonbelted bias-ply tire and a belted radial tire apply to the pavement during a turn.

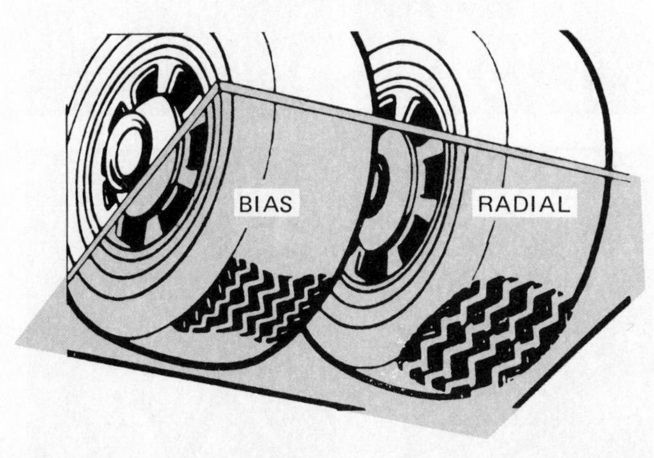

Fig. 20-6 Tread patterns ("footprints") of a nonbelted bias-ply tire and a belted radial tire on a flat surface. The belted radial tire puts more rubber on the road.

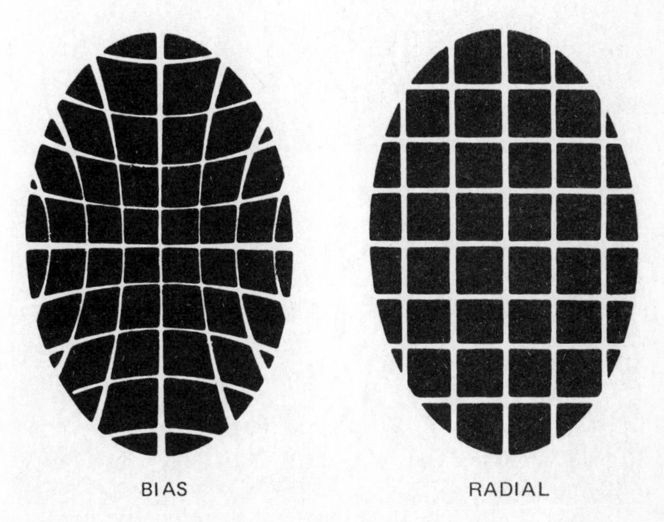

Fig. 20-8 Bias-ply tire tread tends to squirm as it meets the road. Belted radial tire tread tends to remain apart.

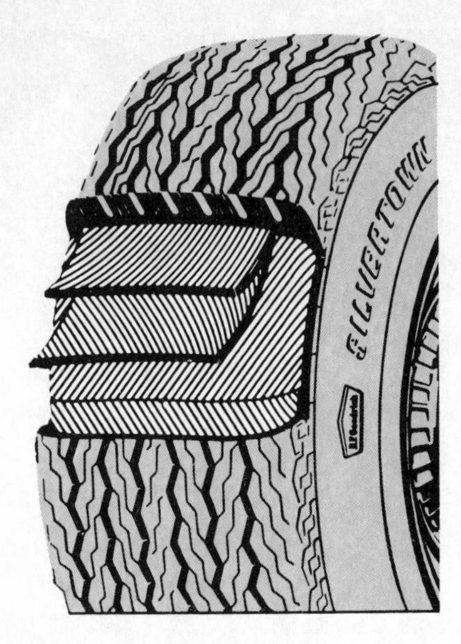

Fig. 20-9 Belted-bias tire, partly cut away to show bias plies and belt. (*B. F. Goodrich Company*)

snow. However, some people claim studded tires shorten the life of the road surface. For this reason, studded tires are banned in some localities.

⚙ 20-5 Tire valve Air is put into the tire, or into the inner tube, through a valve that opens when an air hose is applied to it (Fig. 20-2). The valve is sometimes called a Schrader valve. On a tubed tire, the valve is mounted in the inner tube and sticks out through a hole in the wheel rim. On the tubeless tire, the valve is mounted in the hole in the wheel rim. When the valve is closed, spring force and air pressure inside the tire or tube hold the valve on its seat. Most valves carry a valve cap. The cap is screwed down over the end of the valve. It protects the valve from dirt and acts as an added safeguard against air leaks.

⚙ 20-6 Tire size and markings Tire size is marked on the sidewall of the tire. An older tire might be marked 7.75-14. This means that the tire fits on a wheel that is 14 inches [356 mm] in diameter at the rim where the tire bead rests. The 7.75 means the tire itself is about 7.75 inches [197 mm] wide when it is properly inflated.

Figure 20-11 shows a tire with an explanation of each mark on it. Tires carry several markings on the sidewall. The markings include a letter code to designate the type of car the tire is designed for. D means a lightweight car. F means intermediate. G means a standard car. H, J, and L are for large luxury cars and high-performance vehicles. For example, some cars use a G78-14 tire. The 14 means a rim diameter of 14 inches [356 mm]. The 78 indicates the ratio between the tire height and width (Fig. 20-12). This tire is 78 percent as high as it is wide. The ratio of the height to the width is called the *aspect ratio* or the *profile ratio*. There are four aspect ratios at present: 83, 78, 70, and 60. The lower the number, the wider the tire looks. A 60 tire is only 60 percent as high as it is wide.

The addition of an R to the sidewall marking, such as GR78-14, indicates that the tire is a radial. Also, if a tire is a radial, the word "radial" must be molded into the sidewall. Some radials are marked in the metric system. For example, a tire marked 175R13 is a radial tire which measures 175 mm (6.9 inches) wide. It mounts on a wheel with a rim diameter of 13 inches [330 mm].

Fig. 20-10 Types of tire tread. (*B. F. Goodrich Company; Goodyear Tire and Rubber Company*)

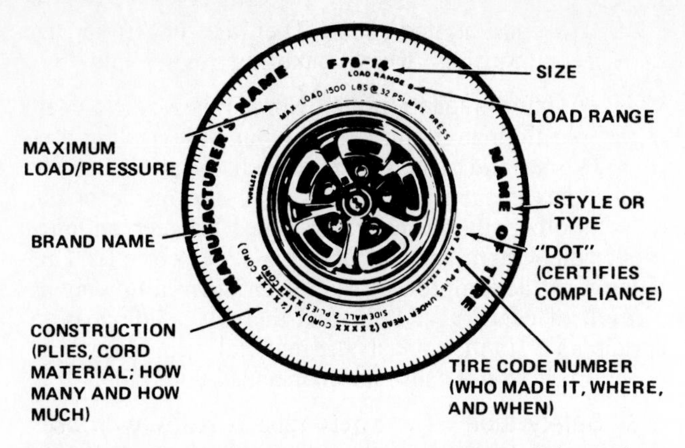

Fig. 20-11 Every marking on a tire is important information. Note the location of tire size, pressure, and load limit.

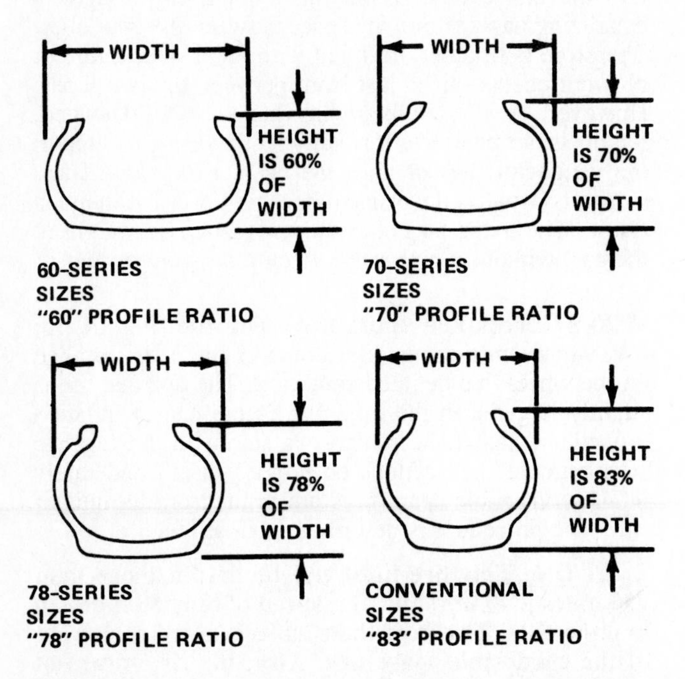

Fig. 20-12 Four aspect ratios of car tires. (*American Motors Corporation*)

Wheel diameter is given in inches in all tire-sizing systems.

Some cars use metric size tires. The meaning of each letter and number of a metric tire is shown in Fig. 20-13. This is the latest size designation for tires. Comparing the two tire-size labels, a tire formerly marked as an ER78-14 is now marked P195/75R14. An SL ("standard load") on the sidewall means a maximum inflation pressure of 35 psi [240 kPa]. When the tires are marked XL ("extra load"), maximum inflation pressure is 41 psi [280 kPa].

To identify the load that an older-type tire can safely carry, each tire is classified by load range. The load range indicates the allowable load for the tire as inflation pressure is increased. The tire in Fig. 20-11 is marked "Load Range B." Most passenger car tires are in load range B. There are three load ranges for passenger-car tires, B, C, and D. Under the oldest system, "ply rat-

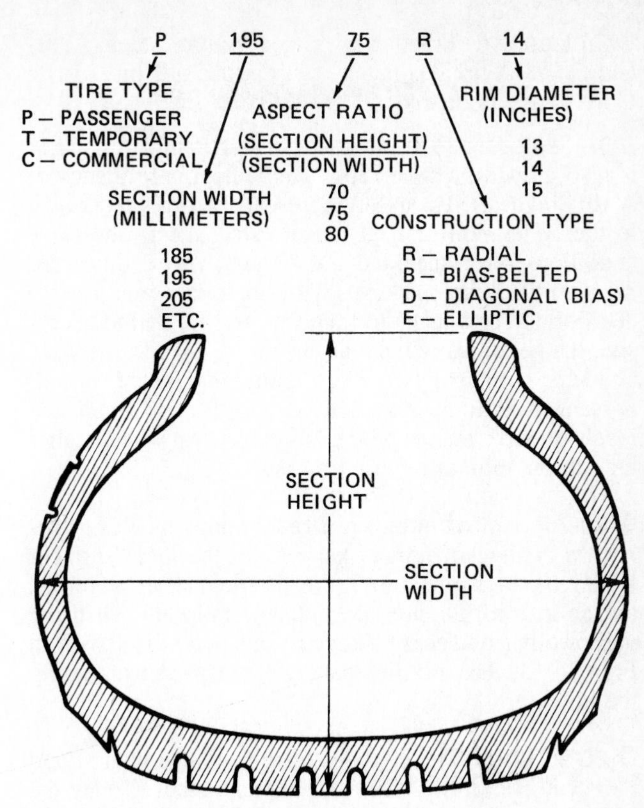

Fig. 20-13 Meanings of the size designations for a metric tire. (*Chevrolet Motor Division of General Motors Corporation*)

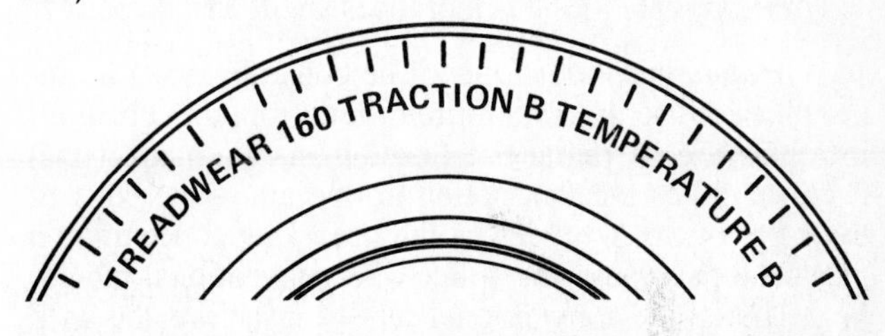

Fig. 20-14 Uniform Tire Quality Grading System markings on the tire sidewall. (*ATW*)

ing" was used to indicate load range. The load range B tire has the same load-carrying capacity as a tire with a 4-ply rating. Load range C equals a 6-ply-rating tire. Load range D equals an 8-ply-rating tire.

In addition, new tires must comply with the Uniform Tire Quality Grading System (UTQGS) of the Department of Transportation (DOT). Under this system, tires are graded for tread wear, traction, and temperature resistance. Each tire is graded with a series of numbers and letters indicating comparative tread wear, traction, and temperature. The grades are molded on the tire sidewall (Fig. 20-14).

1. Tread wear Tread wear is a number grade, such as 90 or 120. A tire with a tread-wear grade of 150 can be expected to last 50 percent longer on the government test course than a 100 tread-wear tire, if the tire is not abused. A tire with a tread-wear grade of 100 should last for about 30,000 miles [48,280 km]. The tire graded 150 should last about 45,000 miles [72,420 km].

2. Traction Traction is graded in three steps, A, B, and C. An A tire has better traction than a B tire. Grade C tires have poor traction on wet roads.

3. Temperature resistance Temperature resistance is also graded in three seps. The temperature grades of A (the highest), B, and C represent the tire's resistance to the generation of heat when tested under laboratory conditions on a specified indoor test wheel. Sustained high temperature can cause the materials in a tire to deteriorate. This leads to reduced tire life, and to excessive temperatures which can lead to sudden tire failure. Grade C corresponds to the minimum standard that all passenger-car tires must meet. Grades B and A represent levels of performance on the laboratory test wheel higher than the minimum required by law.

20-7 Puncture-sealing tires Some tubeless tires have a coating of plastic material in the inner surface. When the tire is punctured, this plastic material is forced by the internal air pressure into the hole left when the nail or other object is removed. This action is shown in Fig. 20-15. The plastic material then hardens, sealing the hole.

20-8 Tubes Three types of rubber, one natural and two synthetic, have been used to make tubes. Today the most common tube material is butyl. A butyl tube can be identified by its blue stripe. The other synthetic rubber type (GR-S) has a red stripe. Natural rubber is not striped.

Three special types of inner tubes are described below.

1. Radial-tire inner tube The construction of an inner tube for use in a radial tire differs from the tube used in a bias tire. A radial tire flexes in such a manner that it concentrates the flex action in one area—the edge of the belts in the shoulder of the tire. This concentration of stress may cause the splice of a conventional tube to fail. Therefore, conventional tubes must never be used in radial tires.

To overcome the problem, the radial tube is made from a special rubber compound that is designed to overcome this concentrated stress. Then the tube is spliced on a special machine which makes a stronger splice.

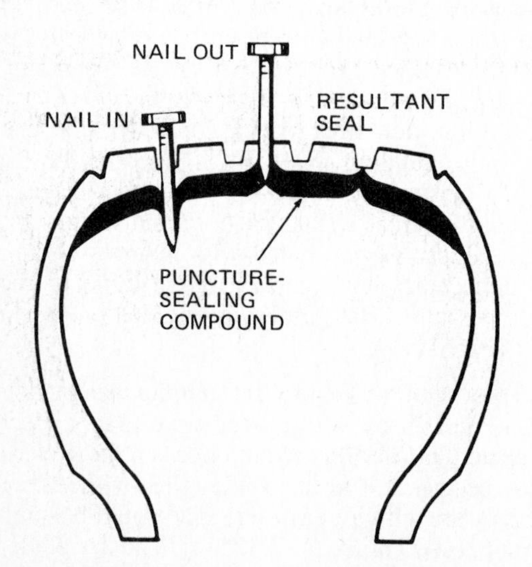

Fig. 20-15 Sealing action in a puncture-sealing tire. (*Pontiac Motor Division of General Motors Corporation*)

2. Puncture-sealing tubes Some tubes have a coating of plastic material used in the puncture-sealing tire. It flows into and seals any holes left by punctures. In some tubes the plastic material coats the inside of the tube. In others the material is retained between an inner rubber diaphragm and the tube in a series of cells. This latter construction prevents the material from flowing as a result of centrifugal force and thereby building up in certain spots in the tube. If the material were allowed to build up, it would cause an unbalanced condition.

3. Safety tube The safety tube is really two tubes in one, one smaller than the other and joined at the rim edge. When the tube is filled with air, the air flows first into the inside tube. From there it passes through an equalizing passage into the space between the two tubes. Therefore both tubes are filled with air. If a puncture or blowout occurs, air is lost from between the two tubes. However, the inside tube, which has not been damaged, retains its air pressure. It is sufficiently strong to support the weight of the car until the car can be slowed and stopped. Usually, the inside tube is reinforced with nylon fabric. The nylon takes the suddenly imposed weight of the car, without giving way, when a blowout occurs.

20-9 Collapsible spare tire This tire (Fig. 20-16) saves space in the luggage compartment. It is installed on the wheel in a deflated condition. The deflated tire is slightly larger than the rim. A pressure can of inflation propellant, called the inflator, is stored in the luggage compartment. Instructions on how to install and safely inflate collapsible spare tires are printed on the inflator can. The procedure is described in Chap. 21.

CAUTION: **This tire must not be driven more than 150 miles [240 km] and at a speed of only 50 mph [80 km/h] or less. To exceed these limits is to risk a blowout of the collapsible spare tire. Also, the tire must not be inflated from the usual air hose at the shop or in a service station. This can cause the tire to explode.**

20-10 Compact spare tire Another tire that saves space in the luggage compartment is the compact spare tire (Fig. 20-17). This spare is lighter and considerably smaller than the standard tire. The compact spare tire can be driven for the 1000- to 3000-mile [1609- to 4800-km] life of the tread. It is mounted on a narrow 51 × 4 wheel. The tire must not be mounted on any other wheel. No other tire, wheel cover, or trim ring should be installed on the special wheel. Also, the compact spare tire should not be used on the rear of a car equipped with a nonslip differential. The collapsible spare tire is smaller in diameter than the tire on the other side of the rear axle. Differential action would have to take place continuously. This may cause damage and failure of the differential.

The compact spare tire is for emergency use only. As soon as the standard tire has been repaired, reinstall it on the car. The compact spare carries an inflation pressure of 60 psi [415 kPa]. It gives a rough and noisy ride.

Fig. 20-16 A collapsible spare tire with inflator. (*ATW*)

Fig. 20-17 A compact spare tire. (*ATW*)

20-11 Wheels Most cars use a pressed steel or disk wheel. This type of wheel also is called a *safety-rim* wheel. Figure 20-18 shows how a disk wheel is made. The outer part, called the rim, is in one piece, and it is welded to the disk. This forms the seamless and airtight wheel that is needed to mount a tubeless tire. The center of the rim is smaller in diameter than the rest. This gives the rim the name "drop center." The center well is necessary to permit removal and installation of the tire. The bead of the tire must be pushed off the bead seat and into the smaller diameter. Only then can the tire beads be worked up over the rim flange. Tire service is covered in Chap. 21.

The 14-inch [356-mm] wheel is used on most cars today. However, some smaller cars have 12-inch [305-mm] or 13-inch [330-mm] wheels. Fourteen-inch [356-mm] wheels are available in three widths. The width

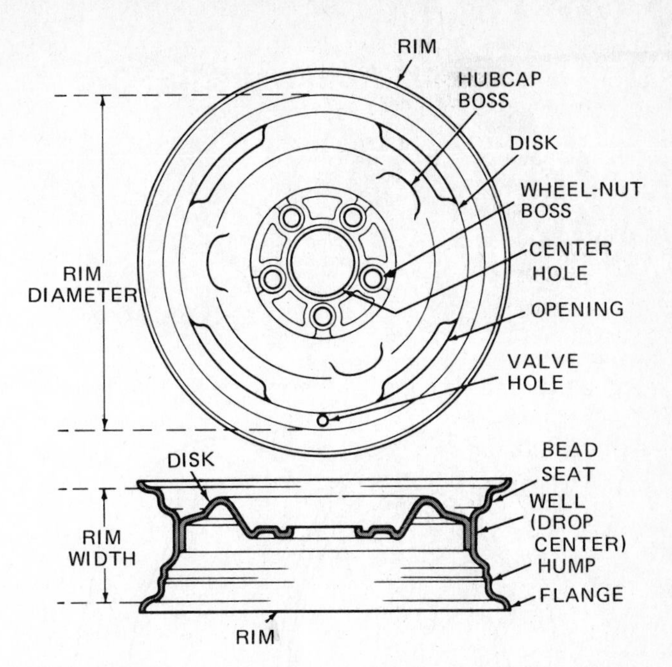

Fig. 20-18 Construction of a car wheel. (*American Motors Corporation*)

used on a car depends on the tire specified for the car. The rim widths are 4.5 inches [114 mm], 5 inches [127 mm], and 6 inches [152 mm]. Optional larger tires usually require wider rims.

Most manufacturers recommend that a wheel be replaced if it is bent or leaks air. The new wheel must be exactly the same as the old wheel. Installation of the wrong wheel could cause the wheel bearing to fail, the brakes to overheat, the speedometer to read inaccurately, and the tire to rub the body and frame.

20-12 Special wheels Plain steel wheels, decorated with hub caps or wheel covers, are used on cars today. A variety of special wheels are available. These special wheels can be classified as styled steel or styled aluminum wheels. The "mag" wheel is very popular. It looks like the magnesium wheels used on some race cars (Fig. 20-19). Magnesium metal is very light. However, for passenger cars, most "mag" wheels are made of aluminum. Actually, the term "mag wheel" can mean almost any chromed, aluminum-offset, or wide-rim wheel of spoke design.

Some aluminum wheels are lighter than the steel wheels they replace. Lighter wheels reduce unsprung weight. This improves handling and performance. Also, some aluminum wheels can improve brake and tire performance by allowing them to run cooler. Aluminum transmits and dissipates heat faster than steel.

20-13 Split-rim wheels Split-rim wheels are used on heavy trucks, trailers, and earthmoving equipment. They are heavily made and require a different method of tire installation. The drop-center rim (Figs. 20-18 and 20-19) is not satisfactory for a heavy-duty tire.

There are many types of two- and three-piece truck rims. One type retains the tire on the rim with a lock ring (Fig. 20-20). Then the lock ring is removed so that

Fig. 20-19 Various designs of "mag" wheels. (*Shelby International, Inc.*)

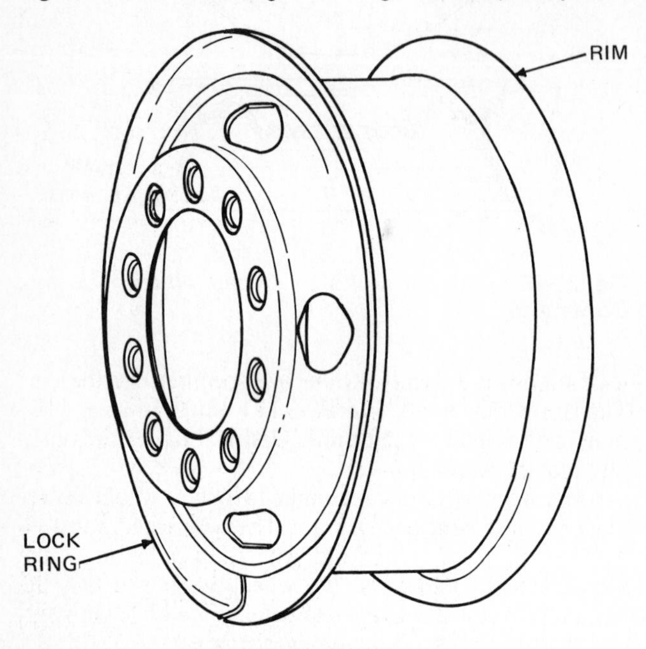

Fig. 20-20 A split-rim wheel. (*ATW*)

the tire can be taken off the rim. These tires require tubes because the rim cannot be made airtight.

CAUTION: **If you ever work on tires mounted on split-rim wheels, make sure all air pressure has been released from the tube before you begin to remove the lock ring or flange. If air pressure is still in the tube, it could blow the tire off the rim when the lock ring or flange is removed and seriously injure or kill anyone nearby. Heavy-duty tires carry high air pressures. Make sure the lock ring or flange is securely in place before attempting to inflate the tire. *Never* stand over the tire while inflating it! It could explode.**

Chapter 20 review questions

Select the *one* correct, best, or most probable answer to each question. Then check your answers against the correct answers given at the end of the book.

1. One purpose of tires is to:
   a. grip the road and provide good traction
   b. substitute for springs
   c. act as brakes
   d. none of the above
2. Two general types of tires are:
   a. tube type and tubeless
   b. solid and tubeless
   c. air and pneumatic
   d. split-rim and drop-center
3. Typical passenger-car tire inflation pressures are from
   a. 12 to 22 psi
   b. 22 to 36 psi
   c. 40 to 100 psi
   d. none of the above
4. Vulcanizing means:
   a. heating rubber under pressure
   b. spraying with a special paint
   c. melting rubber while stirring it
   d. none of the above
5. Bias-ply tires have:
   a. all plies running parallel to one another
   b. belts of steel mesh in the tires
   c. one ply layer that runs diagonally one way and another layer that runs diagonally the other way
   d. all of the above
6. In radial tires:
   a. one ply layer runs diagonally one way, and the other layer runs diagonally the other way
   b. all plies run parallel to one another and vertical to the road
   c. inner tubes are always used
   d. none of the above
7. Different types of tires should never be mixed on the same car, such as
   a. tube and tubeless tires
   b. whitewall and blackwall tires
   c. radial and bias-ply tires
   d. all of the above
8. A tire size given as G78-15 means that the
   a. tire fits a lightweight car
   b. tire is 78 percent as high as it is wide
   c. circumference of the rim is 15 inches
   d. none of the above
9. Mag wheels usually are made of:
   a. magnesium
   b. aluminum
   c. cast iron
   d. steel
10. The wheel diameter for a tire marked P195/75R14 is:
    a. 14 mm
    b. 195 mm
    c. 75 mm
    d. 14 inches

CHAPTER 21

WHEEL AND TIRE SERVICE

After studying this chapter, and with proper instruction and equipment, you should be able to:

1. Diagnose tire wear.
2. Check tire pressure and inflate tires.
3. Rotate tires.
4. Mount a tubeless tire.
5. Service the collapsible spare tire.
6. Repair a punctured tire.
7. Repair a punctured tube.

Tire service

21-1 Tire service Tire service includes periodic checking of the air pressure and addition of air as needed. Failure to maintain correct air pressure can cause rapid tire wear and early tire failure. Incorrect air pressure can also cause handling problems (21-2). Tire service also includes periodic inspection of the tire for abnormal wear, cuts, bruises, and other damage. In addition, tire service includes removal, repair, and replacement of tires.

21-2 Tire inflation and tire wear The driver has more effect on tire life than anything else. Rapid tire wear results from quick starts and stops, heavy braking, high speed, taking corners too fast, and striking or rubbing curbs. Too little air in the tire can cause hard steering, front-wheel shimmy, steering kickback, and tire squeal on turns. A tire with too little air will wear on the shoulders and not in the center of the tread, as shown at the upper left in Fig. 21-1.

Also, the underinflated tire is subject to rim bruises. If the tire strikes a rut or stone, or bumps a curb too hard, it flexes so much that it is pinched against the rim. Any of these kinds of damage can lead to early tire failure.

With overinflation, the tire rides on the center of the tread. Then, only the center wears, as shown at the upper center in Fig. 21-1. In addition, the overinflated tire will not flex normally. As a result, the tire fabric can be weakened or even broken.

Other types of tire wear are discussed below.

1. Toe-in or toe-out tire wear Excessive toe-in or toe-out on turns causes the tire to be dragged sideways as it moves forward. For example, a tire on a front wheel that toes in 1 inch [25.4 mm] from straight ahead will be dragged sideways about 150 feet [46 m] every mile [1.6 km]. This sideways drag scrapes off rubber, as shown at the upper right in Fig. 21-1. Note the feather edges of rubber that appear on one side of the tread. If both sides show this type of wear, the toe is incorrect. If only one tire shows this type of wear, a steering arm probably is bent. This causes one wheel to toe in or out more than the other.

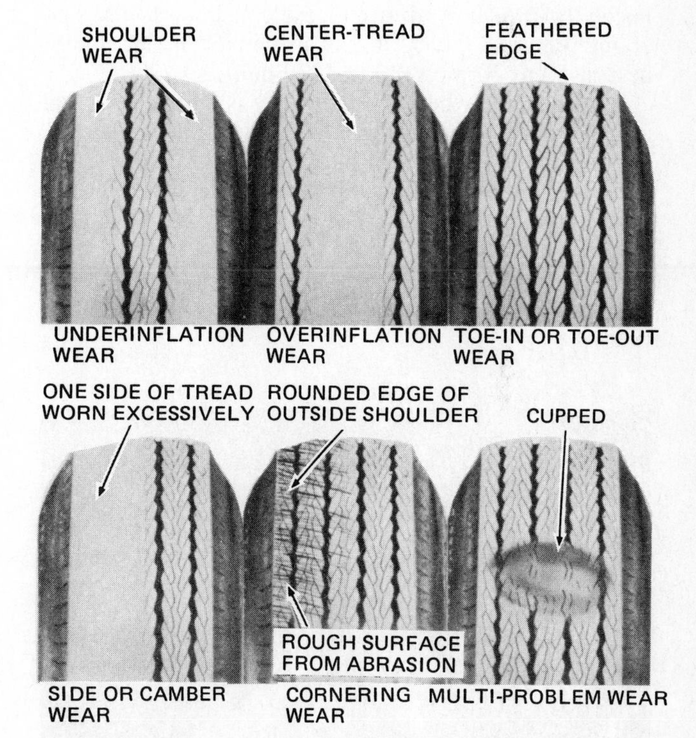

Fig. 21-1 Patterns of abnormal tire-tread wear. (*Buick Motor Division of General motors Corporation*)

2. Camber wear If a wheel has excessive camber, the tire runs more on one shoulder than on the other.

The tread wears excessively on that side, as shown at the lower left in Fig. 21-1.

3. Cornering wear Cornering wear, shown at the lower center in Fig. 21-1, is caused by taking curves at excessive speeds. The tire skids and tends to roll, producing the diagonal type of wear shown. This is one of the more common causes of tire wear. The only remedy is to have the driver slow down around curves.

4. Uneven tire wear Uneven tire wear, with the tread unevenly or spottily worn, is shown at the lower right in Fig. 21-1. It can result from several mechanical problems. These include misaligned wheels, unbalanced wheels, uneven or "grabby" brakes, overinflated tires, and out-of-round brake drums.

5. High-speed wear Tires wear more rapidly at high speed than at low speed. Tires driven consistently at 70 to 80 miles per hour [113 to 129 km/h] will give less than half the life of tires driven at 30 miles per hour [48 km/h].

21-3 Radial-tire waddle *Waddle* is side-to-side movement of the front or rear of the car as it moves forward. It is caused by the steel belt not being straight inside the tire. It is most noticeable at low speeds, 5 to 30 mph [8 to 48 km/h]. It may also be felt as ride roughness at 50 to 70 mph [80 to 113 km/h].

To determine where the faulty tire is on the car (front or rear), make a road test. If the faulty tire is on the rear, the rear end of the car will shake from side to side. From the driver's seat it feels as though someone were pushing on the side of the car.

If the faulty tire is on the front, the waddle is more visual. The front body sections will appear to move back and forth or from side to side.

Fig. 21-2 Checking tire air pressure.

kPa	psi	kPa	psi
140	20	215	31
145	21	220	32
155	22	230	33
160	23	235	34
165	24	240	35
170	25	250	36
180	26	275	40
185	27	310	45
190	28	345	50
200	29	380	55
205	30	415	60

Conversion: 6.9 kPa = 1 psi

Fig. 21-3 Inflation pressure chart, showing the conversion from kilopascals (kPa) to pounds per square inch (psi).

21-4 Checking tire pressure and inflating tires To check tire pressure (Fig. 21-2) and inflate the tires, you must know the correct tire pressure for the tire you are servicing. You find this spec on the tire sidewall (on many tires), in the shop manual, in the owner's manual, or on one of the door jambs (or at some similar place) on the car. Specifications are for cold tires. Tires that are hot from being driven or from sitting in the sun will have an increased air pressure. Air expands when hot. Tires that have just come off an interstate highway may show as much as a 5 to 7 psi [35 to 48 kPa] increase.

As a hot tire cools, it loses pressure. So, never bleed a hot tire to reduce the pressure. If you do this, then when the tire cools, its presure could drop below the specified minimum.

There are times when the tire pressure should be on the high side. For example, one tire manufacturer recommends adding 4 psi [28 kPa] for turnpike speed, trailer pulling, or extra-heavy loads. But never exceed the maximum pressure specified on the tire sidewall.

CAUTION: **Never stand over a tire while inflating it. It could explode, and you would be injured.**

If the tire valve has a cap, always install the cap after checking pressure or adding air.

Inflation specifications are often given in kilopascals (kPa) instead of pounds per square inch (psi). The conversion table in Fig. 21-3 will help you convert from one to the other.

21-5 Tire rotation The amount of wear a tire gets depends on its location on the car. For example, the right rear tire wears about twice as fast as the left front tire. This is because roads are slightly crowned (higher in the center), and also because the right rear tire is driving. The crown causes the car to lean out a little, so that the right tires carry more weight. The combination of this and carrying power through the right rear tire causes it to wear faster. To equalize wear as much as possible, tires should be rotated any time uneven wear is noticed and at the distance specified by the car manufacturer. One manufacturer recommends rotating radial tires after the first 7500 miles [12,000 km] and then every 15,000 miles [24,000 km]. Bias tires should be rotated every 7500 miles [12,000 km].

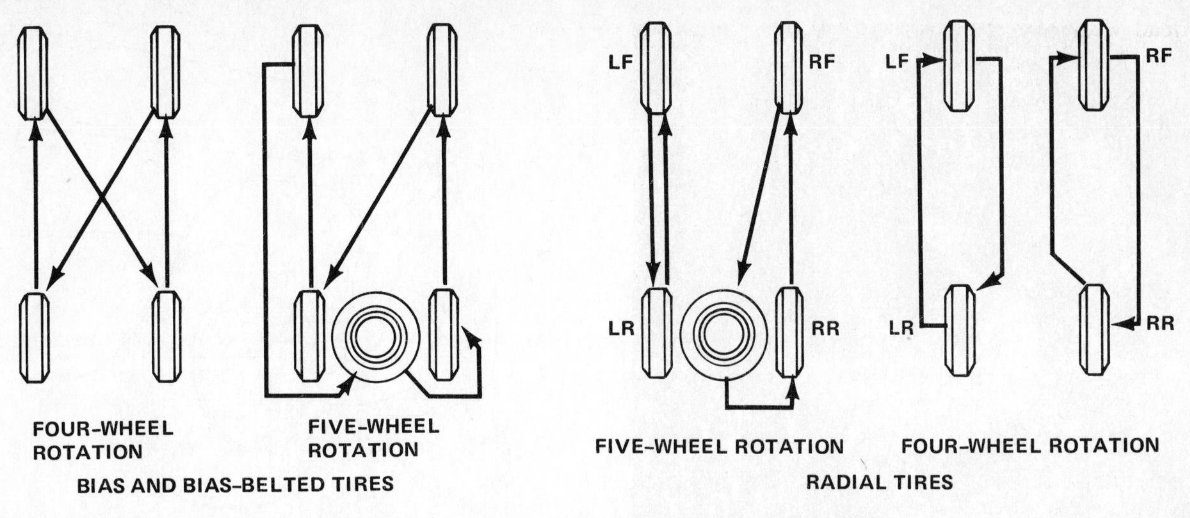

Fig. 21-4 Tire-rotation patterns for cars with and without a rotatable spare tire. (*Chevrolet Motor Division of General Motors Corporation*)

Figure 21-4 shows the recommended rotation pattern for bias, bias-belted, and radial tires. Bias and bias-belted tires can be switched from one side of the car to the other. However, radial tires must not be switched from one side to the other. This would reverse their direction of rotation and cause handling and wear problems.

On cars with a collapsible or a compact spare tire, use the four-wheel rotation pattern shown in Fig. 21-4.

Always check tire pressure after switching tires. Many cars require that the front tires carry different pressures from the rear tires. Therefore, when you switch tires from front to back, the pressure will require adjustment to meet the specifications.

Careful! Studded tires should never be rotated. A studded tire should be put back on the wheel from which it was removed. Before you remove a studded tire, mark its location (LR or RR) on the sidewall. A studded tire should be put back on the same wheel from which it was removed, mounted so that it rolls in the same direction. Reversing its rotation can result in serious handling problems.

⚙ 21-6 Tire inspection The purpose of inspecting the tires is to determine whether they are safe for further use. When an improper wear pattern is found, the technician must know the cause of the abnormal tread wear (Fig. 21-1). The technician must correct the cause or tell the driver what is wrong. When the tires are found to be in good condition, they can be rotated (Fig. 21-4). After the tires are cool, check and adjust inflation pressures.

When inspecting a tire, check for bulges in the sidewalls. A bulge is a danger signal. It can mean that the plies are separated or broken and that the tire is likely to go flat. A tire with a bulge should be removed. If the plies are broken or separated, the tire should be thrown away. To make a complete tire inspection, remove all stones from the tread. This is to make sure that no tire damage is hidden by the stones. Also, any time the tire is to be spin-balanced, remove all stones from the tread. This will ensure that no person is struck and injured by stones thrown from the tread as the tire rotates.

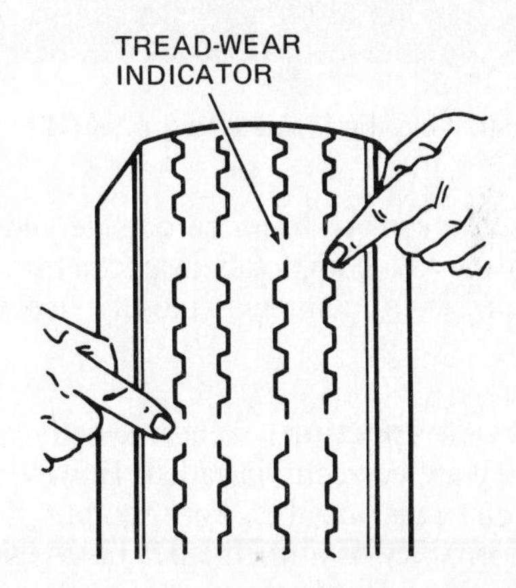

Fig. 21-5 A tire tread worn down so much that the tread-wear indicator shows up.

Many tires have tread-wear indicators, which are filled-in sections of the tread grooves. When the tread has worn down enough to show the indicators (Fig. 21-5), the tire should be replaced. There also are special tire-tread gauges that can be inserted into the tread grooves to measure how much tread remains. A quick way to check tread wear is with a Lincoln penny (Fig. 21-6). If at any point you can see all of Lincoln's head, the tread is excessively worn. Some state laws require a tread depth of at least 1/32 inch [0.79 mm] in any two adjacent grooves at any location on the tire.

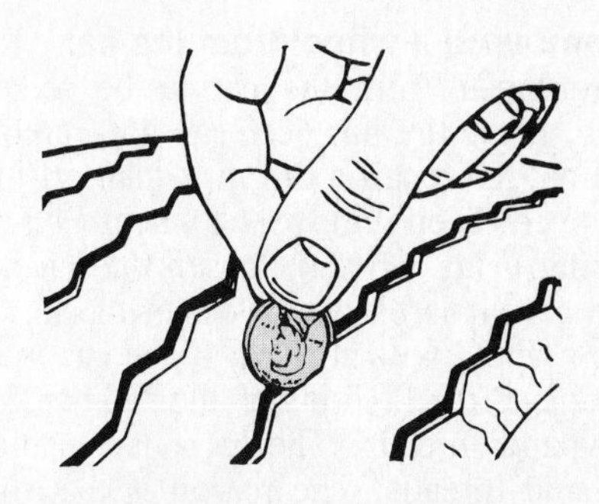

Fig. 21-6 Using a penny to check tread wear.

Fig. 21-7 Inspecting the inside of a tire. (*ATW*)

A tire can look okay from the outside and still have internal damage. To completely inspect a tire, remove it from the rim. Then examine it closely, inside and out (Fig. 21-7).

21-7 Tube inspection Tubes usually give little trouble if they are correctly installed. However, careless installation can cause trouble. For example, if the wheel rim is rough or rusty or if the tire bead is rough, the tube may wear through. Dirt in the casing can cause the same trouble. Another condition that can cause trouble is installing a tube that is too large in the tire. Sometimes an old tube (which may have stretched) is put in a new tire. When a tube that is too large is put into a casing, the tube can overlap at some point. The overlap will rub and wear and possibly cause early tube failure.

A radial tire that is used with a tube must have a special radial-tire inner tube in it. If regular tubes are used in radial tires, the tube splice may come apart.

Always check carefully around the valve stem when inspecting a tube. If the tube has been run flat or at low pressure, the valve stem may be broken or tearing away from the tube. Valve-stem trouble requires installation of a new tube.

21-8 Removing a wheel from the car Radial tires must be removed from the car to be repaired (Fig. 21-8). Also, if the tire has been run flat, remove the tire for inspection. To repair a tire, first take off the hub cap or wheel cover. Then remove the wheel from the car. If you are using a lug wrench, loosen the lug nuts before raising the car. It is easier to loosen the lug nuts first, because the wheel will not turn if the car weight is on it. On some cars the lug nuts on the right side of the car have right-hand threads. The lug nuts on the left side have left-hand threads. The reason is that the forward

Fig. 21-8 Using an impact wrench to remove the lug nuts. (*ATW*)

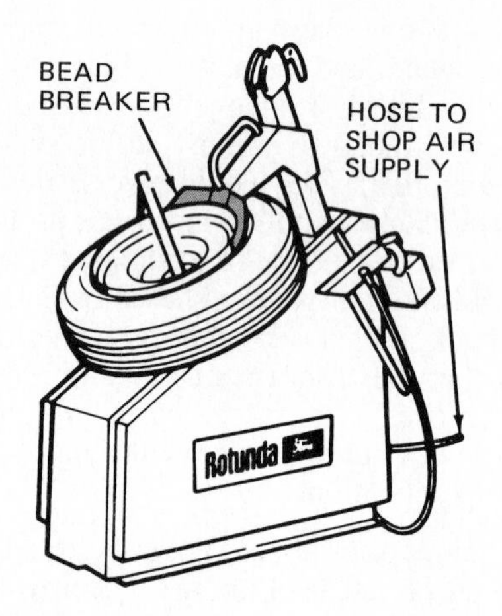

Fig. 21-9 Using an air-powered tire changer to break the bead so that the tire can be removed from the wheel. (*Ford Motor Company*)

rotation of the wheels tends to tighten the nuts, not loosen them.

When using an impact wrench (Fig. 21-8), use an impact socket with it. A regular socket may crack when used on an impact wrench.

21-9 Demounting a tire from a drop-center rim With the wheel off the car, make a chalk mark across the tire and rim so that you can reinstall the tire in the same position. This preserves the balance of the wheel and tire. Next, release the air from the tire. This can be done by holding the tire valve open or removing the valve core.

CAUTION: **The air coming out could shoot dirt particles into your eyes, so protect them with glasses or a shield.**

The tire should then be removed from the rim, using a tire changer (Fig. 21-9). At one time, tire irons—flat strips of steel—were used to remove and install tires.

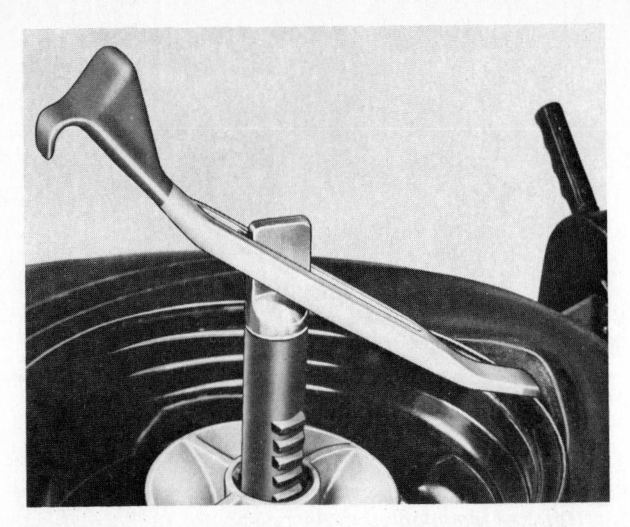

Fig. 21-10 Using the powered tire changer to lift the upper bead above the rim. The center post rotates, carrying the bead-lifting tool around with it.

They can damage the tire bead so that it will not seal, and this ruins the tire. Do not use tire irons!

⚙ 21-10 Using shop tire changers Today many shops have an air-powered tire changer (Fig. 21-9). After the bead is pushed off the rim (this is called "breaking the bead"), a tool is used with the tire changer to lift the bead up over the rim (Fig. 21-10). The powered tire changer also has a tool to remount the tire on the rim.

⚙ 21-11 Remounting the tire on the rim To mount the tire on the rim, use the tire changer. Coat the rim and beads with rubber lubricant or a soap-and-water mixture. This will make the mounting procedure easier. Do not use a nondrying lubricant, such as antifreeze, silicone, grease, or oil. They will allow the tire to "walk around" the rim so that the tire balance is lost. Oil or grease will damage the rubber.

When you are remounting the same tire that was removed from the rim, make sure the chalk marks on the tire and rim align. After the tire is on the rim, reposition the beads against the bead seat. Slowly inflate the tire (Fig. 21-11). If the beads do not hold air, use a tire-mounting band to spread the beads. You usually will hear a "pop" as the beads seat on the rim. Then install the valve core and inflate the tire to the recommended pressure.

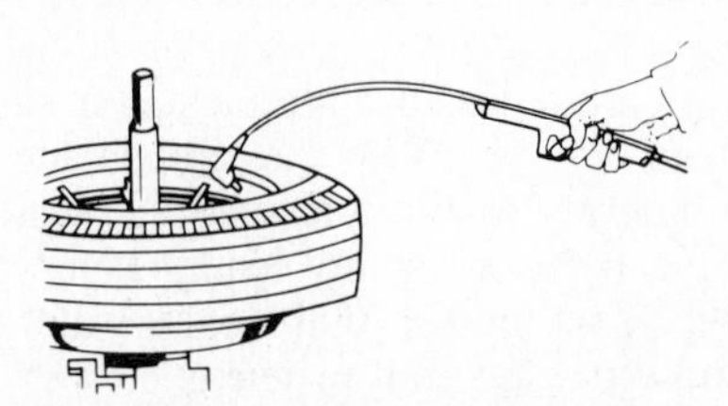

Fig. 21-11 Inflating the tire while seating the beads. Never exceed a pressure of 40 psi [276 kPa] in a passenger-car tire. (*Tire Industry Safety Council*)

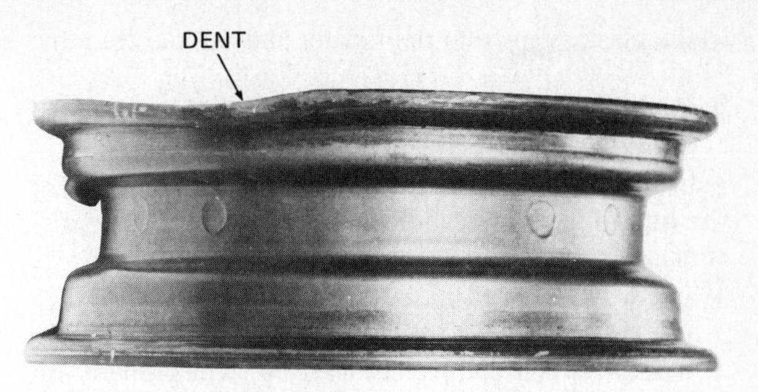

Fig. 21-12 A wheel that is bent or dented should be discarded. (*ATW*)

***CAUTION:* Do not stand over the tire while inflating it. If the tire should explode, you could be injured.**

⚙ 21-12 Checking the wheel When the tire is off the wheel, check the rim for dents (Fig. 21-12) and roughness. Steel wool can be used to clean rust spots from standard steel wheels. Aluminum wheels can be cleaned only with mild soap and water. File off nicks or burrs. Then clean the rim to remove all filings and dirt. A wheel that has been bent should be discarded. A bent wheel may be weakened by heating, welding, or straightening so that it could fail on the highway.

Some wheels now have decorative plastic inserts. The plastic can be cleaned by using a sponge and soap and water.

⚙ 21-13 Tire-valve service If the valve in the wheel requires replacement, remove the old valve and install a new one. There are two types: the snap-in type, and the type that is secured with a nut. To remove the snap-in type, cut off the base of the old valve. Lubricate the new valve with rubber lubricant. Then attach a tire-valve installing tool to the valve and pull the new valve into place.

On the clamp-in type that is secured with a nut, remove the nut to take the old valve out. Be sure to tighten the nut sufficiently when installing the new valve.

⚙ 21-14 Servicing the collapsible spare tire The space-saving collapsible tire (⚙ 20-9) is installed on the wheel deflated (Fig. 20-16). The wheel must be installed on the car before the tire is inflated.

***CAUTION:* Do not inflate the tire before the wheel is mounted on the car. Follow the safety cautions listed on the inflator can.**

The inflator can has detailed instructions on how to install and inflate the tire. Briefly, here is the inflation procedure.

1. If the temperature is 10°F [−12°C] or below, the inflator must be heated. Put the inflator over the defroster outlet of the car. Set the heater at "Defrost" at the highest temperature. Then run the blower at the fasted speed for 10 minutes.
2. Do not inflate the tire off the car. First bolt the wheel to the car with the air valve at the bottom. Remove

the plastic cap from the inflator and the cap from the tire valve.

3. Push the inflator onto the valve stem until you hear the sound of gas entering the tire.

CAUTION: **Keep your hands off the metal parts of the inflator. They become extremely cold during discharge. You could freeze your fingers!**

4. When the sound stops, wait 1 minute. Then remove the inflator and install the valve cap. The gas in the inflator, when completely used up, will properly fill the tire.

After you have inflated the tire and have removed the jack so that the tire rests on the pavement, the tire may look underinflated. This can happen especially in cold weather. Drive slowly for the first mile [1.6 km]. This will warm up the tire and increase the pressure.

5. If the inflator is the nonrefillable type, dispose of it in a safe waste receptacle. Do not burn or puncture the inflator. If it is the refillable type, it can be recharged with the proper equipment.
6. The collapsible spare tire must not be driven farther than necessary. The maximum distance is 150 miles [240 km]. As soon as the regular tire has been repaired and installed in place of the collapsible spare tire, remove the valve core from the collapsible spare tire. This will allow the gas to escape so that the tire collapses. It can then be stored, as before, in the luggage compartment.

Instead of a can of inflator, some manufacturers supply an electric air compressor to inflate the tire. Porsche, for example, supplies an electric air compressor. It is plugged into the cigarette-lighter socket and gets power from the car battery. The manufacturer warns that you should not use any other equipment to inflate the tire.

CAUTION: **This tire must not be driven more than 150 miles [240 km] and at a speed of only 50 mph [80 km/h] or less. To exceed these limits is to risk a blowout of the collapsible spare tire. Also, the tire must not be inflated from the usual air hose at the shop or in a service station. This can cause the tire to explode.**

Tire and tube repair

21-15 Tube repair If a tire has been punctured but has no other damage, it can be repaired with a patch. Remove the tube from the tire to find the leak. Inflate the tube and then submerge it in water. Bubbles will appear where there is a leak. Mark the spot. Then deflate the tube and dry it.

There are two ways to patch a tube leak. They are the cold-patch method and the hot-patch method. With the cold-patch method (also known as *chemical vulcanizing*), first make sure the rubber is clean, dry, and free of oil or grease. Buff, or roughen, the area around the leak. Then cover the area with vulcanizing cement. Let the cement dry until it is tacky. Press the patch into place. Roll it from the center out with a "stitching tool" or with the edge of a patch-kit can.

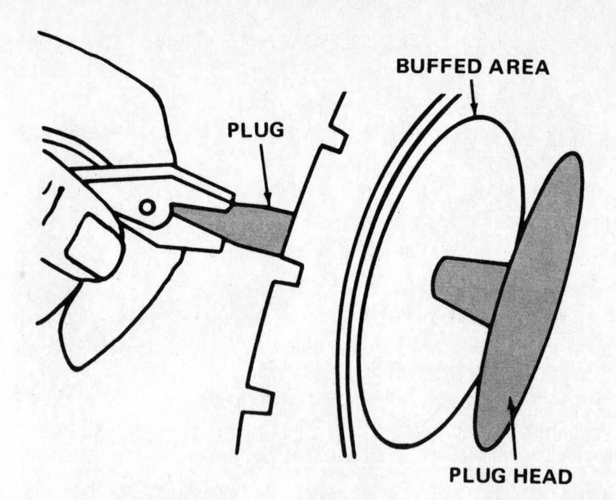

Fig. 21-13 Installing a head-type plug in a radial tubeless tire. (*Rubber Manufacturers Association*)

With the hot-patch method, prepare the tube in the same way as for the cold patch. Put the hot patch into place and clamp it. Then, with a match, light the fuel on the back of the patch. As the fuel burns, the heat vulcanizes the patch to the tube. After the patch has cooled, recheck the tube for leaks by submerging the tube in water.

Another kind of hot patch uses a vulcanizing hot plate. The hot plate supplies the heat required to bond the patch to the tube.

The hot-patch method is preferred to the cold-patch method by many technicians.

21-16 Tire repair No attempt should be made to repair a tire that has been badly damaged. If the plies are torn or have holes in them, the tire should be thrown away. A puncture bigger than ¼ inch [6.35 mm] should not be patched. Instead, the tire should be replaced. Even though you might be able to patch the tire, it would be dangerous to use. The tire might blow out on the highway.

To repair small holes in a tubeless tire, first make sure that the object that caused the hole has been removed. Check the tire for other puncturing objects. Sometimes a tubeless tire can carry a nail for a long distance without losing air.

A radial tire should be removed from the wheel for repair. The plug should be of the head type and applied from inside the tire (Fig. 21-13). Figure 21-14 shows the area of a tire in which a puncture can be repaired. Punctures outside this area require replacement of the tire.

Leaks from a tubeless tire are located in the same way as leaks from a tube. With the tire on the wheel and inflated, submerge the tire and wheel in water. Bubbles will show the location of any leaks. If a water tank is not available, coat the tire with soapy water. Soap bubbles will show the location of leaks.

If air leaks from around the spoke welds of the wheel, you can repair the leaks. Clean the area and apply two coats of cold-patch vulcanizing cement on the inside of the rim. Allow the first coat to dry before applying the

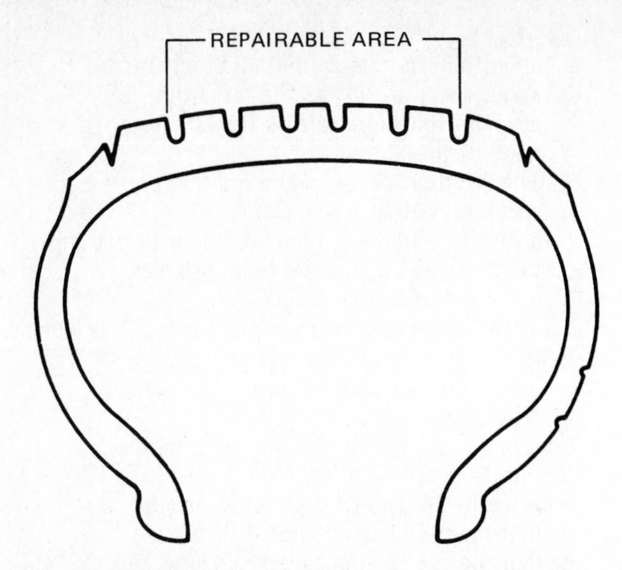

Fig. 21-14 The area of a tire in which a puncture can safely be patched. (*Chrysler Corporation*)

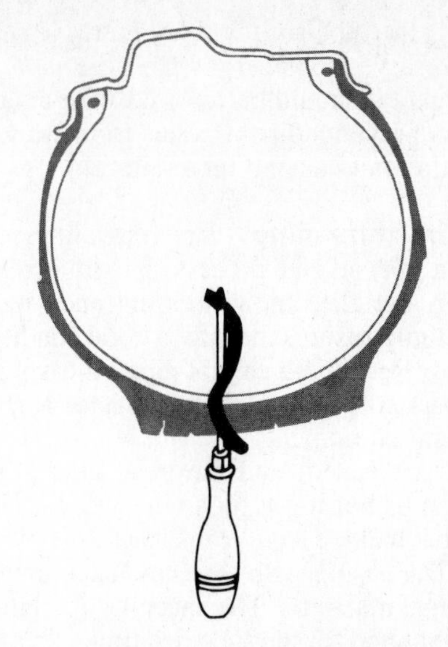

Fig. 21-15 A tire cut away to show a needle being used to insert the rubber plug into a hole in the tire. (*ATW*)

second coat. Then cement a strip of rubber patching material over the area.

❁ 21-17 Plugging a mounted tubeless tire A temporary repair of a small puncture can be made with the tire still mounted on the rim. However, this repair is only a temporary fix. At the first opportunity, the tire must be removed from the rim and repaired from the inside, as explained in ❁ 21-18.

Remove the puncturing object, and clean the hole with a rasp. Apply the special vulcanizing fluid supplied with the repair kit to the outside of the hole. Push the snout of the vulcanizing-fluid can into the hole to get fluid inside the tire. There are different kinds of rubber plugs. One kind is installed with a plug needle.

To use this plug, first cover the hole with vulcanizing fluid. Then select a plug of the right size for the hole. The plug should be at least twice the diameter of the hole. Roll the small end of the plug into the eye of the needle. Dip the plug into vulcanizing fluid. Push the needle and plug through the hole (Fig. 21-15). Then pull the needle out. Trim off the plug ⅛ inch [3.2 mm] above the tire surface. Check for leakage. If there is no leakage, the tire is ready for service after it is inflated.

❁ 21-18 Repairing a demounted tubeless tire- There are three methods of repairing holes in tires: the rubber-plug method, the cold-patch method, and the hot-patch method. Permanent repairs are made from inside the tire—with the tire off the rim.

1. Rubber-plug method Rubber plugs can be used in the same way as explained in ❁ 21-17. The basic difference is that the repair is made from inside the tire, and the inside area around the puncture is buffed and cleaned. Then the plug is installed from inside the tire (❁ 21-16).

2. Cold-patch method In the cold-patch method, first clean and buff the inside area around the puncture. Then pour a small amount of self-vulcanizing fluid around the injury. Allow it to dry for 5 minutes. Next, remove the backing on the patch, and place the patch over the puncture. Stitch it down with the stitching tool. Start stitching at the center and work out, making sure to stitch down the edges.

Careful! Make sure no dirt gets on the fluid or patch during the repair job. Dirt could allow leakage.

3. Hot-patch method The hot-patch method is very similar to the cold-patch method. The difference is that heat is applied after the patch has been put into place over the area. This is done by lighting the patch with a match, or using with an electric hot plate, according to the type of patch being used (Fig. 21-16).

After the repair job is done, mount the tire on the rim. Inflate it and test it for leakage (❁ 21-16).

❁ 21-19 Repairing a demounted tube-type tire If a tire that uses a tube has a small hole, clean out the

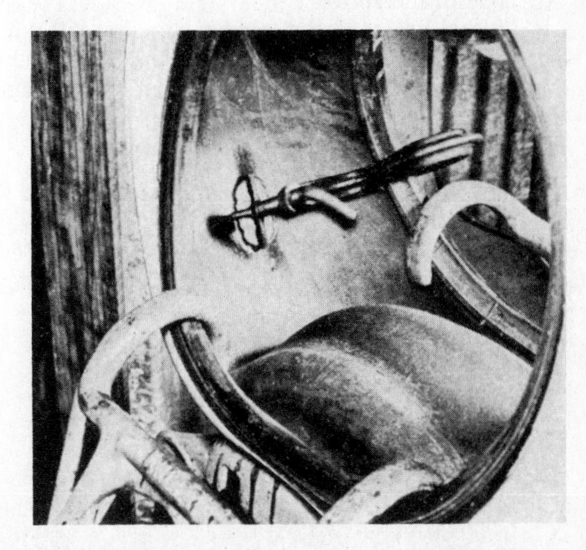

Fig. 21-16 Repairing a tubeless tire with a hot patch.

hole. Repair the tube so it will hold air. No repair to the tire is necessary. However, if the hole is ¼ inch [6.3 mm] or larger, it should be repaired with a patch on the inside. This prevents dirt or water from working in between the tire and causing tube failure.

21-20 Tire retreading Tire retreading, or "recapping," is a specialized process that involves applying new tread material to the old casing and vulcanizing it into place. Only casings that are in good condition should be recapped. Recapping cannot repair a casing with broken or separated plies or other damage. Recapping requires special equipment.

The tire is cleaned, and the tread area is roughened by rasping it or buffing it on a wire wheel. Then a strip of new rubber tread, called "camelback," is placed around the tread. The casing with the camelback then goes into the recapping machine. The machine is clamped shut, and heat is applied for the specified time. This vulcanizes the new tread onto the old casing.

21-21 Balancing the wheel-and-tire assembly After a tire change or repair, the tire-and-wheel assembly should be checked for balance. Wheel balancing on and off the car is covered in 13-5.

Chapter 21 review questions

Select the *one* correct, best, or most probable answer to each question. Then check your answers against the correct answers given at the end of the book.

1. Failure to maintain correct air pressure in tires can cause:
   a. rapid tire wear and early tire failure
   b. no bad effects
   c. failure of the wheel
   d. none of the above
2. If a wheel has excessive camber, the tire will run:
   a. more on one shoulder of the tire than on the other
   b. more on the center of the tire than on the shoulders
   c. slightly sideways to the direction of travel
   d. none of the above
3. The right rear tire wears:
   a. about half as fast as the left front tire
   b. about twice as fast as the left front tire
   c. at about the same rate as the left front tire
   d. none of the above
4. Tires have tread-wear indicators that are:
   a. special types of valve stems
   b. a series of warning lights on the instrument panel
   c. filled-in sections of the tread grooves
   d. all of the above
5. To remove a wheel from a car using a lug wrench, first:
   a. loosen the lug nuts before lifting the car
   b. jack up the car
   c. remove the wheel-bearing grease retainer
   d. raise the car on a lift
6. A bulge in the sidewall of a tire means:
   a. that the plies are separated or broken
   b. that the tire is about ready to blow out
   c. both *a* and *b*
   d. neither *a* nor *b*
7. When removing a tire from a drop-center rim:
   a. make a chalk mark across the tire and rim so that they can be put back in the same position
   b. use tire irons and large hammers
   c. leave the wheel on the car
   d. none of the above
8. Mechanic A says that tire cupping is caused by wheel tramp. Mechanic B says that the cause is loose tie-rod ends.
   Who is right?
   a. A only
   b. B only
   c. either A or B
   d. neither A nor B
9. When rotating tires, cross-switching is recommended for all of the following *except:*
   a. bias tires
   b. belted-bias tires
   c. 60-series belted-bias tires
   d. steel-belted radial tires
10. When checking the tires on a car with bias-belted tires, the tires show that the second rib on both sides of each tire is worn lower than the other ribs. The mechanic should:
    a. allow a camber spread of 1.5°
    b. readjust the caster
    c. change the toe setting
    d. none of the above

GLOSSARY

This glossary of automotive words and phrases provides a ready reference for the automotive technician. The definitions may differ from those given in a standard dictionary. They are not intended to be all-inclusive but to cover only what specifically applies to the automotive service field.

accessories Devices not considered essential to the operation of a vehicle, such as the radio, car heater, and electric window lifts.

adjust To bring the parts of a component or system into a specified relationship, or to a specified dimension or pressure.

adjustments Necessary or desired changes in clearances, fit, or settings.

air brakes A brake system that uses compressed air to supply the force required to apply brakes.

air compressor An engine-driven pump used to supply air under pressure for operating air brakes and air-powered accessories on a vehicle.

air line A hose, pipe, or tube through which air passes.

air pressure Atmospheric pressure; also the pressure produced by an air pump or by compression of air in a cylinder.

air resistance The drag on a vehicle moving through the air; it increases as the square of vehicle speed.

air suspension Any suspension system that uses contained air for vehicle springing.

alignment The act of lining up; also, the state of being in a true line.

analyzer A device used to check the internal functioning of a system or component.

anchor The point in a drum brake where the braking forces are transmitted to the chassis.

antifriction bearing Name given to almost any type ball, roller, or tapered-roller bearing.

antilock brake A system installed with the brakes to prevent wheel lockup during braking, thereby reducing the chance of skidding.

anti-rattle clips or springs Devices that attach to disk-brake pads to keep them from rattling when the brakes are not applied.

asbestos A fiber material that is heat-resistant and nonburning; used for brake linings, clutch facings, and gaskets.

aspect ratio The ratio of tire height to width. For example, a G78 tire is 78 percent as high as it is wide. The lower the number, the wider the tire.

assemble To put together.

assembly A component part, itself made up of assembled pieces which form a self-contained, independently mounted unit. For example, in the automobile, the transmission is an assembly.

atmospheric pressure The weight of the atmosphere per unit area. Atmospheric pressure at sea level is 14.7 psi absolute [101.35 kPa]; it decreases as altitude increases.

attrition Wearing down by rubbing or by friction; abrasion.

automatic adjuster A drum-brake mechanism that adjusts the lining clearance as wear occurs; usually actuated on reverse stops or when the parking brake is set.

automatic level control A suspension system which compensates for variations in load in the rear of the car; it positions the rear at a predesignated level regardless of load.

auxiliary-drum parking brake Incorporates an auxiliary parking-brake drum inside a rear rotor on some four-wheel disk-brake systems.

auxiliary springs Usually rear only, for increased load stability or capacity without affecting light ride. Mounted to act only after regular springs are partially deflected.

axis The center line of a rotating part, a symmetrical part, or a circular bore.

axle A theoretical or actual crossbar supporting a vehicle and on which one or more wheels turn.

backing plate The mounting plate upon which drum-brake components are attached.

backlash In gearing, the clearance between the meshing teeth of two gears.

balanced valve A type of hydraulic valve that produces pressure changes proportional to the movement of a mechanical linkage, or to variations in spring force.

ball-and-nut steering gear See *recirculating-ball-and-nut steering gear*.

ball bearing An antifriction bearing with an inner race and an outer race, and one or more rows of balls between them.

ball check valve A valve consisting of a ball and a seat. Fluid can pass in one direction only; flow in the other direction is checked by the ball seating tightly on the seat.

ball joint A flexible joint consisting of a ball within a socket; used in front-suspension systems.

ball-joint angle The inward tilt of the steering axis from the vertical.

ball-joint suspension A type of front-suspension system in which the wheel spindle is attached directly to the upper and lower suspension arms through ball joints. The ball joints pivot for steering, and carry the vertical load of the vehicle.

ball stud A stud with a ball-shaped end; used in steering linkages to connect the pitman arm to the linkage or to connect tie rods to the steering arms.

bead That part of the tire which is shaped to fit the rim; the bead is made of steel wires, wrapped and reinforced by the plies of the tire.

bearing A part that transmits a load to a support and in so doing absorbs the friction of moving parts.

bearing groove A channel cut in the surface of a bearing to distribute oil.

bearing oil clearance The space purposely provided between a shaft and a bearing through which lubricating oil can flow.

bearing spin A type of bearing failure in which a lack of lubrication overheats the bearing until it seizes on the shaft, shears its locking lip, and rotates in the housing or block.

bell-shaped wear Condition in which an opening (such as a brake drum) is worn mostly at one end, so that the opening flares out like a bell.

belt In a tire, a layer of fiber glass, rayon, or woven steel located under the tread around the circumference of a tire.

belted-bias tire A tire in which the plies are laid on the bias, diagonally crisscrossing each other, with two circumferential belts on top of them. The rubber tread is vulcanized on top of the belts and plies.

belted-radial tire A tire in which the plies run parallel to each other and perpendicular to the tire bead. Belts running parallel to the tire tread are applied over this radial section.

belt tension The tightness of a drive belt.

bench bleeding A procedure used to bleed the air from a new or rebuilt master cylinder before installation in a vehicle.

bevel gear A gear shaped like the lower part of a cone; used to transmit motion through an angle.

bias-belted tire See *belted-bias tire*.

bias-ply tire A tire constructed so that the plies are laid diagonally, crisscrossing each other at an angle of about 30° to 40°.

bleed A process by which air is removed from a hydraulic system (brake or power steering) by draining part of the fluid or by operating the sytem to work out the air.

bleeding sequence Order of bleeding brake-system components.

body On a vehicle, the assembly of formed sections, together with windows, doors, seats, and other parts, that provides enclosures for the passengers, engine, and luggage compartments.

body mounting Putting a car body onto a car chassis. Also, the placing of rubber cushions at strategic points along the chassis to soak up noise and vibration.

body panels Sheets of steel or plastic which are fastened together to form the car body.

boiling Conversion from the liquid to the vapor state, which takes place throughout the liquid. The conversion is accompanied by bubbling as vapor rises from below the surface.

boiling point The temperature at which a liquid begins to boil.

brake An energy-conversion device used to slow, stop, or hold a vehicle or mechanism. A device which changes the kinetic energy of motion into useless and wasted heat energy.

brake disk The parallel-faced rotating member of a disk-brake assembly acted upon by the brake pads to slow or stop a vehicle.

brake drag A constant, relatively light contact between brake linings and drums or disks when the brakes are not applied.

brake drum Rotating cylindrical member of drum-brake assembly which is acted upon by the brake shoes to slow or stop a vehicle.

brake-drum glaze Excessively smooth brake-drum surface that lowers friction and, therefore, braking effectiveness.

brake fade A temporary reduction, or "fading out," of braking effectiveness; caused by overheating from excessively long and hard brake application, or by water reducing the friction between braking surfaces.

brake feel The reaction of the brake pedal against the driver's foot; tells the driver how heavily the brakes are being applied.

brake fluid A special fluid used in hydraulic braking systems which is incompressible and transmits hydraulic force from the master cylinder to the wheel cylinders or calipers.

brake grab A sudden increase in braking at a wheel; usually caused by contaminated brake linings.

brake hose A flexible tubing which conducts brake fluid through the brake system.

brake-light switch A mechanical or hydraulic switch that closes when the brakes are applied, turning on the brake lights.

brake lines The tubes or hoses connecting the master cylinder and wheel cylinders, or calipers, in a hydraulic-brake system.

brake lining A special friction material with which brake shoes or brake pads are lined. It takes the wear when the brake shoe is forced against the drum or rotor.

brake pad The friction-lining-and-shoe assembly that is forced against the rotor to cause braking action in a disk brake.

brake pedal The foot-actuated lever the driver depresses to cause brake application.

brake shoe The curved metal part, faced with brake lining, which is forced against the brake drum to produce braking action. In disk brakes, flat metal parts faced with brake lining which are forced against the rotor face.

brake-shoe hold-downs Spring-loaded retainers that hold brake shoes against the backing plate.

bushing A one-piece sleeve placed in a bore to serve as a bearing surface.

butyl A type of synthetic rubber used in making tire tubes.

bypass A separate passage which permits a liquid, gas, or electric current to take a path other than that normally used.

calibrate To check or correct the initial setting of a test instrument.

caliper A C-shaped housing that fits over the rotor, holds the pads, and contains the hydraulic components that force the pads against the rotors when braking.

caliper seal A caliper-piston seal fixed in the caliper, through which the piston moves.

cam A rotating lobe or eccentric which can be used with a cam follower to change rotary motion to reciprocating motion.

camber The tilt of the top of the wheel from the vertical; when the tilt is outward, the camber is positive.

capacity The ability to perform or to hold.

case hardening The carburizing method used on low-carbon steel or other alloys to make the case or outer layer of the metal harder than its core.

casing The outer part of the tire assembly, made of fabric or cord to which rubber is vulcanized.

caster The tilting of the steering axis forward or backward to provide directional steering stability. Also, the angle which a front-wheel kingpin makes with the vertical.

cc Abbreviation for cubic centimeter.

Celsius A thermometer scale (formerly called centigrade) on which water boils at 100° and freezes at 0°. The formula $°C = \frac{5}{9}(°F - 32)$ converts Fahrenheit readings to Celsius readings.

center of gravity (CG) The point at which the weight of the vehicle and/or body and payload appears to be concentrated, and if suspended at that point would balance front and rear and side to side.

centigrade See *Celsius*.

centimeter (cm) A unit of linear measure in the metric system; 1 centimeter equals approximately 0.390 inch.

centrifugal force The force acting on a rotating body which tends to move it outward and away from the center of rotation. The force increases as rotational speed increases.

ceramic A type of material made from various minerals by baking or firing at high temperatures; can be used as an electric insulator or a filter element.

chassis The assembly of mechanisms that make up the major operating systems of the vehicle; usually assumed to include everything except the car body.

check To verify that a component, system, or measurement complies with specifications.

check valve A valve that opens to permit the passage of air or fluid in one direction only, or operates to prevent (check) some undesirable action.

circuit The complete path of an electric current, including the current source. When the path is continuous, the circuit is closed, and current flows. When the path is broken, the circuit is open, and no current flows. Also used to refer to fluid paths, as in hydraulic systems.

circuit breaker An automatic switch, used as a protective device, that opens an electric circuit to prevent damage when it is overheated by excess current flow. One type contains a thermostatic blade that warps to open the circuit when the maximum safe current is exceeded.

clearance The space between two moving parts, or between a moving and a stationary part, such as a journal and a bearing. The bearing clearance is filled with lubricating oil when the mechanism is running.

cm See *centimeter*.

coil spring A spring made of an elastic metal such as steel, formed into a wire and wound into a coil.

cold patching A method of repairing a punctured tire or tube by gluing a thin rubber patch over the hole.

collapsible spare tire A wheel that mounts a special deflated tire. It is furnished with a can of tire-inflation propellant to use when the tire must be inflated and installed.

collapsible steering column An energy-absorbing steering column designed to collapse if the driver is thrown into it by a severe collision.

combination valve A brake-warning-lamp valve in combination with a proportioning and/or metering valve.

compact spare tire A special high-pressure spare tire, mounted on a narrow wheel, that can be used on cars without a limited-slip differential.

compensating port A passage in each section of the master cylinder through which excess fluid returns to the reservoir when the brakes are released.

component A part of a whole, assembly, system, or unit which may be identified and serviced separately. For example, a bulb is a component of the lighting system.

compressor A pump that is used to increase the pressure of a gas or vapor.

conductor Any material or substance that allows current or heat to flow easily.

control arm A part of the suspension system designed to control wheel movement precisely.

cord The strands forming the plies in a tire.

cornering wear A type of tire-tread wear caused by cornering at excessive speeds.

corrosion Chemical action, usually by an acid, that eats away (decomposes) a metal.

cubic centimeter (cu cm or cc) A unit of volume in the metric system; equal to approximately 0.03 fluid ounce or 0.061 cubic inch.

cu cm See *cubic centimeter*.

cup The hydraulic piston seal in master and wheel cylinders. By design, hydraulic pressure assists the sealing action.

curb weight The weight of an empty vehicle, without payload or driver but including fuel, coolant, oil, and all items of standard equipment.

cylinder hone An expandable rotating tool with abrasive stones turned by an electric motor; used to clean and smooth the inside surface of a cylinder.

dead axle An axle that supports weight and attached parts but does not turn or deliver power to a wheel or other rotating member.

de-energize The removal of the control from a device to allow it to return to its normal "at rest" position.

deflection rate For a spring, the weight required to compress the spring exactly 1 inch [25.4 mm].

degree Part of a circle; 1 degree is 1/360 of a complete circle.

detent A small depression in a shaft, rail, or rod into which a pawl or ball drops when the shaft, rail, or rod is moved, to provide a locking effect.

device A mechanism, tool, or other piece of equipment designed to serve a special purpose or perform a special function.

diagnosis A procedure followed in locating the cause of a malfunction. Also, to specifically identify and answer the question: "What is wrong?"

diagonal-brake system A dual-brake system with separate hydraulic circuits connecting diagonal wheels together (RF to LR and LF to RR).

dial indicator A gauge that has a dial face and a needle to register movement; used to measure variations in dimensions and movements too small to be measured accurately by other means.

diaphragm A flat, flexible material used to separate two areas of a chamber.

differential A gear assembly between axle shafts that permits one wheel to turn at a different speed than the other, while transmitting power from the drive shaft to the wheel axles.

directional signal A device on the car that flashes lights to indicate the direction in which the driver intends to turn.

disassemble To take apart.

disc brake See *disk brake*.

disk In a disk brake, the rotor, or revolving piece of metal, against which brake shoes are pressed to provide braking action.

disk brake A brake in which the frictional forces act upon the faces of a revolving disk to slow or stop it.

disk runout The amount by which a brake disk wobbles during rotation.

double flare A type of flare used on ends of brake lines for extra strength. The flared end of the tubing is doubled over.

dowel A metal pin attached to one object which, when inserted into a hole in another object, ensures proper alignment.

driveability The general operation of an automobile, usually rated from good to poor; based on characteristics of concern to the average driver, such as smoothness of idle, even acceleration, ease of starting, quick warmup, and tendency to overheat at idle.

drive line The driving connection between the transmission and the differential; made up of one or more drive shafts.

drive pinion A rotating shaft with a small gear on one end that transmits torque to another gear.

drive shaft An assembly of one or two universal joints and slip joints connected to a heavy metal tube; used to transmit power from the transmission to the differential. Also called the *propeller shaft*.

drop-center wheel The conventional passenger-car wheel, which has a well or drop in the center for one tire bead to fit into while the other bead is being lifted over the rim flange.

drum brake A brake in which curved brake shoes press against the inner surface of a rotating metal drum to produce the braking action.

drum lathe A special lathe for turning brake drums; some can be used to resurface disk-brake rotors.

dry friction The friction between two dry solids.

dual-brake system A car brake system using two separate hydraulic systems, which may be split front and rear or diagonally.

dual master cylinder A master cylinder with two separate pressure chambers; used in dual-brake systems.

duct A tube or channel used to convey air or liquid from one place to another.

durability The quality of being useful for a long period of time and service.

duration The length of time during which something exists or lasts.

dynamic balance The balance of an object when it is in motion; for example, the dynamic balance of a rotating wheel.

dynamometer A device for measuring power output, or brake horsepower, of an engine. An engine dynamometer measures power output at the flywheel; a chassis dynamometer measures power output at the drive wheels.

eccentric A disk or offset section (of a shaft, for example) used to convert rotary to reciprocating motion. Sometimes called a cam.

ECU See *electronic control unit*.

efficiency The ratio between the power of an effect and the power expended to produce the effect; the ratio between an actual result and the theoretically possible result.

electric brakes A brake system with an armature-electromagnet combination at each wheel; when the electromagnet is energized, the magnetic attraction between the armature and the electromagnet causes the brake shoes to move against the brake drum.

electric current A movement of electrons through a conductor such as a copper wire; measured in amperes.

electric system In the automobile, the system that electrically cranks the engine for starting; it furnishes high-voltage sparks to the engine cylinders to fire the compressed air-fuel charges; lights the lights; and powers the heater motor, radio, and other accessories. It consists, in part, of the starting motor, wiring, battery, alternator, regulator, ignition distributor, and ignition coil.

electromagnet A coil of wire, usually wound around an iron core, which produces magnetism as long as an electric current (dc) flows through the coil.

electromagnetic induction The characteristic of a magnetic field that causes an electric current to be created in a conductor as it passes through the field, or when the field builds and collapses around the conductor.

electromechanical A device whose mechanical movement is dependent upon an electric current.

electron A negatively charged particle that circles the nucleus of an atom. The movement of electrons is an electric current.

electronic control unit A solid-state device that receives information from sensors and is programmed to operate various circuits and systems based on that information.

electronics Electrical assemblies, circuits, and systems that use electronic devices such as diodes and transistors.

element A substance that cannot be further divided into a simpler substance. In a battery, the group of unlike positive and negative plates, separated by insulators, that make up each cell.

energize To activate; to cause movement or action.

energy The capacity or ability to do work. Usually measured in work units of pound-feet [kilogram-meters], but also expressed in heat-energy units (Btus [joules]).

engine A machine that converts heat energy into mechanical energy. A device that burns fuel to produce mechanical power; sometimes referred to as a power plant.

engine speed See *rpm*.

epoxy A plastic compound that can be used to repair some types of cracks in metal.

exhaust system The system through which exhaust gases leave the vehicle. Consists of the exhaust manifold, exhaust pipe, catalytic converter, muffler, tail pipe, and resonator (if used).

extreme-pressure lubricant A special lubricant for use in hypoid-gear differentials; needed because of the heavy wiping loads imposed on the gear teeth.

fatigue failure A type of metal failure resulting from repeated stress which finally alters the character of the metal so that it cracks.

feeler gauge See *thickness gauge*.

filter A device through which air, gases, or liquids are passed to remove impurities.

fins On a radiator or heat exchanger, thin metal projections over which cooling air flows to remove heat from hot liquid flowing through internal passages. On an air-cooled engine, thin metal projections on the cylinder and head which greatly increase the area of the heat-dissipating surfaces and help cool the engine.

fixed-caliper disk brake Disk brake using a caliper which is fixed in position and creates braking force through opposing pistons; the caliper usually has four pistons, two on each side of the disk.

floating-caliper disk brake Disk brake using a caliper mounted through rubber bushings which permit the caliper to float, or move, when the brakes are applied; has a piston in the caliper on only one side of the disk.

fluid Any liquid or gas.

flushing hydraulic system To wash out the hydraulic system and the master and wheel cylinders, or calipers, with clean brake fluid to remove dirt or impurities that have gotten into the system.

FMVSS Federal Motor Vehicle Safety Standard. Established Department of Transportation (DOT) minimum safety standards that apply to the vehicle and certain of its components.

force Any push or pull exerted on an object; in the U.S. Customary System, measured in units of weight, such as pounds and ounces. In the metric system, pounds-force is measured in Newtons (N).

four-wheel drive On a vehicle, driving axles at both front and rear, so that all four wheels can be driven.

frame The assembly of metal structural parts and channel sections that supports the car engine and body and is supported by the wheels.

frame gauges Gauges that may be hung from the car frame to check its alignment.

friction The resistance to motion between two bodies in contact with each other.

friction bearing A bearing in which there is sliding contact between the moving surfaces. Sleeve bearings, such as those used in connecting rods, are friction bearings.

front-end geometry The angular relationship between the front wheels, wheel-attaching parts, and car frame. Includes camber, caster, steering-axis inclination, toe, and turning radius.

front-suspension system The springs, arms, pivots, spindles, and other parts which act together to support the weight of the front end, absorb road shocks, provide steering control, and reduce the strains of braking.

front-wheel drive A vehicle having its drive wheels located on the front axle.

fuel tank The storage tank for fuel on the vehicle.

full-floating rear axle An axle which only transmits driving forces to the rear wheels; the weight of the vehicle (including payload) is supported by the axle housing.

fuse A ribbon of fusible metal, used as a protective device, that burns through and opens an electric circuit to prevent damage when overheated by excess current flow. An open, or "blown," fuse must be replaced after the circuit problem is corrected.

fuse block A boxlike unit that holds the fuses for the various electric circuits in an automobile.

fusible link A type of fuse in which a special wire melts to open the circuit when the current is excessive. An open, or "blown," fusible link must be replaced after the circuit problem is corrected.

fusion Melting; conversion from the solid to the liquid state.

gap The air space between two parts or electrodes.

gas A state in which matter has neither a definite shape nor a definite volume; air is a mixture of several gases.

gasket A layer of material, such as paper, cork, or metal, placed between two parts to prevent leaks.

gasket cement A liquid adhesive material, or sealer, used to install gaskets; sometimes a layer of gasket cement is used as the gasket.

gauge pressure A pressure reading on a scale which ignores atmospheric pressure. Therefore, the atmospheric pressure of 14.7 psi absolute is equivalent to 0 psi gauge.

GCW Abbreviation for *gross combination weight;* the total weight of a tractor and semitrailer or trailer, including payload, fuel, driver, etc.

gear lubricant A type of grease or oil designed especially to lubricate gears.

gear ratio The number of revolutions of a driving gear required to turn a driven gear through one complete revolution. For a pair of gears, the ratio is found by dividing the number of teeth on the driven gear by the number of teeth on the driving gear.

geared speed The theoretical vehicle speed based on engine rpm, transmission gear ratio, drive axle ratio, and tire size.

gear-type pump A pump in which a pair of rotating gears mesh to force oil (or other fluid) from between the teeth to the pump outlet.

gears Toothed wheels that transmit power between shafts.

gram A measurement of mass (weight) in the metric system; 1 ounce equals 28.33 grams.

grease Lubricating oil to which thickening agents have been added.

greasy friction The friction between two solids coated with a thin film of oil.

grommet A device, usually made of hard rubber or similar material, used to encircle or support a component.

groove The space between two adjacent tread ribs of a tire.

gross vehicle weight The total weight of a vehicle, including the body, payload, fuel, driver, etc.

ground The terminal of the battery that is connected to the engine or metal frame of the car. In the United States, the negative terminal is grounded on 12-volt batteries in automotive electrical systems.

ground clearance The distance between the ground and the lowest point under the vehicle.

guide pins Floating calipers slide on guide pins to center over the rotor when braking action occurs.

GVW abbreviation for *gross vehicle weight;* the total weight of the vehicle, including the body, payload, fuel, driver, etc.

hazard system Also called the *emergency signal system;* a driver-controlled system of flashing front and rear lights to warn approaching motorists of a stalled car.

heat A form of energy, released by the burning of fuel or caused by friction.

heat exchanger A device in which heat is transferred from one fluid to another fluid across a tube or other solid surface.

helical gear A gear in which the teeth are cut at an angle to the center line of the gear.

Heli-coil See *threaded insert.*

hone An abrasive stone that is rotated in a bore or bushing to remove material.

hood The part of the car body that fits over and protects the engine.

horsepower A measure of mechanical power, or the rate at which work is done. 1 horsepower equals 33,000 ft-lb (foot-pounds) of work per minute; it is the power necessary to raise 33,000 pounds a distance of one foot in one minute.

Hotchkiss drive The type of rear suspension in which leaf springs absorb the rear-axle-housing torque.

hot patching A method of repairing a tire or tube by using heat to vulcanize a patch onto the damaged surface.

hub The center part of a wheel.

hydraulic brakes A brake system that uses hydraulic pressure to force the brake shoes against the brake drums or rotors as the brake pedal is depressed.

hydraulic-brake system A brake system in which brake operation and control uses hydraulic-brake fluid.

hydraulic circuit The path of hydraulic fluid from the master cylinder to the wheels, through all valves, lines, hoses, and fittings.

hydraulic piston A piston in a cylinder, acted upon by or acting on a hydraulic fluid.

hydraulic pressure The force per unit area exerted in all parts of a hydraulic system by a liquid.

hydraulics The use of a liquid under pressure to transfer force or motion, or to increase an applied force.

hydraulic valve A valve in a hydraulic system that operates on, or controls, the hydraulic pressure in the system. Also, any valve that is operated or controlled by hydraulic pressure.

hypoid gear A type of gear used in the differential (drive pinion and ring gear); it is cut in a spiral form to allow the pinion to be set below the centerline of the ring gear, so that the car floor can be lower.

idle Engine speed when the accelerator pedal is fully released, and there is no load on the engine.

idler arm In the steering system, a link that supports the tie rod and transmits steering motion to both wheels through the tie-rod ends.

idle speed The speed, or rpm, at which the engine runs without load when the accelerator pedal is released.

impact wrench An air-powered or electrically driven hand-held tool that rapidly turns nuts and bolts using a series of sharp, fast blows, or impacts.

impeller A rotating finned disk; used in centrifugal pumps, such as water pumps, and in torque converters.

impulse A wave of energy resulting in a physical activity.

included angle In the front-suspension system, camber angle plus steering-axis inclination angle.

independent front suspension The conventional front-suspension system in which each front wheel has its own spring.

indicator A device used to make some condition known by use of a light or a dial and pointer; for example, the temperature indicator or oil-pressure indicator.

inertia Property of an object that causes it to resist any change in its speed or direction of travel.

in-line steering gear A type of integral power steering; uses a recirculating-ball steering gear to which are added a control valve and an actuating piston.

inner tube See *tire tube*.

inspect To examine a part or system for surface condition or function to answer the question: "Is something wrong?"

install To set up for use on a vehicle any part, accessory, option, or kit.

insulation Material that stops the travel of electricity (electric insulation) or heat (heat insulation).

insulator A poor conductor of electricity or heat.

intake port The passage in the master cylinder through which brake fluid flows from the reservoir to refill the low-pressure area ahead of the cup on the return stroke.

integral Built into, as part of the whole.

interchangeability The manufacture of similar parts to close tolerances so that any one of them can be substituted for another in a device, and the part will fit and operate properly; the basis of mass production.

internal gear A gear with teeth pointing inward, toward the hollow center of the gear.

jack stand See *safety stand*.

jet A calibrated passage through which a fluid flows.

journal The part of a rotating shaft which turns in a bearing.

key A wedgelike metal piece, usually rectangular or semicircular, inserted in grooves to transmit torque while holding two parts in the same relative position. Also, the small strip of metal with coded peaks and grooves used to operate a lock, such as that for the ignition switch.

kg/cm^2 Abbreviation for kilograms per square centimeter, a metric engineering term for the measurement of pressure; 1 kilogram per square centimeter equals 4.22 pounds per square inch.

kilogram (kg) In the metric system, a unit of weight and mass; approximately equal to 2.2 pounds.

kilometer (km) In the metric system, a unit of linear measure; equal to 0.621 mile.

kilowatt (kW) 1000 watts; a unit of power, equal to about 1.34 horsepower.

kinetic energy The energy of motion; the energy stored in a moving body through its momentum; for example, kinetic energy is stored in a rotating flywheel.

kingpin In older cars and trucks, the steel pin on which the steering knuckle pivots; it attaches the steering knuckle to the knuckle support or axle.

kingpin inclination Inward tilt of the kingpin from the vertical. See *steering-axis inclination*.

knuckle A steering knuckle; a front-suspension part that acts as a hinge to support a front wheel and permit it to be turned to steer the car. The knuckle pivots on ball joints or, in earlier cars and trucks, on kingpins.

knurl A series of ridges formed on the outer surfaces of a piston or on the inner surface of a valve guide by a wheel which forces metal above the surface while making indentations below the surface.

kPa Abbreviation for kilopascals, the metric measurement of pressure; 1 kilopascal equals 0.145 pound per square inch.

kW Abbreviation for kilowatt.

laminated Made up of several thin sheets or layers.

lamp A divisible assembly that provides light; contains a bulb or other light source and sometimes a lens and a reflector.

lash The amount of free motion in a gear train, between gears, or in a mechanical assembly, such as the lash in a valve train.

lateral runout A measurement of the lateral change in position of the disk-brake-rotor surface during one revolution.

lead A cable or conductor to carry electric current (pronounced "leed"). A heavy metal; used as wheel weights.

leaf spring A spring made up of a single flat steel plate, or several plates of graduated lengths assembled one on top of another; used on vehicles to absorb road shocks by bending or flexing.

light A gas-filled bulb enclosing a wire that glows brightly when an electric current passes through it; a lamp. Also, any visible radiant energy.

light-duty vehicle Any motor vehicle manufactured primarily for transporting persons or property and having a gross vehicle weight of 6000 pounds [2727.6 kg] or less.

limited-slip differential A differential designed so that when one wheel is slipping, a major portion of the drive torque is supplied to the wheel with better traction; also called a *nonslip differential*.

linear measurement A measurement taken in a straight line; for example, the measurement of shaft end play.

lining See *brake lining*.

linkage An assembly of rods or links used to transmit motion.

linkage-type power steering A type of power steering in which the power-steering units (power cylinder and valve) are part of the steering linkage; frequently a bolt-on type of system.

liter (L) In the metric system, a measure of volume; approximately equal to 0.26 gallon (U.S.) or 61.02 cubic inches (33.8 fluid ounces, or 1 quart 1.8 ounces, or 1000 cubic centimeters).

live axle An axle that drives wheels which are rigidly attached to it.

lobe A projecting part; for example, the rotor lobe or the cam lobe.

locknut A second nut turned down on a holding nut to prevent loosening.

lock washer A type of washer which, when placed under the head of a bolt or nut, prevents the bolt or nut from working loose.

lubricant Any material, usually a petroleum product such as grease or oil, that is placed between two moving parts to reduce friction.

machining The process of using a machine to remove metal from a metal part.

Magna-Flux A process in which an electromagnet and a special magnetic powder are used to detect surface and subsurface cracks in iron and steel which otherwise might not be seen.

magnetic Having the ability to attract iron. This ability may be permanent, or it may depend on a current flow, as in an electromagnet.

magnetic switch A switch with a winding (a coil of wire); when the winding is energized, the switch is moved to open or close a circuit.

mag wheel A magnesium wheel assembly; also used to refer to many types of styled-chrome, aluminum-offset, or wide-rim wheel.

make A distinctive name applied to a group of vehicles produced by one manufacturer; may be further subdivided into car lines, body types, etc.

malfunction Improper or incorrect operation.

manifold vacuum The vacuum in the intake manifold that develops as a result of the vacuum in the cylinders during their intake strokes.

manual bleeding A technique for bleeding hydraulic brakes that requires two people. One pumps the brakes, the other opens and closes the bleeder screws.

manufacturer Any person, firm, or corporation engaged in the production or assembly of motor vehicles or other products.

mass production The manufacture of interchangeable parts and similar products in large quantities.

master cylinder The liquid-filled cylinder in the hydraulic-brake system in which hydraulic pressure is developed when the driver depresses a foot pedal.

matter Anything that has weight and occupies space.

measuring The act of determining the size, capacity, or quantity of an object.

mechanical advantage In a machine, the ratio of the output force to the input force applied to it.

mechanical efficiency In an engine, the ratio between brake horsepower and indicated horsepower.

mechanism A system of interrelated parts that make up a working assembly.

member Any essential part of a machine or assembly.

meshing The mating, or engaging, of the teeth of two gears.

meter (m) A unit of linear measure in the metric sytem, equal to 39.37 inches. Also, the name given to any test instrument that measures a property of a substance passing through it, as an ammeter measures electric current. Also, any device that measures and controls the flow of a substance passing through it, as a carburetor jet meters fuel flow.

metering valve Used in some brake systems to slightly delay the application of front-disk brakes.

millimeter (mm) In the metric system, a unit of linear measure, approximately equal to 0.039 inch.

mm See *millimeter*.

model year The production period for new motor vehicles or new engines, designated by the calendar year in which the period ends.

modification An alteration; a change from the original.

moisture Humidity, dampness, wetness, or very small drops of water.

motor vehicle A vehicle propelled by a means other than muscle power, usually mounted on rubber tires, which does not run on rails or tracks.

mph Abbreviation for miles per hour, a measure of speed.

neck A portion of a shaft that has a smaller diameter than the rest of the shaft.

needle bearing An antifriction bearing of the roller type, in which the rollers are very small in diameter (needle-sized).

needle valve A small, tapered, needle-pointed valve which can move into or out of a valve seat to close or open the passage through the seat.

neoprene A synthetic rubber that is not affected by the various chemicals that are harmful to natural rubber.

neutral In a transmission, the setting in which all gears are disengaged and the output shaft is disconnected from the drive wheels.

nonslip differential See *limited-slip differential*.

nut A removable fastener used with a bolt to lock pieces together; made by threading a hole through the center of a piece of metal which has been shaped to a standard size.

odometer The meter that indicates the total distance a vehicle has traveled, in miles or kilometers; usually located in the speedometer.

OEM Abbreviation for *original-equipment manufacturer*.

oil A liquid lubricant usually made from crude oil; used to provide lubrication between moving parts.

oil cooler A small radiator that lowers the temperature of oil flowing through it.

oil pump The device that delivers oil from the reservoir to the moving parts.

oil seal A seal placed around a rotating shaft or other moving part to prevent leakage of oil.

open circuit In an electric circuit, a break or opening which prevents the passage of current.

orifice A small calibrated hole in a line carrying a liquid or gas.

O ring A type of sealing ring, made of a special rubberlike material; in use, the O ring is compressed into a groove to provide the sealing action.

oscillating Moving back and forth, as a swinging pendulum.

output shaft The main shaft of the transmission; the shaft that delivers torque from the transmission to the drive shaft.

overflow Spilling of the excess of a substance; also, to run or spill over the sides of a container, usually because of overfilling.

overhaul To completely disassemble a unit, clean and inspect all parts, reassemble it with the original or new parts, and make all adjustments necessary for proper operation.

overheat To heat excessively; also, to become excessively hot.

oversteer A built-in characteristic of certain types of rear-suspension systems; causes the rear wheels to turn toward the outside of a turn.

oxidation Burning or combustion or the combining of a material with oxygen; rusting is slow oxidation, and combustion is rapid oxidation.

parallel The quality of two items being the same distance from each other at all points; usually applied to lines and, in automotive work, to machined surfaces.

parallelism A measurement of disk-brake-rotor thickness variation at various points around the rotor.

parallelogram linkage A steering system in which a short idler arm is mounted on the right side, so that it is parallel to the pitman arm.

parking-brake cables Cables that transmit brake actuating force in the parking-brake sytem.

parking-brake equalizer A device to equalize pull between the parking-brake actuator and two wheel brakes.

part A basic mechanical element or piece, which normally cannot be further disassembled, of an assembly, component, system, or unit. Also applied to any separate entry in a parts catalog, or one that has a ''part number.''

particle A very small piece of metal, dirt, or other impurity which may be contained in the air, fuel, or lubricating oil used in an assembly or device.

passage A small hole or gallery in an assembly or casting through which air, coolant, fuel, or oil flows.

passenger car Any four-wheeled motor vehicle manufactured primarily for use on streets and highways and carrying 10 or fewer passengers.

pawl An arm, pivoted so that its free end can fit into a detent, slot, or groove at certain times to hold a part stationary.

payload The weight of the cargo carried by a truck, not including the weight of the body.

pedal reserve The distance from the brake pedal to the floorboard after the brakes are applied.

peen To mushroom, or spread, the end of a pin or rivet.

percent of grade The quotient obtained by dividing the height of a hill by its length; used in computing the power requirements of trucks.

petroleum The crude oil from which gasoline, lubricating oil, and other such products are refined.

pilot shaft A shaft that is used to align parts, and that is removed before final installation of the parts; a dummy shaft.

pinion gear The smaller of two meshing gears.

pintle valve An upright pivot pin on which another part moves.

piston A movable part, fitted in a cylinder, which can receive or transmit motion as a result of pressure changes in a fluid.

piston displacement The cylinder volume displaced by the piston as it moves from the bottom to the top of the cylinder during one complete stroke.

piston ring A split ring that is installed in a groove in the piston.

pitch The number of threads per inch on any threaded part.

pitman arm In the steering system, the arm that is connected between the steering-gear sector shaft and the steering linkage or tie rod; it swings back forth to move the steering arms as the steering wheel is turned.

pivot A pin or shaft upon which another part rests or turns.

plastic gasket compound A plastic paste which can be squeezed out of a tube to make a gasket in any shape.

plies The layers of cord in a tire casing; each of these layers is a ply.

plunger A sliding reciprocating piece driven by an auxiliary power source, having the motion of a ram or piston.

ply In a tire, a layer of rubber-coated parallel cords.

ply rating A measure of the strength of a tire based on the strength of a single ply of designated construction.

pneumatic tire A tire designed so that compressed air supports the load.

port The opening through which a fluid passes.

potential energy Energy stored in a body because of its position. A weight raised to a height has potential energy because it can do work coming down. Likewise, a tensed or compressed spring contains potential energy.

power The rate at which work is done. A common power unit is the horsepower, which is equal to 33,000 foot-pounds per minute.

power cylinder An operating cylinder which produces the power to actuate a mechanism. Both power-brake and power-steering units contain power cylinders.

power hop The loss of traction at the drive wheels caused by an excessive transfer of power, resulting in wheel bounce.

power plant The engine or power source of a vehicle.

power steering A steering system that uses hydraulic pressure from a pump to multiply the driver's steering effort.

power take-off An attachment for connecting the engine to devices or other machinery when its use is required.

power team The combination of an engine, transmission, and specific axle ratio.

power tool A tool whose power source is not muscle power; a tool powered by air or electricity.

power train The mechanisms that carry power from the engine to the drive wheels; these include the clutch, transmission, drive line, differential, and axles.

PR Abbreviation for ply rating.

preload In bearings, the amount of load placed on a bearing before actual operating loads are imposed. Proper preloading requires bearing adjustment and ensures alignment and minimum looseness in the system.

press fit A fit between two parts so tight that one part has to be pressed into the other, usually with a shop press.

pressure Force per unit area, or force divided by area; usually measured in pounds per square inch (psi) and kilopascals (kPa).

pressure bleeder A container with a brake-fluid and an air compartment separated by a diaphragm; used with adapters to supply a contant, clean, pressurized source of brake fluid for bleeding.

pressure bleeding Bleeding the hydraulic-brake system using a pressure bleeder to charge the system; then one person can open and close the bleed screws as required.

pressure-differential valve The valve in a dual-brake system that turns on a warning light if the pressure drops in one section of the hydraulic system.

pressure regulator A device that operates to prevent excessive pressure from developing.

pressure-relief valve A valve in the oil line that opens to relieve excessive pressure.

pressurize To apply more-than-atmospheric pressure to a gas or liquid.

preventive maintenance The systematic inspection of a vehicle to detect and correct failures, either before they occur or before they develop into major defects. A procedure for economically maintaining a vehicle in a satisfactory and dependable operating condition.

primary shoe The forward shoe in a drum brake, which often has shorter lining than the secondary shoe.

propeller shaft See *drive shaft*.

proportioning valve A valve in the brake hydraulic system that reduces pressure to the rear wheels to achieve better brake balance.

psi Abbreviation for *pounds per square inch,* a measurement of pressure.

psig Abbreviation for *pounds per square inch of gauge pressure.*

pull The result of an unbalanced condition. For example, uneven braking at the front brakes or unequal front-wheel alignment will cause a car to swerve (pull) to one side when the brakes are applied.

puller A shop tool used to separate two closely fitted parts without damage. Often contains a screw, or several screws, which can be turned to gradually apply force.

pulley A metal wheel with a V-shaped groove around the rim; drives or is driven by a belt.

pulsation A surge felt in the brake pedal during low-pressure braking.

pump A device that transfers fluid from one place to another.

puncture-sealing tires or tubes Tires or tubes coated on the inside with a plastic material. As air leaks from a puncture, air pressure in the tire or tube forces the sealing material into the hole; it hardens on contact with the air to seal the puncture.

purge To remove, evacuate, or empty trapped substances from a space.

quick-takeup master cylinder A master cylinder using a step piston which has two different diameters. The larger piston diameter takes up system clearances rapidly. The smaller piston builds up braking pressures.

races The metal rings on which ball or roller bearings rotate.

rack-and-pinion steering gear A steering gear in which a pinion on the end of the steering shaft meshes with a rack of gear teeth on the major crossmember of the steering linkage.

radial tire A tire in which the plies are placed radially, or perpendicular to the rim, with a circumferential belt on top of them. The rubber tread is vulcanized on top of the belt and plies.

ratio The relationship in size or quantity of two or more objects; the gear ratio is found by dividing the number of teeth on the driven gear by the number of teeth on the drive gear.

reamer A round metal-cutting tool with a series of sharp cutting edges; it enlarges a hole when turned inside it.

rear-end torque The reaction torque that acts on the rear-axle housing when torque is applied to the wheels; it tends to turn the axle housing in a direction opposite to wheel rotation.

reassembly Putting back together the parts of a device.

rebore To increase the diameter of a cylinder.

recapping A form of tire repair in which a cap of new tread material is placed on the old casing and vulcanized into place.

reciprocating motion Motion of an object between two limiting positions; motion in a straight line either back-and-forth or up-and-down.

recirculating-ball-and-nut steering gear A type of steering gear in which a nut (meshing with a gear sector) is assembled on a worm gear; balls circulate between the nut and the worm threads.

relay An electric device that opens or closes a circuit or circuits in response to a voltage signal.

relief valve A valve that opens when a preset pressure is reached. This relieves or prevents excessive pressures.

remove and reinstall (R and R) To perform a series of servicing procedures on an original part or assembly; includes removal, inspection, lubrication, all necessary adjustments, and reinstallation.

replace To remove a used part or assembly and install a new part or assembly in its place; includes cleaning, lubricating, and adjusting as required.

reservoir-diaphragm gasket The gasket under the master-cylinder reservoir cap that separates the fluid from the atmosphere. Moves with the fluid to allow atmospheric pressure above the fluid.

residual check valve A check valve in some hydraulic circuits that maintains a slight pressure in the brake lines to prevent air leaks.

resistance The opposition to a flow of current through a circuit or electrical device; measured in ohms. A voltage of 1 volt will cause 1 ampere to flow through a resistance of 1 ohm. This is known as Ohm's law, which can be written in three ways: amperes = volts/ohms; ohms = volts/amperes; and volts = amperes × ohms.

retaining key A metal plate used to retain some sliding calipers in their anchor plate.

retaining ring A removable fastener used as a shoulder to retain and position a round bearing in a hole.

rim A metal support around the outside of a wheel for a tire, or a tire-and-tube assembly, upon which the tire beads are seated.

ring expander A special tool used to expand piston rings for installation on the piston.

ring gap The gap between the ends of the piston ring when the ring is in place in the cylinder.

ring gear A large gear carried by the differential case; meshes with and is driven by the drive pinion.

ring grooves Grooves cut in a piston, into which the piston rings are assembled.

rivet A semipermanent fastener used to hold two pieces together.

roadability The steering and handling qualities of a vehicle while it is being driven on the road.

road load The power required to hold constant vehicle speed on a level road.

room temperature 68 to 72°F (20 to 22°C).

rotary The motion of a part that continually rotates or turns.

rotary-valve steering gear A type of power-steering gear.

rotor The rotating part of a device, such as a disk-brake rotor. See *brake disk.*

rotor oil pump A type of oil pump in which a pair of rotors, one inside the other, produce the pressure required to circulate oil to system parts.

rpm Abbreviation for *revolutions per minute,* a measure of rotational speed.

RTV sealer Room-temperature vulcanizing gasket material which cures at room temperature; a plastic paste squeezed from a tube to form a gasket of any shape.

runout Wobble.

SAE Abbreviation for Society of Automotive Engineers. Used to indicate a grade or weight of oil measured according to Society of Automotive Engineers standards.

safety Freedom from injury or danger.

safety rim A wheel rim with a hump on the inner edge of the ledge on which the tire bead rides. The hump helps hold the tire on the rim in case of a blowout.

safety stand A pinned, or locked, type of stand placed under a car to support its weight after the car has been raised with a floor jack. Also called a car stand or jack stand.

schematic A pictorial representation, most often in the form of a line drawing. A systematic positioning of components, showing their relationship to each other or to an overall function.

Schrader valve A spring-loaded valve through which a connection can be made to the air chamber in a tire.

science The understanding of nature.

scoring Irregular grooves in the friction surfaces of brake drums or rotors caused by contamination or worn-out linings.

screens Pieces of fine-mesh metal fabric; used to prevent solid particles from circulating through any liquid or vapor system and damaging vital moving parts.

screw A fastener with threads that can be turned into a threaded hole, usually with a screwdriver. There are many different types and sizes of screws.

scuffing A type of wear in which there is a transfer of material between parts moving against each other; shows up as pits or grooves in the mating surfaces.

seal A material, shaped around a shaft, used to close off the operating compartment of the shaft, preventing oil leakage.

sealer A thick, tacky compound, usually spread with a brush, which may be used as a gasket or sealant to seal small openings or surface irregularities.

seat The surface upon which another part rests, as a valve seat. Also, to wear into a good fit.

secondary shoe The rear shoe in a drum brake; often has longer lining than the other shoe.

section modulus A measure of the strength of the car-frame side rails; depends on the cross-sectional area and shape of the rails.

sector A gear that is not a complete circle. Specifically, the gear sector on the pitman shaft in many steering gears.

sediment The accumulation of matter which settles to the bottom of a liquid.

self-adjuster A mechanism on drum brakes which compensates for shoe wear by automatically keeping the shoe adjusted close to the drum.

self-energizing brakes A drum brake in which braking action pulls the shoe lining tighter against the drum.

self-locking screw A screw that locks itself in place, without the use of a separate nut or lock washer.

self-tapping screw A screw that cuts its own threads as it is turned into an unthreaded hole.

semifloating rear axle An axle that supports the weight of the vehicle on the axle shaft in addition to transmitting driving forces to the rear wheels.

sensor Any device that receives and reacts to a signal, such as a change in voltage, temperature, or pressure.

separator A thin sheet of wood, rubber, plastic, or fiber glass mat that is placed between each positive and negative plate in a battery cell to insulate the plates from each other.

serviceable Parts or systems that can be repaired and maintained to continue in operation.

service-brake system The foot-operated brake system used for retarding, stopping, and controlling the vehicle under normal driving conditions.

service manual A book, usually published annually by each vehicle manufacturer, listing the specifications and service procedures for each make and model of vehicle. Also called a shop manual.

servo A device in a hydraulic system that converts hydraulic pressure to mechanical movement. Consists of a cylinder with a piston which moves as hydraulic pressure acts on it.

servo brake A drum brake in which the action of one shoe reinforces the action of the other shoe.

setscrew A type of fastener that holds a collar or gear on a shaft when its point is turned down into the shaft.

shackle The swinging support by which one end of a leaf spring is attached to the car frame.

shim A slotted strip of metal used as a spacer to adjust the front-wheel alignment on many cars; also used to make small corrections in the position of body sheet metal and other parts.

shimmy Rapid oscillation. For example, in wheel shimmy, the front wheel turns in and out alternately and rapidly, causing the front end of the car to oscillate, or shimmy.

shim stock Sheets of metal of accurately known thicknesses which can be cut into strips and used to measure or correct clearances.

shock absorber A device placed at each vehicle wheel to regulate spring rebound and compression.

shoe In the brake system, a metal plate that supports the brake lining and absorbs and transmits braking forces.

short-arm, long-arm (SALA) suspension The front-suspension system which uses a short upper control arm and a longer lower control arm.

short circuit A defect in an electric circuit which permits current to take a short path, or circuit, instead of following the desired path.

shrink fit A tight fit of one part into another, achieved by heating or cooling one part and then assembling it to the other part. A heated part will shrink on cooling to provide the tight fit; a cooled part will expand on warming to provide the tight fit.

shunt A parallel connection or circuit.

side clearance The clearance between the sides of moving parts when the sides do not serve as load-carrying surfaces.

sidewall That portion of the tire between the tread and the bead.

single brake system A brake system using only one hydraulic circuit for all wheels.

single master cylinder A master cylinder with only one pressure chamber, used in single brake systems.

sintered bronze Tiny particles of bronze pressed tightly together so that they form a solid piece which is highly porous and which is often used as a filter for liquids.

skid-control system A system which responds to a locking wheel by relieving hydraulic pressure to the locking brake.

slick Performance term for a smooth, treadless racing tire.

slip joint In the power train, a variable-length connection that permits the drive shaft to change its effective length.

sluggish The condition in which the engine delivers limited power under load or at high speed, and will not accelerate as fast as normal, loses too much speed on hills, or has a lower top speed than normal.

snap ring A metal fastener, available in two types; the external snap ring fits into a groove in a shaft, and the internal snap ring fits into a groove in a housing. Snap rings must be installed and removed with special snap-ring pliers.

soldering Joining pieces of metal with solder, flux, and heat.

solenoid An electrically operated magnetic device used to mechanically operate some other device through movement of an iron core placed inside a coil. When current flows through the coil, the core attempts to center itself in the coil, thereby exerting a strong force on anything connected to the core.

solenoid relay A relay that connects a solenoid to a current source when its contacts close.

solvent A petroleum product of low volatility used in the cleaning of engine and vehicle parts.

solvent tank In the shop, a tank of cleaning fluid in which most parts are brushed and washed clean.

specifications Information and service procedures provided by the manufacturer for each automotive system and its components, operation, and clearances.

specific heat The quantity of heat (in Btus) required to change the temperature of one pound of a substance by 1°F.

specs Short for specifications.

speed The rate of motion; for vehicles, measured in miles per hour or kilometers per hour.

speedometer An instrument that indicates vehicle speed; usually driven from the transmission.

splines Slots or grooves cut in a shaft or bore; splines on a shaft are matched to splines in a bore to ensure that two parts turn together.

spongy pedal A condition in which the brake pedal is not solid when depressed, but bounces softly, because of air trapped in the hydraulic sytem.

spool valve A rod with indented sections.

spring A device that changes shape under stress or pressure, but returns to its original shape when the stress or pressure is removed; the component of the automotive suspension system that absorbs road shocks by flexing and twisting.

spring capacity at ground The total weight (sprung plus unsprung) which will deflect the spring its maximum normal amount; used for rating springs as installed on the vehicle.

spring rate The load required to move a spring or a suspended wheel a given distance; indicates the softness or firmness of a given spring or suspension.

spring shackle See *shackle*.

sprung weight That part of the car which is supported on springs (includes the engine, frame, body, and payload).

spur gear A gear in which the teeth are parallel to the center line of the gear.

squeak A high-pitched noise of short duration.

squeal A continuous, high-pitched, low-volume noise.

stabilizer bar An interconnecting shaft between the two lower suspension arms; reduces body roll on turns.

stalls The condition in which an engine quits running, at idle or while driving.

star-wheel adjuster A drum-brake mechanism that separates the brake shoes at the bottom; it rotates a nut on a threaded link to adjust brake-lining clearance.

static balance The balance of an object while it is not moving.

static friction The friction between two bodies at rest.

steam cleaner A machine used for cleaning large parts with a spray of steam, often mixed with soap.

steering-and-ignition lock A device that locks the ignition switch in the OFF position and locks the steering wheel so that it cannot be turned.

steering arm The arm, attached to the steering knuckle, that turns the knuckle and wheel for steering.

steering axis The centerline of the ball joints in a front-suspension system.

steering-axis inclination The inward tilt of the steering axis from the vertical.

steering-column shift An arrangement in which the transmission shift lever is mounted on the steering column.

steering gear That part of the steering system that is located at the lower end of the steering shaft; changes the rotary motion of the steering wheel into linear motion of the front wheels for steering.

steering kickback Sharp and rapid movements of the steering wheel as the front wheels encounter obstructions in the road; the shocks of these encounters "kick back" to the steering wheel.

steering knuckle The front-wheel spindle, which is supported by upper and lower ball joints and by the wheel; the part on which a front wheel is mounted, and which is turned for steering.

steering shaft The shaft extending from the steering gear to the steering wheel.

steering system The mechanism that enables the driver to turn the wheels to change the direction of vehicle movement.

steering wheel The wheel, at the top of the steering shaft, which is used by the driver to guide or steer the vehicle.

stepped thickness gauge A thickness gauge which has a thin tip of a known dimension, and is thicker along the rest of the guage; a go/no-go thickness gauge.

stop Complete cessation from movement; a device that prevents travel of a part past a certain point.

stoplights Lights at the rear of a vehicle which indicate that the driver is applying the brakes to slow or stop the vehicle.

stoplight switch The switch that turns the stoplights on and off as the brakes are applied and released.

stopping distance The distance traveled by a vehicle from the point of application of force to the brake control to the point at which the vehicle reaches a full stop.

streamlining The shaping of a car body or truck cab so that it minimizes air resistance and can be moved through the air with less energy.

strut A bar that connects the lower control arm to the car frame; used when the lower control arm is of the type that is attached to the frame at only one point. Also called a brake reaction rod.

stud A headless bolt that is threaded on both ends.

stud extractor A special tool used to remove a broken stud or bolt.

substance Any matter or material; may be a solid, a liquid, or a gas.

surge To occur suddenly to an excessive or abnormal value. The condition in which the engine speed increases and decreases slightly under constant-throttle operation.

suspension arm In the front suspension, one of the arms pivoted on the frame at one end and on the steering-knuckle support at the other end.

suspension system The springs and other parts which support the upper part of a vehicle on its axles and wheels.

sway bar See *stabilizer bar*.

switch A device that opens and closes an electric circuit.

synchronize To make two or more events or operations occur at the same time or at the same speed.

synthetic oil An artificial oil that is manufacured; not a natural mineral oil made from petroleum.

system A combination or grouping of two or more parts or components into a whole which in operation performs some function that cannot be done by the separate parts.

tachometer A device for measuring the speed of an engine in revolutions per minute (rpm).

tandem master cylinder. See *dual master cylinder*.

tap A tool used for cutting threads in a hole.

taper A shaft or hole that gets gradually smaller toward one end.

technology The applications of science.

temperature The measure of heat intensity in degrees. Temperature is *not* a measure of heat quantity.

terminal A connecting point in an electric circuit.

thermal Of or pertaining to heat.

thermistor A heat-sensing device with a negative temperature coefficient of resistance; as its temperature increases, its electric resistance decreases.

thermometer An instrument which measures heat intensity (temperature) by the thermal expansion of a liquid.

thermostat A device for the automatic regulation of temperature; usually contains a temperature-sensitive element that expands or contracts to open or close off fluid flow.

thermostatic gauge An indicating device (for fuel quantity, oil pressure, engine temperature) that contains a thermostatic blade or blades.

thickness gauge Strips of metal made to an exact thickness, used to measure clearances between parts.

thread chaser A device, similar to a die, run over threads to clean them.

threaded insert A threaded coil that is used to restore the original thread size to a hole with damaged threads; the hole is drilled oversize and tapped, and the insert is threaded into the tapped hole.

thread series A designation indicating the pitch, or number of threads per inch, on a threaded part.

tie-rod end A socket and ball stud in a housing. They rotate and tilt to transmit steering action under all conditions.

tie rods In the steering system, the rods that link the pitman arm to the steering-knuckle arms; small steel components that connect the front wheels to the steering mechanism.

tilt steering wheel A type of steering wheel that can be tilted at various angles, through a flex joint in the steering shaft.

tire The rubber-and-cord donut on the wheel rim that is filled with pressurized air and transmits vehicle braking and tractive forces to the road.

tire carcass The plies that constitute the underbody of the tire; the "skeleton" over which the rubber of the sidewalls and the thicker tread area are molded.

tire casing Layers of cord, called plies, shaped in a tire form and impregnated with rubber, to which the tread is applied.

tire rotation The interchanging of the running locations of the tires on a car, to minimize noise and to equalize tire wear.

tire tread See *tread*.

tire tube An inflatable rubber device mounted inside some tires to contain air at sufficient pressure to inflate the casing and support the vehicle weight.

tire-wear indicator Small strips of rubber molded into the bottom of the tire-tread grooves; they appear as narrow strips of smooth rubber across the tire when the tread depth decreases to 1/16 inch [1.6 mm].

toe-in The amount in inches or millimeters by which the front wheels point inward.

toe-out-on-turns See *turning radius*.

tolerance The range of variation in a given dimension.

torque Turning or twisting effort; usually measured in pound-feet or newton-meters. Also, a turning force such as that required to tighten a connection.

torque wrench A wrench that indicates the amount of torque or turning force being applied with the wrench.

torsional vibration Rotary vibration that causes a twist-untwist action on a rotating shaft, so that a part of the shaft repeatedly moves ahead of, or lags behind, the remainder of the shaft.

torsion-bar spring A long, straight bar that is fastened to the vehicle frame at one end and to a suspension part at the other. Spring action is produced by a twisting of the bar.

torsion-bar steering gear A rotary-valve power-steering gear.

tracking Rear wheels following the front wheels in a parallel path when the vehicle is moving straight ahead.

tractive effort The force available at the road surface in contact with the driving wheels of a vehicle. It is determined by engine torque, transmission ratio, axle ratio, tire size, and frictional losses in the drive line. "Rim pull" is also known as tractive effort.

tractor breakaway valve A valve that couples the tractor and trailer emergency-brake systems. Provides air to the trailer emergency-brake system for normal operating conditions. If the trailer brake system fails, the breakaway valve automatically seals off the tractor braking system and activates the trailer emergency brake.

tramp Up-and-down motion (hopping) of the front wheels at higher speeds, due to unbalanced wheels or excessive wheel runout. Also called high-speed shimmy.

transaxle A power transmission device that combines the functions of the transmission and the drive axle (differential) into a single assembly; used in front-wheel-drive cars with front-mounted engines, and in rear-wheel-drive cars with rear-mounted engines.

transducer Any device which converts an input signal of one form into an output signal of a different form. For example, the automobile horn converts an electric signal to sound.

tread (tire) That part of the tire that contacts the road. It is the thickest part of the tire, and is cut with grooves to provide traction for driving and stopping.

tread (vehicle) The distance between the centers of tires at the points where they contact the road surface. On trucks, duals are measured from the center of dual wheels.

tread rib The tread section running circumferentially around the tire.

trouble diagnosis The detective work necessary to find the cause of a trouble.

truck Any motor vehicle primarily designed for the transportation of property which carries the load on its own wheels.

truck tractor Any motor vehicle designed primarily for pulling truck trailers and constructed so as to carry part of the weight and load of a semitrailer.

tube seat An insert or machined face, against which a flared tube end seals.

tubeless tire A tire that holds air without the use of an inner tube.

tuneup A procedure for inspecting, testing, and adjusting an engine, and replacing any worn parts, to restore the engine to its best performance.

turning radius The difference between the angles each of the front wheels makes with the car frame during turns, usually measured with the outside wheel turned 20°.

On a turn, the inner wheel turns, or toes out, more. Also called toe-out-on-turns.

turn signal See *directional signal*.

twin I-beam A type of front-suspension system used on some trucks.

U bolt An iron rod with threads on both ends, bent into the shape of a U and fitted with a nut at each rod.

unit An assembly or device that can perform its function only if it is not further divided into its component parts.

unitized construction A type of automotive construction in which the frame and body parts are welded together to form a single unit.

universal joint In the power train, a jointed connection in the drive shaft that permits the driving angle to change.

unsprung weight The weight of that part of the car which is not supported on springs; for example, the wheels and tires.

vacuum A pressure less than atmospheric pressure; a negative gauge pressure. Vacuum can be measured in pounds per square inch, but is usually measured in inches or millimeters of mercury (Hg); a reading of 30 inches [762 mm] Hg would indicate a perfect vacuum.

vacuum booster A power-brake unit which uses vacuum on one side of a diaphragm as a power source.

vacuum gauge In automotive-engine service, a device that measures intake-manifold vacuum and thereby indicates actions of engine components.

vacuum motor A small motor, powered by intake-manifold vacuum; used for jobs such as raising and lowering headlight doors.

vacuum power unit A device for operating accessory doors and valves using vacuum as a source of power.

vacuum pump A mechanical device used to evacuate, or pump out, a system.

vacuum-suspended power brake A type of power brake in which both sides of the piston are subjected to vacuum; therefore the piston is "suspended" in vacuum.

vacuum switch A switch that closes or opens its contacts in response to changing vacuum conditions.

valve Any device used to control the direction, volume, or pressure of a fluid flowing through a hydraulic system. The word preceding *valve* usually designates the type of valve (*needle* valve) or the function it performs (*check* valve).

valve seat The surface against which a valve comes to rest to provide a seal against leakage.

valve spool A spool-shaped valve, such as in the power-steering unit.

vane A flat, extended surface that is moved around an axis by or in a fluid. Part of the internal revolving portion of vane-type pumps.

vapor A gas; any substance in the gaseous state, as distinguished from the liquid or solid state.

vaporization A change of state from liquid to vapor or gas by evaporation or boiling; a general term including both evaporation and boiling.

variable-ratio power steering Power-steering system in which the response of the car wheels varies according to how much the steering wheel is turned.

vehicle See *motor vehicle*.

vehicle identification number (VIN) The number assigned to each vehicle by its manufacturer, primarily for registration and identification purposes.

vent An opening through which air can leave an enclosed chamber.

ventilation The circulating of fresh air through any space to replace impure air.

vibration A rapid back-and-forth motion; an oscillation.

VIN Abbreviation for *vehicle identification number*.

viscosity The resistance to flow exhibited by a liquid. A thick oil has greater viscosity than a thin oil.

viscosity index A number indicating how much the viscosity of an oil changes with heat.

viscosity rating An indicator of the viscosity of engine oil. There are separate ratings for winter driving and for summer driving. The winter grades are SAE5W, SAE10W, and SAE20W. The summer grades are SAE20, SAE30, SAE40, and SAE50. Many oils have multiple-viscosity ratings, as, for example, SAE10W-30.

viscous Thick; tending to resist flowing.

viscous friction The friction between layers of a liquid.

volatile Evaporating readily.

volatility A measure of the ease with which a liquid vaporizes; has a direct relationship to the flammability of a fuel.

vulcanizing A process of treating raw rubber with heat and pressure; the treatment forms the rubber and gives it toughness and flexibility.

warning blinker See *hazard system*.

wear limit The allowable play set by the manufacturer for a given part.

wear sensor A projection on an inboard brake pad that causes a squealing sound when the brake pads are worn thin.

weight distribution The percentage of a vehicle's total weight that rests on each axle.

weight, sprung See *sprung weight*.

weight, unsprung See *unsprung weight*.

welding The process of joining pieces of metal by fusing them together with heat.

wheel A disk or spokes with a hub at the center, which revolves around an axle, and a rim around the outside for mounting on the tire.

wheel alignment A series of tests and adjustments to ensure that wheels and tires are properly positioned on the vehicle.

wheel balancer A device that checks a wheel-and-tire assembly (statically, dynamically, or both) for balance.

wheelbase The distance between the centerlines of the front and rear axles. For trucks with tandem rear axles, the rear centerline is considered to be midway between the two rear axles.

wheel cylinder A drum-brake device for converting hydraulic-fluid pressure to mechanical force for brake application.

wheel tramp Tendency for a wheel to move up and down so that it repeatedly bears down hard, or "tramps," on the road. Sometimes called high-speed shimmy.

wire thickness gauges A set of round wires of known diameters used to check clearances between parts.

wiring harness A group of two or more individually insulated wires, wrapped together to form a neat, easily installed bundle.

work The changing of the position of an object against an opposing force; measured in foot-pounds or joules. The product of a force and the distance through which it acts.

worm Type of gear in which the teeth resemble threads; used on the lower end of the steering shaft.

zip gun An air-powered cutting tool often used for work on vehicle exhaust systems.

INDEX

ANSWERS TO REVIEW QUESTIONS

CHAPTER 1

1. *c* 2. *d* 3. *b* 4. *c* 5. *c*

CHAPTER 2

1. *b* 2. *a* 3. *c* 4. *a* 5. *c*
6. *c* 7. *c* 8. *a* 9. *b* 10. *c*
11. *c* 12. *a* 13. *a* 14. *a* 15. *b*
16. *b* 17. *c* 18. *b* 19. *c* 20. *c*

CHAPTER 3

1. *b* 2. *c* 3. *a* 4. *b* 5. *c*
6. *b* 7. *a* 8. *b* 9. *b* 10. *a*
11. *b* 12. *c* 13. *a* 14. *d* 15. *b*
16. *c* 17. *c* 18. *b* 19. *d* 20. *a*

CHAPTER 4

1. *d* 2. *c* 3. *c* 4. *b* 5. *d*
6. *a* 7. *a* 8. *a* 9.*b* 10. *d*

CHAPTER 5

1. *a* 2. *d* 3. *c* 4. *d* 5. *a*
6. *b* 7. *a* 8. *d* 9. *c* 10. *d*

CHAPTER 6

1. *d* 2. *c* 3. *d* 4. *d* 5. *a*
6. *c* 7. *b* 8. *c* 9. *d* 10. *c*
11. *d* 12. *d* 13. *b* 14. *a* 15. *b*
16. *c* 17. *c* 18. *b* 19. *a* 20. *d*

CHAPTER 7

1. *d* 2. *a* 3. *a* 4. *c* 5. *b*

CHAPTER 8

1. *c* 2. *d* 3. *c* 4. *c* 5. *c*
6. *c* 7. *a* 8. *c* 9. *b* 10. *c*

CHAPTER 9

1. *c* 2. *b* 3. *c* 4. *b* 5. *a*
6. *a* 7. *c* 8. *b* 9. *d* 10. *c*

CHAPTER 10

1. *d* 2. *d* 3. *a* 4. *c* 5. *b*
6. *c* 7. *a* 8. *a* 9. *a* 10. *a*
11. *c* 12. *a* 13. *c* 14. *b* 15. *b*
16. *a* 17. *a* 18. *c* 19. *c* 20. *b*

CHAPTER 11

1. *c* 2. *d* 3. *c* 4. *b* 5. *c*
6. *c* 7. *b* 8. *c* 9. *a* 10. *b*

CHAPTER 12

1. *b* 2. *c* 3. *b* 4. *c* 5. *a*
6. *c* 7. *a* 8. *c* 9. *a* 10. *b*

CHAPTER 13

1. *b* 2. *d* 3. *d* 4. *a* 5. *b*
6. *b* 7. *c* 8. *c* 9. *a* 10. *c*

CHAPTER 14

1. *d* 2. *c* 3. *a* 4. *c* 5. *b*

CHAPTER 15

1. *b* 2. *b* 3. *c* 4. *b* 5. *c*

CHAPTER 16

1. *c* 2. *a* 3. *c* 4. *a* 5. *b*

CHAPTER 17

1. *a* 2. *b* 3. *c* 4. *b* 5. *c*

CHAPTER 18

1. *b* 2. *a* 3. *c* 4. *d* 5. *d*

CHAPTER 19

1. *c* 2. *d* 3. *a* 4. *c* 5. *b*

CHAPTER 20

1. *a* 2. *a* 3. *b* 4. *a* 5. *c*
6. *b* 7. *c* 8. *b* 9. *b* 10. *d*

CHAPTER 21

1. *a* 2. *a* 3. *b* 4. *c* 5. *a*
6. *c* 7. *a* 8. *c* 9. *d* 10. *d*